Engineering Mathematics

Engineering Mathematics

VOLUME ONE

Third Edition

S.S. SASTRY
Formerly Scientist/Engineer SF
Vikram Sarabhai Space Centre
Trivandrum

Prentice-Hall of India Private Limited
New Delhi-110 001
2008

Rs. 295.00

ENGINEERING MATHEMATICS, Volume One, 3rd ed.
S.S. Sastry

ISBN-978-81-203-2579-1

Thirteenth Printing (Third Edition) ... ... May, 2008

Published by Asoke K. Ghosh, Prentice-Hall of India Private Limited, M-97, Connaught Circus, New Delhi-110001 and Printed by Rajkamal Electric Press, B-35/9, G.T. Karnal Road Industrial Area, Delhi-110033.

To
My Mother and Father

Contents

Preface

During the last few years, there have been several changes in the mathematics curriculum of engineering studies. Keeping in view the present requirements for a first course in engineering mathematics, the following changes were effected in the material of the previous editions:

(a) *Higher Algebra.* The section on determinants has been deleted since this topic forms a part of the higher secondary syllabus. Instead, a new section on theory of equations has been introduced, whereas the section on matrices has been rewritten.

(b) *Differentiation.* Most of the elementary material involving functions, limits, continuity and differentiation have been omitted while new sections on envelopes and evolutes have been incorporated.

(c) *Integration.* Most of the elementary material have been removed and the various methods of integration are illustrated through numerical examples.

(d) *Numerical methods.* This is a new chapter (involving the introductory topics of interpolation, numerical differentiation and integration) included in this volume so as to provide a ready reference of numerical methods to the engineering student. In Chapter 1, an elementary idea of interpolation is given, but Chapter 7 discusses the problem of interpolation in greater detail. Similarly, the idea of numerical integration introduced in the chapter on integration finds detailed discussion in Chapter 7. Most of this material are drawn from the author's book, *Introductory Methods of Numerical Analysis*. It is hoped that this chapter will greatly help the engineering student.

(e) More illustrative examples and exercises have been included in Chapters 1–6. This volume contains about 464 worked examples and over 1200 problems in the exercises.

The author is grateful to Mr. Asoke K. Ghosh, Chairman and Managing Director, Prentice-Hall of India, for his interest and cooperation in bringing out the present edition.

S.S. SASTRY

Preface to the First Edition

This volume is primarily intended to serve as a basic text meeting the requirements of students following introductory courses in engineering and the physical sciences. It also provides a clear, elementary and well-motivated, yet precise, treatment of the essential tools of applicable mathematics required by engineers and scientists.

There are six chapters in the present volume. The first-two chapters contain introductory material on algebra, coordinate geometry, vector algebra and complex numbers which serve as a review for many students. Vectors are designated by means of arrows, instead of the bold-face type, to conform to the handwriting practice. Chapters 3 and 4 are concerned with differentiation and integration and their applications. Functions of several variables are discussed in Chapter 5, which include a brief account of calculus of variations. This is intended to serve as an introduction to the more subtle and advanced branch of mathematics, optimization techniques, to be covered in a future volume. Certain sections and exercises in this volume are marked with an asterisk and these may be omitted while reading. The last chapter, Chapter 6, is on multiple integrals and their applications. The entire subject matter in this volume can be covered in the first year of the engineering course.

The more advanced topics of engineering mathematics, viz., ordinary differential equations and their numerical treatment, Laplace transforms and their applications, Fourier series and integrals, partial differential equations and their numerical solutions, vector and tensor analysis and finally a chapter on modern methods of computation are covered in volume 2 of this book, which is presently in press. To complete the series, it is proposed to bring out volume 3 comprising the topics on statistics and probability, complex variables and applications, special functions, integral transforms, optimization techniques and interval mathematics.

Special attention is drawn to the abundant sets of worked examples and exercises in the present volume—there are about 350 worked examples and over 1250 exercises. A majority of the exercises are routine drills for practice but a few of them are intermediate exercises that extend the material of the text. Whenever it is found desirable, hints are also provided. Answers to selected exercises have also been given.

I owe a great deal to the works of several authors which have been helpful to me in preparing my lecture notes. In particular, I acknowledge, gratefully, the help I derived from: *Calculus and Analytic Geometry* by G.B. Thomas (Addison-Wesley, Massachusetts) and *Mathematics for Engineers and Scientists*, Volumes 1 and 2, by A.C. Bajpai, I.M. Calus and J.A. Fairley (John Wiley & Sons, New York) and other books mentioned in the bibliography.

The book owes its origin to my experience in giving such courses at Birla Institute of Technology, Ranchi for over 14 years and to the lectures I gave to a large number of students who were preparing for AMIE examinations. Their reactions have helped the development of this book to a great extent.

I thank my friends and colleagues at Vikram Sarabhai Space Centre (VSSC), Trivandrum, whose suggestions, criticisms and comments helped me in this long and arduous task. My special gratitude is to Dr. V.R. Gowariker (previously Director of VSSC and presently a Visiting Scientist at Stanford University, California) and Dr. S.C. Gupta, Director, VSSC, for their words of encouragement; Dr. V. Swaminathan, Head, Applied Mathematics Division, VSSC, for his encouragement and advice in the publication of this book; Dr. S.C. Purohit and Dr. P.V. Subba Rao for their critical review of the manuscript; and to Dr. R.K. Sharma for his help in correcting the page-proofs. I gratefully acknowledge the encouraging attitudes of Dr. A.P.J. Abdul Kalam, Director, DRDL, Hyderabad and Dr. V. Adimurty, Head, Aerodynamics Division, VSSC, whose comments improved the first draft of this book. My thanks are also due to Professor R.S. Nanda and Professor S.K. Rao of IIT Kharagpur for the interest they have shown in this book.

Finally, I wish to express my sincere appreciation to Mr. Asoke K. Ghosh, Chairman and Managing Director, Prentice-Hall of India, for his efforts and cooperation in bringing out this volume.

Any suggestions for further improvement in the contents of the book will be gratefully accepted.

S.S. SASTRY

1

Higher Algebra

1.1 PARTIAL FRACTIONS

We frequently come across expressions of the form $\frac{p}{q}$ where p and q are polynomials in x. Such expressions are called *rational fractions.* If p is of lower degree than q, then $\frac{p}{q}$ is called a *proper fraction*; otherwise, it is an *improper fraction.* By division, we can express an improper fraction in terms of an integral function and a proper fraction. Thus, for example,

$$\frac{2x^2 - x + 4}{x - 1} = 2x + 1 + \frac{5}{x - 1}.$$

The theory of partial fractions rests on an important property of polynomials which is given below: "If q and r are two polynomials with no common factor, there exist polynomials x and y such that $qx + ry = 1$, where x and y have no common factor".

To show this, let us suppose that the degree of q is equal to or higher than that of r. Dividing q by r, we obtain

$$q = ry_1 + x_1,$$

where y_1 is the quotient and x_1 is the remainder. It follows that

$$x_1 = q - ry_1.$$

The degree of r is greater than that of x_1. Dividing r by x_1, we get

$$r = y_2 x_1 + x_2,$$

so that

$$x_2 = r - y_2 x_1 = r - y_2(q - ry_1) = (1 + y_1 y_2)r - y_2 q,$$

on using the expression for x_1 obtained above. Proceeding in a similar way, we obtain

$$\begin{aligned} x_3 = x_1 - y_3 x_2 &= q - ry_1 - y_3[(1 + y_1 y_2)r - y_2 q] \\ &= (1 + y_3 y_2)q - r\,[y_1 + y_3(1 + y_1 y_2)], \end{aligned}$$

and so on. It is clear that each member of the sequence $x_1, x_2, x_3,...$ is of the form $qx' + ry'$, where x' and y' are polynomials. Now, since q and r have no common factor, the remainder obtained after a certain stage (say x_n) must be a constant, say k. Hence, we must have

$$k = qx' + ry',$$

where x' and y' are polynomials. Again, since q and r have no common factor, it follows that x' and y' also have no common factor. Dividing throughout by k, we obtain

$$q\frac{x'}{k} + r\frac{y'}{k} = 1 \quad \text{or} \quad qx + ry = 1,$$

where x and y are polynomials, which have no common factor.

Using this property, it is possible to express a proper function $\frac{p}{qr}$ (where q and r are polynomials which have no common factor) as the sum of two proper fractions with q and r as its denominators. Since q and r have no common factor, there exist x and y such that

$$qx + ry = 1.$$

Hence

$$\frac{p}{qr} = \frac{p(qx + ry)}{qr} = \frac{p}{r}x + \frac{p}{q}y.$$

If $\frac{px}{r}$ and $\frac{py}{q}$ are improper fractions, let

$$\frac{px}{r} = x' + \frac{l}{r} \quad \text{and} \quad \frac{py}{q} = y' + \frac{n}{q}$$

where x' and y' are polynomials and $\frac{l}{r}$ and $\frac{n}{q}$ are proper fractions. Hence,

$$\frac{p}{qr} = x' + y' + \frac{l}{r} + \frac{n}{q} = x' + y' + \frac{lq + nr}{qr},$$

where the fraction on the right-hand side is a proper fractions as $lq + nr$ is of degree lower than that of qr. It easily follows that $x' + y' = 0$ and that

$$\frac{p}{qr} = \frac{l}{r} + \frac{n}{q}.$$

In general, if we have polynomials $qrst\ldots$ which do not have a common factor, then it is possible to find $l, m, n\ldots$ such that

$$\frac{p}{qrs\ldots} = \frac{l}{q} + \frac{m}{r} + \frac{n}{s} + \cdots$$

where $l, m, n, \ldots$ are polynomials of *lower* degree than $q, r, s, t, \ldots$ In such a case, we say that the proper fraction $\frac{p}{qrst\ldots}$ is resolved into partial fractions. This resolution into partial fractions, which finds important applications in integration and the theory of Laplace transforms, is done according to certain specified rules which are described below.

In what follows, we consider the proper fraction $\frac{p}{q}$, where p and q are polynomials in x. In order to resolve $\frac{p}{q}$ into partial fractions, we first factorize q, and the following cases arise depending upon the nature of the factors of q.

1.1.1 Linear Factors

Let $q=(ax+b)r$, where $(ax+b)$ and r do not have a common factor. We can now express $\frac{p}{q}$ as

$$\frac{p}{q}=\frac{l}{ax+b}+\frac{x}{r},$$

where since the denominator $(ax+b)$ is linear and the degree of l is lower than that of $(ax+b)$, l is a constant. It follows that to every non-repeated linear factor, there corresponds a fraction of the form $\frac{l}{ax+b}$, where l is a constant.

Example 1.1 Resolve into partial fractions

$$\frac{3x+7}{x^2-3x+2}.$$

Since $x^2-3x+2=(x-1)(x-2)$, corresponding to each linear factor we have a fraction with a constant in the numerator. We therefore write

$$\frac{3x+7}{x^2-3x+2}=\frac{A}{x-2}+\frac{B}{x-1}$$

where A and B are constants. Clearing off the fractions, we obtain

$$3x+7=A(x-1)+B(x-2). \qquad \text{(i)}$$

Now, equating the coefficients of like powers of x on both sides, we get

$$A+B=3 \quad \text{and} \quad -A-2B=7, \qquad \text{(ii)}$$

whose solution is $A=13$ and $B=-10$.

Alternatively, we may proceed, more simply. By putting in succession $x=1$ and $x=2$ in Eq. (i), we obtain $B=-10$ and $A=13$, as before.

Example 1.2 Resolve into partial fractions

$$\frac{x^2+x-3}{(x-1)(x-2)(x+1)}$$

Let

$$\frac{x^2+x-3}{(x-1)(x-2)(x+1)}=\frac{A}{x-1}+\frac{B}{x-2}+\frac{C}{x+1}$$

Clearing off the fractions, we obtain

$$A(x-2)(x+1)+B(x-1)(x+1)+C(x-1)(x-2)=x^2+x-3 \qquad \text{(i)}$$

Setting $x=1$ yields

$$-2A=-1 \quad \text{or} \quad A=\frac{1}{2}$$

setting $x=2$ yields

$$3B=3 \quad \text{or} \quad B=1$$

and setting $x=-1$ yields

$$6C=-3 \quad \text{or} \quad C=-\frac{1}{2}$$

1.1.2 Repeated Linear Factors

We take $q=(ax+b)^n r$, where r and $(ax+b)^n$ do not have a common factor, $\frac{p}{q}$ is now expressed as

$$\frac{p}{q}=\frac{y}{(ax+b)^n}+\frac{z}{r},$$

where y and z are polynomials of degree lower than those of $(ax+b)^n$ and r, respectively. The degree of y is atmost $(n-1)$ and hence y can be expressed as

$$y=\frac{k_0+k_1(ax+b)+k_2(ax+b)^2+\cdots+k_{n-1}(ax+b)^{n-1}}{(ax+b)^n}+\frac{z}{r}$$

$$=\frac{k_{n-1}}{ax+b}+\frac{k_{n-2}}{(ax+b)^2}+\cdots+\frac{k_0}{(ax+b)^n}+\frac{z}{r},$$

which means that corresponding to a linear factor, say $(ax+b)$, repeated n times, there exists a group of n terms given by

$$\frac{k_0}{(ax+b)^n}+\frac{k_1}{(ax+b)^{n-1}}+\cdots+\frac{k_{n-1}}{ax+b}.$$

Example 1.3 Resolve into partial fractions

$$\frac{x+1}{(x-1)^2(x-2)}.$$

Corresponding to $(x-2)$ we write $\frac{k}{x-2}$ and corresponding to $(x-1)^2$, we take

$$\frac{k_0}{x-1}+\frac{k_1}{(x-1)^2}.$$

Hence we write

$$\frac{x+1}{(x-1)^2(x-2)} = \frac{k}{x-2} + \frac{k_0}{x-1} + \frac{k_1}{(x-1)^2},$$

where k, k_0 and k_1 are constants. Clearing off fractions, we obtain

$$x+1 = k(x-1)^2 + k_0(x-2)(x-1) + k_1(x-2). \tag{i}$$

Setting $x=2$, we immediately obtain, $k=3$. Hence, (i) becomes:

$$x+1 = 3(x-1)^2 + k_0(x-2)(x-1) + k_1(x-2). \tag{ii}$$

Equating the coefficients of like powers of x on both sides, we get

$$3+k_0 = 0 \quad \text{or} \quad k_0 = -3,$$

$$-6-3k_0+k_1 = 1 \quad \text{or} \quad k_1 = -2.$$

The student should check that these values satisfy the third equation obtained by equating the constant terms of both sides of (ii). We therefore have

$$\frac{x+1}{(x-2)(x-1)^2} = \frac{3}{x-2} - \frac{3}{x-1} - \frac{2}{(x-1)^2}.$$

Example 1.4 Resolve into partial fractions

$$\frac{x^3-5x^2+8}{(x-2)^4}$$

We write

$$\frac{x^3-5x^2+8}{(x-2)^4} = \frac{A}{x-2} + \frac{B}{(x-2)^2} + \frac{C}{(x-2)^3} + \frac{D}{(x-2)^4}.$$

Proceeding as in Example 1.3, we can determine A, B, C and D. This example can be solved, more simply, in the following way:

Put $x-2=y$ so that $x=y+2$. Hence

$$\frac{x^3-5x^2+8}{(x-2)^4} = \frac{(y+2)^3-5(y+2)^2+8}{y^4}$$

$$= \frac{y^3+y^2-8y-4}{y^4}$$

$$= \frac{1}{y} + \frac{1}{y^2} - \frac{8}{y^3} - \frac{4}{y^4}$$

$$= \frac{1}{x-2} + \frac{1}{(x-2)^2} - \frac{8}{(x-2)^3} - \frac{4}{(x-2)^4}.$$

1.1.3 Nonlinear Factors

Let $q = (ax^2 + bx + c)r$, where $(ax^2 + bx + c)$ and r do not have a common factor. Then we write

$$\frac{p}{q} = \frac{y}{ax^2 + bx + c} + \frac{z}{r},$$

where y and z are polynomials of lower degree than that of $(ax^2 + bx + c)$ and r respectively. It follows that y should be a linear expression of the form $k_0 x + k_1$, where k_0 and k_1 are constants.

Example 1.5 Resolve into partial fractions

$$\frac{x^3 + 2x^2 + 6x + 3}{(x^2 + x + 1)(x^2 - x + 1)}.$$

In this case, we write

$$\frac{x^3 + 2x^2 + 6x + 3}{(x^2 + x + 1)(x^2 - x + 1)} = \frac{k_0 x + k_1}{x^2 + x + 1} + \frac{k_2 x + k_3}{x^2 - x + 1}.$$

Simplifying this, we obtain:

$$x^3 + 2x^2 + 6x + 3 = (k_0 x + k_1)(x^2 - x + 1) + (k_2 x + k_3)(x^2 + x + 1).$$

Equating the coefficients of like powers of x on both sides, we get:

$$k_0 + k_2 = 1, \qquad -k_0 + k_1 + k_2 + k_3 = 2$$

$$k_0 - k_1 + k_2 + k_3 = 6, \qquad k_1 + k_3 = 3.$$

The solution of these equations yields

$$k_0 = 1, \quad k_1 = -1, \quad k_2 = 0, \quad k_3 = 4,$$

so that we have,

$$\frac{x^3 + 2x^2 + 6x + 3}{(x^2 + x + 1)(x^2 - x + 1)} = \frac{x - 1}{x^2 + x + 1} + \frac{4}{x^2 - x + 1}.$$

Example 1.6 Resolve into partial fractions

$$\frac{3x^2 + x - 2}{(x - 1)(x^2 + 1)}.$$

We write

$$\frac{3x^2 + x - 2}{(x - 1)(x^2 + 1)} = \frac{k_0}{x - 1} + \frac{k_1 x + k_2}{x^2 + 1}.$$

Hence

$$k_0(x^2+1)+(x-1)(k_1x+k_2)=3x^2+x-2 \qquad \text{(i)}$$

Setting $x=1$, we get $k_0=1$. Then, Eq. (i) becomes

$$x^2+1+(x-1)(k_1x+k_2)=3x^2+x-2$$

or

$$(k_1x+k_2)(x-1)=2x^2+x-3.$$

Equating the coefficients of like powers of x of both sides, we obtain

$$k_1=2, \qquad -k_1+k_2=1 \quad \text{or} \quad k_2=3.$$

Hence

$$\frac{3x^2+x-2}{(x-1)(x^2+1)}=\frac{1}{x-1}+\frac{2x+3}{x^2+1}.$$

1.1.4 Repeated Nonlinear Factors

Let $q=(ax^2+bx+c)^n r$, so that

$$\frac{p}{q}=\frac{y}{(ax^2+bx+c)^n}+\frac{z}{r}.$$

Since the degree of y is utmost $(2n-1)$, we write

$$y=(l_1x+m_1)(ax^2+bx+c)^{n-1}+(l_2x+m_2)(ax^2+bx+c)^{n-2}+\cdots+(l_nx+m_n).$$

Hence

$$\frac{p}{q}=\frac{l_nx+m_n}{(ax^2+bx+c)^n}+\frac{l_{n-1}x+m_{n-1}}{(ax^2+bx+c)^{n-1}}+\cdots+\frac{l_1x+n_1}{ax^2+bx+c}+\frac{z}{r}.$$

The evaluation of the coefficients can then be effected as in the previous cases. In Sections 1.1.3 and 1.1.4, we have considered quadratic factors only. However, the procedure for cubic and higher degree factors is similar and is therefore omitted here.

Example 1.7 Resolve into partial fractions

$$\frac{x^4+4x^3+7}{(x-1)(x^2+1)^2}.$$

We write

$$\frac{x^4+4x^3+7}{(x-1)(x^2+1)^2}=\frac{k}{x-1}+\frac{k_0x+k_1}{x^2+1}+\frac{k_2x+k_3}{(x^2+1)^2}$$

or

$$x^4 + 4x^3 + 7 = k(x^2+1)^2 + (k_0 x + k_1)(x^2+1)(x-1) + (k_2 x + k_3)(x-1).$$

Setting $x = 1$, we immediately obtain $k = 3$. Hence, the above equation becomes:

$$x^4 + 4x^3 + 7 = 3(x^2+1)^2 + (k_0 x + k_1)(x^2+1)(x-1) + (k_2 x + k_3)(x-1).$$

We easily obtain $k_0 = -2$, $k_1 = 2$, $k_2 = -2$ and $k_3 = -6$ so that

$$\frac{x^4 + 4x^3 + 7}{(x-1)(x^2+1)^2} = \frac{3}{x-1} - \frac{2x-2}{x^2+1} - \frac{2x+6}{(x^2+1)^2}.$$

EXERCISES 1.1

Resolve into partial fractions

1. $\dfrac{4x-11}{x^2-5x+6}$

2. $\dfrac{1}{x^2+x^4}$

3. $\dfrac{3x+7}{x^2-3x+2}$

4. $\dfrac{x^2}{(1-x)^3}$

5. $\dfrac{x^2+x+1}{x^2+2x+1}$

6. $\dfrac{1}{(x^2+1)(x^2+4)}$

7. $\dfrac{5x+1}{x^3-3x+2}$

8. $\dfrac{4x^2-5x+3}{x(x-1)^2}$

9. $\dfrac{3(1+x)}{x^3-1}$

10. $\dfrac{1}{(x-2)(x^2+1)}$

11. $\dfrac{x^2-2}{x^4+x^2}$

12. $\dfrac{x^4+4x^3+7}{x^5-x^4+2x^3-2x^2+x-1}$

13. $\dfrac{x+4}{(x^2-4)(x+1)}$

14. $\dfrac{3x^3-8x^2+10}{(x-1)^4}$

15. $\dfrac{x^2}{(x+1)^2(x^2+1)}$

16. $\dfrac{x+1}{x(x^2+4x+8)}$

17. $\dfrac{2x^5}{(x^2-1)(x^2-4)}$

18. $\dfrac{a}{(x^2+p)(x^2+q)}$

19. $\dfrac{9}{(x-1)(x+2)^2}$

20. $\dfrac{4x^2-2x-8}{x^3-4x^2}$

21. $\dfrac{13-7x}{(x+2)(x-1)^3}$

22. $\dfrac{x^2+2}{x^2(x+1)(x+2)}$

23. $\dfrac{7x^2+4x+9}{(2x+3)(x^2+4)}$ **24.** $\dfrac{x^2-3x+4}{x^3-x^2+x-1}$

25. $\dfrac{3x^4+2x^3+8x^2+x+2}{x^5+2x^3+x}$ **26.** $\dfrac{x^5}{(x-1)^3}$.

27. Find k_1 and k_2 so that $k_1(x+2)^3+k_2(x+3)^2=1$. Hence express $\frac{1}{(x+2)^3(x+3)^2}$ in partial fractions

28. If

$$\frac{1}{(1-ax)(1-bx)}=\frac{k_1}{1-ax}+\frac{k_2}{1-bx},$$

then prove that

$$\frac{1}{(1-ax)^2(1-bx)}=\frac{k_1}{(1-ax)^2}+\frac{k_1k_2}{1-ax}+\frac{k_2^2}{1-bx}.$$

1.2 INFINITE SERIES

An ordered set of numbers

$$a_1, a_2, a_3, a_4, \ldots, a_n, \tag{1.1}$$

is called a *sequence* and is usually denoted by $\{a_n\}$. For example, the set of numbers

$$1, 3, 5, 7, \ldots, 2n-1$$

is a sequence represented by the symbol $\{2n-1\}$. In this case, the number of terms is unlimited and so the sequence is called an *infinite* sequence; otherwise, it is a *finite* sequence.

From the sequence given in (1.1), we can form an expression

$$a_1+a_2+a_3+\cdots+a_n+\cdots \tag{1.2}$$

This consists of an infinite number of terms and is called an *infinite series.* The study of infinite series constitutes an important branch of mathematics, since almost every function that is encountered in applications of mathematics can be represented as an infinite series. If the series in (1.2) is terminated after n terms, we get

$$a_1+a_2+a_3+\cdots+a_n, \tag{1.3}$$

and this is a finite series, a series of n terms. It is easy to sum up these n terms; let us denote it by S_n. Now, what is the sum of the infinite series in (1.2)? Clearly, this is related to the sum, S_n, of the first n terms. In what way? To see this, we consider the infinite geometric series given by

$$1+\frac{1}{2}+\frac{1}{4}+\frac{1}{8}+\frac{1}{16}+\cdots \tag{1.4}$$

Let S_n represents the sum to n terms of this series. Cut off a line OA (see Fig. 1.1) of length 2 units, and let B be its midpoint. Then $OB = OA = 1$ unit and the first term of the series in (1.4) is represented by the length OB. If C is the midpoint of BA, then $BC = CA = 1/2$. Then the sum of the first-two terms is represented by the length OC. Similarly, the length OD, where D is the midpoint of CA, represents the sum of the first-three terms of the geometric series.

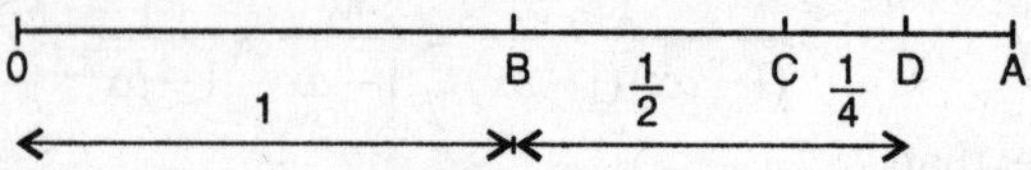

Figure 1.1

In this way, addition of a new term of the sum S_n leads us to a further subdivision of the right-hand portion of the straight line and it is clear that by taking sufficiently large number of terms, we can approach *as close* to A as we desire. We thus obtain

$$S_1 = 1.0, \qquad S_2 = 1.5, \qquad S_4 = 1.875, \qquad S_8 = 1.992, \ldots$$

It is easy to see that the sum S_n becomes and remains arbitrarily close to 2, if n is large enough. We say that S_n tends to 2 as its *limit* as n approaches infinity. Symbolically, this is written as

$$\lim_{n\to\infty} S_n = 2, \tag{1.5}$$

and the series in (1.4) is said to be *convergent.* In fact, as $n \to \infty$, there are four distinct possibilities for S_n:

(i) S_n may tend to a finite limit (say S),
(ii) S_n may tend to plus-infinity,
(iii) S_n may tend to minus-infinity, and
(iv) S_n may tend to more than one limit, i.e. *oscillate.*

If S_n tends to a finite limit S, the infinite series is said to be *convergent* and S is called its *sum to infinity.* Thus, the sum to infinity is not a sum in the ordinary sense, but it is 'the limit of a sum'. For the infinite geometric series, for example,

$$S_n = \frac{1-(1/2^n)}{1-1/2} = 2 - \frac{1}{2^{n-1}}.$$

Now, as $n \to \infty$,

$$\frac{1}{2^{n-1}} \to 0.$$

Hence, $S_n \to 2$ as $n \to \infty$.

If S_n tends to infinity or minus-infinity, the series is said to be *divergent.* For example, the series

$$1+2+3+\cdots+n+\cdots$$

represented by $\Sigma_{n=1}^{\infty} u_r$, where $u_r = r$, is divergent, for in this case,

$$S_n = 1+2+3+\cdots+n = \frac{n(n+1)}{2},$$

which tends to infinity as $n \to \infty$. Hence, the series $\Sigma_1^{\infty} r$ is divergent.

Sometimes, S_n tends to more than one limit. For example, consider the series

$$1-1+1-1+\cdots,$$

represented by

$$\sum_1^{\infty}(-1)^{r+1}$$

Here

$$S_n = 1-1+1-1+\ldots n \text{ terms.}$$

Clearly

$$\begin{aligned} S_n &= +1 \quad \text{if } n \text{ is odd} \\ &= 0 \quad \text{if } n \text{ is even.} \end{aligned}$$

Thus, the value of S_n *oscillates* between 0 and 1 and such a series is said to be an *oscillating series.*

Example 1.8 Discuss the convergence of the series

$$1+x+x^2+\cdots+x^{n-1}+\cdots$$

Here

$$S_n = \frac{1-x^n}{1-x}, \quad \text{if} \quad x \neq 1.$$

If $-1 < x < 1$, then $x^n \to 0$, and hence

$$S_n \to \frac{1}{1-x} \quad \text{as } n \to \infty,$$

and the series is convergent.

If $x = 1$, the series becomes

$$1+1+1+1+\ldots,$$

which is obviously divergent.

If $x > 1$,

$$S_n = \frac{x^n - 1}{x-1},$$

which tends to infinity as $n \to \infty$ and is therefore divergent.

If $x=-1$,

$$S_n = 1 \quad \text{if } n \text{ is odd}$$
$$= 0 \quad \text{if } n \text{ is even.}$$

Hence, for $x=-1$; the series oscillates.

For $x<-1$, $S_n \to \infty$, or $-\infty$, according as n is odd or even. We thus conclude that the geometric series.

$$\sum_{n=0}^{\infty} x^n$$

converges if $-1<x<1$, diverges if $x \geq 1$ and oscillates if $x=-1$ and $x<-1$.

The meaning of the statement

$$\lim_{n\to\infty} S_n = S \tag{1.6}$$

can be made more precise. This will be done by referring once again to the infinite geometric series given in (1.4). We have already found that its sum to infinity is 2. The difference between the limiting value and the sum to n terms is given by

$$2 - S_n = 2 - \frac{1-(1/2^n)}{1-(1/2)} = \frac{1}{2^{n-1}},$$

and this becomes less and less as n increases. Suppose that we want this difference to be less than a very small positive number, say 1/1000. Then we require

$$\frac{1}{2^{n-1}} < \frac{1}{1000}$$

or

$$2^{n-1} > 1000$$

whence

$$n > 10.$$

Therefore, if we take $n>10$, i.e. more than 10 terms of the series, the condition that the difference between the limiting value and the sum to n terms can be made less than 1/1000, is satisfied. Usually, the numerical difference or absolute value is taken and this is indicated by the symbol $|2-S_n|$. In case of the geometric series, this difference is made smaller than a pre-assigned small positive number by choosing n sufficiently large. Thus, more generally, the precise meaning of the statement in (1.6) is that for any pre-assigned positive number, say $\in$, however small, we can find a number N such that

$$|S - S_n| < \varepsilon \text{ for all } n > N. \tag{1.7}$$

We have thus seen that if the sum to n terms of an infinite series can be expressed by means of elementary functions, then the nature of the infinite series can be determined by finding the limit of S_n. But, in many situations, it may not be possible to find S_n and therefore we require methods of determining whether as infinite series is convergent or divergent. Some important tests of convergence are described below.

1.2.1 Comparison Tests

We start by stating, without proof, some general theorems concerning infinite series.

Theorem 1.1 If

$$u_1 + u_2 + u_3 + \cdots + u_r + \cdots$$

is convergent and has the sum S, then

$$u_{m+1} + u_{m+2} + \cdots$$

is convergent and has the sum

$$S - (u_1 + u_2 + \cdots + u_m)$$

where m is a positive integer.

Theorem 1.2 If

$$u_1 + u_2 + u_3 + \cdots + u_n + \cdots$$

is convergent and has the sum S, then

$$ku_1 + ku_2 + ku_3 + \cdots + ku_n + \cdots$$

is convergent and has the sum kS.

Theorem 1.3 If the two series

$$u_1 + u_2 + u_3 + \cdots + u_n + \cdots$$

and

$$v_1 + v_2 + v_3 + \cdots + v_n + \cdots$$

are convergent, then the series $\Sigma_{n=1}^{\infty} (u_n + v_n)$ is convergent and its sum is the sum of the two series.

Theorem 1.4 If

$$u_1 + u_2 + \cdots + u_n + \cdots$$

is convergent, then $\lim_{n\to\infty} u_n = 0$.

As an example, we consider the infinite geometric series given in (1.4). We have already seen that this series is convergent. Further,

$$u_n = \frac{1}{2^{n-1}},$$

and this tends to zero, as $n \to \infty$. The converse, however, is not true, i.e. when $u_n \to 0$ as $n \to \infty$, we cannot say definitely that the series is convergent. To show this, we consider the harmonic series

$$1 + \frac{1}{2} + \frac{1}{3} + \frac{1}{4} + \cdots + \frac{1}{n} + \cdots$$

Here $u_n = 1/n \to 0$ as $n \to \infty$. We now regroup the series in terms of 2, 4, 8, 16, ... terms:

$$1 + \frac{1}{2} + \left(\frac{1}{3} + \frac{1}{4}\right) + \left(\frac{1}{5} + \frac{1}{6} + \frac{1}{7} + \frac{1}{8}\right) + \cdots$$

Then the original series is

$$> 1 + \frac{1}{2} + 2\left(\frac{1}{4}\right) + 4\left(\frac{1}{8}\right) + \cdots$$

$$> \frac{1}{2} + \frac{1}{2} + \frac{1}{2} + \frac{1}{2} + \cdots,$$

which is a divergent series. Hence the original series is also divergent. This example illustrates the idea of comparison of two series which is described below.

Theorem 1.5 (*Comparison test*) Let $\Sigma_{n=1}^{\infty} u_n$ and $\Sigma_{n=1}^{\infty} v_n$ be two series of positive terms and let $\Sigma_1^{\infty} v_n$ be convergent. If $u_n \le v_n$ for all n, then the series $\Sigma_1^{\infty} u_n$ is also convergent.

Since the second series is convergent, we have

$$v_1 + v_2 + \cdots + v_n + \cdots = S, \quad \text{(say)}.$$

Then

$$v_1 + v_2 + \cdots + v_n \le S$$

and so

$$u_1 + u_2 + \cdots + u_n \le S$$

which means that $\Sigma_1^{\infty} u_n$ is convergent.

Conversely, if $\Sigma_1^{\infty} v_n$ is divergent and $u_n \ge v_n$, then $\Sigma_1^{\infty} u_n$ is also divergent.

The following examples illustrate the use of this test.

Example 1.9 Show that the series

$$1 + \frac{1}{1!} + \frac{1}{2!} + \frac{1}{3!} + \cdots$$

is convergent.

Here,

$$u_n = \frac{1}{(n-1)!} = \frac{1}{1\cdot 2\cdot 3\ldots(n-1)} < \frac{1}{1\cdot 2\cdot 2\ldots(n-2)\text{ times}} = \frac{1}{2^{n-2}}.$$

But $\Sigma_1^\infty 1/2^{n-2}$ is the infinite geometric series, which is known to be convergent. Hence, the given series is also convergent.

Example 1.10 Prove that the series

$$\frac{1}{1\cdot 3}+\frac{2}{3\cdot 5}+\frac{3}{5\cdot 7}+\frac{4}{7\cdot 9}+\cdots$$

is divergent

Here,

$$u_n = \frac{n}{(2n-1)(2n+1)} > \frac{n}{2n\cdot 3n} = \frac{1}{6n}.$$

If we choose $v_n = 1/n$, then $u_n > v_n/6$ and $\Sigma_1^\infty v_n$ is divergent. Hence, $\Sigma_1^\infty u_n$ is divergent.

A useful particular case of the comparison test is now given.

Theorem 1.6 If $\Sigma_1^\infty v_n$ is convergent and $\frac{u_n}{v_n}$ tends to a limit other than zero, as n tends to infinity, then $\Sigma_1^\infty u_n$ is convergent. Again, if $\Sigma_1^\infty v_n$ is divergent and $\frac{u_n}{v_n}$ tends to a limit other than zero as $n \to \infty$, then $\Sigma_1^\infty u_n$ is divergent.

Let

$$\lim_{n\to\infty} \frac{u_n}{v_n} = k \quad (k \text{ is positive and } \neq 0).$$

Therefore, after a certain value of n, say m, the ratios

$$\frac{u_n}{v_n}, \frac{u_{n+1}}{v_{n+1}}, \frac{u_{n+2}}{v_{n+2}}, \ldots$$

lie in the interval $(k-\varepsilon)$ to $(k+\varepsilon)$, where ε is a very small finite positive quantity. Thus

$$u_n < v_n(k+\varepsilon), \qquad n \geq m.$$

Since $\Sigma_1^\infty v_n$ is convergent, it follows that $\Sigma_1^\infty u_n$ is also convergent.

The proof of the second part is left as an exercise to the student.

Example 1.11 We consider the series

$$\frac{1}{1+x}+\frac{1}{2+x}+\frac{1}{3+x}+\cdots+\frac{1}{n+x}+\cdots$$

Here

$$u_n = \frac{1}{n+x}$$

We choose

$$v_n = \frac{1}{n}$$

Now

$$\frac{u_n}{v_n} = \frac{n}{n+x} = \frac{1}{1+x/n},$$

and $\frac{1}{1+(x/n)}$ varies from $\frac{1}{1+x}$ to 1 as n changes from 1 to ∞, and is therefore finite. Since $\Sigma_1^\infty v_n$ is divergent, it follows that $\Sigma_1^\infty u_n$ is also divergent.

Example 1.12 Test the convergence of the series

$$\frac{1}{1\cdot 2} + \frac{1}{3\cdot 4} + \frac{1}{5\cdot 6} + \cdots + \frac{1}{(2n-1)\,2n} + \cdots$$

Here

$$u_n = \frac{1}{2n(2n-1)}$$

Let $v_n = 1/n^2$ so that $\Sigma_1^\infty v_n$ is convergent. Now

$$\frac{u_n}{v_n} = \frac{n^2}{2n(2n-1)} = \frac{1}{2\left(2-\frac{1}{n}\right)} \to \frac{1}{4}, \quad \text{as } n \to \infty$$

Hence $\Sigma_1^\infty u_n$ is convergent.

Example 1.13 Test the convergence of the series

$$u_n = \sqrt{n^2+1} - n.$$

We have

$$u_n = \sqrt{n^2+1} - n = \frac{(\sqrt{n^2+1}-n)(\sqrt{n^2+1}+n)}{\sqrt{n^2+1}+n} = \frac{1}{\sqrt{n^2+1}+n}.$$

We choose $v_n = \frac{1}{n}$. Hence

$$\frac{u_n}{v_n} = \frac{n}{\sqrt{n^2+1}+n} = \frac{1}{\sqrt{1+1/n^2}+1} \to \frac{1}{2}, \quad \text{as } n \to \infty.$$

Since $\Sigma_1^\infty v_n$ is divergent, it follows that $\Sigma_1^\infty u_n$ is also divergent.

1.2.2 D'Alembert's Ratio Test

This test is particularly useful for power series and is one of the best known tests for convergence.

Here the ratio used is $\frac{u_{n+1}}{u_n}$, which is the ratio of any term to the preceeding. Suppose that for $n = r, r+1, \ldots$ each ratio is less than a fixed quantity which is less than unity, i.e. let

$$\frac{u_{r+1}}{u_r} < k, \quad \frac{u_{r+2}}{u_{r+1}} < k, \ldots$$

It follows that

$$\begin{aligned} u_{r+1} &< ku_r, \\ u_{r+2} &< ku_{r+1} < k^2 u_r, \\ u_{r+3} &< ku_{r+2} < k^3 u_r, \\ &\vdots \end{aligned}$$

Hence

$$\begin{aligned} u_r + u_{r+1} + u_{r+2} + \cdots &< u_r + ku_r + k^2 u_r + \cdots \\ &= (1 + k + k^2 + \cdots) u_r \\ &= \frac{u_r}{1-k}, \quad \text{since } k < 1. \end{aligned}$$

Hence, the sum of the terms beginning at the *r*th term is finite which shows that the series $\Sigma_1^\infty u_n$ is convergent, as the nature of a series does not change if we omit a finite number of terms in the beginning. On the other hand, if each ratio is greater than *k* which is equal to or greater than unity, it can be shown that the series is divergent. The results we have just now obtained are contained in the following theorems:

Theorem 1.7 If, after a particular term of an infinite series, the ratio of each term to the preceding is less than a fixed quantity (which is itself less than unity), then the series is convergent.

For applications, however, it may be easier to use the test in the form given below.

Theorem 1.8 If

$$\lim_{n\to\infty} \frac{u_{n+1}}{u_n} = k < 1,$$

then $\Sigma_1^\infty u_n$ is convergent.

Theorem 1.9 If, after a particular term of an infinite series, the ratio of each term to the preceding is either equal to or greater than unity, then the series is divergent. Alternatively, we have

Theorem 1.10 If

$$\lim_{n\to\infty} \frac{u_{n+1}}{u_n} = k > 1,$$

then $\Sigma_1^\infty u_n$ is divergent.

Note: If

$$\lim_{n\to\infty}\frac{u_{n+1}}{u_n}=1,$$

the series may be convergent or divergent and we say that the test fails. In this case, therefore, one has to use the comparison test or judge the nature of the series from elementary principles.

The following examples illustrate the use of the ratio test.

Example 1.14 Test for convergence the series

$$\frac{1}{1\cdot 2}+\frac{1}{2\cdot 3}+\frac{1}{3\cdot 4}+\cdots$$

Here

$$u_n=\frac{1}{n(n+1)} \quad \text{and} \quad u_{n+1}=\frac{1}{(n+1)(n+2)}$$

Then

$$\frac{u_{n+1}}{u_n}=\frac{n}{n+2}=\frac{1}{1+2/n}\to 1, \quad \text{as } n\to\infty.$$

Hence the ratio test fails. To use the comparison test, we take $v_n=\frac{1}{n^2}$. Then

$$\frac{u_n}{v_n}=\frac{n^2}{n(n+1)}=\frac{1}{1+(1/n)}\to 1, \quad \text{as } n\to\infty.$$

Since $\Sigma_1^\infty v_n$ is convergent, it follows, by Theorem 1.6, that $\Sigma_1^\infty u_n$ is convergent.

Example 1.15 Examine the convergence of the series

$$1+\frac{1}{1!}+\frac{1}{2!}+\frac{1}{3!}+\cdots+\frac{1}{(n-1)!}+\cdots$$

Here

$$u_n=\frac{1}{(n-1)!}, \quad u_{n+1}=\frac{1}{n!}$$

Then

$$\frac{u_{n+1}}{u_n}=\frac{(n-1)!}{n}=\frac{1}{n}\to 0, \quad \text{as } n\to\infty$$

or

$$\lim_{n\to\infty} \frac{u_{n+1}}{u_n} < 1.$$

Hence, $\Sigma_1^\infty u_n$ is convergent.

Example 1.16 Discuss the convergence of the series

$$\frac{x}{1+x} + \frac{x^2}{1+x^2} + \frac{x^3}{1+x^3} + \cdots + \frac{x^n}{1+x^n} + \cdots$$

Here

$$u_n = \frac{x^n}{1+x^n}, \quad u_{n+1} = \frac{x^{n+1}}{1+x^{n+1}},$$

and

$$\frac{u_{n+1}}{u_n} = \frac{1+x^n}{1+x^{n+1}} x.$$

If $x < 1$, $x^n \to 0$ as $n \to \infty$, we get

$$\lim_{n\to\infty} \frac{u_{n+1}}{u_n} = x < 1.$$

Hence, $\Sigma_1^\infty u_n$ is convergent if $x < 1$.

If $x > 1$,

$$\lim_{n\to\infty} u_n = \lim_{n\to\infty} \frac{x^n}{1+x^n} = \lim_{n\to\infty} \frac{1}{1+1/x^n} = 1$$

Hence $\Sigma_1^\infty u_n$ is divergent for $x > 1$.

Finally, for $x = 1$, the series becomes

$$\frac{1}{2} + \frac{1}{2} + \frac{1}{2} + \cdots,$$

which is divergent. Hence the given series is convergent for $0 < x < 1$.

1.2.3 Cauchy's Root Test

Let $\Sigma_1^\infty u_n$ be a series of positive terms. Then this series is convergent or divergent according as

$$\lim_{n\to\infty} u_n^{1/n} < 1 \quad \text{or} \quad > 1.$$

Let

$$\lim_{n\to\infty} u_n^{1/n} = k < 1.$$

Since $k < 1$, we can choose a sufficiently small positive number, ε, so that $k+\varepsilon < 1$. Again, since $\lim_{n\to\infty} u_n^{1/n} = k$, we can find a large number m such that u_n differs from k by less than ε for all $n \geq m$. That is

$$u_n^{1/n} < k+\varepsilon \quad \text{or} \quad u_n < (k+\varepsilon)^n.$$

Hence the series $\Sigma_{n=m}^{\infty} u_n$ is convergent, since the terms of this series are less than those of the geometric series $(k+\varepsilon)^n$, which is convergent since $k+\varepsilon < 1$.

The proof of case (ii), i.e. when $\lim_{n\to\infty} u_n^{1/n} = k > 1$, is similar and is left as an exercise to the student.

It should be understood that D'Alembert's ratio test is usually easier to apply than Cauchy's root test, but the latter is more powerful.

Example 1.17 Establish the convergence of the series

$$\frac{1}{3} + \left(\frac{2}{5}\right)^2 + \left(\frac{3}{7}\right)^3 + \cdots + \left(\frac{n}{2n+1}\right)^n + \cdots$$

To apply Cauchy's test, we observe that

$$\begin{aligned}\lim_{n\to\infty} (u_n)^{1/n} &= \lim_{n\to\infty} \sqrt[n]{\left(\frac{n}{2n+1}\right)^n} \\ &= \lim_{n\to\infty} \frac{n}{2n+1} \\ &= \lim_{n\to\infty} \frac{1}{2+1/n} \\ &= \frac{1}{2} < 1.\end{aligned}$$

Hence the series converges.

Example 1.18 Discuss the convergence of

$$2\left(\frac{4}{5}\right) + \left(\frac{4}{5}\right)^2 + 2\left(\frac{4}{5}\right)^3 + \cdots + \left(\frac{4}{5}\right)^{2m} + 2\left(\frac{4}{5}\right)^{2m+1} + \cdots$$

If n is even,

$$u_{n-1} = 2\left(\frac{4}{5}\right)^{n-1} \quad \text{and} \quad u_n = \left(\frac{4}{5}\right)^n.$$

Hence

$$\frac{u_n}{u_{n-1}} = \frac{(4/5)^n}{2(4/5)^{n-1}} = \frac{2}{5} \quad \text{and} \quad \sqrt[n]{u_n} = \frac{4}{5}$$

On the other hand, if n is odd, then

$$u_{n-1} = \left(\frac{4}{5}\right)^{n-1} \quad \text{and} \quad u_n = 2\left(\frac{4}{5}\right)^n,$$

Hence

$$\frac{u_n}{u_{n-1}} = \frac{2(4/5)^n}{(4/5)^{n-1}} = \frac{8}{5} \quad \text{and} \quad \sqrt[n]{u_n} = \frac{4}{5}(2)^{1/n}$$

Thus the ratio u_n/u_{n-1} oscillates between 2/5 and 8/5 and the ratio test cannot be applied. On the other hand, since $\lim_{n\to\infty}(2)^{1/n} = 1$, it follows that $\sqrt[n]{u_n}$ tends to the number 4/5 as its limit, as $n \to \infty$. Hence the series is convergent by the root test.

1.2.4 Absolute and Conditional Convergence

The tests for convergence or divergence which we have discussed so far can be applied only to the series whose terms are *all positive* or *all negative*. But these do not work for a series like

$$1 + x + x^2 + x^3 + \cdots + x^n + \cdots$$

which becomes an alternating series when x is negative. For example, a series of the form

$$u_1 - u_2 + u_3 - u_4 + \cdots,$$

where $u_n > 0$ for every n, is an *alternating* series. In order to handle such a series, we introduce the concept of *absolute convergence.*

Absolute convergence. A series

$$\sum_{n=1}^{\infty} u_n = u_1 + u_2 + u_3 + \cdots$$

is said to be absolutely convergent if the series formed by the absolute (or numerical) values of the terms, i.e.

$$\sum_{n=1}^{\infty} |u_n| = |u_1| + |u_2| + \cdots + |u_n| + \cdots,$$

is convergent. For example, the series

$$1 - \frac{1}{2^2} + \frac{1}{3^2} - \frac{1}{4^2} + \cdots,$$

is absolutely convergent since the series

$$1+\frac{1}{2^2}+\frac{1}{3^2}+\frac{1}{4^2}+\cdots$$

is convergent. On the other hand, if a series converges but not absolutely, it is said to be *conditionally convergent.*

The following theorem demonstrates the importance of absolute convergence.

Theorem 1.11 If a series converges absolutely, then it converges *necessarily,* i.e. it converges when the absolute value signs are removed.

Let $\Sigma_1^\infty u_n$ be the given series. Since $\Sigma_1^\infty u_n$ is absolutely convergent, it follows that $\Sigma_1^\infty |u_n|$ is convergent. Now

$$\begin{aligned} u_n+|u_n| &= 2u_n, \quad \text{if } u_n \text{ is positive} \\ &= 0, \quad \text{if } u_n \text{ is negative.} \end{aligned}$$

Hence, every term of the series $\Sigma_1^\infty (u_n+|u_n|)$ is positive and is less than or equal to the corresponding term of the series $\Sigma_1^\infty 2|u_n|$ which is convergent. It follows, therefore, that the series $\Sigma_1^\infty (u_n+|u_n|)$ is convergent. But the series $\Sigma_1^\infty |u_n|$ is convergent. Hence the series $\Sigma_1^\infty u_n$ is also convergent.

(The reader should understand that when we say that the series $\Sigma_1^\infty u_n$ is absolutely convergent, we are asserting the convergence of another series, viz, $\Sigma_1^\infty |u_n|$ and not that of the series $\Sigma_1^\infty u_n$ alone.)

An important consequence of this theorem is that if two series are absolutely convergent, they can be multiplied term by term and the resulting series is also an absolutely convergent series.

In determining the interval of convergence of a power series, we shall apply the ratio test to the series of absolute values. Thus, if

$$\lim_{n\to\infty}\left|\frac{u_{n+1}}{u_n}\right| = k,$$

the series:

(a) converges absolutely if $k<1$
(b) diverges if $k>1$, and
(c) may either converge or diverge if $k=1$.

Example 1.19 Find the interval of convergence of the series

$$1+2x+3x^2+4x^3+\cdots+nx^{n-1}+\cdots$$

We have

$$k=\lim_{n\to\infty}\left|\frac{u_{n+1}}{u_n}\right| = \lim_{n\to\infty}\left|\frac{(n+1)x^n}{nx^{n-1}}\right| = |x|$$

By the ratio test, the series converges absolutely if $|x|<1$ and diverges if $|x|>1$. When x is either $+1$ or -1, the nth term does not approach zero as n becomes infinite. Hence the series is divergent for $x=\pm 1$. The interval of convergence is, therefore, $-1<x<1$.

1.2.5 Alternating Series

Let the alternating series be

$$u_1 - u_2 + u_3 - u_4 + \cdots + (-1)^{n-1} u_n + \cdots$$

where the u_n are all positive, $u_{n+1} < u_n$ for all n, and $\lim_{n\to\infty} u_n = 0$.

If n is an even integer, say $n = 2m$, then the sum of the first n terms is

$$\begin{aligned} S_{2m} &= (u_1 - u_2) + (u_3 - u_4) + \cdots + (u_{2m-1} - u_{2m}) \\ &= u_1 - (u_2 - u_3) - (u_4 - u_5) - \cdots - (u_{2m-2} - u_{2m-1}) - u_{2m}. \end{aligned}$$

In the above, we have expressed S_{2m} in two ways: in the first, S_{2m} is the sum of m non-negative terms and in the second representation, $S_{2m} \le u_1$. Such a sequence $\{S_{2m}\}$ has a limit, say

$$\lim_{m\to\infty} S_{2m} = l. \tag{1.8a}$$

If, on the other hand, we form the sum of the odd number of terms, say $n = 2m+1$, we have

$$S_{2m+1} = u_1 - u_2 + u_3 - \cdots + u_{2m-1} - u_{2m} + u_{2m+1} = S_{2m} + u_{2m+1}$$

Taking limits on both sides, we obtain

$$\lim_{m\to\infty} S_{2m+1} = \lim_{m\to\infty} S_{2m} + \lim_{m\to\infty} u_{2m+1}.$$

By hypotheses, we have

$$\lim_{m\to\infty} u_{2m+1} = 0.$$

Hence, it follows that

$$\lim_{m\to\infty} S_{2m+1} = \lim_{m\to\infty} S_{2m} = l \tag{1.8b}$$

Combining Eqs. (1.8a) and (1.8b), we may write

$$\lim_{n\to\infty} S_n = l,$$

where n may be either even or odd. This shows that any series of this nature must necessarily converge, its sum l lying between $(u_1 - u_2)$ and u_1. The alternating series test may, formally, be stated as follows:

Theorem 1.12 If $u_1 - u_2 + u_3 - u_4 + \cdots$ is a series of terms alternately positive and negative, and if $u_n > u_{n+1}$ for all values of n and if $\lim_{n\to\infty} u_n = 0$, then the series converges.

Example 1.20 The series

$$1-\frac{1}{2}+\frac{1}{3}-\frac{1}{4}+\cdots+\frac{(-1)^n}{n}+\cdots$$

satisfies the three conditions of Theorem 1.12:

(i) it is alternating,

(ii) the terms steadily decrease in numerical value, and

(iii) $\lim_{n\to\infty} u_n = \lim_{n\to\infty} (1/n) = 0.$

The alternating series test, therefore, applies and the series is convergent. We can also conclude that there is a certain number l, where $1/2 \le l \le 1$, such that

$$l = 1-\frac{1}{2}+\frac{1}{3}-\frac{1}{4}+\cdots$$

Since the series of absolute values, i.e. the series

$$1+\frac{1}{2}+\frac{1}{3}+\frac{1}{4}+\cdots$$

is divergent, it follows that the alternating series given above is only conditionally convergent.

Example 1.21 Determine the interval of convergence of the series

$$x+\frac{x^2}{2}+\frac{x^3}{3}+\frac{x^4}{4}+\cdots$$

With

$$u_n = \frac{x^n}{n},$$

we have

$$\left|\frac{u_{n+1}}{u_n}\right| = \left|\frac{x^{n+1}}{n+1}\,\frac{n}{x^n}\right| = \left|\frac{x}{1+1/n}\right|$$

Hence

$$\lim_{n\to\infty}\left|\frac{u_{n+1}}{u_n}\right| = |x|.$$

It therefore follows, by the ratio test, that the series converges absolutely if $|x|<1$ and it diverges if $|x|>1$.

For $x=+1$, the series becomes

$$1+\frac{1}{2}+\frac{1}{3}+\frac{1}{4}+\cdots$$

which diverges to $+\infty$. For $x=-1$, the series becomes

$$-1+\frac{1}{2}-\frac{1}{3}+\frac{1}{4}-\cdots$$

which is convergent by alternating series test. Hence the interval of convergence of the series is given by $-1 \le x < 1$.

EXERCISES 1.2

Discuss the convergence of the following series (Problems 1–15):

1. $\frac{1}{1.2}+\frac{1}{3.4}+\frac{1}{5.6}+\cdots$

2. $\frac{1}{1.3}+\frac{1}{3.5}+\frac{1}{5.7}+\cdots$

3. $\frac{1}{3.7}+\frac{1}{7.11}+\frac{1}{11.15}+\cdots$

4. $\frac{1.2}{3.4.5}+\frac{2.3}{4.5.6}+\frac{3.4}{5.6.7}+\cdots$

5. $1+\frac{1+2}{1+2^2}+\frac{1+3}{1+3^2}+\cdots+\frac{1+n}{1+n^2}+\cdots$

6. $\frac{3}{2}+\left(\frac{3}{2}\right)^2+\left(\frac{3}{2}\right)^3+\cdots$

7. $\sin 1+\sin\frac{1}{2}+\sin\frac{1}{3}+\cdots+\sin\frac{1}{n}+\cdots$

8. $\frac{1}{a}+\frac{1}{a+b}+\frac{1}{a+2b}+\frac{1}{a+3b}+\cdots$

9. $1+\frac{1}{2^2}+\frac{2^2}{3^3}+\frac{3^3}{4^4}+\cdots$

10. $\frac{1}{1.2.3}+\frac{1}{2.3.4}+\frac{1}{3.4.5}+\cdots$

11. $\frac{1}{2}+\frac{3!}{2.4}+\frac{5!}{2.4.6}+\frac{7!}{2.4.6.8}+\cdots$

12. $\frac{1}{x}+\frac{1}{x+1}+\frac{1}{x+2}+\frac{1}{x+3}+\cdots$

13. $\frac{1}{1^2+x^2}+\frac{1}{2^2+x^2}+\frac{1}{3^2+x^2}+\cdots$ for real values of x

14. $\frac{1}{1+x}+\frac{1}{1+2x^2}+\frac{1}{1+3x^3}+\cdots$ for positive values of x

15. $\frac{x}{1^2+1}+\frac{2^3}{2^2+1}x^2+\frac{3^3}{3^2+1}x^3+\cdots$ for real values of x

16. Show that the series

$$\frac{1}{1^p}+\frac{1}{2^p}+\frac{1}{3^p}+\cdots$$

is convergent when $p>1$ and divergent when $p\le 1$.

Examine the convergence of the following series (Problems 17–25):

17. $\sum_1^\infty \frac{\sqrt{n}}{n+1}$

18. $\sum_1^\infty \left(\sqrt{n^4+n}-n^2\right)$

19. $\sum_1^\infty \left(\sqrt[3]{n^3+1}-n\right)$

20. $\sum_1^\infty \frac{\sqrt{n+1}-\sqrt{n}}{n^p}$

21. $\sum_1^\infty \frac{\ln n}{2n^3-1}$

22. $\sum_1^\infty \left(\sqrt{n^4+1}-\sqrt{n^4-1}\right)$

23. $\sum_1^\infty \frac{4n^3-n+3}{n^3+2n}$

24. $\sum_1^\infty \left(\sqrt{n^2+1}-n\right)$

25. $\sum_1^\infty \frac{2n^3-3n^2}{7n^4+100n^3+7}$

Examine the convergence of the following series (Problems 26–29):

26. $\sum_1^\infty \sqrt{\frac{n}{n+1}}\, x^n$

27. $\sum_1^\infty \frac{x^n}{1+x^{2n}}$

28. $\sum_1^\infty \frac{(n+1)^2}{n!}$

29. $\frac{1}{1^p}+\frac{x}{3^p}+\frac{x^2}{5^p}+\cdots+\frac{x^{n-1}}{(2n-1)^p}+\cdots$

30. Use Cauchy's root test to show that the series

$$\sum_1^\infty \frac{[(n+1)r]^n}{n^{n+1}}$$

is convergent if $r<1$ and divergent if $r \ge 1$.

Which of the following series is convergent? If convergent, determine whether the convergence is absolute or conditional (Problems 31–35):

31. $1-\frac{1}{5}+\frac{1}{9}-\frac{1}{13}+\cdots$

32. $1-\frac{3}{2}+\frac{5}{4}-\frac{7}{8}+\frac{9}{16}+\cdots$

33. $1-\frac{1}{2}\sqrt{2}+\frac{1}{3}\sqrt{3}-\frac{1}{4}\sqrt{4}+\cdots$

34. $\frac{1}{6}-\frac{2}{11}+\frac{3}{16}-\frac{4}{21}+\frac{5}{26}-\cdots$

35. $1-\frac{1}{2!}+\frac{1}{3!}-\frac{1}{4!}+\cdots$

Test for the convergence the following series (Problems 36–38):

36. $\sum_1^\infty (-1)^n \left(\sqrt{n+1}-\sqrt{n}\right)$

37. $\sum_1^\infty (-1)^n \frac{n}{(n+1)}$

38. $\sum_{1}^{\infty}(-1)^n \frac{(1+n^2)}{(1+n^3)}$.

Find the interval of convergence of the following series (Problems 39–42):

39. $\sum_{1}^{\infty}(-1)^n \frac{X^n}{\sqrt{n}}$

40. $\sum_{1}^{\infty}(-1)^n \frac{x^n}{\log(n+1)}$

41. $\sum_{1}^{\infty}\frac{(x-2)^n}{n^2}$

42. $\sum_{1}^{\infty}\frac{(x+3)^{n-1}}{n}$.

1.3 BINOMIAL, EXPONENTIAL AND LOGARITHMIC SERIES

Many practical situations occurring in science and engineering deal with approximations concerning binomial, exponential and logarithmic series. We shall discuss these approximations in this section.

1.3.1 The Binomial Theorem

When n is a positive integer, the theorem is

$$(1+x)^n = 1 + nx + \frac{n(n-1)}{2!}x^2 + \frac{n(n-1)(n-2)}{3!}x^3 + \cdots$$

$$+\frac{n(n-1)\ldots(n-r+1)}{r!}x^r + \cdots + x^n, \tag{1.9}$$

for any value of x, the series consisting of $(n+1)$ terms as shown. The expansion given in (1.9) can be proved easily by the method of induction. But when n is a fraction or a negative integer, the series does not terminate as in (1.9) but will be an infinite series:

$$(1+x)^n = 1 + nx + \frac{n(n-1)}{2!}x^2 + \cdots + \frac{n(n-1)(n-2)\ldots(n-r+1)}{r!}x^r + \cdots \tag{1.10}$$

Writing, for brevity, n_r for $n(n-1)(n-2)\ldots(n-r+1)$, Eq. (1.10) can be written as

$$(1+x)^n = 1 + \frac{n_1 x}{1!} + \frac{n_2}{2!}x^2 + \cdots + \frac{n_r}{r!}x^r + \cdots \tag{1.11}$$

To know the range of convergence of this series, we note that

$$u_r = \frac{n(n-1)(n-2)\ldots(n-r+2)}{(r-1)!}x^{r-1}$$

and

$$u_{r+1} = \frac{n(n-1)(n-2)\ldots(n-r+1)}{r!}x^r$$

Applying the ratio test, we have

$$\left|\frac{u_{r+1}}{u_r}\right| = \left|\frac{n-r+1}{r}\right||x| = \left|\frac{n+1}{r}-1\right||x|$$

Therefore,

$$\lim_{r\to\infty} \frac{u_{r+1}}{u_r} = |x|,$$

so that the binomial series is absolutely convergent if $|x|<1$. This means that for this range of values of x, the series is equal to $(1+x)^n$.

Assuming the binomial theorem for a positive integral index, we first prove Vandermonde's theorem which will then be used to prove the binomial theorem for any rational index.

Vandermonde's theorem

If r is a positive integer, then

$$(m+n)_r = m_r + {}^r\mathrm{C}_1\, m_{r-1}\, n_1 + {}^r\mathrm{C}_2\, m_{r-2}\, n_2 + \cdots + n_r,$$

for all values of m and n. If m and n are positive integers, we have

$$(1+x)^m = 1 + \frac{m_1}{1!}x + \frac{m_2}{2!}x^2 + \cdots + \frac{m_r}{r!}x^r + \cdots x^m$$

and

$$(1+x)^n = 1 + \frac{n_1}{1!}x + \frac{n_2}{2!}x^2 + \cdots + \frac{n_r}{r!}x^r + \cdots x^n$$

Multiplication gives

$$(1+x)^{m+n} = \left(1 + \frac{m_1}{1!}x + \frac{m_2}{2!}x^2 + \cdots + \frac{m_r}{r!}x^r + \cdots\right)$$

$$\times\left(1 + \frac{n_1}{1!}x + \frac{n_2}{2!}x^2 + \cdots + \frac{n_r}{r!}x^r + \cdots\right)$$

The coefficient of x^r on the left-hand side is $\frac{(m+n)_r}{r!}$ whereas that on the right-hand side is

$$\frac{m^r}{r!} + \frac{m_{r-1}}{(r-1)!}\frac{n_1}{1!} + \frac{m_{r-2}}{(r-2)!}\frac{n_2}{2!} + \cdots + \frac{m_1}{1!}\frac{n_{r-1}}{(r-1)!} + \frac{n_r}{r!}$$

$$= \frac{1}{r!}\left[m_r + \frac{r!}{(r-1)!(1!)}m_{r-1}\,n_1 + \frac{r!}{(r-2)!2!}m_{r-2}\,n_2 + \cdots + \frac{r!}{(r-1)!(1!)}m_1 n_{r-1} + n_r\right]$$

$$= \frac{1}{r!}\left(m_r + {}^r\mathrm{C}_1\, m_{r-1}n_1 + {}^r\mathrm{C}_2\, m_{r-2}n_2 + \cdots + {}^r\mathrm{C}_{r-1}\, m_1 n_{r-1} + {}^r\mathrm{C}_r\, n_r\right)$$

Equating the coefficients of x^r on both sides, we obtain the desired result, i.e.

$$(m+n)_r = m_r + {}^rC_1\, m_{r-1}n_1 + {}^rC_2\, m_{r-2}n_2 + \cdots + n_r,$$

where each side is a polynomial of degree r in m and n. This result is true for all integral values of m and n, i.e. it is true for more than r values of m and n. Hence, the polynomials are identically equal, i.e. the result is true for all values of m and n. Using this theorem, we can now prove binomial theorem for a rational index.

Let

$$F(m) = 1 + \frac{m_1}{1!}x + \frac{m_2}{2!}x^2 + \cdots + \frac{m_r}{r!}x^r + \cdots \tag{1.12}$$

We have already seen that this series will be absolutely convergent when $|x| < 1$, i.e. when $-1 < x < 1$. Similarly, the two series

$$F(n) = 1 + \frac{n_1}{1!}x + \frac{n_2}{2!}x^2 + \cdots + \frac{n_r}{r!}x^r + \cdots \tag{1.13}$$

and

$$F(m+n) = 1 + \frac{(m+n)_1}{1!}x + \frac{(m+n)_2}{2!}x^2 + \cdots + \frac{(m+n)_r}{r!}x^r + \cdots \tag{1.14}$$

are both absolutely convergent when $-1 < x < 1$. Hence, $F(m)$ and $F(n)$ can be multiplied and the resulting series will be absolutely convergent. The coefficient of x^r in the product of $F(m)$ and $F(n)$ is

$$\frac{m_r}{r!} + \frac{m_{r-1}}{(r-1)!}\frac{n_1}{1!} + \frac{m_{r-2}}{(r-2)!}\frac{n_2}{2!} + \cdots + \frac{n_r}{r!}$$

or

$$\frac{1}{r!}[m_r + {}^rC_1 m_{r-1}n_1 + {}^rC_2 m_{r-2}n_2 + \cdots + n_r]$$

By Vandermonde's theorem, this expression is equal to $\frac{(m+n)_r}{r!}$, which is the coefficient of x^r in $F(m+n)$. Therefore,

$$F(m)F(n) = F(m+n) \tag{1.15}$$

for all values of m and n, provided that $|x| < 1$. The result can obviously be extended to any number of factors so that we can write

$$F(m)F(n)F(p)F(q)\cdots = F(m+n+p+q+\cdots) \tag{1.16}$$

Case (i): n is a positive fraction

Put $n = r/s$, where r and s are positive integers. We have

$$F(r) = (1+x)^r,$$

as r is a positive integer. It follows that

$$F(0) = 1 \quad \text{and} \quad F(1) = 1 + x.$$

Therefore,

$$F\left(\frac{r}{s}\right) \cdot F\left(\frac{r}{s}\right) \cdots s \text{ factors } = F\left(\frac{r}{s} + \frac{r}{s} + \cdots s \text{ term}\right)$$

or

$$\left[F\left(\frac{r}{s}\right)\right]^s = F(r) = (1+x)^r,$$

as r is a positive integer. Hence

$$F\left(\frac{r}{s}\right) = (1+x)^{r/s} \quad \text{or} \quad F(n) = (1+x)^n,$$

which proves the binomial theorem for a positive fraction.

Case (*ii*) n is a negative rational number:

Let $n = -m$, where m is a positive rational number. We have

$$F(m)\,F(n) = F(m+n) = F(0) = 1$$

Therefore,

$$F(n) = \frac{1}{F(m)} = \frac{1}{(1+x)^m} = (1+x)^{-m} = (1+x)^n,$$

where m is positive. This proves the theorem for any negative index.

An important particular case is given by

$$(1-x)^{-n} = 1 + nx + \frac{n(n+1)}{2!}x^2 + \frac{n(n+1)(n+2)}{3!}x^3 + \cdots \tag{1.17}$$

so that we have the expansions:

$$(1-x)^{-1} = 1 + x + x^2 + x^3 + \cdots,$$

$$(1-x)^{-2} = 1 + 2x + 3x^2 + 4x^3 + \cdots,$$

$$(1-x)^{-1/2} = 1 + \frac{1}{2}x + \frac{1.3}{2.4}x^2 + \frac{1.3.5}{2.4.6}x^3 + \cdots,$$

$$(1-x)^{-1/3} = 1 + \frac{1}{3}x + \frac{1.4}{3.6}x^2 + \frac{1.4.7}{3.6.9}x^3 + \cdots,$$

and so on.

Example 1.22 Expand

$$\frac{1}{(1-2x)(1+3x)}$$

in a series of ascending powers of x and find when the expansion is valid.

On splitting into partial fractions, we have

$$\frac{1}{(1-2x)(1+3x)} = \frac{2}{5(1-2x)} + \frac{3}{5(1+3x)}$$

$$= \frac{2}{5}(1-2x)^{-1} + \frac{3}{5}(1+3x)^{-1}$$

$$= \frac{2}{5}(1+2x+4x^2+8x^3+\cdots+2^n x^n+\cdots)$$

$$+\frac{3}{5}[1-3x+9x^2-27x^3+\cdots+(-1)^n 3^n x^n+\cdots]$$

$$= 1 - x + 7x^2 - 13x^3 + \cdots + \frac{1}{5}[2^{n+1} + (-1)^n 3^{n+1}]\, x^n$$

By ratio test, it follows that the series is convergent, if and only if $|2x|<1$ and $|3x|<1$. It is easy to see that both these conditions are satisfied if $|x|<1/3$, which is therefore the condition of validity of the expansion.

Example 1.23 Given positive integers $r>1$, $n>2$ and that the coefficients of $(3r)$th and $(r+2)$th terms in the binomial expansion of $(1+x)^{2n}$ are equal, show that $n=2r$.

The given expression is $(1+x)^{2n}$. Therefore,

$$\text{Coefficient of } (3r)\text{th term} = {}^{2n}C_{3r-1},$$

and

$$\text{Coefficient of } (r+2)\text{th term} = {}^{2n}C_{r+1}.$$

We therefore, have

$${}^{2n}C_{3r-1} = {}^{2n}C_{r+1},$$

or

$$\frac{(2n)!}{(3r-1)!(2n-3r+1)!} = \frac{(2n)!}{(r+1)!(2n-r-1)!},$$

from which follows the required result on simplification.

Example 1.24 Show that

$$({}^{2n}C_0)^2 - ({}^{2n}C_1)^2 + ({}^{2n}C_2)^2 - \cdots + ({}^{2n}C_{2n})^2 = (-1)^n({}^{2n}C_n)$$

We have

$$(1+x)^{2n} = {}^{2n}C_0 + {}^{2n}C_1\, x + {}^{2n}C_2\, x^2 + \cdots + {}^{2n}C_{2n}\, x^{2n},$$

and

$$(1-x)^{2n} = {}^{2n}C_0 - {}^{2n}C_1\, x + {}^{2n}C_2\, x^2 + \cdots + (-1)^{2n}\, {}^{2n}C_{2n}\, x^{2n}.$$

Multiplying the two series, we obtain

$$(1-x^2)^{2n} = ({}^{2n}C_0 + {}^{2n}C_1 x + {}^{2n}C_2 x^2 + \cdots + {}^{2n}C_{2n} x^{2n})$$

$$\times ({}^{2n}C_0 - {}^{2n}C_1 x + {}^{2n}C_2 x^2 - \cdots + {}^{2n}C_{2n} x^{2n})$$

$$= ({}^{2n}C_0)^2 - {}^{2n}C_0 {}^{2n}C_1 x + {}^{2n}C_0 {}^{2n}C_2 x^2 + \cdots + {}^{2n}C_0 {}^{2n}C_{2n} x^{2n}$$

$$+ {}^{2n}C_1 {}^{2n}C_0 x - ({}^{2n}C_1)^2 x^2 + {}^{2n}C_1 {}^{2n}C_2 x^3 - \cdots$$

$$+ {}^{2n}C_1 {}^{2n}C_{2n} x^{2n+1} + {}^{2n}C_2 {}^{2n}C_0 x^2 - {}^{2n}C_2 {}^{2n}C_1 x^3$$

$$+({}^{2n}C_2)^2 x^4 + \cdots + {}^{2n}C_2 {}^{2n}C_{2n} x^{2n+2} + \cdots + {}^{2n}C_{2n} {}^{2n}C_0 x^{2n}$$

$$- {}^{2n}C_{2n} {}^{2n}C_1 x^{2n+1} + {}^{2n}C_{2n} {}^{2n}C_2 x^{2n+2} - \cdots + ({}^{2n}C_{2n})^2 x^{4n}.$$

Now, the coefficient of x^{2n} on the left-hand side is $(-1)^n \, {}^{2n}C_n$. Similarly, the coefficient of x^{2n} on the right-hand side is given by

$${}^{2n}C_0 {}^{2n}C_{2n} - {}^{2n}C_1 {}^{2n}C_{2n-1} + {}^{2n}C_2 {}^{2n}C_{2n-2} - \cdots + {}^{2n}C_{2n} {}^{2n}C_0$$

$$= ({}^{2n}C_0)^2 - ({}^{2n}C_1)^2 + ({}^{2n}C_2)^2 - \cdots + ({}^{2n}C_{2n})^2.$$

Equating the coefficients of x^{2n} on both sides, we obtain

$$({}^{2n}C_0)^2 - ({}^{2n}C_1)^2 + ({}^{2n}C_2)^2 - \cdots + ({}^{2n}C_{2n})^2 = (-1)^n ({}^{2n}C_n)$$

which is the required result.

1.3.2 Applications of Binomial Theorem to Approximation and Interpolation

Approximations

A useful application of the binomial theorem occurs when x is quite small compared with unity so that successive powers of x, namely $x^2, x^3, \ldots$ diminish quite rapidly. We can then obtain approximations to the sum of the series by retaining a few terms only, the number of terms so retained being dependent on the degree of accuracy required. In many applications, however, the first and second approximations are quite sufficient. The following examples illustrate the use of binomial series to approximations.

Example 1.25 Find the first and second approximations for $(130)^{1/3}$.

We write

$$(130)^{1/3} = (125+5)^{1/3}$$

$$= 5\left(1+\frac{1}{25}\right)^{1/3}$$

$$= 5(1+0.04)^{1/3}$$

$$= 5(1+0.01333-0.000177+\cdots)$$

The first approximation is given by

$$(130)^{1/3} = 5(1+0.01333) = 5.06665 = 5.07$$

The above result is correct to two decimal places, since the third term has three zeros after the decimal point.

Similarly, the second approximation is given by

$$(130)^{1/3} = 5(1+0.01333-0.000177) = 5.065780 = 5.066,$$

correct to three decimal places.

Example 1.26 Calculate the value of

$$\sqrt{1.01}-\sqrt{0.99},$$

correct to six decimal places.

We have

$$(1+x)^{1/2} = 1+\frac{1}{2}x+\frac{\frac{1}{2}\left(-\frac{1}{2}\right)}{2}x^2+\frac{\frac{1}{2}\left(-\frac{1}{2}\right)\left(-\frac{3}{2}\right)}{6}x^3+\cdots$$

and

$$(1-x)^{1/2} = 1-\frac{1}{2}x+\frac{\frac{1}{2}\left(-\frac{1}{2}\right)}{2}x^2-\frac{\frac{1}{2}\left(-\frac{1}{2}\right)\left(-\frac{3}{2}\right)}{6}x^3+\cdots$$

so that

$$(1+x)^{1/2}-(1-x)^{1/2} = x+\frac{1}{8}x^3+\cdots$$

Putting $x = 0.01$, we obtain

$$(1.01)^{1/2}-(0.99)^{1/2} = 0.01+\frac{1}{8}(0.01)^3+\cdots$$

$$= 0.01+\frac{1}{8}(0.000001)+\cdots$$

$$= 0.01 \text{ (correct to six decimal places).}$$

Example 1.27 Obtain the first-four terms in the expansion of $(N^5+a)^{1/5}$ by the binomial theorem when $a < N^5$, and hence show that

$$N\frac{5N^5+3a}{5N^5+2a}$$

is an approximation to $(N^5+a)^{1/5}$. Deduce that the difference between $(100100)^{1/5}$ and 50030/50020 is of the order $1/10^{10}$.

We have

$$(N^5+a)^{1/5} = N\left(1+\frac{a}{N^5}\right)^{1/5}$$

$$= N\left[1+\frac{1}{5}\frac{a}{N^5}+\frac{\frac{1}{5}\left(\frac{1}{5}-1\right)}{2}\frac{a^2}{N^{10}}+\frac{\frac{1}{5}\left(\frac{1}{5}-1\right)\left(\frac{1}{5}-2\right)}{6}\frac{a^3}{N^{15}}+\cdots\right]$$

$$= N+\frac{a}{5N^4}-\frac{2}{25}\frac{a^2}{N^9}+\frac{6}{125}\frac{a^3}{N^{14}}-\cdots \qquad \text{(i)}$$

Now

$$N\frac{5N^5+3a}{5N^5+2a} = N\left(1+\frac{a}{5N^5+2a}\right)$$

$$= N\left[1+\frac{a}{5n^5}\left(1+\frac{2a}{5N^5}\right)^{-1}\right]$$

$$= N\left[1+\frac{a}{5N^5}\left(1-\frac{2a}{5N^5}+\frac{4a^2}{25N^{10}}-\frac{8a^3}{125N^{15}}+\cdots\right)\right]$$

$$= N\left(1+\frac{a}{5N^5}-\frac{2a^2}{25N^{10}}+\frac{4a^3}{125N^{15}}-\cdots\right)$$

$$= N+\frac{a}{5N^4}-\frac{2a^2}{25N^9}+\frac{4a^3}{125N^{14}}-\cdots \qquad \text{(ii)}$$

Neglecting the terms containing $[a/(5N^5)]^3$, etc. it is clear, from (i) and (ii), that

$$N\frac{5N^5+3a}{5N^5+2a}$$

is an approximate value of $(N^5+a)^{1/5}$. If we take the fourth term into consideration, it will be seen that

$$(N^5+a)^{1/5}-N\frac{5N^5+3a}{5N^5+2a}=\frac{2}{125}\frac{a^3}{N^{14}}$$

Putting $N = 10$ and $a = 100$, we have

$$\frac{2a^3}{125N^{14}} = \frac{2(100)^3}{125(10)^{14}} = \frac{2}{125}\frac{2}{10^8} = O\left(\frac{1}{10^{10}}\right)$$

Example 1.28 If $\sqrt{n} = a + x$ when x is very small, then show that

$$\sqrt{n} = a\frac{3n + a^2}{n + 3a^2}, \qquad \text{(approximately).}$$

Since $\sqrt{n} = a + x$, we have $n = (a + x)^2$. Hence

$$a\frac{3n + a^2}{n + 3a^2} = a\frac{3(a + x)^2 + a^2}{(a + x)^2 + 3a^2}$$

$$= a\frac{4a^2 + 6ax}{4a^2 + 2ax} \qquad \text{(neglecting } x^2\text{)}$$

$$= a\frac{1 + (3x/2a)}{1 + x/(2a)}$$

$$= a\left(1 + \frac{3x}{2a}\right)\left(1 + \frac{x}{2a}\right)^{-1}$$

$$= a\left(1 + \frac{3x}{2a}\right)\left(1 - \frac{x}{2a}\right) \qquad \text{(approximately)}$$

$$= a\left(1 + \frac{x}{a}\right) \qquad \text{(again neglecting } x^2\text{)}$$

$$= a + x$$

$$= \sqrt{n}.$$

Interpolation

The binomial expansion can be used for interpolating across a given table of values.

Let $(x_0, y_0), (x_1, y_1), (x_2, y_2), \ldots, (x_n, y_n)$ be a given set of $(n+1)$ values of x and y with $x_n = x_0 + nh$, where h is constant. We define *forward differences* in the following manner. The difference $y_{n+1} - y_n$ is called the *first forward difference* of y_n and is denoted Δy_n so that

$$y_{n+1} - y_n = \Delta y_n \tag{1.18}$$

In (1.18), Δ is called the *forward difference operator.* We thus have

$$y_1 - y_0 = \Delta y_0, \quad y_2 - y_1 = \Delta y_1, \quad y_3 - y_2 = \Delta y_2.$$

and so on. In a similar way, higher-order forward differences can also be defined. Thus,

$$\Delta^2 y_0 = \Delta y_1 - \Delta y_0 = y_2 - y_1 - (y_1 - y_0) = y_2 - 2y_1 + y_0,$$

$$\Delta^3 y_0 = \Delta^2 y_1 - \Delta^2 y_0 = y_3 - 2y_2 + y_1 - (y_2 - 2y_1 + y_0) = y_3 - 3y_2 + 3y_1 - y_0,$$

and so on. A table of forward differences is easily formed. This is given in Table 1.1 for a set of 5 tabular values.

Table 1.1 Forward Differences

x	y	Δ	Δ^2	Δ^3	Δ^4
x_0	y_0				
		Δy_0			
x_1	y_1		$\Delta^2 y_0$		
		Δy_1		$\Delta^3 y_0$	
x_2	y_2		$\Delta^2 y_1$		$\Delta^4 y_0$
		Δy_2		$\Delta^3 y_1$	
x_3	y_3		$\Delta^2 y_2$		
		Δy_3			
x_4	y_4				

Using forward differences, an important formula for interpolation can be derived in the following way. Since $y_1 - y_0 = \Delta y_0$, we have $y_1 = y_0 + \Delta y_0$. Again, since $y_2 - y_1 = \Delta y_1$, we have

$$\begin{aligned} y_2 &= y_1 + \Delta y_1 \\ &= y_0 + \Delta y_0 + \Delta(y_0 + \Delta y_0) \\ &= y_0 + \Delta y_0 + \Delta^2 y_0 = (1 + \Delta + \Delta^2) y_0 \\ &= (1+\Delta)^2 y_0. \end{aligned}$$

We similarly obtain

$$\begin{aligned} y_3 &= y_2 + \Delta y_2 \\ &= y_0 + 2\Delta y_0 + \Delta^2 y_0 + \Delta(y_0 + 2\Delta y_0 + \Delta^2 y_0) \\ &= y_0 + 3\Delta y_0 + 3\Delta^2 y_0 + \Delta^3 y_0 \\ &= (1+\Delta)^3 y_0. \end{aligned}$$

It therefore follows that

$$\begin{aligned} y_n &= (1+\Delta)^n y_0 \\ &= y_0 + n\Delta y_0 + \frac{n(n-1)}{2!}\Delta^2 y_0 + \frac{n(n-1)(n-2)}{3!}\Delta^3 y_0 + \cdots, \end{aligned} \tag{1.19}$$

where the coefficients on the right-hand side are the normal binomial coefficients and Δy_0, $\Delta^2 y_0$, etc., are the forward differences of y_0 which lie on a line diagonally downwards from y_0. Formula (1.19) is called *Newton's forward difference interpolation formula.*

In a similar manner, one can define *backward* differences. We thus have

$$\left.\begin{aligned} y_1 - y_0 &= \nabla y_1, & y_2 - y_1 &= \nabla y_2, \\ \nabla y_1 - \nabla y_0 &= \nabla^2 y_1, & \nabla y_2 - \nabla y_1 &= \nabla^2 y_2, \\ \nabla^2 y_1 - \nabla^2 y_0 &= \nabla^3 y_1, & \nabla^2 y_2 - \nabla^2 y_1 &= \nabla^3 y_2 \end{aligned}\right\} \tag{1.20}$$

and so on. In (1.20), ∇ is called the *backward difference operator.*

Using backward differences and proceeding in the same way as above, the following formula is easily derived:

$$\begin{aligned} y_0 &= (1-\nabla)^n y_n \\ &= y_n - n\nabla y_n + \frac{n(n-1)}{2!}\nabla^2 y_n - \frac{n(n-1)(n-2)}{3!}\nabla^3 y_n + \cdots \end{aligned} \tag{1.21}$$

It is seen that the coefficients are again the normal binomial coefficients and that ∇y_n, $\nabla^2 y_n$, etc., are the backward differences which lie along a line diagonally upwards from y_n. Formula (1.21) is called *Newton's backward difference interpolation formula.*

In the above, we have derived two simple interpolation formulae. There are, however, many other formulae for interpolation.[1] The use of formulae (1.19) and (1.21) is illustrated by the following numerical examples.

Example 1.29 The following table of values are given. Find $f(1)$ and $f(5)$.

x	0	2	4	6
$y = f(x)$	−3	−5	1	15

We first form the table of differences;

x	$f(x)$	Δ	Δ^2	Δ^3
0	−3			
		−2		
2	−5		8	
		6		0
4	1		8	
		14		
6	15			

1. Refer to Sastry, S.S., *Introductory Methods of Numerical Analysis,* 3rd ed., Prentice-Hall of India, New Delhi, 2001.

To find $f(1)$, we choose $x_n = 1$, $x_0 = 0$. With $h = 2$, we obtain

$$n = \frac{x_n - x_0}{h} = \frac{1}{2}.$$

Using formula (1.19), we obtain

$$f(1) = (1+\Delta)^{1/2} y_0 = y_0 + \frac{1}{2}\Delta y_0 + \frac{\frac{1}{2}\left(\frac{1}{2}-1\right)}{2}\Delta^2 y_0,$$

since the third- and higher-order differences vanish.

Here $y_0 = -3$ and the differences Δy_0 and $\Delta^2 y_0$ are those lying on a line *diagonally downwards* from y_0 in the table of differences given above. Hence, we obtain

$$f(1) = -3 + \frac{1}{2}(-2) - \frac{1}{8}(8) = -3 - 1 - 1 = -5.$$

It should be noted that the given table of values is obtained from $f(x) = x^2 - 3x - 3$, so that $f(1) = -5$. The calculated value of $f(1)$ is therefore exact.

To find $f(5)$, we choose $x_0 = 5$ and $x_n = 6$. With $h = 2$, we now obtain

$$n = \frac{x_n - x_0}{h} = \frac{1}{2}.$$

Hence, formula (1.21) gives:

$$f(5) = (1-\nabla)^{1/2} y_n = y_n - \frac{1}{2}\nabla y_n + \frac{\frac{1}{2}\left(\frac{1}{2}-1\right)}{2}\nabla^2 y_n,$$

since the third- and higher-order backward differences vanish.

Here $y_n = 15$ and the backward differences ∇y_n and $\nabla^2 y_n$ are those lying on a line *diagonally upwards* from y_n. We then obtain

$$f(5) = 15 - \frac{1}{2}(14) - \frac{1}{8}(8) = 15 - 7 - 1 = 7,$$

which is found to be exact on comparison with the functional relation given above.

Example 1.30 The following is a table of values taken from cube root tables. Determine the values of $\sqrt[3]{7}$ and $\sqrt[3]{13}$.

x	6	8	10	12	14
$\sqrt[3]{x}$	1.817	2.000	2.154	2.289	2.410

The table of differences is shown below:

x	$\sqrt[3]{x}$	Δ	Δ^2	Δ^3	Δ^4
6	1.817				
		0.183			
8	2.000		−0.029		
		0.154		0.010	
10	2.154		−0.019		−0.005
		0.135		0.005	
12	2.289		−0.014		
		0.121			
14	2.410				

To find $\sqrt[3]{7}$, we choose $x_n = 7$ and $x_0 = 6$. With $h = 2$ this gives $n = 1/2$. We therefore obtain

$$\sqrt[3]{7} = 1.817 + \frac{1}{2}(0.183) + \frac{\frac{1}{2}\left(\frac{1}{2}-1\right)}{2}(-0.029)$$

$$+\frac{\frac{1}{2}\left(\frac{1}{2}-1\right)\left(\frac{1}{2}-2\right)}{6}(0.010) + \frac{\frac{1}{2}\left(\frac{1}{2}-1\right)\left(\frac{1}{2}-2\right)\left(\frac{1}{2}-3\right)}{24}(-0.005)$$

$$= 1.817 + 0.0915 + 0.0036 + 0.0006 + 0.0002$$

$$= 1.913.$$

The determination of $\sqrt[3]{13}$ using the backward difference formula is left as an exercise to the reader.

1.3.3 The Exponential Series

We start with the expansion of $(1+x)^n$:

$$(1+x)^n = 1 + \frac{nx}{1!} + \frac{n(n-1)}{2!}x^2 + \frac{n(n-1)(n-2)}{3!}x^3$$

$$+\frac{n(n-1)(n-2)(n-3)}{4!}x^4 + \cdots$$

Putting $x = 1/n$, this gives

$$\left(1+\frac{1}{n}\right)^n = 1 + \frac{1}{1!} + \frac{1-\frac{1}{n}}{2!} + \frac{\left(1-\frac{1}{n}\right)\left(1-\frac{2}{n}\right)}{3!}$$

$$+\frac{\left(1-\frac{1}{n}\right)\left(1-\frac{2}{n}\right)\cdots\left(1-\frac{r-1}{n}\right)}{r!} + \cdots$$

Now suppose that n becomes infinitely great; then each of the fraction $1/n$, $2/n$, $3/n$,... tends to zero and all the numerators become unity. The series on the right becomes an infinite series.

$$1+\frac{1}{1!}+\frac{1}{2!}+\frac{1}{3!}+\cdots+\frac{1}{r!}+\cdots$$

This is a convergent series and we write

$$\lim_{n\to\infty}\left(1+\frac{1}{n}\right)^n=1+\frac{1}{1!}+\frac{1}{2!}+\frac{1}{3!}+\cdots+\frac{1}{r!}+\cdots$$

The successive terms of this series diminish quite rapidly, and the sum of the series can be found to any desired degree of accuracy. This sum to infinity, which is a finite number, is usually denoted by the letter e, so that

$$\lim_{n\to\infty}\left(1+\frac{1}{n}\right)^n=e=1+\frac{1}{1!}+\frac{1}{2!}+\frac{1}{3!}+\cdots+\frac{1}{r!}+\cdots \qquad (1.22)$$

It is easy to see that

$$e>1+\frac{1}{1!}=2,$$

as all the terms are positive. Further,

$$e=1+\frac{1}{2}+\frac{1}{1.2}+\frac{1}{1.2.3}+\frac{1}{1.2.3.4}+\cdots$$

$$<1+1+\frac{1}{2}+\frac{1}{2.2}+\frac{1}{2.2.2}+\cdots$$

$$=1+\left(1+\frac{1}{2}+\frac{1}{2^2}+\frac{1}{2^3}+\cdots\right)$$

$$=1+\frac{1}{1-(1/2)}=3.$$

Thus, e lies between 2 and 3. In fact, e is an incommensurable number, i.e. e cannot be expressed as a fraction a/b and its value has been calculated to more than 500 decimal places ($e=2.7182818284\ldots$).

The exponential theorem

This states that for all values of x,

$$e^x=1+x+\frac{x^2}{2!}+\frac{x^3}{3!}+\cdots+\frac{x^n}{n!}+\cdots \qquad (1.23)$$

Let

$$F(x) = 1 + x + \frac{x^2}{2!} + \frac{x^3}{3!} + \cdots + \frac{x^n}{n!} + \cdots$$

so that $F(1) = e$. Since

$$\frac{u_{r+1}}{u_r} = \frac{x^r}{r!} \frac{(r-1)!}{x^{r-1}} = \frac{x}{r} \to 0 \qquad \text{as } r \to \infty,$$

the series for $F(x)$ is convergent for all values of x. Hence the series

$$F(m) = 1 + \frac{m}{1!} + \frac{m^2}{2!} + \frac{m^3}{3!} + \cdots + \frac{m^r}{r!} + \cdots$$

$$F(n) = 1 + \frac{n}{1!} + \frac{n^2}{2!} + \frac{n^3}{3!} + \cdots + \frac{n^r}{r!} + \cdots$$

and

$$F(m+n) = 1 + \frac{(m+n)}{1!} + \frac{(m+n)^2}{2!} + \cdots + \frac{(m+n)^r}{r!} + \cdots$$

are convergent for all values of m and n. We can therefore multiply the series for $F(m)$ and $F(n)$. This gives us:

$$F(m)F(n) = \left(1 + \frac{m}{1!} + \frac{m^2}{2!} + \cdots + \frac{m^r}{r!} + \cdots\right)\left(1 + \frac{n}{1!} + \frac{n^2}{2!} + \cdots + \frac{n^r}{r!} + \cdots\right)$$

$$= 1 + \left(\frac{m}{1!} + \frac{n}{1!}\right) + \left(\frac{m^2}{2!} + \frac{m}{1!}\frac{n}{11} + \frac{n^2}{2!}\right) + \left(\frac{m^3}{3!} + \frac{m^2}{2!}\frac{n}{1!} + \frac{m}{1!}\frac{n^2}{2!} + \frac{n^3}{3!}\right) + \cdots$$

The $(r+1)$th term in the product is

$$\frac{m^r}{r!} + \frac{m^{r-1}}{(r-1)!}\frac{n}{1!} + \frac{m^{r-2}}{(r-2)!}\frac{n^2}{2!} + \cdots + \frac{m^2}{2!}\frac{n^{r-2}}{(r-2)!} + \frac{m}{1!}\frac{n^{r-1}}{(r-1)!} + \frac{n^r}{r!}$$

$$= \frac{1}{r!}(m^r + {}^rC_1 m^{r-1} n + {}^rC_2 m^{r-2} n^2 + \cdots + {}^rC_{r-1} m n^{r-1} + n^r)$$

$$= \frac{1}{r!}(m+n)^r.$$

Hence we have

$$F(m)F(n) = 1 + \frac{m+n}{1!} + \frac{(m+n)^2}{2!} + \cdots + \frac{(m+n)^r}{r!} + \cdots = F(m+n).$$

Case (i): Let x be a positive integer. Then from the result proved earlier, we have

$$F(m)F(n)F(p)F(q)\cdots = F(m+n+p+q+\cdots)$$

If we now set

$$m = n = p = q = \cdots = 1,$$

we obtain

$$F(1)F(1)\cdots = F(1+1+\cdots)$$

If we consider x such factors, we obtain

$$[F(1)]^x = F(x)$$

or

$$e^x = F(x) = 1 + x + \frac{x^2}{2!} + \frac{x^3}{3!} + \cdots + \frac{x^r}{r!} + \cdots$$

Case (ii): Let x be a negative integer, say $-t$, where t is a positive integer. We have

$$F(t)F(-t) = F(t-t) = F(0) = 1.$$

Therefore,

$$F(-t) = \frac{1}{F(t)}$$

or

$$F(x) = \frac{1}{F(t)} = \frac{1}{[F(1)]^t} = [F(1)]^{-t} = [F(1)]^x$$

Hence for all x, we have

$$[F(1)]^x = F(x).$$

i.e.

$$e^x = 1 + x + \frac{x^2}{2!} + \frac{x^3}{3!} + \cdots + \frac{x^r}{r!} + \cdots$$

Also since

$$a^x = e^{x \log_e a}$$

We deduce from the above

$$a^x = 1 + x \log_e a + \frac{x^2}{2!}(\log_e a)^2 + \cdots + \frac{x^r}{r!}(\log_e a)^r + \cdots \qquad (1.24)$$

Example 1.31 Find the coefficient of x^n in the expansion of

$$\frac{1+2x+3x^2}{e^x}.$$

Now

$$\frac{1+2x+3x^2}{e^x} = (1+2x+3x^2)\,e^{-x}$$

$$= (1+2x+3x^2)\left[1-x+\frac{x^2}{2!}-\frac{x^3}{3!}+\cdots(-1)^n\frac{x^n}{n!}+\cdots\right]$$

The coefficient of x^n is

$$(-1)^n\frac{1}{n!}+2(-1)^{n-1}\frac{1}{(n-1)!}+3\,(-1)^{n-2}\frac{1}{(n-2)!}=\frac{(-1)^n}{n!}[-2n+3n\,(n-1)+1]$$

$$=\frac{(-1)^n}{n!}(1-5n+3n^2).$$

Example 1.32 Expand $x/(e^x-1)$ in powers of x as far as x^4.

We have

$$\frac{x}{e^x-1}=\frac{x}{1+x+\dfrac{x^2}{2!}+\dfrac{x^3}{3!}+\cdots-1}$$

$$=\frac{1}{1+\dfrac{x}{2}+\dfrac{x^2}{6}+\dfrac{x^3}{24}+\cdots}$$

$$=\left[1+\left(\frac{x}{2}+\frac{x^2}{6}+\frac{x^3}{24}+\frac{x^4}{120}+\cdots\right)\right]^{-1}$$

$$=1-\left(\frac{x}{2}+\frac{x^2}{6}+\frac{x^3}{24}+\frac{x^4}{120}+\cdots\right)+\left(\frac{x}{2}+\frac{x^2}{6}+\frac{x^3}{24}+\frac{x^4}{120}\cdots\right)^2$$

$$-\left(\frac{x}{2}+\frac{x^2}{6}+\frac{x^3}{24}+\frac{x^4}{120}+\cdots\right)^3+\left(\frac{x}{2}+\frac{x^2}{6}+\frac{x^3}{24}+\frac{x^4}{120}+\cdots\right)^4$$

$$=1-\left(\frac{x}{2}+\frac{x^2}{6}+\frac{x^3}{24}+\frac{x^4}{120}\right)+\left(\frac{x^2}{4}+\frac{x^4}{36}+\frac{x^3}{6}+\frac{x^4}{24}\right)-\left(\frac{x^3}{8}+\frac{x^4}{8}\right)+\frac{x^4}{16}.$$

Neglecting terms containing x^5 and higher powers, we get

$$\frac{x}{e^x-1}=1-\frac{x}{2}+\frac{1}{12}x^2-\frac{1}{720}x^4.$$

Example 1.33 If

$$\frac{e^x}{1-x} = B_0 + B_1x + B_2x^2 + \cdots + B_nx^n + \cdots$$

find the value of $B_n - B_{n-1}$.

We are given

$$\frac{e^x}{1-x} = B_0 + B_1x + B_2x^2 + \cdots + B_nx^n + \cdots$$

or

$$e^x = (1-x)(B_0 + B_1x + B_2x^2 + \cdots + B_nx^n + \cdots)$$

Using the expansion of e^x, we obtain

$$1 + x + \frac{x^2}{2!} + \frac{x^3}{3!} + \cdots + \frac{x^n}{n!} + \cdots = (1-x)(B_0 + B_1x + B_2x^2 + \cdots + B_{n-1}x^{n-1} + B_nx^n + \cdots)$$

Equating the coefficients of x^n on both sides, we get

$$B_n - B_{n-1} = \frac{1}{n!}.$$

Example 1.34 Find the value of

$$S = \frac{1^3}{1!} + \frac{2^3}{2!} + \frac{3^3}{3!} + \frac{4^3}{4!} + \cdots$$

Let u_n be the nth term. Then, we have

$$u_n = \frac{n^3}{n!} = \frac{n^2}{(n-1)!}$$

$$= \frac{n^2 - 1 + 1}{(n-1)!}$$

$$= \frac{n+1}{(n-2)!} + \frac{1}{(n-1)!}$$

$$= \frac{n-2+3}{(n-2)!} + \frac{1}{(n-1)!}$$

$$= \frac{1}{(n-3)!} + \frac{3}{(n-2)!} + \frac{1}{(n-1)!}$$

$$= \frac{1}{(n-1)!} + \frac{3}{(n-2)!} + \frac{1}{(n-3)!}$$

Putting $n = 1, 2, 3, \ldots$, we get

$$u_1 = 1,$$

$$u_2 = \frac{1}{1!} + 3,$$

$$u_3 = \frac{1}{2!} + \frac{3}{1!} + 1,$$

$$u_4 = \frac{1}{3!} + \frac{3}{2!} + \frac{1}{1!},$$

$$u_5 = \frac{1}{4!} + \frac{3}{3!} + \frac{1}{2!}, \text{ etc.}$$

Hence

$$S = \left(1 + \frac{1}{1!} + \frac{1}{2!} + \frac{1}{3!} + \cdots\right) + 3\left(1 + \frac{1}{1!} + \frac{1}{2!} + \cdots\right) + \left(1 + \frac{1}{1!} + \frac{1}{2!} + \cdots\right) = 5e$$

Example 1.35 If $e^x = y + 1 - 2x$, where x and y are small, show that

$$x = \frac{1}{54}(18y - y^2)$$

Let

$$e^x = y + 1 - 2x \tag{i}$$

We know that

$$e^x = 1 + x + \frac{x^2}{2!} + \frac{x^3}{3!} + \cdots = 1 + x + \frac{x^2}{2} + \frac{x^3}{6} + \cdots$$

or

$$e^x + 2x = 1 + 3x + \frac{x^2}{2} + \frac{x^3}{6} + \cdots \tag{ii}$$

From (i) and (ii), it follows that

$$y + 1 = 1 + 3x + \frac{x^2}{2} + \frac{x^3}{6} + \cdots$$

or

$$y = 3x + \frac{x^2}{2} + \frac{x^3}{6} + \cdots \tag{iii}$$

If $x^2, x^3, \ldots$ are neglected, (iii) gives us:

$$y = 3x \quad \text{or} \quad x = \frac{1}{3}y,$$

as a first approximation to x. As a better approximation, we put

$$x = \frac{1}{3} y + u \tag{iv}$$

Then (iii) gives:

$$y = 3\left(\frac{1}{3} y + u\right) + \frac{1}{2}\left(\frac{1}{3} y + u\right)^2 + \cdots = y + 3u + \frac{1}{2}\left(\frac{1}{9} y^2 + u^2 + \frac{2}{3} yu\right) + \cdots$$

Neglecting terms containing u^2, $yu, \ldots,$ we obtain:

$$3u = -\frac{1}{18} y^2 \quad \text{or} \quad u = -\frac{1}{54} y^2 \qquad \text{(approx.)}$$

Hence (iv) gives:

$$x = \frac{1}{3} y - \frac{1}{54} y^2 = \frac{1}{54}(18y - y^2).$$

Example 1.36 If x and y are small, show that

$$\frac{(1+x)^y}{(1+y)^x} = 1 + \frac{1}{2} xy(y - x).$$

We have

$$(1+x)^y = 1 + yx + \frac{y(y-1)}{2} x^2 + \cdots$$

and

$$\frac{1}{(1+y)^x} = (1+y)^{-x} = 1 - xy + \frac{x(x+1)}{2} y^2 + \cdots$$

Hence

$$\frac{(1+x)^y}{(1+y)^x} = \left[1 + yx + \frac{y(y-1)}{2} x^2 + \cdots\right]\left[1 - xy + \frac{x(x+1)}{2} y^2 - \cdots\right]$$

$$= 1 - xy + \frac{x(x+1)}{2} y^2 + yx - y^2x^2 + \frac{x^2(x+1)}{2} y^3 + \frac{y(y-1)}{2} x^2$$

$$- \frac{y^2(y-1)}{2} x^3 + \cdots$$

$$= 1 + \frac{x^2y^2 + xy^2}{2} - y^2x^2 + \frac{y^2x^2 - yx^2}{2} + \cdots$$

Neglecting the other terms, we have

$$\frac{(1+x)^y}{(1+y)^x}=1+\frac{1}{2}xy^2-\frac{1}{2}yx^2=1+\frac{1}{2}xy(y-x).$$

1.3.4 The Logarithmic Series

We have from (1.24):

$$a^y=1+y\log_e a+\frac{y^2}{2!}(\log_e a)^2+\cdots$$

Putting $a=1+x$, this gives

$$(1+x)^y=1+y\log_e(1+x)+\frac{y^2}{2!}[\log_e(1+x)]^2+\cdots \quad (1.25)$$

If $|x|<1$, the left-hand side can be expanded by the binomial theorem, and we have

$$(1+x)^y=1+yx+\frac{y(y-1)}{2!}x^2+\frac{y(y-1)(y-2)}{3!}x^3+\cdots \quad (1.26)$$

Since the series is absolutely convergent, it can be rearranged in ascending powers of y. Then the coefficients of corresponding powers of y in (1.25) and (1.26) are equal. Equating the coefficients of y, we obtain

$$\log_e(1+x)=x-\frac{x^2}{2}+\frac{x^3}{3}-\frac{x^4}{5}+\cdots+(-1)^{n-1}\frac{x^n}{n}+\cdots \quad (1.27)$$

which is called the *logarithmic series* and is valid for $-1<x<1$. This series can be used to calculate the natural logarithms of numbers. Thus, putting $x=1$ in both sides of (1.27), we obtain

$$\log_e 2=1-\frac{1}{2}+\frac{1}{3}-\frac{1}{4}+\cdots \quad (1.28)$$

The above series converges slowly, i.e. the successive terms decrease slowly and therefore a large number of terms must be taken to evaluate the logarithm to a required degree of accuracy. We must therefore obtain an alternate series which converges more rapidly.

From (1.27), we obtain

$$\log_e(1-x)=-x-\frac{x^2}{2}-\frac{x^3}{3}-\frac{x^4}{4}-\cdots \quad (1.29)$$

so that we get by subtraction:

$$\log_e\frac{1+x}{1-x}=2\left(x+\frac{x^3}{3}+\frac{x^5}{5}+\cdots\right)$$

Now, putting

$$\frac{1+x}{1-x}=\frac{m}{n} \quad \text{or} \quad x=\frac{m-n}{m+n},$$

we obtain

$$\log_e \frac{m}{n}=2\left[\frac{m-n}{m+n}+\frac{1}{3}\left(\frac{m-n}{m+n}\right)^3+\frac{1}{5}\left(\frac{m-n}{m+n}\right)^5+\cdots\right] \tag{1.30}$$

This series converges quite rapidly and is therefore suitable for the computation of logarithms. Thus, setting $m=2$ and $n=1$ in (1.30), we obtain

$$\log_e 2=2\left[\frac{1}{3}+\frac{1}{3}\left(\frac{1}{3}\right)^2+\frac{1}{5}\left(\frac{1}{3}\right)^5+\cdots\right]$$

$$=2\left[0.33333+\frac{1}{3}(0.03704)+\frac{1}{5}(0.00411)+\frac{1}{7}(0.00046)+\cdots\right]$$

$$=0.6931 \quad \text{(correct to four decimal places).}$$

Example 1.37 Assuming that x is small, show that

$$\log[(1+x)^{1/3}+(1-x)^{1/3}]\approx \log 2-\frac{1}{9}x^2.$$

Now

$$(1+x)^{1/3}=1+\frac{1}{3}x-\frac{1}{9}x^2+\cdots$$

$$(1-x)^{1/3}=1-\frac{1}{3}x-\frac{1}{9}x^2+\cdots$$

and

$$(1+x)^{1/3}+(1-x)^{1/3}=2\left(1-\frac{1}{9}x^2+\cdots\right)$$

Hence

$$\log\,[(1+x)^{1/3}+(1-x)^{1/3}]=\log\left[2\left(1-\frac{1}{9}x^2+\cdots\right)\right]$$

$$=\log\,2+\log\left(1-\frac{1}{9}x^2+\cdots\right)$$

$$=\log\,2-\frac{1}{9}x^2+\cdots$$

Thus

$$\log\,[(1+x)^{1/3}+(1-x)^{1/3}]\approx\log 2-\frac{1}{9}x^2.$$

Example 1.38 If m and n are the roots of the equation $x^2+bx+c=0$, prove that

$$\log\,(1-bx+cx^2)=(m+n)x-\frac{m^2+n^2}{2}x^2+\frac{m^3+n^3}{3}x^3-\cdots$$

It is given that m and n are the roots of the equation $x^2+bx+c=0$. Hence

$$m+n=-b \quad\text{and}\quad mn=c.$$

We therefore write

$$1-bx+cx^2=1+(m+n)x+mnx^2=(1+mx)(1+nx).$$

Hence

$$\begin{aligned}\log\,(1-bx+cx^2)&=\log\,[(1+mx)(1+nx)]\\&=\log\,(1+mx)+\log\,(1+nx)\\&=mx-\frac{1}{2}m^2x^2+\frac{1}{3}m^3x^3-\frac{1}{4}m^4x^4+\cdots\\&\quad+nx-\frac{1}{2}n^2x^2+\frac{1}{3}n^3x^3-\frac{1}{4}n^4x^4+\cdots\\&=(m+n)x-\frac{1}{2}(m^2+n^2)x^2+\frac{1}{3}(m^3+n^3)x^3-\cdots,\end{aligned}$$

which is the required result.

Example 1.39 Show that

$$\log_3 e-\log_9 e+\log_{27} e-\log_{81} e+\cdots=\frac{\log_e 2}{\log_e 3}.$$

Using the formula

$$\log_a x\log_x a=1,$$

we write

$$\log_3 e=\frac{1}{\log_e 3},\quad \log_9 e=\frac{1}{\log_e 9},\ \text{etc.}$$

Hence

$$\log_3 e - \log_9 e + \log_{27} e - \log_{81} e + \cdots = \frac{1}{\log_e 3} - \frac{1}{\log_e 9} + \frac{1}{\log_e 27} - \frac{1}{\log_e 81} + \cdots$$

$$= \frac{1}{\log_e 3} - \frac{1}{2\log_e 3} + \frac{1}{3\log_e 3} - \frac{1}{4\log_e 3} + \cdots$$

$$= \frac{1}{\log_e 3}\left(1 - \frac{1}{2} + \frac{1}{3} - \frac{1}{4} + \cdots\right)$$

Using (1.27), we obtain

$$\log_3 e - \log_9 e + \log_{27} e - \log_{81} e + \cdots = \frac{\log_e 2}{\log_e 3}.$$

Example 1.40 Show that

$$(1+x)^{1+x} = 1 + x + x^2 + \frac{1}{2}x^3,$$

when x^4 and higher powers of x are neglected. Hence find an approximate value of $(1.02)^{1.02}$.

Let $u = (1+x)^{1+x}$. Then

$$\ln u = (1+x)\ \ln(1+x).$$

Neglecting terms containing x^4 and higher powers of x, we have

$$\ln\ u = (1+x)\left(x - \frac{x^2}{2} + \frac{x^3}{3}\right),$$

or

$$\ln u = x - \frac{x^2}{2} + \frac{x^3}{3} + x^2 - \frac{x^3}{2} = x + \frac{x^2}{2} - \frac{x^3}{6}.$$

Hence

$$u = (1+x)^{1+x} = \exp\left(x + \frac{x^2}{2} - \frac{x^3}{6}\right)$$

$$= e^x \exp\left(\frac{x^2}{2}\right) \exp\left(-\frac{x^3}{6}\right)$$

$$= \left(1 + x + \frac{x^2}{2} + \frac{x^3}{6}\right)\left(1 + \frac{x^2}{2}\right)\left(1 - \frac{x^3}{6}\right)$$

$$= \left(1 + x + \frac{x^2}{2} + \frac{x^3}{6}\right)\left(1 + \frac{x^2}{2} - \frac{x^3}{6}\right)$$

$$= 1 + x + \frac{x^2}{2} + \frac{x^3}{6} + \frac{x^2}{2} + \frac{x^3}{2} - \frac{x^3}{6}.$$

$$= 1 + x + x^2 + \frac{x^3}{2}, \qquad \text{(nearly)}.$$

Putting $x = 0.02$, we obtain

$$(1.02)^{1.02} = 1 + 0.02 + (0.02)^2 + \frac{1}{2}(0.02)^3$$

$$= 1 + 0.02 + 0.0004 + \frac{1}{2}(0.000008)$$

$$= 1.020404.$$

Example 1.41 If p, q, r denote three consecutive integers, prove that

$$\log_e q = \frac{1}{2}\log_e (pr) + \frac{1}{2pr+1} + \frac{1}{3}\frac{1}{(2pr+1)^3} + \cdots$$

Let

$$\frac{1}{2pr+1} = x.$$

Then

$$\frac{1}{2pr+1} + \frac{1}{3}\frac{1}{(2pr+1)^3} + \cdots = x + \frac{1}{3}x^3 + \cdots$$

$$= \frac{1}{2}\ln\frac{1+x}{1-x}$$

$$= \frac{1}{2}\ln\frac{1+1/(2pr+1)}{1-1/(2pr+1)}$$

$$= \frac{1}{2}\ln\frac{2pr+2}{2pr}$$

$$= \frac{1}{2}\ln\frac{pr+1}{pr} \qquad \text{(i)}$$

Hence

$$\frac{1}{2}\log_e(pr)+\frac{1}{2pr+1}+\frac{1}{3}\frac{1}{(2pr+1)^3}+\cdots=\frac{1}{2}\log_e(pr)+\frac{1}{2}\log_e\frac{pr+1}{pr}, \quad \text{[using (i)]}$$

$$=\frac{1}{2}\log_e(pr+1). \qquad \text{(ii)}$$

Since p, q, r are consecutive integers, we have

$$p=q-1 \quad \text{and} \quad r=q+1.$$

Therefore,

$$pr=q^2-1 \quad \text{or} \quad pr+1=q^2. \qquad \text{(iii)}$$

Hence (ii) gives:

$$\frac{1}{2}\log_e(pr)+\frac{1}{2pr+1}+\frac{1}{3}\frac{1}{(2pr+1)^3}+\cdots=\frac{1}{2}\log_e(q)^2, \qquad \text{[using (iii)]}$$

$$=\log_e q.$$

EXERCISES 1.3

1. Find the following correct to four decimal places:

   (a) $(138)^{1/2}$ (b) $(734)^{1/3}$ (c) $(630)^{1/4}$.

2. Expand $1/(4-3x)^3$ to four terms. Compare the value of the function with that of its approximation when $x=0.0625$.

3. Compute $\sqrt{18}$ to four significant figures without tables. If x is so small that the second and higher powers of x may be neglected, find an approximation for

$$(2-x)^3\frac{(1+x/2)^{1/2}}{4-3x}.$$

4. Write down the first-four terms in the expansion of $\sqrt{(a^2+x^2)}$ in ascending powers of x. If x is small compared with a, show that $a+x^2/(2a)$ exceeds $\sqrt{(a^2+x^2)}$ by an amount which is less than $x^4/(8a^3)$. Deduce that an approximate value of the square root of 4.01 is 2.0025, and that this approximation is correct is five decimal places.

5. If c is small in comparison with l, show that

$$\left(\frac{l}{l+c}\right)^{1/2}+\left(\frac{l}{l-c}\right)^{1/2}=2+\frac{3c^2}{4l^2}, \quad \text{(approx.)}$$

6. If n is large, show that $\sqrt{n}$ exceeds

$$\frac{\sqrt{n+1}+\sqrt{n-1}}{2}$$

by $1/(8\sqrt{n^3})$ approximately. Hence show that

$$10-\frac{1}{2}\left(\sqrt{101}+\sqrt{99}\right)=0.000125 \quad \text{(nearly)}$$

7. When x is small, prove that

$$\frac{(1-3x)^{-2/3}+(1-4x)^{-3/4}}{(1-3x)^{-1/3}+(1-4x)^{-1/4}}=1+\frac{3}{2}x+4x^2, \quad \text{(approx.)}$$

8. Find the coefficient of x^7 in the expansion of

$$\left(\frac{x^2}{2}-\frac{2}{x}\right)^8.$$

9. Find the coefficient of x^r in the expansion of $(1+x+x^2)/e^x$.

10. If $C_0, C_1, C_2, C_3, \ldots, C_n$ be the coefficients in the expansion of $(1+x)^n$, prove that

(a) $C_0^2+C_1^2+C_2^2+\cdots+C_n^2=\dfrac{(2n)!}{(n!)^2}$

(b) $C_0C_r+C_1C_{r+1}+C_2C_{r+2}+\cdots+C_{n-r}C_n=\dfrac{n!}{(n-r)!(n+r)!}$

11. Given that

$$C_1+2C_2x+3C_3x^2+\cdots+2nC_{2n}x^{2n-1}=2n(1+x)^{2n-1}$$

where

$$C_r=\frac{(2n)!}{r!(2n-r)!}, \quad r=0, 2, \ldots, 2n.$$

Prove that

$$C_1^2-2C_2^2+3C_3^2-\cdots-2nC_{2n}^2=(-1)^n C_n^n.$$

12. If the coefficients of x^{r-1}, x^r and x^{r+1} in the binomial expansion of $(1+x)^n$ are in arithmetic progression, prove that

$$n^2-n(4r+1)+4r^2-2=0.$$

13. If $p=q$ nearly and $n>1$, show that

$$\frac{(n+1)p+(n-1)q}{(n-1)p+(n+1)q}=\left(\frac{p}{q}\right)^{1/n}$$

Hence, find the 7th root of 131/133.

14. If

$$y = -\left(x^3 + \frac{x^6}{2} + \frac{x^9}{3} + \cdots\right),$$

show that $x^3 = 1 - e^y$.

15. If x is so small that terms of higher order than x^2 may be neglected, show that

$$e^x\sqrt{1+x} + e^{-x}\sqrt{1-x} \approx \frac{7}{4}x^2 + 2.$$

16. For the function given by $f(x) = x^4 - 5x^3 + 6$, where $x = -1, 0, 1, 2, 3$, form the table of differences.

17. Given that $f_x = x^2 - 2x + 3$ and $h = 1$, find analytical expressions for Δf_x, $\Delta^2 f_x$ and $\Delta^3 f_x$.

18. The following table of values of x and y are given. Find $y(0.5)$ and $y(3.6)$.

x	0	1	2	3	4
y	–5	–10	–9	4	35

19. Values of $y(x) = \tan x$ are given below for $x = 35°, 37°, 39°, 41°, 43°$ and $45°$. Compute $\tan 36°$ and $\tan 44°$.

$x°$	35	37	39	41	43	45
$\tan x$	0.7002	0.7536	0.8098	0.8693	0.9325	1.0000

20. The following set of values of x and y was obtained from an experiment. Find values of y for $x = 0.5$, 1.5 and 3.8.

x	0	1	2	3	4
y	6.9897	7.4036	7.7815	8.1291	8.4510

21. The following set of values is taken from a table of cubes. Estimate, as accurately as possible, $\sqrt[3]{51}$ and $\sqrt[3]{59}$.

x	50	52	54	56	58	60
$\sqrt[3]{x}$	3.684	3.733	3.780	3.826	3.871	3.915

22. Show that e is an incommensurable number.

23. Expand $(1 + 2x + 3x^2)/e^x$ as a power series in x and write down the coefficient of x^n.

24. Expand $e^{a+bx+cx^2}$ in ascending powers of x up to and including the term in x^3. If the first-three terms of the expansion are $p+qx+rx^2$, where p, q, r are supposed to be given, show that

$$a = \log p, \quad b = \frac{q}{p}, \quad c = \frac{2pr - q^2}{2p^2}.$$

25. When n is large, show that

$$e\left(1 - \frac{1}{2n} + \frac{11}{24n^2}\right)$$

is an approximate value of $(1+1/n)^n$.

26. Prove that, if x is sufficiently small,

$$\log_e (1 + x + x^2) = x + \frac{1}{2}x^2 - \frac{2}{3}x^3 + \frac{1}{4}x^4 + \frac{1}{5}x^5 - \frac{1}{3}x^6 + \cdots$$

Prove that the coefficient of x^n is $1/n$ if n is not a multiple of 3, and is $-2/n$ if n is a multiple of 3.

27. Given that $\log_{10} 3 = 0.47712$ and $\log_e 10 = 2.30259$, find $\log_{10} 29$ correct to five decimal places.

28. Expand

$$\log_e \frac{1}{1 + x + x^2 + x^3}$$

in ascending powers of x and find the coefficient of x^n when n is a multiple of 4.

29. Show that if

$$\frac{1}{2} \log \frac{1+x}{1-x},$$

where $|x| < 1$, is replaced by the first-n terms of the series

$$x + \frac{1}{3}x^3 + \frac{1}{5}x^5 + \cdots$$

the error is less than the numerical value of

$$\frac{x^{2n+1}}{(2n+1)(1-x^2)}.$$

30. Show that

$$\log \sqrt{12} = 1 + \left(\frac{1}{2} + \frac{1}{3}\right)\frac{1}{4} + \left(\frac{1}{4} + \frac{1}{5}\right)\frac{1}{4^2} + \left(\frac{1}{6} + \frac{1}{7}\right)\frac{1}{4^3} + \cdots$$

1.4 CONTINUED FRACTIONS

The expression

$$a_0 + \cfrac{b_1}{a_1 + \cfrac{b_2}{a_2 + \cfrac{b_3}{a_3 + \cdots}}} \tag{1.31}$$

is called a *continued fraction* and is usually written, more conveniently, as

$$a_0 + \frac{b_1}{a_1+}\,\frac{b_2}{a_2+}\,\frac{b_3}{a_3+}\cdots \tag{1.32}$$

In particular, the continued fraction

$$a_0 + \frac{1}{a_1+}\,\frac{1}{a_2+}\,\frac{1}{a_3+}\cdots$$

is called a *simple continued fraction.*

In general, the elements a_0, a_k, b_k $(k = 1, 2, \ldots)$ are real or complex numbers, or functions of one or more variables. The fractions

$$\frac{a_0}{1}, \qquad \frac{b_k}{a_k}\ (k = 1, 2, \ldots)$$

are called its *components.* If the continued fraction contains an infinite number of components, it is called an *infinite continued fraction;* otherwise, it is a finite continued fraction.

Any positive rational number can be expressed uniquely as a simple continued fraction. For example,

$$\frac{73}{43} = 1 + \frac{30}{43} = 1 + \frac{1}{43/30} = 1 + \frac{1}{1 + 13/30}$$

$$= 1 + \cfrac{1}{1 + \cfrac{1}{\cfrac{30}{13}}} = 1 + \cfrac{1}{1 + \cfrac{1}{2 + \cfrac{4}{13}}} = 1 + \cfrac{1}{1 + \cfrac{1}{2 + \cfrac{1}{\cfrac{13}{4}}}}$$

$$= 1 + \cfrac{1}{1 + \cfrac{1}{2 + \cfrac{1}{3 + \cfrac{1}{4}}}} = 1 + \frac{1}{1+}\,\frac{1}{2+}\,\frac{1}{3+}\,\frac{1}{4}.$$

Conversely, any finite continued fraction can easily be converted to a simple fraction by simplifying the fractions in succession starting with the lowest. For example,

$$2+\frac{1}{2+}\,\frac{1}{1+}\,\frac{1}{4}=2+\cfrac{1}{2+\cfrac{1}{1+\cfrac{1}{4}}}=2+\cfrac{1}{2+\cfrac{4}{5}}=2+\frac{5}{14}=\frac{33}{14}.$$

1.4.1 Convergents of a Continued Fraction

If the infinite continued fraction is terminated by setting $b_{n+1}=0$, we would be left with

$$a_0+\frac{b_1}{a_1+}\,\frac{b_2}{a_2+}\,\frac{b_3}{a_3+}\cdots\frac{+b_n}{a_n}, \tag{1.33}$$

which is called the *nth convergent* of the continued fraction. For example,

a_0 is the 0th convergent,

$a_0+\dfrac{b_1}{a_1}$ is the 1st convergent,

$a_0+\dfrac{b_1}{a_1+}\,\dfrac{b_2}{a_2}$ is the 2nd convergent, etc.

Let

$$C_n=a_0+\frac{b_1}{a_1+}\,\frac{b_2}{a_2+}\cdots\frac{+b_n}{a_n}=\frac{B_n}{A_n}.$$

Then, with

$$B_0=a_0,\quad A_0=1,\quad C_0=a_0,\quad C_1=a_0+\frac{b_1}{a_1}=\frac{a_0a_1+b_1}{a_1}$$

Hence

$$B_1=a_0a_1+b_1,\qquad A_1=a_1.$$

Again

$$C_2=a_0+\frac{b_1}{a_1+}\,\frac{b_2}{a_2}=a_0+\frac{b_1a_2}{a_1a_2+b_2}=\frac{a_0a_1a_2+a_0b_2+b_1a_2}{a_1a_2+b_2}.$$

Therefore,

$$B_2=a_0a_1a_2+a_0b_2+b_1a_2=a_2(a_0a_1+b_1)+a_0b_2=a_2B_1+b_2B_0$$

$$A_2=a_1a_2+b_2=a_2A_1+b_2A_0.$$

The above expressions for B_2 and A_2 suggest the law of formation of successive convergents as

$$B_n=a_nB_{n-1}+b_nB_{n-2},\quad A_n=a_nA_{n-1}+b_nA_{n-2},\quad n=2,3,\ldots \tag{1.34}$$

The relations given in (1.34) are called the *recurrence formulae* for the calculation of the successive convergents of a continued fraction. These formulae can be proved by the method of mathematical induction.

Proof Let the relations (1.34) hold good for $n = 2, 3, \ldots, N$. Then

$$C_{N+1} = \frac{B_{N+1}}{A_{N+1}}$$

$$= a_0 + \frac{b_1}{a_1+} \frac{b_2}{a_2+} \cdots \frac{+b_N}{a_N+} \frac{b_{N+1}}{a_{N+1}}$$

$$= a_0 + \frac{b_1}{a_1+} \frac{b_2}{a_2+} \cdots \frac{+b_{N-1}}{a_{N-1}+} \frac{b_N a_{N+1}}{a_N a_{N+1} + b_{N+1}}$$

This is now an Nth convergent of the continued fraction

$$c'_n = a'_0 + \frac{b'_1}{a'_1+} \frac{b'_2}{a'_2+} \cdots \frac{+b'_n}{a'_n} = \frac{B'_n}{A'_n}, \qquad n = 0, 1, \ldots, N.$$

Then

$$B'_n = a'_n B'_{n-1} + b'_n B'_{n-2}.$$

But

$$B'_n = B'_n \quad \text{for } n = 1, 2, \ldots, N-1$$

and so

$$\begin{aligned} B'_N &= (a_N a_{N+1} + b_{N+1})\, B_{N-1} + b_N a_{N+1} B_{N-2} \\ &= a_{N+1}\,(a_N B_{N-1} + b_N B_{N-2}) + b_{N+1} B_{N-1} \\ &= a_{N+1} B_N + b_{N+1} B_{N-1} \\ &= B_{N+1}, \end{aligned}$$

which completes the proof.

Example 1.42 Calculate

$$\sqrt{2} = 1 + \frac{1}{2+} \frac{1}{2+} \frac{1}{2+} \cdots$$

correct to six decimal places.

We have

$$\frac{B_n}{A_n} = a_0 + \frac{b_1}{a_1+} \frac{b_2}{a_2+} \cdots \frac{b_n}{+a_n},$$

where

$$\begin{aligned} A_0 &= 1, \quad B_0 = a_0, \\ A_1 &= a_1, \quad B_1 = a_0 a_1 + b_1, \end{aligned}$$

and

$$B_n = a_n B_{n-1} + b_n B_{n-2}, \quad A_n = a_n A_{n-1} + b_n A_{n-2}, \qquad n > 2.$$

Now

$$\frac{B_0}{A_0} = 1,$$

$$\frac{B_1}{A_1} = \frac{3}{2},$$

$$\frac{B_2}{A_2} = \frac{7}{5};$$

$$\frac{B_3}{A_3} = \frac{a_3 B_2 + b_3 B_1}{a_3 A_2 + b_3 A_1} = \frac{2(7)+1(3)}{2(5)+1(2)} = \frac{17}{12} = 1.4167$$

$$\frac{B_4}{A_4} = \frac{a_4 B_3 + b_4 B_2}{a_4 A_3 + b_4 A_2} = \frac{2(17)+1(7)}{2(12)+1(5)} = \frac{41}{29} = 1.4137931$$

$$\frac{B_5}{A_5} = \frac{99}{70} = 1.4142857$$

$$\frac{B_6}{A_6} = \frac{239}{169} = 1.4142011$$

$$\frac{B_7}{A_7} = \frac{577}{408} = 1.4142156$$

$$\frac{B_8}{A_8} = \frac{1393}{985} = 1.4142131$$

$$\frac{B_9}{A_9} = \frac{3363}{2378} = 1.4142136$$

$$\frac{B_{10}}{A_{10}} = \frac{8119}{5741} = 1.4142135$$

Hence

$$\sqrt{2} = 1 + \frac{1}{2+}\,\frac{1}{2+}\,\frac{1}{2+}\cdots = 1.414214,$$

correct to six decimal places.

Example 1.43 Prove that

$$a + \frac{1}{b+}\,\frac{1}{2a+}\,\frac{1}{b+}\,\frac{1}{2a+}\cdots = \sqrt{\frac{a}{b}(ab+2)},$$

where a and b are positive.

Let

$$z = a + \frac{1}{b+}\,\frac{1}{2a+}\,\frac{1}{b+}\,\frac{1}{2a+}\cdots\infty,$$

and

$$x = \frac{1}{b+}\,\frac{1}{2a+}\,\frac{1}{b+}\,\frac{1}{2a+}\cdots\infty,$$

so that

$$z = a + x.$$

Clearly

$$x = \frac{1}{b+}\,\frac{1}{2a+x} = \frac{2a+x}{(2a+x)b+1},$$

whence $bx^2 + 2abx - 2a = 0$. Solving this quadratic equation, we get

$$x = -a \pm \sqrt{a^2 + \frac{2a}{b}}.$$

Hence

$$z = x + a = \pm\sqrt{a^2 + \frac{2a}{b}} = \pm\sqrt{\frac{a}{b}\,(ab+2)}.$$

Since a and b are positive, the value of the given continued fraction is also positive and hence we have

$$z = \sqrt{\frac{a}{b}(ab+2)}.$$

1.4.2 Estimation of Error

Theorem 1.13 Let S be the value of the simple continued fraction

$$a_0 + \frac{1}{a_1+}\,\frac{1}{a_2+}\cdots\frac{1}{+a_n+}\cdots$$

whose elements are positive, and let B_K/A_K be its successive convergents. Then

$$\left|S - \frac{B_K}{A_K}\right| \le \frac{1}{A_K A_{K+1}} \tag{1.35}$$

Since all the elements are positive, we have

$$S > \frac{B_0}{A_0} = a_0, \quad S < \frac{B_1}{A_1}, \quad S > \frac{B_2}{A_2}, \cdots$$

i.e.

$$\frac{B_K}{A_K} < S < \frac{B_{K+1}}{A_{K+1}}, \qquad \text{(for } k \text{ even)}$$

and

$$\frac{B_K}{A_K} > S > \frac{B_{K+1}}{A_{K+1}}, \qquad \text{(for } k \text{ odd)}$$

Hence

$$\left| S - \frac{B_K}{A_K} \right| \le \left| \frac{B_{K+1}}{A_{K+1}} - \frac{B_K}{A_K} \right| \tag{1.36}$$

Now

$$\frac{B_{K+1}}{A_{K+1}} - \frac{B_K}{A_K} = \frac{B_{K+1}A_K - B_K A_{K+1}}{A_K A_{K+1}}$$

$$= \frac{(a_{K+1}B_K + b_{K+1}B_{K-1})A_K - B_K(a_{K+1}A_K + b_{K+1}A_{K-1})}{A_K A_{K+1}}$$

$$= -\frac{B_K A_{K-1} - B_{K-1}A_K}{A_K A_{K+1}}, \qquad \text{(since } b_{K+1} = 1\text{)}.$$

$$= (-1)^2 \frac{B_{K-1}A_{K-2} - B_{K-2}A_{K-1}}{A_K A_{K+1}}$$

$$= (-1)^K \frac{B_1 A_0 - B_0 A_1}{A_K A_{K+1}}.$$

But $B_0 = a_0$, $A_0 = 1$, $B_1 = a_0 a_1 + b_1$, $A_1 = a_1$. Therefore,

$$B_1 A_0 - B_0 A_1 = a_0 a_1 + 1 - a_0 a_1 = 1,$$

and

$$A_K A_{K+1} = A_K(a_{K+1}A_K + A_{K-1})$$

Hence,

$$\frac{B_{K+1}}{A_{K+1}} - \frac{B_K}{A_K} = \frac{(-1)^K}{A_K(a_{K+1}A_K + A_{K-1})}$$

and

$$\left| \frac{B_{K+1}}{A_{K+1}} - \frac{B_K}{A_K} \right| \le \frac{1}{A_K(a_{K+1}A_K + A_{K-1})}.$$

We therefore obtain from (1.36):

$$\left| S - \frac{B_K}{A_K} \right| \le \frac{1}{A_K(a_{K+1}A_K + A_{K-1})}. \tag{1.37}$$

Application of this estimate is illustrated in the following example.

Example 1.44 Express the number $\sqrt{41}$ as a continued fraction and hence find its approximate value.

We write

$$\sqrt{41} = 6 + \frac{1}{a_1}$$

so that

$$a_1 = \frac{1}{\sqrt{41}-6} = \frac{\sqrt{41}+6}{5} = 2 + \frac{1}{a_2}, \quad \text{(say)}$$

Therefore,

$$a_2 = \frac{1}{a_1 - 2} = \frac{5}{\sqrt{41}-4} = \frac{\sqrt{41}+4}{5} = 2 + \frac{1}{a_3},$$

so that

$$a_3 = \frac{1}{a_2 - 2} = \frac{5}{\sqrt{41}-6} = \sqrt{41} + 6 = 12 + \frac{1}{a_4}.$$

We then have

$$a_4 = \frac{1}{a_3 - 12} = \frac{1}{\sqrt{41}-6} = \frac{\sqrt{41}+6}{5} = 2 + \frac{1}{a_5}.$$

This gives

$$a_5 = \frac{5}{\sqrt{41}-4}.$$

It is clear, at this stage, that $a_4 = a_1$, $a_5 = a_2$, etc. Hence the elements are repeating which means that the continued fraction is a recurring one, and we have

$$\sqrt{41} = 6 + \frac{1}{2+}\,\frac{1}{2+}\,\frac{1}{12+}\,\frac{1}{2+}\,\frac{1}{2+}\,\frac{1}{12+}\cdots$$

Here

$B_0 = 6, \quad A_0 = 1; \quad B_1 = 13, \quad A_1 = 2; \quad B_2 = 32, \quad A_2 = 5.$

Therefore,

$$B_3 = 12(32) + 1(13) = 397,\ A_3 = 12(5) + 1(2) = 62,\ B_4 = 826,\ A_4 = 129, \text{ etc.}$$

If we terminate the computations here, then the value S of the continued fraction is given by

$$S \approx \frac{B_4}{A_4} = \frac{826}{129} = 6.4031$$

Then

$$\text{The absolute error} \leq \frac{1}{A_K A_{K+1}} = \frac{1}{A_K (a_{K+1} A_K + A_{K-1})}$$

$$= \frac{1}{129[2(129)+62]}$$

$$= \frac{1}{(129)(320)},$$

which shows that S is correct to four decimal places.

1.4.3 Application to Interpolation

Suppose we are given a set of data values of x and $f(x)$, viz., $x_0, f(x_0)$; $x_1, f(x_1)$; ...; $x_n, f(x_n)$, where $f(x)$ is an unknown function of x. We are required to compute the value of $f(x)$ for some non-tabular value of x, such that $x_0 < x < x_n$. Such a process, viz., the process of 'reading' across a table of values, is called *interpolation.* In practice, however, it would be convenient to determine a function, say $\phi_0(x)$, which agrees with $f(x)$ at the set of tabulated points. In what follows, we determine a continued fraction which agrees with $f(x)$ at $x_0, x_1, x_2, \ldots, x_n$.

We first define quantities $\phi_0[x]$, $\phi_1[x_0, x]$, $\phi_2[x_0, x_1, x]$, $\phi_3[x_0, x_1, x_2, x]$, ... in the following way:

$$\phi_0[x] = f(x),$$

$$\phi_1[x_0, x] = \frac{x - x_0}{\phi_0[x] - \phi_0[x_0]} = \frac{x - x_0}{f(x) - f(x_0)},$$

$$\phi_2[x_0, x_1, x] = \frac{x - x_1}{\phi_1[x_0, x] - \phi_1[x_0, x_1]}$$

$$\vdots$$

$$\phi_n[x_0, x_1, \ldots, x_{n-1}, x] = \frac{x - x_{n-1}}{\phi_{n-1}[x_0, x_1, \ldots, x_{n-2}, x] - \phi_{n-1}[x_0, x_1, \ldots, x_{n-1}]}$$

These quantities are called *reciprocal differences.* It is easily seen that these are symmetric with respect to their arguments. For example,

$$\phi_1[x_0, x] = \frac{x - x_0}{\phi_0[x] - \phi_0[x_0]} = \frac{x_0 - x}{\phi_0[x_0] - \phi_0[x]} = \phi_1[x, x_0].$$

For the given set of data values, i.e. $x_0, f(x_0)$; $x_1, f(x_1)$;...; $x_n, f(x_n)$, the reciprocal differences can be computed and arranged in Table 1.2.

Table 1.2 Reciprocal Differences

x_0	$f(x_0)$			
x_1	$f(x_1)$	$\phi_1[x_0, x_1]$		
x_2	$f(x_2)$	$\phi_1[x_0, x_2]$	$\phi_2[x_0, x_1, x_2]$	
x_3	$f(x_3)$	$\phi_1[x_0, x_3]$	$\phi_2[x_0, x_1, x_3]$	$\phi_3[x_0, x_1, x_2, x_3]$
$\vdots$	$\vdots$	$\vdots$	$\vdots$	$\vdots$
x_n	$f(x_n)$	$\phi_1[x_0, x_n]$	$\phi_2[x_0, x_1, x_n]$	$\phi_3[x_0, x_1, x_2, x_n]$

An important advantage of the table given above is that the reciprocal differences along a downward diagonal line can be used to construct a continued fraction approximating the given data values. This is shown below:

We have

$$\phi_1[x_0, x] = \frac{x - x_0}{\phi_0[x] - \phi_0[x_0]}$$

From the above, we obtain

$$\phi_0[x] = \phi_0[x_0] + \frac{x - x_0}{\phi_1[x_0, x]} = f(x_0) + \frac{x - x_0}{\phi_1[x_0, x]} \tag{1.38}$$

Now

$$\phi_2[x_0, x_1, x] = \frac{x - x_1}{\phi_1[x_0, x] - \phi_1[x_0, x_1]},$$

which gives

$$\phi_1[x_0, x] = \phi_1[x_0, x_1] + \frac{x - x_1}{\phi_2[x_0, x_1, x]}.$$

Substituting the above in (1.38), we obtain

$$\phi_0[x] = f(x_0) + \frac{x - x_0}{\phi_1[x_0, x_1]} + \frac{x - x_1}{\phi_2[x_0, x_1, x]}.$$

It is clear, at this stage, that the above procedure can be repeated to derive the general formula:

$$\phi_0[x] = f(x_0) + \cfrac{x - x_0}{\phi_1[x_0, x_1] + \cfrac{x - x_1}{\phi_2[x_0, x_1, x_2] + \cfrac{x - x_2}{\phi_3[x_0, x_1, x_2, x_3] + \cdots}}} \tag{1.39}$$

which is the continued fraction approximation to the given set of data values. It can be seen that the reciprocal differences $\phi_1[x_0, x_1]$, $\phi_2[x_0, x_1, x_2]$,

... in (1.39) are all on a downward diagonal line in the table (see Table 1.2) of reciprocal differences. When once the approximating function $f(x)$ is obtained, the value of $f(x)$ for any other value of x in the range can be computed by directed substitution in (1.38). The computational procedure for obtaining the continued fraction is illustrated in the following example.

Example 1.45 Obtain a continued fraction approximation for the following set of points: (1, 1), (2, 4), (3, 9) and (4, 16).

x	$f(x)$			
1	1			
2	4	$\frac{1}{3}$		
3	9	$\frac{1}{4}$	−12	
4	16	$\frac{1}{5}$	−15	$\frac{-1}{3}$

We have

$$\phi_1[x_1, x_2] = \frac{x_2 - x_1}{f(x_2) - f(x_1)} = \frac{1}{3},$$

$$\phi_1[x_1, x_3] = \frac{2}{8} = \frac{1}{4},$$

$$\phi_1[x_1, x_4] = \frac{3}{15} = \frac{1}{5}.$$

Again,

$$\phi_2[x_1, x_2, x_3] = \frac{x_3 - x_2}{\phi_1[x_1, x_3] - \phi_1[x_1, x_2]} = \frac{1}{(1/4)-(1/3)} = -12,$$

$$\phi_2[x_1, x_2, x_4] = \frac{x_4 - x_2}{\phi_1[x_1, x_4] - \phi_1[x_1, x_2]} = \frac{2}{(1/5)-(1/3)} = -15,$$

Finally,

$$\phi_3[x_1, x_2, x_3, x_4] = \frac{x_4 - x_3}{\phi_2[x_1, x_2, x_4] - \phi_2[x_1, x_2, x_3]} = \frac{1}{-15+12} = -\frac{1}{3}.$$

These values are arranged in the table given.

The continued fraction is therefore given by

$$f(x) = 1 + \cfrac{x-1}{\cfrac{1}{3} + \cfrac{x-2}{-12 + \cfrac{x-3}{-1/3}}}$$

The tabulated function is $f(x) = x^2$ and the reader should verify the correctness of the result obtained.

EXERCISES 1.4

1. (a) Convert the following fractions into continued fractions:

$$\frac{29}{140}, \frac{674}{313}, \frac{613}{1497}.$$

(b) Show that

$$\frac{1-x}{1-5x+6x^2} = \frac{1}{1-}\,\frac{4x}{1-}\,\frac{2x}{-4+6x}.$$

2. Calculate the successive convergents to:

(a) $1 + \frac{1}{3+}\,\frac{1}{5+}\,\frac{1}{7+}\,\frac{1}{9+}\,\frac{1}{11}$,

(b) $1 + \frac{1}{2+}\,\frac{1}{3+}\,\frac{1}{2+}\,\frac{1}{3+}.$

3. Evaluate the continued fraction

$$y = 1 + \cfrac{x}{-3 + \cfrac{x-1}{-2/3}} \quad \text{at } x = 0, 1 \text{ and } 2.$$

4. Given the continued fraction

$$a_1 + \frac{1}{a_2+}\,\frac{1}{a_3+}\cdots\frac{1}{+a_n+}\cdots,$$

show that

$$\frac{B_n}{B_{n-1}} = a_n + \frac{1}{a_{n-1}+} + \frac{1}{a_{n-2}+}\cdots\frac{1}{+a_1}.$$

5. If B_n/A_n denotes the nth convergent of the continued fraction given by

$$\frac{1}{a+}\,\frac{1}{b+}\,\frac{1}{a+}\,\frac{1}{b+}\cdots$$

show that $B_nA_{n-2} - B_{n-2}A_n = +a$ or $-b$ according as n is even or odd.

6. If

$$S = 1 + \frac{1}{2+}\,\frac{1}{3+}\,\frac{1}{4+}\,\frac{1}{5+}\cdots,$$

find the error made when the fourth convergent is taken as an approximation to S.

7. Prove that the reciprocal differences are symmetric with respect to the arguments.

8. Form the table of reciprocal differences for

   (i) x^3 for $x = 0, 1, 2, 3, 4, 5, 6$

   (ii) $\dfrac{1}{1+x}$ for $x = 0, 1, 2, 3$

   (iii) $\dfrac{1}{1+x^2}$ for $x = 0, 1, 2, \ldots, 6$

   (iv) $\dfrac{1-x^2}{1+x^2}$ for $x = 1, 2, \ldots, 6$

   (v) $\tan x$ for $x = 65°, 66°, 67°, 68°, 69°$

   (vi) $\sec x$ for $x = 65°, 66°, 67°, 68°, 69°$.

9. Calculate a continued fraction which agrees with $x(t)$ at the values given in the table:

t	0	1	2	3	4	5
$x(t)$	2	3/2	4/5	1/2	6/17	7/26

10. Apply reciprocal differences to recover the function $f(x) = 1/(1+x^2)$ from the following data:

x	0	1	2	3	4	5
$f(x)$	1	1/2	1/5	1/10	1/17	1/26

11. The following values of x and cosec x are given:

x	1°	2°	3°	4°
cosec x	57.30	28.65	19.11	14.34

Calculate cosec 1°30′ using a continued fraction approximation.

1.5 THEORY OF EQUATIONS

Let $a_0 \neq 0$ and

$$f(x) = a_0x^n + a_1x^{n-1} + a_2x^{n-2} + \cdots + a_n = 0 \tag{1.40}$$

where a_i, $i = 0, 1, 2, \ldots, n$ are constants and n is a positive integer. Then $f(x)$ is called a *polynomial* in x of degree n and $f(x) = 0$ is a *polynomial equation* of the nth degree. If $f(c) = 0$, then c is called a *root* of the equation $f(x) = 0$. Some questions which often arise in practice are:

(i) How many roots does an equation $f(x) = 0$ of nth degree has?
(ii) How many of these roots are real and how many imaginary?
(iii) What are the methods for determining the roots of $f(x) = 0$ to a given accuracy?

1.5.1 The Fundamental Theorem of Algebra

The first question is answered by *fundamental theorem of algebra* which states that:

Every polynomial in x has a root, real or imaginary.

The proof of this theorem is beyond the scope of this book but we assume its truth in proving an important corollary of it which is stated below.

Theorem 1.14 A polynomial equation of the nth degree has n and only n roots.

Proof Let $f(x) = 0$ be a polynomial equation of nth degree. By the fundamental theorem of algebra, this equation has at least one root, say $x = \alpha_1$. Then it follows that $(x - \alpha_1)$ is a factor of $f(x)$ and therefore we can write

$$f(x) = (x - \alpha_1)\, f_1(x),$$

where $f_1(x)$ is a polynomial in x of degree $(n-1)$. Now, $f_1(x) = 0$ has at least one root, say $x = \alpha_2$, and so we write

$$f_1(x) = (x - \alpha_2)\, f_2(x),$$

where $f_2(x)$ is a polynomial in x of degree $(n-2)$. Hence we have

$$f(x) = (x - \alpha_1)(x - \alpha_2)\, f_2(x)$$

Again, since $f_2(x) = 0$ has at least one root, say $x = \alpha_3$, we can write

$$f_2(x) = (x - \alpha_3)\, f_3(x),$$

where $f_3(x)$ is a polynomial in x of degree $(n-3)$. Hence we have

$$f(x) = (x - \alpha_1)(x - \alpha_2)(x - \alpha_3)\, f_3(x).$$

Proceeding in this way, we can write

$$f(x) = (x - \alpha_1)(x - \alpha_2)(x - \alpha_3)\ldots(x - \alpha_n)\, f_n(x),$$

where $f_n(x)$ is a constant. We see that $f(x)$ vanishes at $x = \alpha_1, \alpha_2, \ldots, \alpha_n$. Hence $f(x) = 0$ has n roots. Also for values of x other than α_1, α_2, $\ldots, \alpha_n$, $f(x) \neq 0$. Therefore, $f(x) = 0$ has n and only n roots.

If $\alpha_1 = \alpha_2 = \alpha$, then $x = \alpha$ is said to be a *double root* of $f(x) = 0$. Similarly, $x = \alpha$ is called *r-multiple root* of the equation $f(x) = 0$ if $f(x) = (x - \alpha)^r \phi(x)$, where $\phi(\alpha) \neq 0$.

1.5.2 Relations between the Roots and Coefficients

We consider Eq. (1.40) in the form

$$x^n + a_1 x^{n-1} + a_2 x^{n-2} + \cdots + a_n = 0 \tag{1.41}$$

and let the roots of (1.41) be given by $\alpha_1, \alpha_2, \ldots, \alpha_n$. In terms of these roots, Eq. (1.40) can also be written as

$$(x - \alpha_1)(x - \alpha_2)(x - \alpha_3) \cdots (x - \alpha_n) = 0 \tag{1.42}$$

or as

$$x^n - \left(\sum \alpha_1\right) x^{n-1} + \left(\sum \alpha_1 \alpha_2\right) x^{n-2} + \cdots + (-1)^n \alpha_1 \alpha_2 \ldots \alpha_n = 0 \tag{1.43}$$

Comparing equations (1.41) and (1.43), we obtain:

$$\left.\begin{aligned} \sum \alpha_1 &= -a_1 \\ \sum \alpha_1 \alpha_2 &= a_2 \\ \sum \alpha_1 \alpha_2 \alpha_3 &= -a_3 \\ &\vdots \\ \alpha_1 \alpha_2 \alpha_3 \ldots \alpha_n &= (-1)^n a_n \end{aligned}\right\} \tag{1.44}$$

For example, if $\alpha_1, \alpha_2, \alpha_3$ are the roots of the cubic equation

$$a_0 x^3 + a_1 x^2 + a_2 x + a_3 = 0,$$

Then we obtain

$$\alpha_1 + \alpha_2 + \alpha_3 = \frac{-a_1}{a_0}$$

$$\alpha_1 \alpha_2 + \alpha_1 \alpha_3 + \alpha_2 \alpha_3 = \frac{a_2}{a_0}$$

$$\alpha_1 \alpha_2 \alpha_3 = \frac{-a_3}{a_0}.$$

Example 1.46 Solve the equation $4x^3 - 24x^2 + 23x + 18 = 0$ given that its roots are in arithmetic progression.

Let the roots be $\alpha - d$, α and $\alpha + d$. Then we obtain

$$(\alpha - d) + \alpha + (\alpha + d) = \frac{24}{4} = 6 \tag{i}$$

$$(\alpha - d)\alpha + (\alpha - d)(\alpha + d) + \alpha(\alpha + d) = \frac{23}{4} \qquad \text{(ii)}$$

$$(\alpha - d)\alpha(\alpha + d) = -\frac{18}{4} = -\frac{9}{2} \qquad \text{(iii)}$$

From (i), we get:

$$3\alpha = 6 \text{ or } \alpha = 2.$$

Also from (ii),

$$3\alpha^2 - d^2 = \frac{23}{4}$$

or

$$d^2 = 3\alpha^2 - \frac{23}{4} = 12 - \frac{23}{4} = \frac{25}{4}.$$

Therefore, $d = 5/2$. The three roots are therefore $-1/2$, 2 and 9/2.

Example 1.47 If $\alpha_1, \alpha_2, \alpha_3$ are the roots of $x^3 - 3ax + b = 0$, find the value of $\Sigma(\alpha_1 - \alpha_2)(\alpha_1 - \alpha_3)$.

Let

$$x^3 - 3ax + b = 0 \qquad \text{(i)}$$

Since $\alpha_1, \alpha_2, \alpha_3$ are the roots of (i), we have

$$\alpha_1 + \alpha_2 + \alpha_3 = 0 \qquad \text{(ii)}$$

$$\alpha_1\alpha_2 + \alpha_1\alpha_3 + \alpha_2\alpha_3 = -3a \qquad \text{(iii)}$$

$$\alpha_1\alpha_2\alpha_3 = -b \qquad \text{(iv)}$$

Now,

$$\begin{aligned}\sum(\alpha_1 - \alpha_2)(\alpha_1 - \alpha_3) &= (\alpha_1 - \alpha_2)(\alpha_1 - \alpha_3) + (\alpha_2 - \alpha_3)(\alpha_2 - \alpha_1) \\ &\quad + (\alpha_3 - \alpha_1)(\alpha_3 - \alpha_2) \\ &= \alpha_1^2 + \alpha_2^2 + \alpha_3^2 - \alpha_1\alpha_2 - \alpha_1\alpha_3 - \alpha_2\alpha_3 \\ &= (\alpha_1 + \alpha_2 + \alpha_3)^2 - 3(\alpha_1\alpha_2 + \alpha_1\alpha_3 + \alpha_2\alpha_3) \\ &= 0 - 3(-3a) \\ &= 9a.\end{aligned}$$

Example 1.48 If a, b, c are the roots of $x^3 - x - 1 = 0$, find the value of

$$\frac{1+a}{1-a} + \frac{1+b}{1-b} + \frac{1+c}{1-c}.$$

We shall first find the equation whose roots are

$$\frac{1+a}{1-a}, \qquad \frac{1+b}{1-b}, \qquad \frac{1+c}{1-c}.$$

For this we set

$$\frac{1+a}{1-a}=u \quad \text{or} \quad a=\frac{u-1}{u+1}$$

Since $x=a$ is a root of the equation $x^3-x-1=0$, we have

$$\left(\frac{u-1}{u+1}\right)^3-\left(\frac{u-1}{u+1}\right)-1=0 \quad \text{or} \quad u^3+7u^2+u+1=0,$$

which is the equation with roots

$$\frac{1+a}{1-a}, \qquad \frac{1+b}{1-b}, \qquad \frac{1+c}{1-c}.$$

Then

$$\text{Sum of the roots} = \frac{1+a}{1-a}+\frac{1+b}{1-b}+\frac{1+c}{1-c}=-7.$$

Example 1.49 If two roots of the equation $x^4+ax^3+bx^2+cx+d=0$ are equal in value but different in sign, show that $c^2=a(bc-ad)$.

Let α, β, $-\alpha$ and γ be the roots of the equation

$$x^4+ax^3+bx^2+cx+d=0 \tag{i}$$

Then we have

$$\alpha+\beta-\alpha+\gamma=\beta+\gamma=-a \tag{ii}$$

$$\alpha\beta-\alpha^2+\alpha\gamma-\alpha\beta+\beta\gamma-\gamma\alpha=\beta\gamma-\alpha^2=b \tag{iii}$$

$$-\alpha^2\beta+\alpha\beta\gamma-\beta\alpha\gamma-\alpha^2\gamma=-\alpha^2(\beta+\gamma)=-c \tag{iv}$$

$$-\alpha^2\beta\gamma=d \tag{v}$$

From (ii) and (iv),

$$-\alpha^2(-a)=-c \quad \text{or} \quad \alpha^2=-\frac{c}{a}.$$

From (iii) and (v),

$$-\alpha^2(\alpha^2+b)=d \Rightarrow \frac{c}{a}\left(-\frac{c}{a}+b\right)=d$$

$$\Rightarrow c(ab-c)=a^2d$$

$$\Rightarrow -c^2=-abc+a^2d$$

$$\Rightarrow c^2=a(bc-ad).$$

1.5.3 Imaginary and Irrational Roots

Theorem 1.15 If $\alpha + i\beta$ is a root of the equation

$$f(x) = a_0x^n + a_1x^{n-1} + a_2x^{n-2} + \cdots + a_n = 0,$$

then $\alpha - i\beta$ is also a root of $f(x) = 0$.

Let

$$f(x) = a_0x^n + a_1x^{n-1} + a_2x^{n-2} + \cdots + a_n = 0 \tag{1.45}$$

Now,

$$[x - (\alpha + i\beta)][x - (\alpha - i\beta)] = (x - \alpha)^2 + \beta^2.$$

We can therefore write

$$f(x) = [(x - \alpha)^2 + \beta^2]\,\phi(x) + c_1x + c_2$$

$$= [x - (\alpha + i\beta)][x - (\alpha - i\beta)]\,\phi(x) + c_1x + c_2 \tag{1.46}$$

Putting $x = \alpha + i\beta$ in (1.46), we obtain

$$f(\alpha + i\beta) = c_1(\alpha + i\beta) + c_2.$$

But $f(\alpha + i\beta) = 0$ since $\alpha + i\beta$ is a root of $f(x) = 0$, we get

$$c_1(\alpha + i\beta) + c_2 = 0 \;\Rightarrow\; c_1\alpha + c_2 = 0 \text{ and } c_1 = 0 \text{ since } \beta \neq 0$$

Therefore, $c_2 = 0$. Hence (1.46) gives:

$$f(x) = [x - (\alpha + i\beta)][x - (\alpha - i\beta)]\,\phi(x) \tag{1.47}$$

Therefore, $f(\alpha - i\beta) = 0$, which shows that $(\alpha - i\beta)$ is a root of $f(x) = 0$.

Theorem 1.16 Let $f(x) = a_0x^n + a_1x^{n-1} + a_2x^{n-2} + \cdots + a_n = 0$, where the a_i are rational. Then the irrational roots occur in conjugate pairs.

We know that

$$\left[x - \left(\alpha + \sqrt{\beta}\right)\right]\left[x - \left(\alpha - \sqrt{\beta}\right)\right] = (x - \alpha)^2 - \beta \tag{1.48}$$

Let

$$f(x) = \left[(x - \alpha)^2 - \beta\right]\phi(x) + c_1x + c_2$$

$$= \left[x - (\alpha + \sqrt{\beta})\right]\left[x - (\alpha - \sqrt{\beta})\right]\phi(x) + c_1x + c_2 \tag{1.49}$$

Putting $x = \alpha + \sqrt{\beta}$ in (1.49), we get

$$f(\alpha + \sqrt{\beta}) = c_1(\alpha + \sqrt{\beta}) + c_2$$

But $f(\alpha + \sqrt{\beta}) = 0$ as $\alpha + \sqrt{\beta}$ is a root of $f(x) = 0$. Therefore,

$$c_1(\alpha + \sqrt{\beta}) + c_2 = 0 \Rightarrow c_1 = 0 \text{ and } c_2 = 0.$$

Hence (1.49) becomes:

$$f(x)=\left[x-\left(\alpha+\sqrt{\beta}\right)\right]\left[x-\left(\alpha-\sqrt{\beta}\right)\right]\phi(x) \qquad (1.50)$$

Putting $x=\alpha-\sqrt{\beta}$ in (1.49), we get $f(x)=0$, which shows that $x=\alpha-\sqrt{\beta}$ is also a root of $f(x)=0$.

Example 1.50 If $1-\sqrt{5}$ is a root of the equation $x^4-5x^3+4x^2+8x-8=0$, find the other roots.

Let

$$f(x)=x^4-5x^3+4x^2+8x-8=0 \qquad \text{(i)}$$

Since $1-\sqrt{5}$ is a root of $f(x)=0$, it follows that $1+\sqrt{5}$ is also a root. Let

$$f(x)=\left[x-\left(1+\sqrt{5}\right)\right]\left[x-\left(1-\sqrt{5}\right)\right]\phi(x)$$

$$=\left[\left(x-1\right)^2-5\right]\phi(x)$$

$$=(x^2-2x-4)\,\phi(x)$$

Since $f(x)=x^4-5x^3+4x^2+8x-8$, we find, by division, that $\phi(x)=x^2-3x+2$. Hence

$$f(x)=(x^2-2x-4)\,(x^2-3x+2)$$

$$=\left[x-\left(1+\sqrt{5}\right)\right]\left[x-\left(1-\sqrt{5}\right)\right](x-1)\,(x-2).$$

Therefore, the roots of $f(x)=0$ are 1, 2 and $1\pm\sqrt{5}$.

Example 1.51 If a, b, c are all positive, show that the roots of the equation

$$\frac{1}{x-a}+\frac{1}{x-b}+\frac{1}{x-c}-\frac{1}{x}=0$$

are all real.

Suppose that the given equation has an imaginary root and it is $(\alpha+i\beta)$. Then $(\alpha-i\beta)$ is also a root.

Substituting $x=\alpha+i\beta$ and $\alpha-i\beta$ in succession in the given equation, we get

$$\frac{1}{\alpha+i\beta-a}+\frac{1}{\alpha+i\beta-b}+\frac{1}{\alpha+i\beta-c}-\frac{1}{\alpha+i\beta}=0 \qquad \text{(i)}$$

and

$$\frac{1}{\alpha-i\beta-a}+\frac{1}{\alpha-i\beta-b}+\frac{1}{\alpha-i\beta-c}-\frac{1}{\alpha-i\beta}=0 \qquad \text{(ii)}$$

Subtracting (ii) from (i) and simplifying, we obtain

$$-2i\beta\left[\frac{1}{(\alpha-a)^2+\beta^2}+\frac{1}{(\alpha-b)^2+\beta^2}+\frac{1}{(\alpha-c)^2+\beta^2}+\frac{1}{\alpha^2+\beta^2}\right]=0$$

Since the expression within the brackets is non-zero, it follows that $\beta = 0$, which shows that the roots are all real.

1.5.4 Symmetric Functions of the Roots

Let α, β, γ be the roots of a given equation. Then the functions $\alpha+\beta+\gamma$, $\alpha^2+\beta^2+\gamma^2$, $\alpha\beta+\beta\gamma+\gamma\alpha$, etc. are *symmetric functions* as the function remains unaltered if any two of α, β, γ are interchanged.

Example 1.52 If a, b, c are the roots of the cubic equation $x^3+px^2+qx+r=0$, find the value of:

(a) $a^3+b^3+c^3$

(b) $\dfrac{1}{a^2b^2}+\dfrac{1}{b^2c^2}+\dfrac{1}{c^2a^2}$

(c) $\dfrac{b^2+c^2}{bc}+\dfrac{c^2+a^2}{ca}+\dfrac{a^2+b^2}{ab}$.

Since a, b, c are the roots of $x^3+px^2+qx+r=0$, we have

$$a+b+c=-p \tag{i}$$

$$ab+bc+ca=q \tag{ii}$$

$$abc=-r \tag{iii}$$

We know that

$$\begin{aligned} a^3+b^3+c^3-3abc &= (a+b+c)(a^2+b^2+c^2-ab-bc-ca) \\ &= -p[(a+b+c)^2-3(ab+bc+ca)] \\ &= -p(p^2-3q) \end{aligned}$$

or

$$a^3+b^3+c^3=-3r-p^3+3pq.$$

Also

$$\begin{aligned} \frac{1}{a^2b^2}+\frac{1}{b^2c^2}+\frac{1}{c^2a^2} &= \frac{1}{a^2b^2c^2}(c^2+a^2+b^2) \\ &= \frac{1}{r^2}[(a+b+c)^2-2(ab+bc+ca)] \\ &= \frac{1}{r^2}(p^2-2q). \end{aligned}$$

Finally,

$$\frac{b^2+c^2}{bc}+\frac{c^2+a^2}{ca}+\frac{a^2+b^2}{ab}=\frac{1}{abc}[a(b^2+c^2)+b(c^2+a^2)+c(a^2+b^2)]$$

$$=\frac{1}{abc}[(a+b+c)(ab+bc+ca)-3abc]$$

$$=-\frac{1}{r}(-pq+3r)$$

$$=\frac{pq}{r}-3.$$

Example 1.53 If a, b, c are the roots of the equation $x^3+px+q=0$, form the equation whose roots are $(b-c)^2, (c-a)^2$ and $(a-b)^2$. Hence find the value of $(b-c)^2(c-a)^2(a-b)^2$.

Let

$$x^3+px+q=0 \tag{i}$$

Since a, b, c are the roots of (i), we have

$$a+b+c=0 \tag{ii}$$

$$ab+bc+ca=p \tag{iii}$$

$$abc=-q \tag{iv}$$

Now,

$$(b-c)^2=(b+c)^2-4bc$$

$$=(a+b+c-a)^2-\frac{4abc}{a}$$

$$=a^2+\frac{4q}{a}, \qquad \text{[using (ii) and (iv)]}$$

Putting $a^2+(4q/u)=u$, we get

$$a^3+4q=au \quad \text{or} \quad a^3-au+4q=0 \tag{v}$$

From (i) and (v), we obtain, by subtraction:

$$(x^3-a^3)+(px+au)-3q=0$$

or

$$pa+au=3q, \qquad (\text{since } x=a)$$

or

$$a=\frac{3q}{p+u}=x.$$

Substitution in (i) gives:

$$\left(\frac{3q}{p+u}\right)^3+p\left(\frac{3q}{p+u}\right)+q=0,$$

which on simplification gives the equation:

$$u^3+6u^2p+9up^2+4p^3+27q^2=0,$$

Therefore,

$$\text{Product of the roots}=(b-c)^2(c-a)^2(a-b)^2=-4p^3-27q^2.$$

Example 1.54 If a, b, c are the roots of the equation, $x^3-px^2+qx-r=0$, find the equation whose roots are

$$\frac{a}{b+c-a},\quad \frac{b}{c+a-b},\quad \frac{c}{a+b-c}$$

we have

$$a+b+c=p \tag{i}$$

$$ab+bc+ca=q \tag{ii}$$

$$abc=r \tag{iii}$$

Let $a/(b+c-a)=u$ so that $a/(a+b+c-2a)=u$. Therefore,

$$\frac{a}{p-2a}=u \quad \text{or} \quad a=\frac{pu}{2u+1}=x$$

Substituting for x in the given equation, we obtain

$$\left(\frac{pu}{2u+1}\right)^3-p\left(\frac{pu}{2u+1}\right)^2+q\left(\frac{pu}{2u+1}\right)-r=0$$

Simplifying, we get

$$(p^3-4pq+8r)u^3+(p^3-4pq+12r)u^2-(pq-6r)u+r=0,$$

which is the required equation.

Example 1.55 If a, b, c are the roots of the equation $x^3+px+q=0$, find the value of

$$\left(\frac{1}{a^2}-\frac{1}{bc}\right)\left(\frac{1}{b^2}-\frac{1}{ca}\right)\left(\frac{1}{c^2}-\frac{1}{ab}\right).$$

Since a, b, c are the roots of $x^3+px+q=0$, we have

$$a+b+c=0 \tag{i}$$

$$ab+bc+ca=p \tag{ii}$$

$$abc=-q \tag{iii}$$

Putting

$$\frac{1}{a^2}-\frac{1}{bc}=u,$$

we obtain

$$\frac{1}{a^2}-\frac{a}{abc}=u \quad \text{or} \quad \frac{1}{a^2}-\frac{a}{(-q)}=u.$$

Then

$$a^3-a^2uq+q=0 \tag{iv}$$

Subtracting (iv) from the given equation, we get

$$(x^3-a^3)+(px+a^2uq)=0$$

when $x=a$, we have

$$x=-\frac{p}{uq}.$$

Substituting $x=-p/(uq)$ in the given equation, we obtain

$$-\frac{p^3}{u^3q^3}-\frac{p^2}{uq}+q=0 \quad \text{or} \quad u^3q^4-u^2p^2q^2-p^3=0.$$

Therefore,

$$\text{Product of the roots} =\left(\frac{1}{a^2}-\frac{1}{bc}\right)\left(\frac{1}{b^2}-\frac{1}{ca}\right)\left(\frac{1}{c^2}-\frac{1}{ab}\right)=\frac{p^3}{q^4}.$$

Example 1.56 If a, b, c are the roots of the equation, $x^3+2x^2-x-3=0$, find the value of

$$\frac{a+3}{a-2}+\frac{b+3}{b-2}+\frac{c+3}{c-2}.$$

we have

$$a+b+c=-2 \tag{i}$$

$$ab+bc+ca=-1 \tag{ii}$$

$$abc=3 \tag{iii}$$

Let

$$\frac{a+3}{a-2}=u$$

Then

$$a=\frac{2u+3}{u-1}=x$$

Substituting $x=(2u+3)/(u-1)$ in the given equation, we obtain

$$\left(\frac{2u+3}{u-1}\right)^3+2\left(\frac{2u+3}{u-1}\right)^2-\left(\frac{2u+3}{u-1}\right)-3=0$$

or

$$11u^3+62u^2+43u+3=0.$$

Hence

$$\text{Sum of the roots}=\frac{a+3}{a-2}+\frac{b+3}{b-2}+\frac{c+3}{c-2}=-\frac{62}{11}.$$

1.5.5 Reciprocal Equations

An equation in which the reciprocal of every root is also a root of the equation, is called a *reciprocal equation.*

Let the equation given by

$$f(x)=x^n+a_1x^{n-1}+a_2x^{n-2}+\cdots+a_n=0 \tag{1.41}$$

be a reciprocal equation. Putting $x=1/x$ in Eq. (1.41), we get

$$\frac{1}{x^n}+\frac{a_1}{x^{n-1}}+\frac{a_2}{x^{n-2}}+\cdots+a_n=0$$

or

$$x^n+\frac{a_{n-1}}{a_n}x^{n-1}+\frac{a_{n-2}}{a_n}x^{n-2}+\cdots+\frac{a_1}{a_n}x+\frac{1}{a_n}=0 \tag{1.51}$$

Since (1.41) is given to be a reciprocal equation, it follows that (1.41) and (1.50) are the same. Hence comparing the coefficients of like powers of x in both the equations, we obtain

$$a_1=\frac{a_{n-1}}{a_n},\qquad a_2=\frac{a_{n-2}}{a_n},\ \ldots,\qquad a_n=\frac{1}{a_n}.$$

The last equation gives: $a_n^2=1$ or $a_n=\pm1$.

Case (i): Let $a_n=1$. Then $a_1=a_{n-1}$, $a_2=a_{n-2},\ldots,$ That is the coefficients of the terms equidistant from the first and last terms are equal in magnitude and are also of the *same* sign. These are called *reciprocal equations with like signs.*

Case (ii): Let $a_n=-1$. Then $a_1=-a_{n-1}, a_2=-a_{n-2},\ldots$. That is the coefficients of the terms equidistant from the first and last terms are equal in magnitude but have *opposite* signs. These are called *reciprocal equations with unlike signs.*

We consider four types of reciprocal equations:

Type-1: Equations of degree 4 with like or unlike signs.

Type-2: Equations of *odd* degree with like signs $x = -1$ is a root of such equations.

Type-3: Equations of *odd* degree with unlike signs. $x = 1$ is a root of such equation.

Type-4: Equations of *even* degree with unlike signs and the middle term absent. Both $x = 1$ and $x = -1$ are the roots of such equations.

The following examples demonstrate the method of solution in each case.

Example 1.57 Solve the reciprocal equation

$$6x^4 - 35x^3 + 62x^2 - 35x + 6 = 0$$

This is an equation of degree 4 with like signs and therefore it is of Type-1. In this case, we divide all the terms by x^2 and regroup them as follows:

$$6\left(x^2 + \frac{1}{x^2}\right) - 35\left(x + \frac{1}{x}\right) + 62 = 0 \tag{i}$$

Putting $x + 1/x = y$, we obtain

$$x^2 + \frac{1}{x^2} = \left(x + \frac{1}{x}\right)^2 - 2 = y^2 - 2.$$

Hence equation (i) becomes:

$$\begin{aligned} 6y^2 - 12 - 35y + 62 = 0 &\Rightarrow 6y^2 - 35y + 50 = 0 \\ &\Rightarrow (2y - 5)(3y - 10) = 0 \\ &\Rightarrow y = 5/2 \text{ or } y = 10/3. \end{aligned}$$

Hence we have

$$\begin{aligned} x + \frac{1}{x} = \frac{5}{2} &\Rightarrow 2x^2 - 5x + 2 = 0 \\ &\Rightarrow (2x - 1)(x - 2) = 0 \\ &\Rightarrow x = 1/2 \text{ or } x = 2. \end{aligned}$$

Again

$$\begin{aligned} x + \frac{1}{x} = \frac{10}{3} &\Rightarrow 3x^2 - 10x + 3 = 0 \\ &\Rightarrow (3x - 1)(x - 3) = 0 \\ &\Rightarrow x = 1/3 \text{ or } x = 3. \end{aligned}$$

The required solutions are therefore given by

$$x = \frac{1}{2},\ x = 2,\ x = \frac{1}{3} \text{ and } x = 3.$$

Example 1.58 Solve: $2x^6 - 9x^5 + 10x^4 - 3x^3 + 10x^2 - 9x + 2 = 0$.

This is an equation with even degree and like signs. Hence we divide the equation throughout by x^3. We obtain

$$2\left(x^3 + \frac{1}{x^3}\right) - 9\left(x^2 + \frac{1}{x^2}\right) + 10\left(x + \frac{1}{x}\right) - 3 = 0 \qquad \text{(i)}$$

Putting $x + (1/x) = y$, we get

$$x^2 + \frac{1}{x^2} = \left(x + \frac{1}{x}\right)^2 - 2 = y^2 - 2$$

and

$$x^3 + \frac{1}{x^3} = \left(x + \frac{1}{x}\right)^3 - 3\left(x + \frac{1}{x}\right) = y^3 - 3y.$$

Hence equation (i) becomes:

$$2(y^3 - 3y) - 9(y^2 - 2) + 10y - 3 = 0$$

or

$$2y^3 - 9y^2 + 4y + 15 = 0 \qquad \text{(ii)}$$

The roots of (ii) are easily seen to be -1, $5/2$ and 3.

Now

$$x + \frac{1}{x} = -1 \Rightarrow x^2 + x + 1 = 0$$

$$\Rightarrow x = \frac{-1 \pm \sqrt{1-4}}{2}$$

$$\Rightarrow x = \frac{-1 \pm i\sqrt{3}}{2}$$

Also

$$x + \frac{1}{x} = \frac{5}{2} \Rightarrow 2x^2 - 5x + 2 = 0 \Rightarrow x = \frac{1}{2},\ 2.$$

Finally,

$$x + \frac{1}{x} = 3 \Rightarrow x^2 - 3x + 1 = 0$$

$$\Rightarrow x = \frac{3 \pm \sqrt{9-4}}{2}$$

$$\Rightarrow x = \frac{3 \pm \sqrt{5}}{2}.$$

Hence the roots of the given equation are

$$\frac{1}{2},\ 2,\ \frac{3 \pm \sqrt{5}}{2},\ \frac{-1 \pm i\sqrt{3}}{2}.$$

Example 1.59 Solve: $6x^5 - 41x^4 + 97x^3 - 97x^2 + 41x - 6 = 0$.

This is a reciprocal equation of odd degree with unlike signs. Hence $x = 1$ is a root of this equation. Taking out the factor $x - 1$, we find that the given equation simplifies to

$$(x-1)\,f(x) = 0$$

where

$$f(x) = 6x^4 - 35x^3 + 62x^2 - 35x + 6.$$

It can be seen that $f(x) = 0$ is a reciprocal equation of Type-1. This equation was solved in Example 1.56 and its roots are $x = 1/2,\ 2,\ 1/3,\ 3$. Hence the roots of the given equation are $x = 1,\ 1/2,\ 2,\ 1/3$ and 3.

Example 1.60 Solve: $6x^6 - 35x^5 + 56x^4 - 56x^2 + 35x - 6 = 0$.

This is a reciprocal equation of even degree with unlike signs and middle term absent. It is of Type-4 with $x = -1$ and $x = 1$ as its roots. The given equation can be written as

$$(x-1)(x+1)\,f(x) = 0$$

where

$$f(x) = 6x^4 - 35x^3 + 62x^2 - 35x + 6.$$

Since the roots of $f(x) = 0$ (see Example 1.56) are 1/2, 2, 1/3, 3, it follows that the roots of the given equation are $x = -1,\ 1,\ 1/3,\ 3,\ 1/2$ and 2.

1.5.6 Transformation of Equations

We shall consider again Eq. (1.40), viz.,

$$f(x) = a_0 x^n + a_1 x^{n-1} + a_2 x^{n-2} + \cdots + a_n = 0 \tag{1.40}$$

and let its roots be $\alpha_1, \alpha_2, \ldots, \alpha_n$. It would often be convenient to transform Eq. (1.40) into another equation whose roots are related to $\alpha_1, \alpha_2, \ldots, \alpha_n$. in some way. Such a transformation may enable us to solve the given equation more easily. In what follows, we list some of these transformations and illustrate them with examples.

(a) *To transform Eq. (1.40) into another equation, whose roots are* α_1, α_2, ..., α_n

Let $y = -\alpha_1 = -x$, (say). Then $x = -y$. Putting $x = -y$ in Eq. (1.40), we obtain

$$a_0 (-y)^n + a_1 (-y)^{n-1} + a_2 (-y)^{n-2} + \cdots + a_{n-1} (-y) + a_n = 0.$$

or

$$a_0 y^n - a_1 y^{n-1} + a_2 y^{n-2} - \cdots + (-1)^n a_n = 0$$

or

$$a_0 x^n - a_1 x^{n-1} + a_2 x^{n-2} - \cdots + (-1)^n a_n = 0, \qquad (1.52)$$

which is the required equation.

(b) *To transform Eq. (1.40) into another equation, whose roots are* $k\alpha_1$, $k\alpha_2$, ..., $k\alpha_n$.

Let $y = k\alpha_1 = kx$, (say). This gives: $x = y/k$. Putting $x = y/k$ In Eq. (1.40), we obtain

$$a_0 \frac{1}{k^n} y^n + a_1 \frac{1}{k^{n-1}} y^{n-1} + a_2 \frac{1}{k^{n-2}} y^{n-2} + \cdots + a_n = 0.$$

or

$$a_0 y^n + a_1 k y^{n-1} + a_2 k^2 y^{n-2} + \cdots + k^n a_n = 0,$$

or

$$a_0 x^n + a_1 k x^{n-1} + a_2 k^2 x^{n-2} + \cdots + k^n a_n = 0, \qquad (1.53)$$

which is the required equation with roots $k\alpha_1$, $k\alpha_2$, ..., $k\alpha_n$.

(c) *To transform Eq. (1.40) into another equation, whose roots are* $1/\alpha_1$, $1/\alpha_2$, ..., $1/\alpha_n$

Let $y = 1/\alpha_1 = 1/x$, (say). This gives $x = 1/y$. Putting $x = 1/y$ in Eq. (1.40), we obtain:

$$a_0 \frac{1}{y_n} + a_1 \frac{1}{y^{n-1}} + a_2 \frac{1}{y^{n-2}} + \cdots + a_n = 0$$

or

$$a_0 + a_1 y + a_2 y^2 + \cdots + a_n y^n = 0$$

or

$$a_n x^n + a_{n-1} x^{n-1} + a_{n-2} x^{n-2} + \cdots + a_1 x + a_0 = 0, \qquad (1.54)$$

which is the required equation with roots $1/\alpha_1$, $1/\alpha_2$, ..., $1/\alpha_n$.

(d) *To transform Eq. (1.40) into another equation, whose roots are* $\alpha_1 - h, \alpha_2 - h, \ldots, \alpha_n - h$.

Let $y = \alpha_1 - h = x - h$, (say). This gives: $x = y + h$. Putting $x = y + h$ in Eq. (1.40), we obtain

$$a_0 (y+h)^n + a_1 (y+h)^{n-1} + a_2 (y+h)^{n-2} + \cdots + a_n = 0$$

or

$$A_0 y^n + A_1 y^{n-1} + A_2 y^{n-2} + \cdots + A_n = 0,$$

where $A_0, A_1, A_2, \ldots, A_n$ are functions of $a_0, a_1, a_2, \ldots, a_n$.

Putting back $y = x - h$, the above equation becomes:

$$A_0 (x-h)^n + A_1 (x-h)^{n-1} + A_2 (x-h)^{n-2} + \cdots + A_n = 0, \qquad (1.55)$$

which is identical to Eq. (1.40). The coefficients $A_0, A_1, \ldots, A_n$ in Eq. (1.55) may easily be evaluated by the following method.

We observe that

$$A_0 (x-h)^n + A_1(x-h)^{n-1} + \cdots + A_{n-1}(x-h) + A_n$$
$$= [A_0 (x-h)^{n-1} + A_1 (x-h)^{n-2} + \cdots + A_{n-1}](x-h) + A_n,$$

or

$$A_0(x-h)^{n-1} + A_1 (x-h)^{n-2} + \cdots + A_{n-1} = [A_0 (x-h)^{n-2} + A_1 (x-h)^{n-3}$$
$$+ \cdots + A_{n-2}](x-h) + A_{n-1}, \text{ etc.}$$

Thus, $A_n, A_{n-1}, \ldots, A_1$ are the remainders obtained by successive divisions of the left side of (1.54) by $x - h$, A_0 being the quotient in the last division.

Similarly, to increase the roots by h, we must decrease the roots by $(-h)$ and obtain the new equation.

(e) *To transform Eq. (1.40) into another equation with a certain term removed.*

Suppose we want to transform Eq. (1.40) into another equation with the second term removed. For this, we first transform Eq. (1.40) into another equation whose roots are diminished by h. Putting $x = y + h$ in (1.40), we obtain:

$$a_0 (y+h)^n + a_1 (y+h)^{n-1} + a_2 (y+h)^{n-2} + \cdots + a_n = 0$$

or

$$a_0 y^n + n a_0 y^{n-1} h + a_1 y^{n-1} + \cdots + a_n = 0$$

or

$$a_0 y^n + (n a_0 h + a_1) y^{n-1} + \cdots + a_n = 0.$$

Since the term containing y^{n-1} is to be removed, we set

$$na_0h + a_1 = 0 \quad \text{or} \quad h = -\frac{a_1}{na_0}.$$

Thus, we can remove the second term by diminishing the roots of Eq. (1.40) by h.

Similarly, to remove the third term, we set

$$\frac{n(n-1)}{2}a_0h^2 + (n-1)a_1h + a_2 = 0,$$

which gives two values for h, and therefore we obtain two equations in which the third term is removed.

The following examples illustrate the different transformations.

Example 1.61 Find the equation whose roots are the negatives of the roots of the equation $x^3 + 6x^2 + 8x + 9 = 0$.

Putting $x = -y$ in the given equation, we obtain

$$-y^3 + 6y^2 - 8y + 9 = 0 \quad \text{or} \quad y^3 - 6y^2 + 8y - 9 = 0$$

Hence the required equation is $x^3 - 6x^2 + 8x - 9 = 0$.

Example 1.62 Find the equation whose roots are 3 times the roots of the equation $x^3 + x^2 - x + 1 = 0$.

Putting $y = 3x$, we obtain $x = y/3$. The given equation then becomes:

$$\left(\frac{1}{3}y\right)^3 + \left(\frac{1}{3}y\right)^2 - \left(\frac{1}{3}y\right) + 1 = 0$$

or

$$\frac{y^3}{27} + \frac{1}{9}y^2 - \frac{1}{3}y + 1 = 0$$

or

$$y^3 + 3y^2 - 9y + 27 = 0$$

Hence the required equation is $x^3 + 3x^2 - 9x + 27 = 0$.

Example 1.63 Multiply the roots of the equation $x^3 - 5x + 1 = 0$ by 10.

The given equation is

$$x^3 + 0x^2 - 5x + 1 = 0 \tag{i}$$

Let $y = 10x$ so that $x = y/10$. Putting $x = y/10$ in (i), we obtain:

$$\left(\frac{1}{10}y\right)^3 + 0\left(\frac{1}{10}y\right)^2 - 5\left(\frac{1}{10}y\right) + 1 = 0$$

or

$$y^3 + 0(10)y^2 - 5(10^2)y + 10^3 = 0$$

Hence the required equation is $x^3 - 500x + 1000 = 0$.

Example 1.64 Remove the fractional coefficients from the equation

$$x^5 - \frac{1}{4}x^4 + \frac{25}{64}x^2 + \frac{7}{128}x - \frac{1}{32} = 0$$

such that the coefficient of x^5 is unity.

We transform the given equation so that the roots of the transformed equation are those of the given equation multiplied by 4. The required equation is therefore

$$x^5 - \frac{1}{4}(4x^4) + 0(4^2 x^3) + \frac{25}{64}(4^3 x^2) + \frac{7}{128}(4^4 x) - \frac{1}{32}(4^5) = 0$$

or

$$x^5 - x^4 + 25x^2 + 14x - 32 = 0.$$

Example 1.65 Find the equation whose roots are the roots of the equation $x^4 - 5x^3 + 7x^2 - 17x + 11 = 0$ diminished by 2.

Let $y = x - 2$ so that $x = y + 2$. Substituting $x = y + 2$ in the given equation, we obtain

$$(y+2)^4 - 5(y+2)^3 + 7(y+2)^2 - 17(y+2) + 11 = 0.$$

or

$$(y^4 + 8y^3 + 24y^2 + 32y + 16) - 5(y^3 + 6y^2 + 12y + 8)$$
$$+ 7(y^2 + 4y + 4) - 17y - 34 + 11 = 0$$

or

$$y^4 + 3y^3 + y^2 - 17y - 19 = 0.$$

The required equation is therefore given by $x^4 + 3x^2 + x^2 - 17x - 19 = 0$.

Example 1.66 Find the equation whose roots exceed the roots of the equation $x^4 - x^3 - 10x^2 + 4x + 24 = 0$ by 2.

Putting $x = y - 2$ in the given equation, we obtain

$$(y-2)^4 - (y-2)^3 - 10(y-2)^2 + 4(y-2) + 24 = 0.$$

or

$$(y^4 - 8y^3 + 24y^2 - 32y + 16) - (y^3 - 6y^2 + 12y - 8)$$
$$- 10(y^2 - 4y + 4) + 4y - 8 + 24 = 0$$

or

$$y^4 - 9y^3 + 20y^2 = 0$$

Simplifying, we get

$$y^2(y^2 - 9y + 20) = 0 \quad \text{or} \quad y^2(y-4)(y-5) = 0.$$

Therefore, the roots of the above equation are 0, 0, 4 and 5. Hence the roots of the given equation are –2, –2, 2 and 3.

1.5.7 Descartes' Rule of Signs

In a polynomial equation $f(x) = 0$, the number of real positive roots cannot exceed the number of changes of sign in its coefficients. Similarly, the number of real negative roots cannot exceed the number of changes of sign in the coefficients of $f(-x)$.

Example 1.67 Show that the equation $x^3 + 2x + 3 = 0$ has no real positive roots and has one real negative root.

Here $f(x) = x^3 + 2x + 3$. There are no changes in sign in $f(x)$ and hence $f(x) = 0$ has no real positive roots. Further,

$$f(-x) = -x^3 - 2x + 3$$

Since there is one change of sign in $f(-x)$, the equation $f(x) = 0$ has therefore one negative root, and two imaginary roots.

Example 1.68 Show that the equation $6x^7 - 5x^4 + 10x^3 - 280 = 0$ has at least four imaginary roots.

Let $f(x) = 6x^7 - 5x^4 + 10x^3 - 280$. There are three changes in sign in $f(x)$ and therefore $f(x) = 0$ has atmost three real positive roots.

Now let $f(-x) = -6x^7 - 5x^4 - 10x^3 - 280$. There are no changes in sign in $f(-x)$ and hence $f(x) = 0$ cannot have any real negative roots. Since the equation is of the 7th degree, it must have 7 roots. It follows that $f(x) = 0$ has at least four imaginary roots.

1.5.8 Horner's Method for Finding a Real Root

Let $f(x) = 0$ be a polynomial equation whose real root is required. Suppose we have found, by Descartes' rule of signs, that a real root lies between two consecutive integers a and $a+1$. Let this root be written as $a.k_1k_2k_3\ldots$, where k_1, k_2, k_3 are the numbers in the first, second, third, ... decimal places in the value of the root.

Now, we transform the given equation to another equation whose roots are diminished by 'a'. Let the transformed equation be $\phi_1(x) = 0$, whose real root is $0.k_1k_2k_3\ldots$ Now, we can transform the equation $\phi_1(x) = 0$ to another equation whose root is 10 times the root of $\phi_1(x) = 0$. Let this transformed equation be $\phi_2(x)=0$, whose real root is $k_1.k_2k_3\ldots$ Thus we have obtained k_1, the first number after the decimal. Repeating this procedure we can find $k_2, k_3\ldots$

To find a negative real root, we proceed as above with the equation $f(-x)=0$. For example, consider the polynomial equation given by

$$f(x)=x^3-3x+1=0 \tag{i}$$

There are two changes in sign in $f(x)$ and therefore there are two positive real roots.

Again $f(-x)=-x^3+3x+1$ and we observe that there is one change in sign. Hence there is one negative real root. Therefore, $f(x)=0$ has two positive real roots and one negative real root.

To determine the limits between which these roots lie, we form the following table:

x	sign of $f(x)=x^3-3x+1$
$-\infty$	–ve
–3	–ve
–2	–ve
–1	+ve
0	+ve
1	–ve
2	+ve.

Hence the positive roots lie between 0 and 1, and between 1 and 2, whereas the negative root lies between –2 and –1. We first determine the root between 1 and 2 to two decimal places.

Let the positive root between 1 and 2 be $1.k_1k_2k_3\ldots$ We start by diminishing the root of $f(x)=0$ by 1. Putting $x=y+1$ in $f(x)=0$, we obtain:

$$(y+1)^3-3(y+1)+1=0 \Rightarrow y^3+3y^2+3y+1-3y-3+1=0$$

$$\Rightarrow y^3+3y^2-1=0$$

$$\Rightarrow x^3+3x^2-1=0$$

whose root is $.k_1k_2k_3\ldots$ We now multiply the root by 10 so that the transformed equation is

$$x^3+3(10)x^2-1(10^3)=0$$

or

$$x^3+30x^2-1000=0, \text{ i.e. } \phi_1(x)=0 \quad \text{(say)}.$$

Now

$$\phi_1(5) = 125 + 750 - 1000 = -\text{ve}$$

and

$$\phi_1(6) = 216 + 1080 - 1000 = +\text{ve}.$$

Then, $k_1 = 5$. We therefore diminish the root of $\phi_1(x) = 0$ by 5. For this we put $x = y + 5$ in $\phi_1(x) = 0$. We obtain

$$(y+5)^3 + 30(y+5)^2 - 1000 = 0$$

or

$$y^3 + 15y^2 + 75y + 125 + 30y^2 + 300y + 750 - 1000 = 0$$

or

$$y^3 + 45y^2 + 375y - 125 = 0$$

or

$$x^3 + 45x^2 + 375x - 125 = 0$$

whose root is $0.k_2k_3$.

Multiplying the roots of the above equation by 10, we obtain:

$$x^3 + 450x^2 + 37500x - 125000 = 0 = \phi_2(x), \qquad \text{(say)}$$

Now, considering only dominant terms, we get

$$\phi_2(3) = 112500 - 125000 = -\text{ve}$$

and

$$\phi_2(4) = 150000 - 125000 = +\text{ve}$$

Therefore, $k_2 = 3$. To diminish the root of $\phi_2(x) = 0$ by 3, we put $x = y + 3$ and get:

$$(y+3)^3 + 450(y+3)^2 + 37500(y+3) - 125000 = 0$$

or

$$y^3 + 459y^2 + 40227y - 8423 = 0$$

or

$$x^3 + 459x^2 + 40227x - 8423 = 0,$$

whose root is $0.k_3\ldots$

Multiplying the roots of the above equation by 10, we get the equation

$$x^3 + 4590x^2 + 4022700x - 8423000 = 0 = \phi_3(x), \qquad \text{(say)}$$

Now, considering only the dominant terms, we have

$$\phi_3(2) = 8045400 - 8423000 = -\text{ve}$$

and

$$\phi_3(3) = 12068100 - 8423000 = +\text{ve}.$$

Hence $k_3 = 2$. The positive root between 1 and 2 is therefore 1.532, i.e. 1.53, correct to two decimal places.

Next, we find the negative real root of $f(x) = 0$. For this, we find the positive real root of $f(-x) = 0$.

Now

$$f(-x) = -x^3 + 3x + 1 = 0 \quad \text{or} \quad x^3 - 3x - 1 = 0$$

Let $F(x) = x^3 - 3x - 1 = 0$. We find

$$F(1) = 1 - 3 - 1 = -\text{ve} \quad \text{and} \quad F(2) = 8 - 6 - 1 = +\text{ve}.$$

Let the root be $1.k_1 k_2 k_3$. Now, we diminish the root of $F(x) = 0$ by 1. For this, we put $x = y + 1$ in $F(x) = 0$. We obtain

$$(y+1)^3 - 3(y+1) - 1 = 0 \Rightarrow y^3 + 3y^2 + 3y + 1 - 3y - 3 - 1 = 0$$

$$\Rightarrow y^3 + 3y^2 - 3 = 0$$

$$\Rightarrow x^3 + 3x^2 - 3 = 0,$$

whose root is $0.k_1 k_2 k_3$. We therefore multiply the roots of this equation by 10. The transformed equation is

$$x^3 + 30x^2 - 3000 = 0 = \phi_1(x), \qquad \text{(say)}.$$

Now

$$\phi_1(8) = 512 + 1920 - 3000 = -\text{ve}$$

and

$$\phi_1(9) = 729 + 2430 - 3000 = +\text{ve}.$$

Then $k_1 = 8$. To diminish the root of $\phi_1(x) = 0$ by 8, we put $x = y + 8$ and obtain

$$(y+8)^3 + 30(y+8)^2 - 3000 = 0$$

or

$$y^3 + 24y^2 + 192y + 512 + 30y^2 + 480y + 1920 - 3000 = 0$$

or

$$y^3 + 54y^2 + 672y - 568 = 0$$

or

$$x^3 + 54x^2 + 672x - 568 = 0$$

The equation whose roots are 10 times the roots of the above equation is

$$x^3 + 540x^2 + 67200x - 568000 = 0 = \phi_2(x), \qquad \text{(say)}.$$

Now, considering the dominant terms only, we get

$$\phi_2(7) = 470400 - 568000 = -\text{ve}$$

and

$$\phi_2(8) = 537600 - 568000 + 512 + 34560 = +\text{ve}.$$

Then $k_2 = 7$. Diminishing the roots of $\phi_2(x) = 0$ by 7, we obtain the equation

$$(y+7)^3 + 540(y+7)^2 + 67200(y+7) - 568000 = 0$$

or

$$y^3 + 561y^2 + 74907y - 70797 = 0,$$

whose root is $0.k_3$. Multiplying the roots of the above equation by 10, we obtain

$$y^3 + 5610y^2 + 7490700y - 70797 = 0$$

or

$$x^3 + 5610y^2 + 7490700x - 70797000 = 0 \quad \text{or} \quad = \phi_3(x), \qquad \text{(say)}.$$

Since $\phi_3(0)$ is negative and $\phi_3(10)$ is positive, it follows that the positive root of $F(x) = 0$ is $1.879 = 1.88$, correct to two decimal places. Hence the negative root of $f(x) = 0$ is -1.88, correct to two decimal places.

EXERCISES 1.5

Solve the following equations when one of the roots is given (Problems 1–5):

1. $x^3 + 6x + 20 = 0; \quad x = -2.$
2. $x^3 - 6x^2 + 13x - 10 = 0; \quad x = 2 - i.$
3. $x^4 - 5x^3 + 4x^2 + 8x - 8 = 0; \quad x = 1 + \sqrt{5}.$
4. $3x^5 - 4x^4 - 42x^3 + 56x^2 + 27x - 36 = 0; \quad x = \sqrt{2} + \sqrt{5}.$
5. $x^4 + 2x^3 - 5x^2 + 6x + 2 = 0; \quad x = 1 + i.$

Solve the following equations (Problems 6–10):

6. $x^3 - 9x^2 + 26x - 24 = 0$, whose roots are in arithmetic progression.
7. $3x^3 - 26x^2 + 52x - 24 = 0$, whose roots are in geometric progression.
8. $40x^4 + 22x^3 - 21x^2 + 2x + 1 = 0$, whose roots are in harmonic progression.
9. $x^3 - 12x^2 + 39x - 28 = 0$, whose roots are in arithmetic progression.
10. $x^4 - 10x^3 - 120x^2 + 320x + 1024 = 0$, whose roots are in geometric progression.
11. If the roots of the equation $x^3 + ax^2 + bx + c = 0$ are in geometric progression, show that $2a^3 - 9ab + 27c = 0$.
12. Let $\alpha, \beta, \gamma, \delta$ be the roots of the equation $ax^4 + bx^3 + cx^2 + dx + e = 0$, so that $\alpha + \beta = \gamma + \delta$. Show that $4abc - b^3 - 8a^2d = 0$.
13. Solve the equation $x^3 - 4x^2 - 20x + 48 = 0$ given that two of its roots α, β satisfy the condition $\alpha + 2\beta = 0$.

14. If the roots of $x^3+3ax^2+3bx+c=0$ are in harmonic progression, show that $2b^3+c^2=3abc$.

15. Let $\alpha, \beta, \gamma, \delta$ be the roots of the equation $x^4-4x^3-17x^2+24x+36=0$, such that $\alpha\beta=12$. Solve the equation completely.

16. If α, β, γ are the roots of the equation $x^3+px^2+qx+r=0$, find the value of:

(i) $\sum\frac{1}{\alpha\beta}$ (ii) $\sum\frac{1}{\alpha}$ (iii) $\sum\alpha^2$.

(iv) $(\alpha^2+1)(\beta^2+1)(\gamma^2+1)$ (v) $(\alpha^2-\beta\gamma)(\beta^2-\gamma\alpha)(\gamma^2-\alpha\beta)$

(vi) $\sum\alpha^2\beta$ (vii) $(\alpha+\beta-\gamma)^3+(\beta+\gamma-\alpha)^3+(\gamma+\alpha-\beta)^3$.

17. If p, q, r, s are the roots of $x^4+ax^3+bx^2+cx+d=0$, find the value of: (i) $\Sigma\alpha^2\beta\gamma$ (ii) $\Sigma\alpha^4$.

18. If a, b, c are the roots of $x^3+px+q=0$, then show that

$$\frac{a^7+b^7+c^7}{7}=\frac{1}{10}(a^5+b^5+c^5)(a^2+b^2+c^2).$$

19. If a, b, c are the roots of $x^3-x-1=0$, form the equation whose roots are

$$\frac{1+a}{1-a}, \quad \frac{1+b}{1-b}, \quad \frac{1+c}{1-c}.$$

20. If p is a root of $x^2(x+1)^2-k(x-1)(2x^2+x+1)=0$, prove that $(p+1)/(p-1)$ is also a root.

21. If α, β, γ are the roots of $x^3+px^2+qx+r=0$, show that $(1+\alpha^2)(1+\beta^2)(1+\gamma^2)=r^2(q-1)^2+(p-q)^2$.

22. If a, b, c are the roots of $x^3+px+q=0$, find the equation whose roots are

$$\frac{b^2+c^2}{a}, \quad \frac{c^2+a^2}{b}, \quad \frac{a^2+b^2}{c}.$$

23. The roots of the equation $x^4-5x^3+7x^2-17x+11=0$ are diminished by 2. Find the new equation.

24. The roots of the equation $3x^4+7x^3-15x^2+x-2=0$ are increased by 7. Find the new equation.

25. Find the equation whose roots are those of $x^4-2x^3+3x-5=0$ diminished by 2.

26. Solve the equation $x^4-8x^3+19x^2-12x+2=0$ by removing the term containing x^3.

Solve the following equations (Problems 27–30):

27. $4x^4-20x^3+33x^2-20x+4=0$.

28. $6x^5 + 11x^4 - 33x^3 - 33x^2 + 11x + 6 = 0.$

29. $x^5 - 5x^4 + 9x^3 - 9x^2 + 5x - 1 = 0.$

30. $6x^6 - 35x^5 + 56x^4 - 56x^2 + 35x - 6 = 0.$

Solve the following equations by Horner's method (Problems 31–35):

31. $x^3 - 3x - 4 = 0$, a positive root to 2 decimal places.

32. $2x^3 - 3x + 6 = 0$, a negative root to 2 decimal places.

33. $x^3 + x^2 + 12x - 24 = 0$, a positive root to 3 decimal places.

34. $x^3 - 3x - 5 = 0$, positive root lying between 2 and 3, correct to 3 decimal places.

35. $x^4 + 12x + 7 = 0$, a negative root lying between –2 and –3, correct to 3 decimal places.

1.6 MATRICES

First discovered by the French mathematician, Cayley, matrices occur in a wide variety of subjects—advanced mathematics, applied sciences, quantum mechanics, nuclear physics, probability and statistics, mathematical economics and electrical circuits. The matrix notation is powerful and is convenient in expressing basic relationships in engineering problems. In this section, we give a brief account of matrices, including some special matrices, and their applications to the solution of linear algebraic equations and also the algebraic eigenvalue problem.

1.6.1 Basic Definitions

A *matrix* is an array of *mn* elements (where *m* and *n* are integers) arranged in *m* rows and *n* columns. Such a matrix *A* is usually denoted by

$$A = \begin{bmatrix} a_{11} & a_{12} & a_{13} & \cdots & a_{1n} \\ a_{21} & a_{22} & a_{23} & \cdots & a_{2n} \\ a_{31} & a_{32} & a_{33} & \cdots & a_{3n} \\ \vdots & \vdots & \vdots & & \vdots \\ a_{m1} & a_{m2} & a_{m3} & \cdots & a_{mn} \end{bmatrix} = [a_{ij}], \tag{1.56}$$

where $a_{11}, a_{12}, \ldots, a_{mn}$ are called its *elements* and may be either real or complex. The matrix A is said to be of size $(m \times n)$. If $m = n$, the matrix is said to be a *square* matrix of order n. For example,

$$B = \begin{bmatrix} b_{11} & b_{12} & b_{13} \\ b_{21} & b_{22} & b_{23} \\ b_{31} & b_{32} & b_{33} \end{bmatrix}$$

is a square matrix of order 3. We may also have single row or single column matrices. Thus,

$$[a_1\ a_2\ a_3\ \dots\ a_n]$$

is a single row matrix, and

$$\begin{bmatrix} b_1 \\ b_2 \\ \vdots \\ b_n \end{bmatrix}$$

is a single column matrix.

The elements a_{ii} in a square matrix form the *principal* diagonal (or main diagonal). Their sum $a_{11} + a_{22} + \cdots + a_{mn}$ is called the *trace of A*. If all the elements of a square matrix are zero, then the matrix is called a *null* matrix. On the other hand, if only the elements on the main diagonal are non-zero, then the matrix is said to be a *diagonal* matrix. For example,

$$C = \begin{bmatrix} 3 & 0 & 0 \\ 0 & 4 & 0 \\ 0 & 0 & 5 \end{bmatrix}$$

is a diagonal matrix. In particular, if all the diagonal elements are equal to 1, it is called a *unit* matrix, and is usually denoted by *I*. Thus,

$$I = \begin{bmatrix} 1 & 0 & 0 \\ 0 & 1 & 0 \\ 0 & 0 & 1 \end{bmatrix}$$

is a unit matrix of order 3.

A square matrix is said to be an *upper-triangular* matrix if $a_{ij} = 0$ for $i > j$, and a *lower-triangular* matrix if $a_{ij} = 0$ for $i < j$. The matrices U and L defined by

$$U = \begin{bmatrix} u_{11} & u_{12} & u_{13} \\ 0 & u_{22} & u_{23} \\ 0 & 0 & u_{33} \end{bmatrix} \quad \text{and} \quad L = \begin{bmatrix} l_{11} & 0 & 0 \\ l_{21} & l_{22} & 0 \\ l_{31} & l_{32} & l_{33} \end{bmatrix}$$

are respectively, upper- and and lower-triangular matrices. In particular, if the diagonal elements are unity, the matrices are said to be *unit lower-triangular* or *unit upper-triangular* as the case may be.

A square matrix A is said to be *banded* if

$$a_{ij} = 0, \quad \text{when } |i - j| > k,$$

in which case only $(2k + 1)$ elements in every row are non-zero. In particular, matrices of the type

$$A=\begin{bmatrix} a_{11} & a_{12} & 0 & 0 \\ a_{21} & a_{22} & a_{23} & 0 \\ 0 & a_{32} & a_{33} & a_{34} \\ 0 & 0 & a_{43} & a_{44} \end{bmatrix}$$

are called *tridiagonal* matrices.

Every square matrix A is associated with a number called its *determinant*[1] and is written as

$$|A|=\begin{vmatrix} a_{11} & a_{12} & a_{13} & \cdots & a_{1n} \\ a_{21} & a_{22} & a_{23} & \cdots & a_{2n} \\ \vdots & \vdots & \vdots & & \vdots \\ a_{n1} & a_{n2} & a_{n3} & \cdots & a_{nn} \end{vmatrix}$$

If $|A| \neq 0$, the matrix is said to be *non-singular*; otherwise it is a *singular* matrix. As an example, the matrix

$$A=\begin{bmatrix} 2 & 1 & -2 \\ 3 & 0 & 5 \\ 4 & 2 & -4 \end{bmatrix}$$

is singular, since $|A|=0$.

A square matrix A in which $a_{ij}=a_{ji}$ is said to be *symmetric*. If $a_{ij}=-a_{ji}$, it is said to be *skew-symmetric*. For a skew-symmetric matrix, therefore, we have

$$a_{ii}=-a_{ii},$$

whence $a_{ii}=0$. For example, the matrix

$$\begin{bmatrix} 0 & 2 & -3 \\ -2 & 0 & 5 \\ 3 & -5 & 0 \end{bmatrix}$$

is a skew-symmetric matrix.

1.6.2 Matrix Operations

Equality of two matrices. Two matrices are said to be equal if they are of the same size and if their corresponding elements are equal.

Addition and subtraction of matrices. Two matrices of the *same* size can be added or subtracted by adding or subtracting their corresponding elements.

---

1. In this book, we assume that the reader is familiar with the theory of determinants and their properties.

Thus, if

$$A = \begin{bmatrix} 5 & 4 \\ 3 & 2 \end{bmatrix} \quad \text{and} \quad B = \begin{bmatrix} 2 & 3 \\ 1 & -1 \end{bmatrix}$$

then

$$A + B = \begin{bmatrix} 7 & 7 \\ 4 & 1 \end{bmatrix} \quad \text{and} \quad A - B = \begin{bmatrix} 3 & 1 \\ 2 & 3 \end{bmatrix}$$

Scalar multiplication of a matrix. If all the elements of a matrix A are multiplied by a constant k, then the resulting matrix is kA, i.e. if

$$A = \begin{bmatrix} 2 & -1 & 3 \\ 1 & 2 & -1 \end{bmatrix}$$

then

$$2A = \begin{bmatrix} 4 & -2 & 6 \\ 2 & 4 & -2 \end{bmatrix}$$

Example 1.69 If

$$A = \begin{bmatrix} 1 & 2 & 3 \\ 0 & 1 & 2 \\ 3 & 0 & -4 \end{bmatrix} \quad \text{and} \quad B = \begin{bmatrix} 17 & 0 & 13 \\ 0 & -8 & 4 \\ 7 & 6 & -5 \end{bmatrix}$$

find $2A - B$.

Here

$$2A = \begin{bmatrix} 2 & 4 & 6 \\ 0 & 2 & 4 \\ 6 & 0 & -8 \end{bmatrix}$$

Hence

$$2A - B = \begin{bmatrix} 2 & 4 & 6 \\ 0 & 2 & 4 \\ 6 & 0 & -8 \end{bmatrix} - \begin{bmatrix} 17 & 0 & 13 \\ 0 & -8 & 4 \\ 7 & 6 & -5 \end{bmatrix} = \begin{bmatrix} -15 & 4 & -7 \\ 0 & 10 & 0 \\ -1 & -6 & -3 \end{bmatrix}$$

Associativity, commutativity and distributivity of matrices. The following properties follow easily for three matrices A, B and C:

(a) $A + (B + C) = (A + B) + C$,

(b) $A + B = B + A$,

(c) $k(A + B) = kA + kB$, (k being a scalar)

(d) $(k_1 + k_2)A = k_1A + k_2A$, ($k_1$ and k_2 being constants).

Multiplication of a matrix by another matrix. Two matrices A and B can be multiplied only if the number of columns of A equals the number of rows of B. Thus, if A and B are of sizes (3×4) and (4×2) respectively, then their product AB is defined and will be of size (3×2). If C is the product, then the element C_{ij} of C is equal to the sum of the products of the corresponding elements of the ith row of A by those of the jth column of B, i.e.

$$C_{ij} = a_{i1}b_{1j} + a_{i2}b_{2j} + \ldots + a_{in}b_{nj} = \sum_{k=1}^{n} a_{ik}b_{kj}. \tag{1.57}$$

Consider an example, if

$$A = \begin{bmatrix} 3 & 4 & 2 & 1 \\ 1 & 2 & 3 & 1 \\ 0 & 1 & 2 & 3 \end{bmatrix}, \quad B = \begin{bmatrix} 1 & 2 \\ 3 & 4 \\ 1 & 0 \\ -1 & 1 \end{bmatrix}$$

then

$$AB = \begin{bmatrix} 3+12+2-1 & 6+16+0+1 \\ 1+6+3-1 & 2+8+0+1 \\ 0+3+2-3 & 0+4+0+3 \end{bmatrix} = \begin{bmatrix} 16 & 23 \\ 9 & 11 \\ 2 & 7 \end{bmatrix}$$

We note that the number of columns in B is 2 and that the number of rows in A is 3. Since these are not the same, the product BA is not defined.

In general, $AB \neq BA$. For example,

$$\begin{bmatrix} 1 & 2 \\ -1 & 1 \end{bmatrix}\begin{bmatrix} 3 & 4 \\ 2 & 1 \end{bmatrix} = \begin{bmatrix} 7 & 6 \\ -1 & -3 \end{bmatrix},$$

but

$$\begin{bmatrix} 3 & 4 \\ 2 & 1 \end{bmatrix}\begin{bmatrix} 1 & 2 \\ -1 & 1 \end{bmatrix} = \begin{bmatrix} -1 & 10 \\ 1 & 5 \end{bmatrix},$$

which is a different matrix.

Example 1.70

(a) If

$$A = \begin{bmatrix} a_{11} & a_{12} & a_{13} \\ a_{21} & a_{22} & a_{23} \\ a_{31} & a_{32} & a_{33} \end{bmatrix} \quad \text{and} \quad X = \begin{bmatrix} x_1 \\ x_2 \\ x_3 \end{bmatrix}$$

then

$$AX = \begin{bmatrix} a_{11}x_1 + a_{12}x_2 + a_{13}x_3 \\ a_{21}x_1 + a_{22}x_2 + a_{23}x_3 \\ a_{31}x_1 + a_{32}x_2 + a_{33}x_3 \end{bmatrix}$$

(b) If

$$A = \begin{bmatrix} 0 & 1 \\ 1 & 0 \end{bmatrix}$$

then

$$A^2 = AA = \begin{bmatrix} 0 & 1 \\ 1 & 0 \end{bmatrix} \begin{bmatrix} 0 & 1 \\ 1 & 0 \end{bmatrix} = \begin{bmatrix} 1 & 0 \\ 0 & 1 \end{bmatrix} = I$$

(c) If

$$A = \begin{bmatrix} 1 & 0 \\ 0 & 0 \end{bmatrix} \quad \text{and} \quad B = \begin{bmatrix} 0 & 0 \\ 1 & 1 \end{bmatrix}$$

then

$$AB = \begin{bmatrix} 0 & 0 \\ 0 & 0 \end{bmatrix},$$

which shows that the product AB may be a null matrix with neither A nor B being null matrices.

(d) If

$$A = \begin{bmatrix} 1 & 0 & 2 \\ 3 & 4 & 1 \\ 5 & 6 & 7 \end{bmatrix} \quad \text{and} \quad I = \begin{bmatrix} 1 & 0 & 0 \\ 0 & 1 & 0 \\ 0 & 0 & 1 \end{bmatrix}$$

then $AI = IA = A$, which shows that the cummutative law holds good in this case.

(e) If we define the matrices:

$$A = \begin{bmatrix} 0 & 1 & 3 \\ 1 & 3 & 0 \\ 0 & 0 & 4 \end{bmatrix}, \quad X = \begin{bmatrix} x_1 \\ x_2 \\ x_3 \end{bmatrix}, \quad B = \begin{bmatrix} 1 \\ 0 \\ 2 \end{bmatrix},$$

then the matrix equation $AX = B$ represents the system of equations:

$$x_2 + 3x_3 = 1$$

$$x_1 + 3x_2 = 0$$

$$x_3 = 2.$$

Example 1.71 If

$$A = \begin{bmatrix} 1 & 2 \\ 3 & 4 \end{bmatrix},$$

show that $A^2 - 5A = 2I$, where I is the unit matrix of order 2. Hence, or otherwise, determine A^4.

We have

$$A^2 = \begin{bmatrix} 1 & 2 \\ 3 & 4 \end{bmatrix}\begin{bmatrix} 1 & 2 \\ 3 & 4 \end{bmatrix} = \begin{bmatrix} 7 & 10 \\ 15 & 22 \end{bmatrix}$$

and

$$5A = 5\begin{bmatrix} 1 & 2 \\ 3 & 4 \end{bmatrix} = \begin{bmatrix} 5 & 10 \\ 15 & 20 \end{bmatrix}$$

Hence,

$$A^2 - 5A = \begin{bmatrix} 7 & 10 \\ 15 & 22 \end{bmatrix} - \begin{bmatrix} 5 & 10 \\ 15 & 20 \end{bmatrix} = \begin{bmatrix} 2 & 0 \\ 0 & 2 \end{bmatrix} = 2\begin{bmatrix} 1 & 0 \\ 0 & 1 \end{bmatrix} = 2I.$$

From the above result, we have

$$A^2 = 5A + 2I.$$

Hence,

$$\begin{aligned} A^4 &= (5A+2I)(5A+2I) \\ &= 25A^2 + 10AI + 10IA + 4I^2 \\ &= 25A^2 + 20AI + 4I, \text{ since } AI = IA \text{ and } I^2 = I. \\ &= 25(5A+2I) + 20AI + 4I, \text{ since } A^2 = 5A + 2I. \\ &= 125A + 50I + 20AI + 4I. \\ &= 145A + 54I, \text{ since } AI = A \end{aligned}$$

We get

$$\begin{aligned} A^4 &= 145\begin{bmatrix} 1 & 2 \\ 3 & 4 \end{bmatrix} + 54\begin{bmatrix} 1 & 0 \\ 0 & 1 \end{bmatrix} \\ &= \begin{bmatrix} 145 & 290 \\ 435 & 580 \end{bmatrix} + \begin{bmatrix} 54 & 0 \\ 0 & 54 \end{bmatrix} \\ &= \begin{bmatrix} 199 & 290 \\ 435 & 634 \end{bmatrix} \end{aligned}$$

1.6.3 Transpose, Adjoint and Inverse of a Matrix

The matrix obtained by interchanging the rows and columns of a square matrix A is defined as the *transpose* of A and is denoted by A^{T}.

Thus, if

$$A = \begin{bmatrix} a_{11} & a_{12} & a_{13} \\ a_{21} & a_{22} & a_{23} \\ a_{31} & a_{32} & a_{33} \end{bmatrix}$$

then

$$A^{\mathrm{T}} = \begin{bmatrix} a_{11} & a_{21} & a_{31} \\ a_{12} & a_{22} & a_{32} \\ a_{13} & a_{23} & a_{33} \end{bmatrix}$$

It can easily be shown that

(i) If $C = A + B$, then $C^{\mathrm{T}} = A^{\mathrm{T}} + B^{\mathrm{T}}$, and

(ii) If A and B are two square matrices, then the transpose of the product AB is the product of the transposes taken in the reverse order, i.e.

$$(AB)^{\mathrm{T}} = B^{\mathrm{T}} A^{\mathrm{T}}.$$

More generally,

$$(ABCD \ldots)^{\mathrm{T}} = \cdots D^{\mathrm{T}} C^{\mathrm{T}} B^{\mathrm{T}} A^{\mathrm{T}}$$

The proofs are left as exercises to the reader.

Example 1.72 Let

$$A = \begin{bmatrix} 1 & 2 & 3 \\ 4 & 5 & 6 \\ 7 & 8 & 9 \end{bmatrix}, \quad B = \begin{bmatrix} 1 & 1 & 1 \\ 1 & 1 & 1 \\ 1 & 1 & 1 \end{bmatrix}$$

then

$$AB = \begin{bmatrix} 6 & 6 & 6 \\ 15 & 15 & 15 \\ 24 & 24 & 24 \end{bmatrix} \quad \text{and} \quad (AB)^{\mathrm{T}} = \begin{bmatrix} 6 & 15 & 24 \\ 6 & 15 & 24 \\ 6 & 15 & 24 \end{bmatrix}$$

Also

$$B^{\mathrm{T}} A^{\mathrm{T}} = \begin{bmatrix} 1 & 1 & 1 \\ 1 & 1 & 1 \\ 1 & 1 & 1 \end{bmatrix} \begin{bmatrix} 1 & 4 & 7 \\ 2 & 5 & 8 \\ 3 & 6 & 9 \end{bmatrix} = \begin{bmatrix} 6 & 15 & 24 \\ 6 & 15 & 24 \\ 6 & 15 & 24 \end{bmatrix} = (AB)^{\mathrm{T}}.$$

Example 1.73 Given

$$A = \begin{bmatrix} 1 & 0 & 1 \\ 0 & 1 & 0 \\ 3 & 0 & 4 \end{bmatrix},$$

show that

$$(A^2)^{\mathrm{T}} = (A^{\mathrm{T}})^2.$$

Since

$$A = \begin{bmatrix} 1 & 0 & 1 \\ 0 & 1 & 0 \\ 3 & 0 & 4 \end{bmatrix},$$

we have

$$A^2 = \begin{bmatrix} 1 & 0 & 1 \\ 0 & 1 & 0 \\ 3 & 0 & 4 \end{bmatrix} \begin{bmatrix} 1 & 0 & 1 \\ 0 & 1 & 0 \\ 3 & 0 & 4 \end{bmatrix} = \begin{bmatrix} 4 & 0 & 5 \\ 0 & 1 & 0 \\ 15 & 0 & 19 \end{bmatrix}$$

Hence,

$$(A^2)^{\mathrm{T}} = \begin{bmatrix} 4 & 0 & 15 \\ 0 & 1 & 0 \\ 5 & 0 & 19 \end{bmatrix}$$

Therefore,

$$(A^{\mathrm{T}})^2 = \begin{bmatrix} 1 & 0 & 3 \\ 0 & 1 & 0 \\ 1 & 0 & 4 \end{bmatrix} \begin{bmatrix} 1 & 0 & 3 \\ 0 & 1 & 0 \\ 1 & 0 & 4 \end{bmatrix} = \begin{bmatrix} 4 & 0 & 15 \\ 0 & 1 & 0 \\ 5 & 0 & 19 \end{bmatrix}$$

which is the same as $(A^2)^{\mathrm{T}}$.

Let A be a square matrix given by

$$A = \begin{bmatrix} a_{11} & a_{12} & a_{13} & \cdots & a_{1n} \\ a_{21} & a_{22} & a_{23} & \cdots & a_{2n} \\ \vdots & \vdots & \vdots & & \vdots \\ a_{n1} & a_{n2} & a_{n3} & \cdots & a_{nn} \end{bmatrix}$$

If $|A|$ is the determinant of the matrix A, then

$$|A| = \begin{bmatrix} a_{11} & a_{12} & a_{13} & \cdots & a_{1n} \\ a_{21} & a_{22} & a_{23} & \cdots & a_{2n} \\ \vdots & \vdots & \vdots & & \vdots \\ a_{n1} & a_{n2} & a_{n3} & \cdots & a_{nn} \end{bmatrix}$$

Let $A_{11}, A_{12}, \ldots$ be respectively the cofactors of the elements $a_{11}, a_{12}, \ldots$ in $|A|$. Then the transpose of the matrix

$$\begin{bmatrix} A_{11} & A_{12} & A_{13} & \cdots & A_{1n} \\ A_{21} & A_{22} & A_{23} & \cdots & A_{2n} \\ \vdots & \vdots & \vdots & & \vdots \\ A_{n1} & A_{n2} & A_{n3} & \cdots & A_{nn} \end{bmatrix}$$

i.e the matrix

$$\begin{bmatrix} A_{11} & A_{21} & \cdots & A_{n1} \\ A_{12} & A_{22} & \cdots & A_{n2} \\ A_{13} & A_{23} & \cdots & A_{n3} \\ \vdots & \vdots & & \vdots \\ A_{1n} & A_{2n} & \cdots & A_{nn} \end{bmatrix}$$

is defined as the *adjoint* of the matrix A and is denoted by adj A.

Example 1.74 Find the adjoint of

$$A = \begin{bmatrix} 1 & 2 & -1 \\ 3 & 8 & 2 \\ 4 & 9 & -1 \end{bmatrix}$$

We find

$$A_{11} = \begin{vmatrix} 8 & 2 \\ 9 & -1 \end{vmatrix} = -26, \quad A_{12} = -\begin{vmatrix} 3 & 2 \\ 4 & -1 \end{vmatrix} = 11,$$

$$A_{13} = -5, \quad A_{21} = -7, \quad A_{22} = 3,$$

$$A_{23} = -1, \quad A_{31} = 12, \quad A_{32} = -5,$$

$$A_{33} = 2.$$

Hence, adj A is given by

$$\text{adj } A = \begin{bmatrix} -26 & -7 & 12 \\ 11 & 3 & -5 \\ -5 & -1 & 2 \end{bmatrix}$$

Let A be a non-singular square matrix of order n. Let B be another square matrix of the same order such that

$$BA = I,$$

where I is the unit matrix of order n. Then B is said to be the *inverse of* A which is written as A^{-1} so that

$$AA^{-1} = A^{-1}A = I$$

It follows that A and A^{-1} are square matrices of the same order.

Theorem 1.17 If an inverse to A exists, then it is unique.

Let B and C be the inverse matrices to A. Then

$$AB = BA = I \quad \text{and} \quad AC = CA = I$$

Premultiplying AB by C, we get

$$CAB = CI \quad \text{or} \quad IB = CI \quad \text{or} \quad B = C,$$

which proves that the inverse is *unique*.

Theorem 1.18 The inverse A^{-1} exists only when $|A| \neq 0$ and then

$$A^{-1} = \frac{1}{|A|} \text{ adj } A. \tag{1.58}$$

We have

$$A\,(\text{adj } A) = \begin{bmatrix} A_{11} & A_{21} & \cdots & A_{n1} \\ A_{12} & A_{22} & \cdots & A_{n2} \\ \vdots & \vdots & & \vdots \\ A_{1n} & A_{2n} & \cdots & A_{nn} \end{bmatrix} = \begin{bmatrix} |A| & 0 & 0 & \cdots & 0 \\ 0 & |A| & 0 & \cdots & 0 \\ \vdots & \vdots & \vdots & & \vdots \\ 0 & 0 & 0 & \cdots & |A| \end{bmatrix}$$

since

$$a_{11}A_{11} + a_{12}A_{12} + a_{13}A_{13} + \cdots + a_{1n}A_{1n} = |A|,$$
$$a_{11}A_{21} + a_{12}A_{22} + a_{13}A_{23} + \cdots + a_{1n}A_{2n} = 0, \text{ etc.}$$

We therefore obtain:

$$A\ (\text{adj } A) = |A|\,I,$$

where I is the unit matrix of order n. Hence,

$$\frac{A(\text{adj } A)}{|A|} A = I.$$

Similarly, it can be shown that

$$\frac{(\text{adj } A)}{|A|} A = I.$$

It follows therefore that

$$\frac{\text{adj } A}{|A|}$$

is the inverse of A so that

$$A^{-1} = \frac{\text{adj } A}{|A|}.$$

Clearly, the inverse A^{-1} exists only when $|A| \neq 0$, i.e. when A is non-singular.

If A and B are non-singular square matrices, then

$$(AB)^{-1} = B^{-1}A^{-1}$$

and

$$(A^T)^{-1} = (A^{-1})^T. \tag{1.59}$$

Since A and B are non-singular, we have $|A| \neq 0$ and $|B| \neq 0$, so that $|AB| \neq 0$. Now,

$$AB(B^{-1}A^{-1}) = A(BB^{-1})A^{-1} = AIA^{-1} = AA^{-1} = I$$

Similarly,

$$(B^{-1}A^{-1})AB = B^{-1}(A^{-1}A)B = B^{-1}IB = B^{-1}B = I.$$

Hence, it follows that $B^{-1}A^{-1}$ is the inverse of the product AB. The proof of the second result is left as an exercise to the reader.

Example 1.75 Show that

$$\begin{bmatrix} 1 & -\tan\dfrac{\theta}{2} \\ \tan\dfrac{\theta}{2} & 1 \end{bmatrix} \begin{bmatrix} 1 & \tan\dfrac{\theta}{2} \\ -\tan\dfrac{\theta}{2} & 1 \end{bmatrix}^{-1} = \begin{bmatrix} \cos\theta & -\sin\theta \\ \sin\theta & \cos\theta \end{bmatrix}$$

Let

$$A = \begin{bmatrix} 1 & -\tan\dfrac{\theta}{2} \\ \tan\dfrac{\theta}{2} & 1 \end{bmatrix} \quad \text{and} \quad B = \begin{bmatrix} 1 & \tan\dfrac{\theta}{2} \\ -\tan\dfrac{\theta}{2} & 1 \end{bmatrix}$$

To compute the inverse of B, we first observe that

$$|B| = 1 + \tan^2\frac{\theta}{2} = \sec^2\frac{\theta}{2}.$$

Now,

$$\text{adj } B = \begin{bmatrix} 1 & -\tan\dfrac{\theta}{2} \\ \tan\dfrac{\theta}{2} & 1 \end{bmatrix},$$

so that

$$B^{-1} = \frac{1}{\sec^2(\theta/2)} \begin{bmatrix} 1 & -\tan\dfrac{\theta}{2} \\ \tan\dfrac{\theta}{2} & 1 \end{bmatrix}$$

Hence,

$$AB^{-1} = \frac{1}{\sec^2(\theta/2)} \begin{bmatrix} 1 & -\tan\frac{\theta}{2} \\ \tan\frac{\theta}{2} & 1 \end{bmatrix} \begin{bmatrix} 1 & -\tan\frac{\theta}{2} \\ \tan\frac{\theta}{2} & 1 \end{bmatrix}$$

$$= \frac{1}{\sec^2(\theta/2)} \begin{bmatrix} 1-\tan^2\frac{\theta}{2} & -2\tan\frac{\theta}{2} \\ 2\tan\frac{\theta}{2} & 1-\tan^2\frac{\theta}{2} \end{bmatrix}$$

$$= \begin{bmatrix} \cos^2\frac{\theta}{2}-\sin^2\frac{\theta}{2} & -2\sin\frac{\theta}{2}\cos\frac{\theta}{2} \\ 2\sin\frac{\theta}{2}\cos\frac{\theta}{2} & \cos^2\frac{\theta}{2}-\sin^2\frac{\theta}{2} \end{bmatrix}$$

$$= \begin{bmatrix} \cos\theta & -\sin\theta \\ \sin\theta & \cos\theta \end{bmatrix},$$

which is the required result.

Example 1.76 Find the inverse of the matrix

$$A = \begin{bmatrix} 5 & -2 & 4 \\ -2 & 1 & 1 \\ 4 & 1 & 0 \end{bmatrix}$$

We have

$$|A| = 5(-1) + 2(-4) + 4(-2-4) = -37,$$

and

$$A^{\mathrm{T}} = \begin{bmatrix} 5 & -2 & 4 \\ -2 & 1 & 1 \\ 4 & 1 & 0 \end{bmatrix}$$

We find

$$A_{11} = \begin{vmatrix} 1 & 0 \\ 1 & 1 \end{vmatrix} = -1; \quad A_{12} = -\begin{vmatrix} -2 & 1 \\ 4 & 0 \end{vmatrix} = 4; \quad A_{13} = \begin{vmatrix} -2 & 1 \\ 4 & 1 \end{vmatrix} = -6;$$

$$A_{21} = -\begin{vmatrix} -2 & 4 \\ 1 & 0 \end{vmatrix} = 4; \quad A_{22} = \begin{vmatrix} 5 & 4 \\ 4 & 0 \end{vmatrix} = -16; \quad A_{23} = -\begin{vmatrix} 5 & -2 \\ 4 & 1 \end{vmatrix} = -13;$$

$$A_{31} = \begin{vmatrix} -2 & 4 \\ 1 & 1 \end{vmatrix} = -6; \quad A_{32} = -\begin{vmatrix} 5 & 4 \\ -2 & 1 \end{vmatrix} = -13; \quad A_{33} = \begin{vmatrix} 5 & -2 \\ -2 & 1 \end{vmatrix} = 1.$$

Therefore,

$$\text{adj } A = \begin{bmatrix} -1 & 4 & -6 \\ 4 & -16 & -13 \\ -6 & -13 & 1 \end{bmatrix}$$

and hence

$$A^{-1} = -\frac{1}{37}\begin{bmatrix} -1 & 4 & -6 \\ 4 & -16 & -13 \\ -6 & -13 & 1 \end{bmatrix}$$

The reader may verify that $AA^{-1} = I$.

1.6.4 Some Special Matrices

Orthogonal matrix

Let $A = [a_{ij}]$ be a square matrix of order n. If $AA^{\mathrm{T}} = I$, then A is said to be an *orthogonal* matrix.

For example, the matrices

$$A = \begin{bmatrix} \cos\theta & -\sin\theta \\ \sin\theta & \cos\theta \end{bmatrix} \quad \text{and} \quad B = \begin{bmatrix} -\frac{2}{3} & \frac{1}{3} & \frac{2}{3} \\ \frac{2}{3} & \frac{2}{3} & \frac{1}{3} \\ \frac{1}{3} & -\frac{2}{3} & \frac{2}{3} \end{bmatrix}$$

are orthogonal matrices.

As we have

$$A^{\mathrm{T}} = \begin{bmatrix} \cos\theta & \sin\theta \\ -\sin\theta & \cos\theta \end{bmatrix}$$

and therefore

$$AA^{\mathrm{T}} = \begin{bmatrix} \cos\theta & -\sin\theta \\ \sin\theta & \cos\theta \end{bmatrix}\begin{bmatrix} \cos\theta & \sin\theta \\ -\sin\theta & \cos\theta \end{bmatrix} = \begin{bmatrix} 1 & 0 \\ 0 & 1 \end{bmatrix} = I.$$

Hence, A is orthogonal. In a similar manner, B can also be shown to be orthogonal.

Hermitian matrix

The *Hermitian* matrix is a generalization of the symmetric matrix which is often more useful when some of the elements are complex numbers. If A is a square matrix with its elements a_{ij} as complex numbers, then the Hermitian matrix is defined to be the matrix which satisfies

$$A^{\mathrm{T}} = \overline{A} \tag{1.60}$$

where $\bar{A}$ denotes the matrix with elements which are complex conjugates of those of A. In terms of general elements, the above equation implies

$$a_{ji} = \bar{a}_{ij} \quad \text{or} \quad a_{ii} = \bar{a}_{ii},$$

which shows that all the diagonal elements are real. For example, the matrix

$$\begin{bmatrix} 2 & 1+i & 2+3i \\ 1-i & 3 & 3+4i \\ 2-3i & 3-4i & 5 \end{bmatrix}$$

is a typical third-order Hermitian matrix.

The matrix $(\bar{A})^{\text{T}}$, which is the transpose of the conjugate of A is called the *transposed conjugate* of A and is denoted by A^*.

Skew-Hermitian matrix

If a complex matrix A satisfies the condition

$$(\bar{A})^{\text{T}} = -\bar{A}, \quad \text{i.e.} \quad A^* = -\bar{A}, \tag{1.61}$$

then A is called a *skew-Hermitian matrix*. In this case,

$$a_{ji} = -\bar{a}_{ij} \quad \text{and} \quad a_{ii} = -a_{ii},$$

which means that the diagonal elements are pure imaginary numbers. Thus the matrix

$$B = \begin{bmatrix} 2i & 2+3i & 1+i \\ -2+3i & 3i & 3+4i \\ -1+i & -3+4i & 4i \end{bmatrix}$$

is a skew-Hermitian matrix.

If A and B are Hermitian, then $AB + BA$ is also Hermitian but $AB - BA$ is skew-Hermitian. For,

$$\begin{aligned} (AB + BA)^* &= (AB)^* + (BA)^* \\ &= B^*A^* + A^*B^* \\ &= BA + AB \\ &= AB + BA \end{aligned}$$

Similarly, it can be shown that $AB - BA$ is skew-Hermitian. Some results concerning Hermitian and skew-Hermitian matrices are stated below without proof:

(a) If A is a square matrix, then $A - A^*$, A^*A are Hermitian, whereas $A - A^*$ is skew-Hermitian.

(b) B^*AB is Hermitian or skew-Hermitian according as A is Hermitian or skew-Hermitian.

(c) Every square matrix is uniquely expressible as the sum of a Hermitian matrix and a skew-Hermitian matrix.

(d) $\bar{A}$ is Hermitian or skew-Hermitian according as A is Hermitian or skew-Hermitian.

(e) Every square matrix can be uniquely expressed as $P+iQ$, where P and Q are Hermitian matrices.

Unitary matrices

A square matrix is said to be *unitary* if $A^*A = I$.

An important result concerning unitary matrices is that if A and B are unitary matrices, then AB and BA are also unitary matrices, for we have

$$A^*A = AA^* = I \quad \text{and} \quad B^*B = BB^* = I,$$

since A and B are unitary matrices.

Now,

$$(AB)(AB)^* = (AB)B^*A^* = ABB^*A^* = AIA^* = AA^* = I.$$

Therefore, AB is a unitary matrix.

Similarly, it can be shown that BA is also unitary.

Example 1.77 Show that

$$\begin{bmatrix} \frac{1+i}{2} & \frac{-1+i}{2} \\ \frac{1+i}{2} & \frac{1-i}{2} \end{bmatrix}$$

is unitary.

Let

$$A = \begin{bmatrix} \frac{1+i}{2} & \frac{-1+i}{2} \\ \frac{1+i}{2} & \frac{1-i}{2} \end{bmatrix}$$

Then

$$\bar{A} = \begin{bmatrix} \frac{1-i}{2} & \frac{-1-i}{2} \\ \frac{1-i}{2} & \frac{1+i}{2} \end{bmatrix}$$

Therefore,

$$(\bar{A})^{\mathrm{T}} = \begin{bmatrix} \frac{1-i}{2} & \frac{1-i}{2} \\ \frac{-1-i}{2} & \frac{1+i}{2} \end{bmatrix} = A^*.$$

Hence

$$AA^* = \begin{bmatrix} \frac{1+i}{2} & \frac{-1+i}{2} \\ \frac{1+i}{2} & \frac{1-i}{2} \end{bmatrix} \begin{bmatrix} \frac{1-i}{2} & \frac{1-i}{2} \\ \frac{-1-i}{2} & \frac{1+i}{2} \end{bmatrix} = \begin{bmatrix} \frac{1-i^2}{4} + \frac{1-i^2}{4} & \frac{1-i^2}{4} - \frac{1-i^2}{4} \\ \frac{1-i^2}{4} - \frac{1-i^2}{4} & \frac{1-i^2}{4} + \frac{1-i^2}{4} \end{bmatrix}$$

$$= \begin{bmatrix} 1 & 0 \\ 0 & 1 \end{bmatrix} = I.$$

It follows that A is a unitary matrix.

1.6.5 Useful Relationships Involving Matrices

Theorem 1.19 Any real square matrix can be expressed as the sum of a symmetric matrix and a skew-symmetric matrix.

Let A, B, C be three square matrices such that

$$A = B + C.$$

Taking transposes, we obtain

$$A^{\mathrm{T}} = B^{\mathrm{T}} + C^{\mathrm{T}}.$$

Since B and C are symmetric and skew-symmetric matrices, we have

$$B^{\mathrm{T}} = B \quad \text{and} \quad C^{\mathrm{T}} = -C.$$

Hence we obtain $A^{\mathrm{T}} = B - C$. It follows that

$$\frac{A + A^{\mathrm{T}}}{2} = B \quad \text{and} \quad \frac{A - A^{\mathrm{T}}}{2} = C.$$

In terms of their elements, these relationships are

$$b_{ij} = \frac{a_{ij} + a_{ji}}{2} \quad \text{and} \quad c_{ij} = \frac{a_{ij} - a_{ji}}{2} \tag{1.62}$$

In a similar way, any square complex matrix can be expressed as the sum of a Hermitian and a skew-Hermitian matrix.

Example 1.78 Express the matrix

$$A = \begin{bmatrix} 1 & 6 & 5 \\ 8 & 9 & 3 \\ 7 & 2 & 4 \end{bmatrix}$$

as the sum of a symmetric and a skew-symmetric matrix.

Here,

$$A^{\mathrm{T}} = \begin{bmatrix} 1 & 8 & 7 \\ 6 & 9 & 2 \\ 5 & 3 & 4 \end{bmatrix}$$

Hence,

$$B = \begin{bmatrix} 1 & 7 & 6 \\ 7 & 9 & 2.5 \\ 6 & 2.5 & 4 \end{bmatrix} \quad \text{and} \quad C = \begin{bmatrix} 0 & -1 & -1 \\ 1 & 0 & 0.5 \\ 1 & -0.5 & 0 \end{bmatrix},$$

which are symmetric and skew-symmetric, respectively.

Theorem 1.20 Any square matrix can be expressed as the product of a unit lower triangular matrix and an upper triangular matrix.

If A is a square matrix, then A can be factorized into the form LU, where L is unit lower triangular and U is upper triangular, if all the principal minors of A are non-singular, i.e. if

$$a_{11} \neq 0, \qquad \begin{vmatrix} a_{11} & a_{12} \\ a_{21} & a_{22} \end{vmatrix} \neq 0, \qquad \begin{vmatrix} a_{11} & a_{12} & a_{13} \\ a_{21} & a_{22} & a_{23} \\ a_{31} & a_{32} & a_{33} \end{vmatrix} \neq 0, \text{ etc.} \tag{1.63}$$

It can be shown that such a factorization, when it exists, is unique.[2]

We demonstrate the method by considering the simple linear system:

$$\left.\begin{aligned} a_{11}x_1 + a_{12}x_2 + a_{13}x_3 &= b_1 \\ a_{21}x_1 + a_{22}x_2 + a_{23}x_3 &= b_2 \\ a_{31}x_1 + a_{32}x_2 + a_{33}x_3 &= b_3 \end{aligned}\right\} \tag{1.64a}$$

which can be written in the matrix form

$$AX = B, \tag{1.64b}$$

where

$$A = \begin{bmatrix} a_{11} & a_{12} & a_{13} \\ a_{21} & a_{22} & a_{23} \\ a_{31} & a_{32} & a_{33} \end{bmatrix}, \qquad X = \begin{bmatrix} x_1 \\ x_2 \\ x_3 \end{bmatrix}, \qquad B = \begin{bmatrix} b_1 \\ b_2 \\ b_3 \end{bmatrix}.$$

Let

$$A = LU \tag{1.65}$$

where

$$U = \begin{bmatrix} u_{11} & u_{12} & u_{13} \\ 0 & u_{22} & u_{23} \\ 0 & 0 & u_{33} \end{bmatrix} \tag{1.66}$$

2. See Issacson and Keller (1967).

and

$$L = \begin{bmatrix} 1 & 0 & 0 \\ l_{21} & 1 & 0 \\ l_{31} & l_{32} & 1 \end{bmatrix} \tag{1.67}$$

Hence, (1.64b) becomes:

$$LUX = B \tag{1.68}$$

If we set

$$UX = Y, \tag{1.69}$$

then (1.68) may be written as

$$LY = B, \tag{1.70}$$

which is equivalent to the system

$$\begin{aligned} y_1 &= b_1, \\ l_{21}y_1 + y_2 &= b_2 \\ l_{31}y_1 + l_{32}y_2 + y_3 &= b_3, \end{aligned}$$

and can therefore be solved for y_1, y_2, y_3 by the forward substitution. When Y is known, the system (1.69) becomes:

$$\begin{aligned} u_{11}x_1 + u_{12}x_2 + u_{13}x_3 &= y_1 \\ u_{22}x_2 + u_{23}x_3 &= y_2 \\ u_{33}x_3 &= y_3, \end{aligned}$$

which can be solved for x_3, x_2 and x_1, in that order, by the backward substitution process.

We shall now describe a scheme for computing the matrices L and U, and illustrate the procedure with a matrix of order 3. From the relation (1.65), we obtain

$$\begin{bmatrix} 1 & 0 & 0 \\ l_{21} & 1 & 0 \\ l_{31} & l_{32} & 1 \end{bmatrix} \begin{bmatrix} u_{11} & u_{12} & u_{13} \\ 0 & u_{22} & u_{23} \\ 0 & 0 & u_{33} \end{bmatrix} = \begin{bmatrix} a_{11} & a_{12} & a_{13} \\ a_{21} & a_{22} & a_{23} \\ a_{31} & a_{32} & a_{33} \end{bmatrix}$$

Multiplying the matrices on the left and equating the corresponding elements of both sides, we get:

$$\left.\begin{aligned} u_{11} &= a_{11}, & u_{12} &= a_{12}, & u_{13} &= a_{13} \\ l_{21}u_{11} &= a_{21}, & l_{21}u_{12} + u_{22} &= a_{22}, & l_{21}u_{13} + u_{23} &= a_{23} \\ l_{31}u_{11} &= a_{31}, & l_{31}u_{12} + l_{32}u_{22} &= a_{32}, & l_{31}u_{13} + l_{32}u_{23} + u_{33} &= a_{33} \end{aligned}\right\} \tag{1.71}$$

from which u_{33} can be computed.

Thus, we have a systematic procedure to evaluate the elements of L and U. First, we determine the first row of U and the first column of L, then the second row of U and the second column of L, and finally we compute the third row of U. The procedure can easily be generalized. When the factorization is effected, the inverse of A can be computed from the formula

$$A^{-1} = (LU)^{-1} = U^{-1}L^{-1} \tag{1.72}$$

Example 1.79 Solve the equations

$$2x + 3y + z = 9$$

$$x + 2y + 3z = 6$$

$$3x + y + 2z = 8$$

by the factorization method.

We have

$$A = \begin{bmatrix} 2 & 3 & 1 \\ 1 & 2 & 3 \\ 3 & 1 & 2 \end{bmatrix}$$

Let

$$\begin{bmatrix} 1 & 0 & 0 \\ l_{21} & 1 & 0 \\ l_{31} & l_{32} & 1 \end{bmatrix} \begin{bmatrix} u_{11} & u_{12} & u_{13} \\ 0 & u_{22} & u_{23} \\ 0 & 0 & u_{33} \end{bmatrix} = \begin{bmatrix} 2 & 3 & 1 \\ 1 & 2 & 3 \\ 3 & 1 & 2 \end{bmatrix}$$

Clearly,

$$u_{11} = 2, \quad u_{12} = 3, \quad u_{13} = 1.$$

Also,

$$l_{21}u_{11} = 1, \quad \text{so that} \quad l_{21} = \frac{1}{2},$$

and

$$l_{31}u_{11} = 3, \quad \text{so that} \quad l_{31} = \frac{3}{2}.$$

For u_{22} and u_{23}, we have the equations

$$l_{21}u_{12} + u_{22} = 2 \quad \text{and} \quad l_{21}u_{13} + u_{23} = 3,$$

from which we obtain

$$u_{22} = \frac{1}{2} \quad \text{and} \quad u_{23} = \frac{5}{2}.$$

Finally, the equations

$$l_{31}u_{12} + l_{32}u_{22} = 1 \quad \text{and} \quad l_{31}u_{13} + l_{32}u_{23} + u_{33} = 2,$$

yield

$$l_{32} = -7 \quad \text{and} \quad u_{33} = 18.$$

We therefore have

$$A = \begin{bmatrix} 1 & 0 & 0 \\ \frac{1}{2} & 1 & 0 \\ \frac{3}{2} & -7 & 1 \end{bmatrix} \begin{bmatrix} 2 & 3 & 1 \\ 0 & \frac{1}{2} & \frac{5}{2} \\ 0 & 0 & 18 \end{bmatrix}$$

Writing $UX = Y$, the given system becomes:

$$\begin{bmatrix} 1 & 0 & 0 \\ \frac{1}{2} & 1 & 0 \\ \frac{3}{2} & -7 & 1 \end{bmatrix} \begin{bmatrix} y_1 \\ y_2 \\ y_3 \end{bmatrix} = \begin{bmatrix} 9 \\ 6 \\ 8 \end{bmatrix}$$

Solving this system by forward substitution, we get

$$y_1 = 9,$$

$$\frac{1}{2}y_1 + y_2 = 6 \quad \text{or} \quad y_2 = \frac{3}{2}.$$

$$\frac{3}{2}y_1 - 7y_2 + y_3 = 8 \quad \text{or} \quad y_3 = 5.$$

Hence, the solution of the original system is given by

$$\begin{bmatrix} 2 & 3 & 1 \\ 0 & \frac{1}{2} & \frac{5}{2} \\ 0 & 0 & 18 \end{bmatrix} \begin{bmatrix} x_1 \\ x_2 \\ x_3 \end{bmatrix} = \begin{bmatrix} 9 \\ \frac{3}{2} \\ 5 \end{bmatrix},$$

which, when solved by the back-substitution process, gives:

$$x = \frac{35}{18}, \qquad y = \frac{29}{18}, \qquad z = \frac{5}{18}.$$

1.6.6 Rank of a Matrix

We consider a square matrix of order n. Of the n rows and n columns, if there are at least k rows and k columns which must be deleted in order to obtain a non-vanishing determinant, then the order of the highest ordered non-vanishing determinant in A is given by $r = n - k$, and this number is defined as the *rank* of A and is written as $r(A)$. Hence, the rank of a matrix is equal to the order of the highest ordered non-vanishing determinant in A. It follows, therefore, that for a nonsingular square matrix of order n, the rank is equal to n. To determine The rank of a matrix, we have to find the order of the highest ordered non-vanishing determinant. This method, although general, would be tedious when applied to matrices of higher order, for which the Gaussian elimination method, to be described later in this section, would be particularly suitable.

Consider the following examples:

(a) $A = \begin{bmatrix} 0 & 0 \\ 0 & 0 \end{bmatrix}$, $r(A) = 0$, since all the elements are zero.

(b) $A = \begin{bmatrix} 1 & 2 \\ 3 & 4 \end{bmatrix}$, $r(A) = 2$, since $|A| \neq 0$.

(c) $B = \begin{bmatrix} 2 & 1 \\ 4 & 2 \end{bmatrix}$, $r(B) = 1$, since $|A| = 0$ and there is at least one non-vanishing element.

(d) $A = \begin{bmatrix} 1 & 2 & 3 \\ 4 & 5 & 6 \\ 3 & 2 & 1 \end{bmatrix}$. Here $|A| = 0$; and $\begin{vmatrix} 1 & 2 \\ 4 & 5 \end{vmatrix} \neq 0$; hence $r(A) = 2$.

(e) $A = \begin{bmatrix} 1 & 2 & 3 \\ 4 & 5 & 8 \\ 3 & 2 & 1 \end{bmatrix}$. Here $|A| \neq 0$; hence $r(A) = 3$.

1.6.7 Elementary Transformations

The following are the elementary transformations (or operations) in a matrix:

(a) Interchange of any two rows (or columns);

(b) Multiplication of the elements of any row (or column) by a nonzero number;

(c) Addition to the elements of any row (or column) the corresponding elements of any other row (or column) multiplied by a number.

Elementary transformations are thus either row or column transformations. In this text, we use the following symbols for elementary transformations:

r_{ij} indicates that the ith and jth rows are interchanged; $r_i(k)$ stands for the multiplication, by k, of the elements of the ith row; $r_{ij}(k)$ stands for the addition of k times the elements of the jth row to the corresponding elements of the ith row; similarly, c_{ij}, $c_i(k)$ and $c_{ij}(k)$ stand for the corresponding column transformations.

If we subject a unit matrix to a number of elementary transformations, we obtain a new matrix which is called an *elementary matrix*. For example,

$$\begin{bmatrix} 0 & 0 & 1 \\ 0 & 1 & 0 \\ 1 & 0 & 0 \end{bmatrix}$$

is an elementary matrix obtained by interchanging the first and third columns of a unit matrix.

If we transform a matrix A to another matrix B by means of elementary transformations, then the two matrices A and B are said to be *equivalent* and we write $A \sim B$. An important property of the elementary transformations is that the rank of a matrix remains unaltered if we subject it to elementary transformations. This means that *equivalent matrices have the same rank.*

By using elementary transformations, every nonzero matrix A of rank r can be reduced to the form

$$\begin{bmatrix} I_k & 0 \\ 0 & 0 \end{bmatrix},$$

where I_k is the unit matrix of order k. Such a form is called canonical or normal form. Now, the rank of I_k is k and therefore the rank of the above canonical form is also k. It follows that the rank of A is k. The following examples demonstrate the use of elementary transformations.

Example 1.80 Find the rank of the matrix

$$A = \begin{bmatrix} 1 & -1 & 2 & -3 \\ 4 & 1 & 0 & 2 \\ 0 & 3 & 1 & 4 \\ 0 & 1 & 0 & 2 \end{bmatrix}$$

Applying $r_{21}(-4)$, we get

$$A \sim \begin{bmatrix} 1 & -1 & 2 & -3 \\ 0 & 5 & -8 & 14 \\ 0 & 3 & 1 & 4 \\ 0 & 1 & 0 & 2 \end{bmatrix}$$

Applying $C_{21}(-1)$, $C_{31}(-2)$, $C_{41}(3)$, we obtain

$$A \sim \begin{bmatrix} 1 & 0 & 0 & 0 \\ 0 & 5 & -8 & 14 \\ 0 & 3 & 1 & 4 \\ 0 & 1 & 0 & 2 \end{bmatrix}$$

Now, applying R_{24}, we find

$$A \sim \begin{bmatrix} 1 & 0 & 0 & 0 \\ 0 & 1 & 0 & 2 \\ 0 & 3 & 1 & 4 \\ 0 & 5 & -8 & 14 \end{bmatrix}$$

Applying $C_{42}(-1)$, we get

$$A \sim \begin{bmatrix} 1 & 0 & 0 & 0 \\ 0 & 1 & 0 & 0 \\ 0 & 3 & 1 & -2 \\ 0 & 5 & -8 & 4 \end{bmatrix}$$

Applying $R_{32}(-3)$, $R_{42}(-5)$, we obtain

$$A \sim \begin{bmatrix} 1 & 0 & 0 & 0 \\ 0 & 1 & 0 & 0 \\ 0 & 0 & 1 & -2 \\ 0 & 0 & 8 & 4 \end{bmatrix}$$

Applying $C_{43}(2)$, we get

$$A \sim \begin{bmatrix} 1 & 0 & 0 & 0 \\ 0 & 1 & 0 & 0 \\ 0 & 0 & 1 & 0 \\ 0 & 0 & -8 & -12 \end{bmatrix}$$

Hence, we get $r(A) = 4$.

Example 1.81 Find the rank of the matrix

$$A = \begin{bmatrix} 1 & 2 & 1 & 2 \\ 3 & 1 & -2 & 1 \\ 4 & -3 & -1 & 3 \\ 2 & 4 & 2 & 4 \end{bmatrix}$$

Applying $R_{21}(-3)$, $R_{31}(-4)$ and $R_{41}(-2)$, we get

$$A \sim \begin{bmatrix} 1 & 2 & 1 & 2 \\ 0 & -5 & -5 & -5 \\ 0 & -11 & -5 & -5 \\ 0 & 0 & 0 & 0 \end{bmatrix}$$

Applying $C_{21}(-2)$, $C_{31}(-1)$ and $C_{41}(-2)$, we obtain

$$A \sim \begin{bmatrix} 1 & 0 & 0 & 0 \\ 0 & -5 & -5 & -5 \\ 0 & -11 & -5 & -5 \\ 0 & 0 & 0 & 0 \end{bmatrix}$$

Applying $R_{23}(-1)$, we find

$$A \sim \begin{bmatrix} 1 & 0 & 0 & 0 \\ 0 & 6 & 0 & 0 \\ 0 & -11 & -5 & -5 \\ 0 & 0 & 0 & 0 \end{bmatrix}$$

Applying $R_2(1/6)$, we get

$$A \sim \begin{bmatrix} 1 & 0 & 0 & 0 \\ 0 & 1 & 0 & 0 \\ 0 & -11 & -5 & -5 \\ 0 & 0 & 0 & 0 \end{bmatrix}$$

Applying $R_{32}(11)$, we get

$$A \sim \begin{bmatrix} 1 & 0 & 0 & 0 \\ 0 & 1 & 0 & 0 \\ 0 & 0 & -5 & -5 \\ 0 & 0 & 0 & 0 \end{bmatrix}$$

Applying $R_3(-1/5)$, we have

$$A \sim \begin{bmatrix} 1 & 0 & 0 & 0 \\ 0 & 1 & 0 & 0 \\ 0 & 0 & 1 & 1 \\ 0 & 0 & 0 & 0 \end{bmatrix}$$

Here $r(A) = 3$.

1.6.8 Linear Dependence and Independence of Vectors

A vector is an ordered set of numbers called its components. If there are n components, the vector is an *n-dimensional vector.* There are *row-vectors* and *column-vectors.* A row-vector of order n is a $(1\times n)$ matrix and a column-vector of order n is an $(n\times 1)$ matrix.

Let $X_1, X_2, \dots, X_m$ be a set of m vectors. If $k_1, k_2, \dots, k_m$ are m constants (not all zero) such that

$$k_1x_1 + k_2x_2 + \cdots + k_mx_m = 0,$$

then we say that the vectors $X_1, X_2, \dots, X_m$ are *linearly dependent.* If this is not the case, then the vectors are said to be *linearly independent.* Hence $X_1, X_2, \dots, X_m$ are linearly independent if $k_1x_1 + k_2x_2 + \cdots + k_mx_m = 0$ implies $k_1 = k_2 = \cdots = k_m = 0$. These definitions may directly be used to test for the linear dependence or independence of vectors.

Let $A_1, A_2, \dots, A_n$ be n column vectors defined by

$$A_1 = \begin{bmatrix} a_{11} \\ a_{21} \\ \vdots \\ a_{m1} \end{bmatrix}, \quad A_2 = \begin{bmatrix} a_{12} \\ a_{22} \\ \vdots \\ a_{m2} \end{bmatrix}, \quad \dots, \quad A_n = \begin{bmatrix} a_{1n} \\ a_{2n} \\ \vdots \\ a_{mn} \end{bmatrix}.$$

Let constants $k_1, k_2, \dots, k_n$ exist such that

$$k_1A_1 + k_2A_2 + \cdots + k_nA_n = 0 \tag{1.73}$$

Equation (1.73) implies that

$$\left.\begin{aligned} a_{11}k_1 + a_{12}k_2 + \cdots + a_{1n}k_n &= 0 \\ a_{21}k_1 + a_{22}k_2 + \cdots + a_{2n}k_n &= 0 \\ &\vdots \\ a_{m1}k_1 + a_{m2}k_2 + \cdots + a_{mn}k_n &= 0 \end{aligned}\right\} \tag{1.74}$$

The system (1.74) can be written in the matrix form

$$Ak = 0, \tag{1.75}$$

which is a homogeneous system. This system will have a nontrivial solution only if the columns of A are linearly dependent. In particular, if A is a square matrix of order n, the columns of A will be linearly dependent only if $|A| = 0$. Since $|A| = |A^{\mathrm{T}}| = 0$, it follows that the columns of A^{T}, i.e. the rows of A are linearly dependent. It follows that the rows or columns of A will be linearly independent if $|A| \neq 0$.

Now, let $A = [a_{ij}]$ be an $(m\times n)$ matrix of rank r. Then, A will have at least one minor A_r of order r for which $|A_r| \neq 0$. Hence the columns of A_r are linearly independent. It therefore follows that the rank of a matrix is equal to the number of linearly independent rows or columns.

Example 1.82 Show that the vectors (1, 7, 5), (–7, 20, –2) and (3, –2, 4) are linearly dependent.

Let $x_1 = (1, 7, 5)$, $x_2 = (-7, 20, -2)$ and $x_3 = (3, -2, 4)$. We have to show that there exist constants k_1, k_2, k_3 (not all zero) such that $k_1x_1 + k_2x_2 + k_3x_3 = 0$. Using the given vectors, the above relation gives:

$$k_1(1, 7, 5) + k_2(-7, 20, -2) + k_3(3, -2, 4) = 0.$$

This implies that

$$k_1 - 7k_2 + 3k_3 = 0 \quad \text{(i)}$$

$$7k_1 + 20k_2 - 2k_3 = 0 \quad \text{(ii)}$$

$$5k_1 - 2k_2 + 4k_3 = 0 \quad \text{(iii)}$$

From (i) and (ii), we obtain $k_3 = 3k_2$. Putting this value in (i), we get $k_1 = -2k_2$. Substituting these values of k_1 and k_3 in (iii), we obtain

$$5(-2k_2) - 2k_2 + 4(3k_2) = 0,$$

which means that (iii) is satisfied. Hence the vectors x_1, x_2 and x_3 are linearly dependent.

Aliter: It is also possible to combine the vectors x_1, x_2 and x_3 in a suitable manner and find a relation of the type $k_1x_1 + k_2x_2 + k_3x_3 = 0$, where k_1, k_2, k_3 are not all zero

We have

$$7x_1 + x_2 = (7, 49, 35) + (-7, 20, -2) = (0, 69, 33) \quad \text{(i)}$$

Also

$$3x_1 - x_3 = (3, 21, 15) - (3, -2, 4) = (0, 23, 11) \quad \text{(ii)}$$

From (i) and (ii), we obtain

$$7x_1 + x_2 - 3(3x_1 - x_3) = 0 \quad \text{or} \quad -2x_1 + x_2 + 3x_3 = 0.$$

Example 1.83 Show that the vectors: $x_1 = (1, 2, 1, 0)$, $x_2 = (1, 3, 1, 2)$, $x_3 = (4, 2, 1, 0)$ and $x_4 = (6, 1, 0, 1)$ are linearly independent.

Let

$$k_1(1, 2, 1, 0) + k_2(1, 3, 1, 2) + k_3(4, 2, 1, 0) + k_4(6, 1, 0, 1) = 0 \quad \text{(i)}$$

Then (i) implies:

$$k_1 + k_2 + 4k_3 + 6k_4 = 0 \quad \text{(ii)}$$

$$2k_1 + 3k_2 + 2k_3 + k_4 = 0 \quad \text{(iii)}$$

$$k_1 + k_2 + k_3 = 0 \quad \text{(iv)}$$

$$2k_2 + k_4 = 0 \quad \text{(v)}$$

From (v),

$$k_4 = -2k_2 \tag{vi}$$

From (ii) and (iv)

$$3k_3 + 6k_4 = 0$$

Using (vi), we get

$$k_3 = -2k_4 = 4k_2, \tag{vii}$$

Then (iv) gives;

$$k_1 = -k_2 - k_3 = -k_2 - 4k_2 = -5k_2 \tag{viii}$$

Finally, substitution for k_1, k_3 and k_4 in (iii) gives:

$$2(-5k_2) + 3k_2 + 8k_2 - 2k_2 = 0 \quad \text{or} \quad k_2 = 0.$$

Therefore, $k_1 = k_3 = k_4 = 0$. It follows therefore that the given vectors are linearly independent.

Aliter: We can also combine the given vectors in a suitable manner to show that they are linearly independent. We have

$$x_1 - x_2 = (0, -1, 0, -2);$$
$$4x_1 - x_3 = (0, 6, 3, 0);$$
$$6x_2 - x_4 = (6, 18, 6, 12) - (6, 1, 0, 1) = (0, 17, 6, 11)$$

To reduce the second components to zero, we obtain

$$6(x_1 - x_2) + (4x_1 - x_3) = (0, -6, 0, -12) + (0, 6, 3, 0) = (0, 0, 3, -12)$$

and

$$17(x_1 - x_2) + (6x_2 - x_4) = (0, -17, 0, -34) + (0, 17, 6, 11) = (0, 0, 6, -23)$$

To reduce the third component to zero, we finally obtain:

$$\begin{aligned} 2[6(x_1 - x_2) + (4x_1 - x_3)] - [17(x_1 - x_2) + (6x_2 - x_4)] &= (0, 0, 6, -24) \\ &\quad -(0, 0, 6, -23) \\ &= (0, 0, 0, -1), \end{aligned}$$

in which the fourth component is not zero. It follows that the given vectors are linearly independent.

1.6.9 Consistency of a Nonhomogeneous Linear System

Let the system of m linear equations in n unknowns be:

$$\left.\begin{aligned} a_{11}x_1 + a_{12}x_2 + \cdots + a_{1n}x_n &= b_1 \\ a_{21}x_1 + a_{22}x_2 + \cdots + a_{2n}x_n &= b_2 \\ &\vdots \\ a_{m1}x_1 + a_{m2}x_2 + \cdots + a_{mn}x_n &= b_m \end{aligned}\right\} \tag{1.76}$$

The matrix A defined by

$$A = \begin{bmatrix} a_{11} & a_{12} & \cdots & a_{1n} \\ a_{21} & a_{22} & \cdots & a_{2n} \\ \vdots & \vdots & & \vdots \\ a_{m1} & a_{m2} & \cdots & a_{mn} \end{bmatrix}$$

is called the coefficient matrix, and the matrix (A, b) defined by

$$(A, b) = \begin{bmatrix} a_{11} & a_{12} & \cdots & a_{1n} & b_1 \\ a_{21} & a_{22} & \cdots & a_{2n} & b_2 \\ \vdots & \vdots & & \vdots & \vdots \\ a_{m1} & a_{m2} & \cdots & a_{mn} & b_m \end{bmatrix}$$

is called the *augmented matrix*. Let $r(A)$ and $r(A, b)$ be the ranks of A and (A, b), respectively. Then we have the following theorem for consistency of the system (1.76).

Theorem 1.21 The necessary and sufficient condition that the system of equations (1.76) is consistent is that $r(A) = r(A, b)$.

Proof Let $A_1, A_2, \ldots, A_n$ be the column vectors of the matrix A. Then the system (1.76) can be written as

$$[A_1, A_2, \ldots, A_n] \begin{bmatrix} x_1 \\ x_2 \\ \vdots \\ x_n \end{bmatrix} = \mathbf{b}$$

or

$$A_1 x_1 + A_2 x_2 + \cdots + A_n x_n = \mathbf{b}$$

or

$$x_1 A_1 + x_2 A_2 + \cdots + x_n A_n = \mathbf{b} \tag{1.77}$$

For convenience, let $r(A) = r$. Since r is the rank of A, the matrix A has r linearly independent columns. For definiteness, let these columns be the first r columns of A, viz., $A_1, A_2, \ldots, A_r$. It follows that each of the columns $A_{r+1}, A_{r+2}, \ldots, A_n$ will be a linear combination of $A_1, A_2, \ldots, A_r$.

Now, let the given system (1.76) be consistent, i.e. possesses a solution. Then there must exist scalars $k_1, k_2, \ldots, k_n$ such that

$$k_1 A_1 + k_2 A_2 + \ldots + k_n A_n = \mathbf{b},$$

which shows that $\mathbf{b}$ is also a linear combination of $A_1, A_2, \ldots, A_r$. This means that the maximum number of linearly independent columns of the augmented matrix $(A, \mathbf{b})$ is also r. Therefore, $r(A, \mathbf{b}) = r$. Hence $r(A) = r(A, \mathbf{b})$. This proves that the condition is necessary.

Next, let $r(A) = r(A, b) = r$, be given. Then the maximum number of linearly independent columns of (A, b) is r. But the first r columns, viz., $A_1, A_2, \ldots, A_r$ of $(A, \mathbf{b})$ are linearly independent. Then the column vector $\mathbf{b}$ can be expressed as a linear combination of $A_1, A_2, \ldots, A_r$. This means that there exist scalars $k_1, k_2, \ldots, k_r$ such that

$$k_1 A_1 + k_2 A_2 + \cdots + k_r A_r = \mathbf{b} \tag{1.78}$$

Equation (1.78) can be written as

$$k_1 A_1 + k_2 A_2 + \cdots + k_r A_r + 0.A_{r+1} + 0.A_{r+2} + \cdots + 0.A_n = \mathbf{b}. \tag{1.79}$$

Comparing (1.77) and (1.78), we see that

$$x_1 = k_1,\ x_2 = k_2,\ \ldots,\ x_r = k_r,\ x_{r+1} = x_{r+2} \cdots = x_n = 0;$$

which is a solution of the system (1.76). Hence the system (1.76) is consistent and the condition is sufficient.

This completes the proof of Theorem 1.21.

Note:

(i) If $r(A) \neq r(A, b)$ then the equations are inconsistent.

(ii) If $r(A) = r(A, b) = n$, the equations are consistent and possess a unique solution.

(iii) If $r(A) = r(A, b) < n$, the equations are consistent and have an infinite number of solutions.

Example 1.84 Examine for consistency the equations

$$2x - 3y + 5z = 1$$
$$3x + y - z = 2$$
$$x + 4y - 6z = 1$$

We have

$$A = \begin{bmatrix} 2 & -3 & 5 \\ 3 & 1 & -1 \\ 1 & 4 & -6 \end{bmatrix} \quad \text{and} \quad (A, b) = \begin{bmatrix} 2 & -3 & 5 & 1 \\ 3 & 1 & -1 & 2 \\ 1 & 4 & -6 & 1 \end{bmatrix}$$

It can be seen that

$$|A| = 0 \quad \text{and} \quad \begin{vmatrix} 2 & -3 \\ 3 & 1 \end{vmatrix} \neq 0;$$

hence $r(A) = 2$.

Similarly, it may be verified that $r(A, b) = 2$. As all the third-order determinants formed from (A, b) are zero, and

$$\begin{vmatrix} 2 & -3 \\ 3 & 1 \end{vmatrix} \neq 0.$$

Since $r(A) = 2$ and $r(A, b) = 2$, it follows that the equations are consistent.

Example 1.85 Find whether the following system is consistent:

$$x - 4y + 5z = 8$$
$$3x + 7y - z = 3$$
$$x + 15y - 11z = 14.$$

We have

$$A = \begin{bmatrix} 1 & -4 & 5 \\ 3 & 7 & -1 \\ 1 & 15 & -11 \end{bmatrix};$$

then

$$|A| = 0 \quad \text{and} \quad \begin{vmatrix} 1 & -4 \\ 3 & 7 \end{vmatrix} \neq 0.$$

Hence, $r(A) = 2$. Further,

$$(A, b) = \begin{bmatrix} 1 & -4 & 5 & 8 \\ 3 & 7 & -1 & 3 \\ 1 & 15 & -11 & -14 \end{bmatrix},$$

where it can be verified that

$$\begin{vmatrix} -4 & 5 & 8 \\ 7 & -1 & 3 \\ 15 & -11 & -14 \end{vmatrix} \neq 0.$$

Hence, $r(A, b) = 3$.

It follows that the equations are inconsistent and that there exists no solution.

Example 1.86 Show that the equations:

$$x + y + z = 6,$$
$$x - y + 2z = 5,$$
$$3x + y + z = 8$$
$$2x - 2y + 3z = 7$$

are consistent and solve them.

We have

$$(A,\ b) = \begin{bmatrix} 1 & 1 & 1 & 6 \\ 1 & -1 & 2 & 5 \\ 3 & 1 & 1 & 8 \\ 2 & -2 & 3 & 7 \end{bmatrix}$$

Applying $R_{21}(-1)$, $R_{31}(-3)$, $R_{41}(-2)$, we get

$$(A, b) \sim \begin{bmatrix} 1 & 1 & 1 & 6 \\ 0 & -2 & 1 & -1 \\ 0 & -2 & -2 & -10 \\ 0 & -4 & 1 & -5 \end{bmatrix}$$

Applying $R_{32}(-1)$, $R_{42}(-2)$, we find

$$(A, b) \sim \begin{bmatrix} 1 & 1 & 1 & 6 \\ 0 & -2 & 1 & -1 \\ 0 & 0 & -3 & -9 \\ 0 & 0 & -1 & -3 \end{bmatrix}$$

Applying $R_3(1/3)$, we have

$$(A,\ b) \sim \begin{bmatrix} 1 & 1 & 1 & 6 \\ 0 & -2 & 1 & -1 \\ 0 & 0 & -1 & -3 \\ 0 & 0 & -1 & -3 \end{bmatrix}$$

Applying $R_{43}(-1)$, we get

$$(A,\ b) \sim \begin{bmatrix} 1 & 1 & 1 & 6 \\ 0 & -2 & 1 & -1 \\ 0 & 0 & -1 & -3 \\ 0 & 0 & 0 & 0 \end{bmatrix}$$

It is clear that $R(A) = R(A,\ b) = 3$. For the solution, we have the equations:

$$\begin{aligned} x + y + z &= 6 \\ -2y + z &= -1 \\ -z &= -3 \end{aligned}$$

Using the backward substitution, we obtain $x = 1,\ y = 2,\ z = 3$.

Example 1.87 Solve the system of equations:

$$x+2y-3z-4u=6$$
$$x+3y+z-2u=4$$
$$2x+5y-2z-5u=10.$$

We have

$$(A,\ b)=\begin{bmatrix} 1 & 2 & -3 & -4 & 6 \\ 1 & 3 & 1 & -2 & 4 \\ 2 & 5 & -2 & -5 & 10 \end{bmatrix}$$

Applying $R_{21}(-1)$, $R_{31}(-2)$, we get

$$(A,\ b)\sim\begin{bmatrix} 1 & 2 & -3 & -4 & 6 \\ 0 & 1 & 4 & 2 & -2 \\ 0 & 1 & 4 & 3 & -2 \end{bmatrix}$$

Applying $R_{12}(-2)$, $R_{32}(-1)$, we find

$$(A,\ b)\sim\begin{bmatrix} 1 & 0 & -11 & -8 & 10 \\ 0 & 1 & 4 & 2 & -2 \\ 0 & 0 & 0 & 1 & 0 \end{bmatrix}$$

Applying $R_{13}(8)$, $R_{23}(-2)$, we have

$$(A,\ b)\sim\begin{bmatrix} 1 & 0 & -11 & 0 & 10 \\ 0 & 1 & 4 & 0 & -2 \\ 0 & 0 & 0 & 1 & 0 \end{bmatrix}$$

It is easily seen that $r(A)=r(A,b)=3$. Hence the equations are consistent. For the solution, we observe that the given equations are equivalent to

$$x-11z=10 \qquad \text{(i)}$$

$$y+4z=-2 \qquad \text{(ii)}$$

$$u=0. \qquad \text{(iii)}$$

From (i), $x=11z+10$ and from (ii), $y=-4z-2$. Hence the infinite number of solutions are given by

$$x=11z+10$$

$$y=-4z-2$$

$$u=0,$$

where z is arbitrary.

1.6.10 Homogeneous System of Equations

Let the system of equations be

$$\left.\begin{aligned} a_{11}x_1 + a_{12}x_2 + \cdots + a_{1n}x_n &= 0 \\ a_{21}x_1 + a_{22}x_2 + \cdots + a_{2n}x_n &= 0 \\ &\vdots \\ a_{m1}x_1 + a_{m2}x_2 + \cdots + a_{mn}x_n &= 0 \end{aligned}\right\} \tag{1.80}$$

The system (1.80) may be written in the matrix form:

$$A\mathbf{X} = \mathbf{0},$$

where

$$A = \begin{bmatrix} a_{11} & a_{12} & \cdots & a_{1n} \\ a_{21} & a_{22} & \cdots & a_{2n} \\ \vdots & \vdots & & \vdots \\ a_{m1} & a_{m2} & \cdots & a_{mn} \end{bmatrix} \tag{1.81}$$

Since the augmented matrix is $(A, \mathbf{0})$, we have

$$r(A) = r(A, \mathbf{0}) = r, \quad \text{(say)}$$

Hence the system (1.80) is always consistent. If $r = n$, the number of unknowns, the system has only the trivial solution given by $x_1 = x_2 = \cdots = x_n = 0$. But if $r < n$, the system will have an infinite number of solutions (nontrivial).

Example 1.88 Solve the system:

$$x_1 + 3x_2 - 2x_3 = 0 \qquad \text{(i)}$$

$$2x_1 - x_2 + 4x_3 = 0 \qquad \text{(ii)}$$

$$x_1 - 11x_2 + 14x_3 = 0 \qquad \text{(iii)}$$

We have

$$\begin{bmatrix} 1 & 3 & -2 \\ 2 & -1 & 4 \\ 1 & -11 & 14 \end{bmatrix} \begin{bmatrix} x_1 \\ x_2 \\ x_3 \end{bmatrix} = \begin{bmatrix} 0 \\ 0 \\ 0 \end{bmatrix},$$

where

$$A = \begin{bmatrix} 1 & 3 & -2 \\ 2 & -1 & 4 \\ 1 & -11 & 14 \end{bmatrix}$$

Applying $R_{21}(-2)$, $R_{31}(-1)$, we get

$$A \sim \begin{bmatrix} 1 & 3 & -2 \\ 0 & -7 & 8 \\ 0 & -14 & 16 \end{bmatrix}$$

Applying $R_{32}(-2)$, we find

$$A \sim \begin{bmatrix} 1 & 3 & -2 \\ 0 & -7 & 8 \\ 0 & 0 & 0 \end{bmatrix} \tag{i}$$

Clearly $r(A) = 2$ and $|A| = 0$. Since $r(A) < n$, the number of unknowns, the given system will have an infinite number of solutions. For the solution, we write from (i), the equivalent system:

$$x_1 + 3x_2 - 2x_3 = 0 \tag{ii}$$

$$-7x_2 + 8x_3 = 0 \tag{iii}$$

From (iii), we have

$$x_3 = \frac{7}{8}x_2$$

Equation (ii) gives:

$$x_1 = 2x_3 - 3x_2 = \frac{14}{8}x_2 - 3x_2 = -\frac{10}{8}x_2 = -\frac{5}{4}x_2.$$

Hence the infinite number of nontrivial solutions are given by

$$x_1 = -\frac{5}{4}x_2 \quad \text{and} \quad x_3 = \frac{7}{8}x_2$$

where x_2 is arbitrary.

Example 1.89 Solve the system of equations:

$$x_1 - 2x_2 + x_3 - x_4 = 0$$
$$x_1 + x_2 - 2x_3 + 3x_4 = 0$$
$$4x_1 + x_2 - 5x_3 + 8x_4 = 0$$
$$5x_1 - 7x_2 + 2x_3 - x_4 = 0$$

We have

$$A = \begin{bmatrix} 1 & -2 & 1 & -1 \\ 1 & 1 & -2 & 3 \\ 4 & 1 & -5 & 8 \\ 5 & -7 & 2 & -1 \end{bmatrix}$$

Applying $R_{21}(-1)$, $R_{31}(-4)$, $R_{41}(-5)$, we get

$$A \sim \begin{bmatrix} 1 & -2 & 1 & -1 \\ 0 & 3 & -3 & 4 \\ 0 & 9 & -9 & 12 \\ 0 & 3 & -3 & 4 \end{bmatrix}$$

Applying $R_3(1/3)$, we have

$$A \sim \begin{bmatrix} 1 & -2 & 1 & -1 \\ 0 & 3 & -3 & 4 \\ 0 & 3 & -3 & 4 \\ 0 & 3 & -3 & 4 \end{bmatrix}$$

Applying $R_{23}(-1)$, $R_{34}(-1)$, we find

$$A \sim \begin{bmatrix} 1 & -2 & 1 & -1 \\ 0 & 0 & 0 & 0 \\ 0 & 0 & 0 & 0 \\ 0 & 3 & -3 & 4 \end{bmatrix}$$

Applying $R_{14}(1/3)$, we have

$$A \sim \begin{bmatrix} 1 & -1 & 0 & \frac{1}{3} \\ 0 & 0 & 0 & 0 \\ 0 & 0 & 0 & 0 \\ 0 & 3 & -3 & 4 \end{bmatrix}$$

Since $r(A) = 2 < n$, the system will have an infinite number of solutions. For the solution, we write

$$x_1 - x_2 + \frac{1}{3} x_4 = 0 \qquad \text{(i)}$$

and

$$3x_2 - 3x_3 + 4x_4 = 0 \quad \text{or} \quad x_2 - x_3 + \frac{4}{3} x_4 = 0 \qquad \text{(ii)}$$

From (ii):

$$x_2 = x_3 - \frac{4}{3} x_4.$$

and (i) gives

$$x_1 = x_2 - \frac{1}{3} x_4 = x_3 - \frac{5}{3} x_4.$$

Hence the infinite number of solutions are given by

$$x_1 = x_3 - \frac{5}{3}x_4 \quad \text{and} \quad x_2 = x_3 - \frac{4}{3}x_4$$

where x_3 and x_4 are arbitrary.

1.6.11 Solution of Linear Systems

We have already discussed, in Section 1.6.5, the factorization method for the solution of a linear system of equations. Two more methods: the matrix inversion method and the Gaussian elimination method, are described below.

Matrix inversion method

We consider the system of n linear equations in n unknowns:

$$\left.\begin{aligned} a_{11}x_1 + a_{12}x_2 + \ldots + a_{1n}x_n &= b_1 \\ a_{21}x_1 + a_{22}x_2 + \cdots + a_{2n}x_n &= b_2 \\ &\vdots \\ a_{n1}x_1 + a_{n2}x_2 + \cdots + a_{nn}x_n &= b_n \end{aligned}\right\} \tag{1.82}$$

which may be written in the matrix form

$$AX = B, \tag{1.83}$$

where

$$A = \begin{bmatrix} a_{11} & a_{12} & \cdots & a_{1n} \\ a_{21} & a_{22} & \cdots & a_{2n} \\ \vdots & \vdots & & \vdots \\ a_{n1} & a_{n2} & \cdots & a_{nn} \end{bmatrix}, \quad X = \begin{bmatrix} x_1 \\ x_2 \\ \vdots \\ x_n \end{bmatrix}, \quad B = \begin{bmatrix} b_1 \\ b_2 \\ \vdots \\ b_n \end{bmatrix}$$

Let A be nonsingular so that A^{-1} exists. Then, premultiplying both sides of (1.83) by A^{-1}, we obtain:

$$A^{-1}AX = A^{-1}B \quad \text{or} \quad X = A^{-1}B, \tag{1.84}$$

since

$$A^{-1}A = I \quad \text{and} \quad IX = X.$$

If A^{-1} is known, the solution vector X can then be found out from (1.84).

Example 1.90 Solve the equations

$$3x + y + 2z = 3$$

$$2x - 3y - z = -3$$

$$x + 2y + z = 4$$

Here

$$|A| = \begin{bmatrix} 3 & 1 & 2 \\ 2 & -3 & -1 \\ 1 & 2 & 1 \end{bmatrix} = 8.$$

Also,

$$A^{\mathrm{T}} = \begin{bmatrix} 3 & 2 & 1 \\ 1 & -3 & 2 \\ 2 & -1 & 1 \end{bmatrix}$$

and

$$\text{adj } A = \begin{bmatrix} -1 & 3 & 5 \\ -3 & 1 & 7 \\ 7 & -5 & -11 \end{bmatrix}$$

Hence,

$$A^{-1} = \frac{1}{8}\begin{bmatrix} -1 & 3 & 5 \\ -3 & 1 & 7 \\ 7 & -5 & -11 \end{bmatrix}$$

It follows that

$$x = \begin{bmatrix} x \\ y \\ z \end{bmatrix} = A^{-1}B = \frac{1}{8}\begin{bmatrix} -1 & 3 & 5 \\ -3 & 1 & 7 \\ 7 & -5 & -11 \end{bmatrix}\begin{bmatrix} 3 \\ -3 \\ 4 \end{bmatrix} = \frac{1}{8}\begin{bmatrix} 8 \\ 16 \\ -8 \end{bmatrix} = \begin{bmatrix} 1 \\ 2 \\ -1 \end{bmatrix},$$

which gives $x = 1$, $x = 2$ and $z = -1$.

Gaussian elimination method

This is the elementary elimination method and it reduces the system of equations to an equivalent upper triangular system which can then be solved by *back substitution.* Although quite general, we shall describe this method by considering a system of three equations for the sake of clarity and simplicity.

Let the system be

$$\left.\begin{aligned} a_{11}x_1 + a_{12}x_2 + a_{13}x_3 = b_1 \\ a_{21}x_1 + a_{22}x_2 + a_{23}x_3 = b_2 \\ a_{31}x_1 + a_{32}x_2 + a_{33}x_3 = b_3 \end{aligned}\right\} \tag{1.85}$$

We first form the augmented matrix

$$\begin{bmatrix} a_{11} & a_{12} & a_{13} & b_1 \\ a_{21} & a_{22} & a_{23} & b_2 \\ a_{31} & a_{32} & a_{33} & b_3 \end{bmatrix}$$

To eliminate x_1 from the second equation, we multiply the first equation by $-a_{21}/a_{11}$ and then add it to the second equation. Similarly, to eliminate x_1 from the third equation, we multiply the first equation by $-a_{31}/a_{11}$ and add it to the third. This procedure can be shown thus:

$$\begin{matrix} \\ -\dfrac{a_{21}}{a_{11}} \\ -\dfrac{a_{31}}{a_{11}} \end{matrix} \begin{bmatrix} a_{11} & a_{12} & a_{13} & b_1 \\ a_{21} & a_{22} & a_{23} & b_2 \\ a_{31} & a_{32} & a_{33} & b_3 \end{bmatrix} \tag{1.86}$$

where $-a_{21}/a_{11}$ and $-a_{31}/a_{11}$ are called the *multipliers* for the *first* stage of elimination. In this stage, we have assumed that $a_{11} \neq 0$. The first equation is called the *pivotal* equation and a_{11} is called the *first pivot.* At the end of the first stage, the augmented matrix (1.86) becomes:

$$\begin{matrix} \\ \\ -\dfrac{a'_{32}}{a'_{22}} \end{matrix} \begin{bmatrix} a_{11} & a_{12} & a_{13} & b_1 \\ 0 & a'_{22} & a'_{23} & b'_2 \\ 0 & a'_{32} & a'_{33} & b'_3 \end{bmatrix} \tag{1.87}$$

where $a'_{22}, a'_{23}, \ldots$ are all changed elements. a'_{22} is the new pivot and the multiplier is $-a'_{32}/a'_{22}$. At the end of the second stage, we have the upper triangular system:

$$\begin{bmatrix} a_{11} & a_{12} & a_{13} & b_1 \\ 0 & a'_{22} & a'_{23} & b'_2 \\ 0 & 0 & a''_{33} & b''_3 \end{bmatrix} \tag{1.88}$$

from which the values of x_1, x_2 and x_3 can be obtained by back substitution.

It is clear that this method will fail if one of the pivots a_{11}, a'_{22} or a''_{33} vanishes. In such a case, the method can be modified by rearranging the rows so that the pivot is nonzero. This procedure is called *partial pivoting* and can easily be implemented on a computer. If this is impossible, then the matrix is singular and the equations have no solution. In particular, the number of nonzero diagonal elements in (1.88) will represent the rank of the original matrix.

A modification of this procedure, called the *Gauss–Jordan method*, consists in obtaining a unit matrix of the third order in the first three columns of the augmented matrix by elementary transformations as above. The resulting fourth column represents the solution of the given system. Both the methods are efficient numerical procedures and can be implemented on high-speed digital computers. The following examples illustrate the two methods.

Example 1.91 Solve the following system:

$$2x + y + z = 0$$
$$3x + 2y + 3z = 18$$
$$x + 4y + 9z = 16$$

In the first stage, the multipliers are –3/2 and –1/2. We multiply the first equation successively by –3/2 and –1/2 and add it to the second and third equations respectively to obtain and equations

$$\frac{1}{2}y + \frac{3}{2}z = 3$$

and

$$\frac{7}{2}y + \frac{17}{2}z = 11.$$

The augmented matrix therefore becomes

$$\begin{array}{c} \\ -\frac{3}{2} \\ -\frac{1}{2} \end{array}\left[\begin{array}{cccc} 2 & 1 & 1 & 10 \\ 0 & \frac{1}{2} & \frac{3}{2} & 3 \\ 0 & \frac{7}{2} & \frac{17}{2} & 11 \end{array}\right]$$

At the second stage, we eliminate y from the third equation by multiplying the second equation by –7 and adding it to the third. The resulting system will be upper triangular:

$$2x + y + z = 10$$
$$\frac{1}{2}y + \frac{3}{2}z = 3$$
$$-2z = -10$$

Back substitution gives the solution: $x = 7,\ y = -9$ and $z = 5$.

Example 1.92 Using the Gauss–Jordan procedure, solve the following system:

$$x+2y-z=-1$$
$$3x+8y+2z=28$$
$$4x+9y-z=14$$

The augmented matrix is

$$\begin{bmatrix} 1 & 2 & -1 & -1 \\ 3 & 8 & 2 & 28 \\ 4 & 9 & -1 & 14 \end{bmatrix}$$

We successively reduce the first-three columns of the above matrix to the unit matrix of order 3. The elementary transformations are indicated below.

We first use $R_{21}(-3)$ and $R_{31}(-4)$ indicating the transformations for the second and third rows. These produce the matrix

$$\begin{bmatrix} 1 & 2 & -1 & -1 \\ 0 & 2 & 5 & 31 \\ 0 & 1 & 3 & 18 \end{bmatrix}$$

The successive transformations are shown below:

The operation $R_{23}(-1)$ implies

$$A \sim \begin{bmatrix} 1 & 2 & -1 & -1 \\ 0 & 1 & 2 & 13 \\ 0 & 1 & 3 & 18 \end{bmatrix}$$

Applying $R_{12}(-2)$, we get

$$A \sim \begin{bmatrix} 1 & 0 & -5 & -27 \\ 0 & 1 & 2 & 13 \\ 0 & 1 & 3 & 18 \end{bmatrix}$$

Applying $R_{32}(-1)$, we get

$$A \sim \begin{bmatrix} 1 & 0 & -5 & -27 \\ 0 & 1 & 2 & 13 \\ 0 & 0 & 1 & 5 \end{bmatrix}$$

Applying $R_{23}(-2)$, we have

$$A \sim \begin{bmatrix} 1 & 0 & -5 & -27 \\ 0 & 1 & 0 & 3 \\ 0 & 0 & 1 & 5 \end{bmatrix}$$

On applying $R_{13}(5)$, we get

$$A \sim \begin{bmatrix} 1 & 0 & 0 & -2 \\ 0 & 1 & 0 & 3 \\ 0 & 0 & 1 & 5 \end{bmatrix}$$

It follows that the solution of the given system is $x = -2$, $y = 3$ and $z = 5$.

1.6.12 The Algebraic Eigenvalue Problem

Let $A = [a_{ij}]$ be a square matrix of order n and I be a unit matrix of the same order. Then the matrix $A - \lambda I$, where λ is an indeterminate, is called the *characteristic matrix* of A and the determinant $|A - \lambda I|$, which is a *polynomial of degree n in* λ, is called the *characteristic polynomial* of A. The equation $|A - \lambda I| = 0$ is called the *characteristic equation* of A and its roots are called the *characteristic roots* or *latent* roots or *eigenvalues* of A. In what follows we shall refer to them as *eigenvalues*. Now,

$$\begin{aligned}
|A - \lambda I| &= |AI - \lambda I| \\
&= |AI - \lambda AA^{-1}| \\
&= |A(I - \lambda A^{-1})| \\
&= \left| A\lambda\left(\frac{1}{\lambda} - A^{-1}\right)\right| \\
&= \left| -A\lambda\left(A^{-1} - \frac{1}{\lambda}\right)\right| \\
&= -\lambda |A| \left| A^{-1} - \frac{1}{\lambda}\right|
\end{aligned}$$

Since $\lambda \neq 0$ and $|A| \neq 0$, it follows that the characteristic equation of A is

$$\left| A^{-1} - \frac{1}{\lambda}\right| = 0,$$

which shows that $1/\lambda$ is an eigenvalue of A^{-1}.

Next, we consider the matrix equation $AX = \lambda X$, where $A = [a_{ij}]$ is a matrix of order n and

$$X = \begin{bmatrix} x_1 \\ x_2 \\ \vdots \\ x_n \end{bmatrix}$$

is a column vector. The equation

$$AX = \lambda X \tag{1.89}$$

represents the set of homogeneous equations:

$$\left.\begin{aligned} (a_{11} - \lambda)x_1 + a_{12}x_2 + \cdots + a_{1n}x_n &= 0 \\ a_{21}x_1 + (a_{22} - \lambda)x_2 + \cdots + a_{2n}x_n &= 0 \\ &\vdots \\ a_{n1}x_1 + a_{n2}x_2 + \cdots + (a_{nn} - \lambda)x_n &= 0 \end{aligned}\right\} \tag{1.90}$$

For a nontrivial solution of the system (1.90), we have

$$\begin{vmatrix} a_{11} - \lambda & a_{12} & \cdots & a_{1n} \\ a_{21} & a_{22-\lambda} & \cdots & a_{2n} \\ \vdots & \vdots & & \vdots \\ a_{n1} & a_{n2} & \cdots & a_{nn-\lambda} \end{vmatrix} = 0, \tag{1.91}$$

which is the *characteristic equation* of A. The characteristic polynomial $|A - \lambda I| = 0$ has n roots, in general. These are the eigenvalues of A, say $\lambda_1, \lambda_2, \ldots, \lambda_n$. Corresponding to each eigenvalue, the equation $AX = \lambda X$ possesses a nonzero solution vector X, called the *eigenvector* of A.

Note: The number of linearly independent solutions of $AX = \lambda X$ is equal to n-rank of $|A - \lambda I|$.

1.6.13 Cayley–Hamilton Theorem

This theorem states that every square matrix satisfies its own characteristic equation.

Let the characteristic polynomial be

$$f(\lambda) = |A - \lambda I| = 0,$$

i.e.

$$p_0 + p_1\lambda + p_2\lambda^2 + \cdots + p_n\lambda^n = 0 \tag{1.92}$$

Then the Cayley–Hamilton theorem states that

$$f(A) = p_0 I + p_1 A + p_2 A^2 + \cdots + p_{n-1}A^{n-1} + p_n A^n = 0 \tag{1.93}$$

Proof The elements of $|A - \lambda I|$ are of degree n or less in l. Hence the elements of adj $[A - \lambda I]$ are atmost of degree $(n - 1)$ in l. We therefore write

$$\text{adj } [A - \lambda I] = Q_0 + Q_1\lambda + Q_2\lambda^2 + \cdots + Q_{n-1}\lambda^{n-1} \tag{1.94}$$

where the Q_i's are square matrices of order $(n - 1)$.

Since

$$[A-\lambda I] \text{ adj } [A-\lambda I]=|A-\lambda I|I,$$

we have

$$(A-\lambda I)(Q_0+Q_1\lambda)+Q_2\lambda^2+\cdots+Q_{n-1}\lambda^{n-1}=(p_0+p_1\lambda+p_2\lambda^2+\cdots+p_n\lambda^n)I \tag{1.95}$$

Equating the coefficients of like powers of λ on both the sides of (1.95), we get:

$$\left.\begin{aligned} AQ_0 &= p_0 I \\ AQ_1 - Q_0 &= p_1 I \\ AQ_2 - Q_1 &= p_2 I \\ &\vdots \\ -Q^{n-1} &= p_n I. \end{aligned}\right\} \tag{1.96}$$

Premultiplying these equations respectively by I, A, A^2, ..., A^n and adding, we obtain

$$IAQ_0 + A^2Q_1 - AQ_0 + A^3Q_2 - A^2Q_1 + \cdots - A^nQ^{n-1}$$
$$= p_0 I + Ap_1 I + A^2 p_2 I + \cdots + A^n p_n I$$

or

$$p_0 I + p_1 A + p_2 A^2 + \cdots + p_n A^n = 0,$$

since the terms on the left side cancel each other.

This proves Cayley–Hamilton theorem.

We can write the above statement as:

$$p_0 I = -p_1 A - p_2 A^2 - \cdots - p_n A^n \tag{1.97}$$

Premultiplying both sides of (1.97) by A^{-1}, we obtain:

$$A^{-1}p_0 I = -p_1 A^{-1}A - p_2 A^{-1}A^2 - \cdots - p_n A^{-1}A^n$$

or

$$A^{-1} = -\frac{p_1}{p_0}I - \frac{p_2}{p_0}A - \frac{p_3}{p_0}A^2 - \cdots - \frac{p_n}{p_0}A^{n-1} \tag{1.98}$$

Properties of eigenvalues of eigenvectors

The following properties of eigenvalues and eigenvectors are useful in applications:

(i) If $\lambda_1, \lambda_2, \ldots, \lambda_n$ are the eigenvalues of A, then

(a) $\frac{1}{\lambda_1}, \frac{1}{\lambda_2}, \ldots, \frac{1}{\lambda_n}$ are the eigenvalues of A^{-1}.

(b) $k\lambda_1, k\lambda_2, \ldots, k\lambda_n$ are the eigenvalues of kA, $k \neq 0$.

(c) $\lambda_1^m, \lambda_2^m, \ldots, \lambda_n^m$ are the eigenvalues of A^m.

(d) $\lambda_1\lambda_2, \ldots, \lambda_n = |A|$.

(ii) The matrices A and A^T have the same eigenvalues.

(iii) If two or more eigenvalues are equal, then the eigenvectors may be linearly independent or linearly dependent.

(iv) If X_1 and X_2 are eigenvectors of a matrix A corresponding to eigenvalues λ_1 and λ_2 respectively, and if $\lambda_1 \neq \lambda_2$, then X_1 and X_2 are linearly independent.

Proof Let

$$\alpha_1 X_1 + \alpha_2 X_2 = 0 \tag{1.99}$$

Premultiplying both sides of (1.99) by A, we obtain:

$$A(\alpha_1 X_1 + \alpha_2 X_2) = 0$$

$$A\alpha_1 X_1 + A\alpha_2 X_2 = 0 \tag{1.100}$$

Since $AX_1 = \lambda X_1$ and $AX_2 = \lambda X_2$, Eq. (1.100) gives:

$$\lambda_1 \alpha_1 X_1 + \lambda_2 \alpha_2 X_2 = 0. \tag{1.101}$$

But from (1.98), $\alpha_1 X_1 = -\alpha_2 X_2$. Thus (1.100) becomes:

$$\alpha_2 X_2 (\lambda_2 - \lambda_1) = 0$$

Since $\lambda_1 \neq \lambda_2$, we obtain immediately:

$$\alpha_2 X_2 = 0,$$

which yields $\alpha_2 = 0$, as X_2 is a nonzero vector. It follows that $\alpha_1 = 0$.

Hence equation (1.99) implies that $\alpha_1 = 0$ and $\alpha_2 = 0$, which means that X_1 and X_2 are linearly independent.

(v) The eigenvectors corresponding to distinct eigenvalues of a real symmetric matrix are orthogonal.

(vi) The eigenvalues of a real symmetric matrix are real.

(vii) Similar matrices have the same eigenvalues.

Two matrices A and B are said to be *similar* if there exists a non-singular matrix P such that $P^{-1}AP = B$.

Proof Let A and B be two similar matrices. We have

$$P^{-1}AP = B \tag{1.102}$$

Then

$$B - \lambda I = P^{-1}AP - \lambda I = P^{-1}AP - P^{-1}\lambda IP = P^{-1}(A - \lambda I)P$$

Hence

$$|B - \lambda I| = \left|P^{-1}\right| |A - \lambda I| |P| = |A - \lambda I| \left|P^{-1}P\right| = |A - \lambda I|,$$

which shows that the eigenvalues of A and B are the same.

(viii) If A and D are similar matrices, where D is a diagonal matrix, then the diagonal elements of D are distinct eigenvalues of A.

Proof Since the eigenvalues are distinct, the corresponding eigenvalues are linearly independent. Let P be a matrix such that

$$P = [X_1, X_2, \ldots, X_n].$$

where $X_1, X_2, \ldots, X_n$ are the eigenvectors corresponding to the eigenvalues $\lambda_1, \lambda_2, \ldots, \lambda_n$ of A. It follows that A is non-singular.

Now,

$$\begin{aligned} AP &= A[X_1, X_2, \ldots, X_n] \\ &= [AX_1, AX_2, \ldots, AX_n] \\ &= [\lambda_1 X_1, \lambda_2 X_2, \ldots, \lambda_n X_n] \end{aligned}$$

Also

$$PD = [X_1, X_2, \ldots, X_n]\begin{bmatrix} \lambda_1 & 0 & 0 & \cdots & 0 \\ 0 & \lambda_2 & 0 & \cdots & 0 \\ \vdots & \vdots & \vdots & & \vdots \\ 0 & 0 & 0 & \cdots & \lambda_n \end{bmatrix} = [\lambda_1 X_1, \lambda_2 X_2, \ldots, \lambda_n X_n],$$

which is the same as AP. Hence $AP = PD$. Therefore, $P^{-1}AP = D$, where D is a diagonal matrix.

Example 1.93 Given

$$A = \begin{bmatrix} 1 & 2 \\ 4 & 3 \end{bmatrix},$$

find A^3 and A^{-1}.

The characteristic equation is

$$\begin{vmatrix} 1-\lambda & 2 \\ 4 & 3-\lambda \end{vmatrix} = 0 \quad \text{or} \quad \lambda^2 - 4\lambda - 5 = 0$$

Hence by Cayley–Hamilton theorem, we have

$$A^2 - 4A - 5I = 0 \tag{i}$$

or

$$A^2 = 4A + 5I \tag{ii}$$

Therefore,

$$\begin{aligned} A^3 = AA^2 &= A(4A + 5I) \qquad \text{[using (ii)]} \\ &= 4A^2 + 5A \\ &= 4(4A + 5I) + 5A, \qquad \text{[using (ii) again]} \end{aligned}$$

$$= 16A + 20I + 5A$$
$$= 16A + 20I$$
$$= 21A + 20I$$
$$= 21\begin{bmatrix} 1 & 2 \\ 4 & 3 \end{bmatrix} + 20\begin{bmatrix} 1 & 0 \\ 0 & 1 \end{bmatrix}$$
$$= \begin{bmatrix} 41 & 42 \\ 84 & 83 \end{bmatrix}$$

To find A^{-1}, we rewrite (i) as:

$$5I = A^2 - 4A. \qquad \text{(iii)}$$

Premultiplying both sides of (iii) by A^{-1}, we get

$$5A^{-1} = A^{-1}A^2 - 4A^{-1}A = A - 4I.$$

or

$$A^{-1} = \frac{1}{5}(A - 4I)$$
$$= \frac{1}{5}\begin{bmatrix} 1 & 2 \\ 4 & 3 \end{bmatrix} - \frac{4}{5}\begin{bmatrix} 1 & 0 \\ 0 & 1 \end{bmatrix}$$
$$= \begin{bmatrix} -\frac{3}{5} & \frac{2}{5} \\ \frac{4}{5} & -\frac{1}{5} \end{bmatrix}.$$

Example 1.94 If

$$A = \begin{bmatrix} 4 & -1 & -1 \\ -1 & 4 & -1 \\ -1 & -1 & 4 \end{bmatrix} \quad \text{and} \quad B = I - \frac{1}{4}A,$$

then show that $\mu_i = 1 - \frac{1}{4}\lambda_i$, where λ_i and μ_i are the eigenvalues of A and B respectively.

Eigenvalues of A are given by

$$\begin{vmatrix} 4-\lambda & -1 & -1 \\ -1 & 4-\lambda & -1 \\ -1 & -1 & 4-\lambda \end{vmatrix} = 0$$

or

$$\begin{vmatrix} 2-\lambda & -1 & -1 \\ 2-\lambda & 4-\lambda & -1 \\ 2-\lambda & -1 & 4-\lambda \end{vmatrix} = 0$$

or

$$(2-\lambda)\begin{vmatrix} 1 & -1 & -1 \\ 1 & 4-\lambda & -1 \\ 1 & -1 & 4-\lambda \end{vmatrix} = 0$$

or

$$(2-\lambda)(5-\lambda)^2 = 0$$

Hence $\lambda_1 = 2$, $\lambda_2 = 5$ and $\lambda_3 = 5$.

Now,

$$B = I - \frac{1}{4}A = \begin{bmatrix} 0 & \frac{1}{4} & \frac{1}{4} \\ \frac{1}{4} & 0 & \frac{1}{4} \\ \frac{1}{4} & \frac{1}{4} & 0 \end{bmatrix}.$$

Therefore, eigenvalues of B are given by

$$\begin{vmatrix} -\mu & \frac{1}{4} & \frac{1}{4} \\ \frac{1}{4} & -\mu & \frac{1}{4} \\ \frac{1}{4} & \frac{1}{4} & -\mu \end{vmatrix} = 0$$

or

$$\begin{vmatrix} -\mu+\frac{1}{2} & \frac{1}{4} & \frac{1}{4} \\ -\mu+\frac{1}{2} & -\mu & \frac{1}{4} \\ \frac{1}{2}-\mu & \frac{1}{4} & -\mu \end{vmatrix} = 0$$

or

$$\left(\frac{1}{2}-\mu\right)\begin{vmatrix} 1 & \frac{1}{4} & \frac{1}{4} \\ 1 & -\mu & \frac{1}{4} \\ 1 & \frac{1}{4} & -\mu \end{vmatrix} = 0$$

or

$$\left(\frac{1}{2}-\mu\right)\begin{vmatrix} 1 & \frac{1}{4} & \frac{1}{4} \\ 0 & -\mu-\frac{1}{4} & 0 \\ 0 & \frac{1}{4}+\mu & -\mu-\frac{1}{4} \end{vmatrix} = 0$$

or

$$\left(\frac{1}{2}-\mu\right)\left(\mu+\frac{1}{4}\right)^2 = 0$$

Therefore, $\mu_1 = 1/2$, $\mu_2 = -1/4$, $\mu_3 = -1/4$. The relation $\mu_i = 1-(1/4)\lambda_i$ for all i, is easily verified.

Example 1.95 Find the eigenvalues and eigenvectors of the matrix

$$A = \begin{bmatrix} 8 & -6 & 2 \\ -6 & 7 & -4 \\ 2 & -4 & 3 \end{bmatrix}$$

For the eigenvalues of A, we have

$$\begin{vmatrix} 8-\lambda & -6 & 2 \\ -6 & 7-\lambda & -4 \\ 2 & -4 & 3-\lambda \end{vmatrix} = 0 \quad \text{or} \quad \lambda^3 - 18\lambda^2 + 45\lambda = 0$$

Solving, we get $\lambda_1 = 0$, $\lambda_2 = 3$ and $\lambda_3 = 15$.

We shall find the eigenvectors corresponding to each of the above eigenvalues.

(a) When $\lambda_1 = 0$, we have

$$\begin{bmatrix} 8 & -6 & 2 \\ -6 & 7 & -4 \\ 2 & -4 & 3 \end{bmatrix}\begin{bmatrix} x_1 \\ x_2 \\ x_3 \end{bmatrix} = 0, \qquad \text{where } X_1 = \begin{bmatrix} x_1 \\ x_2 \\ x_3 \end{bmatrix}$$

is the eigenvector corresponding to $\lambda_1 = 0$. Then

$$8x_1 - 6x_2 + 2x_3 = 0 \qquad \text{(i)}$$

$$-6x_1 + 7x_2 - 4x_3 = 0 \qquad \text{(ii)}$$

$$2x_1 - 4x_2 + 3x_3 = 0 \qquad \text{(iii)}$$

Since equation (ii) can be obtained from equations (i) and (iii), there are only 2 independent equations. Since the rank is 2, there is only one independent solution. To obtain this, we use equations (ii) and (iii), of which the solution is $x_2 = x_3 = 2x_1$. Hence

$$X_1 = \begin{bmatrix} a \\ 2a \\ 2a \end{bmatrix},$$

where a is a constant.

(b) When $\lambda_2 = 3$, we have

$$\begin{bmatrix} 8 & -6 & 2 \\ -6 & 7 & -4 \\ 2 & -4 & 3 \end{bmatrix} \begin{bmatrix} x_1' \\ x_2' \\ x_3' \end{bmatrix} = 3 \begin{bmatrix} x_1' \\ x_2' \\ x_3' \end{bmatrix},$$

where

$$X_2 = \begin{bmatrix} x_1' \\ x_2' \\ x_3' \end{bmatrix}$$

is the eigenvector corresponding to $\lambda_2 = 3$.

The above matrix equation implies:

$$5x_1' - 6x_2' + 2x_3' = 0 \qquad \text{(iv)}$$

$$-6x_1' + 4x_2' + 4x_3' = 0 \qquad \text{(v)}$$

$$2x_1' - 4x_2' = 0 \qquad \text{(vi)}$$

From (vi), we get: $x_1' = 2x_2'$, and from (v) $x_1' = -x_3'$.

It may be verified that these values satisfy (iv). Taking $x_2 = b$, we write

$$X_2 = \begin{bmatrix} 2b \\ b \\ -2b \end{bmatrix},$$

where b is a constant.

(c) When $\lambda_3 = 15$, we have

$$\begin{bmatrix} 8 & -6 & 2 \\ -6 & 7 & -4 \\ 2 & -4 & 3 \end{bmatrix} \begin{bmatrix} x_1'' \\ x_2'' \\ x_3'' \end{bmatrix} = 15 \begin{bmatrix} x_1'' \\ x_2'' \\ x_3'' \end{bmatrix},$$

where

$$X_3 = \begin{bmatrix} x_1'' \\ x_2'' \\ x_3'' \end{bmatrix}$$

is the eigenvector corresponding to $\lambda_3 = 15$. The above matrix equation implies:

$$-7x_1'' - 6x_2'' + 2x_3'' = 0 \quad \text{(i)}$$

$$-6x_1'' - 8x_2'' - 4x_3'' = 0 \quad \text{(ii)}$$

$$2x_1'' - 4x_2'' - 12x_3'' = 0 \quad \text{(iii)}$$

Since the last equation can be obtained from the first-two, there is only one independent solution. From the equations, we obtain

$$x_1'' = -x_2'' = 2x_3''.$$

Taking $x_3'' = a'$, we obtain

$$X_3 = \begin{bmatrix} 2a' \\ -2a' \\ a^1 \end{bmatrix}$$

as the eigenvector corresponding to the eigenvalue $\lambda_3 = 15$.

Example 1.96 Determine the eigenvalues and eigenvectors of the matrix

$$A = \begin{bmatrix} 6 & -2 & 2 \\ -2 & 3 & -1 \\ 2 & -1 & 3 \end{bmatrix}$$

We have

$$\begin{vmatrix} 6-\lambda & -2 & 2 \\ -2 & 3-\lambda & -1 \\ 2 & -1 & 3-\lambda \end{vmatrix} = 0.$$

Applying $C_{32}(1)$, we get

$$\begin{vmatrix} 6-\lambda & -2 & 0 \\ -2 & 3-\lambda & 2-\lambda \\ 2 & -1 & 2-\lambda \end{vmatrix} = 0 \quad \text{or} \quad (2-\lambda)\begin{vmatrix} 6-\lambda & -2 & 0 \\ -2 & 3-\lambda & 1 \\ 2 & -1 & 1 \end{vmatrix} = 0$$

on simplification, we find

$$(2-\lambda)(\lambda^2 - 10\lambda + 16) = 0 \quad \text{or} \quad (2-\lambda)(\lambda-2)(\lambda-8) = 0$$

Thus, $\lambda_1 = 8$, $\lambda_2 = 2$ and $\lambda_3 = 2$.

(a) When $\lambda_1 = 8$, we have

$$\begin{bmatrix} 6 & -2 & 2 \\ -2 & 3 & -1 \\ 2 & -1 & 3 \end{bmatrix} \begin{bmatrix} x_1 \\ x_2 \\ x_3 \end{bmatrix} = 8 \begin{bmatrix} x_1 \\ x_2 \\ x_3 \end{bmatrix},$$

where

$$X_1 = \begin{bmatrix} x_1 \\ x_2 \\ x_3 \end{bmatrix},$$

is the eigenvector corresponding to the eigenvalue $\lambda_1 = 8$. The above matrix equation gives:

$$-2x_1 - 2x_2 + 2x_3 = 0 \quad \text{(i)}$$

$$-2x_1 - 5x_2 - x_3 = 0 \quad \text{(ii)}$$

$$2x_1 - x_2 - 5x_3 = 0 \quad \text{(iii)}$$

There are two independent equations and the rank is 2. Hence there is only one independent solution in terms of one variable. We obtain $x_1 = -2x_2 = 2x_3$. Putting $x_3 = a$, we obtain

$$X_1 = \begin{bmatrix} 2a \\ -a \\ a \end{bmatrix}.$$

(b) When $\lambda_2 = \lambda_3 = 2$, we have

$$\begin{bmatrix} 6 & -2 & 2 \\ -2 & 3 & -1 \\ 2 & -1 & 3 \end{bmatrix} \begin{bmatrix} x_1' \\ x_2' \\ x_3' \end{bmatrix} = 2 \begin{bmatrix} x_1' \\ x_2' \\ x_3' \end{bmatrix},$$

where

$$X_2 = \begin{bmatrix} x_1' \\ x_2' \\ x_3' \end{bmatrix}$$

is the eigenvector corresponding to the eigenvalue $\lambda_2 = \lambda_3 = 2$. The above matrix equation implies:

$$4x_1' - 2x_2' + 2x_3' = 0$$

$$-2x_1' + x_2' - x_3' = 0$$

$$2x_1' - x_2' + x_3' = 0$$

or

$$2x_1' - x_2' + x_3' = 0$$

$$2x_1' - x_2' + x_3' = 0$$

$$2x_1' - x_2' + x_3' = 0$$

Thus all the equations are same and rank = 1. Hence two variables have to be assigned values. Taking $x_1' = k_1$ and $x_2' = k_2$, we obtain $x_3' = k_2 - 2k_1$. Hence

$$X_2 = X_3 = \begin{bmatrix} k_1 \\ k_2 \\ k_2 - 2k_1 \end{bmatrix},$$

where k_1 and k_2 are constants.

Example 1.97 Diagonalize the matrix given by

$$A = \begin{bmatrix} 1 & 1 & 3 \\ 1 & 5 & 1 \\ 3 & 1 & 1 \end{bmatrix}$$

For the eigenvalues, we have

$$\begin{vmatrix} 1-\lambda & 1 & 3 \\ 1 & 5-\lambda & 1 \\ 3 & 1 & 1-\lambda \end{vmatrix} = 0 \quad \text{or} \quad \lambda^3 - 7\lambda^2 + 36 = 0$$

Solving, we get $\lambda_1 = -2$, $\lambda_2 = 3$ and $\lambda_3 = 6$. Corresponding to $\lambda_1 = -2$, let the eigenvector be

$$X_1 = \begin{bmatrix} x_1 \\ x_2 \\ x_3 \end{bmatrix}$$

Then we have

$$\begin{bmatrix} 1 & 1 & 3 \\ 1 & 5 & 1 \\ 3 & 1 & 1 \end{bmatrix} \begin{bmatrix} x_1 \\ x_2 \\ x_3 \end{bmatrix} = -2 \begin{bmatrix} x_1 \\ x_2 \\ x_3 \end{bmatrix}.$$

The above matrix equation gives:

$$3x_1 + x_2 + 3x_3 = 0 \qquad \text{(i)}$$

$$x_1 + 7x_2 + x_3 = 0 \qquad \text{(ii)}$$

$$3x_1 + x_2 + 3x_3 = 0 \qquad \text{(iii)}$$

Since (i) and (iii) are the same and rank = 2, we have to assign an arbitrary value to one variable. Solving (i) and (ii), we obtain $x_2 = 0$ and $x_1 = -x_3$. Hence We take

$$x_1 = \begin{bmatrix} 1 \\ 0 \\ -1 \end{bmatrix}.$$

Again, corresponding to $\lambda_2 = 3$, let the eigenvector be

$$x_2 = \begin{bmatrix} x_1' \\ x_2' \\ x_3' \end{bmatrix}.$$

We then have

$$\begin{bmatrix} 1 & 1 & 3 \\ 1 & 5 & 1 \\ 3 & 1 & 1 \end{bmatrix} \begin{bmatrix} x_1' \\ x_2' \\ x_3' \end{bmatrix} = 3 \begin{bmatrix} x_1' \\ x_2' \\ x_3' \end{bmatrix}.$$

The above matrix equation implies:

$$-2x_1' + x_2' + 3x_3' = 0 \tag{iv}$$

$$x_1' + 2x_2' + x_3' = 0 \tag{v}$$

$$3x_1' + x_2' - 2x_3' = 0. \tag{vi}$$

Since (iv) can be obtained from (v) and (vi) and rank = 2, we obtain one independent solution in terms of one variable. From the first-two equations, we obtain

$$\frac{x_1'}{-5} = \frac{x_2'}{5} = \frac{x_3'}{-5} \Rightarrow \frac{x_1'}{-1} = \frac{x_2'}{1} = \frac{x_3'}{-1}.$$

We therefore take

$$X_2 = \begin{bmatrix} 1 \\ -1 \\ 1 \end{bmatrix}.$$

Following the same procedure as above, we obtain

$$X_3 = \begin{bmatrix} 1 \\ 2 \\ 1 \end{bmatrix},$$

corresponding to the eigenvalue $\lambda_3 = 6$. Hence

$$P = \begin{bmatrix} 1 & 1 & 1 \\ 0 & -1 & 2 \\ -1 & 1 & 1 \end{bmatrix}$$

and it can be verified that

$$P^{-1}AP = \begin{bmatrix} 2 & 0 & 0 \\ 0 & 3 & 0 \\ 0 & 0 & 6 \end{bmatrix}.$$

EXERCISES 1.6

1. Evaluate $AB - BA$ when

$$A = \begin{bmatrix} 1 & 0 & -2 \\ 0 & -3 & 1 \\ 2 & 1 & 0 \end{bmatrix} \text{ and } B = \begin{bmatrix} 3 & 1 & -2 \\ 3 & -2 & 4 \\ -3 & 5 & -1 \end{bmatrix}.$$

2. If

$$A = \begin{bmatrix} 1 & 2 \\ 3 & 4 \end{bmatrix},$$

compute A^2, A^3 and A^4.

3. If

$$A = \begin{bmatrix} 1 & -1 \\ 4 & 0 \end{bmatrix}, \quad B = \begin{bmatrix} 1 & 2 \\ 3 & 4 \end{bmatrix}, \quad C = \begin{bmatrix} 1 & -1 \\ 4 & 0 \end{bmatrix}$$

verify that $(AB)C = A(BC)$.

4. If

$$A = \begin{bmatrix} \cos\theta & \sin\theta \\ -\sin\theta & \cos\theta \end{bmatrix},$$

prove that

$$A^2 = \begin{bmatrix} \cos 2\theta & \sin 2\theta \\ -\sin 2\theta & \cos 2\theta \end{bmatrix} \text{ and } A^3 = \begin{bmatrix} \cos 3\theta & \sin 3\theta \\ -\sin 3\theta & \cos 3\theta \end{bmatrix}$$

What is A^n?

5. If

$$A = \begin{bmatrix} 2 & -4 & 1 \\ 3 & -2 & 5 \\ 1 & 6 & 1 \end{bmatrix}, \quad B = \begin{bmatrix} -1 & 2 & 3 \\ 1 & 3 & -1 \\ 4 & 5 & 1 \end{bmatrix}$$

show that $|AB| = |BA| = |A||B|$.

6. Find the matrices $A+B$, $A-B$, AB and BA, if they are possible.

(a) $A=\begin{bmatrix} 1 & 2 \\ 3 & 4 \end{bmatrix}$, $B=\begin{bmatrix} 3 & 0 \\ 1 & 2 \end{bmatrix}$

(b) $A=\begin{bmatrix} 2 & 1 \\ 3 & -4 \end{bmatrix}$, $B=\begin{bmatrix} 2 & 1 \\ 1 & 0 \\ 3 & 1 \end{bmatrix}$

(c) $A=[1 \quad 2 \quad 3 \quad 4]$, $B=\begin{bmatrix} -1 \\ 2 \\ 4 \\ 1 \end{bmatrix}$

7. If

$$U_1=\begin{bmatrix} 1 & 2 & 3 \\ 0 & 1 & 2 \\ 0 & 0 & 1 \end{bmatrix} \quad \text{and} \quad U_2=\begin{bmatrix} 2 & 1 & 2 \\ 0 & 1 & 3 \\ 0 & 0 & 2 \end{bmatrix},$$

show that U_1U_2 is an upper triangular matrix.

Express the following matrices as the sum of a symmetric and a skew-symmetric matrix (Problems 8–11):

8. $\begin{bmatrix} 2 & 1 & 3 \\ -1 & 4 & 1 \\ 0 & 2 & -2 \end{bmatrix}$ **9.** $\begin{bmatrix} 3 & 4 & 0 \\ 7 & 2 & 0 \\ 0 & 3 & 4 \end{bmatrix}$

10. $\begin{bmatrix} 2 & 3 & 4 \\ 5 & 0 & 3 \\ 1 & 2 & 3 \end{bmatrix}$ **11.** $\begin{bmatrix} 0 & 1 & 2 \\ 3 & 0 & 0 \\ 0 & 5 & 7 \end{bmatrix}$

Which of the following matrices (Problems 12–16) are orthogonal?

12. $\begin{bmatrix} 1 & -2 \\ 2 & 1 \end{bmatrix}$ **13.** $\begin{bmatrix} \frac{\sqrt{3}}{2} & \frac{1}{2} \\ -\frac{1}{2} & \frac{\sqrt{3}}{2} \end{bmatrix}$

14. $\begin{bmatrix} \frac{1}{9} & \frac{\sqrt{8}}{9} \\ \frac{\sqrt{5}}{9} & \frac{2}{9} \end{bmatrix}$ **15.** $\begin{bmatrix} \cos\theta & 0 & \sin\theta \\ 0 & 1 & 0 \\ -\sin\theta & 0 & \cos\theta \end{bmatrix}$

16. $\begin{bmatrix} -\frac{2}{3} & \frac{1}{3} & \frac{2}{3} \\ \frac{2}{3} & \frac{2}{3} & \frac{1}{3} \\ \frac{1}{3} & -\frac{2}{3} & \frac{2}{3} \end{bmatrix}$.

17. Show that

$$\begin{bmatrix} 1 & -\tan\theta \\ \tan\theta & 1 \end{bmatrix}\begin{bmatrix} 1 & \tan\theta \\ -\tan\theta & 1 \end{bmatrix}^{-1} = \begin{bmatrix} \cos 2\theta & -\sin 2\theta \\ \sin 2\theta & \cos 2\theta \end{bmatrix}$$

18. If

$$A = \begin{bmatrix} 1 & 1 & -2 \\ -1 & 2 & 1 \\ 0 & 1 & -1 \end{bmatrix}$$

show that $(A+I)(A-2I)(A-I) = 0$. Hence, find expressions for A^{-1} and $(2I+A)^{-1}$ in terms of I, A and A^2.

19. Factorize the matrix

$$A = \begin{bmatrix} 1 & 3 & 1 \\ 1 & 4 & 2 \\ -1 & -2 & 3 \end{bmatrix}$$

into the product LU and hence solve the system of equations

$$x + 3y + z = 3$$
$$x + 4y + 2z = 3$$
$$-x - 2y + 3z = -6.$$

20. Given that

$$B = \begin{bmatrix} 1 & 2 & 1 \\ -1 & -1 & -1 \\ 0 & -2 & -2 \end{bmatrix},$$

calculate $B^{-1}AB$.

21. If

$$A = \begin{bmatrix} \cosh x & \sinh x \\ \sinh x & \cosh x \end{bmatrix},$$

prove that

$$A^n = \begin{bmatrix} \cosh nx & \sinh nx \\ \sinh nx & \cosh nx \end{bmatrix}$$

22. Prove that the product of two upper (lower) triangular matrices is an upper (lower) triangular matrix.

23. Prove that (a) the product of two orthogonal matrices is orthogonal, (b) the inverse of an orthogonal matrix is orthogonal.

Write the following system of equations in the matrix form $AX = B$ (Problems 24–27):

24.
$$\begin{aligned} x_1 + 2x_2 + 3x_3 &= 5 \\ 2x_1 + 5x_3 &= 10 \\ x_2 + 3x_3 &= 6. \end{aligned}$$

25.
$$\begin{aligned} x + 2y + z &= 0 \\ 7x - y + z &= 0 \\ 2x - 13y - z &= 0. \end{aligned}$$

26.
$$\begin{aligned} x + z &= 1 \\ 7x + 2y &= 0 \\ 6x + z &= 2. \end{aligned}$$

27.
$$\begin{aligned} x_1 + x_3 &= 2 \\ 2x_1 - x_2 &= 7 \\ 6x_1 + 3x_2 + x_3 &= 8. \end{aligned}$$

28. If

$$A = \begin{bmatrix} 1 & 2 \\ 3 & 4 \end{bmatrix}$$

compute A^{-1} and check your result by direct multiplication with A.

Find inverse in the following cases (Problems 29–33):

29. $\begin{bmatrix} 2 & 4 & 3 \\ 0 & 1 & 1 \\ 2 & 2 & -1 \end{bmatrix}$

30. $\begin{bmatrix} 1 & 6 & 4 \\ 0 & 2 & 3 \\ 0 & 1 & 2 \end{bmatrix}$

31. $\begin{bmatrix} 1 & 2 & -1 \\ 3 & 8 & 2 \\ 4 & 9 & -1 \end{bmatrix}$

32. $\begin{bmatrix} 2 & 4 & 8 \\ 3 & 7 & 12 \\ 1 & 3 & 5 \end{bmatrix}$

33. $\begin{bmatrix} 1 & 3 & 12 \\ 0 & 4 & 8 \\ 2 & 6 & 13 \end{bmatrix}$

34. Show that

$$A = \begin{bmatrix} 1 & 2 & 2 \\ 2 & 1 & 2 \\ 2 & 2 & 1 \end{bmatrix}$$

satisfies the equation $A^2 - 4A - 5I = 0$. Hence, determine A^{-1}.

35. Compute the inverse of the matrix

$$\begin{bmatrix} 3 & 2 & 4 \\ 2 & 1 & 1 \\ 1 & 3 & 5 \end{bmatrix}$$

and use the result to solve the system of equations:

$$3x + 2y + 4z = 7$$

$$2x + y + z = 4$$

$$x + 3y + 5z = 2.$$

Find the ranks of the following matrices (Problems 36–37):

36. $\begin{bmatrix} 1 & -1 & 2 \\ 2 & 6 & 3 \\ 3 & 13 & 4 \end{bmatrix}$

37. $\begin{bmatrix} 3 & 2 & 1 \\ 4 & -1 & -2 \\ -6 & 7 & 8 \end{bmatrix}$

Find whether the following systems are consistent or not; if consistent, solve them by the Gaussian elimination method (Problems 38–43):

38. $x + 2y - z = -1$
$3x + 8y + 2z = 28$
$4x + 9y - z = 14$

39. $5x - 3y + 7z = 4$
$3x + 26y + 2z = 9$
$7x + 2y + 10z = 5$

40. $3x - y + 2z = 1$
$x - y + z = -1$
$2x - 2y + 3z = 2$

41. $x + 2y - 3z + \omega = 1$
$2x - y + 2z - \omega = 1$
$4x + 3y - 4z + \omega = 2$

42. $x + 2y + 3z = 1$
$y + 2z = 3$
$3x + 5y + 7z = 15$

43. $x + 2y + 3z = 1$
$2x + 3y + 4z = 3$
$3x + 5y + 8z = 2$

44. Find x so that

$$[1 \quad x \quad 1]\begin{bmatrix} 1 & 3 & 2 \\ 0 & 5 & 1 \\ 0 & 3 & 2 \end{bmatrix}\begin{bmatrix} 1 \\ 1 \\ x \end{bmatrix} = 0.$$

45. Find the LU decomposition of the matrix

$$A = \begin{bmatrix} 1 & 2 & -6 \\ 2 & 4 & 3 \\ -3 & 4 & 1 \end{bmatrix}$$

by Gaussian elimination and verify that in fact $A = LU$.

46. Find the LU decomposition of the matrix

$$A = \begin{bmatrix} 2 & 2 \\ 4 & -1 \end{bmatrix}.$$

47. The following matrices

$$A = \begin{bmatrix} 0 & 1 \\ 1 & 0 \end{bmatrix}, \quad B = \begin{bmatrix} 0 & -i \\ i & 0 \end{bmatrix}, \quad C = \begin{bmatrix} 1 & 0 \\ 0 & -1 \end{bmatrix}$$

are called Pauli spin matrices and occur in atomic physics. Find whether

(i) A, B, C commute, and

(ii) A, B, C are Hermitian.

48. Given that

$$A = \begin{bmatrix} a & b & c \\ b & c & a \\ c & a & b \end{bmatrix}$$

where a, b, c are the roots of $x^3 + x^2 + k = 0$ (k is a constant), prove that A is orthogonal.

49. If

$$A = \begin{bmatrix} 1 & 2 & 4 \\ 0 & 2 & 3 \\ 0 & 0 & 3 \end{bmatrix},$$

find A^{-1} by solving the matrix equation $AX = I$.

Solve the following systems by the Gauss–Jordan method (Problems 50–54):

50.
$$x + 2y - z = -1$$
$$3x - y - 2z = 5$$
$$x - y - 3z = 0$$

51.
$$x + 2y + 3z = 6$$
$$2x + 4y + z = 7$$
$$3x + 2y + 9z = 14$$

52.
$$6x + 2y - 2z = 6$$
$$-2x + 2y + 2z = 2$$
$$2x + 2y + 2z = 6$$

53.
$$x + y + 2z = -1$$
$$2x - y + 3z = -4$$
$$3x - y - z = 0$$

54.
$$x + y - z = 0$$
$$2x - y + 3z = 9$$
$$-3x + 4y + 2z = 11$$

Find the ranks of the following matrices by reducing them to the canonical form (Problems 55–58):

55. $\begin{bmatrix} 2 & 3 & 4 & -1 \\ 5 & 2 & 0 & -1 \\ -4 & 5 & 12 & -1 \end{bmatrix}$.

56. $\begin{bmatrix} 2 & 3 & -1 & -1 \\ 1 & -1 & -2 & -4 \\ 3 & 1 & 3 & -2 \\ 6 & 3 & 0 & -7 \end{bmatrix}$.

57. $\begin{bmatrix} 1 & 3 & 2 & 2 \\ 1 & 2 & 1 & 2 \\ 3 & 7 & 4 & 6 \\ 2 & 4 & 3 & 4 \end{bmatrix}$

58. $\begin{bmatrix} 1 & 2 & 1 & 2 \\ 3 & 1 & -2 & 1 \\ 4 & -3 & -1 & 3 \\ 2 & 4 & 2 & 4 \end{bmatrix}$

59. Determine the values of k_1 and k_2 so that the system of equations:

$$x + y + z = 6 \qquad x + 2y + 3z = 10 \qquad x + 2y + k_1 z = k_2$$

have (i) a unique solution, (ii) no solution, (iii) an infinite number of solutions.

Determine the eigenvalues and eigenvectors of the following matrices (Problems 60–63):

60. $\begin{bmatrix} -2 & -1 \\ 5 & 4 \end{bmatrix}$

61. $\begin{bmatrix} 2 & 0 & -2 \\ 0 & 4 & 0 \\ -2 & 0 & 5 \end{bmatrix}$

62. $\begin{bmatrix} 6 & -2 & 2 \\ -2 & 3 & -1 \\ 2 & -1 & 3 \end{bmatrix}$

63. $\begin{bmatrix} 5 & 0 & 1 \\ 0 & -2 & 0 \\ 1 & 0 & 5 \end{bmatrix}$

64. Let A and B be two similar matrices. If X is an eigenvector corresponding to the eigenvalue λ, then show that $P^{-1}X$ is an eigenvector of B corresponding to the eigenvalue l, where $B = P^{-1}AP$.

65. Given that

$$A = \begin{bmatrix} 6 & -2 & 2 \\ -2 & 3 & -1 \\ 2 & -1 & 3 \end{bmatrix}$$

find a matrix P such that $P^{-1}AP$ is a diagonal matrix.

2

Geometry, Vectors and Complex Numbers

2.1 COORDINATE SYSTEMS

The philosopher–mathematician Rene Descartes was the first to realize that geometrical ideas can be translated into algebraic relations and this had enabled him to write his book *la Géometrie* (1637) in which geometry was studied systematically by the use of algebra.

The name *coordinate* geometry or analytic geometry was given because of the fact that numbers (called coordinates) which are associated with points of some 'space', are employed in this study. In this chapter, we shall be concerned with a brief review of geometry as also an introduction to vectors and complex numbers.

As we have already remarked, the study of analytic geometry consists in using the mathematical ideas developed in algebra for the analysis of geometric situations and this therefore requires the establishment of a link between geometry and algebra. A basic necessity for this is the introduction of a *coordinate system* and to define 'coordinates' in the concerned space.

2.1.1 Rectangular Coordinates in a Plane

Two mutually perpendicular lines, with their intersection at the origin (0, 0) of coordinates, are known as *rectangular cartesian axes,* named after Descartes for his work in analytic geometry. Such a system of coordinates, called the *rectangular cartesian coordinate system* in two dimensions is depicted in Fig. 2.1. Where the horizontal line is called the x-axis and the vertical line the y-axis. All points to the right of the origin O are positive, while those to the left are negative. Similarly, all points above the origin are positive while all those below are negative.

It may be seen from Fig. 2.1 that the two axes divide the entire plane into four *quadrants* and these are numbered in the customary way, beginning with the first quadrant in the upper right hand quarter and proceeding in the counter-clockwise direction. Using the coordinate system specified above, we can now specify any point on a plane by its distance from the origin along both the x and y directions. In what follows, $P(x, y)$ denotes the position of a point P having coordinates x and y. The value x is called the *abscissa* of the point, while the value y is the *ordinate*. If we are given a set of values for x and the corresponding values of y, we say that there exists a functional relationship between x and y. If we can plot these points, then the curve obtained by joining them is called the *graph* of the function.

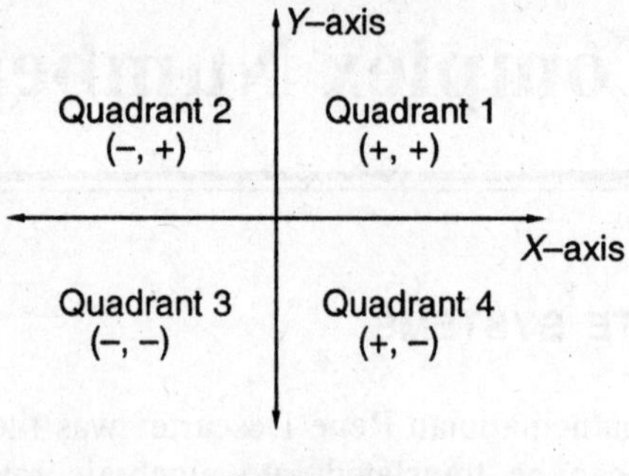

Figure 2.1

It should be mentioned that, in addition to rectangular cartesian axes, we have oblique coordinate system in which the axes are inclined to each other at an angle not equal to a right angle. We also have polar coordinates and cylindrical coordinates, and these will be discussed in later sections.

2.1.2 Distance between Two Points

Once a coordinate system is established, it will be relatively easy to find the distance between two points, say $P(x_1, y_1)$ and $Q(x_2, y_2)$. Let these points be, as shown in Fig. 2.2. We draw PM and QN perpendiculars to the x-axis and also PL perpendicular to NQ meeting it at L. Since x_1 and y_1 are the coordinates of P, we have $OM = x_1$ and $MP = y_1$. Similarly, $ON = x_2$ and $NQ = y_2$. It follows that $PL = MN = x_2 - x_1$, and $LQ = NQ - NL = NQ - MP = y_2 - y_1$.

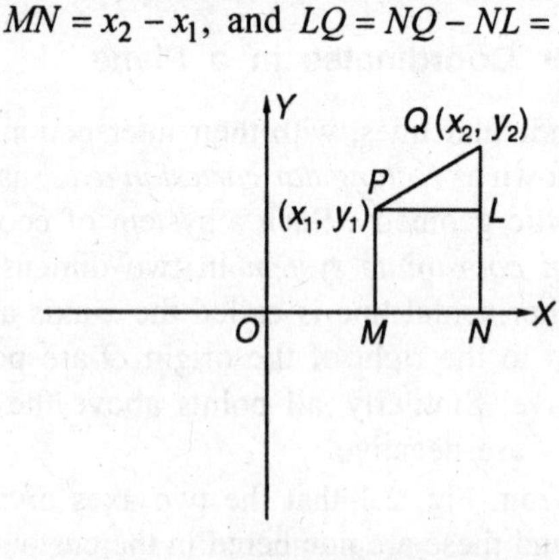

Figure 2.2

Since the triangle QPL is a right triangle, we have by Pythagorous theorem:

$$PQ^2 = PL^2 + LQ^2 = (x_2 - x_1)^2 + (y_2 - y_1)^2 \tag{2.1}$$

It is easy to deduce from this that

$$OP = \sqrt{x_1^2 + y_1^2} \quad \text{and} \quad OQ = \sqrt{x_2^2 + y_2^2}$$

The following results will be useful exercises:

(i) If $P(x_1, y_1)$ and $Q(x_2, y_2)$ are two given points, then the point

$$R\left(\frac{mx_2 + nx_1}{m+n}, \frac{my_2 + ny_1}{m+n}\right)$$

divides PQ *internally* in the ratio $m : n$.

(ii) If $A\ (x_1, y_1)$, $B(x_2, y_2)$ and $C(x_3, y_3)$ represent the three vertices of a triangle ABC, then

$$G\left(\frac{x_1 + x_2 + x_3}{3}, \frac{y_1 + y_2 + y_3}{3}\right)$$

represents its *centroid.*

(iii) The area, Δ, of the triangle ABC is given by

$$\Delta = \frac{1}{2}\begin{vmatrix} x_1 & y_1 & 1 \\ x_2 & y_2 & 1 \\ x_3 & y_3 & 1 \end{vmatrix}$$

when the three vertices A, B, C are given as in (ii) above.

Example 2.1 The vertices of a triangle are $(-3, -4)$, $(2, 6)$ and $(-6, 10)$. Show that the triangle is a right-angled triangle.

Let $A(-3, -4)$, $B(2, 6)$ and $C(-6, 10)$ be the three given points. Then, $AB^2 = 25 + 100 = 125$, $BC^2 = 64 + 16 = 80$ and $CA^2 = 9 + 196 = 205$.

Since $AB^2 + BC^2 = CA^2$, it follows that the triangle ABC is a right-angled triangle.

Example 2.2 Show that the three points, whose coordinates are $(-1, 1)$, $(-5, -7)$ and $(10, 23)$, are collinear.

If the area of the triangle formed by the three points is zero, then the three points lie on a straight line. We have here

$$\text{Area} = \frac{1}{2}\begin{vmatrix} -1 & 1 & 1 \\ -5 & -7 & 1 \\ 10 & 23 & 1 \end{vmatrix}$$

$$= \frac{1}{2}[-1(-7-23)-1(-5-10)+1(-115+70)]$$

$$= \frac{1}{2}[30+15-45]$$

$$= 0.$$

Hence the given points are collinear.

2.1.3 Polar Coordinates

The rectangular coordinate system, described in Section 2.2.1, is the most frequently used coordinate system. However, a system involving *polar coordinates* will be found to be particularly useful and convenient in certain applications, e.g. in locating objects in surveying or navigation.

The polar coordinate system is described with reference to (i) a point O called the origin or *pole*, and (ii) a horizontal line, originating at O, called the *polar axis* or the *radius vector.* In this system, a point P is located by means of two quantities: (a) the distance OP usually denoted by $r(r \geq 0)$, and (b) the angle θ (say, in degrees) formed by the radius vector OP and the polar axis, reckoned positive if the angle is measured from the polar axis in a counter-clockwise direction. If the direction is clockwise, then the angle will be taken as negative. Therefore, in the polar system, the coordinates of a point P are written as $P(r, \theta)$. If $P(x, y)$ represents the *same* point in the cartesian system, we then have:

$$P(x, y) = P(r, \theta) \tag{2.2}$$

The pair of values (r, θ) are referred to as the polar coordinates of the point P, and Eq. (2.2) has geometrical significance since it says that the two points are one and the same whatever might be the coordinate system employed. Due to the equivalence, it is possible to find coordinate values in one system if their values in the other system are given.

In Fig. 2.3, we give the rectangular and polar coordinates of a point P and we find easily

$$x = r\cos\theta \quad \text{and} \quad y = r\sin\theta \tag{2.3}$$

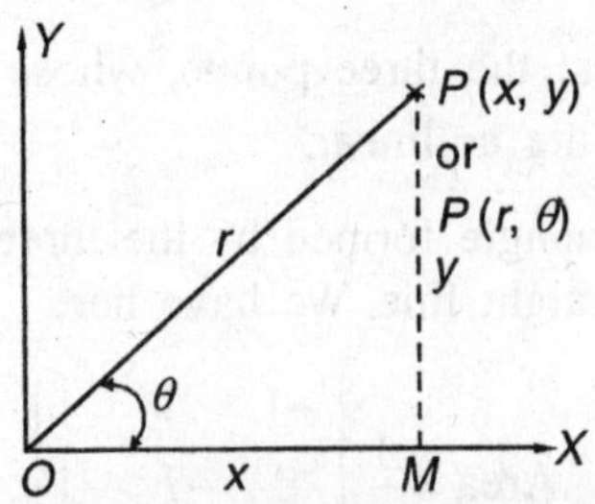

Figure 2.3

Equation (2.3) enables us to calculate the rectangular coordinates if the polar coordinates are given. Conversely, if the rectangular coordinates are given, the polar coordinates can be calculated from

$$r = \sqrt{x^2 + y^2} \quad \text{and} \quad \theta = \tan^{-1}\frac{y}{x} \tag{2.4}$$

Let $P(r_1, \theta_1)$ and $Q(r_2, \theta_2)$ be two points, as given in Fig. 2.4. Then, we have

$$OP = r_1, \; OQ = r_2 \; \text{and} \; P\hat{O}Q = \theta_2 - \theta_1.$$

From $\Delta\, POQ$,

$$PQ^2 = OP^2 + OQ^2 - 2(OP)(OQ)\cos P\hat{O}Q = r_1^2 + r_2^2 - 2r_1r_2\cos(\theta_2 - \theta_1),$$

which is the formula for calculating the distance between two points whose polar coordinates are given.

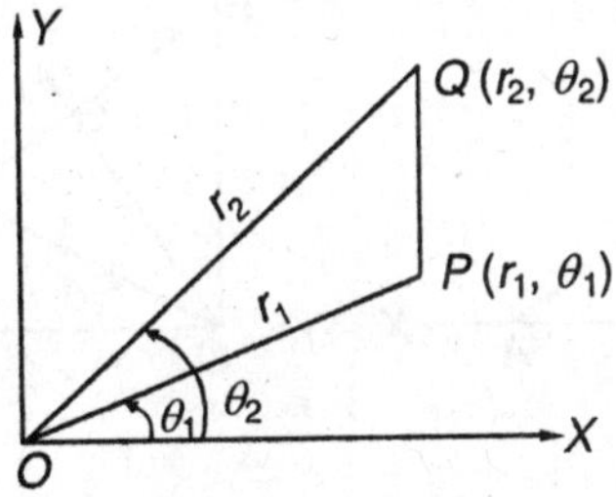

Figure 2.4

If $P(r_1, \theta_1)$, $Q(r_2, \theta_2)$ and $R(r_3, \theta_3)$ are the vertices of a triangle PQR, then its area Δ is given by

$$\Delta = \frac{1}{2}[r_1r_2\sin(\theta_2 - \theta_1) + r_2r_3\sin(\theta_3 - \theta_2) + r_3r_1\sin(\theta_1 - \theta_3)] \tag{2.5}$$

The proof of this result is left as an exercise.

2.1.4 Equation for a Line in Polar Coordinates

In determining the equation of a line in polar form, two cases arise :

(i) the line passes through the pole,
(ii) the line does not pass through the pole.

When the line passes through the pole, θ remains constant, say θ_0. If $n\pi$ is added to or subtracted from θ_0, $\theta_0 \pm n\pi$ determines the same line. Hence, in general, the equation of a line passing through the pole may be written as

$$\theta = \theta_0 \pm n\pi, \qquad n = 0, 1, 2, \ldots \tag{2.6}$$

As an example, let the equation of the line be

$$\theta = \frac{\pi}{4} \pm n\pi, \qquad n = 0, 1, 2, \ldots$$

Then, it is easy to see that the point $(3, 9\pi/4)$ satisfies the equation $\theta = (\pi/4) + 2\pi$. Hence the point $(3, 9\pi/4)$ is on the line given by $\theta = (\pi/4) + n\pi$.

If the line does not pass through the pole (e.g. the line AB in Fig. 2.5(a)), let $P(r, \theta)$ be a point on this line. Let OM, whose length is p, be the perpendicular from O on AB, and let $\angle XOM$ be equal to α. Then, it follows that $\angle MOP = \theta - \alpha$. Hence the equation

$$r \cos(\theta - \alpha) = p \tag{2.7}$$

represents the polar equation of AB.

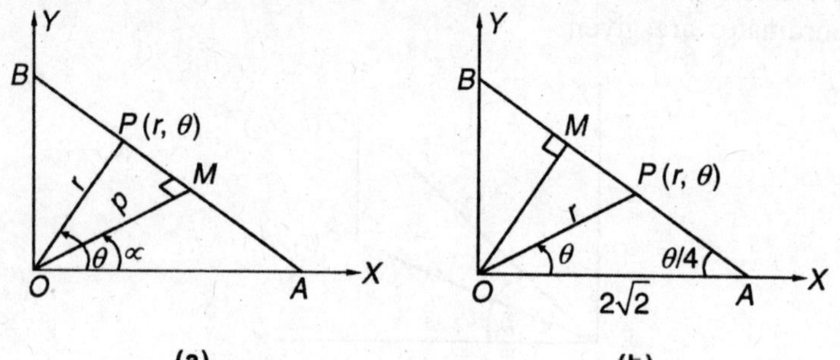

Figure 2.5

Example 2.3 Find the equation of the line AB given in Fig. 2.5(b). From the figure,

$$\angle AOM = \frac{\pi}{4} \quad \text{and} \quad \angle POM = \frac{\pi}{4} - \theta.$$

From $\Delta\, AOM$,

$$OM = 2\sqrt{2}\cos\frac{\pi}{4} = 2.$$

Therefore, from $\Delta\, POM$,

$$r\cos\left(\frac{\pi}{4} - \theta\right) = 2$$

which is the required of equation the line AB.

2.1.5 Rectangular Coordinates in Space

To establish a three-dimensional rectangular cartesian coordinate system, we proceed analogously as in Section 2.2.1. Instead of two mutually perpendicular lines, we now need three, viz., OX, OY, OZ through the origin O. It is easy to visualize such a coordinate system by looking at the corner of a room where the intersections of two walls and the floor provide three mutually perpendicular lines. We have, however, two types of rectangular coordinate systems in three-dimensional space: the *right-handed* or the *left-handed* system. A three-dimensional rectangular coordinate system is said to be right-handed if a rotation from OX towards OY advances a right-handed corkscrew along OZ; similarly, a rotation from OY to OZ takes a right-handed corkscrew along OX, and so on. Otherwise, it is said to be a left-handed system. We shall use right-handed cartesian axes only, unless otherwise stated.

The three coordinate axes OX, OY and OZ define three *coordinate planes OXY, OYZ* and *OZX,* which divide the three-dimensional space surrounding the origin into eight sections called *octants.* The position of a point P is specified by its perpendicular distances from each of the coordinate planes, i.e. distances measured from the planes to P parallel to the axes. In Fig. 2.6, a three-dimensional cartesian coordinate system is shown.

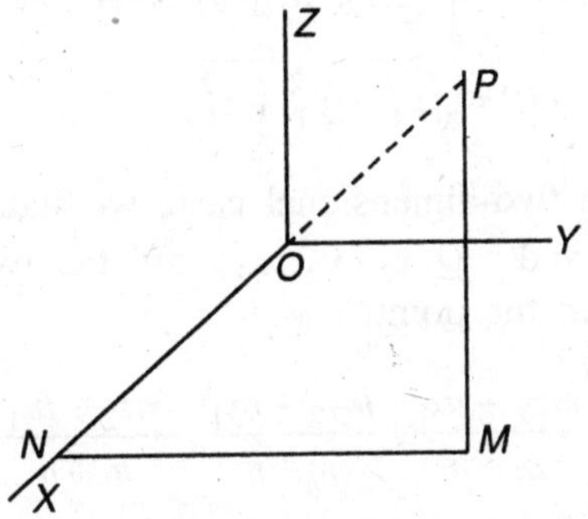

Figure 2.6

Let PM be drawn perpendicular to the plane XOY and from M in this plane, we draw MN perpendicular to OX meeting it at N. Let

$$ON = x, \quad NM = y \quad \text{and} \quad MP = z.$$

Then, (x, y, z) are the cartesian coordinates of P with reference to this coordinate system. Clearly x, y and z are the distances of P from the three coordinate planes YOZ, ZOX and XOY, respectively.

Let $P(x_1, y_1, z_1)$ and $Q(x_2, y_2, z_2)$ be two points in space, as shown in Fig. 2.7. Through the points P, Q we draw planes parallel to the coordinate planes to form a rectangular parallelopiped whose one diagonal is PQ. It is easily seen that the coordinates of N are x_2, y_2 and z_1, and that the angle PNQ is a right angle. The Pythagorean theorem therefore yields

$$PQ^2 = PN^2 + NQ^2 = (x_2 - x_1)^2 + (y_2 - y_1)^2 + (z_2 - z_1)^2,$$

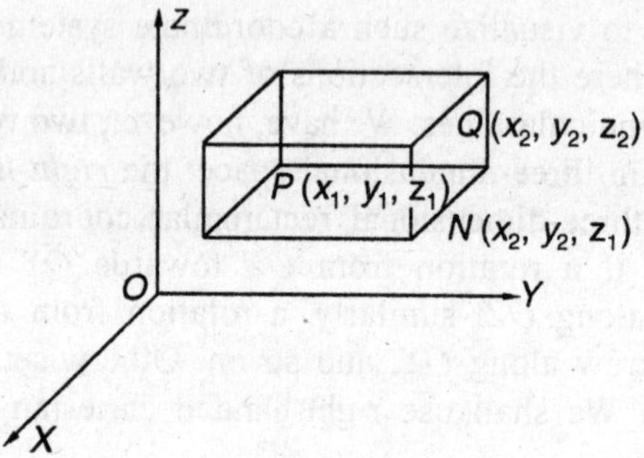

Figure 2.7

or

$$PQ = \sqrt{(x_2 - x_1)^2 + (y_2 - y_1)^2 + (z_2 - z_1)^2} \qquad (2.8)$$

Formula (2.8) is seen to be an extension to three dimensions of the two dimensional formula given in (2.1). In particular, we have

$$OP = \sqrt{x_1^2 + y_1^2 + z_1^2}, \text{ etc.} \qquad (2.9)$$

Analogously, as in the two-dimensional case, we state the result:

If $P(x_1, y_1, z_1)$ and $Q(x_2, y_2, z_2)$ are the two points in a three-dimensional space, then the point

$$R\left(\frac{mx_2 + nx_1}{m+n}, \frac{my_2 + ny_1}{m+n}, \frac{mz_2 + nz_1}{m+n}\right)$$

divides the line PQ in the ratio $m:n$.

Example 2.4 Find the ratio in which the line joining the points $A(1, 2, 3)$ and $B(2, -3, 4)$ is divided by the plane $x = 0$.

Let $P(x, y, z)$ be a point on AB dividing it in the ratio $m:n$. Then the coordinates of P are given by

$$x = \frac{2m+n}{m+n}, \quad y = \frac{-3m+2n}{m+n}, \quad z = \frac{4m+3n}{m+n},$$

where $m + n \neq 0$. If this point lies on the plane $x = 0$, we have

$$2m + n = 0 \quad \text{or} \quad m:n = -1:2.$$

Hence the coordinates of P are (0, 7, 2).

2.1.6 Cylindrical Polar Coordinates

The polar coordinate system described in Section 2.1.3 can be generalized to three dimensions by choosing as the first-two coordinates of a point the polar coordinates of the projection of the point on the polar plane, and as the third coordinate the number indicating the distance of the point from the polar plane. Let P be a point in space and let M be its projection in the polar plane XOY. Let (r, θ) be the polar coordinates of M and let $MP = z$ (see Fig. 2.8). Then the cylindrical polar coordinates of P are given by (r, θ, z).

If x, y, z are the cartesian coordinates of P referred to OX, OY and OZ as axes, the following formulae give the relations between x, y, z and r, θ, z:

$$x = r\cos\theta, \quad y = r\sin\theta, \quad z = z \tag{2.10}$$

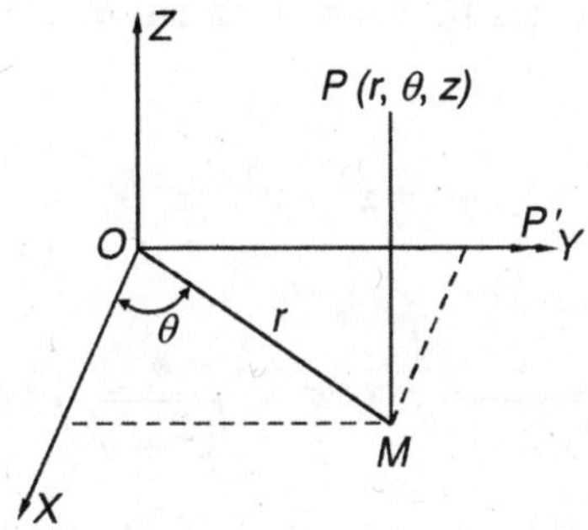

Figure 2.8

2.1.7 Spherical Polar Coordinates

Let P be the point in space and let M be the foot of the perpendicular from P on the plane XOY. Let $OP = r$, $X\hat{O}M = \theta$ and $Z\hat{O}P = \phi$ (see Fig. 2.9). Then r, θ, ϕ are called the *spherical polar coordinates* of P. It is clear that

$$OM = r\sin\phi.$$

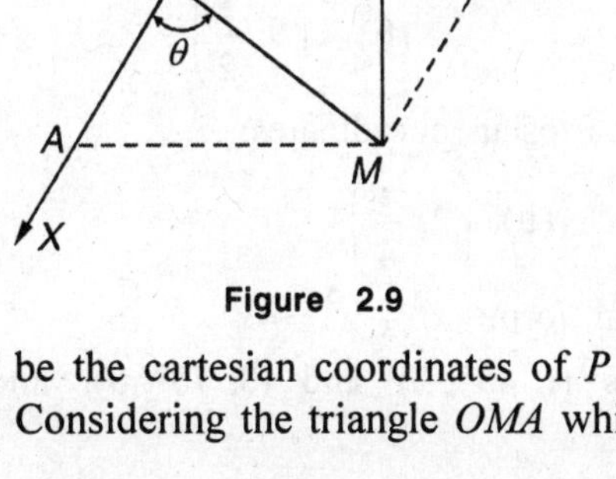

Figure 2.9

Now, let x, y and z be the cartesian coordinates of P with respect to OX, OY and OZ as axes. Considering the triangle OMA which is right angled at A, we obtain

$$OA = x = r \sin \phi \cos \theta.$$

Similarly, from $\Delta\, OMB$, we have

$$OB = y = r \sin \phi \sin \theta.$$

Finally,

$$MP = z = ON = r \cos \phi.$$

Hence, we have the following formulae of transformation:

$$x = r \sin \phi \cos \theta,\ y = r \sin \phi \sin \theta,\ z = r \cos \phi \tag{2.11}$$

It is easily seen that

$$r^2 = x^2 + y^2 + z^2$$

where

$$\sin \theta = \frac{y}{\sqrt{x^2 + y^2}},\ \cos \theta = \frac{x}{\sqrt{x^2 + y^2}},\ \cos \phi = \frac{z}{r} \tag{2.12}$$

It may be verified that if a point (r, θ, ϕ) varies in the interior of a sphere whose centre is at the origin and radius a, then r varies from 0 to a, θ varies from 0 to π and ϕ varies from 0 to 2π.

EXERCISES 2.1

1. Find the area of the triangle formed by the points:
   (a) (2, 3), (3, − 1) and (−4, 2)
   (b) (1, 4), (3, − 2) and (−3, 16)
   (c) $(a, b + c)$, $(c, b - c)$ and $(-a, c)$.
2. Find the area of the quadrilateral whose vertices are (1, 1), (3, 4), (5, −2) and (4, −7).
3. Show that the three points (4, 2), (7, 5) and (9, 7) lie on a straight line.
4. Transform to polar coordinates:

   (a) (5, 4) (b) $\left(9, \frac{5}{2}\right)$ (c) (6, −1).
5. Transform to cartesian coordinates:

   (a) (10, 30°) (b) $\left(2, \frac{\pi}{4}\right)$ (c) (10, 75°).
6. Change to polar form: $x^2 + y^2 = 4x$.
7. Plot the points (1, 45°, 2) and (3, 75°, 5), and find the distance between them.

8. Derive equations for transferring from rectangular to cylindrical polar coordinates.

9. Write rectangular coordinates for the points whose spherical polar coordinates are:

(a) $\left(1, \frac{\pi}{4}, \frac{\pi}{3}\right)$ (b) $(3, \pi, \pi)$.

10. Transform the equations: (a) $x^2 + y^2 + z^2 = 2z$ (b) $x^2 + y^2 - 3z^2 = 0$, by using spherical polar coordinates.

11. Change to spherical polar and cylindrical polar coordinates:

(a) $x^2 + y^2 = 5$ (b) $3x - 4y + 5z = 1$

(c) $x^2 + y^2 + 2z^2 = 4$ (d) $x^2 - z^2 = 4$.

12. Prove that the distance between two points in a plane with polar coordinates (r_1, θ_1) and (r_2, θ_2) is given by

$$d^2 = r_1^2 + r_2^2 - 2r_1r_2 \cos (\theta_1 - \theta_2).$$

13. Transform the equation $x^2 + y^2 = x$ to cylindrical coordinates.

14. Prove that $r = \sqrt{(x^2 + y^2)}$ is the shortest distance from the point (x, y, z) to the z-axis.

15. Describe the locus of the equation $x^2 + 6xy + y^2 + z^2 = 1$ in cylindrical and spherical polar coordinates.

16. In spherical polar coordinates, a certain surface is described by the equation $\rho = 2 \cos \phi$. Find its equation in cartesian coordinates and cylindrical coordinates.

17. Write down the rectangular coordinates of the two unit vectors with spherical polar angles (θ_1, ϕ_1) and (θ_2, ϕ_2). If α is the angle between the two unit vectors, show that

$$\cos \alpha = \cos \theta_1 \cos \theta_2 + \sin \theta_1 \sin \theta_2 \cos (\phi_1 - \phi_2).$$

18. Two points have spherical polar coordinates (r_1, θ_1, ϕ_1) and (r_2, θ_2, ϕ_2). Show that the distance d between the points is given by

$$d^2 = r_1^2 + r_2^2 - 2r_1r_2 [\cos \theta_1 \cos\theta_2 + \sin \theta_1 \sin \theta_2 \cos (\phi_1 - \phi_2)].$$

Simplify this formula by using the result of Q.17.

2.2 ELEMENTARY COORDINATE GEOMETRY

We assume that the reader is already familiar with some of the standard results in elementary coordinate geometry. We, however, collect here a few of them for ready reference and revision.

2.2.1 The Straight Line

The equation of a straight line can be written in several forms:

(i) *Line through two points* (x_1, y_1) and (x_2, y_2):

$$y - y_1 = \frac{y_2 - y_1}{x_2 - x_1}(x - x_1)$$

(ii) *Line through a given point* (x_1, y_1) *with a given slope m:*

$$y - y_1 = m(x - x_1)$$

(iii) *Intercept form:*

$$\frac{x}{a} + \frac{y}{b} = 1,$$

where a and b are the lengths of the intercepts made by the line on OX and OY respectively.

(iv) *Slope-intercept form:*

$$y = mx + c;$$

where m is the slope and c is the intercept on the y-axis.

(v) *Perpendicular form:*

$$x \cos\alpha + y \sin\alpha = p,$$

where α is the angle made by the line with the x-axis and p is the length of the perpendicular on it from the origin.

(vi) *General form:* The equation $ax + by + c = 0$ always represents a straight line.

To prove this, we assume two points $P(x_1, y_1)$ and $Q(x_2, y_2)$ on the locus so that

$$ax_1 + by_1 + c = 0$$

$$ax_2 + by_2 + c = 0$$

Let the point R divide PQ in the ratio $\lambda:1$. Then the coordinates of R are given by

$$x = \frac{x_1 + \lambda x_2}{1 + \lambda}, \quad y = \frac{y_1 + \lambda y_2}{1 + \lambda}$$

This point lies on the locus if

$$a\frac{x_1 + \lambda x_2}{1 + \lambda} + b\frac{y_1 + \lambda y_2}{1 + \lambda} + c = 0$$

i.e. if

$$ax_1 + by_1 + c + \lambda(ax_2 + by_2 + c) = 0,$$

which is true for all values of λ. It follows that if any two points P and Q are taken on the locus, then every point R on the straight line PQ is also a point on the locus. Therefore, the locus is a straight line.

It is easily seen that all the above forms are equivalent. Thus, if the general form given by

$$ax + by + c = 0 \tag{2.13}$$

is divided by $-c$, we get

$$-\frac{a}{c}x - \frac{b}{c}y = 1 \quad \text{or} \quad \frac{x}{-c/a} + \frac{y}{-c/b} = 1$$

which is the intercept form.

Again, if we divide Eq. (2.13) by $\sqrt{(a^2 + b^2)}$, we obtain

$$\frac{a}{\sqrt{a^2 + b^2}}x + \frac{b}{\sqrt{a^2 + b^2}}y + \frac{c}{\sqrt{a^2 + b^2}} = 0.$$

If we put

$$\frac{a}{\sqrt{a^2 + b^2}} = \cos\alpha, \quad \frac{b}{\sqrt{a^2 + b^2}} = \sin\alpha \quad \text{and} \quad p = -\frac{c}{\sqrt{a^2 + b^2}},$$

we obtain the perpendicular form.

Angle between two lines

Let the two lines be given by $y = m_1x + c_1$ and $y = m_2x + c_2$ and let they make angles θ_1 and θ_2 with x-axis (Fig. 2.10). Now,

$$m_1 = \tan\theta_1 \quad \text{and} \quad m_2 = \tan\theta_2$$

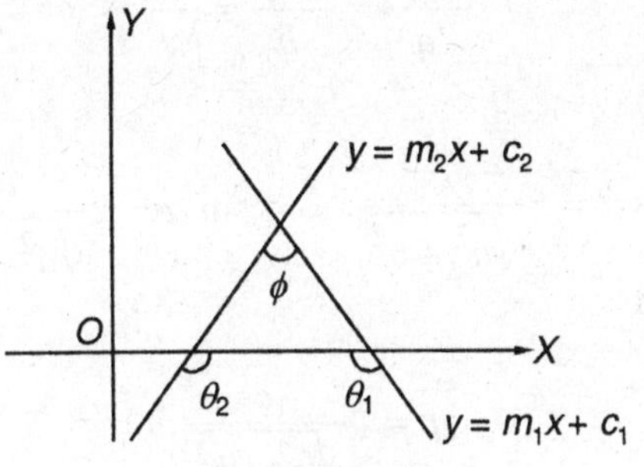

Figure 2.10

If ϕ is the angle between the two lines, we have

$$\phi = \theta_1 - \theta_2 \quad \text{(see Fig. 2.10)},$$

and so,

$$\tan\phi = \tan(\theta_1 - \theta_2) = \frac{\tan\theta_1 - \tan\theta_2}{1 + \tan\theta_1 \tan\theta_2} = \frac{m_1 - m_2}{1 + m_1m_2}$$

It easily follows that if the two straight lines are parallel, $m_1 = m_2$ and if they are perpendicular to each other $m_1 m_2 = -1$.

Length of the perpendicular from the point (x_1, y_1) to the line given by $ax + by + c = 0$

Let P be the given point (x_1, y_1) and AB be the given line (Fig. 2.11). Let OS be the perpendicular from O on the line and let $X\hat{O}S$ be α.

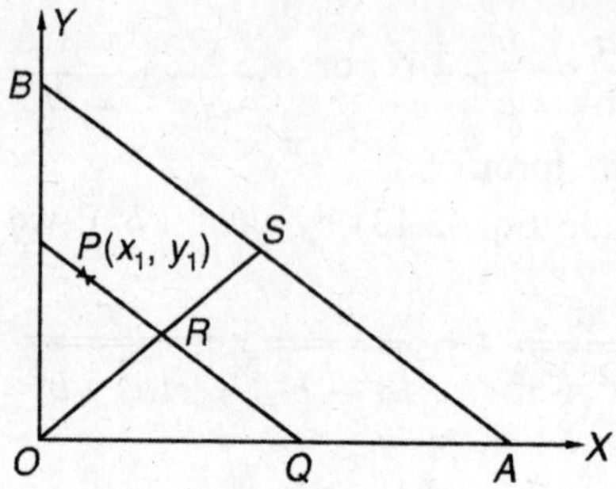

Figure 2.11

Through P draw PQ parallel to AB meeting OS in R. Let $OS = p$ and $OR = p'$. Then the equation of AB is

$$x \cos\alpha + y \sin\alpha - p = 0.$$

But its equation is given as

$$ax + by + c = 0.$$

Hence,

$$\frac{\cos\alpha}{a} = \frac{\sin\alpha}{b} = \frac{-p}{c}$$

or,

$$\cos\alpha = \pm\frac{a}{\sqrt{a^2 + b^2}}, \qquad \sin\alpha = \pm\frac{b}{\sqrt{a^2 + b^2}}$$

and

$$p = \mp\frac{c}{\sqrt{a^2 + b^2}}.$$

Now, the equation of PQ is

$$x \cos\alpha + y \sin\alpha - p' = 0.$$

Since it passes through (x_1, y_1),

$$x_1 \cos\alpha + y_1 \sin\alpha - p' = 0$$

or,

$$p' = x_1 \cos \alpha + y_1 \sin \alpha.$$

Let d be the length of the perpendicular from (x_1, y_1) on AB. Clearly,

$$d = p - p'$$

$$= \mp \frac{c}{\sqrt{a^2 + b^2}} \mp \frac{a}{\sqrt{a^2 + b^2}} x_1 \mp \frac{b}{\sqrt{a^2 + b^2}} y_1$$

$$= \mp \frac{ax_1 + by_1 + c}{\sqrt{a^2 + b^2}} \tag{2.14}$$

Pair of straight lines

We now deduce the condition that the general equation of the second degree should represent a pair of straight lines. Let the general equation of the second degree be

$$ax^2 + 2hxy + by^2 + 2gx + 2fy + c = 0 \tag{2.15}$$

Arranging the equation as a quadratic in x, we get

$$ax^2 + 2(hy + g)x + (by^2 + 2fy + c) = 0,$$

whose solution is given by

$$x = \frac{-(hy + g) \pm \sqrt{(hy + g)^2 - a(by^2 + 2fy + c)}}{a}$$

or

$$ax + hy + g = \pm\sqrt{(hy + g)^2 - a(by^2 + 2fy + c)}$$

$$= \pm\sqrt{(h^2 - ab)\,y^2 + 2(gh - af)\,y + (g^2 - ac)}$$

If this is to represent a pair of straight lines, the expression under the radical sign should be a perfect square and the condition for this is

$$(h^2 - ab)(g^2 - ac) = (gh - af)^2$$

Simplifying and dividing by a, we get

$$abc + 2fgh - af^2 - bg^2 - ch^2 = 0 \tag{2.16}$$

which is the required condition. It is easily seen that the above condition may be written as

$$\begin{vmatrix} a & h & g \\ h & b & f \\ g & f & c \end{vmatrix} = 0 \tag{2.17}$$

The angle θ between the pair of straight lines given by $ax^2 + 2hxy + by^2 = 0$ may easily be obtained as:

$$\tan\theta = \pm\frac{2\sqrt{h^2 - ab}}{a+b},$$

where the positive sign gives the acute angle between the lines.

Example 2.5 Find the equation of a straight line which passes through the point (–2, –7) and has the length 3 intercepted on it by the parallel straight lines $4x + 3y = 12$ and $4x + 3y = 3$. How many such lines are there?

Let one such line be given by

$$\frac{x+2}{\cos\theta} = \frac{y+7}{\sin\theta} = r,$$

so that the coordinates of any point on this line are of the form

$$x = -2 + r\cos\theta \quad \text{and} \quad y = -7 + r\sin\theta.$$

If this straight line meets the parallel lines (say at A and B) at distances r_1 and r_2 from (–2, –7), we have

$$r_1 - r_2 = 3,$$

and A and B are given by

$$(-2 + r_1\cos\theta,\ -7 + r_1\sin\theta) \quad \text{and} \quad (-2 + r_2\cos\theta,\ -7 + r_2\sin\theta),$$

respectively. Since A lies on the line $4x + 3y = 12$, we must have

$$4(-2 + r_1\cos\theta) + 3(-7 + r_1\sin\theta) = 12 \tag{i}$$

Again, since B lies on the line $4x + 3y = 3$, we have

$$4(-2 + r_2\cos\theta) + 3(-7 + r_2\sin\theta) = 3 \tag{ii}$$

Subtracting (ii) from (i), we obtain

$$4(r_1 - r_2)\cos\theta + 3(r_1 - r_2)\sin\theta = 9$$

Since $r_1 - r_2 = 3$, the above equation becomes

$$12\cos\theta + 9\sin\theta = 9 \quad \text{or} \quad 4\cos\theta + 3\sin\theta = 3,$$

the solution to which is $\theta = \pi/2$ or $\tan\theta = -7/24$.

Hence the two lines are given by

$$x + 2 = 0 \quad \text{and} \quad y + 7 = -\frac{7}{24}(x+2)$$

or

$$x + 2 = 0 \quad \text{and} \quad 7x + 24y + 182 = 0.$$

Example 2.6 Find the coordinates of the orthocentre of the triangle whose vertices are the points (1, − 2), (3, 1) and (−2, 3).

The orthocentre is the point of intersection of the altitudes *AD*, *BE* and *CF* (Fig. 2.12). The slope of *BC* is 2/(−5). Hence the slope of *AD* is 5/2.

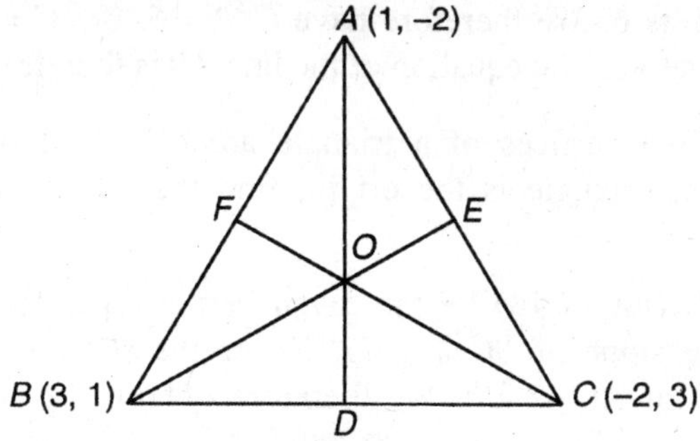

Figure 2.12

The equation of *AD* is given by

$$y+2=\frac{5}{2}(x-1)$$

or

$$5x-2y=9 \tag{i}$$

Similarly, the equation of *BE* is obtained as

$$y-1=\frac{3}{5}(x-3)$$

or

$$3x-5y=4 \tag{ii}$$

The point of intersection of *AD* and *BE* is then obtained by solving the two equations (i) and (ii). This point is found to be (37/19, 7/19) which is the required orthocentre.

Example 2.7 A straight line *AB* is perpendicular to the line $3x-y=5$. The area of the triangle formed by the line *AB* and the coordinate axes is 6. Find the equation of the line *AB*.

The given line is $3x-y=5$. Equation of any line *AB* which is perpendicular to this line may be written as

$$y=-\frac{1}{3}x+C^1,$$

or

$$x+3y=C \tag{i}$$

The intercepts of (i) are C and $C/3$. Hence the area A of the triangle formed by these intercepts and the line AB is

$$A = \frac{1}{2} \times C \times \frac{C}{3} = \frac{C^2}{6}.$$

But this is given as 6. We therefore have $C^2/6 = 6$, or, $C = -6$, as both the intercepts are negative. The equation of the line AB is therefore $x + 3y + 6 = 0$.

Example 2.8 Two vertices of a triangle are (3, –1) and (–2, 2). If the orthocentre of the triangle is the origin, find the coordinates of the third vertex.

Let B and C (Fig. 2.13) be the given vertices and let $A(h, k)$ be the third vertex. The slope of BC is $-3/5$. Hence the slope of AD is 5/3. But the slope of AD or AO is k/h. We therefore obtain

$$\frac{k}{h} = \frac{5}{3},$$

or

$$5h - 3k = 0 \tag{i}$$

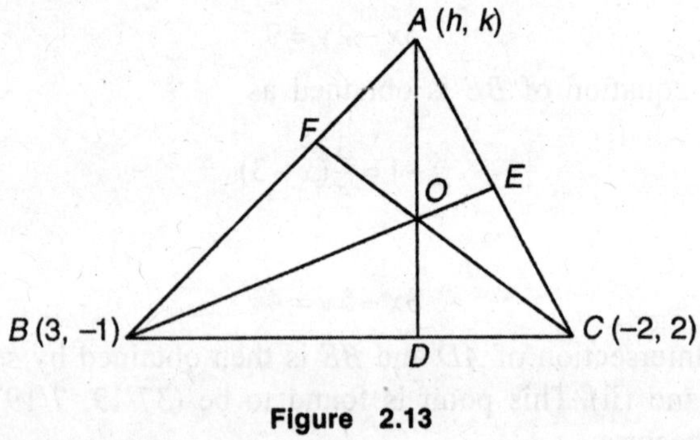

Figure 2.13

Again, slope of BO or BE is $-1/3$, and slope of CA is $(k-2)/(h+2)$. Hence

$$\left(\frac{k-2}{h+2}\right)\left(-\frac{1}{3}\right) = -1,$$

which gives

$$k - 2 = 3(h + 2) \tag{ii}$$

From (i) and (ii), we obtain $h = -6$ and $k = -10$. Hence the third vertex is $(-6, -10)$.

Example 2.9 A variable straight line, drawn through the point of intersection of the straight lines

$$\frac{x}{a} + \frac{y}{b} = 1 \quad \text{and} \quad \frac{x}{b} + \frac{y}{a} = 1,$$

meets the coordinate axes at A and B. Find the locus of the midpoint of AB.
Let C be the point of intersection of the lines. Then,

$$C \equiv \left(\frac{ab}{a+b}, \frac{ab}{a+b} \right)$$

Let M (h, k) be the midpoint of AB (see Fig. 2.14). Then, the equation of AB may be written as

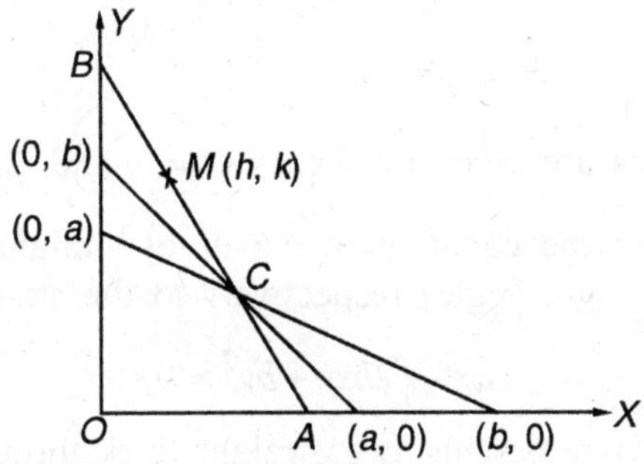

Figure 2.14

$$\frac{x}{2h} + \frac{y}{2k} = 1.$$

Since it passes through C, we have

$$\frac{ab}{2h(a+b)} + \frac{ab}{2k(a+b)} = 1 \quad \text{or} \quad \frac{1}{h} + \frac{1}{k} = \frac{2(a+b)}{ab}.$$

The locus of M (h, k) is therefore

$$\frac{1}{x} + \frac{1}{y} = \frac{2(a+b)}{ab} \quad \text{or} \quad 2xy(a+b) = ab(x+y).$$

Example 2.10 Show that $4x^2 + 4xy - 3y^2 - 8x + 8y - 5 = 0$ represents a pair of straight lines and find the equations of the lines.

In this case, $a = 4$, $h = 2$, $b = -3$, $g = -4$, $f = 4$ and $C = -5$. The condition (2.17) gives:

$$\begin{vmatrix} 4 & 2 & -4 \\ 2 & -3 & 4 \\ -4 & 4 & -5 \end{vmatrix} = 0,$$

which is satisfied, since the value of the determinant is found to be zero. Hence, the given equation represents a pair of lines. We now write the equation as

$$4x^2 + 4x(y-2) - (3y^2 - 8y + 5) = 0.$$

Hence,

$$x = \frac{-4(y-2) \pm \sqrt{16(y-2)^2 + 16(3y^2 - 8y + 5)}}{8}$$

or

$$2x + y - 2 = \pm\sqrt{y^2 - 4y + 4 + 3y^2 - 8y + 5}$$

$$= \pm\sqrt{4y^2 - 12y + 9}$$

$$= \pm(2y - 3).$$

Hence, the two lines are given by $2x - y + 1 = 0$ and $2x + 3y - 5 = 0$.

Example 2.11 Find the equation of a pair of straight lines through the origin which are at right angles respectively to the straight lines:

$$ax^2 + 2hxy + by^2 = 0. \qquad \text{(i)}$$

The equation (i) represents two straight lines through the origin. Let these lines be given by the equations:

$$y = m_1 x \quad \text{and} \quad y = m_2 x. \qquad \text{(ii)}$$

Then, we have

$$(y - m_1 x)(y - m_2 x) = 0$$

or

$$y^2 - (m_1 + m_2)x + m_1 m_2 x^2 = 0 \qquad \text{(iii)}$$

Comparing (i) and (iii), we obtain

$$m_1 + m_2 = -\frac{2h}{b} \quad \text{and} \quad m_1 m_2 = \frac{a}{b} \qquad \text{(iv)}$$

Now, the equations of the lines passing through the origin and perpendicular to the lines in (ii) are given by

$$x + m_1 y = 0 \quad \text{and} \quad x + m_2 y = 0.$$

or

$$x^2 + (m_1 + m_2)xy + m_1 m_2 y^2 = 0 \qquad \text{(v)}$$

Substituting for $(m_1 + m_2)$ and $m_1 m_2$ from (iv) in (v), we obtain

$$x^2 - \frac{2h}{b}xy + \frac{a}{b}y^2 = 0, \quad \text{or} \quad ay^2 - 2hxy + bx^2 = 0,$$

which is the required equation of the pair of lines.

Example 2.12 Show that the straight line $ax + by + c = 0$ and the straight lines $(ax + by)^2 - 3(bx - ay)^2 = 0$ form the sides of an equilateral triangle.

The given lines are

$$(ax+by)^2-3(bx-ay)^2=0 \tag{i}$$

and

$$ax+by+c=0 \tag{ii}$$

Equation (i) may be rewritten as:

$$x^2(a^2-3b^2)+8abxy+y^2(b^2-3a^2)=0 \tag{iii}$$

If θ is the angle between the lines in (iii), then we have

$$\tan\theta=\pm\frac{\sqrt{64a^2b^2-4(a^2-3b^2)(b^2-3a^2)}}{-2(a^2+b^2)}$$

$$=\pm\frac{\sqrt{3}\sqrt{2a^2b^2+b^4+a^4}}{a^2+b^2}$$

$$=\pm\sqrt{3},$$

which gives $\theta=60°$ or $120°$. Thus the acute angle between the lines (i) is $60°$. Now, the separate equations of the lines in (iii) may be written as:

$$(a+\sqrt{3}b)x+(b-\sqrt{3}a)y=0 \tag{iv}$$

and

$$(a-\sqrt{3}b)x+(b+\sqrt{3}a)y=0 \tag{v}$$

Let the angle between (ii) and (iv) be θ_1. Then

$$\tan\theta_1=\pm\frac{a(b-\sqrt{3}a)-b(a+\sqrt{3}b)}{a(\sqrt{3}b+a)+b(b-\sqrt{3}a)}=\pm\frac{-\sqrt{3}(a^2+b^2)}{a^2+b^2}$$

Therefore, the acute angle between the lines (ii) and (iv) is $60°$. It follows that the three lines form the sides of an equilateral triangle.

2.2.2 The Circle

The simplest equation of the circle is given by

$$x^2+y^2=a^2$$

with radius a and centre at the origin. The equation of the circle with radius a and centre at (g, f) is given by

$$(x-g)^2+(y-f)^2=a^2,$$

which is often written as

$$x^2+y^2-2gx-2fy+c=0, \quad \text{where} \quad c=g^2+f^2-a^2.$$

Thus, for example, the circle $x^2+y^2+2x-4y-4=0$ can be written as

$$(x+1)^2+(y-2)^2=9.$$

Hence, its centre is (–1, 2) and radius 3 units. The general equation of the second degree given by Eq. (2.16) represents a circle if $a=b$ and $h=0$.

Tangent at a point

Let the straight line

$$y=mx+c$$

touches the circle

$$x^2+y^2=a^2.$$

The points of intersection are given by

$$x^2+(mx+c)^2=a^2 \quad \text{or} \quad x^2(1+m^2)+2mxc+(c^2-a^2)=0,$$

the roots of which are

$$x=\frac{-2mc\pm\sqrt{4m^2c^2-4(1+m^2)(c^2-a^2)}}{2(1+m^2)}$$

Since the line touches the circle, the two points of intersection will coincide and therefore we must have from the above

$$m^2c^2-(1+m^2)(c^2-a^2)=0 \quad \text{or} \quad c^2=a^2(1+m^2)$$

Hence,

$$x=-\frac{mc}{1+m^2} \quad \text{and} \quad y=\frac{c}{1+m^2},$$

If we require the tangent at (x_1, y_1), we have then

$$x_1=-\frac{mc}{1+m^2} \quad \text{and} \quad y_1=\frac{c}{1+m^2},$$

which give

$$x_1=-my_1 \quad \text{or} \quad m=-\frac{x_1}{y_1},$$

and so

$$y_1=\frac{c}{1+x_1^2/y_1^2}=\frac{y_1^2c}{a^2},$$

as $x_1^2+y_1^2=a^2$. Hence,

$$c = \frac{a^2}{y_1}.$$

The equation of the tangent at (x_1, y_1) is therefore given by

$$y = -\frac{x_1}{y_1}x + \frac{a^2}{y_1} \quad \text{or} \quad xx_1 + yy_1 = a^2.$$

Again, the slope of the normal is y_1/x_1 and hence, equation of the normal at (x_1, y_1) is given by

$$y - y_1 = \frac{y_1}{x_1}(x - x_1) = \frac{y_1}{x_1}x - y_1$$

or

$$y = \frac{y_1}{x_1}x$$

Pair of tangents

Let the equation of the circle be given by

$$S = x^2 + y^2 + 2gx + 2fy + c = 0$$

and let $O(x_1, y_1)$ be any point. We wish to find the equation of the pair of tangents to the circle from O. Now, any line through the point O may be written as

$$\frac{x - x_1}{\cos\theta} = \frac{y - y_1}{\sin\theta} = r \tag{2.18}$$

and any point on this line is $(x_1 + r\cos\theta,\ y_1 + r\sin\theta)$. The points P and Q, where this line meets the circle, are given by

$$(x_1 + r\cos\theta)^2 + (y_1 + r\sin\theta)^2 + 2g(x_1 + r\cos\theta) + 2f(y_1 + r\sin\theta) + c = 0$$

or

$$r^2 + 2r[(x_1 + g)\cos\theta + (y_1 + f)\sin\theta] + x_1^2 + y_1^2 + 2gx_1 + 2fy_1 + c = 0$$

If OPQ is a tangent to the circle, then the above equation must have equal roots. The condition for this is:

$$[(x_1 + g)\cos\theta + (y_1 + f)\sin\theta]^2 = x_1^2 + y_1^2 + 2gx_1 + 2fy_1 + c$$

or

$$[(x_1 + g) + (y_1 + f)\tan\theta]^2 = (x_1^2 + y_1^2 + 2gx_1 + 2fy_1 + c)\sec^2\theta.$$

Substituting for $\tan\theta$ from Eq. (2.18) and simplifying, the above equation becomes:

$$[xx_1 + yy_1 + g(x + x_1) + f(y + y_1) + c]^2 = (x^2 + y^2 + 2gx + 2fy + c)$$
$$\times (x_1^2 + y_1^2 + 2gx_1 + 2fy_1 + c)$$

This equation is usually written as

$$T^2 = SS_1 \tag{2.19a}$$

where

$$T = xx_1 + yy_1 + g(x + x_1) + f(y + y_1) + c,$$
$$S = x^2 + y^2 + 2gx + 2fy + c$$
$$S_1 = x_1^2 + y_1^2 + 2gx_1 + 2fy_1 + c.$$

Radical axis of two circles

This is defined as the locus of a point from which the tangents to the circles are of equal length. Let the two circles be given by

$$S_1 = x^2 + y^2 + 2g_1x + 2f_1y + c_1 = 0$$

and

$$S_2 = x^2 + y^2 + 2g_2x + 2f_2y + c_2 = 0$$

Let $P(x_1, y_1)$ be any point from which the lengths of the tangents to the two circles are equal. Now,

$$(PT_1)^2 = x_1^2 + y_1^2 + 2g_1x_1 + 2f_1y_1 + c_1$$

and

$$(PT_1)^2 = x_1^2 + y_1^2 + 2g_1x_1 + 2f_1y_1 + c_2.$$

Since $PT_1 = PT_2$, we have

$$x_1^2 + y_1^2 + 2g_1x_1 + 2f_1y_1 + c_1 - (x_1^2 + y_1^2 + 2g_2x_1 + 2f_2y_1 + c_2) = 0,$$

which is linear. This shows that the point (x_1, y_1) lies on the line given by

$$S_1 - S_2 = 0 \tag{2.19b}$$

where

$$S_1 = x^2 + y^2 + 2g_1x + 2f_1y + c_1$$
$$S_2 = x^2 + y^2 + 2g_2x + 2f_2y + c_2.$$

This line is called the *radical axis* of the circles and it is easy to see that the radical axis is perpendicular to the line joining the centres of the circles.

Orthogonal circles

When two circles intersect at a point such that their tangents at this point are at right angles to each other, then the circles are called *orthogonal circles* and they are said to cut one another orthogonally.

To find the condition of orthogonality, let the equations of the circles be

$$x^2 + y^2 + 2g_1x + 2f_1y + c_1 = 0$$

and

$$x^2 + y^2 + 2g_2x + 2f_2y + c_2 = 0.$$

The centres of the circles are $(-g_1, -f_1)$ and $(-g_2, -f_2)$, and their radii are $\sqrt{(g_1^2 + f_1^2 - c_1)}$ and $\sqrt{(g_2^2 + f_2^2 - c_2)}$ respectively. Since the circles cut orthogonally, the tangents at the point of intersection are at right angles, i.e. the sum of the squares of the radii is equal to the square on the line joining their centres. This gives:

$$(-g_1 + g_2)^2 + (-f_1 + f_2)^2 = g_1^2 + f_1^2 - c_1 + g_2^2 + f_2^2 - c_2;$$

which becomes on simplification

$$2g_1g_2 + 2f_1f_2 = c_1 + c_2 \tag{2.20}$$

Coaxal circles

A system of circles in which every pair of circles has the same radical axis is called a *coaxal system.* We know that the line joining the centres of two circles is perpendicular to the radical axis. From this, it follows that the centres of all the circles of the system lie on a straight line which is perpendicular to the common radical axis. Hence the equations of the circles of the system may be written as

$$\left.\begin{aligned} x^2 + y^2 + 2g_1x + c_1 = 0 \\ x^2 + y^2 + 2g_2x + c_2 = 0 \end{aligned}\right\} \tag{2.21}$$

Such a system is represented by the equation

$$x^2 + y^2 + 2\lambda x + c = 0, \tag{2.22}$$

where c is a constant and λ is a variable.

Example 2.13 Find the equation of the circle that passes through the three points $A(x_1, y_1)$, $B(x_2, y_2)$ and $C(x_3, y_3)$.

Let the required equation of the circle be

$$(x^2 + y^2) + 2gx + 2fy + c = 0 \tag{i}$$

Since it passes through A, B and C, we must have

$$(x_1^2 + y_1^2) + 2gx_1 + 2fy_1 + c = 0 \tag{ii}$$

$$(x_2^2 + y_2^2) + 2gx_2 + 2fy_2 + c = 0 \tag{iii}$$

$$(x_3^2 + y_3^2) + 2gx_3 + 2fy_3 + c = 0 \tag{iv}$$

Eliminating g, f and c from equations (i) to (iv), we obtain

$$\begin{vmatrix} x^2+y^2 & x & y & 1 \\ x_1^2+y_1^2 & x_1 & y_1 & 1 \\ x_2^2+y_2^2 & x_2 & y_2 & 1 \\ x_3^2+y_3^2 & x_3 & y_3 & 1 \end{vmatrix} = 0,$$

which is the required equation of the circle.

Example 2.14 A circle passing through the points (1, 1) and (1, 9) touches the x-axis and coordinates of its centre are both positive. Find its equation and the length of its intercept on the y-axis.

Let the equation of the circle be

$$x^2 + y^2 + 2gx + 2fy + c = 0 \tag{i}$$

Since it passes through (1, 1) and (1, 9), we have

$$2 + 2g + 2f + c = 0 \tag{ii}$$

and

$$82 + 2g + 18f + c = 0 \tag{iii}$$

The points of intersection of the circle with the axis are given by

$$x^2 + 2gx + c = 0.$$

Since the circle touches the x-axis, the two roots of this equation must be equal and therefore

$$g^2 = c. \tag{iv}$$

Solving equations (ii), (iii) and (iv), we obtain

$$f = -5, \quad g = 2, \quad c = 4$$

or

$$f = -5, \quad g = -4, \quad c = 16.$$

The equation of the circle is therefore given by

$$x^2 + y^2 - 8x - 10y + 16 = 0.$$

To find its intercept on the y-axis, we put $x = 0$ and obtain:

$$y^2 - 10y + 16 = 0,$$

which gives $y = 2$ or $y = 8$. Hence the required intercept on the y-axis $= 8 - 2 = 6$.

Example 2.15 Tangents are drawn to the circles $x^2 + y^2 = a^2$, $x^2 + y^2 = b^2$ at right angles to each another. Find the locus of their point of intersection.

The equation of any tangent to the circle $x^2 + y^2 = a^2$ is given by

$$y = m_1 x + a\sqrt{1 + m_1^2} \tag{i}$$

Similarly, the equation of any tangent to the circle $x^2 + y^2 = b^2$ is given by

$$y = m_2 x + b\sqrt{1 + m_2^2} \tag{ii}$$

Since the two tangents are at right angles, we have

$$m_1 m_2 = -1 \quad \text{or} \quad m_2 = -\frac{1}{m_1}$$

Equation (ii) therefore becomes:

$$y = -\frac{1}{m_1}x + b\sqrt{1 + \frac{1}{m_1^2}}$$

Let (x_1, y_1) be the point of intersection of the two tangents, so that we have

$$y_1 = m_1 x_1 + a\sqrt{1 + m_1^2}$$

and

$$y_1 = -\frac{1}{m_1}x_1 + b\sqrt{1 + \frac{1}{m_1^2}}$$

The above equations can be written as

$$y_1 - m_1 x = a\sqrt{1 + m_1^2}$$

and

$$m_1 y_1 = x_1 = b\sqrt{1 + m_1^2}$$

Squaring and adding, we obtain $x_1^2 + y_1^2 = a^2 + b^2$. Hence, the locus of (x_1, y_1) is given by $x^2 + y^2 = a^2 + b^2$, which is a circle with radius $\sqrt{(a^2 + b^2)}$ and centre at the origin.

Example 2.16 Find the equations of the circles passing through $(-4, 3)$ and touching the lines $x + y = 2$ and $x - y = 2$.

Let AB and AC be the straight lines $x + y = 2$ and $x - y = 2$, respectively. Let $P(h, 0)$ be the centre of the circle touching AB and AC (Fig. 2.15). Then, the radius of the circle is given by

$$r = \sqrt{(h + 4)^2 + 9},$$

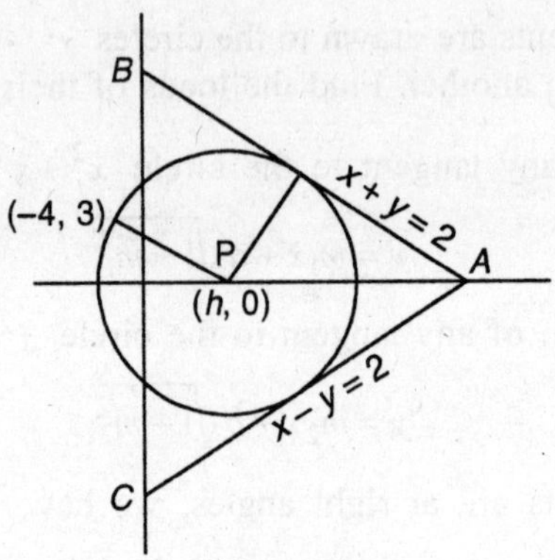

Figure 2.15

since this is the distance between the centre P and the point (–4, 3). Hence, the equation of the circle is given by

$$(x-h)^2 + y^2 = r^2 = (h+4)^2 + 9$$

Further, the perpendicular distance from P upon the line AB is given by

$$\frac{h-2}{\sqrt{2}}.$$

Since this is equal to the radius r, we must have

$$\frac{(h-2)^2}{2} = (h+4)^2 + 9.$$

Simplifying, we obtain

$$h = -10 \pm 3\sqrt{6}$$

Hence, the equation of the circles with centre $(h, 0)$ and radius r are given by

$$(x-h)^2 + y^2 = (h+4)^2 + 9, \quad \text{where } h = -10 \pm 3\sqrt{6}.$$

Example 2.17 Find the locus of the middle points of the chords of the circle $x^2 + y^2 = a^2$ which subtend a right angle at the point $(d, 0)$.

Let M (h, k) be the middle point of the chord AB, and let O be the centre of the circle (see Fig. 2.16) Then $AM = MB = MD$, where D is the point $(d, 0)$.

Also,

$$AM^2 = OA^2 - OM^2 = a^2 - (h^2 + k^2),$$

and

$$MD^2 = (h-d)^2 + k^2$$

Since $AM^2 = MD^2$, it follows that

$$a^2 - (h^2 + k^2) = (h-d)^2 + k^2,$$

which simplifies to

$$2(h^2 + k^2) - 2hd + d^2 - a^2 = 0.$$

It follows that the locus of (h, k) is given by

$$2(x^2 + y^2) - 2xd + d^2 - a^2 = 0.$$

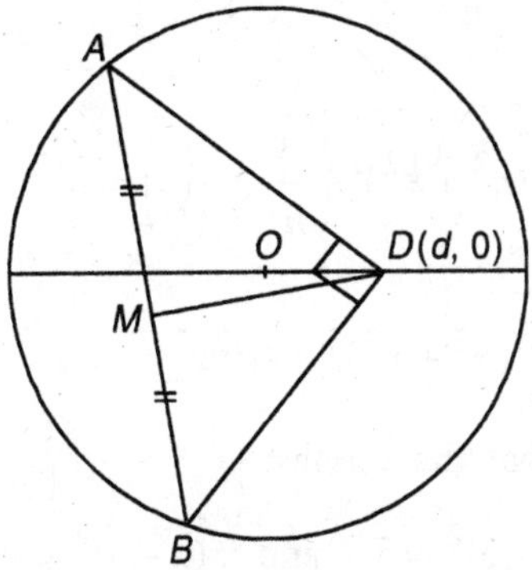

Figure 2.16

Example 2.18 A circle of constant radius d passes through the origin O and cuts the axes in two points A and B. Show that the locus of the foot of the perpendicular from O to AB is $(x^2 + y^2)^3 - 4d^2x^2y^2 = 0$.

Let the two points A and B be $(a, 0)$ and $(0, b)$, respectively. Then the equation of the circle passing through O, A and B is given by

$$x^2 + y^2 - ax - by = 0$$

Since the radius of the above circle is d, we have

$$a^2 + b^2 = 4d^2 \tag{i}$$

Now, equation of AB is $(x/a) + (y/b) = 1$

Let $M(h, k)$ be the foot of the perpendicular from O on AB. Since OM is perpendicular to AB, we have

$$\frac{k}{h}\left(-\frac{b}{a}\right) = -1 \quad \text{or} \quad \frac{h}{1/a} = \frac{k}{1/b} = r \qquad \text{(say).}$$

Hence

$$h = \frac{r}{a} \quad \text{and} \quad k = \frac{r}{b}$$

Since (h, k) lies on AB, we have

$$\frac{h}{a} + \frac{k}{b} = 1,$$

which implies that $h^2 + k^2 = r$. Hence,

$$a = \frac{h^2 + k^2}{h} \quad \text{and} \quad b = \frac{h^2 + k^2}{k}$$

Putting these values in (i), we get

$$\left(\frac{h^2 + k^2}{h}\right)^2 + \left(\frac{h^2 + k^2}{k}\right)^2 = 4d^2,$$

or

$$(h^2 + k^2)^2 \left(\frac{1}{h^2} + \frac{1}{k^2}\right) = 4d^2$$

Hence the locus is

$$(x^2 + y^2)^3 - 4d^2x^2y^2 = 0.$$

Example 2.19 Show that the circles:

$$(x-2)^2 + (y+3)^2 = 5 \quad \text{and} \quad (x-5)^2 + (y-3)^2 = 20$$

touch each other and find the coordinates of the point of contact.

The radical axis of the two circles is given by

$$6x + 12y - 6 = 0$$

or

$$x + 2y - 1 = 0 \tag{i}$$

The centre of the first circle is (2, –3) and its radius is $\sqrt{5}$ units. The perpendicular distance from (2, –3) to (i) is

$$\left|\frac{2-6-1}{\sqrt{5}}\right| = \sqrt{5} \text{ units.}$$

Since this is equal to the radius of the first circle, it follows that the radical axis touches the first circle. Hence the two circles touch each other and let (x_1, y_1) be the point of contact. The tangent to the first circle at (x_1, y_1) is

$$xx_1 + yy_1 - 2(x + x_1) + 3(y + y_1) + 8 = 0$$

or

$$(x_1 - 2)x + (y_1 + 3)y - 2x_1 + 3y_1 + 8 = 0 \tag{ii}$$

Since this is the radical axis, we compare (i) and (ii), and obtain

$$\frac{x_1 - 2}{1} = \frac{y_1 + 3}{2} = \frac{2x_1 - 3y_1 - 8}{1}.$$

We therefore have the equations:

$$x_1 - 2 = (2x_1 - 3y_1 - 8) \quad \text{and} \quad y_1 + 3 = 4x_1 - 6y_1 - 16.$$

Solving the above two equations, we obtain the point of contact: $x_1 = 3$ and $y_1 = -1$.

Example 2.20 Prove that the two circles which pass through the points $(0, d)$ and $(0, -d)$ and touch the line $y = mx + c$ will cut orthogonally if $c^2 = d^2(2 + m^2)$.

Let the equation of the circle be

$$x^2 + y^2 + 2gx + 2fy + h = 0 \qquad \text{(i)}$$

Since the circle passes through $(0, d)$ and $(0, -d)$, we have

$$d^2 + 2fd + h = 0 \quad \text{and} \quad d^2 - 2fd + h = 0.$$

Solving the above equations, we get: $h = -d^2$ and $f = 0$. Hence Eq. (i) becomes:

$$x^2 + y^2 + 2gx - d^2 = 0 \qquad \text{(ii)}$$

This circle will touch the line $y = mx + c$ if

$$\frac{-mg + c}{\sqrt{m^2 + 1}} = \sqrt{g^2 + d^2}$$

or

$$g^2 + 2mcg + d^2(1 + m^2) - c^2 = 0 \qquad \text{(iii)}$$

If g_1 and g_2 are the two roots of Eq. (iii), we have

$$g_1 + g_2 = -2mc \quad \text{and} \quad g_1 g_2 = d^2(1 + m^2) - c^2 \qquad \text{(iv)}$$

The equations of the two circles are

$$x^2 + y^2 + 2g_1 x - d^2 = 0 \quad \text{and} \quad x^2 + y^2 + 2g_2 x - d^2 = 0.$$

These two circles cut each other orthogonally if $2g_1 g_2 = -2d^2$. Using the second equation in (iv), the above becomes:

$$d^2(1 + m^2) - c^2 = -d^2 \quad \text{or} \quad c^2 = d^2(2 + m^2),$$

which is the required condition.

2.2.3 The Conic Sections

If a point moves so that its distance from a fixed point bears a constant ratio to its distance from a fixed line, the curve described is called a *conic*. Thus, if S is a fixed point (Fig. 2.17), MN is a fixed line and P is a point moving in such a way that

$$\frac{SP}{MP} = \text{constant} \qquad (2.23)$$

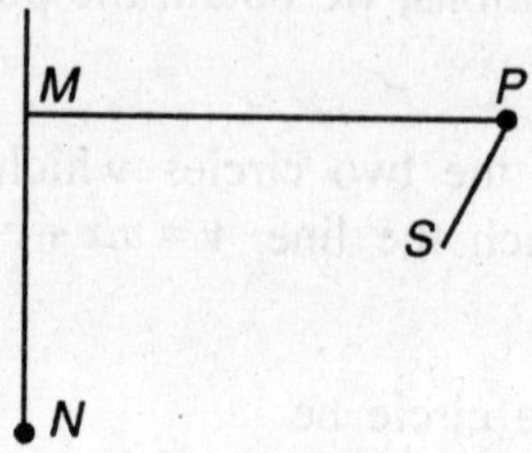

Figure 2.17

the curve traced by P is a conic. This constant is known as the *eccentricity* of the conic and is denoted by e.

If $e=1$, the conic is called a *parabola.* If $e<1$, the conic is called an *ellipse*, and if $e>1$, it is called a *hyperbola*. The fixed point S is called the *focus* of the conic and the fixed line MN is called the *directrix*.

These well-known curves were first studied as the plane sections of a right circular cone and, hence, are called *conic sections.* The equation of any conic can be obtained without difficulty. Thus, if (a, b) is the focus S and $lx+my+n=0$ is the directrix, e is the eccentricity and $P(x, y)$ is any point on the conic, we have

$$(x-a)^2+(y-b)^2=e^2\,\frac{(lx+my+n)^2}{l^2+m^2}$$

or

$$\begin{aligned}&x^2(l^2+m^2-e^2l^2)-2lme^2xy+y^2(l^2+m^2-e^2m^2)\\&-2x[a(l^2+m^2)+lne^2]-2y[b(l^2+m^2)+mne^2]\\&+[(a^2+b^2)(l^2+m^2)-e^2n^2]=0\end{aligned} \tag{2.24}$$

This is an equation of the second degree; it can be proved that every equation of the second degree can be expressed in this form and will represent a conic.

We shall now discuss the three curves in turn.

The parabola

In its simplest form, the parabola has the equation

$$y^2=4ax \tag{2.25}$$

where the axes are, as shown in Fig. 2.18(a). The focus S has coordinates $(a, 0)$ and the chord LSL' passing through S is called the *latus rectum.* The line $x+a=0$ is the *directrix*. Putting $x=a$ in (2.25), we obtain

$$y^2=4a^2 \quad \text{or} \quad y=\pm\,2a.$$

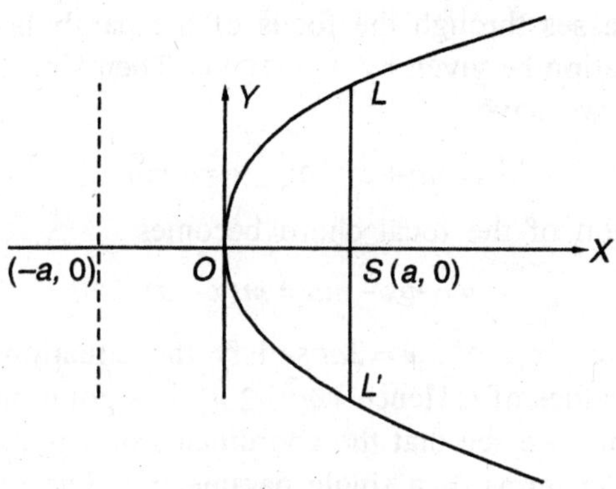

Figure 2.18(a)

Hence the length of the latus rectum LL' is $4a$.

If the line $y = mx + c$ intersects the parabola $y^2 = 4ax$, we have

$$(mx + c)^2 = 4ax$$

i.e.

$$m^2x^2 + 2x(mc - 2a) + c^2 = 0 \tag{2.26}$$

If the line is a tangent to the parabola, then the two roots of (2.26) must coincide, and hence

$$(mc - 2a)^2 = m^2c^2,$$

which gives on simplification:

$$c = \frac{a}{m}. \tag{2.27}$$

The equation of the tangent is therefore given by

$$y = mx + \frac{a}{m} \tag{2.28a}$$

and the point of intersection is given by

$$x_1 = \frac{a}{m^2} \quad \text{and} \quad y_1 = \frac{2a}{m}.$$

Therefore,

$$m = \frac{2a}{y_1},$$

and the equation of the tangent becomes

$$y = \frac{2a}{y_1}x + x_1\frac{2a}{y_1} = \frac{2a}{y_1}(x + x_1)$$

i.e.

$$yy_1 = 2a(x + x_1). \tag{2.28b}$$

A chord which passes through the focus of the parabola is called a *focal chord.* Let its equation be given by $y = mx + c$. Then since this chord passes through $S(a, 0)$, we have

$$0 = ma + c \quad \text{or} \quad c = -ma.$$

Hence, the equation of the focal chord becomes

$$y = mx - ma = m(x - a). \tag{2.29}$$

The coordinates $x = at^2$, $y = 2at$ satisfy the equation of the parabola $y^2 = 4ax$ for all values of t. Hence $(at^2, 2at)$ is a point on the parabola for all values of t. Thus we see that the coordinates of a point on the parabola can be expressed in terms of a single parameter t. The representation of a point by means of single parameter is of great importance.

If we consider two points $(at_1^2, 2at_1)$ and $(at_2^2, 2at_2)$ on the parabola $y^2 = 4ax$, then the chord joining these two points is given by

$$\frac{y - 2at_1}{2at_1 - 2at_2} = \frac{x - at_1^2}{at_1^2 - at_2^2} \quad \text{or} \quad y(t_1 + t_2) = 2x + 2at_1t_2.$$

As t_1 and t_2 coincide with the point t, the chord joining the points t_1 and t_2 will become the tangent at t. Thus the equation of the tangent at t becomes:

$$yt = x + at^2 \quad \text{or} \quad y = \frac{x}{t} + at.$$

Using the parametric representation, some important properties of the parabola can be deduced and the reader is referred to the examples and exercises given at the end of this section.

Some useful results for the parabola $y^2 = 4ax$ are given below:

(i) Equation of the normal at $(at^2, 2at)$ is $y + xt = 2at + at^3$.

(ii) Point of intersection of the tangents at $(at_1^2, 2at_1)$ and $(at_2^2, 2at_2)$ is $[at_1t_2, a(t_1 + t_2)]$.

(iii) Equation of the chord of contact of tangents from (x_1, y_1) is $yy_1 = 2a(x + x_1)$.

(iv) Equation of the chord in terms of its middle point (x_1, y_1) is $T = S_1$: $yy_1 - 2a(x + x_1) = y_1^2 - 4ax_1$.

(v) Equation of the pair of tangents from (x_1, y_1) to the parabola is $T^2 = SS_1$: $[yy_1 - 2a(x + x_1)]^2 = (y^2 - 4ax)(y_1^2 - 4ax_1)$.

(vi) Polar of a point (x_1, y_1) with respect to the parabola is $yy_1 = 2a(x + x_1)$.

(vii) Pole of the line $lx + my + n = 0$ with respect to the parabola is $(n/l, -2am/l)$.

(viii) In general, three normals can be drawn from a given point to a parabola and the sum of the ordinates of the feet of the normals is zero. The points of concurrence of the normals with the parabola are called *conormal points.*

(ix) A circle and a parabola will always cut at four points and the sum of the ordinates of the four points of intersection is zero.

(x) The equation of the circle passing through the feet of the normals from a point (h, k) is

$$x^2 + y^2 - (2a + h)x - \frac{k}{2}y = 0.$$

Different forms of the parabola If the focus is situated on the left side of the directrix, the equation of the parabola with the vertex as orgin and the axis as x-axis is $y^2 = -4ax$, since in this case the focus S is $(-a, 0)$. The latus rectum, which is a positive quantity, is equal to $4a$, as before [see Fig. 2.18(b)]. If we interchange the roles of x and y in the equations $y^2 = 4ax$, and $y^2 = 4ax$, we obtain the equations $x^2 = 4ay$, and $x^2 = -4ay$ respectively. Figures 2.18(c) and 2.18(d) show the graphs of these equations.

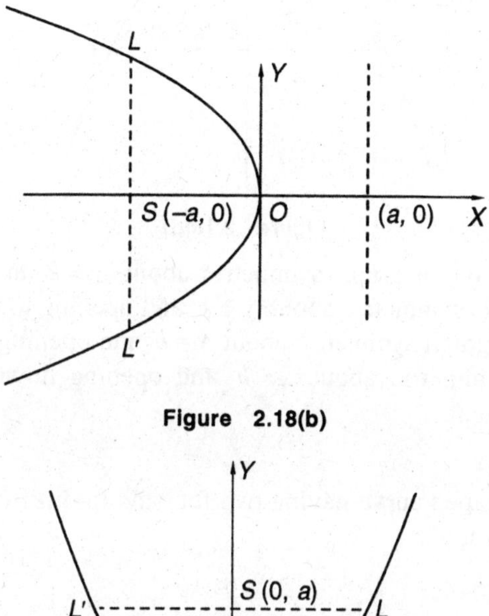

Figure 2.18(b)

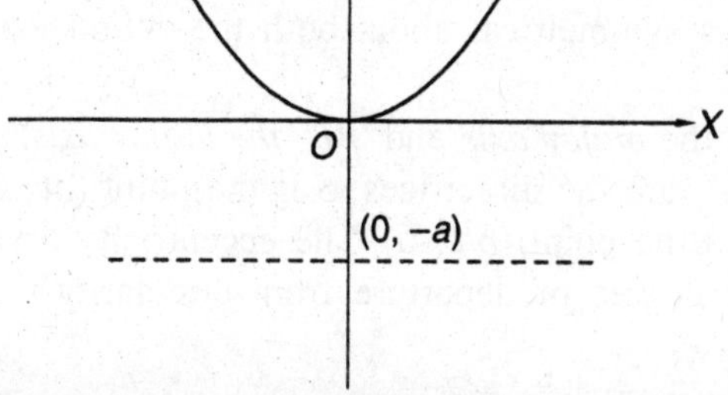

Figure 2.18(c)

Other forms of equations of parabolas are:

$$(y-k)^2 = 4a\,(x-h) \tag{2.30a}$$

$$(y-k)^2 = -4a\,(x-h) \tag{2.30b}$$

$$(x-h)^2 = 4a\,(y-k) \tag{2.30c}$$

$$(x-h)^2 = -4a\,(y-k) \tag{2.30d}$$

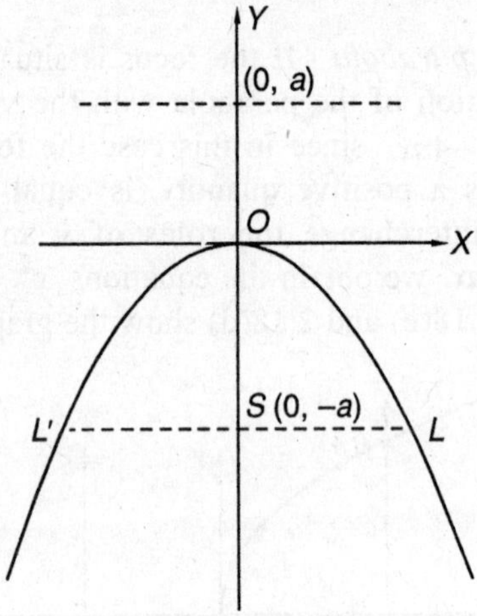

Figure 2.18(d)

Equation (2.30a) has a graph symmetric about $y = k$ and opening to the right and (2.30b) is symmetric about $y = k$ and opening to the left. Similarly, Eq. (2.30c) has a graph symmetric about $x = h$ and opening upward, whereas Eq. (2.30d) is symmetric about $x = h$ and opening downward.

The ellipse

This is an oval-shaped curve having two foci and to directrices. The simplest form of the equation is

$$\frac{x^2}{a^2} + \frac{y^2}{b^2} = 1 \tag{2.31}$$

where the curve is symmetrical about both the x- and y-axes, as shown in Fig. 2.19.

AA' is called the *major axis* and *BB'* *the minor axis*, S and S' are the foci, DC and $D'C'$ are the directrices. S is the point $(ae, 0)$ S' is the point $(-ae, 0)$ and C is the point $(a/e, 0)$. The eccentricity e varies from 0 to 1 and indicates the degree of departure from circularity.

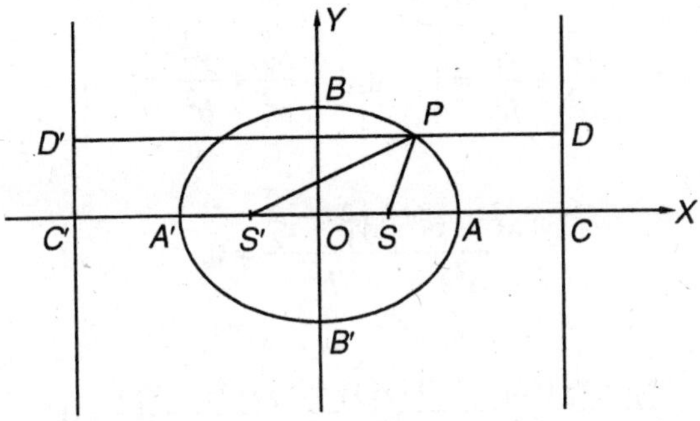

Figure 2.19

For convenience we list out some properties of the ellipse with reference to Fig. 2.20:

(i) $S'P + PS = 2a$, where P is any point on the ellipse.

(ii) $e = \sqrt{1 - b^2/a^2}$.

(iii) $PM \cdot OT = b^2$, where PM is the perpendicular from P on OX and PT is the tangent to the ellipse at P.

(iv) $OM \cdot OT = a^2$

(v) $S'N' \cdot SN = b^2$.

(vi) The locus of the point of intersection of a pair of perpendicular tangents to the ellipse is a circle called the *director circle* whose equation is given by $x^2 + y^2 = a^2 + b^2$.

To find the equation of the tangent to the ellipse at the point $P(x_1, y_1)$ on it, let $Q(x_2, y_2)$ be another point on the ellipse and is very close to P.

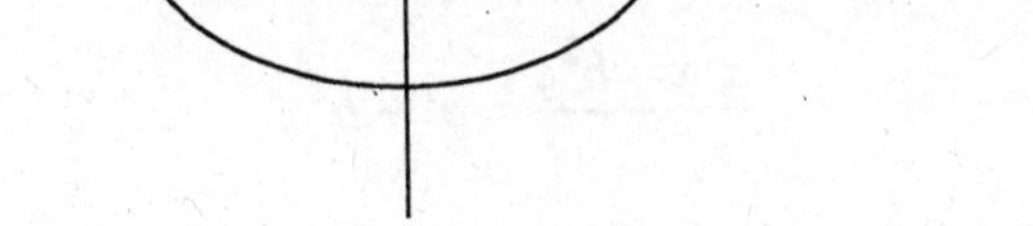

Figure 2.20

The equation to PQ is

$$y - y_1 = \frac{y_2 - y_1}{x_2 - x_1}(x - x_1).$$

Since P and Q lie on the ellipse, we have

$$\frac{x_1^2}{a^2} + \frac{y_1^2}{b^2} = 1 \quad \text{and} \quad \frac{x_2^2}{a^2} + \frac{y_2^2}{b^2} = 1.$$

By subtraction,

$$\frac{x_1^2 - x_2^2}{a^2} + \frac{y_1^2 - y_2^2}{b^2} = 0,$$

i.e.

$$\frac{(x_1 + x_2)(x_1 - x_2)}{a^2} + \frac{(y_1 + y_2)(y_1 - y_2)}{b^2} = 0,$$

from which we obtain

$$\frac{x_1 - y_2}{x_1 - x_2} = \frac{b^2}{a^2} \frac{x_1 + x_2}{y_1 + y_2}$$

Substituting the values in the equation of PQ, we get

$$y - y_1 = -\frac{b^2}{a^2} \frac{x_1 + x_2}{y_1 + y_2}(x - x_1).$$

When Q tends to P, in the limiting position, the secant PQ becomes the tangent at P and y_2 becomes equal to y_1. Hence, putting $x_2 = x_1$ and $y_2 = y_1$ in the above, we obtain:

$$y - y_1 = -\frac{b^2}{a^2} \frac{x_1}{y_1}(x - x_1),$$

Which simplifies to:

$$\frac{xx_1}{a^2} + \frac{yy_1}{b^2} = 1, \tag{2.32}$$

which is the equation of the tangent.

We list below some well-known results for the ellipse:

(i) Equation of the normal at (x_1, y_1) is

$$\frac{a^2 x}{x_1} - \frac{b^2 y}{y_1} = a^2 - b^2.$$

(ii) Equation of the chord of contact of tangents from (x_1, y_1) is

$$\frac{xx_1}{a^2} + \frac{yy_1}{b^2} = 1.$$

(iii) The straight line $y = mx + c$ is a tangent to the ellipse if $c^2 = a^2m^2 + b^2$.

(iv) Equation of the chord of the ellipse with (x_1, y_1) as its midpoint is $T = S_1$:

$$\frac{xx_1}{a^2}+\frac{yy_1}{b^2}-1=\frac{x_1^2}{a^2}+\frac{y_1^2}{b^2}-1.$$

(v) Equation of the pair of tangents from (x_1, y_1) is $T^2 = SS_1$:

$$\left(\frac{xx_1}{a^2}+\frac{yy_1}{b^2}-1\right)^2=\left(\frac{x^2}{a^2}+\frac{y^2}{b^2}-1\right)\left(\frac{x_1^2}{a^2}+\frac{y_1^2}{b^2}-1\right).$$

(vi) The point $x = a\cos\theta$, $y = b\sin\theta$ satisfies the equation

$$\frac{x^2}{a^2}+\frac{y^2}{b^2}=1$$

and is usually represented by θ.

(vii) Equation of the chord joining θ_1 and θ_2 is

$$\frac{x}{a}\cos\frac{\theta_1+\theta_2}{2}+\frac{y}{b}\sin\frac{\theta_1+\theta_2}{2}=\cos\frac{\theta_1-\theta_2}{2}. \tag{2.33}$$

(viii) Equation of the tangent at θ is

$$\frac{x\cos\theta}{a}+\frac{y\sin\theta}{b}=1. \tag{2.34}$$

(ix) Equation of the normal at θ is

$$\frac{ax}{\cos\theta}-\frac{by}{\sin\theta}=a^2-b^2. \tag{2.35}$$

(x) In general, four normals can be drawn to an ellipse from any point and if θ_1, θ_2, θ_3 and θ_4 are the eccentric angles of these four conormal points, then $\theta_1+\theta_2+\theta_3+\theta_4$ is an odd multiple of π.

(xi) A circle and an ellipse will cut at four points (called concyclic points) and the sum of the eccentric angles of these four points is an even multiple of π.

(xii) Two diameters $y = m_1x$ and $y = m_2x$ are said to be conjugate if $m_1m_2 = -b^2/a^2$.

(xiii) The tangents at the extremities of a pair of conjugate diameters of an ellipse form a parallelogram of constant area.

The hyperbola

In its simplest form, the equation of the hyperbola (Fig. 2.21) is given by

$$\frac{x^2}{a^2}-\frac{y^2}{b^2}=1, \tag{2.36}$$

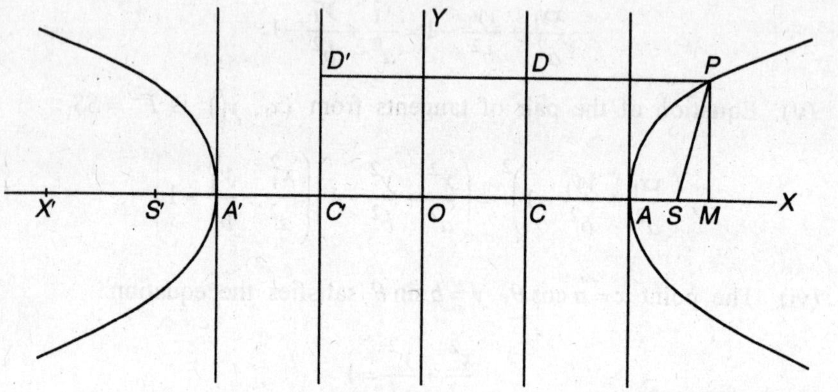

Figure 2.21

S is the focus $(ae, 0)$, S is the focus $(-ae, 0)$, DC and $D'C'$ are the directrices, C is the point $(a/e, 0)$ and C' is the point $(-a/e, 0)$. The x-axis is called the *transverse* axis of the hyperbola given by (2.36). If we take two points B and B' on YOY' such that $BO = B'O = b = \sqrt{a(e^2-1)/B}$ BOB' is called the *conjugate* axis of the hyperbola. If P (x_1, y_1) is any point on the hyperbola, we have

$$\begin{aligned} SP &= ePD \\ &= eMC \\ &= e(OM - OC) \\ &= e\left(x_1 - \frac{a}{e}\right) \\ &= ex_1 - a, \end{aligned} \tag{2.37}$$

and

$$\begin{aligned} S'P &= ePD' \\ &= eMC' \\ &= e(OM + MC') + e\left(x_1 + \frac{a}{e}\right) \\ &= ex_1 + a. \end{aligned} \tag{2.38}$$

Hence,

$$S'P - SP = 2a \tag{2.39}$$

If $b = a$, i.e. if the transverse axis is equal to the conjugate axis, the equation of the hyperbola becomes:

$$x^2 - y^2 = a^2, \tag{2.40}$$

in which case it is called a *rectangular hyperbola.* It is easy to deduce that its eccenntricity is $\sqrt{2}$.

The following properties can easily be proved:

(i) The line

$$y = mx \pm \sqrt{a^2m^2 - b^2} \tag{2.41}$$

is a tangent to the hyperbola given by (2.36).

(ii) The equation of the tangent at $(x_1 y_1)$ is given by

$$\frac{xx_1}{a^2} - \frac{yy_1}{b^2} = 1. \tag{2.42}$$

(iii) The equation of the direction circle

$$x^2 + y^2 = a^2 - b^2 \tag{2.43}$$

An important concept is that of the *asymptote* of the hyperbola, and this is discussed below:

An *asymptote* is the straight line which touches the hyperbola at infinity but does not lie altogether at infinity.

The coordinates of the point of intersection of the line $y = mx + c$ and the hyperbola $(x^2/a^2) - (y^2/b^2) = 1$ are given by

$$\frac{x^2}{a^2} - \frac{(mx+c)^2}{b^2} = 1$$

or

$$x^2(b^2 - a^2m^2) - 2ma^2cx - a^2(c^2 + b^2) = 0 \tag{2.44}$$

Since both the roots of Eq. (2.44) are at infinity, we set $x = 1/z$ and the above equation becomes:

$$a^2(c^2 + b^2)z^2 + 2ma^2cz - (b^2 - a^2m^2) = 0, \tag{2.45}$$

for which both the roots must vanish. This requires that

$$2ma^2c = 0 \quad \text{and} \quad b^2 - a^2m^2 = 0,$$

from which we obtain:

$$c = 0 \quad \text{and} \quad m = \pm\frac{b}{a} \tag{2.46}$$

Hence, the asymptotes exits and are given by

$$y = \frac{b}{a}x \quad \text{and} \quad y = -\frac{b}{a}x. \tag{2.47}$$

If the asymptotes are at right angles, we have

$$\frac{b^2}{a^2} = 1. \tag{2.48}$$

Hence, the equation of the hyperbola in this case, is

$$x^2 - y^2 = a^2 \tag{2.49}$$

which is the *rectangular hyperbola.*

Finally, for the hyperbola

$$\frac{x^2}{a^2} - \frac{y^2}{b^2} = 1, \tag{2.36}$$

we see that the point $(a \sec\theta, b \tan\theta)$, which is briefly written as q, is a point on the hyperbola for all values of q. The tangent at q is

$$\frac{x}{a}\sec\theta - \frac{y}{b}\tan\theta = 1 \tag{2.50}$$

and the equation of the normal is

$$(x - a\sec\theta)\frac{a}{\sec\theta} + (y - b\tan\theta)\frac{b}{\tan\theta} = 0,$$

or

$$\frac{ax}{\sec\theta} + \frac{by}{\tan\theta} = a^2 + b^2. \tag{2.51}$$

The equation $xy = c^2$ is the equation of a rectangular hyperbola referred to its asymptotes as axes. It may be verified that the coordinates $x = ct$ and $y = c/t$ are the parametric coordinates of the rectangular hyperbola $xy = c^2$.

Some important results concerning the hyperbola [Eq. (2.36)] are listed below:

(i) Equation of the normal at (x_1, y_1) is

$$\frac{a^2x}{x_1} + \frac{b^2y}{y_1} = a^2 + b^2.$$

(ii) Equation of the chord of contact of tangents from (x_1, y_1) is

$$\frac{xx_1}{a^2} - \frac{yy_1}{b^2} = 1.$$

(iii) Equation of the chord with (x_1, y_1) as its midpoint is $T = S_1$:

$$\frac{xx_1}{a^2} - \frac{yy_1}{b^2} - 1 = \frac{x_1^2}{a^2} - \frac{y_1^2}{b^2} - 1.$$

(iv) Equation of the pair of tangents from (x_1, y_1) is $T^2 = SS_1$:

$$\left(\frac{xx_1}{a^2} - \frac{yy_1}{b^2} - 1\right)^2 = \left(\frac{x^2}{a^2} - \frac{y^2}{b^2} - 1\right)\left(\frac{x_1^2}{a^2} - \frac{y_1^2}{b^2} - 1\right).$$

For the rectangular hyperbola $xy = c^2$:

(i) Equation of the tangent at $(ct, c/t)$ is

$$x + yt^2 = 2ct.$$

(ii) Equation of the normal at $(ct, c/t)$ is

$$xt - \frac{y'}{t} = c\left(t^2 - \frac{1}{t^2}\right).$$

Example 2.21 The chain of a suspension bridge hangs in the form of a parabola, whose axis is vertical. In the case of a certain bridge, the chain hangs symmetrically with a span of 170 m and a dip of 13 m. Find the latus rectum of the parabola and the angle of inclination to the horizon at each end of the chain.

From Fig. 2.22, we have $x = 85$ m and $y = 13$ m. Hence,

$$4a = \frac{x^2}{y} = \frac{85 \times 85}{13} = \frac{7225}{13} = 555.77 \text{ m},$$

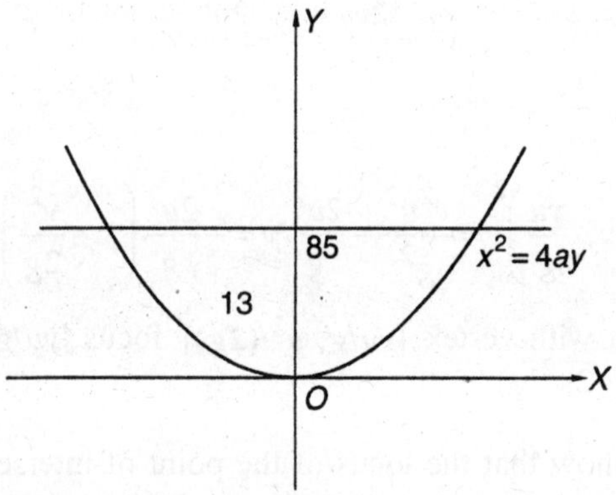

Figure 2.22

which is the latus rectum of the parabola.

Also, the tangent at the point (x_1, y_1) to the parabola $x^2 = 4ay$ is given by

$$xx_1 = 2a(y + y_1), \quad \text{or} \quad y = \frac{x_1}{2a}x - y_1.$$

If θ is the angle of inclination to the horizon, then we have

$$\tan\theta = \frac{x_1}{2a} = \frac{2x_1}{4a} = \frac{170 \times 13}{7225} = 0.3059,$$

from which we obtain $\theta = 17°$ (nearly).

Example 2.22 A body is projected from a point O with a velocity whose horizontal and vertical components are u and v, respectively. Neglecting air resistance, show that the equation of its path is a parabola and find its vertex, focus and latus rectum.

Let the origin be at O, the point of projection and let the axes be the x-axis and y-axis. Let the position of the body be $P(x, y)$ after t secs. Then, at time t, the position of the body is defined by

$$x = ut \quad \text{and} \quad y = vt - \frac{1}{2}gt^2,$$

where g is the acceleration due to gravity. To find the equation of the path, we eliminate t from the above equations. From the first equation, we write

$$t = \frac{x}{u}.$$

Substituting this in the second equation, we obtain

$$y = v\frac{x}{u} - \frac{1}{2}g\left(\frac{x}{u}\right)^2,$$

which can be written as

$$x^2 - \frac{2vu}{g}x = -\frac{2u^2}{g}y$$

or

$$\left(x - \frac{vu}{g}\right)^2 = \frac{v^2u^2}{g^2} - \frac{2u^2}{g}y = -\frac{2u^2}{g}\left(y - \frac{v^2}{2g}\right).$$

This is a parabola with vertex $[vu/g, u^2/(2g)]$ focus $[vu/g, (v^2 - u^2)/(2g)]$ and latus rectum $(2u^2/g)$.

Example 2.23 Show that the locus of the point of intersection of tangents to $y^2 = 4ax$ which intercept a constant length d on the directrix is

$$(y^2 - 4ax)(x + a)^2 = d^2x^2.$$

The equation of any line through $P(x_1, y_1)$ is

$$\frac{x - x_1}{\cos\theta} = \frac{y - y_1}{\sin\theta} = r \tag{i}$$

and therefore any point on this line is $(x_1 + r\cos\theta, y_1 + r\sin\theta)$. If this line intersects the parabola, we obtain

$$(y_1 + r\sin\theta)^2 = 4a(x_1 + r\cos\theta)$$

or

$$r^2 \sin^2\theta + 2ry_1 \sin\theta + y_1^2 = 4ax_1 + 4ar\cos\theta,$$

or

$$r^2 \sin^2\theta + 2r(y_1 \sin\theta - 2a\cos\theta) = -(y_1^2 - 4ax_1).$$

If this is to be a tangent to the parabola, then the two roots for r must coincide, and hence we have

$$(y_1 \sin\theta - 2a\cos\theta)^2 = \sin^2\theta\,(y_1^2 - 4ax_1),$$

or

$$(y_1 \tan\theta - 2a)^2 = \tan^2\theta\,(y_1^2 - 4ax_1).$$

But from (i),

$$\tan\theta = \frac{y - y_1}{x - x_1}.$$

Substituting this value in the above equation, we obtain

$$\left[\frac{y_1(y - y_1)}{x - x_1} - 2a\right]^2 = \left(\frac{y - y_1}{x - x_1}\right)^2 (y_1^2 - 4ax_1).$$

Simplification gives:

$$(y^2 - 4ax)(y_1^2 - 4ax_1) = [yy_1 - 2a(x + x_1)]^2, \tag{ii}$$

which is the equation of the pair of tangents from (x_1, y_1) to the parabola (see Fig. 2.23).

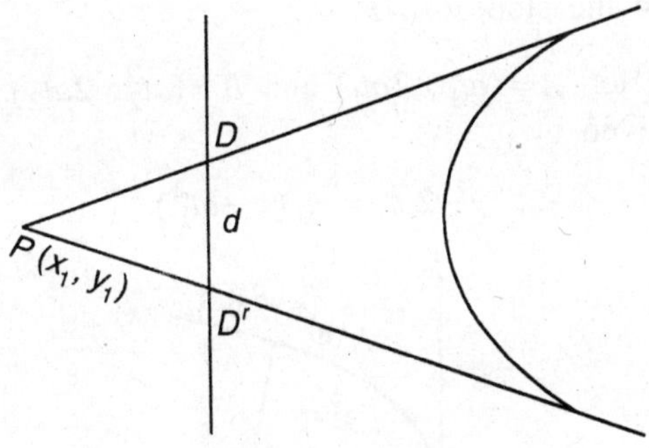

Figure 2.23

The equation of the directrix is $x + a = 0$, and hence the points of intersection of (ii) with the directrix are given by substituting $x = -a$ in (ii). We then obtain:

$$(y^2 + 4a^2)(y_1^2 - 4ax_1) = [yy_1 - 2a(x_1 - a)]^2,$$

which yields on simplification:

$$x_1 y^2 - y_1 (x_1 - a) y + a[(x_1 + a)^2 - y_1^2] = 0 \tag{iii}$$

Let the roots of (iii) be y_1 and y_2 so that

$$y_1 + y_2 = \frac{y_1 (x_1 - a)}{x_1} \quad \text{and} \quad y_1 y_2 = \frac{a[(x_1 + a)^2 - y_1^2]}{x_1}$$

Since $d = y_1 - y_2$, we have

$$d^2 = (y_1 - y_2)^2 = (y_1 + y_2)^2 - 4y_1 y_2 = \frac{y_1^2 (x_1 - a)^2}{x_1^2} - \frac{4a}{x_1}[(x_1 + a)^1 - y_1^2]$$

or

$$d^2 x_1^2 = y_1^2 (x_1 - a)^2 - 4ax_1 [(x_1 + a)^2 - y_1^2]$$

$$= y_1^2 [(x_1 + a)^2 - 4ax_1] - 4ax_1 [(x_1 + a)^2 - y_1^2]$$

$$= y_1^2 (x_1 + a)^2 - 4ax_1 (x_1 + a)^2$$

$$= (y_1^2 - 4ax_1)(x_1 + a)^2,$$

Hence the locus of (x_1, y_1) is given by

$$d^2 x^2 = (y^2 - 4ax)(x + a)^2.$$

Example 2.24 A is a point on the parabola $y^2 = 4ax$. The normal at A cuts the parabola again at point B. If AB subtends a right angle at the vertex of the parabola, find the slope of AB.

In Fig. 2.24, let $A = (at_1^2, 2at_1)$ and $B = (at_2^2, 2at_2)$. Equation of the normal at t_1 is given by

$$y - 2at_1 = -t_1 (x - at_1^2).$$

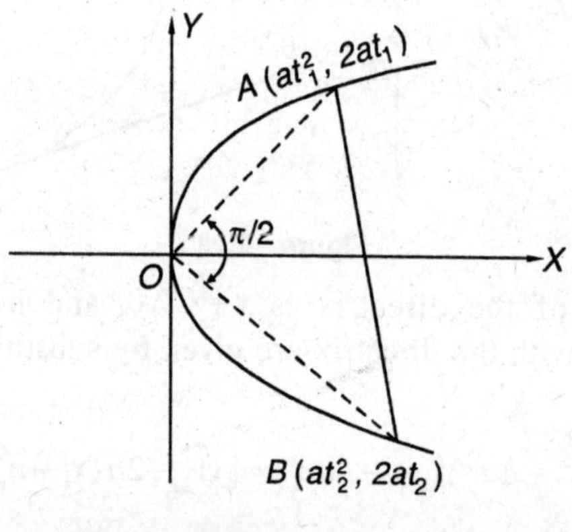

Figure 2.24

Since it passes through B, we have

$$2at_2 - 2at_1 = -t_1\,(at_2^2 - at_1^2)$$

or

$$2a(t_2 - t_1) = -at_1\,(t_2^2 - t_1^2)$$

or

$$2 = -t_1\,(t_2 + t_1) \qquad \text{(i)}$$

Since OA and OB are perpendicular to each other, we must have

$$\frac{2}{t_1}\frac{2}{t_2} = -1$$

Then

$$t_1t_2 = -4 \quad \text{or} \quad t_2 = -\frac{4}{t_1}.$$

Putting this value of t_2 in (i), we get

$$2 = -t_1\left(-\frac{4}{t_1} + t_1\right) = 4 - t_1^2 \quad \text{or} \quad t_1^2 = 2.$$

Hence, using (i)

$$\text{Slope of } AB = \frac{2a(t_2 - t_1)}{a(t_2^2 - t_1^2)} = \frac{2}{t_1 + t_2} = -t_1 = \pm\sqrt{2}.$$

Example 2.25 Prove that the tangents at the ends of a focal chord intersect at right angles on the directrix.

Let $P(at_1^2,\, 2at_1)$ and $Q(at_2^2,\, 2at_2)$ be the extremities of a focal chord. Then the equation of PQ is given by

$$\frac{y - 2at_1}{x - at_1^2} = \frac{2a\,(t_2 - t_1)}{a\,(t_2^2 - t_1^2)} \quad \text{or} \quad y\,(t_1 + t_2) = 2x + 2at_1t_2.$$

Since this passes through the focus $(a, 0)$, we obtain:

$$2a + 2at_1t_2 = 0$$

or

$$t_1t_2 = -1. \qquad \text{(i)}$$

Now, the equations of the tangents at P and Q are given by

$$y = \frac{1}{t_1}x + at_1 \quad \text{and} \quad y = \frac{1}{t_2}x + at_2,$$

respectively. Since $t_1t_2 = -1$, it follows that the two tangents are at right angles.

Further, the point of intersection of the two tangents is given by

$$x = at_1t_2 \quad \text{and} \quad y = a(t_1 + t_2).$$

Since $t_1t_2 = -1$, it follows that $x = -a$, or, $x + a = 0$. This means that the tangents meet on the directrix $x + a = 0$.

Example 2.26 Prove that three normals can be drawn to a parabola from any point in its plane and that the sum of the ordinates of the feet of the normals is zero.

Let (h, k) be any point in the plane of the parabola $y^2 = 4ax$ and let 't' be the foot of the normal drawn from (h, k). Equation of the normal at 't' is

$$y + xt = 2at + at^3 \tag{i}$$

Since (i) passes through (h, k), we have

$$k + ht = 2at + at^3$$

or

$$at^3 + (2a - h)t - k = 0 \tag{ii}$$

Equation (ii) is a cubic in t and therefore three normals, real or imaginary, can be drawn to the parabola. Now let t_1, t_2, t_3 be the roots of (ii). Then

$$t_1 + t_2 + t_3 = 0 \tag{iii}$$

But

$$\begin{aligned}\text{Sum of the ordinates of the feet of the normals} &= 2at_1 + 2at_2 + 2at_3\\ &= 2a(t_1 + t_2 + t_3)\\ &= 0, \quad \text{using (iii)}\end{aligned}$$

Example 2.27 Show that the locus of the midpoints of chords of a parabola which subtend a right angle at the vertex is another parabola whose latus rectum is half of the latus rectum of the original parabola.

Let $y^2 = 4ax$ be the equation of the given parabola. Then the equation of the chord whose middle point is (x_1, y_1) is:

$$yy_1 - 2a(x + x_1) = y_1^2 - 4ax_1$$

or

$$yy_1 - 2ax = y_1^2 - 2ax_1 \tag{i}$$

Now, the equation of the pair of lines joining the vertex to the points of intersection of (i) and the parabola is obtained by homogenising the equation $y^2 = 4ax$ using (i). Therefore, the equation of the pair of lines is given by

$$y^2 = 4ax\,\frac{yy_1 - 2ax}{y_1^2 - 2ax_1}$$

or

$$y^2(y_1^2 - 2ax_1) = 4ax(yy_1 - 2ax). \quad \text{(ii)}$$

Since the chord subtends a right angle at the vertex, the two lines in (ii) must be perpendicular to each other. The condition for this is

Coefficient of x^2 + coefficient of $y^2 = 0$.

i.e.

$$8a^2 + y_1^2 - 2ax_1 = 0$$

or

$$y_1^2 = 2ax_1 - 8a^2 = 2a(x_1 - 4a).$$

Hence the locus of (x_1, y_1) is $y^2 = 2a(x - 4a)$.

Example 2.28 Normal chords are drawn to the parabola $y^2 = 4ax$. Show that the locus of their poles is $y^2(x + 2a) + 4a^3 = 0$.

Let $P(at^2, 2at)$ be a point on the parabola and PQ be a normal chord. Then the equation of PQ is

$$y + xt = 2at + at^3 \quad \text{(i)}$$

If (x_1, y_1) is the pole of PQ, then PQ will be the polar of (x_1, y_1). Hence, equation of PQ is

$$yy_1 = 2a(x + x_1) \quad \text{(ii)}$$

Since (i) and (ii) represent the same line, we obtain:

$$\frac{1}{y_1} = \frac{t}{-2a} = \frac{2at + at^3}{2ax_1}.$$

Therefore,

$$t = -\frac{2a}{y_1} \quad \text{and} \quad 2ax_1 = y_1(2at + at^3)$$

Eliminating t, we get

$$2ax_1 = y_1\left[2a\left(-\frac{2a}{y_1}\right) + a\left(-\frac{2a}{y_1}\right)^3\right] = -4a^2 - \frac{8a^4}{y_1^2}.$$

or

$$x_1 y_1^2 = -2ay_1^2 - 4a^3$$

or

$$y_1^2(x_1 + 2a) + 4a^3 = 0.$$

Hence the locus of (x_1, y_1) is $y^2(x + 2a) + 4a^3 = 0$.

Example 2.29 Tangents P_1Q_1 and P_2Q_2 are drawn to the parabola $y^2 = 4ax$ at P_1 and P_2, respectively. If P_1Q_1, P_2Q_2 and the tangent at the vertex form a triangle of constant area k^2, show that the locus of the point of intersection of P_1Q_1 and P_2Q_2 is $x^2(y^2 - 4ax) - 4k^4 = 0$.

Let $P_1(at_1^2, 2at_1)$ and $P_2(at_2^2, 2at_2)$ be two points on the given parabola. Then the equations of P_1Q_1 and P_2Q_2 are given by

$$yt_1 = x + at_1^2 \tag{i}$$

and

$$yt_2 = x + at_2^2 \tag{ii}$$

The tangent at the vertex is

$$x = 0 \tag{iii}$$

The points of intersection of (i) and (ii) with (iii) are $(0, at_1)$ and $(0, at_2)$, respectively.

Let (x_1, y_1) be the point of intersection of (i) and (ii). Then

$$y_1 = a(t_1 + t_2)$$

and

$$x_1 = at_1t_2. \tag{iv}$$

Hence we have

$$k^2 = \frac{1}{2}\begin{vmatrix} x_1 & y_1 & 1 \\ 0 & at_1 & 1 \\ 0 & at_2 & 1 \end{vmatrix} \tag{v}$$

Now, (v) simplifies to:

$$2k^2 = ax_1(t_1 - t_2).$$

$$= ax_1\sqrt{(t_1 + t_2)^2 - 4t_1t_2}$$

$$= ax_1\sqrt{\frac{y_1^2}{a^2} - 4\frac{x_1}{a}}, \quad \text{[using (iv)]}$$

$$= x_1\sqrt{y_1^2 - 4ax_1}$$

or

$$x_1^2(y_1^2 - 4ax_1) = 4k^4.$$

Hence the required locus is given by

$$x^2(y^2 - 4ax) - 4k^4 = 0.$$

Example 2.30 If the normals to the parabola $y^2 = 4ax$ at three points P, Q and R meet at a point M, prove that

$$SP \cdot SQ \cdot SR = a(SM)^2,$$

where S is the focus of the parabola.

Let M be the point (x_1, y_1). Then P, Q, R are the feet of the normals from $M(x_1, y_1)$. Now the equation of the normal at 't' is given by

$$y + xt = 2at + at^3 \quad \text{or} \quad at^3 + (2a - x)t - y = 0$$

Since this normal passes through $M(x_1, y_1)$, we have

$$at^3 + (2a - x_1)t - y_1 = 0 \tag{i}$$

If the feet of P, Q, R are t_1, t_2, t_3, respectively, then these are given by the roots of (i). We have

$$t_1 + t_2 + t_3 = 0 \tag{i}$$

$$t_1t_2 + t_2t_3 + t_3t_1 = \frac{2a - x_1}{a} \tag{ii}$$

$$t_1t_2t_3 = \frac{y_1}{a} \tag{iii}$$

Now

$$SP^2 = (a - at_1^2)^2 + (2at_1)^2 = a^2(1 - t_1^2)^2 + 4a^2t_1^2 = a^2(1 + t_1^2)^2.$$

or

$$SP = a(1 + t_1^2).$$

Similarly,

$$SQ = a(1 + t_2^2) \quad \text{and} \quad SR = a(1 + t_3^2).$$

Hence

$$\begin{aligned} SP \cdot SQ \cdot SR &= a^3(1 + t_1^2)(1 + t_2^2)(1 + t_3^2) \\ &= a^3[1 + (t_1 + t_2 + t_3)^2 - 2(t_1t_2 + t_3t_1 + t_2t_3) + (t_1t_2 + t_2t_3 + t_3t_1)^2 \\ &\quad -2t_1t_2t_3(t_1 + t_2 + t_3) + (t_1t_2t_3)^2]. \end{aligned}$$

Using (ii), (iii) and (iv), we get

$$\begin{aligned} SP \cdot SQ \cdot SR &= a^3\left[1 - 2\left(\frac{2a - x_1}{a}\right) + \left(\frac{2a - x_1}{a}\right)^2 + \frac{y_1^2}{a^2}\right] \\ &= a^3\left[\left(1 - \frac{2a - x_1}{a}\right)^2 + \frac{y_1^2}{a^2}\right] \end{aligned}$$

$$= a^3\left[\frac{(x_1 - a)^2}{a^2} + \frac{y_1^2}{a^2}\right]$$

$$= a\left[(x_1 - a)^2 + y_1^2\right]$$

$$= a\,(SM)^2.$$

Example 2.31 The line $x + y = 1$ meets the ellipse $(x^2/4) + (y^2/9) = 1$, in the points $A(x_1, y_1)$ and $B(x_2, y_2)$. Prove that x_1 and x_2 are the roots of the equation $13x^2 - 8x - 32 = 0$. Deduce the length of PQ.

The ellipse is given by

$$9x^2 + 4y^2 = 36$$

The x-coordinates of the point of intersection of the ellipse with the line $x + y = 1$ are given by

$$9x^2 + 4(1 - x)^2 = 36.$$

i.e.

$$13x^2 - 8x - 32 = 0. \qquad \text{(i)}$$

Now, the slope of PQ is

$$\frac{y_1 - y_2}{x_1 - x_2}.$$

Since the equation of the line is $x + y = 1$, it follows that its slope is -1. Hence, we have

$$\frac{y_1 - y_2}{x_1 - x_2} = -1,$$

or

$$y_1 - y_2 = -(x_1 - x_2) \qquad \text{(ii)}$$

Also, using (ii), we have

$$PQ = \sqrt{(x_2 - x_1)^2 + (y_2 - y_1)^2} = \sqrt{(x_2 - x_1)^2 + (x_2 - x_1)^2} = \sqrt{2(x_1 - x_2)^2}.$$

From (i), we obtain

$$x_1 + x_2 = \frac{8}{13} \quad \text{and} \quad x_1x_2 = -\frac{32}{13}.$$

Also,

$$(x_1 - x_2)^2 = (x_1 + x_2)^2 - 4x_1x_2 = \frac{64}{169} + \frac{128}{13} = \frac{1728}{169},$$

and so,

$$PQ = \sqrt{2\left(\frac{1728}{169}\right)} = \frac{24\sqrt{6}}{13}.$$

Example 2.32 If the normal at an end of a latus rectum of an ellipse passes through one extremity of the minor axis, prove that the eccentricity of the curve is given by the equation

$$e^4 + e^2 - 1 = 0.$$

Let the equation of the ellipse be

$$\frac{x^2}{a^2} + \frac{y^2}{b^2} = 1$$

and let PL (Fig. 2.25) be the latus rectum. Then the coordinates of P are $(ae, b^2/a)$ and the slope of the tangent at P is $-e$. Hence the slope of the normal at P is $1/e$. The equation of the normal at P is therefore given by

$$y - \frac{b^2}{a} = \frac{1}{e}(x - ae)$$

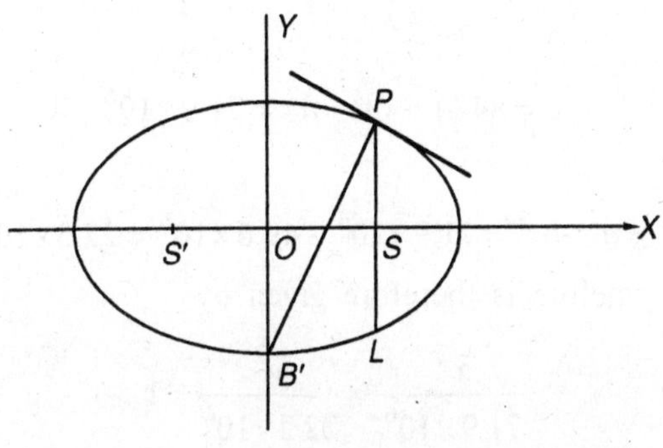

Figure 2.25

Since this passes through $B'(0, -b)$, we must have

$$-b - \frac{b^2}{a} = \frac{1}{e}(-ae) = -a \quad \text{or} \quad b^2 + ab - a^2 = 0.$$

Since $b^2 = a^2(1 - e^2)$, the above equation becomes:

$$a^2(1 - e^2) + a^2\sqrt{1 - e^2} = a^2 \quad \text{or} \quad \sqrt{1 - e^2} = e^2,$$

from which we obtain the equation

$$e^4 + e^2 - 1 = 0.$$

Example 2.33 A satellite is to be placed into an elliptical orbit about the earth having a minimum altitude of 640 km and a maximum altitude of 3520 km. Assuming that the centre of the earth is located at one focus and that the radius of the earth is 6400 km, find the equation describing the path followed by the satellite.

Let the orbit of the satellite be as shown in Fig. 2.26, and let S' be the centre of the earth. Then the distance between S' and the closest vertex A' is

$$c = 6400 + 640 = 7040 \text{ km} \quad \text{or} \quad c^2 = 49.6 \times 10^6 \text{ km}$$

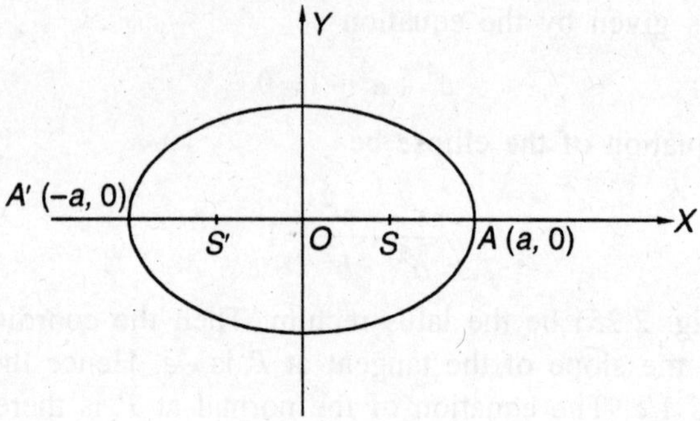

Figure 2.26

Also,

$$2a = \text{major axis of the ellipse} = 640 + 12800 + 3520 = 16960.$$

So,

$$a = 8480 \quad \text{or} \quad a^2 = 71.9 \times 10^6$$

Hence,

$$b^2 = a^2 - c^2 = 71.9 \times 10^6 - 49.6 \times 10^6 = 22.3 \times 10^6.$$

The path of the satellite is therefore given by

$$\frac{x^2}{71.9 \times 10^6} + \frac{y^2}{22.3 \times 10^6} = 1.$$

Example 2.34 Show that the condition for the line $lx + my + n = 0$ to be a normal to the ellipse

$$\frac{x^2}{a^2} + \frac{y^2}{b^2} = 1$$

is

$$\frac{a^2}{l^2} + \frac{b^2}{m^2} = \left(\frac{a^2 - b^2}{n}\right)^2.$$

Let the line

$$lx + my + n = 0 \tag{i}$$

be a normal to the given ellipse at the point 'θ', i.e. $(a \cos \theta, b \sin \theta)$.

Equation of the normal at 'θ' is

$$\frac{ax}{\cos\theta} - \frac{by}{\sin\theta} = a^2 - b^2. \qquad \text{(ii)}$$

Comparing (i) and (ii), we get

$$\frac{l}{a\cos\theta} = \frac{m}{-b\sin\theta} = \frac{-n}{a^2 - b^2},$$

and hence

$$\cos\theta = -\frac{an}{l(a^2 - b^2)}, \qquad \sin\theta = \frac{nb}{m(a^2 - b^2)}.$$

Since $\cos^2\theta + \sin^2\theta = 1$, we must have

$$\frac{a^2n^2}{l^2(a^2 - b^2)^2} + \frac{n^2b^2}{m^2(a^2 - b^2)^2} = 1,$$

or,

$$\frac{a^2}{l^2} + \frac{b^2}{m^2} = \frac{(a^2 - b^2)^2}{n^2},$$

which is the required equation.

Example 2.35 If α and β are the eccentric angles of two points P and Q on the ellipse

$$\frac{x^2}{a^2} + \frac{y^2}{b^2} = 1$$

and if $\alpha - \beta$ is a constant, prove that the chord PQ touches a fixed ellipse.

The equation of PQ is given by

$$\frac{x - a\cos\alpha}{a(\cos\beta - \cos\alpha)} = \frac{y - b\sin\alpha}{b(\sin\beta - \sin\alpha)}$$

or

$$\frac{x - a\cos\alpha}{2a\sin\frac{\alpha+\beta}{2}\sin\frac{\alpha-\beta}{2}} = \frac{y - b\sin\alpha}{2b\cos\frac{\alpha+\beta}{2}\sin\frac{\beta-\alpha}{2}}$$

or

$$\frac{x - a\cos\alpha}{a\sin\frac{\alpha+\beta}{2}} = -\frac{y - b\sin\alpha}{b\cos\frac{\alpha+\beta}{2}},$$

or

$$\frac{x}{a}\cos\frac{\alpha+\beta}{2}+\frac{y}{b}\sin\frac{\alpha+\beta}{2}=\cos\alpha\cos\frac{\alpha+\beta}{2}+\sin\alpha\sin\frac{\alpha+\beta}{2}.$$

$$=\cos\left(\alpha-\frac{\alpha+\beta}{2}\right)$$

$$=\cos\frac{\alpha-\beta}{2}.$$

Hence, the equation of PQ becomes

$$\frac{x}{a}\cos\frac{\alpha+\beta}{2}+\frac{y}{b}\sin\frac{\alpha+\beta}{2}=\cos\frac{\alpha-\beta}{2} \qquad \text{(i)}$$

Let $(\alpha-\beta)/2=k$. Then (i) becomes:

$$\frac{x}{a}\cos\frac{\alpha+\beta}{2}+\frac{y}{b}\sin\frac{\alpha+\beta}{2}=\cos k,$$

or

$$\frac{x}{a}\cos\theta+\frac{y}{b}\sin\theta=\cos k,$$

where $\theta=(\alpha+\beta)/2$. Then

$$\frac{x}{a\cos k}\cos\theta+\frac{y}{b\cos k}\sin\theta=1,$$

which is a straight line touching the ellipse

$$\frac{x^2}{(a\cos k)^2}+\frac{y^2}{(b\cos k)^2}=1.$$

Example 2.36 If the normals at the points θ_1, θ_2, θ_3 of the ellipse $(x^2/a^2)+(y^2/b^2)=1$ are concurrent, prove that

$$\begin{vmatrix} \sec\theta_1 & \operatorname{cosec}\theta_1 & 1 \\ \sec\theta_2 & \operatorname{cosec}\theta_2 & 1 \\ \sec\theta_3 & \operatorname{cosec}\theta_3 & 1 \end{vmatrix}=0$$

The equation of the normal at 'θ' is given by

$$\frac{ax}{\cos\theta}-\frac{by}{\sin\theta}=a^2-b^2 \quad \text{or} \quad ax\sec\theta-by\operatorname{cosec}\theta=a^2-b^2.$$

Since θ_1, θ_2 and θ_3 are points on the ellipse, we have

$$ax \sec\theta_1 - by \operatorname{cosec}\theta_1 = a^2 - b^2,$$
$$ax \sec\theta_2 - by \operatorname{cosec}\theta_2 = a^2 - b^2,$$
$$ax \sec\theta_3 - by \operatorname{cosec}\theta_3 = a^2 - b^2.$$

The eliminant of these equations is

$$\begin{vmatrix} \sec\theta_1 & \operatorname{cosec}\theta_1 & 1 \\ \sec\theta_2 & \operatorname{cosec}\theta_2 & 1 \\ \sec\theta_3 & \operatorname{cosec}\theta_3 & 1 \end{vmatrix} = 0.$$

Example 2.37 If the normals at the four points (x_1, y_1), (x_2, y_2), (x_3, y_3) and (x_4, y_4) on the ellipse $(x^2/a^2)+(y^2/b^2)=1$ are concurrent, prove that

$$(x_1 + x_2 + x_3 + x_4)\left(\frac{1}{x_1}+\frac{1}{x_2}+\frac{1}{x_3}+\frac{1}{x_4}\right) = 4.$$

Equation of the normal at (x, y) is given by

$$\frac{X-x}{x/a^2} = \frac{Y-y}{y/b^2} \quad \text{or} \quad \frac{a^2(X-x)}{x} = \frac{b^2(Y-y)}{y}. \tag{i}$$

If this passes through the point of concurrence (h, k) we obtain

$$\frac{a^2(h-x)}{x} = \frac{b^2(k-y)}{y}.$$

or

$$a^2 y(h-x) = b^2 x(k-y).$$

or

$$y^2[a^2(h-x)+b^2x]^2 = b^4x^2k^2. \tag{ii}$$

But

$$\frac{y^2}{b^2} = 1 - \frac{x^2}{a^2} \quad \text{or} \quad y^2 = b^2\left(1-\frac{x^2}{a^2}\right)$$

Substituting for y^2 in (ii), we get

$$b^2\left(1-\frac{x^2}{a^2}\right)[a^2(h-x)+b^2x]^2 = b^4x^2k^2$$

or

$$(a^2 - x^2)[a^4(h-x)^2 + b^4x^2 + 2a^2b^2x(h-x)] = a^2b^2x^2k^2.$$

The above equation can be simplified and written in powers of x as:

$$x^4(a^2-b^2)^2 - 2a^2h(a^2-b^2)x^3 - x^2[a^4h^2 - a^2(a^2-b^2)^2 - a^2b^2k^2]$$
$$+2a^4h(a^2-b^2)x - a^6h^2 = 0.$$

If x_1, x_2, x_3 and x_4 are the roots of (iii), then we have

$$x_1 + x_2 + x_3 + x_4 = \frac{2a^2h}{a^2-b^2} \qquad \text{(vi)}$$

$$\Sigma x_1x_2x_3 = -\frac{2a^4h}{a^2-b^2} \qquad \text{(v)}$$

$$x_1x_2x_3x_4 = -\frac{a^6h^2}{(a^2-b^2)^2} \qquad \text{(vi)}$$

Now,

$$\frac{1}{x_1}+\frac{1}{x_2}+\frac{1}{x_3}+\frac{1}{x_4} = \frac{\Sigma x_1x_2x_3}{x_1x_2x_3x_4}$$
$$= \frac{2a^4h(a^2-b^2)}{a^6h^2}$$
$$= \frac{2(a^2-b^2)}{a^2h} \qquad \text{(vii)}$$

Hence, using (iv) and (vii), we get

$$(x_1+x_2+x_3+x_4)\left(\frac{1}{x_1}+\frac{1}{x_2}+\frac{1}{x_3}+\frac{1}{x_4}\right) = 4.$$

Example 2.38 Let d be the perpendicular distance from the centre of the ellipse to the tangent drawn at a point P on the ellipse. If F_1 and F_2 are two foci of the ellipse, show that

$$(PF_1 - PF_2)^2 = 4a^2\left(1 - \frac{b^2}{d^2}\right).$$

Let P be the point $(a\cos\theta, b\sin\theta)$. The equation of the tangent at P is given by

$$\frac{x}{a}\cos\theta + \frac{y}{b}\sin\theta = 1 \qquad \text{(i)}$$

The two foci are $F_1 = (ae, 0)$ and $F_2 = (-ae, 0)$. Hence

d = Perpendicular distance of (i) from the centre (0, 0) of the ellipse

$$= \frac{1}{\sqrt{\dfrac{\cos^2\theta}{a^2} + \dfrac{\sin^2\theta}{b^2}}}$$

$$= \frac{ab}{\sqrt{a^2 \sin^2\theta + b^2 \cos^2\theta}}.$$

or

$$\frac{b^2}{d^2} = \frac{a^2 \sin^2\theta + b^2 \cos^2\theta}{a^2}$$

or

$$1 - \frac{b^2}{d^2} = 1 - \frac{a^2 \sin^2\theta + b^2 \cos^2\theta}{a^2} = \frac{(a^2 - b^2)\cos^2\theta}{a^2} = e^2 \cos^2\theta,$$

as $e = \sqrt{1 - b^2/a^2}$. Then

$$4a^2 \left(1 - \frac{b^2}{d^2}\right) = 4e^2 \cos^2\theta \, a^2 = 4a^2 e^2 \cos^2\theta \qquad \text{(ii)}$$

Again,

$$(PF_1 - PF_2)^2 = [(a - ae \cos \theta) - (a + ae \cos \theta)]^2$$

$$= (2\, ae \cos \theta)^2$$

$$= 4a^2 e^2 \cos^2\theta \qquad \text{(iii)}$$

From (ii) and (iii), it follows that

$$(PF_1 - PF_2)^2 = 4a^2 \left(1 - \frac{b^2}{d^2}\right).$$

Example 2.39 Let AB be a chord of an ellipse with θ_1 and θ_2 as the eccentric angles of A and B, respectively. If this chord cuts the major axis of the ellipse at a distance d from the centre, prove that

$$\tan \frac{\theta_1}{2} \tan \frac{\theta_2}{2} = \frac{d - a}{d + a},$$

where $2a$ is the length of the major axis.

Let the equation of the ellipse be

$$\frac{x^2}{a^2}+\frac{y^2}{b^2}=1.$$

Equation of the chord AB is

$$\frac{x}{a}\cos\frac{\theta_1+\theta_2}{2}+\frac{y}{b}\sin\frac{\theta_1+\theta_2}{2}=\cos\frac{\theta_1-\theta_2}{2} \tag{i}$$

Since this chord passes through $(d, 0)$, we obtain

$$\frac{d}{a}\cos\frac{\theta_1+\theta_2}{2}=\cos\frac{\theta_1-\theta_2}{2}.$$

or

$$\frac{d}{a}=\frac{\cos\frac{\theta_1-\theta_2}{2}}{\cos\frac{\theta_1+\theta_2}{2}}.$$

or

$$\frac{d-a}{d+a}=\frac{\cos\frac{\theta_1-\theta_2}{2}-\cos\frac{\theta_1+\theta_2}{2}}{\cos\frac{\theta_1-\theta_2}{2}+\cos\frac{\theta_1+\theta_2}{2}}$$

$$=\frac{2\sin\frac{\theta_1}{2}\sin\frac{\theta_2}{2}}{2\cos\frac{\theta_1}{2}\cos\frac{\theta_2}{2}}$$

$$=\tan\frac{\theta_1}{2}\tan\frac{\theta_2}{2}.$$

Example 2.40 If the equation of a hyperbola $+2a^4h(a^2-b^2)x-a^6h^2=0.$ is $4x^2-9y^2=36$, find its foci, the eccentricity and the latus rectum.

The given equation may be written as

$$\frac{x^2}{9}-\frac{y^2}{4}=1. \tag{i}$$

Comparing with the standard equation of the hyperbola

$$\frac{x^2}{a^2}-\frac{y^2}{b^2}=1,$$

we obtain $a=3,\ b=2$. Further, since $b^2=a^2(e^2-1)$, we have $4=9(e^2-1)$, which gives

$$e^2 = \frac{13}{9} \quad \text{or} \quad e = \frac{\sqrt{13}}{3}.$$

Hence the coordinates of the foci, which are given by $(ae, 0)$ and $(-ae, 0)$ are $(\sqrt{13}, 0)$ and $(-\sqrt{13}, 0)$. Also,

$$\text{Latus rectum} = \frac{2b^2}{a} = \frac{8}{3}.$$

Example 2.41 Show that the product of the perpendiculars drawn from any point on a hyperbola to its asymptotes is constant.

Let (x_1, y_1) be a point on the hyperbola $(x^2/a^2) - (y^2/b^2) = 1$. The asymptotes are given by $(x/a) \pm (y/b) = 0$. Then the lengths of the perpendiculars drawn from (x_1, y_1) to the asymptotes are given by

$$p_1 = \frac{(x_1/a) + (y_1/b)}{\sqrt{(1/a^2) + (1/b^2)}} \quad \text{and} \quad p_2 = \frac{(x_1/a) - (y_1/b)}{\sqrt{(1/a^2) + (1/b^2)}}$$

Therefore,

$$p_1 p_2 = \frac{(x_1^2/a^2) - (y_1^2/b^2)}{(1/a^2) + (1/b^2)}$$

Since (x_1, y_1) is a point on the hyperbola, we have

$$\frac{x_1^2}{a^2} - \frac{y_1^2}{b^2} = 1.$$

Hence,

$$p_1 p_2 = \frac{1}{(1/a^2) + (1/b^2)}, \quad \text{(a constant)}.$$

Example 2.42 Find the asymptotes of the hyperbola

$$3x^2 - 5xy - 2y^2 + 17x + y + 14 = 0.$$

Since the equations of the asymptotes differ only in the constant term, we assume that the asymptotes are given by the equation

$$3x^2 - 5xy - 2y^2 + 17x + y + k = 0 \qquad \text{(i)}$$

where k is a constant. Since (i) represents a pair of straight lines, we use condition (2.17) and obtain:

$$3(-2)k + 2\left(\frac{1}{2}\right)\left(\frac{17}{2}\right)\left(-\frac{5}{2}\right) - 3\left(\frac{1}{2}\right)^2 + 2\left(\frac{17}{2}\right)^2 - k\left(-\frac{5}{2}\right)^2 = 0,$$

which gives on simplification $k=10$. Hence the asymptotes are given by

$$3x^2-5xy-2y^2+17x+y+10=0,$$

or

$$(3x+y+2)(x-2y+5)=0.$$

Example 2.43 Find the condition for the straight line $lx+my+n=0$ to touch the hyperbola $(x^2/a^2)-(y^2/b^2)=1$.

Let the straight line

$$lx+my+n=0 \tag{i}$$

be a tangent to the hyperbola $(x^2/a^2)+(y^2/b^2)=1$ at the point $(a\sec\theta, b\tan\theta)$. The equation of this tangent is given by

$$\frac{x}{a}\sec\theta-\frac{y}{b}\tan\theta=1$$

Comparing (i) and (ii), we obtain

$$\frac{\sec\theta}{al}=-\frac{\tan\theta}{bm}=-\frac{1}{n}$$

which gives

$$\sec\theta=-\frac{al}{n} \quad \text{and} \quad \tan\theta=\frac{bm}{n}.$$

Since $\sec^2\theta-\tan^2\theta=1$, we have

$$\frac{a^2l^2}{n^2}-\frac{b^2m^2}{n^2}=1 \quad \text{or} \quad a^2l^2-b^2m^2=n^2,$$

which is the required condition.

Example 2.44 If a rectangular hyperbola circumscribes a triangle, prove that it also passes through the orthocentre of the triangle.

Let the equation of the rectangular hyperbola be $xy=c^2$ and let the three vertices of the triangle be given by

$$P\left(ct_1, \frac{c}{t_1}\right), \quad Q\left(ct_2, \frac{c}{t_2}\right), \quad R\left(ct_3, \frac{c}{t_3}\right).$$

Now, equation of the side PQ is

$$x+yt_1t_2=C(t_1+t_2)$$

and the equation of the perpendicular from R on PQ is

$$xt_1t_2-y=ct_1t_2t_3-\frac{c}{t_3} \tag{i}$$

Similarly, the equation of the perpendicular from P on QR is

$$x t_2 t_3 - y = c\left(t_1 t_2 t_3 - \frac{1}{t_1}\right) \quad \text{(ii)}$$

Solving (i) and (ii), we obtain

$$x = -\frac{c}{t_1 t_2 t_3} \quad \text{and} \quad y = -c t_1 t_2 t_3,$$

which are the coordinates of the orthocentre. It is easily verified that the orthocentre lies on the hyperbola.

Example 2.45 From points on the circle $x^2 + y^2 = a^2$, tangents are drawn to the hyperbola $x^2 - y^2 = a^2$. Prove that the locus of the middle points of the chords of contact is the curve

$$(x^2 - y^2)^2 = a^2 (x^2 + y^2).$$

Let $P(a\cos\theta, a\sin\theta)$ be any point on the given circle. Equation of the chord of contact of the hyperbola with respect to P is

$$x\cos\theta - y\sin\theta = a \quad \text{(i)}$$

Also equation of the chord of contact with middle point (x_1, y_1) is given by $T = S_1$:

$$xx_1 - yy_1 - a^2 = x_1^2 - y_1^2 - a^2$$

or

$$xx_1 - yy_1 = x_1^2 - y_1^2 \quad \text{(ii)}$$

Since (i) and (ii) represent the same chord, we obtain

$$\frac{\cos\theta}{x_1} = \frac{\sin\theta}{y_1} = \frac{a}{x_1^2 - y_1^2}.$$

Then

$$\cos\theta = \frac{ax_1}{x_1^2 - y_1^2} \quad \text{and} \quad \sin\theta = \frac{ay_1}{x_1^2 - y_1^2}.$$

Eliminating θ, we have

$$\left(\frac{ax_1}{x_1^2 - y_1^2}\right)^2 + \left(\frac{ay_1}{x_1^2 - y_1^2}\right)^2 = 1,$$

which shows that the locus of (x_1, y_1) is

$$(x^2 - y^2) = a^2(x^2 + y^2).$$

2.2.4 Translation of Axes

The coordinates of a point or the equations of a curve are given with reference to any two axes of coordinates in the plane. The Coordinates of the same point or the equations of the same curve can also be obtained with reference to two other axes in the same plane. This process is called the *transformation of coordinates* and formulae for such transformations are given below.

Change of origin without changing the directions of axes

Let (x, y) be the coordinates of P referred to OX and OY as axes. Let O be shifted to the new origin O' (Fig. 2.27) whose coordinates are (h, k), and draw rectangular axes $O'X'$ and $O'Y'$ parallel to OX and OY, respectively. Further, let (x', y') be the coordinates of P referred to $O'X'$ and $O'Y'$. Draw PM perpendicular to OX meeting OX and $O'X'$ at M and M', respectively. Let $O'N$ be perpendicular to OX. Then

$$x = OM = ON + NM = h + x' = x' + h,$$

and

$$y = MP = MM' + M'P = k + y' = y' + k.$$

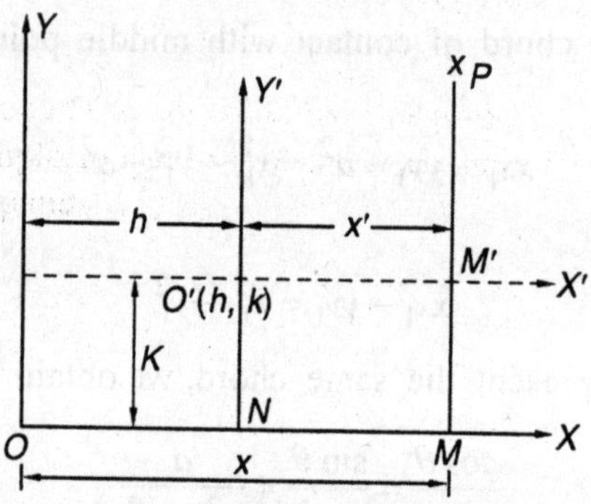

Figure 2.27

The transformation equations are therefore given by

$$x = x' + h, \qquad y = y' + k \tag{2.52}$$

Thus, an equation given in terms of (x', y') coordinates can be transformed to one in terms of (x', y') coordinates by using relations (2.52). The graph of the given equation in the (x, y) system must coincide with that obtained in the (x', y') system. We can therefore reduce a given equation to a standard form by changing the origin, i.e. by translation of axes, and then draw its graph. This is illustrated in the following example.

Example 2.46 Transform the equation $x^2 - 6x - 4y + 1 = 0$ by a proper translation of axes and then draw its graph.

The given equation is

$$x^2 - 6x - 4y + 1 = 0. \tag{i}$$

This is rewritten as

$$(x-3)^2 - 4(y+2) = 0,$$

or,

$$(x-3)^2 = 4(y+2) \tag{ii}$$

Now, let $x - 3 = x'$ and $y + 2 = y'$, so that (ii) reduces to the standard form:

$$x'^2 = 4y', \tag{iii}$$

which is a parabola relative to the new axes (x', y'), as shown in Fig. 2.28.

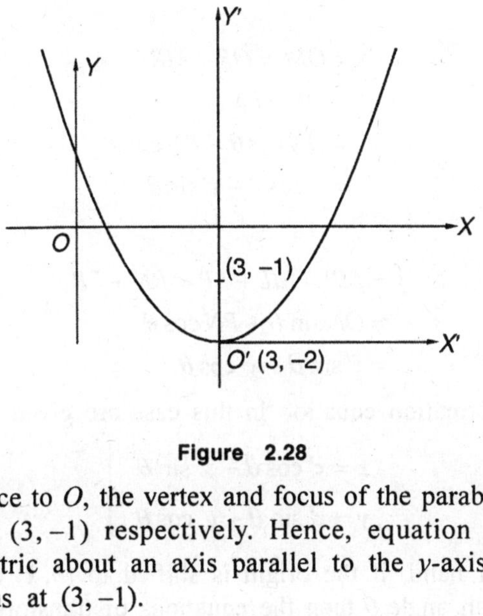

Figure 2.28

With reference to O, the vertex and focus of the parabola given by (iii) are $(3, -2)$ and $(3, -1)$ respectively. Hence, equation (i) represents a parabola symmetric about an axis parallel to the y-axis, with vertex at $(3, -2)$ and focus at $(3, -1)$.

This example illustrates an important use of translation of axes, viz., to reduce a given equation to a simpler form so as to facilitate the sketching of its graph.

Rotation of axes without changing the origin

Let OX and OY be the original set of axes and let (x, y) be the coordinates of a point P with respect to OX and OY. Let OX' and OY' be the new axes inclined at an angle θ to the old axes and let (x', y') be the coordinates of P referred to OX' and OY' (Fig. 2.29).

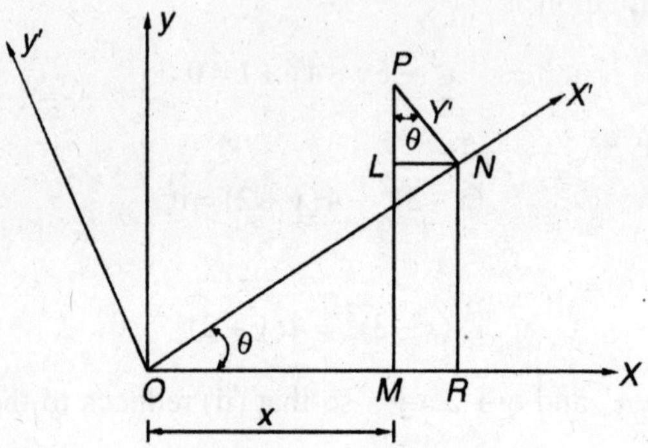

Figure 2.29

Through P draw PM and PN perpendiculars to OX and OX' respectively. Also through N draw NL and NR perpendiculars to PM and OX respectively. Then,

$$\begin{aligned} x &= OM = OR - MR \\ &= OR - LN \\ &= ON\cos\theta - PN\sin\theta \\ &= x'\cos\theta - y'\sin\theta. \end{aligned}$$

Again

$$\begin{aligned} y &= MP = ML + LP = RN + LP \\ &= ON\sin\theta + PN\cos\theta \\ &= x'\sin\theta + y'\cos\theta. \end{aligned}$$

Thus, the transformation equation in this case are given by

$$\left.\begin{aligned} x &= x'\cos\theta - y'\sin\theta \\ y &= x'\sin\theta + y'\cos\theta \end{aligned}\right\} \quad (2.53)$$

On the other hand, if the origin is shifted to (h, k) and the axes are rotated through an angle θ then the equations of transformation are easily obtained as:

$$\left.\begin{aligned} x &= x'\cos\theta - y'\sin\theta + h \\ y &= x'\sin\theta + y'\cos\theta + k \end{aligned}\right\} \quad (2.54)$$

Example 2.47 Transform the equation $x^2 + 4xy + y^2 + 2x = 12$ by rotating the axes through an angle of 45°.

Using the transformation equations (2.53), we obtain

$$x = \frac{1}{\sqrt{2}}(x' - y') \quad \text{and} \quad y = \frac{1}{\sqrt{2}}(x' + y').$$

Hence the given equation transforms into:

$$\frac{1}{2}(x'-y')+4\left(\frac{1}{\sqrt{2}}\right)\left(\frac{1}{\sqrt{2}}\right)(x'^2-y'^2)+\frac{1}{2}(x'+y')^2+2\left(\frac{1}{\sqrt{2}}\right)(x'-y')=12,$$

which becomes $3x'2-y'^2+\sqrt{2}(x'-y')-12=0$, on simplification.

Example 2.48 Transform the expression $ax^2+2hxy+by^2+2gx+2fy+c$ into another of the same degree but without the xy term.

Let the axes be rotated through an angle θ without changing the origin, so that the transformation equations are

$$x=x'\cos\theta-y\sin'\theta$$

$$y=x'\sin\theta+y'\cos\theta.$$

Then, the given expression becomes:

$$a(x'\cos\theta-y'\sin\theta)^2+2h(x'\cos\theta-y'\sin\theta)(x'\sin\theta+y'\cos\theta)$$

$$+b(x'\sin\theta+y'\cos\theta)^2+2g(x'\cos-y'\sin\theta)$$

$$+2f(x'\sin\theta+y'\cos\theta)+c$$

$$=x'^2(a^2\cos^2\theta+2h\cos\theta\sin\theta+b\sin^2\theta)$$

$$+y'2(a\sin^2\theta-2h\cos\theta\sin\theta+b\cos^2\theta)$$

$$+x'y'(-2a\sin\theta\cos\theta+2h\cos^2\theta-2h\sin^2\theta+2b\sin\theta\cos\theta)$$

$$+x'(2g\cos\theta+2f\sin\theta)+y'(-2g\sin\theta+2f\cos\theta)+c.$$

The $x'y'$ term will disappear if

$$(b-a)\sin\theta\cos\theta+h\cos 2\theta=0,$$

i.e. if

$$\frac{1}{2}\tan 2\theta=\frac{h}{a-b}$$

or if

$$\theta=\frac{1}{2}\tan^{-1}\frac{2h}{a-b}.$$

Thus, by rotating the axes through an angle

$$\frac{1}{2}\tan^{-1}\frac{2h}{a-b}$$

without changing the origin, a general second degree expression is transformed into another second-degree expression without the xy term.

2.2.5 Equation of a Conic in Polar Coordinates

The definition of a conic given in section 2.2.3 can be used to find its equation in polar coordinates.

Let Sx be the initial line, the focus S in the pole, and (r, θ) the polar coordinates of a point P (Fig. 2.30). Let DL be the directrix, $D = (-2p, 0)$,

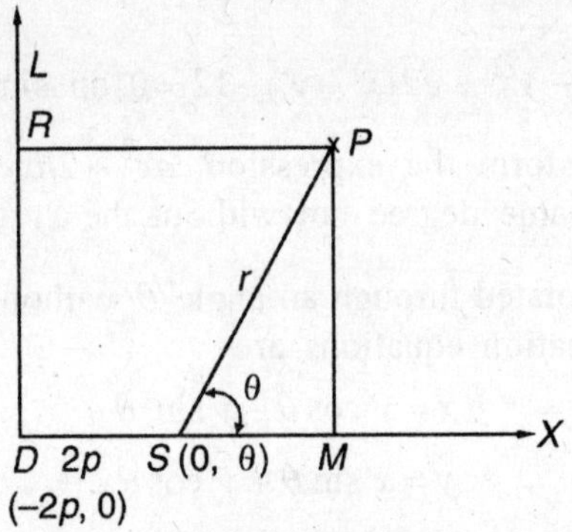

Figure 2.30

and let e be the eccentricity of the conic. We suppose that $e \le 1$. By the definition of a conic, we have

$$\frac{|SP|}{|RP|} = e.$$

since $e \le 1$, the point P is necessarily on the same side of the directrix as the focus S. Now, $|SP| = r$ and $|RP| = |DM| = r \cos\theta + 2p$. Therefore, the point $P(r, \theta)$ satisfies the equation

$$\frac{r}{2p + r\cos\theta} = e.$$

Solving the above equation for r, we get

$$r = \frac{2pe}{1 - e\cos\theta}, \tag{2.55}$$

which is the polar equation of the conic, as this equation is satisfied by every point $P(r, \theta), r > 0$.

For $e > 1$, it can be proved that the above equation represents a conic, but this is left as an exercise to the reader.

Equation (2.25) may be written as

$$r - re\cos\theta = 2pe \quad \text{or} \quad r = e(r\cos\theta + 2p).$$

Since $r = \sqrt{x^2 + y^2}$ and $r\cos\theta = x$, this gives:

$$\sqrt{x^2 + y^2} = e(x + 2p)$$

or,

$$x^2 + y^2 = e^2(x + 2p)^2 = e^2(x^2 + 4px + 4p^2)$$

or,

$$x^2(1-e^2)+y^2=4e^2px+4e^2p^2. \qquad (2.56)$$

If $e=1$, Eq. (2.56) gives:

$$y^2=4px+4p^2=4p(x+p),$$

which is a parabola with focus at (0, 0) and directrix $x=-2p$. On the other hand, if $e<1$, Eq. (2.56) is written as

$$x^2+\frac{y^2}{1-e^2}-\frac{4e^2p}{1-e^2}x=\frac{4e^2p^2}{1-e^2},$$

or

$$\left(x-\frac{2e^2p}{1-e^2}\right)^2+\frac{y^2}{1-e^2}=\frac{4e^2p^2}{(1-e^2)^2},$$

which is the equation of an ellipse.

In a similar way, it can be shown that Eq. (2.55) represents a hyperbola if $e>1$.

Example 2.49 Sketch the graph of the equation

$$r=\frac{10}{3\cos\theta+4\sin\theta+5}$$

The given equation may be written as

$$r=\frac{2}{(3/5)\cos\theta+(4/5)\sin\theta+1}$$

or

$$r=\frac{2}{1+\cos(\theta-\alpha)}, \qquad \text{(i)}$$

where

$$\cos\alpha=\frac{3}{5},\quad \sin\alpha=\frac{4}{5}\quad \text{and}\quad \alpha=53°8' \text{ (approx.)}.$$

Comparison of (i) with Eq. (2.55) shows that $e=1$ and hence the given equation represents a parabola with focus in the pole and its axis inclined at an angle α with the initial line. Further, $p=1$.

If l is the semi latus rectum of the parabola, then the definition of the conic gives:

$$l=2pe=2, \qquad \text{since } p=1 \text{ and } e=1.$$

Hence, the latus rectum is 4. If A is the vertex of the parabola, then $AS=1$.

When $\theta = 0$, Eq. (i) gives:

$$r = \frac{10}{3+5} = \frac{10}{8} = \frac{5}{4},$$

and when $\theta = \pi$, we obtain:

$$r = \frac{10}{-3+5} = \frac{10}{2} = 5.$$

Hence, the parabola meets the initial line SX at C (say) where $SC = 5/4$ and SX' at D (say) where $DS = 5$. The graph of the parabola is given in Fig. 2.31.

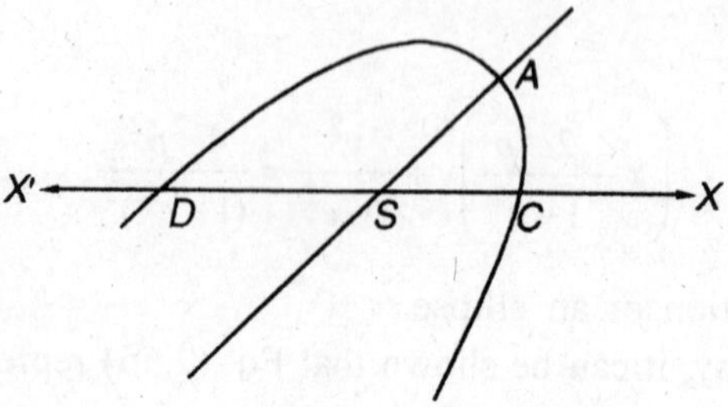

Figure 2.31

EXERCISES 2.2

1. Find the equation of the straight line which passes through the point (1, 2) and is such that the given point bisects the part intercepted between the axes.
2. Find the area of the triangle formed by the lines whose equations are $y = m_1x + c_1$, $y = m_2x + c_2$ and $x = 0$.
3. Find the area of the triangle formed by the straight lines whose equations are $2y + x - 5 = 0$, $y + 2x - 7 = 0$ and $x - y + 1 = 0$.
4. Find the equations of the sides of a square the coordinates of two opposite angular points of which are (3, 4) and (1, – 1).
5. Find the equation of the line which passes through the point of intersection of the lines $5x - 6y = 1$ and $3x + 2y + 5 = 0$ and is perpendicular to the line $3x - 5y + 11 = 0$.
6. Find the equations of the straight lines bisecting the angles between $3x + 4y = 5$, $12x - 5y = 41$.
7. Show that the area of the parallelogram formed by the lines $4y - 3x = a$, $4x - 3y = a$, $4y - 3x = 3a$ and $4x - 3y = 2a$ is $(2/7)a^2$.
8. Find the equation of the line which bisects the obtuse angle between the lines $x - 2y + 4 = 0$ and $4x - 3y + 2 = 0$.
9. The area of a triangle is 5. Two of the vertices are (2, 1) and (3, – 2). The third vertex lines on $y = x + 3$. Find the third vertex.

10. One side of a rectangle lines along the line $4x+7y+5=0$. Two of its vertices are (–3, 1) and (1, 1). Find the equations to the other three sides.

11. Show that the straight lines $7x-2y+10=0$, $7x+2y-10=0$ and $y+2=0$ form an isoceles triangle and find its area.

12. Find the equations of the straight lines passing through (–2, –7) and having an intercept of length 3 between the straight lines $4x+3y=12$, $4x+3y-3=0$.

13. A straight line moves so that the sum of the reciprocals of its intercepts on the axes of coordinates is constant *k*. Show that it passes through a fixed point.

14. Show that the origin is within the triangle whose vertices are (2, 1), (3, –2) and (–4, –1).

15. Find the incentre of the triangle formed by the lines $x+1=0$, $3x-4y=5$ and $5x+12y-27=0$.

16. Find the value of *c* so that the equation $12x^2-10xy+2y^2+14x-5y+c=0$ represents two straight lines. Find their point of intersection and the angle between them.

17. For what value of *k* does the equation $12x^2+7xy+ky^2+13x-y+3=0$ represents a pair of straight lines? Write down the separate equations of the lines and find the angle between them.

18. The equation $ax^2+3xy-2y^2-5x+5y+c=0$ represents two straight lines perpendicular to each other. Find *a* and *c*.

19. If the equation $ax^2+2hxy+by^2+2gx+2fy+c=0$ represents two straight lines, prove that the product of the lengths of the perpendiculars from the origin on these straight lines is

$$\frac{c}{\sqrt{(a-b)^2+4h^2}}$$

20. Show that the equation $x^2+4xy-2y^2+6x-12y-15=0$ represents a pair of straight lines and that these lines together with the pair $x^2+4xy-2y^2=0$ form a rhombus.

21. The straight lines joining the origin to the points of intersection of $3x^2-xy+3y^2+2x-3y+4=0$ and $2x+3y=\lambda$ are at right angles. Find *l*.

22. Show that the straight lines $x^2-4xy+y^2=0$ and $x+y=3$ form an equilateral triangle.

23. Show that the equation $4x^2+4xy+y^2-6x-3y-4=0$ represents a pair of parallel straight lines and find the distance between them.

24. The equation $ax^2+2hxy+by^2+2gx+2fy+c=0$ represents a pair of straight lines at right angles to each other. Prove that the distance of their point of intersection from the origin is

$$\sqrt{\frac{f^2+g^2}{b^2+h^2}}$$

25. Find the value of k so that $2x^2-5xy+2y^2-7x+11y+k=0$ represents two straight lines. For this value of k, show that the two lines intersect at a point on the circle $x^2+y^2-12x+6y+20=0$. Find the equation of the tangent to the circle at this point.

26. A circle touches each of the circles $x^2+y^2=2ax$ and $x^2+y^2=6ax$, one internally and the other extremely. Show that it cuts orthogonally the circle $x^2+y^2=3ax$.

27. Find the equation of the circle which passes through the origin and cuts off chords of lengths 4 and 6 on the positive side of the x-axis and y-axis, respectively.

28. Three circles are given by the equations $x^2+y^2=1$, $x^2+y^2-8x+15=0$ and $x^2+y^2+10y+24=0$, respectively. Find the coordinates of a point such that the tangents drawn from it to the three are equal in length.

29. Show that the circles $x^2+y^2=400$ and $x^2+y^2-10x-24y+120=0$ touch each other internally. Find the coordinates of the point of contact.

30. A point moves so that the square of its distance from the base of an isoceles triangle is equal to the rectangle contained by its distances from the equal sides. Prove that its locus is a circle.

31. Find the equation of the circle which passes through the points $P(1, 0)$, $Q(0, 1)$ and $(2, 2)$.

32. Obtain the equation of the circumcircle of the triangle whose sides are $5x+y=103$, $2x=3y+31$ and $x=y+11$ and find its radius.

33. Two rods whose lengths are 5 and 3 units slide along two rectangular axes in such a way that their extremities are always concyclic. Find the locus of the centre of the circle.

34. Show that the points (3, 4), (0, 5), (–5, 0) and (–3, –4) are concylic and find the radius of the circle.

35. Is the point (7, –11) inside or outside the circle $x^2+y^2-10x=0$?

36. Find the locus of a point the tangents from which to the circles $x^2+y^2=9/4$ and $x^2+y^2=16/9$ are in the ratio 3:4.

37. A chord of length $2b$ is cut off from the line $y=mx+c$ by the circle $x^2+y^2=a^2$. Show that $c^2=(1+m)^2\,(a^2-b^2)$.

38. Tangents are drawn to the circles $x^2+y^2=a^2$, $x^2+y^2=b^2$ at right angles to each other. Find the locus of their point of intersection.

39. $ABCD$ is square of side $2a$ units. Taking AB and AD as axes of coordinates, find the equations to the circle which touches the sides of the square.

40. Prove that the orthocentre of the triangle whose vertices are $(a\cos\alpha, a\sin\alpha)$, $(a\cos\beta, a\sin\beta)$ and $(a\cos\gamma, a\sin\gamma)$, is the point $[a(\cos\alpha + \cos\beta + \cos\gamma), a(\sin\alpha + \sin\beta + \sin\gamma)]$. Hence prove that the centroid of any triangle divides the join of the circumcentre and orthocentre in the ratio 1:2.

41. A parabolic arch erected across a road 30-metre wide is 6.25 metre high. Find (i) the equation of this parabola choosing proper axes of coordinates, (ii) the focus and (iii) the length of its latus rectum.

42. Find the length of the chord cut off the straight line $y = 2x - 1$ by the parabola $2y^2 = 3x$.

43. The line $y = mx + c$ meets the parabola $y^2 = 4ax$ in the points $P(x_1, y_1)$ and $P(x_2, y_2)$. Show that x_1, x_2 are the roots of the equation $mx^2 + 2x(mc - 2a) + c^2 = 0$. Deduce that the line is a tangent if $a = mc$, and that, in general, the length of the chord PQ is

$$\frac{4}{m^2}\sqrt{a(a - mc)(1 + m^2)}.$$

44. If the tangent at a point P on the parabola meets the directrix at L, show that LP subtends a right angle at the focus.

45. Find the equation of the tangent to the parabola $y^2 = 4ax$ at the point $(at^2, 2at)$. Show that the tangents drawn to this parabola from a point on the line $x = c$ meet the line $x = -c$ in points which are equidistant from the axis of the parabola.

46. Show that the locus of the middle points of all parallel chords of the parabola $y^2 = 4ax$, which have gradient m is the line $my = 2a$.

47. A variable straight line, whose direction is fixed, cuts a parabola in P and Q. Prove that the locus of points of intersection of the normals at P and Q to the parabola is a straight line which is itself normal to the parabola.

48. Prove that normals at the ends of a focal chord of the parabola $y^2 = 4ax$ intersect on the parabola $y^2 = ax - 3a^2$.

49. A ray of light comes from a distant light source along the line $y = c$ and strikes the surface of the parabola $y^2 = 4ax$ at P. It is then reflected through the focus to strike the surface again at Q. Along what line will it leave the parabola at Q?

50. Prove that the tangents at the ends of a focal chord of a parabola $y^2 = 4ax$ intersect at right angles on the directrix.

51. If the normal at the point $(at_1^2, 2at_1)$ on a parabola $y^2 = 4ax$ meets it again at $(at_2^2, 2at_2)$, prove that

$$t_2 = -t_1 - \frac{2}{t_1}.$$

52. A parabolic antenna is to be built for a satellite tracking station. It is to have a focal length of 40 cm and a depth of 25 cm. Find its equation. What is the aperture of the antenna?

53. From the points on the line $2x-3y+4=0$, tangents are drawn to the parabola $y^2=4ax$. Show that the chord of contact passes through a fixed point.

54. Show that the parabolas $y^2=8ax$ and $x^2=27ay$ intersect at the origin at right angles and at another point at an angle whose tangent is 9/13.

55. Prove the following properties of the parabola $y^2=4ax$:

(i) P is a point on the parabola, the tangent at P meets the axis at T and PN is the ordinate at P. Then, $AN=AT$ and $SP=ST$, where A is the vertex and S is its focus.

(ii) If the tangent at P meets the tangent at the vertex in Y, then SY is perpendicular to TP and $(SY)^2=AS\cdot SP$.

(iii) The circle circumscribing the tangle formed by any three tangents to a parabola passes through the focus.

(iv) The orthocentre of the triangle formed by three tangents to a parabola is on the directrix.

56. Find the centre and eccentricity of the ellipse

$$9x^2+25y^2-18x-100y-116=0.$$

57. An Earth satellite moves in an ellipse with the earth at one of its foci and is such that the ratio of the shortest distance from the satellite to the earth (perigee) to its largest distance (apogee) is 29:30. Determine the eccentricity of the orbit of the satellite.

58. The apogee and perigee of an Earth satellite with an elliptical orbit are 1720 km and 264 km, respectively. What is the eccentricity of the orbit? (Radius of the earth = 6400 km).

59. Any tangent to an ellipse cuts the tangents at the vertices A and A' at points V and V'. Prove that VV' subtends a right angle at either focus.

60. Obtain the equation of the normal to the ellipse

$$\frac{x^2}{a^2}+\frac{y^2}{b^2}=1 \qquad (a>b>0)$$

at the point $(a\cos\theta,\, b\sin\theta)$. Express $\cos\theta$ and $\sin\theta$ in terms of $t=\tan\frac{1}{2}\theta$ and deduce that, in general, four normals can be drawn to an ellipse from a given point.

61. If ϕ_1, ϕ_2 are the eccentric angles of the extremities of a focal chord of an ellipse of eccentricity e, prove that

$$\cos\frac{1}{2}(\phi_1-\phi_2)=e\cos\frac{1}{2}(\phi_1+\phi_2).$$

62. If the chord joining the variable points 'θ', 'ϕ' on the ellipse $(x^2/a^2) + (y^2/b^2) = 1$ subtends a right angle at the point (a, 0), show that

$$\tan\frac{\theta}{2}\ \tan\frac{\phi}{2} = -\frac{b^2}{a^2}.$$

63. An astronaut is to be fired into an elliptical orbit about the Earth having a minimum altitude of 960 km and a maximum altitude of 5440 km. Assuming the Earth has a radius of approximately 6400 km, determine the equation for describing the path followed by the astronaut's vehicle.

64. Show that the equation of the chord joining the points on the ellipse $(x^2/a^2) + (y^2/b^2) = 1$ whose eccentric angles are θ and ϕ is

$$bx \cos\phi\ \frac{1}{2}(\theta+\phi) + ay \sin\frac{1}{2}(\theta+\phi) = ab \cos\frac{1}{2}(\theta-\phi).$$

65. A diameter of an ellipse is any chord through its centre. Show that the minor axis is the shortest and the major axis is the longest diameter if the axes are of unequal lengths.

66. Find the centre, eccentricity, foci and directrices of the hyperbola $9x^2 - 16y^2 + 72x - 32y - 16 = 0$.

67. Prove that the equation of the normal at the point $(ct, c/t)$ on the rectangular hyperbola $xy = c^2$ is $xt^3 - yt + c(1-t^4) = 0$. Prove also that this normal meets the hyperbola again at the point with parameter $-1/t^3$.

68. The tangent to the hyperbola $xy = c^2$ at the point $(ct, c/t)$ intersects the hyperbola $xy = -c^2$ at the points M and N. Show that the tangents at M and N to the hyperbola $xy = -c^2$ intersect on the hyperbola $xy = c^2$.

69. Prove that the equation of the tangent to the hyperbola $(x^2/a^2) - (y^2/b^2) = 1$ at the point

$$\left[\frac{1}{2}a\left(t+\frac{1}{t}\right), \frac{1}{2}b\left(t-\frac{1}{t}\right)\right]$$

is $bx(t^2+1) - ay(t^2-1) = 2abt$.

70. Find the asymptotes of the hyperbola $3x^2 - 5xy - 2y^2 + 17x + y + 14 = 0$.

71. Show that the normal to the rectangular hyperbola $xy = c^2$ at the point $P(ct, c/t)$ meets the curve again at the point $Q(-c/t^3, -ct^3)$.

72. The rectangular hyperbola $xy = c^2$ is cut by a circle in four points. Prove that the sum of the squares of the distances of these four points from the centre of the hyperbola is equal to the square on the diameter of the circle.

73. The hyperbola which has, for its transverse and conjugate axes, the conjugate and transverse axes of another hyperbola, is called the *conjugate* hyperbola. If e denotes the eccentricity of a hyperbola and e' that of its conjugate, show that $(1/e^2) + (1/e'^2) = 1$.

74. Find the equation of the hyperbola with foci $(\pm 5, 0)$ and vertices $(\pm 4, 0)$.

75. Show that $(xx_1/a^2) - (yy_1/b^2) = 1$ is an equation of the tangent line to the hyperbola $(x^2/a^2) - (y^2/b^2) = 1$ at the point (x_1, y_1).

76. If A and B are the point at which the tangent line to a hyperbola at any point P intersects the asymptotes, show that P is the midpoint of AB.

77. Find the locus of a point which moves so that its distance from the point $(3, 0)$ is always equal to e times its distance to the y-axis. Sketch the graph of the equation if (i) $e = 1/2$, (ii) $e = 1$ and (iii) $e = 3/2$.

78. Show that the curves $xy = a^2$ and $x^2 - y^2 = c^2$ intersect orthogonally, provided $a \neq 0$.

79. Transform the equation $y^2 - x^2 = 4$ by rotating the coordinate axes x and y through an angle of $45°$.

80. If (x, y) and (X, Y) are the coordinates of a point P referred to rectangular axes OX, OY and OX, OY, and if the expression $ax^2 + 2hxy + by^2$ transforms into $AX^2 + 2HXY + BY^2$, prove that (i) $a + b = A + B$ and (ii) $ab - h^2 = AB - H^2$.

81. Show that the graph of the equation: $ax^2 + bxy + cy^2 + dx + ey + f = 0$ is a hyperbola, a parabola, or an ellipse depending on whether $b^2 - 4ac$ is positive, zero, or negative, respectively.

82. Sketch the graphs of the following equations:

(i) $x^2 - 2x - 2y - 3 = 0$ (ii) $y^2 - 6y - 3x + 3 = 0$

(iii) $4x^2 + 3y^2 - 16x + 18y + 31 = 0$ (iv) $y^2 + 8x - 6 = 0$

(v) $x^2 - y^2 + 4x - 4y + 1 = 0$ (vi) $y^2 + 2y + 8x - 15 = 0$.

83. Reduce the following equations to a standard form by rotation of axes and sketch the graph:

(i) $x^2 + y^2 + 4xy + 2x - 12 = 0$ (ii) $3x^2 + 4xy - 4 = 0$

(iii) $x^2 - y^2 = 10$ (iv) $12x^2 + 7xy - 12y^2 - 1 = 0$.

84. If a straight line cuts a hyperbola in P and Q and its asymptotes in R and S, show that $PR = QS$.

85. Show that the equation $l = r(1 + e\cos\theta)$ represents a conic with semilatus rectum l and eccentricity e.

86. Sketch the graphs of the following equations:

(i) $r = \dfrac{2}{1 - \cos\theta}$ (ii) $r = \dfrac{2}{1 + \cos\theta}$

(iii) $r = \dfrac{2}{1+2\cos\theta}$ (iv) $r = \dfrac{16}{5+3\sin\theta}$

(v) $r = \dfrac{12}{3+\cos\theta}$ (vi) $r = 3\cos\theta + 4\sin\theta$

(vii) $r = \dfrac{2}{1+\cos\theta+\sin\theta}$ (viii) $r = \dfrac{3}{1+\sqrt{2}\cos\theta}$

2.3 VECTORS AND SOLID GEOMETRY

In the study of higher mathematics, some knowledge of vectors is almost indispensable. In this section, an introduction to vector algebra and its applications to mechanics and solid geometry is presented. The treatment although brief, is quite adequate for the use of vectors in physics and engineering problems.

2.3.1 The Concept of a Vector

The idea of *vector* is an abstract one denoting the combination of a magnitude and a specified direction, which is usually exemplified in physical quantities such as displacement, velocity, acceleration, force, etc. Typically, a vector is a *directed line segment*, say $\overrightarrow{PQ}$ from one point P called the *initial point,* to another point Q, called the *terminal point.* We also denote vectors by single bold faced letters. Thus $\overrightarrow{PQ}$ is denoted by **a**, as in Fig. 2.32. The *magnitude* or *length* of the vector is then denoted by $|\overrightarrow{PQ}|$ or $|\mathbf{a}|$. We shall consider properties of this segment and assume that these properties also hold good for physical quantities of the type listed above.

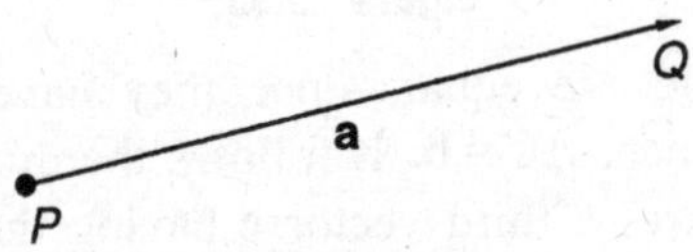

Figure 2.32

We also come across quantities which are characterized by magnitude only, such as mass, length, temperature, etc. Such quantities are called *scalars* to distinguish them from vectors. We denote scalars by ordinary letters as usual. The magnitude of a vector **a** is denoted by a. If $a = 0$, then **a** is a *zero vector.* The vector $\overrightarrow{QP}$ is the *negative* of $\overrightarrow{PQ}$ i.e. the negative of **a**. A *unit* vector is one whose magnitude is unity. We shall denote the unit vector in the direction of **a** by $\hat{a}$. It is easily seen that

$$\mathbf{a} = a\hat{a} \tag{2.57}$$

Two vectors **a** and **b** are said to be *equal* if they have equal magnitudes and the same direction, i.e. if $a = b$ and $\hat{a} = \hat{b}$. If we multiply a vector **a** by a scalar quantity m, then the product $m\mathbf{a}$ is defined to be a vector whose magnitude is m times the magnitude of **a** and whose direction is the same as the direction of **a**, i.e.

$$m\mathbf{a} = ma\hat{a} = (ma)\hat{a}.$$

2.3.2 Addition and Subtraction of Vectors

Two vectors are added according to the parallelogram or triangle law of addition. This is illustrated in Fig. 2.33. Let two vectors **a** and **b** be represented by $\overrightarrow{\mathrm{OA}}$ and $\overrightarrow{\mathrm{OB}}$ respectively, as in Fig. 2.33(a). Then the vector $(\overrightarrow{\mathrm{OA}}+\overrightarrow{\mathrm{OB}})$ is represented by $\overrightarrow{\mathrm{OC}}$ the diagonal of a parallelogram *OACB*. Thus,

$$\mathbf{c} = \mathbf{a} + \mathbf{b}.$$

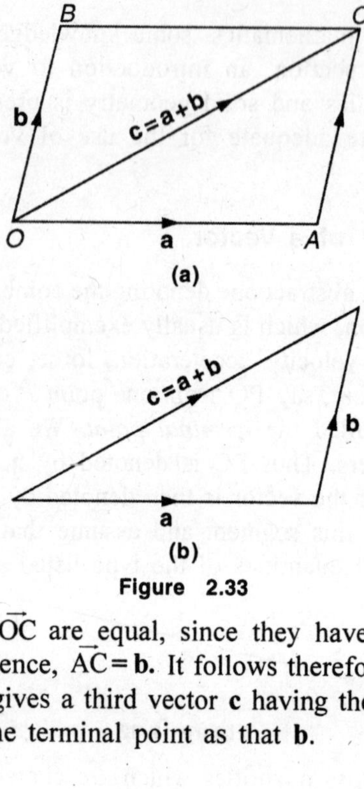

Figure 2.33

The vectors $\overrightarrow{\mathrm{OB}}$ and $\overrightarrow{\mathrm{OC}}$ are equal, since they have the same magnitude and same direction. Hence, $\overrightarrow{\mathrm{AC}}=\mathbf{b}.$ It follows therefore that the addition of two vectors **a** and **b** gives a third vector **c** having the same initial point as that of **a** and the same terminal point as that **b**.

Similarly,

$$\overrightarrow{\mathrm{OA}}+\overrightarrow{\mathrm{AB}}=\overrightarrow{\mathrm{OB}},$$

or

$$a+\overrightarrow{\mathrm{AB}}=b,$$

which means that

$$\overrightarrow{\mathrm{AB}}=\mathbf{b}-\mathbf{a}.$$

For any vector **a**, *a* vector $-\mathbf{a}$ is defined as the vector having the same magnitude as **a** but acting in the opposite direction. The vector $\mathbf{a}-\mathbf{b}$ is then defined as $\mathbf{a}+(-\mathbf{b})$. It follows, therefore, that (i) $-(-\mathbf{a})=\mathbf{a}$ and (ii) $\mathrm{a}+(-\mathbf{a})=0$.

Since $\overrightarrow{\mathrm{BC}}=\overrightarrow{\mathrm{OA}}=a,$ we have

$$\overrightarrow{\mathrm{OB}}+\overrightarrow{\mathrm{BC}}=\mathbf{b}+\mathbf{a}=\overrightarrow{\mathrm{OC}}=\mathbf{c}=\mathbf{a}+\mathbf{b},$$

which shows that the commutative law of addition holds for the vectors. Similarly, it can be shown that the distributive law

$$m(\mathbf{a}+\mathbf{b}) = m\mathbf{a} + m\mathbf{b}$$

also holds good for the vectors.

The magnitude of a vector has some of the properties of the absolute value of a real number. These are listed below:

(i) $|\mathbf{a}| \geq 0$; $|\mathbf{a}| = 0$ if and only if $\mathbf{a} = \mathbf{0}$.

(ii) $|x\mathbf{a}| = |x||\mathbf{a}|$.

(iii) $\left|\dfrac{\mathbf{a}}{x}\right| = \left|\dfrac{\mathbf{a}}{x}\right|$.

(iv) $|\mathbf{a}+\mathbf{b}| \leq |\mathbf{a}| + |\mathbf{b}|$ (The triangle inequality).

(v) $|-\mathbf{a}| = |\mathbf{a}|$.

(vi) $|\mathbf{a}-\mathbf{b}| \leq |\mathbf{a}| + |\mathbf{b}|$.

(vii) $||\mathbf{a}|-|\mathbf{b}|| \leq |\mathbf{a}-\mathbf{b}|$.

2.3.3 Resolution of Vectors

We consider three mutually perpendicular axes *OX*, *OY* and *OZ*, as in Fig. 2.34. Let $P(x, y, z)$ be a point in space and let **i, j, k** denote unit vectors along *OX*, *OY* and *OZ*, respectively. Let *M* be the foot of the perpendicular from *P* on the plane *XOY* and let *MN* be drawn perpendicular to *OX*. If $\vec{\mathrm{OP}} = \mathbf{r}$, then

$$\vec{\mathrm{OP}} = \mathbf{r} = \vec{\mathrm{OM}} + \vec{\mathrm{MP}} = \vec{\mathrm{ON}} + \vec{\mathrm{NM}} + \vec{\mathrm{MP}},$$

since

$$\vec{\mathrm{OM}} = \vec{\mathrm{ON}} + \vec{\mathrm{NM}}.$$

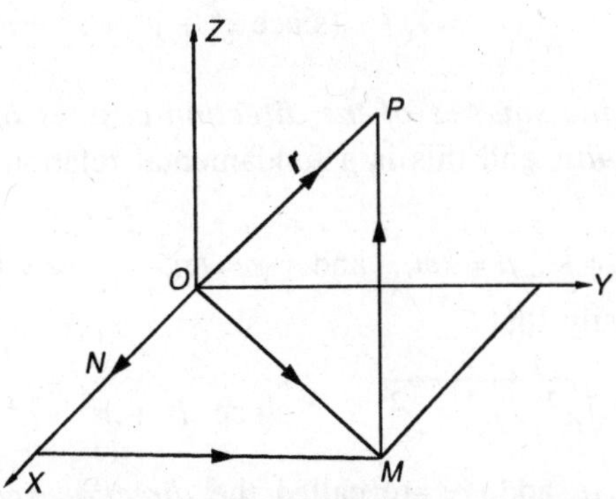

Figure 2.34

Here,

$$\overrightarrow{\mathrm{ON}} = x\mathbf{i}, \qquad \overrightarrow{\mathrm{NM}} = y\mathbf{j}, \qquad \overrightarrow{\mathrm{MP}} = z\mathbf{k},$$

Hence, we have

$$\mathbf{r} = x\mathbf{i} + y\mathbf{j} + z\mathbf{k} \tag{2.58}$$

In (2.58), x, y and z are called the *components* of the vector $\mathbf{r}$ along the coordinate axes; they also represent the projections of the vector $\mathbf{r}$ on OX, OY and OZ, respectively. The vector $\mathbf{r}$ is called the *position vector* of P relative to O as origin.

Let the angles made by $\overrightarrow{\mathrm{OP}}$ with OX, OY and OZ be α, β and γ, respectively. These are usually called the *direction angles of* $\overrightarrow{\mathrm{OP}}$. Their cosines, denoted by l, m, n respectively, are called the *direction cosines* of $\overrightarrow{\mathrm{OP}}$ and are important quantities in the study of three-dimensional geometry. Clearly,

$$l = \cos\alpha = \frac{x}{r}, \qquad m = \cos\beta = \frac{y}{r}, \qquad n = \cos\gamma = \frac{z}{r},$$

where $r = OP$. These relations yield $x = rl$, $y = rm$ and $z = rn$, respectively, whence (2.58) gives:

$$\mathbf{r} = r\,(l\mathbf{i} + m\mathbf{j} + n\mathbf{k}) \tag{2.59}$$

But since $\mathbf{r} = r\hat{r}$, we have

$$\hat{r} = l\mathbf{i} + m\mathbf{j} + n\mathbf{k}, \tag{2.60}$$

which is an important relation since it establishes a connection between the direction cosines of a straight line and a unit vector in its direction. Further, we see that

$$\begin{aligned} l^2 + m^2 + n^2 &= \cos^2\alpha + \cos^2\beta + \cos^2\gamma \\ &= \frac{x^2}{r^2} + \frac{y^2}{r^2} + \frac{z^2}{r^2} \\ &= 1, \qquad (\text{since } x^2 + y^2 + z^2 = r^2). \end{aligned} \tag{2.61}$$

Thus, *the sum of the squares of the direction cosines of a straight line is always equal to unity,* and this is a fundamental relation in solid geometry. If we set

$$\lambda = kl,\ \mu = km \quad \text{and} \quad \gamma = kn, \qquad (k \neq 0)$$

then it follows easily that

$$k = \sqrt{\lambda^2 + \mu^2 + \gamma^2}, \qquad \text{since } l^2 + m^2 + n^2 = 1.$$

The quantities λ, μ and γ are called the *direction ratios* of $\overrightarrow{\mathrm{OP}}$. It is a simple exercise to show that if $P\,(x_1, y_1, z_1)$ and $Q\,(x_2, y_2, z_2)$ are two given points, then the direction ratios of $\overrightarrow{PQ}$ are $x_2 - x_1$, $y_2 - y_1$ and $z_2 - z_1$ respectively, and that its direction cosines are

$$\frac{x_2 - x_1}{d}, \quad \frac{y_2 - y_1}{d}, \quad \frac{z_2 - z_1}{d},$$

where

$$d = PQ = \sqrt{(x_2 - x_1)^2 + (y_2 - y_1)^2 + (z_2 - z_1)^2}.$$

Example 2.50 The points $A(2, 4, -1)$, $B(4, 5, 1)$ and $C(3, 6, -3)$ are the vertices of a triangle ABC. Find AB, BC and CA, and show that the triangle ABC is a right-angled triangle. Find the direction cosines of $\overrightarrow{AB}$.

Let O be an arbitrary fixed origin. Then the position vectors of A, B and C with respect to the origin O are given by

$$\overrightarrow{OA} = 2\mathbf{i} + 4\mathbf{j} - \mathbf{k}, \quad \overrightarrow{OB} = 4\mathbf{i} + 5\mathbf{j} + \mathbf{k}, \quad \overrightarrow{OC} = 3\mathbf{i} + 6\mathbf{j} - 3\mathbf{k},$$

Hence,

$$\overrightarrow{AB} = \overrightarrow{OB} - \overrightarrow{OA} = 2\mathbf{i} + \mathbf{j} + 2\mathbf{k} \quad \text{or} \quad AB = |\overrightarrow{AB}| = \sqrt{4 + 1 + 4} = 3,$$

$$\overrightarrow{BC} = \overrightarrow{OC} - \overrightarrow{OB} = -\mathbf{i} + \mathbf{j} - 4\mathbf{k} \quad \text{or} \quad BC = \sqrt{18},$$

$$\overrightarrow{CA} = \overrightarrow{OA} - \overrightarrow{OC} = -\mathbf{i} - 2\mathbf{j} + 2\mathbf{k} \quad \text{or} \quad CA = 3,$$

Since $AB^2 + CA^2 = BC^2$, it follows that the triangle ABC is a right-angled triangle. The direction cosines of $\overrightarrow{AB}$ are 2/3, 1/3 and 2/3.

Example 2.51 Find all values of λ such that $(x, y, z) \neq (0, 0, 0)$ and

$$(\mathbf{i} + \mathbf{j} + 3\mathbf{k})x + (3\mathbf{i} - 3\mathbf{j} + \mathbf{k})y + (-4\mathbf{i} + 5\mathbf{j})z = \lambda(\mathbf{i}x + \mathbf{j}y + \mathbf{k}z)$$

where **i**, **j**, **k** are unit vectors along the coordinate axes.

We are given that

$$(\mathbf{i} + \mathbf{j} + 3\mathbf{k})x + (3\mathbf{i} - 3\mathbf{j} + \mathbf{k})y + (-4\mathbf{i} + 5\mathbf{j})z = \lambda(\mathbf{i}x + \mathbf{j}y + \mathbf{k}z)$$

or

$$[(1 - \lambda)x + 3y - 4z]\mathbf{i} + [x - (3 + \lambda)y + 5z]\mathbf{j} + [3x + y - \lambda z]\mathbf{k} = 0.$$

It follows therefore that

$$(1 - \lambda)x + 3y - 4z = 0$$

$$x - (3 + \lambda)y + 5z = 0$$

$$3x + y - \lambda z = 0.$$

For a nontrivial, solution, we must have

$$\begin{vmatrix} 1 - \lambda & 3 & -4 \\ 1 & -(3 + \lambda) & 5 \\ 3 & 1 & -\lambda \end{vmatrix} = 0$$

Simplification gives:

$$\lambda = -1 \text{ and } \lambda = 0.$$

Example 2.52 Two points A and B have position vectors $\mathbf{a}$ and $\mathbf{b}$ relative to O as origin. If a point R divides the distance AB in the ratio $m : n$, find the position vector of R. Hence, find the condition of collinearity of three points.

Let $\mathbf{r}$ be the position vector of R relative to O, as origin (Fig. 2.35). We have

$$\frac{\overrightarrow{\text{AR}}}{\overrightarrow{\text{RM}}} = \frac{m}{n}.$$

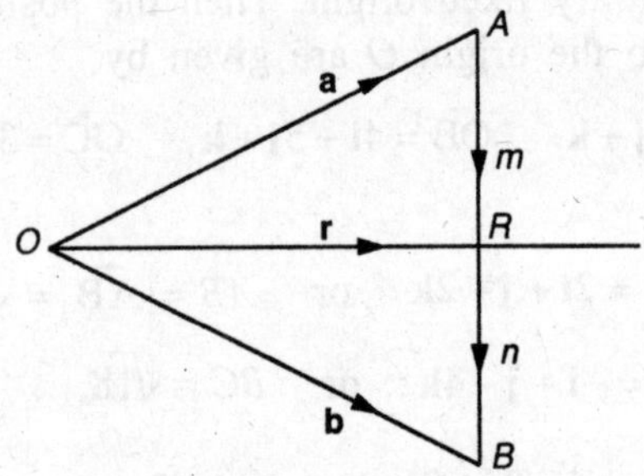

Figure 2.35

But

$$\overrightarrow{\text{AR}} = \overrightarrow{\text{OR}} - \overrightarrow{\text{OA}} = \mathbf{r} - \mathbf{a} \quad \text{and} \quad \overrightarrow{\text{RB}} = \mathbf{b} - \mathbf{r}.$$

Hence,

$$\frac{\mathbf{r} - \mathbf{a}}{\mathbf{b} - \mathbf{r}} = \frac{m}{n},$$

from which we obtain

$$(m+n)\mathbf{r} = n\mathbf{a} + m\mathbf{b}$$

or

$$(m+n)\mathbf{r} - n\mathbf{a} - m\mathbf{b} = 0. \qquad \text{(i)}$$

where it should be noted that the sum of the coefficients of $\mathbf{a}$, $\mathbf{b}$ and $\mathbf{r}$ is zero. Hence, it follows that the three points A, R and B (with position vectors $\mathbf{a}$, $\mathbf{r}$ and $\mathbf{b}$ respectively) are collinear if their position vectors satisfy a relation of the type

$$\lambda\mathbf{r} + \mu\mathbf{a} - \gamma\mathbf{b} = 0 \qquad \text{(ii)}$$

where

$$\lambda + \mu + \gamma = 0 \qquad \text{(iii)}$$

Example 2.53 Forces P, Q act at O and have a resultant R. If any transversal cuts their lines of action at A, B, C, respectively, show that

$$\frac{P}{OA}+\frac{Q}{OB}=\frac{R}{OC}.$$

Let $\vec{OA}=\mathbf{a}$, $\vec{OB}=\mathbf{b}$ and $\vec{OC}=\mathbf{c}$ (see Fig 2.36). Then

$$\frac{\mathbf{a}}{OA},\qquad \frac{\mathbf{b}}{OB},\qquad \frac{\mathbf{c}}{OC}$$

are the unit vectors along OA, OB and OC respectively. Since P, Q and R are their magnitudes, the three forces, **P**, **Q** and **R** can now be obtained:

$$\mathbf{P}=\frac{P}{OA}\,\mathbf{a},\quad \mathbf{Q}=\frac{Q}{OB}\,\mathbf{b},\qquad \mathbf{R}=\frac{R}{OC}\,\mathbf{c}$$

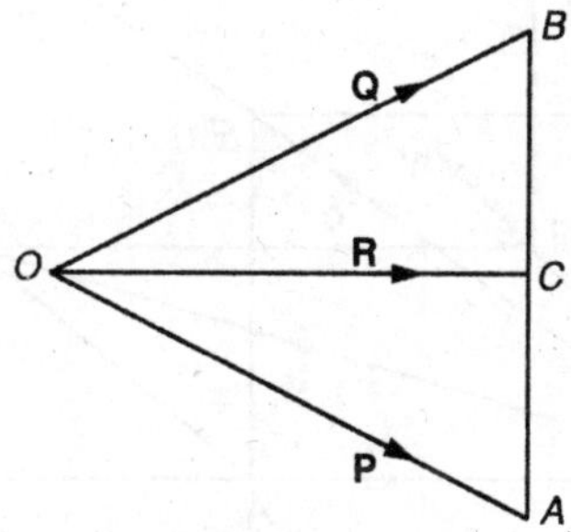

Figure 2.36

But **R** is the resultant of **P** and **Q** and hence we have

$$\frac{P}{OA}\mathbf{a}+\frac{Q}{OB}\mathbf{b}=\frac{R}{OC}\mathbf{c}.$$

or

$$\frac{P}{OA}\mathbf{a}+\frac{Q}{OB}\mathbf{b}-\frac{R}{OC}\mathbf{c}=\mathbf{0}.$$

Since A, B, C are collinear, the sum of the coefficients of their position vectors in the above relation should be zero. Hence, we have

$$\frac{P}{OA}+\frac{Q}{OB}-\frac{R}{OC}=0,$$

or

$$\frac{P}{OA}+\frac{Q}{OB}=\frac{R}{OC},$$

which is the required relation.

Example 2.54 Three vectors of lengths 1, 2, 3 units meeting at the corner of a cube are directed along the diagonals of its three faces meeting at the corner. Find their resultant.

Let the axes be chosen as in Fig 2.37. The three faces of the cube meeting at *O* are seen to be *OADB*, *OAEC* and *OBGC*. Their diagonals are respectively *OD*, *OE* and *OG* and these be represented by the vectors **a**, **c** and **b**, respectively. As before, let **i**, **j** and **k** be unit vectors along *OX*, *OY* and *OZ*, respectively. Since *OADB* is a square, its diagonal *OD* makes an angle of 45° with *OX* and *OY*. Hence, the vector **a** can be written as

$$\mathbf{a} = \cos 45°\,\mathbf{i} + \cos 45°\,\mathbf{j} = \frac{1}{\sqrt{2}}(\mathbf{i}+\mathbf{j}),$$

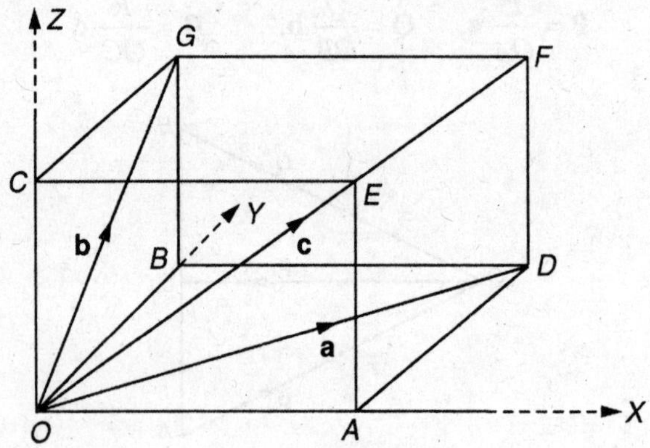

Figure 2.37

taking the magnitude of **a** as unity. Taking $|\mathbf{b}| = 2$ and $|\mathbf{c}| = 3$, we can similarly write:

$$\mathbf{b} = \frac{2}{\sqrt{2}}(\mathbf{j}+\mathbf{k}) \quad \text{and} \quad \mathbf{c} = \frac{3}{\sqrt{2}}(\mathbf{k}+\mathbf{i}).$$

If their resultant is **r**, then we have

$$\mathbf{r} = \mathbf{a}+\mathbf{b}+\mathbf{c} = \frac{1}{\sqrt{2}}(\mathbf{i}+\mathbf{j}) + \frac{2}{\sqrt{2}}(\mathbf{j}+\mathbf{k}) + \frac{3}{\sqrt{2}}(\mathbf{k}+\mathbf{i})$$

$$= \frac{1}{\sqrt{2}}(4\mathbf{i}+3\mathbf{j}+5\mathbf{k})$$

and

$$r = |\mathbf{r}| = \frac{1}{\sqrt{2}}\sqrt{16+9+25} = \frac{1}{\sqrt{2}}(5\sqrt{2}) = \mathbf{5}\ \text{units}.$$

Example 2.55 If $\mathbf{t}_1 = 2\mathbf{i} - \mathbf{j} + \mathbf{k}$, $\mathbf{t}_2 = \mathbf{i} + 3\mathbf{j} - 2\mathbf{k}$, $\mathbf{t}_3 = -2\mathbf{i} + \mathbf{j} - 3\mathbf{k}$ and $\mathbf{t}_4 = 3\mathbf{i} + 2\mathbf{j} - 5\mathbf{k}$, find scalars *p*, *q*, *r* such that $\mathbf{t}_4 = p\mathbf{t}_1 + q\mathbf{t}_2 + r\mathbf{t}_3$, and also show that $q + r = p$.

We have

$$\mathbf{t}_4 = p\mathbf{t}_1 + q\mathbf{t}_2 + r\mathbf{t}_3$$

i.e.,

$$3\mathbf{i} + 2\mathbf{j} - 5\mathbf{k} = p(2\mathbf{i} - \mathbf{j} + \mathbf{k}) + q(\mathbf{i} + 3\mathbf{j} - 2\mathbf{k}) + r(-2\mathbf{i} + \mathbf{j} - 3\mathbf{k}).$$
$$= (2p + q - 2r)\mathbf{i} + (-p + 3q + r)\mathbf{j} + (p - 2q - 3r)\mathbf{k}$$

Comparing the coefficients of **i**, **j** and **k** on both sides, we get

$$2p + q - 2r = 3$$
$$-p + 3q + r = 2$$
$$p - 2q - 3r = -5$$

Solving the above three equations, we obtain $p = -3$, $q = 1$ and $r = -4$. It also follows that $q + r = p$.

Example 2.56 *ABCD* is a quadrilateral. If the mid points of the consecutive sides *AB*, *BC*, *CD* and *DA* are connected by straight lines, show that the resulting quasilateral is a parallelogram.

Let **a**, **b**, **c** and **d** be the position vectors of the points *A*, *B*, *C*, *D*, respectively. Then the position vectors of *P*, *Q*, *R* and *S* are respectively (see Fig. 2.38):

$$\frac{\mathbf{a}+\mathbf{b}}{2}, \quad \frac{\mathbf{b}+\mathbf{c}}{2}, \quad \frac{\mathbf{c}+\mathbf{d}}{2}, \quad \frac{\mathbf{d}+\mathbf{a}}{2}$$

Hence

$$\overrightarrow{PQ} = \frac{\mathbf{c}-\mathbf{a}}{2}, \quad \overrightarrow{SR} = \frac{\mathbf{c}-\mathbf{a}}{2}, \quad \overrightarrow{PS} = \frac{\mathbf{d}-\mathbf{b}}{2}, \quad \text{and} \quad \overrightarrow{QR} = \frac{\mathbf{d}-\mathbf{b}}{2}.$$

It follows that *PQRS* is a parallelogram.

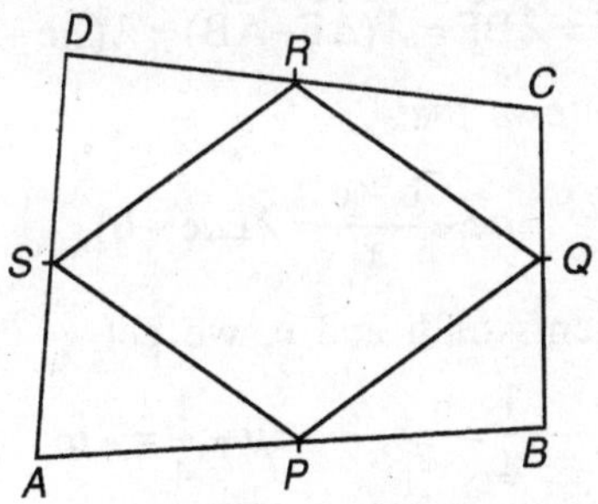

Figure 2.38

Example 2.57 The median *AD* of a triangle *ABC* is bisected at *E* and *BE* is produced to meet the side *AC* in *F*. Using vector methods, show that

$$\overrightarrow{AF} = \frac{1}{3}\overrightarrow{AC} \quad \text{and} \quad \overrightarrow{EF} = \frac{1}{4}\overrightarrow{BF}.$$

In Fig. 2.39, let A be chosen as origin and let $\mathbf{b}$ and $\mathbf{c}$ be the position vectors of B and C. Then the position vector of D is $(\mathbf{b}+\mathbf{c})/2$ and that of E is $(\mathbf{b}+\mathbf{c})/4$.

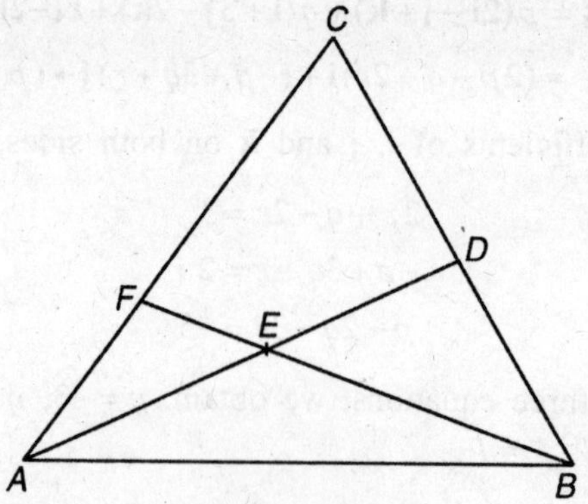

Figure 2.39

Let

$$\frac{\overrightarrow{AF}}{\overrightarrow{AC}}=\mu \quad \text{and} \quad \frac{\overrightarrow{EF}}{\overrightarrow{BF}}=\lambda$$

Now,

$$\overrightarrow{AE}+\overrightarrow{EF}=\overrightarrow{AF}$$

Hence,

$$\overrightarrow{EF}=\overrightarrow{AF}-\overrightarrow{AE}=\mu\overrightarrow{AC}-\frac{\mathbf{b}+\mathbf{c}}{4}=\mu\mathbf{c}-\frac{\mathbf{b}+\mathbf{c}}{4} \qquad \text{(i)}$$

Also,

$$\overrightarrow{EF}=\lambda\overrightarrow{BF}=\lambda(\overrightarrow{AF}-\overrightarrow{AB})=\lambda(\mu\mathbf{c}-\mathbf{b}) \qquad \text{(ii)}$$

From (i) and (ii), it follows that

$$\mu\mathbf{c}-\frac{\mathbf{b}+\mathbf{c}}{4}=\lambda(\mu\mathbf{c}-\mathbf{b})$$

Comparing the coefficients of $\mathbf{b}$ and $\mathbf{c}$, we get

$$-\frac{1}{4}=-\lambda, \qquad \mu-\frac{1}{4}=\lambda\mu$$

Solving them, we find

$$\lambda=\frac{1}{4}, \qquad \mu=\frac{1}{3}$$

or

$$\overrightarrow{EF}=\frac{1}{4}\overrightarrow{BF}, \qquad \overrightarrow{AF}=\frac{1}{3}\overrightarrow{AC}.$$

2.3.4 Scalar or Dot Product of Two Vectors

The multiplication of a vector by a scalar was already discussed in Section 2.3.1. When we multiply a vector by another vector, we must define precisely what we mean. There are, in general, two different ways in which vectors can be multiplied. These are the *scalar* and *vector* products and are discussed below.

The *scalar* or the *dot* product of two vectors **a** and **b** is written as $\mathbf{a} \cdot \mathbf{b}$ (read as '**a** dot **b**') and is defined as:

$$\mathbf{a} \cdot \mathbf{b} = ab \cos \theta \tag{2.62}$$

where θ is the angle between **a** and **b**. It is clear that the dot product is a scalar quantity. If the two vectors are perpendicular, then $\cos \theta = 0$, i.e. $\mathbf{a} \cdot \mathbf{b} = 0$. Conversely, if $\mathbf{a} \cdot \mathbf{b} = 0$, then the vectors **a** and **b** are perpendicular to each other. If the two vectors are parallel, then $\theta = 0$ and $\mathbf{a} \cdot \mathbf{b} = ab$, which means that $\mathbf{a} \cdot \mathbf{a} = a^2$. It is also easy to see that

$$\left.\begin{aligned} \mathbf{i} \cdot \mathbf{i} = \mathbf{j} \cdot \mathbf{j} = \mathbf{k} \cdot \mathbf{k} = 1 \\ \mathbf{i} \cdot \mathbf{j} = \mathbf{j} \cdot \mathbf{k} = \mathbf{k} \cdot \mathbf{i} = 0 \end{aligned}\right\} \tag{2.63}$$

From the definition of the dot product, we have

$$\mathbf{a} \cdot \mathbf{b} = ab \cos \theta = ba \cos \theta = \mathbf{b} \cdot \mathbf{a},$$

which shows that the *commutative law* holds for the scalar products. Similarly, the *distributive law*

$$\mathbf{a} \cdot (\mathbf{b} + \mathbf{c}) = \mathbf{a} \cdot \mathbf{b} + \mathbf{a} \cdot \mathbf{c}$$

also holds.

The dot product can also be written in terms of the vector components in the rectangular coordinate system. If

$$\mathbf{a} = a_1\mathbf{i} + a_2\mathbf{j} + a_3\mathbf{k}, \quad \text{and} \quad \mathbf{b} = b_1\mathbf{i} + b_2\mathbf{j} + b_3\mathbf{k},$$

then

$$|\mathbf{a}| = a = \sqrt{a_1^2 + a_2^2 + a_3^2}, \qquad |\mathbf{b}| = b = \sqrt{b_1^2 + b_2^2 + b_3^2},$$

and

$$\begin{aligned} \mathbf{a} \cdot \mathbf{b} &= (a_1\mathbf{i} + a_2\mathbf{j} + a_3\mathbf{k}) \cdot (b_1\mathbf{i} + b_2\mathbf{j} + b_3\mathbf{k}) \\ &= a_1\mathbf{i} \cdot (b_1\mathbf{i} + b_2\mathbf{j} + b_3\mathbf{k}) + a_2\mathbf{j} \cdot (b_1\mathbf{i} + b_2\mathbf{j} + b_3\mathbf{k}) + a_3\mathbf{k} \cdot (b_1\mathbf{i} + b_2\mathbf{j} + b_3\mathbf{k}) \\ &= a_1b_1 + a_2b_2 + a_3b_3, \end{aligned} \tag{2.64}$$

on using the relations (2.63).

It follows that

$$ab \cos \theta = a_1b_1 + a_2b_2 + a_3b_3$$

Hence, we have

$$\cos\theta = \frac{a_1b_1 + a_2b_2 + a_3b_3}{\sqrt{a_1^2 + a_2^2 + a_3^2}\ \sqrt{b_1^2 + b_2^2 + b_3^2}} \tag{2.65}$$

An important application of the scalar product concept is in mechanics, where the work done by a force in an elementary displacement is defined as the product of the elementary displacement with the resolved part of the force in the direction of the elementary displacement of the point of application of the force, i.e.

$$W = dF\cos\theta = \mathbf{F}\cdot\mathbf{d}, \tag{2.66}$$

where W is the work done due to displacement, $\mathbf{F}$ and $\mathbf{d}$ are the vectors representing the force and the displacement of its point of application.

Example 2.58 If $\mathbf{a}\cdot\mathbf{b} = \mathbf{a}\cdot\mathbf{c}$, show that either $a = 0$ or $b = c$ or $\mathbf{a}$ is orthogonal to $(\mathbf{b} - \mathbf{c})$.

We have

$$\mathbf{a}\cdot\mathbf{b} = \mathbf{a}\cdot\mathbf{c}$$

or

$$\mathbf{a}\cdot(\mathbf{b} - \mathbf{c}) = 0$$

or

$$|\mathbf{a}||\mathbf{b} - \mathbf{c}|\cos\theta = 0, \tag{i}$$

where θ is the angle between the vectors $\mathbf{a}$ and $(\mathbf{b} - \mathbf{c})$. From (i), it follows that either $a = 0$ or $b - c = 0$, or $\cos\theta = 0$, i.e. either $a = 0$, or $b = c$, or the vectors $\mathbf{a}$ and $(\mathbf{b} - \mathbf{c})$ are orthogonal to each other.

Example 2.59 Find the unit vector, which is perpendicular to the two vectors

$$\mathbf{a} = 2\mathbf{i} + \mathbf{j} + 2\mathbf{k}, \qquad \mathbf{b} = 3\mathbf{i} - \mathbf{j} - \mathbf{k}.$$

Let $\mathbf{S} = p\mathbf{i} + q\mathbf{j} + r\mathbf{k}$ be a vector, which is perpendicular to both $\mathbf{a}$ and $\mathbf{b}$. Then, we have

$$\mathbf{s}\cdot\mathbf{a} = \mathbf{s}\cdot\mathbf{b} = 0.$$

But

$$\mathbf{s}\cdot\mathbf{a} = (p\mathbf{i} + q\mathbf{j} + r\mathbf{k})\cdot(2\mathbf{i} + \mathbf{j} + 2\mathbf{k}) = 2p + q + 2r$$

Hence,

$$2p + q + 2r = 0 \tag{i}$$

Since $\mathbf{s}\cdot\mathbf{b} = 0$, we similarly have:

$$3p - q - r = 0 \tag{ii}$$

From (i) and (ii), we have

$$\frac{p}{\begin{vmatrix} 1 & 2 \\ -1 & -1 \end{vmatrix}} = \frac{q}{\begin{vmatrix} 2 & 2 \\ -1 & 3 \end{vmatrix}} = \frac{r}{\begin{vmatrix} 2 & 1 \\ 3 & -1 \end{vmatrix}}$$

or

$$\frac{p}{1} = \frac{q}{8} = \frac{r}{-5} = m, \qquad \text{(say)}$$

Therefore,

$$p = m, \qquad q = 8m, \qquad r = -5m.$$

Hence,

$$\mathbf{s} = m\mathbf{i} + 8m\mathbf{j} - 5m\mathbf{k}.$$

The unit vector is therefore given by

$$\frac{\mathbf{s}}{|\mathbf{s}|} = \frac{m\mathbf{i} + 8m\mathbf{j} - 5m\mathbf{k}}{\sqrt{m^2 + 64m^2 + 25m^2}} = \frac{1}{\sqrt{90}}(\mathbf{i} + 8\mathbf{j} - 5\mathbf{k})$$

Example 2.60 Calculate the work done by a force of 30 kg, whose line of action has direction cosines 2/3, 1/3 and 2/3 in a displacement from the point (1, 3, 5) to the point (7, 9, 2), the unit of length being a metre.

We have

$$\mathbf{F} = 30\left(\frac{2}{3}\mathbf{i} + \frac{1}{3}\mathbf{j} + \frac{2}{3}\mathbf{k}\right) = 10(2\mathbf{i} + \mathbf{j} + 2\mathbf{k})$$

and

$$\mathbf{d} = (7\mathbf{i} + 9\mathbf{j} - 2\mathbf{k}) - (\mathbf{i} + 3\mathbf{j} + 5\mathbf{k}) = 6\mathbf{i} + 6\mathbf{j} - 3\mathbf{k}$$

Hence,

$$\begin{aligned} W = \mathbf{F}\cdot\mathbf{d} &= 10\,(2\mathbf{i} + \mathbf{j} + 2\mathbf{k})\cdot(6\mathbf{i} + 6\mathbf{j} - 3\mathbf{k}) \\ &= 10\,(12 + 6 - 6) \\ &= 120 \text{ kg-force-metre.} \end{aligned}$$

Example 2.61 Using the dot product, prove the law of cosines

$$c^2 = a^2 + b^2 - 2ab\cos\theta.$$

Let a vector triangle be drawn as shown in Fig. 2.40. Since a $\mathbf{a} + \mathbf{c} = \mathbf{b}$, we have $\mathbf{c} = \mathbf{b} - \mathbf{a}$. Hence,

$$\mathbf{c}\cdot\mathbf{c} = (\mathbf{b} - \mathbf{a})\cdot(\mathbf{b} - \mathbf{a}) = \mathbf{b}\cdot\mathbf{b} - \mathbf{b}\cdot\mathbf{a} - \mathbf{a}\cdot\mathbf{b} + \mathbf{a}\cdot\mathbf{a} \qquad \text{(i)}$$

But

$$\mathbf{a}\cdot\mathbf{a} = a^2, \qquad \mathbf{b}\cdot\mathbf{b} = b^2, \qquad \mathbf{a}\cdot\mathbf{b} = \mathbf{b}\cdot\mathbf{a} = ab\cos\theta.$$

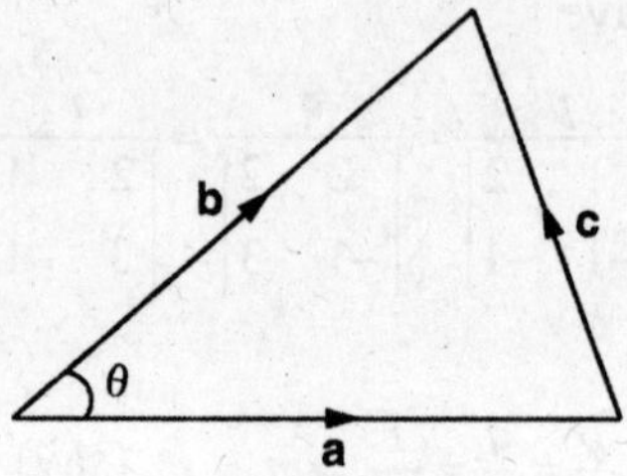

Figure 2.40

Therefore (i) gives:

$$c^2 = b^2 - 2ab\cos\theta + a^2$$

or

$$c^2 = a^2 + b^2 - 2ab\cos\theta,$$

which is the law of cosines.

Example 2.62 The vectors $\mathbf{a}$ and $\mathbf{b}$ make an angle of $2\pi/3$, $a = 2$ and $b = 5$. Find the value of l for which the vectors

$$\mathbf{p} = l\mathbf{a} + 17\mathbf{b} \quad \text{and} \quad \mathbf{q} = 3\mathbf{a} - \mathbf{b}$$

are perpendicular.

Since $\mathbf{p}\cdot\mathbf{q} = 0$, we have

$$(l\mathbf{b} + 17\mathbf{b})\cdot(3\mathbf{a} - \mathbf{b}) = 0$$

or

$$3la^2 - 17b^2 + 51\,\mathbf{b}\cdot\mathbf{a} - l\mathbf{a}\cdot\mathbf{b} = 0.$$

Here $a = 2$, $b = 5$ and

$$\mathbf{a}\cdot\mathbf{b} = \mathbf{b}\cdot\mathbf{a} = 10\cos\frac{2\pi}{3} = -5.$$

Hence the above equation gives:

$$12l - 425 - 255 + 5l = 0$$

which gives $l = 40$.

Example 2.63 If vectors $\mathbf{p}$, $\mathbf{q}$ and $\mathbf{r}$ are coplanar, show that

$$\begin{vmatrix} \mathbf{p} & \mathbf{q} & \mathbf{r} \\ \mathbf{p}\cdot\mathbf{p} & \mathbf{p}\cdot\mathbf{q} & \mathbf{p}\cdot\mathbf{r} \\ \mathbf{q}\cdot\mathbf{p} & \mathbf{q}\cdot\mathbf{q} & \mathbf{q}\cdot\mathbf{r} \end{vmatrix} = 0.$$

Since $\mathbf{p}$, $\mathbf{q}$ and $\mathbf{r}$ coplanar, we can write

$$\mathbf{p} = l\mathbf{q} + m\mathbf{r} \tag{i}$$

Multiplying scalarly with $\mathbf{p}$ and $\mathbf{q}$, we obtain

$$\mathbf{p}\cdot\mathbf{p} = \mathbf{p}\cdot(l\mathbf{q} + m\mathbf{r}) = l(\mathbf{p}\cdot\mathbf{q}) + m(\mathbf{p}\cdot\mathbf{r}) \tag{ii}$$

and

$$\mathbf{q}\cdot\mathbf{p}=\mathbf{q}\cdot(l\mathbf{q}+m\mathbf{r})=l(\mathbf{q}\cdot\mathbf{q})+m(\mathbf{q}\cdot\mathbf{r}) \quad \text{(iii)}$$

Eliminating l and m from (i), (ii) and (iii), we obtain:

$$\begin{vmatrix} \mathbf{p} & \mathbf{q} & \mathbf{r} \\ \mathbf{p}\cdot\mathbf{p} & \mathbf{p}\cdot\mathbf{q} & \mathbf{p}\cdot\mathbf{r} \\ \mathbf{q}\cdot\mathbf{p} & \mathbf{q}\cdot\mathbf{q} & \mathbf{q}\cdot\mathbf{r} \end{vmatrix}=0,$$

which is the required result.

2.3.5 Vector or Cross Product of Two Vectors

We have seen that the scalar product of two vectors is a sclar quanity which is maximum when the two vectors are parallel and zero when they are perpendicular to each other. The vector product of two vectors, on the other hand, is a vector quanity which is maximum when the two vectors are perpendicular to each other and zero when they are parapllel. If **a** and **b** are the two vectors, then their vector product, which is written as $\mathbf{a}\times\mathbf{b}$ (read as '**a** cross **b**') is defined thus:

$$\mathbf{a}\times\mathbf{b}=ab\sin\theta\,\mathbf{n} \quad (2.67)$$

where **n** is a unit vector whose direction is perpendicular to both **a** and **b** such that **a**, **b** and **n** form a right-handed system, as shown in Fig. 2.41. The quanity $ab\sin\theta$ is the magnitude of $\mathbf{a}\times\mathbf{b}$, i.e.

$$|\mathbf{a}\times\mathbf{b}|=ab\sin\theta.$$

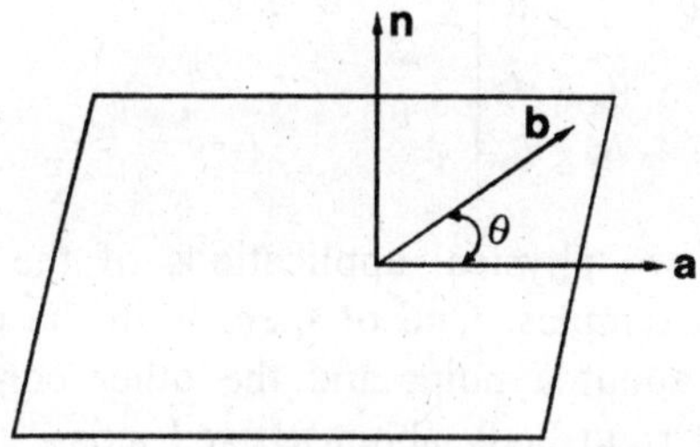

Figure 2.41

Similarly, the vector product $\mathbf{b}\times\mathbf{a}$ has the same magnitude, i.e. $ab\sin\theta$, but its direction woulld be opposite to that of **n**. Hence,

$$\mathbf{a}\times\mathbf{b}=-(\mathbf{b}\times\mathbf{a}).$$

which shows that the commutative law does not hold for the cross product of two vectors. However, the distributive law holds for the cross products, i.e.

$$\mathbf{a}\times(\mathbf{b}+\mathbf{c})=\mathbf{a}\times\mathbf{b}+\mathbf{a}\times\mathbf{c} \quad (2.68)$$

It is also easy to see that $ab \sin\theta$ represents the area of the parallelogram with **a** and **b** its sides. From the definition, its follows easily that

$$\left.\begin{aligned}\mathbf{i}\times\mathbf{j}=-\mathbf{j}\times\mathbf{i}=\mathbf{k}\\ \mathbf{j}\times\mathbf{k}=-\mathbf{k}\times\mathbf{j}=\mathbf{i}\\ \mathbf{k}\times\mathbf{i}=-\mathbf{i}\times\mathbf{k}=\mathbf{j}\end{aligned}\right\} \tag{2.69}$$

and also

$$\mathbf{i}\times\mathbf{i}=\mathbf{j}\times\mathbf{j}=\mathbf{k}\times\mathbf{k}=0 \tag{2.70}$$

The relations (2.69) and (2.70) enable us to compute an explicit expression for $\mathbf{a}\times\mathbf{b}$ in terms of the components of **a** and **b**. Thus, if

$$\mathbf{a}=a\mathbf{i}+a_2\mathbf{j}+a_3\mathbf{k} \quad \text{and} \quad \mathbf{b}=b_1\mathbf{i}+b_2\mathbf{j}+b_3\mathbf{k},$$

then

$$\begin{aligned}\mathbf{a}\times\mathbf{b}&=(a_1\mathbf{i}+a_2\mathbf{j}+a_3\mathbf{k})\times(b_1\mathbf{i}+b_2\mathbf{j}+b_3\mathbf{k})\\ &=a_1b_1(\mathbf{i}\times\mathbf{i})+a_1b_2\,(\mathbf{i}\times\mathbf{j})+a_1b_3\,(\mathbf{i}\times\mathbf{k})\\ &\quad+a_2b_1(\mathbf{j}\times\mathbf{i})+a_2b_2(\mathbf{j}\times\mathbf{j})+a_2b_3(\mathbf{j}\times\mathbf{k})\\ &\quad+a_3b_1(\mathbf{k}\times\mathbf{i})+a_3b_2(\mathbf{k}\times\mathbf{j})+a_3b_3(\mathbf{k}\times\mathbf{k})\\ &=a_1b_2\mathbf{k}-a_1b_3\mathbf{j}-a_1b_1\mathbf{k}-a_2b_3\mathbf{i}+a_3b_1\mathbf{j}-a_3b_2\mathbf{i},\end{aligned}$$

on using the relations (2.69) and (2.70). Hence,

$$\mathbf{a}\times\mathbf{b}=(a_2b_3-a_3b_2)\mathbf{i}-(a_1b_3-a_3b_1)\mathbf{j}+(a_1b_2-a_2b_1)\mathbf{k}$$

$$=\begin{vmatrix}\mathbf{i} & \mathbf{j} & \mathbf{k}\\ a_1 & a_2 & a_3\\ b_1 & b_2 & b_3\end{vmatrix} \tag{2.71}$$

We give beolw two physical applications of the vector product and illustrate them with examples. One of them is in the determination of the moment of a force about a point and the other consists in finding the angular velocity of a rigid body about a fixed axis.

Moment of a force about a point

Let **F** be a force acting along a line through P (see Fig. 2.42), where P is any point whose position vector (with respect to the origin O) is **r**. Let ON be perpendicular from O on the line of action of **F**. We wish to find the vector form for the moment of **F** about the point O. By definition, moment m of **F** about $O=\mathbf{F}\cdot\overrightarrow{\mathrm{ON}}=\mathbf{F}r\sin\theta$, (in magnitude), where F is the magnitude of **F**.

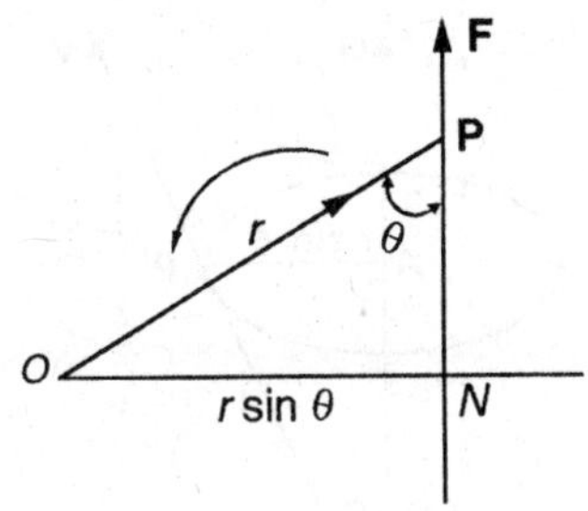

Figure 2.42

and is in the sense indicated in the figure. Hence we write

$$\mathbf{m} = \mathbf{r} \times \mathbf{F} \tag{2.72}$$

This vector, also the *torque* of **F** about O, is perpendicular to the plane of **r** and **F**, i.e. to the plane containing O and the line action of **F**. It therefore follows that if both **F** and **m** are given, then the force **F** is specified in magnitude, direction and line of action.

If we have a number of forces $\mathbf{F}_1, \mathbf{F}_2, \ldots,$ acting through the same point P, whose position vector is **r**, then

$$\begin{aligned}\text{Sum of their moments about } O &= \mathbf{r} \times \mathbf{F}_1 + \mathbf{r} \times \mathbf{F}_2 + \cdots \\ &= \mathbf{r} \times (\mathbf{F}_1 + \mathbf{F}_2 + \cdots) \\ &= \mathbf{r} \times \mathbf{R},\end{aligned}$$

which is the moment of the resultant **R**.

In addition, if we draw a line L through O, then it can be shown that the moment of **F** about L is equal to the projection on L of the moment of **F** about O. This is left as an exercise to the reader.

Angular velocity of a rigid body

We now consider a rigid body rotating about the fixed axis OA with an angular speed of **w** radians per second, **w** being a vector with magnitude w and a direction parallel to OA (see Fig. 2.43). Let P be any point of the body with position vector $\overrightarrow{\mathrm{OP}} = \mathbf{r}$ and PA perpendiculat to the axits of rotation. It follows then that the particle at P is moving in a circular path with centre at A and radius $PA = r \sin\theta$. The velocity of P is therfore equal to $\omega r \sin\theta$, and is represented by the vector $\boldsymbol{\omega} \times \mathbf{r}$, since such a vector has a magnitude $\omega r \sin\theta$, and a direction which is perpendicular to the plane OPA. Hence, we have

$$\mathbf{v} = \boldsymbol{\omega} \times \mathbf{r} \tag{2.73}$$

Example 2.64 Using the cross product of two vector, find a unit vector perpendicular to **a** and **b**, where

$$\mathbf{a} = 2\mathbf{i} + \mathbf{j} + 2\mathbf{k} \quad \text{and} \quad \mathbf{b} = 3\mathbf{i} - \mathbf{j} - \mathbf{k}.$$

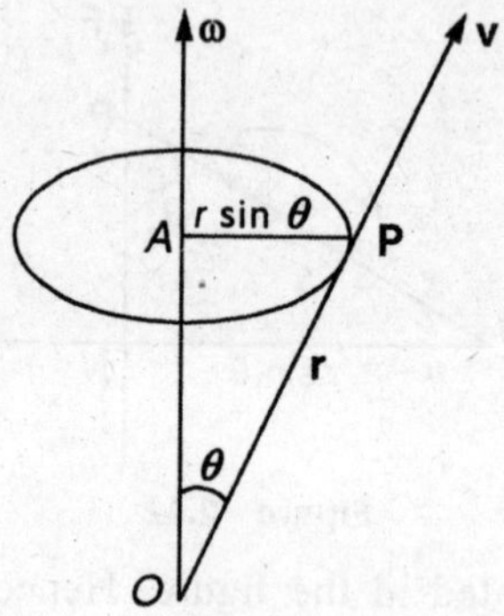

Figure 2.43

From relation (2.71), we obtain

$$\mathbf{a}\times\mathbf{b} = \begin{vmatrix} \mathbf{i} & \mathbf{j} & \mathbf{k} \\ 2 & 1 & 2 \\ 3 & -1 & -1 \end{vmatrix} = \mathbf{i} + 8\mathbf{j} - 5\mathbf{k},$$

on simplification. Hence the unit vector in the direction of $\mathbf{a}\times\mathbf{b}$ is given by

$$\frac{\mathbf{a}\times\mathbf{b}}{|\mathbf{a}\times\mathbf{b}|} = \frac{\mathbf{i}+8\mathbf{j}-5\mathbf{k}}{\sqrt{90}}$$

The reader should verify that this unit vector is indeed perpendicular to both **a** and **b**.

Example 2.65 A rigid body is rotating with a speed of 1.5 radian per second about an axis OR, where R is the point $2\mathbf{i}-2\mathbf{j}+\mathbf{k}$ relative to O. Find the velocity of the particle of the body at the point $4\mathbf{i}+\mathbf{j}+2\mathbf{k}$.

Here,

$$\overrightarrow{OR} = 2\mathbf{i}-2\mathbf{j}+\mathbf{k}.$$

Hence a unit vector in the direction of $\overrightarrow{OR}$ is given by

$$\frac{2\mathbf{i}-2\mathbf{j}+\mathbf{k}}{\sqrt{4+4+1}} = \frac{2}{3}\mathbf{i} - \frac{2}{3}\mathbf{j} + \frac{1}{3}\mathbf{k}.$$

The angular velocity vector $\boldsymbol{\omega}$ is then given by

$$\boldsymbol{\omega} = \frac{3}{2}\left(\frac{2}{3}\mathbf{i} - \frac{2}{3}\mathbf{j} + \frac{1}{3}\mathbf{k}\right) = \frac{1}{2}(2\mathbf{i}-2\mathbf{j}+\mathbf{k}).$$

Also,

$$\mathbf{r} = 4\mathbf{i}+\mathbf{j}+2\mathbf{k}.$$

Hence,

$$\mathbf{v} = \boldsymbol{\omega}\times\mathbf{r} = \begin{vmatrix} \mathbf{i} & \mathbf{j} & \mathbf{k} \\ 1 & -1 & 1/2 \\ 4 & 1 & 2 \end{vmatrix} = -\frac{5}{2}\mathbf{i} + 5\mathbf{k},$$

on simplification. The magnitude of the velocity is therfore given by

$$v = |v| = \sqrt{\frac{25}{4} + 25} = \frac{5\sqrt{5}}{2} \text{ units.}$$

Example 2.66 Prove that the moment about the point $A(3, -1, 3)$ of a force $\mathbf{F} = 2\mathbf{i} + \mathbf{j} + 4\mathbf{k}$ through the point $B(5, 1, 4)$ is given by $7\mathbf{i} - 6\mathbf{j} - 2\mathbf{k}$. Find the moments of $\mathbf{F}$ about axes through A parallel to the coordinate axes.

Here

$$\vec{\mathrm{AB}} = \mathbf{r} = 5\mathbf{i} + \mathbf{j} + 4\mathbf{k} - (3\mathbf{i} - \mathbf{j} + 3\mathbf{k}) = 2\mathbf{i} + 2\mathbf{j} + \mathbf{k}$$

Hence

$$\mathbf{m} = \mathbf{r} \times \mathbf{F} \begin{vmatrix} \mathbf{i} & \mathbf{j} & \mathbf{k} \\ 2 & 2 & 1 \\ 2 & 1 & 4 \end{vmatrix} = 7\mathbf{i} - 6\mathbf{j} = 2\mathbf{k}.$$

The cooefficent of **i**, **j**, **k** in **m** give, respectively, the moments of **F** about axes through A parallel to the coordinate axes.

2.3.6 Equations of Lines and Planes

Let L be a line in space passing through a given point $P_1(x_1, y_1, z_1)$ and is parallel to a given nonzero vector **v**:

$$\mathbf{v} = a\mathbf{i} + b\mathbf{j} + c\mathbf{k} \tag{2.74}$$

See Fig. 2.44. If we now consider any point $P(x, y, z)$ on L, then since the two vector $\vec{\mathrm{P_1P}}$ and **v** are parallel, we can write

$$\vec{\mathrm{P_1P}} = s\mathbf{v} \tag{2.75}$$

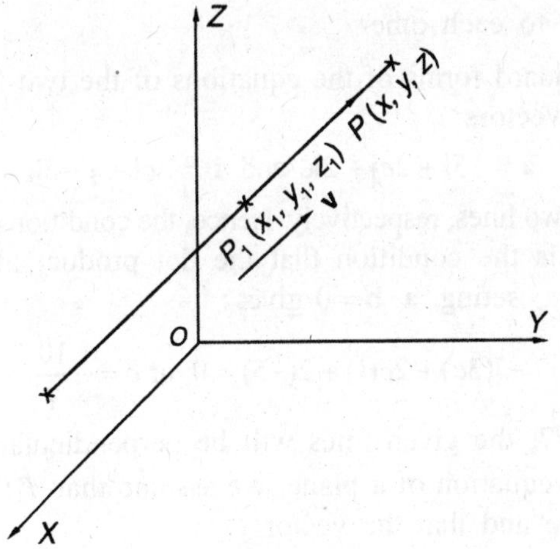

Figure 2.44

where s is a parameter. Since

$$\overrightarrow{P_1P} = (x - x_1)\mathbf{i} + (y - y_1)\mathbf{j} + (z - z_1)\mathbf{k}, \tag{2.76}$$

we must have

$$(x - x_1) = as,\ y - y_1 = bs,\ z - z_1 = cs \tag{2.77}$$

Equations (2.77) are called the *parametric equations* of the line L in space. For each value of s, we get a different point on the line, i.e. different values for x, y and z. Thus, the whole line is obtained by permitting s to take on real values. On eliminating s, we obtain

$$\frac{x - x_1}{a} = \frac{y - y_1}{b} = \frac{z - z_1}{c}, \tag{2.78}$$

which is the *standard form* of the equations of the line.

Example 2.67 Find the parametic equations of the line joining the points $P_1(1, 1, 0)$ and $P_2(0, 2, 3)$.

We have

$$\overrightarrow{P_1P_2} = -\mathbf{i} + \mathbf{j} + 3\mathbf{k}$$

Since the required line is parallel to the vector $\overrightarrow{P_1P_2}$, we must have

$$x - 1 = s, \qquad y - 1 = s, \qquad z = 3s, \tag{i}$$

where $P(x, y, z)$ is any point on the line. Equations (i) are the parametric equation of the line.

Example 2.68 Find the value of c so that the lines

$$\frac{x-1}{-3} = \frac{y-1}{2c} = \frac{z-3}{2} \text{ and } \frac{x-1}{3c} = \frac{y-5}{1} = \frac{z-6}{-5}$$

are perpendicular to each other.

From the standard forms of the equations of the two lines given above, we see that the vectors

$$\mathbf{a} = -3\mathbf{i} + 2c\mathbf{j} + 2\mathbf{k} \text{ and } \mathbf{b} = 3c\mathbf{i} + \mathbf{j} - 5\mathbf{k}$$

are parallel to the two lines, respectively. Hence, the condition of perpendicularity of the two lines is the condition that the dot product of the two vectors vanishes. Therfore, seting $\mathbf{a} \cdot \mathbf{b} = 0$ gives:

$$-3(3c) + 2c(1) + 2(-5) = 0 \text{ or } c = -\frac{10}{7}$$

Thus for $c = -10/7$, the given lines will be perpendicular to each other.

To derive the equation of a plane, we assume that $P_1(x_1, y_1, z_1)$ is given point on the plane and that the vector

$$\mathbf{n} = a\mathbf{i} + b\mathbf{j} + c\mathbf{k} \tag{2.79}$$

is perpendicular to the plane (Fig. 2.45).

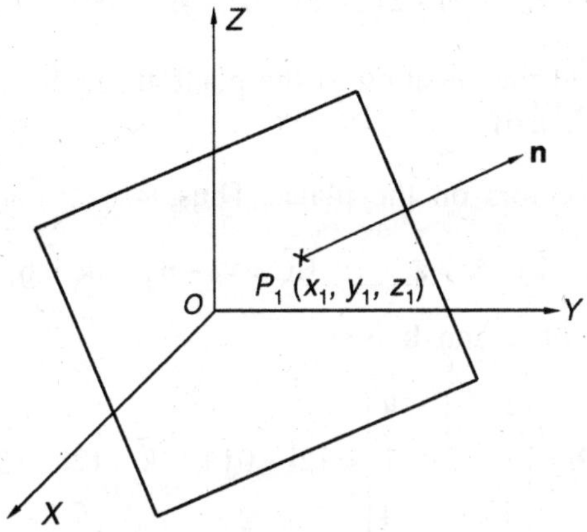

Figure 2.45

If we now consider *any* point $P(x, y, z)$ on the plane, then the two vectors $\overrightarrow{P_1P}$ and **n** are perpendicular to each other. It therfore follows that their dot product, viz., $\overrightarrow{P_1P}$ vanishes:

$$a(x - x_1) + b(y - y_1) + c(z - z_1) = 0, \tag{2.80}$$

which is the equation of the plane.

This is often written in the more general form:

$$ax + by + cz + = d = 0, \tag{2.81a}$$

where the vector $a\mathbf{i} + b\mathbf{j} + c\mathbf{k}$ is normal to the plane.

On the other hand, if we are given a linear equation such as Eq. (2.81a), then a vector perpendicular to the plane is given by

$$\mathbf{n} = a\mathbf{i} + b\mathbf{j} + c\mathbf{k}$$

with the same coefficients which x, y and z have in the equation of the plane. It follows that the coefficients x, y and z in the general equation of the plane denote the direction ratio of the normals to the plane.

Equation (2.81a) can be expressed in the intercept form,

$$\frac{x}{A} + \frac{y}{B} + \frac{z}{C} = 1 \tag{2.81b}$$

where A, B, C are the intercepts made by the plane with the three coordinate planes.

Example 2.69 Find the equation of a plane containing the point $P(1, 2, 3)$ and perpendicular to the line containing $A(5, 2, 1)$ and $(2, 1, -1)$.

Direction ratios for the line containing the points A and B are -3, -1 and -2. Therefore, the equation of the plane containing the point $P(1, 2, 3)$ can be written as

$$-3(x-1)-1(y-2)-2(z-3)=0 \quad \text{or} \quad 3x+y+2z=11.$$

Example 2.70 Find the equation of the plane through the points $A(2, 2, 1)$ $B(3, 4, 2)$ and $C(7, 0, 6)$.

We first two vectors on the plane. Thus,

$$\overrightarrow{\text{AB}} = \mathbf{i}+2\mathbf{j}+\mathbf{k}=\mathbf{a}, \qquad \overrightarrow{\text{BC}} = 4\mathbf{i}-4\mathbf{j}-4\mathbf{k}=\mathbf{b}, \text{ (say)}$$

The cross product of **a** and **b** is

$$\mathbf{a}\times\mathbf{b} = \begin{vmatrix} \mathbf{i} & \mathbf{j} & \mathbf{k} \\ 1 & 2 & 1 \\ 4 & -4 & 4 \end{vmatrix} = 12\mathbf{i}+0\mathbf{j}+12\mathbf{k} = 12\mathbf{i}-12\mathbf{k}.$$

We know that this vector is normal to the plane containing **a** and **b,** and that its direction ratios are $(12, 0, -12)$. Hence the equation of the plane through the point $(2, 2, 1)$ may be written as

$$12(x-2)+0(y-2)-12(z-1)=0 \text{ or, } x-z=1,$$

which is the required equation.

Alternatively, the equation of the plane may be obtained by using the fact that the coordinates of the given points must satisfy the general equation of the plane given by

$$ax+by+cz+d=0.$$

This gives the relations

$$2a+2b+c+d=0$$

$$3a+4b+2c+d=0$$

$$7a+6c+d=0.$$

The above system has the solution

$$a=-d, \qquad b=0, \qquad c=d.$$

Hence, the equation of the plane becomes:

$$-dx+dz=-d, \quad \text{or} \quad x-z=1, \text{ (as before).}$$

Example 2.71 Find the distance of the point $(2, 3, 4)$ from the plane

$$3x-6y+2z+11=0.$$

Let A be any point on the plane (Fig. 2.46). The distance of $P(2, 3, 4)$ from the plane is clearly the projection of $\overrightarrow{\text{AP}}$ on the normal to the plane at the point A.

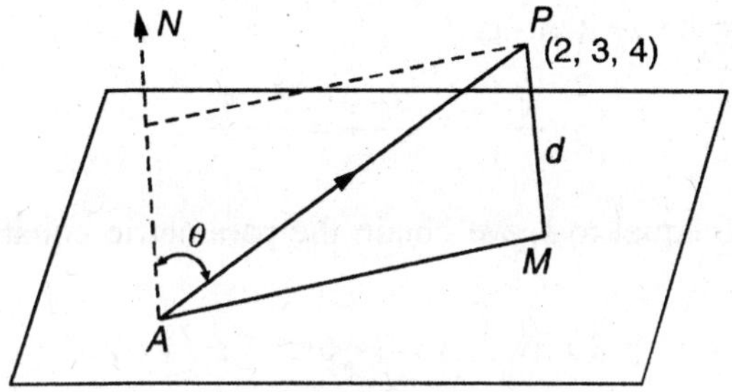

Figure 2.46

Here

$$\vec{AN} = 3\mathbf{i} - 6\mathbf{j} + 2\mathbf{k}$$

and the point A may be chosen as the intersection of the given plane with the x-axis. Clearly, $A = (-11/3, 0, 0)$ and hence

$$\vec{AP} = \frac{17}{3}\mathbf{i} - 3\mathbf{j} + 4\mathbf{k}.$$

Then

$$\text{Projection of } AP \text{ on } AN = |AP| \cos\theta$$

$$= \frac{|\vec{AP}||\vec{AN}| \cos\theta}{|\vec{AN}|}$$

$$= \frac{\vec{AP}.\vec{AN}}{\vec{AN}}$$

Now,

$$\vec{AP} \cdot \vec{AN} = 17 - 18 + 8 = 7,$$

and

$$|\vec{AN}| = \sqrt{9 + 36 + 4} = 7.$$

Hence, the distance of $P(2, 3, 4)$ from the plane $= 7/7 = 1$. Refer to Fig. 2.47.

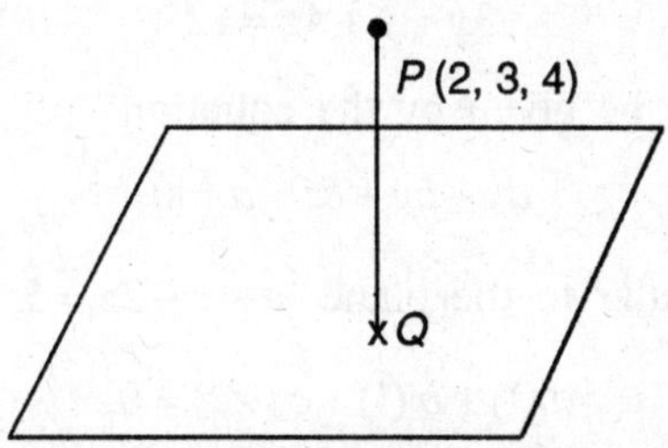

Figure 2.47

Alternatively, the equations of any line through $P(2, 3, 4)$ and perpendicular to the plane may be written as

$$\frac{x-2}{3}=\frac{y-3}{-6}=\frac{z-4}{2}. \tag{i}$$

Setting each ratio equal to s, we obtain the parametric equations of the line as

$$x=2+3s, \quad y=3-6s, \quad z=4+2s. \tag{ii}$$

For a certain value of s, these will represent the coordinates of Q, any point on the plane. Substituting (ii) in the equation of the plane, we obtain

$$3(2+3s)-6(3-6s)+2(4+2s)+11=0,$$

which gives $s=-1/7$.

The coordinates of Q are therefore given by

$$x=2+3\left(-\frac{1}{7}\right), \quad y=3-6\left(-\frac{1}{7}\right), \quad z=4+2\left(-\frac{1}{7}\right).$$

or

$$x=\frac{11}{7}, \quad y=\frac{27}{7}, \quad z=\frac{26}{7}.$$

The perpendicular distance of P from the plane is therefore given by

$$PQ=\sqrt{\left(2-\frac{11}{7}\right)^2+\left(3-\frac{27}{7}\right)^2+\left(4-\frac{26}{7}\right)^2}$$

$$=\sqrt{\frac{9}{49}+\frac{36}{49}+\frac{4}{49}}$$

$$=1, \quad \text{(as before.)}$$

Example 2.72 Find the equation of the plane passing through the point $(1, 2, -1)$ and perpendicular to the planes

$$x+y-2z=5$$

$$3x-y+4z=12$$

Let the required plane be given by the equation

$$ax+by+cz+d=0 \tag{i}$$

Since (i) is perpendicular to the plane $x+y-2z=5$, we must have

$$a(1)+b(1)+c(-2)=0$$

or

$$a + b - 2c = 0 \tag{ii}$$

Again, since (i) is perpendicular to the plane $3x - y + 4z = 12$, we must have

$$a(3) + b - 1 + c(4) = 0$$

i.e.

$$3a - b + 4c = 0. \tag{iii}$$

From (ii) and (iii), we obtain

$$\frac{a}{2} = \frac{b}{-10} = \frac{c}{-4} = s$$

i.e.

$$a = 2s, \quad b = -10s, \quad c = -4s.$$

Now, any plane passing through the point (1, 2, –1) is given by

$$a(x-1) + b(y-2) + c(z+1) = 0.$$

Substituting the values of a, b and c, this gives:

$$2s(x-1) - 10s(y-2) - 4s(z+1) = 0,$$

i.e.

$$x - 5y - 2z + 7 = 0,$$

which is the equation of the plane.

Example 2.73 Find the line of the intersection of the two planes

$$x - 4y + 2z + 7 = 0 \quad \text{and} \quad 3x + 3y - z - 2 = 0$$

Any normal to the plane $x - 4y + 2z + 7 = 0$ is given by

$$\mathbf{N}_1 = \mathbf{i} - 4\mathbf{j} + 2\mathbf{k} = 0$$

Similarly, any normal to the plane $3x + 3y - z - 2 = 0$ is given by

$$\mathbf{N}_2 = 3\mathbf{i} - 3\mathbf{j} - \mathbf{k} = 0$$

Since the line of intersection lies in both the planes, it should be normal to both $\mathbf{N}_1$ and $\mathbf{N}_2$ or equivalently, parallel to $\mathbf{N}_1 \times \mathbf{N}_2$. Now,

$$\mathbf{N}_1 \times \mathbf{N}_2 = \begin{bmatrix} \mathbf{i} & \mathbf{j} & \mathbf{k} \\ 1 & -4 & 2 \\ 3 & 3 & -1 \end{bmatrix} = -2\mathbf{i} + 7\mathbf{j} + 15\mathbf{k}.$$

Any point on the line of intersection of the two planes may be obtained by setting to zero one of the variables and then solving the equations for the other variables. If we set, for example, $z = 0$, then we have

$$x - 4y + 7 = 0 \quad \text{and} \quad 3x + 3y - 2 = 0.$$

These equations give

$$x = -\frac{13}{15} \quad \text{and} \quad y = \frac{23}{15}.$$

A point on the line of intersection is therefore given by

$$P_1\left(-\frac{13}{15}, \frac{23}{15}, 0\right).$$

Then,

$$\overrightarrow{P_1P} = \left(x + \frac{13}{15}\right)\mathbf{i} + \left(y - \frac{23}{15}\right)\mathbf{j} + z\mathbf{k},$$

where P has coordinates x, y and z.

Since $\overrightarrow{P_1P}$ is parallel to $\mathbf{N}_1 \times \mathbf{N}_2$, we have

$$\left(x + \frac{13}{15}\right)\mathbf{i} + \left(y - \frac{23}{15}\right)\mathbf{j} + 2\mathbf{k} = (-2\mathbf{i} + 7\mathbf{j} + 15\mathbf{k})\, s.$$

Equating the coefficients of **i, j** and **k**, we obtain:

$$x + \frac{13}{15} = -2s, \quad y - \frac{23}{15} = 7s, \quad z = 15s.$$

Hence, the line of intersection of the two planes is given by

$$\frac{x + (13/15)}{-2} = y = \frac{23/15}{7} = \frac{z}{15}.$$

Example 2.74 Find the shortest distance between the two skew lines

$$\frac{x - 8}{3} = \frac{y + 9}{-16} = \frac{z - 10}{7}$$

and

$$\frac{x - 15}{3} = \frac{y - 29}{8} = \frac{z - 5}{-5}.$$

Two lines in space are said to be *skew* lines if they do not intersect. In such a case, the length of the line segment which is perpendicular to both the lines is said to be the *shortest distance* between them.

Let PR be the shortest distance between the given lines. Then PR is perpendicular to both the lines and is the projection of QS on PR (Fig. 2.48).

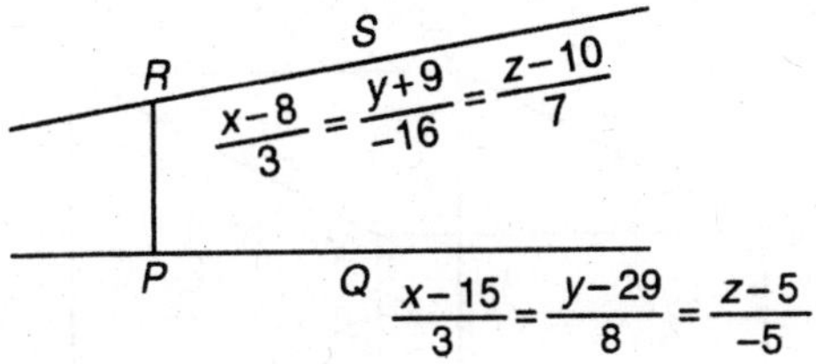

Figure 2.48

The vectors:

$$\mathbf{a} = 3\mathbf{i} - 16\mathbf{j} + 7\mathbf{k} \quad \text{and} \quad \mathbf{b} = 3\mathbf{i} + 8\mathbf{j} - 5\mathbf{k}$$

are parallel to *RS* and *PQ*, respectively.

Now, $\mathbf{a} \times \mathbf{b}$ is perpendicular to both **a** and **b**. Similarly, *PR* is also perpendicular to both **a** and **b**. Hence, it follows that $\mathbf{a} \times \mathbf{b}$ is parallel to $\overrightarrow{PR}$. Now, we have

$$\mathbf{a} \times \mathbf{b} = \begin{bmatrix} \mathbf{i} & \mathbf{j} & \mathbf{k} \\ 3 & -16 & 7 \\ 3 & 8 & -5 \end{bmatrix} = 12(2\mathbf{i} + 3\mathbf{j} + 6\mathbf{k}).$$

and

$$\overrightarrow{QS} = -7\mathbf{i} - 38\mathbf{j} + 5\mathbf{k}$$

Then

$$\text{Projection of } \overrightarrow{QS} \text{ on } \overrightarrow{PR} = \left| \frac{(\mathbf{a} \times \mathbf{b}) \cdot \overrightarrow{QS}}{|\mathbf{a} \times \mathbf{b}|} \right| = 14.$$

2.3.7 Products of Three or More Vectors

When we consider products of three vectors **a, b** and **c**, only three cases arise. These are:

$$\text{(i) } (\mathbf{a} \cdot \mathbf{b})\,\mathbf{c}, \quad \text{(ii) } \mathbf{a} \cdot (\mathbf{b} \times \mathbf{c}), \quad \text{and} \quad \text{(iii) } \mathbf{a} \times (\mathbf{b} \times \mathbf{c}).$$

In (i), we simply have a vector whose direction is that of **c** and whose magnitude is $(\mathbf{a} \cdot \mathbf{b})$ times that of **c**. Cases (ii) and (iii) yield scalar and vector quantities, respectively, and these are, therefore, called *scalar* and *vector triple products*.

The scalar triple product $\mathbf{a} \cdot (\mathbf{b} \times \mathbf{c})$ is simply a number. To interpret it geometrically, we construct a parallelepiped with **a, b** and **c** as its three concurrent edges, as shown in Fig. 2.49. Let $\mathbf{b} \times \mathbf{c} = \mathbf{n}$, where **n** is a vector whose magnitude is the area of the parallelogram *OBEC* and whose directions perpendicular to this face. Hence, it follows that

$$\mathbf{a} \cdot (\mathbf{b} \times \mathbf{c}) = an \cos \theta,$$

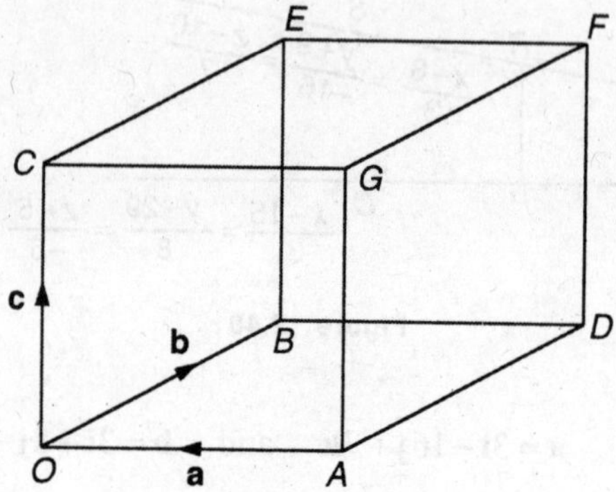

Figure 2.49

where $n = |\mathbf{b} \times \mathbf{c}|$, and θ is the angle between the directions of **a** and **n**. But this being exactly the volume of the parallelopiped, we have the result that the scalar triple product $\mathbf{a} \cdot (\mathbf{b} \times \mathbf{c})$ geometrically, represents the volume of the parallelopiped which is constructed with **a**, **b** and **c** as its three concurrent edges. Similarly, it can be shown that each of the scalar triple products $\mathbf{b} \cdot (\mathbf{c} \times \mathbf{a})$ and $\mathbf{c} \cdot (\mathbf{a} \times \mathbf{b})$ also represent the same volume. Hence, we have

$$\mathbf{a} \cdot (\mathbf{b} \times \mathbf{c}) = \mathbf{b} \cdot (\mathbf{c} \times \mathbf{a}) = \mathbf{c} \cdot (\mathbf{a} \times \mathbf{b}) \tag{2.82}$$

Further, since the scalar triple product $(\mathbf{a} \times \mathbf{b}) \cdot \mathbf{c}$ gives the same value as $\mathbf{a} \cdot (\mathbf{b} \times \mathbf{c})$, it follows that the dot and cross in a scalar triple product, say $\mathbf{a} \cdot (\mathbf{b} \times \mathbf{c})$ can be interchanged at pleasure, provided that the cyclic order **a**, **b**, **c** is maintained. It is also customary to denote $\mathbf{a} \cdot (\mathbf{b} \times \mathbf{c})$ as $(\mathbf{a}\ \mathbf{b}\ \mathbf{c})$. It follows from the above discussion that a scalar triple product of the type $(\mathbf{a}\ \mathbf{b}\ \mathbf{a})$ vanishes identically. The following deductions are immediate consequences of the definition: (a) If **a**, **b** and **c** are on the same plane, then the volume of the parallelopiped is zero. Hence, it follows that $\mathbf{a} \cdot (\mathbf{b} \times \mathbf{c}) = 0$, which is therefore the condition of coplanarity of three vectors **a**, **b** and **c** (b) If **a**, **b** and **c** are mutually perpendicular to each other, then

$$\mathbf{a} \cdot (\mathbf{b} \times \mathbf{c}) = abc.$$

In particular,

$$\mathbf{i} \cdot (\mathbf{j} \times \mathbf{k}) = \mathbf{j} \cdot (\mathbf{k} \times \mathbf{i}) = \mathbf{k} \cdot (\mathbf{i} \times \mathbf{j}) = 1 \tag{2.83}$$

If **a**, **b**, **c** are given in their component form as

$$\mathbf{a} = a_1\mathbf{i} + a_2\mathbf{j} + a_3\mathbf{k}$$

$$\mathbf{b} = b_1\mathbf{i} + b_2\mathbf{j} + b_3\mathbf{k}$$

$$\mathbf{c} = c_1\mathbf{i} + c_2\mathbf{j} + c_3\mathbf{k}$$

Then it is easy to deduce that

$$\mathbf{a}\cdot(\mathbf{b}\times\mathbf{c}) = \begin{bmatrix} a_1 & a_2 & a_3 \\ b_1 & b_2 & b_3 \\ c_1 & c_2 & c_3 \end{bmatrix} \tag{2.84}$$

We now consider the vector triple product $\mathbf{a} \times (\mathbf{b} \times \mathbf{c})$, which is a vector perpendicular to $\mathbf{a}$ and $\mathbf{b} \times \mathbf{c}$. But $\mathbf{b} \times \mathbf{c}$ is a vector perpendicular to the plane containing $\mathbf{b}$ and $\mathbf{c}$. It therefore follows that $\mathbf{a} \times (\mathbf{b} \times \mathbf{c})$ lies in the plane of $\mathbf{b}$ and $\mathbf{c}$ and hence we write

$$\mathbf{a}\times(\mathbf{b}\times\mathbf{c}) = l\mathbf{b} + m\mathbf{c} \tag{2.85}$$

where l and m are scalars to be determined. For this, we chose rectangular axes such that

$$\left.\begin{aligned} &\mathbf{c} = c_1\mathbf{i} \;\; \text{(i.e. } \mathbf{c} \text{ is along an axis parallel to } OX) \\ &\mathbf{b} = b_1\mathbf{i} + b_2\mathbf{j} \;\; \text{(i.e. } \mathbf{b} \text{ is in the } (x,\ y) \text{ plane)} \\ &\mathbf{a} = a_1\mathbf{i} + a_2\mathbf{j} + a_3\mathbf{k} \end{aligned}\right\} \tag{2.86}$$

Then,

$$\mathbf{b}\times\mathbf{c} = (b_1\mathbf{i} + b_2\mathbf{j})\times c_1\mathbf{i} = -b_2c_1\mathbf{k}$$

and hence

$$\mathbf{a}\times(\mathbf{b}\times\mathbf{c}) = a_1\mathbf{i} + a_2\mathbf{j} + a_3\mathbf{k})\times(-b_2c_1\mathbf{k}) = a_1b_2c_1\mathbf{j} - a_2b_2c_1\mathbf{i} \tag{2.87}$$

From (2.85), we have also:

$$\mathbf{a}\times(\mathbf{b}\times\mathbf{c}) = l\,(b_1\mathbf{i} + b_2\mathbf{j}) + mc_1\mathbf{i} = (lb_1 + mc_1)\,\mathbf{i} + lb_2\mathbf{j} \tag{2.88}$$

We now compare the coefficients of $\mathbf{i}$ and $\mathbf{j}$ on the right-hand sides of (2.87) and (2.88), and obtain the equations:

$$lb_1 + mc_1 = a_2b_2c_1 \tag{2.89}$$

$$lb_2 = a_1b_2c_1 \tag{2.90}$$

Equations (2.90) and (2.86) give:

$$l = a_1c_1 = \mathbf{a}\cdot\mathbf{c}$$

Hence, (2.89) gives:

$$mc_1 = -a_2b_2c_1 - lb_1 = -a_2b_2c_1 - a_1c_1b_1,$$

From Eq. (2.86), we get

$$m = -a_2b_2 - a_1b_1 = -(\mathbf{a}\cdot\mathbf{b}).$$

Substituting for l and m in (2.85) gives the formula for $\mathbf{a} \times (\mathbf{b} \times \mathbf{c})$:

$$\mathbf{a}\times(\mathbf{b}\times\mathbf{c}) = (\mathbf{a}\cdot\mathbf{c})\mathbf{b} - (\mathbf{a}\cdot\mathbf{b})\mathbf{c} \tag{2.91}$$

This formula is important and should be memorized.

From (2.91), it follows that

$$\mathbf{b}\times(\mathbf{c}\times\mathbf{a}) = (\mathbf{b}\cdot\mathbf{a})\mathbf{c} - (\mathbf{b}\cdot\mathbf{c})\mathbf{a} \tag{2.92}$$

and

$$\mathbf{c}\times(\mathbf{a}\times\mathbf{b}) = (\mathbf{c}\cdot\mathbf{b})\mathbf{a} - (\mathbf{c}\cdot\mathbf{a})\mathbf{b} \tag{2.93}$$

Since **ab** = **b** · **a**, etc., we deduce from (2.91), (2.92) and (2.93) that

$$\mathbf{a}\times(\mathbf{b}\times\mathbf{c}) + \mathbf{b}\times(\mathbf{c}\times\mathbf{a}) + \mathbf{c}\times(\mathbf{a}\times\mathbf{b}) = 0 \tag{2.94}$$

We now consider four vectors **a**, **b**, **c**, **d** and products of the type

$$(\mathbf{a}\times\mathbf{b})\times(\mathbf{c}\times\mathbf{d}).$$

Let

$$\begin{aligned}(\mathbf{a}\times\mathbf{b})\times(\mathbf{c}\times\mathbf{d}) &= \mathbf{n}\times(\mathbf{c}\times\mathbf{d}), \qquad \text{where } \mathbf{n} = \mathbf{a}\times\mathbf{b}\\ &= (n\cdot d)c - (n\cdot c)d\\ &= [(a\times b)\cdot d]\,c - [(a\times b)\cdot c]\,d\\ &= (a\,b\,d)c - (a\,b\,c)d\end{aligned}$$

The following examples illustrate the use of scalar and vector triple products.

Example 2.75 Show that

$$(\mathbf{b}\times\mathbf{c})\cdot(\mathbf{a}\times\mathbf{d}) + (\mathbf{c}\times\mathbf{a})\cdot(\mathbf{b}\times\mathbf{d}) + (\mathbf{a}\times\mathbf{d})\cdot(\mathbf{c}\times\mathbf{d}) = 0$$

Now,

$$(\mathbf{a}\times\mathbf{b})\cdot(\mathbf{c}\times\mathbf{d}) = \mathbf{n}\cdot(\mathbf{c}\times\mathbf{d}), \quad \text{where } \mathbf{n} = \mathbf{a}\times\mathbf{b}$$

Since the dot and cross can be interchanged, we get

$$\begin{aligned}(\mathbf{a}\times\mathbf{b})\cdot(\mathbf{c}\times\mathbf{d}) &= (\mathbf{n}\times\mathbf{c})\cdot\mathbf{d},\\ &= [(\mathbf{a}\times\mathbf{b})\times\mathbf{c}]\cdot\mathbf{d}\\ &= -[\mathbf{c}\times(\mathbf{a}\times\mathbf{d})]\cdot\mathbf{d}\\ &= -[(\mathbf{c}\cdot\mathbf{b})\mathbf{a} - (\mathbf{c}\cdot\mathbf{a})\mathbf{b}]\cdot\mathbf{d}\\ &= (\mathbf{a}\cdot\mathbf{c})(\mathbf{b}\cdot\mathbf{d}) - (\mathbf{a}\cdot\mathbf{d})(\mathbf{b}\cdot\mathbf{c}).\end{aligned}$$

Similarly,

$$(\mathbf{b}\times\mathbf{c})\cdot(\mathbf{a}\times\mathbf{d}) = (\mathbf{b}\cdot\mathbf{a})(\mathbf{c}\cdot\mathbf{d}) - (\mathbf{b}\cdot\mathbf{d})(\mathbf{c}\cdot\mathbf{a})$$

and,

$$(\mathbf{c}\times\mathbf{a})\cdot(\mathbf{b}\times\mathbf{d}) = (\mathbf{c}\cdot\mathbf{b})(\mathbf{a}\cdot\mathbf{d}) - (\mathbf{c}\cdot\mathbf{d})(\mathbf{a}\cdot\mathbf{b}).$$

The result follows immediately.

Example 2.76 Show that

$$(\mathbf{b}-\mathbf{c})\cdot[(\mathbf{c}-\mathbf{a})\times(\mathbf{a}-\mathbf{b})] = 0.$$

Now,

$$(\mathbf{c}-\mathbf{a})\times(\mathbf{a}-\mathbf{b}) = \mathbf{c}\times\mathbf{a}-\mathbf{c}\times\mathbf{b}-\mathbf{a}\times\mathbf{a}+\mathbf{a}\times\mathbf{b}$$

$$= \mathbf{c}\times\mathbf{a}-\mathbf{c}\times\mathbf{b}+\mathbf{a}\times\mathbf{b} \quad (\text{since } \mathbf{a}\times\mathbf{a}=0)$$

Therefore,

$$(\mathbf{b}-\mathbf{c})\cdot[(\mathbf{c}-\mathbf{a})\times(\mathbf{a}-\mathbf{b})] = (\mathbf{b}-\mathbf{c})\cdot[\mathbf{c}\times\mathbf{a}-\mathbf{c}\times\mathbf{b}+\mathbf{a}\times\mathbf{b}]$$

$$= (\mathbf{b}\,\mathbf{c}\,\mathbf{a})-(\mathbf{b}\,\mathbf{c}\,\mathbf{b})+(\mathbf{b}\,\mathbf{a}\,\mathbf{b})-(\mathbf{c}\,\mathbf{c}\,\mathbf{a})+(\mathbf{c}\,\mathbf{c}\,\mathbf{b})+(\mathbf{c}\,\mathbf{a}\,\mathbf{b})$$

$$= 0,$$

as the triple products of the type $(\mathbf{b}\ \mathbf{c}\ \mathbf{b})$ vanish identically and $(\mathbf{b}\ \mathbf{c}\ \mathbf{a}) = (\mathbf{c}\ \mathbf{a}\ \mathbf{b})$.

Example 2.77 Express the vector **r** in terms of three given non-coplanar vectors **a**, **b** and **c**.

Let

$$\mathbf{r} = l\mathbf{a}+m\mathbf{b}+n\mathbf{c}, \tag{i}$$

where l, m, n are scalars to be determined.

Scalar multiplication of both side of (i) by $(\mathbf{b}\times\mathbf{c})$, gives

$$(\mathbf{b}\ \mathbf{c}\ \mathbf{r}) = l\,(\mathbf{b}\ \mathbf{c}\ \mathbf{a})+m\,(\mathbf{b}\ \mathbf{c}\ \mathbf{b})+n\,(\mathbf{b}\ \mathbf{c}\ \mathbf{c})$$

$$= l(\mathbf{b}\ \mathbf{c}\ \mathbf{a}), \qquad [\text{since } (\mathbf{b}\ \mathbf{c}\ \mathbf{b}) = (\mathbf{b}\ \mathbf{c}\ \mathbf{c}) = 0]$$

$$= l(\mathbf{a}\ \mathbf{b}\ \mathbf{c}).$$

Hence,

$$l = \frac{(\mathbf{b}\ \mathbf{c}\ \mathbf{r})}{(\mathbf{a}\ \mathbf{b}\ \mathbf{c})} = \frac{(\mathbf{r}\ \mathbf{b}\ \mathbf{c})}{(\mathbf{a}\ \mathbf{b}\ \mathbf{c})}$$

Similarly, we obtain

$$m = \frac{(\mathbf{r}\ \mathbf{c}\ \mathbf{a})}{(\mathbf{a}\ \mathbf{b}\ \mathbf{c})} \quad \text{and} \quad n = \frac{(\mathbf{r}\ \mathbf{a}\ \mathbf{b})}{(\mathbf{a}\ \mathbf{b}\ \mathbf{c})}$$

Hence (i) gives:

$$\mathbf{r} = \frac{(\mathbf{r}\ \mathbf{b}\ \mathbf{c})}{(\mathbf{a}\ \mathbf{b}\ \mathbf{c})}\mathbf{a}+\frac{(\mathbf{r}\ \mathbf{c}\ \mathbf{a})}{(\mathbf{a}\ \mathbf{b}\ \mathbf{c})}\mathbf{b}+\frac{(\mathbf{r}\ \mathbf{a}\ \mathbf{b})}{(\mathbf{a}\ \mathbf{b}\ \mathbf{c})}\mathbf{c}.$$

Example 2.78 If $\mathbf{p} = \mathbf{a}\times\mathbf{b}$, $\mathbf{q} = \mathbf{b}\times\mathbf{c}$, $\mathbf{r} = \mathbf{c}\times\mathbf{a}$, show that $(\mathbf{p}\ \mathbf{q}\ \mathbf{r}) = (\mathbf{a}\ \mathbf{b}\ \mathbf{c})^2$.

Now,

$$(\mathbf{p}\mathbf{q}\mathbf{r}) = \mathbf{p}\cdot\mathbf{q}\times\mathbf{r} = \mathbf{p}\cdot\mathbf{q}\times(\mathbf{c}\times\mathbf{a})$$

$$= \mathbf{p}\cdot(\mathbf{q}\cdot\mathbf{a}\mathbf{c}-\mathbf{q}\cdot\mathbf{c}\mathbf{a})$$

$$= \mathbf{p}\,[(\mathbf{b}\ \mathbf{c}\ \mathbf{a})\,\mathbf{c}-(\mathbf{b}\ \mathbf{c}\ \mathbf{c})\,\mathbf{a}], \quad (\text{since } \mathbf{q} = \mathbf{b}\times\mathbf{c})$$

$$= (\mathbf{a}\times\mathbf{b})\cdot\{(\mathbf{b}\ \mathbf{c}\ \mathbf{a})\,\mathbf{c}\} \qquad [\text{since} \quad (\mathbf{b}\ \mathbf{c}\ \mathbf{c}) = 0]$$

$$= [(\mathbf{a}\times\mathbf{b})\cdot\mathbf{c}]\,(\mathbf{b}\ \mathbf{c}\ \mathbf{a})$$

$$= (\mathbf{a\,b\,c}) \cdot (\mathbf{b\,c\,a})$$
$$= (\mathbf{a\,b\,c})^2.$$

2.3.8 Sphere, Cylinder and Cone

The *sphere* is the locus of points in space that are equidistant from one fixed point, the centre. If the centre is $P_0\,(x_0, y_0, z_0)$, then the equation

$$(x - x_0)^2 + (y - y_0)^2 + (z - z_0)^2 = R^2 \tag{2.95}$$

is satisfied by the coordinates of all points that are at a distance R from P_0. Thus, Eq. (2.95) is the equation of a sphere with centre at $P_0\,(x_0, y_0, z_0)$ and radius equal to R (Fig. 2.50).

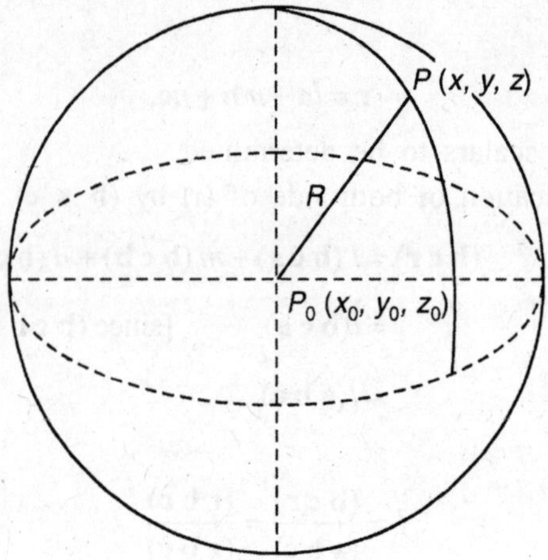

Figure 2.50

On the other hand, if we have any equation of the form

$$x^2 + y^2 + z^2 + 2gx + 2fy + 2hz + c = 0, \tag{2.96}$$

then by completing the squares we can write it in the form

$$(x + g)^2 + (y + f)^2 + (z + h)^2 = g^2 + f^2 + h^2 - c \tag{2.97}$$

Comparing (2.95) and (2.97), it is easy to see that the point $(-g, -f, -h)$ is the centre of a sphere with radius $\sqrt{(g^2 + f^2 + h^2 - c)}$.

Example 2.79 Determine the centre and radius of the sphere whose equation is $x^2 + y^2 + z^2 = 4z$.

Completing the square on z, we obtain

$$x + y^2 + (z - 2)^2 = 4.$$

Therefore, the centre has coordinates (0, 0, 2) and the radius is 2 units.

A *cylinder* is any surface generated by a line which is always parallel to a fixed line and passes through a given curve. The fixed line is called the *axis* of the cylinder and the given curve is called the *guiding* curve.

If the guiding curve consists of one or more straight lines, the cylinder is one or more planes. In other cases, the cylinder is a curved surface and specifically named after the guiding curve. Thus, if the guiding curve is a circle the cylinder is called a *right circular cylinder.* In the same way, we define parabolic, elliptic and hyperbolic cylinders.

We consider below only those cylinders that are parallel to a coordinate axis and whose guiding curve is in a coordinate plane. For such cylinders, the following theorem serves as a guiding principle:

Theorem 2.1 An equation in cartesian coordinates, from which a variable is missing, represents a cylinder parallel to the axis of the missing variable. The curve corresponding to this equation is the guiding curve of the cylinder.

Proof For definiteness, let

$$f(x, y) = 0 \tag{2.98}$$

be the given equation in which the variable z is missing. In three dimensions, Eq. (2.98) may be written in the form

$$f(x, y) = 0, \quad z = 0 \tag{2.99}$$

In either form, the equation imposes restrictions on x and y only, but none on z. Thus, if the point $P_0(x, y, 0)$ lies on the curve (2.98), the point $P(x, y, z)$, having the same values of x and y and any value of z, also lies on the surface. In other words, if a point P is such that its x and y coordinates satisfy Eq. (2.98), then P lies on the cylinder for all values of z. Therefore, corresponding to any point (x_1, y_1) on the curve (2.99), the line

$$x = x_1, \quad y = y_1 \tag{2.100}$$

which is parallel to the z-axis, lies in the surface $f(x, y) = 0$. This surface, therefore, represents a cylinder parallel to the z-axis which is the missing variable here. The curve (2.99) is the guiding curve of the cylinder.

If two variables are missing, the cylinder consists of a collection of planes parallel to the axes of both the missing variables.

Example 2.80 We consider the surface represented by

$$x^2 + 4y^2 = 16 \tag{i}$$

In two dimensions, Eq. (i) represents an ellipse with centre at the origin and with semi-axes 4 and 2. The missing variable is z and therefore the axis of the cylinder is the z-axis. Hence, the surface in (i) represents an *elliptic cylinder* parallel to the z-axis (Fig. 2.51).

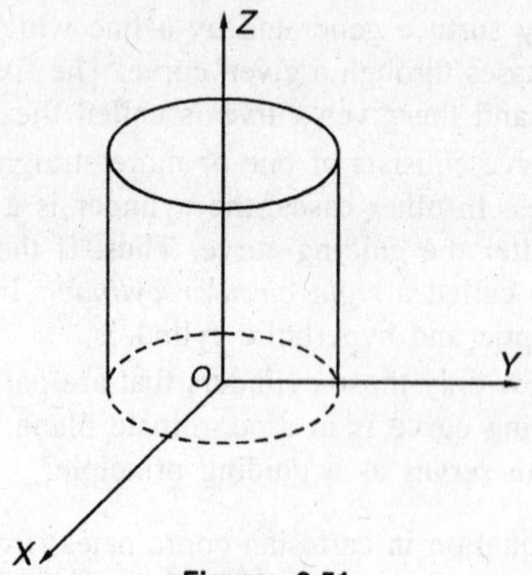

Figure 2.51

Example 2.81 The surface $y^2 + 4z^2 = 16$ is an elliptic cylinder parallel to the x-axis. The cross-section of the cylinder are ellipses with centres on the x-axis and hence the x-axis is the axis of the cylinder (see Fig. 2.52).

A *cone* is a surface generated by a straight line which passes through a fixed point and touches a given curve. The fixed point is called the *vertex* and the given curve is the *guiding curve* of the cone.

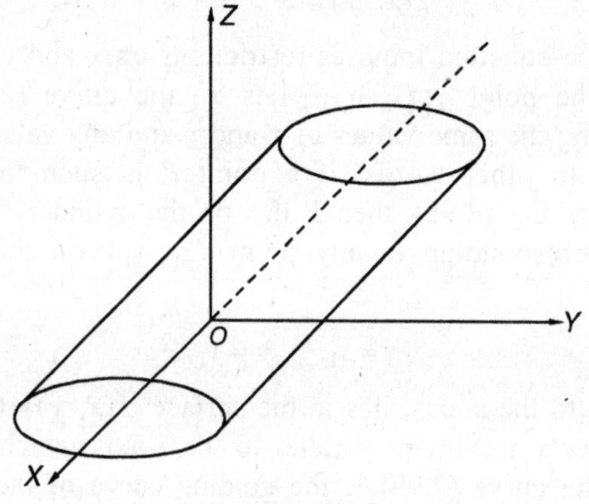

Figure 2.52

If we take the circle $x^2 + y^2 = a^2$ in the xy-plane as the guiding curve and any point on the z-axis, say (0, 0, h) as the vertex, the cone obtained is called the *right-circular cone* (Fig. 2.53). The name right circular is justified from the fact that every section of the cone by a plane perpendicular to the z-axis is a circle. The moving line is called the *generator* of the cone. The vertex V separates the surface into two parts called *nappes.* The generator VP can be so chosen that the angle between VP and VZ is an angle of measure α where $0 < \alpha < \pi/2$. Such an angle when associated with the generating figure is called the *generating angle* (or the *semi-vertical angle).* The z-axis is the axis of the cone.

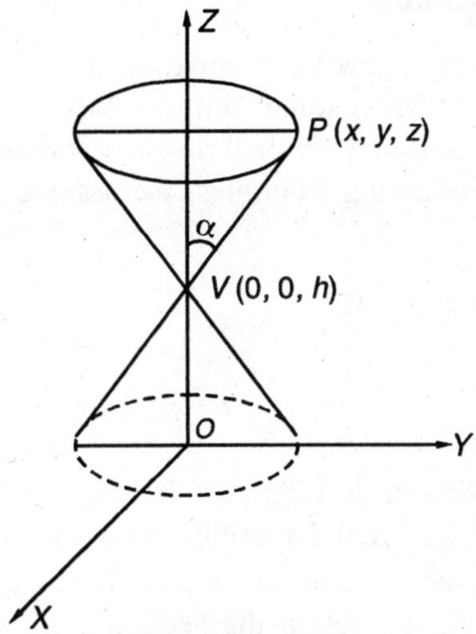

Figure 2.53

We now derive the equation of a right circular cone whose vertex is the point $V(\alpha, \beta, \gamma)$ and whose axis is the line

$$\frac{x-\alpha}{l} = \frac{y-\beta}{m} = \frac{z-\gamma}{n} \tag{i}$$

and semi-vertical angle θ.

Let $P(x, y, z)$ be any point on the cone. Then the equation of the cone may be obtained by using the fact that the line VP makes an angle θ with the axis. Now, the direction ratios of VP are $(x-\alpha)$, $(y-\beta)$ and $(z-\gamma)$, and those of the axis are l, m and n. Hence, we have

$$\cos\theta = \frac{l(x-\alpha)+m(y-\beta)+n(z-\gamma)}{\sqrt{l^2+m^2+n^2}\sqrt{(x-\alpha)^2+(y-\beta)^2+(z-\gamma)^2}} \tag{ii}$$

or

$$[l(x-\alpha)+m(y-\beta)+n(z-\gamma)]^2 = \cos^2\theta\,(l^2+m^2+n^2)$$
$$\times[(x-\alpha)^2+(y-\beta)^2+(z-\gamma)^2]$$

which is the required equation of the right circular cone.

In particular, if we take the origin as the vertex and the z-axis as the axis of the cone, we have

$$\alpha=\beta=\gamma=0 \quad \text{and} \quad l=m=0,\ n=1 \tag{iii}$$

The above equation then simplifies to

$$z^2 = \cos^2\theta\,(x^2+y^2+z^2) \tag{iv}$$

or

$$x^2+y^2 = z^2\tan^2\theta.$$

2.3.9 Quadric Surfaces

A surface represented by a quadratic equation in x, y, z is called *a quadric* surface or a *conicoid.* Only some simpler forms of these surfaces are encountered in applications and we shall restrict ourselves to a brief discussion of them here. In the following examples, we assume a, b, c to be positive constants.

Example 2.82 The equation

$$\frac{x^2}{a^2}+\frac{y^2}{b^2}+\frac{z^2}{c^2}=1 \tag{i}$$

represents a surface known as *ellipsoid.* It is symmetrical with respect to each coordinate plane, axis and the origin. It cuts the coordinate axes at the points $(\pm a, 0, 0)$ $(0, \pm b, 0)$ and $(0, 0, \pm c)$. If we cut the surface by the coordinate plane $x = 0$, we obtain the section

$$\frac{y^2}{b^2}+\frac{z^2}{c^2}=1 \tag{ii}$$

which is an ellipse. Similarly, the sections obtained by cutting the surface with the coordinate planes $y = 0$ and $z = 0$ are also ellipses.

Again, the section of the surface by the plane $z = k$ which is parallel to the XOY plane is an ellipse having the equations

$$\frac{x^2}{a^2(1-k^2/c^2)}+\frac{y^2}{b^2(1-k^2/c^2)}=1, \quad z=k, \quad -c<k<c \tag{iii}$$

Similarly, it can be shown that the sections of the surface by planes parallel to the other coordinate planes are also ellipses. With these ideas, it would be a simple matter to visualize the surface. Figure 2.54 gives a sketch of the surface.

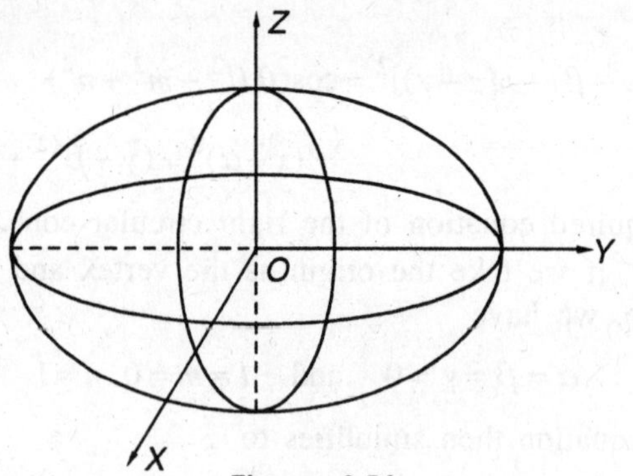

Figure 2.54

When all the three semi-axes are equal, i.e. if $a = b = c$, the equation becomes

$$x^2 + y^2 + z^2 = a^2 \tag{iv}$$

which is the equation of a sphere. When, however, any two of the semi-axes are equal, say $a = b$, the equation becomes

$$\frac{x^2}{a^2} + \frac{y^2}{a^2} + \frac{z^2}{c^2} = 1 \tag{v}$$

Such surfaces, which frequently occur in application, are called *ellipsoids of revolution* and are obtained by revolving a certain ellipse about an axis. For example, the above surface is obtained by revolving the ellipse

$$\frac{y^2}{a^2} + \frac{z^2}{c^2} = 1 \tag{vi}$$

about the z-axis. This surface, with a minor axis as an axis of revolution, is called an *oblate spheroid*. Similarly the surface

$$\frac{x^2}{b^2} + \frac{y^2}{a^2} + \frac{z^2}{b^2} = 1 \tag{vii}$$

is called a *prolate spheroid* since it is obtained by revolving an ellipse about a major axis.

Example 2.83 We next consider the *elliptic paraboloid*

$$\frac{x^2}{a^2} + \frac{y^2}{b^2} = 2z \tag{i}$$

This surface is symmetrical with respect to the planes $x = 0$ and $y = 0$ and the z-axis. The vertex of the paraboloid is at the origin. Since the left side of Eq. (i) is non-negative, the surface lies above the xy-plane (Fig. 2.55).

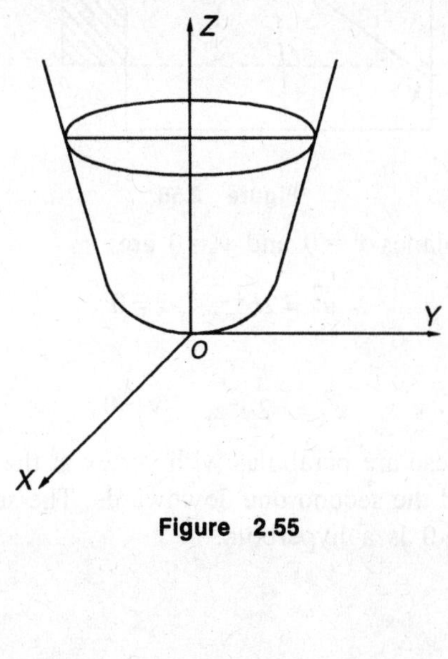

Figure 2.55

For $x = 0$, we obtain from the above equation

$$y^2 = 2b^2 z \tag{i}$$

which is a parabola with vertex at the origin and opening upward. Similarly, when $y = 0$ we have

$$x^2 = 2a^2 z \tag{iii}$$

which is a parabola in the xz-plane.

Finally, the section by a plane $z = k$ is an ellipse. The paraboloid extends upwards indefinitely. When $a = b$, the surface is a *paraboloid of revolution.*

Example 2.84 The equation

$$\frac{x^2}{a^2} - \frac{y^2}{b^2} + 2z = 0 \tag{i}$$

represents a *hyperbolic paraboloid.* This is a saddle-shaped surface and is symmetrical with respect to the planes $x = 0$, $y = 0$ and the z-axis (Fig. 2.56). Its vertex is at origin.

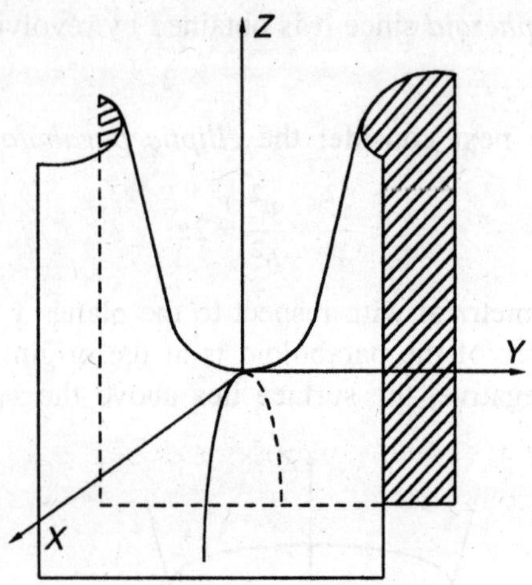

Figure 2.56

The section in the planes $x = 0$ and $y = 0$ are:

$$y^2 = 2b^2 z, \quad x = 0 \tag{ii}$$

and

$$x^2 = -2a^2 z, \quad y = 0 \tag{iii}$$

respectively. Both these are parabolas with vertex at the origin, the first one opening upwards and the second one downwards. The section of the surface by the plane $z = k > 0$ is a hyperbola.

$$\frac{x^2}{a^2} - \frac{y^2}{b^2} + 2k = 0 \tag{iv}$$

When $z = 0$ the surface reduces to a pair of interesting lines and when $z = k < 0$, the section is a hyperbola again.

Example 2.85 Next, we consider the *hyperboloid of one sheet* given by

$$\frac{x^2}{a^2} + \frac{y^2}{b^2} - \frac{z^2}{c^2} = 1 \tag{A}$$

From Fig. 2.57, since the powers of x, y, z are all even, the surface is symmetric with respect to the three coordinate planes. The section by $z = 0$ is the ellipse

$$\frac{x^2}{a^2} + \frac{y^2}{b^2} = 1 \tag{i}$$

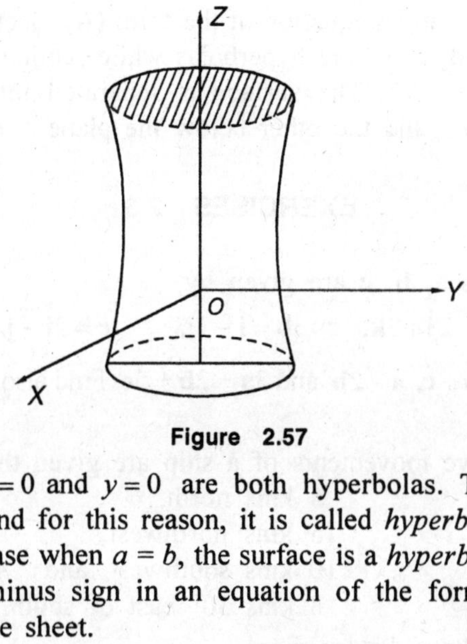

Figure 2.57

and section by $x = 0$ and $y = 0$ are both hyperbolas. The surface is one connected sheet and for this reason, it is called *hyperboloid of one sheet*. In the particular case when $a = b$, the surface is a *hyperboloid of revolution*. In general, one minus sign in an equation of the form of (A) indicates hyperboloid of one sheet.

Example 2.86 We now, finally, consider the *hyperboloid of two sheets* (see Fig. 2.58)

$$\frac{x^2}{a^2} + \frac{y^2}{b^2} - \frac{z^2}{c^2} = -1 \tag{i}$$

or

$$-\frac{x^2}{a^2} - \frac{y^2}{b^2} + \frac{z^2}{c^2} = 1,$$

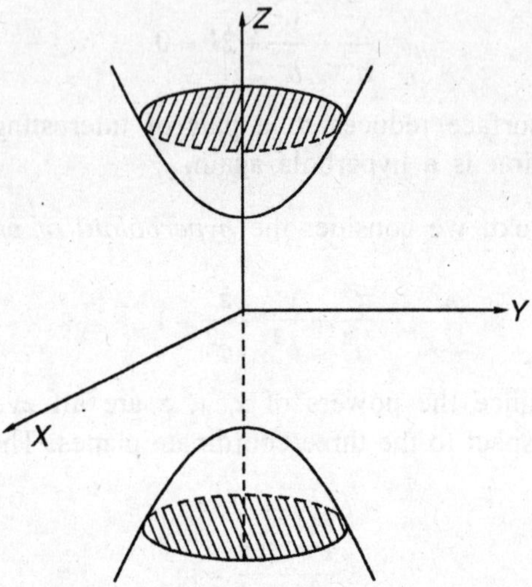

Figure 2.58

with two minus signs in an equation of the form (A). Sections by the planes parallel to $x = 0$ and $y = 0$ are hyperbolas while sections parallel to $z = 0$ are ellipses when $z^2 > c^2$. The surface is separated into two sheets one above the plane $z = c$ and the other below the plane $z = -c$.

EXERCISES 2.3

1. Three vectors **a**, **b**, **c** are given by

$$\mathbf{a} = \mathbf{i} - 2\mathbf{j} + 3\mathbf{k}, \qquad \mathbf{b} = \mathbf{i} - 3\mathbf{j}, \qquad \mathbf{c} = 2\mathbf{i} - \mathbf{j} + 5\mathbf{k}$$

   evaluate $\mathbf{a} + \mathbf{b} + \mathbf{c}$, $\mathbf{a} - 2\mathbf{b}$ and $3\mathbf{a} + 2\mathbf{b} + 5\mathbf{c}$. Find also the the unit vectors **a**, **b** and **c**.

2. The successive movements of a ship are given thus:

   5 kms north,
   15 kms northwest,
   10 kms southwest, and
   8 kms 30° east of south.

   Find its resultant displacement.

3. Find the vector which is perpendicular to the two vectors $\mathbf{a} = \mathbf{i} - 3\mathbf{j} + 2\mathbf{k}$ and $\mathbf{b} = 2\mathbf{i} - \mathbf{j} + \mathbf{k}$ and is of length 5.

4. The vertices of a quadrilateral are $A(-1, -2, 1)$, $B(4, -2, 2,)$, $C(-4, -1, 5)$ and $D(-2, 1, -3)$. At the point A, forces of magnitudes 1, 3, 5 dyne act along the lines, AB, AC and AD respectively. Find their resultant.

5. A line passes through the centre of a cube, making angles α, β, γ, δ with the four diagonals of the cube. Show that

$$\cos^2\alpha + \cos^2\beta + \cos^2\gamma + \cos^2\delta = \frac{4}{3}.$$

6. *OABC* is a parallelogram with $\vec{OA} = \mathbf{a}$, $\vec{OB} = \mathbf{b}$, and $\vec{OC} = \mathbf{c}$. Show that

$$\mathbf{a} + \frac{1}{2}(\mathbf{c} - \mathbf{a}) = \frac{1}{2}\mathbf{b}.$$

7. Using vectors, show that the diagonals of a rhombus are perpendicular.
8. If $\mathbf{a} = \mathbf{i}\cos\alpha + \mathbf{j}\sin\alpha$ and $\mathbf{b} = \mathbf{i}\cos\beta + \mathbf{j}\sin\beta$. Find $\mathbf{a}\cdot\mathbf{b}$ and hence show that

$$\cos(\alpha - \beta) = \cos(\beta - \alpha) = \cos\alpha\cos\beta + \sin\alpha\sin\beta.$$

9. Given $\mathbf{a} = \mathbf{i} - 2\mathbf{j} + 3\mathbf{k}$ and $\mathbf{b} = 2\mathbf{i} + \mathbf{j} - \mathbf{k}$, find $\mathbf{a}\cdot\mathbf{b}$, $\mathbf{a}\times\mathbf{b}$ and $\mathbf{b}\times\mathbf{a}$.
10. A particle acted on by constant forces $2\mathbf{i} + \mathbf{j} - \mathbf{k}$, $\mathbf{i} - 2\mathbf{j} + 3\mathbf{k}$ and $3\mathbf{i} + \mathbf{j} + 5\mathbf{k}$ is displaced from the point $\mathbf{i} + 2\mathbf{j} + 3\mathbf{k}$ to the point $(6\mathbf{i} + 3\mathbf{j} + \mathbf{k})$. Find the word done.
11. The point of application of the force $5\mathbf{i} + 10\mathbf{j} + \mathbf{k}$ is displaced from the point (2, 1, 3) to the point (4, 0, −5). Find the work done by the force.
12. A rigid body is rotating at the rate of 5 revolutions per second about an axis through the origin whose direction ratios are 2, −1, 3. Find the velocity (in cm/sec) of the point of the body whose coordinates are (2, 4, 3), the unit of length being 1 cm.
13. If $\mathbf{a} + \mathbf{b} + \mathbf{c} = 0$, prove that $\mathbf{a}\times\mathbf{b} = \mathbf{b}\times\mathbf{c} = \mathbf{c}\times\mathbf{a}$ and interpret the result geometrically.
14. A force **F** acts through *P* whose position vector relative to *O*, the origin, is **r**. *OL* is any axis through *O* and a unit vector along *OL* is **a**. Show that the moment of *F* about *OL* is given by the scalar triple product $\mathbf{r}\times\mathbf{F}\cdot\mathbf{a}$.
15. *OA*, *OB* and *OC* are adjacent edges of a cube of edge *a* and *OD* is a diagonal of the cube. If the cube rotates round *OD* with an angular speed of magnitude ω, show that the speed of the corner *C* is given by $\omega\sqrt{2/3}$.
16. Show that $\mathbf{a}\times\mathbf{b}\cdot\mathbf{a} = \mathbf{a}\times\mathbf{b}\cdot\mathbf{b} = 0$.
17. Derive a condition for the coplanarity of three vectors **a**, **b** and **c**. Show that the four points $A(4, 5, 1)$, $B(0, -1, -1)$, $C(3, 9, 4)$ and $D(-4, 4, 4)$ are coplanar.
18. If **a**, **b**, **c** and **d** are coplanar, show that $(\mathbf{a}\times\mathbf{b})\times(\mathbf{c}\times\mathbf{d}) = 0$.
19. If $\mathbf{a} = \mathbf{i} - 2\mathbf{j} + \mathbf{k}$, $\mathbf{b} = \mathbf{i} + 2\mathbf{j} - 4\mathbf{k}$, $\mathbf{c} = 2\mathbf{i} - 3\mathbf{j}$, evaluate the following expressions:

    (i) $\mathbf{a}\cdot\mathbf{b}\times\mathbf{c}$ (ii) $\mathbf{a}\times\mathbf{b}\cdot\mathbf{c}$

    (iii) $\mathbf{a}\times(\mathbf{b}\times\mathbf{c})$ (iv) $(\mathbf{a}\times\mathbf{b})\cdot\mathbf{c}$

    (v) $(\mathbf{a}\times\mathbf{b})\times(\mathbf{a}\times\mathbf{c})$.

20. If $\mathbf{a} = a_1\mathbf{i} + a_2\mathbf{j} + a_3\mathbf{k}$, $\mathbf{b} = b_1\mathbf{i} + b_2\mathbf{j} + b_3\mathbf{k}$ and $\mathbf{c} = c_1\mathbf{i} + c_2\mathbf{j} + c_3\mathbf{k}$, show that

$$\mathbf{a} \cdot (\mathbf{b} \times \mathbf{c}) = \begin{vmatrix} a_1 & a_2 & a_3 \\ b_1 & b_2 & b_3 \\ c_1 & c_2 & c_3 \end{vmatrix}$$

21. Given $A = (-1, 1, 2)$, $B = (0, 1, 3)$, $C = (2, 3, 4)$ and $D = (-1, 3, 3)$, find the volume of the parallelopiped with $\overrightarrow{AB}, \overrightarrow{AC}$ and $\overrightarrow{AD}$ as three of its edges.

22. For all vectors **a**, **b**, **c**, prove that

$$\mathbf{a} \times (\mathbf{b} \times \mathbf{c}) + \mathbf{b} \times (\mathbf{c} \times \mathbf{a}) + \mathbf{c} \times (\mathbf{a} \times \mathbf{b}) = 0.$$

23. If a line L is directed from $P(2, 3, 1)$ to $Q(-1, 2, 3)$, find (i) a unit vector in the direction of L, and (ii) parametric equations for L.

24. Determine whether the lines

$$\frac{x-3}{-1} = \frac{y-7}{+2} = \frac{z-1}{3} \quad \text{and} \quad \frac{x-9}{4} = \frac{y-4}{1} = \frac{z-3}{8}$$

intersect.

25. Find parametric equations for the line passing through the points (2, 1, 0) and (3, –2, 1).

26. Find the perpendicular distance from the point $P(-2, 2, 4)$ to the line L whose equations are

$$\frac{x+1}{2} = \frac{y-6}{1} = \frac{z+5}{-4}$$

27. Find the equation of the plane passing through the point (3, 3, 5) and having normal vector $2\mathbf{i} + \mathbf{j} + 2\mathbf{k}$.

28. Find the equation of the plane passing through the three poings (–1, 1, 2), (2, 0, –3) and (5, 1, 2).

29. Find the line of intersection of the planes $2x - y + z = 4$ and $x + 3y - z = 2$.

30. Find the equation of a plane whose perpendicular distance to the origin is 1/3 and which passes through the points (1, 0, 0) and (0, 0, 2).

31. Find the equation of the plane through the three points (1, 1, 1), (1, –1, 1,), (–7, –3, –5) and show that it is perpendicular to the *xz*-plane.

32. Find the equation of the plane passing through the point (–2, –2, 2) and containing the line joining the points (1, 1, 1) and (1, –1, 2).

33. Show that (–1, 4, –3) is the circumcentre of the triangle formed by the points (–3, 2, 1), (–3, 8, –5) and (3, 2, –5).

34. A plane meets the coordinate axes at A, B, C such that the centroid of the triangle ABC is the point (a, b, c). Show that the equation of the plane is

$$\frac{x}{a}+\frac{y}{b}+\frac{z}{c}-3=0.$$

35. A variable plane passes through a fixed point (a, b, c) and meets the coordinate axes in A, B, C. Show that the locus of the point common to the planes through A, B, C parallel to the coordinate planes is

$$\frac{a}{x}+\frac{b}{y}+\frac{c}{z}=1.$$

36. Find the point where the line joining (2, –3, 1), (3, –4, –5) cuts the plane $2x+y+z=7$.

37. Find the image of the point (1, 2, 5) in the plane $2x-y-z+3=0$.

38. Show that the line $x=3+3s$, $y=2-4s$, $z=s-1$ intersects the line $x+2y+3z=0$, $2x+4y+3z=-3$, and find the point of intersection.

39. A variable plane makes intercepts on the coordinate axes the sum of whose squares is constant and equal to k^2. Find the locus of the foot of the perpendicular from the origin to the plane.

40. Prove that the lines

$$\frac{x+1}{3}=\frac{y+3}{5}=\frac{z+5}{7} \quad \text{and} \quad \frac{x-2}{1}=\frac{y-4}{3}=\frac{z-6}{3}$$

intersect. Find their point of intersection and the plane in which they lie.

41. Find the shortest distance between the lines

$$x=2s+1, \qquad y=3s+2, \qquad z=4s+3$$

and

$$x=3s+2, \qquad y=4s+3, \qquad z=5s+4.$$

Show also that the lines are coplanar.

42. Show that the shortest distance between the lines

$$x+c=2y=-12z \quad \text{and} \quad x=y+2c=6z-6c$$

is $2c$.

43. The equations for a line are

$$x=2s+3, \qquad y=s-2, \qquad z=6s+4.$$

Find an equation of a plane perpendicular to this line and containing the point P (3, –2, 1).

44. Determine whether the points $A(1, 5, -2)$, $B(4, 6, 5)$, $C(3, 11, 1)$ and $D(2, 4, 1)$ lie in the same plane.

45. The intercepts of a certain plane on the three coordinates planes are A, B and C. If R is the perpendicular distance from the origin to the plane, then show that

$$\frac{1}{R^2} = \frac{1}{A^2} + \frac{1}{B^2} + \frac{1}{C^2}.$$

46. Find the equation of the sphere passing through the origin and making intercepts a, b, c on the axes.

Discuss and sketch the graphs of the following surfaces (Problems 47–70):

47. $z = 4 - x^2$

48. $z = |y|$

49. $y = e^x$

50. $x^2 + y^2 = 9$

51. $x^2 = 2y$

52. $x^2 + 9z^2 = 9$

53. $y^2 - 9z^2 = 49$

54. $xy = 1$

55. $x^2 - y^2 = 1$

56. $x^2 = y^2$

57. $x^2 + y^2 = 4z$

58. $z = 4y^2 - x^2$

59. $x^2 + y^2 + \frac{z^2}{4} = 1$

60. $x^2 - y^2 - z^2 - 1 = 0$

61. $x^2 + y^2 + z^2 + 2xy + 2yz + 2zx = 1$

62. $4x^2 + y^2 - 9z^2 = 36$

63. $x^2 + 1 = y^2 + z^2$

64. $x^2 + 16y^2 = 9z^2$

65. $x^2 + 4y^2 - 4z^2 + 4z - 1 = 0$

66. $\frac{x^2}{4} + \frac{y^2}{9} - \frac{z^2}{16} = 1$

67. $\frac{x^2}{9} + \frac{y^2}{16} = 8z$

68. $\frac{x^2}{36} + \frac{y^2}{9} + \frac{z^2}{4} = 1$

69. $z^2 - 1 = x^2 + y^2$

70. $x^2 - 4y^2 + 9z^2 + 36 = 0.$

2.4 COMPLEX NUMBERS

This topic requires an explanation of our number system. Generally speaking, the numbers fall into one of two classes: (a) the class of real numbers, (b) the class of imaginary numbers.

The real number system consists of two groups:

(a) *Rational* numbers such as 6, 3/4, 0.75, etc.
(b) *Irrational* numbers such as π, $\sqrt{2}$, $-\sqrt{5}$, etc.

It was found that the real numbers alone are not sufficient to solve all problems we come across. Thus, for example, the problem of finding $\sqrt{-25}$ or equivalently, the problem of solving the algebraic equation $x^2 + 25 = 0$. The solution of such problems requires the introduction to new numbers, the *imaginary numbers*, such as $\sqrt{-25}$, $\sqrt{-1}$, $\sqrt{-9}$, etc. It is clear that all imaginary numbers can be written in terms of $\sqrt{-1}$, which is therefore given a special symbol 'i'; thus $i \equiv \sqrt{-1}$. (It is also customary to use the symbol j instead of i). It is easy to see that $i^2 = -1$, $i^3 = -i$, $i^4 = +1$, $i^5 = i$, $i^6 = -1$, etc., which shows that the powers of i are cyclical in nature. This fact enables us to simplify higher powers of i. Thus, for example,

$$i^{39} = i^{36} i^3 = (i^6) i^3 = (-1)^6 (-i) = -i.$$

In the solution of certain problems, we come across numbers which contain both real and imaginary numbers, e.g. $3+4i$, $-2-3i$, etc. Such numbers are called *complex numbers*. In general, if x and y are real numbers, $x+iy$ is a complex number with x as *real part* and y as *imaginary part*. If $y=0$ we obtain a pure real number and if $x=0$ we obtain a pure imaginary number. The complex number $x-iy$, which is closely related to $x+iy$, is called the *complex conjugate* of $x+iy$. Thus, the complex numbers $2-3i$, $-3i$, $-3+4i$ are, respectively, the complex conjugates of $2+3i$, $3i$, $-3-4i$.

Let $x_1 + iy_1$ and $x_2 + iy_2$ be two complex numbers. If these two are equal, we have

$$x_1 + iy_1 = x_2 + iy_2$$

or

$$x_1 - x_2 = -i(y_1 - y_2)$$

or

$$(x_1 - x_2)^2 = i^2 (y_1 - y_2)^2 = -1(y_1 - y_2)^2,$$

whence

$$(x_1 - x_2)^2 + (y_1 - y_2)^2 = 0,$$

from which follows $x_1 = x_2$ and $y_1 = y_2$. This shows that two complex numbers will be equal if and only if

(i) Their real parts are equal, and
(ii) Their imaginary parts are equal.

Properties (i) and (ii) are very frequently used in applications.

Two complex numbers, say $a+ib$ and $c+id$, can be added by adding separately their real and imaginary parts. Thus,

$$(2+3i) + (4+5i) = (2+4) + (3+5)i = 6+8i.$$

Similarly, they can be subtracted by subtracting their real and imaginary parts separately. For example,

$$(2+3i) - (4+5i) = (2-4) + (3-5)i = -2-2i.$$

The multiplication of two complex numbers can be accomplished by making use of the fact that $i^2 = -1$. Thus

$$\begin{aligned}(2+3i)(4+5i) &= 2(4+5i)+3i(4+5i)\\ &= 8+10i+12i+15i^2\\ &= 8+22i-15\\ &= -7+22i\end{aligned}$$

and

$$(2+3i)(2-3i) = 4-9i^2 = 4+9 = 13.$$

If z is the conjugate of the complex number $z = x+iy$, then

$$z\bar{z} = (x+iy)(x-iy) = x^2+y^2. \tag{2.107}$$

If we denote $\sqrt{(x^2+y^2)}$ by $|z|$, then Eq. (2.107) states that

$$\sqrt{z\bar{z}} = |z|$$

The number $|z|$ is called the *absolute value* or *modulus* of $z = x+iy$. Equation (2.107) may also be written as

$$\frac{z\bar{z}}{|z|^2} = 1,$$

which shows that there exists a number z^{-1} such that

$$z^{-1} = \frac{\bar{z}}{|z|^2}, \tag{2.108}$$

provided that $z \neq 0$. Equation (2.108) suggests a possibility of computing the quotient z_1/z_2 (where $z_2 \neq 0$) of two complex numbers using the conjugate $\bar{z}_2$, for if we let $z_1 = x_1+iy_1$ and $z_2 = x_2+iy_2$, then

$$\begin{aligned}\frac{z_1}{z_2} &= \frac{x_1+iy_1}{x_2+iy_2}\\ &= \frac{(x_1+iy_1)(x_2-iy_2)}{(x_2+iy_2)(x_2-iy_2)}\\ &= \frac{x_1x_2+y_1y_2+i(y_1x_2-y_2x_1)}{x_2^2+y_2^2},\end{aligned}$$

which is a complex number of the form $a+bi$. Thus

$$\frac{2+3i}{4+5i} = \frac{(2+3i)(4-5i)}{(4+5i)(4-5i)}$$

$$= \frac{8+15+12i-10i}{16+25}$$

$$= \frac{23+2i}{41}$$

$$= \frac{23}{41} + \frac{2}{41}i.$$

For all complex numbers z_1 and z_2 and their conjugates $\bar{z}_1$ and $\bar{z}_2$, the following rules hold good:

(i) $\overline{z_1 + z_2} = \bar{z}_1 + \bar{z}_2$

(ii) $\overline{z_1 z_2} = \bar{z}_1 \bar{z}_2$

(iii) $\overline{\left(\frac{z_1}{z_2}\right)} = \frac{\bar{z}_1}{\bar{z}_2}$, where $z_2 \neq 0$.

Further, if $|z_1|$ and $|z_2|$ are the absolute values of z_1 and z_2 respectively, then the following rules are satisfied:

(i) $|z_1 z_2| = |z_1||z_2|$

(ii) $\left|\frac{z_1}{z_2}\right| = \frac{|z_1|}{|z_2|}$, where $z_2 \neq 0$,

(iii) $|z_1 + z_2| \leq |z_1| + |z_2|$.

These rules can easily be derived using the definitions. However, the inequality in (iii) is called, the *triangle inequality*, follows by noting that the points 0, z_1 and $z_1 + z_2$ are the vertices of a triangle (Fig. 2.59) with sides $|z_1|$, $|z_2|$ and $|z_1 + z_2|$, and one side cannot exceed the sum of the other two sides.

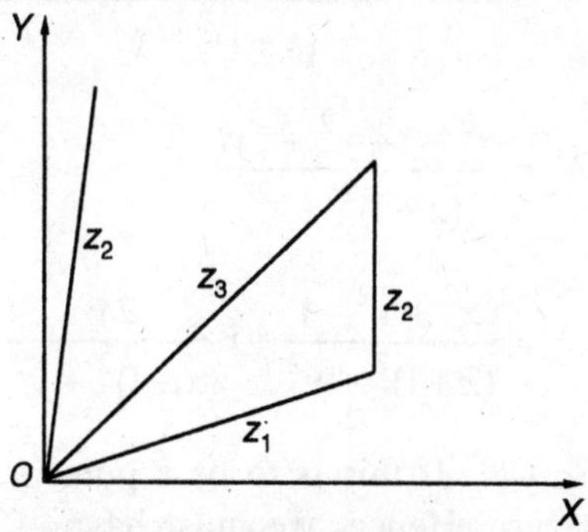

Figure 2.59

Example 2.87 If $z = 2+3i$, find $z^2 + 2z + 3$.

We have $z = 2+3i$. Then

$$z^2 = (2+3i)(2+3i) = 4 - 9 + 12i = -5 + 12i$$

Hence,

$$z^2 + 2z + 3 = -5 + 12i + 2(2+3i) + 3 = 2 + 18i.$$

Example 2.88 If $z_1 = 2+3i$ and $z_2 = 3+4i$, find z_1/z_2.

We have

$$\frac{z_1}{z_2} = \frac{2+3i}{3+4i} = \frac{(2+3i)(3-4i)}{(3+4i)(3-4i)}$$

$$= \frac{6+12+9i-8i}{9+16}$$

$$= \frac{18+i}{25}$$

$$= \frac{18}{25} + \frac{1}{25}i.$$

Example 2.89 Express the complex number

$$\frac{x-1+iy}{x+1+iy}$$

in the form $a+bi$. If this is a pure imaginary number, prove that $x^2 + y^2 = 1$.

Now,

$$\frac{x-1+iy}{x+1+iy} = \frac{(x-1+iy)(x+1-iy)}{(x+1+iy)(x+1-iy)}$$

$$= \frac{x^2 - 1 + y^2 + [y(x+1) - y(x-1)]i}{(x+1)^2 + y^2}$$

$$= \frac{x^2 - 1 + y^2 + 2yi}{(x+1)^2 + y^2}$$

$$= \frac{x^2 + y^2 - 1}{(x+1)^2 + y^2} + i\frac{2y}{(x+1)^2 + y^2}$$

which is of the form $a+bi$. If this is to be a pure imaginary number, then the real part must be zero. Hence, we must have

$$\frac{x^2 + y^2 - 1}{(x+1)^2 + y^2} = 0, \qquad \text{whence } x^2 + y^2 = 1.$$

2.4.1 The Argand Diagram

It is known that we can represent a real number by means of a point on a straight line. But in the case of a complex number, say $x + iy$, we require a two-dimensional coordinate system, the *complex plane*, consisting of the *real axis* (or the axis of real numbers) and the *imaginary axis* (or the axis of imaginary numbers) whose coordinates are x and iy respectively. Thus, in Fig. 2.60, the complex numbers $2+3i$, $1-3i$, $-2+2i$ and $-3-3i$ are represented by the points A, B, C and D, respectively.

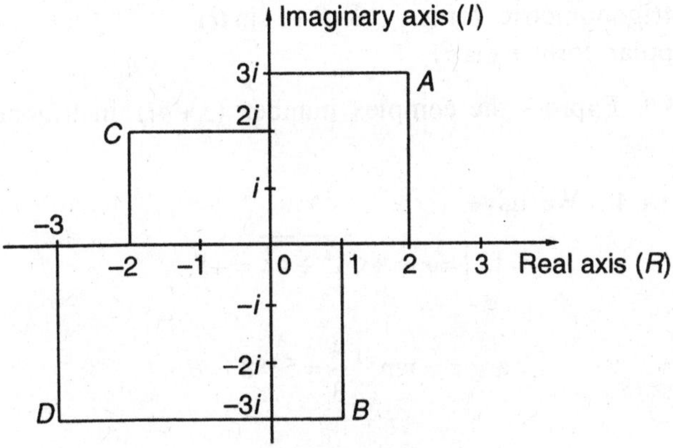

Figure 2.60

We can thus represent any complex number graphically and such a representation is called an *Argand diagram* after its originator. If the complex number $x + iy$ is represented by a point P on the Argand diagram (Fig. 2.61), then the length $OP = r$ (always taken as positive) is the modulus (or absolute value) of the complex number $x + iy$ and is written $|x + iy|$, while the angle θ is called its *argument* (or *amplitude*) and is written arg $(x + iy)$.

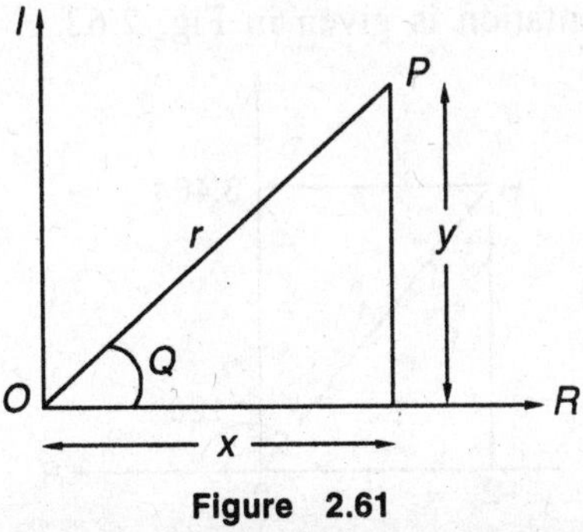

Figure 2.61

It is easily seen that $x = r\cos\theta$ and $y = r\sin\theta$ which gives

$$r = +\sqrt{x^2 + y^2}, \qquad \tan\theta = \frac{y}{x},$$

and

$$\cos\theta = \frac{x}{\sqrt{x^2 + y^2}}, \qquad \sin\theta = \frac{y}{\sqrt{x^2 + y^2}}.$$

[*Note:* Care must be exercised in selecting the correct value of θ.]

Thus, in terms of polar coordinates, we have

$$x + iy = r\cos\theta + ir\sin\theta = r(\cos\theta + i\sin\theta) = r \text{ cis } \theta,$$

where $r \text{ cis } \theta$ is the polar representation of the complex number $x + iy$. We have therefore three representations of a complex number and these are:

(i) the rectangular form $x + iy$
(ii) the trigonometric form $r(\cos\theta + i\sin\theta)$
(iii) the polar form $r \text{ cis } \theta$.

Example 2.90 Express the complex number $(3 + 4i)$ in trigonometric and polar forms.

Let $z = 3 + 4i$. We have

$$|z| = r = +\sqrt{3^2 + 4^2} = +5,$$

and

$$\arg z = \tan^{-1}\frac{4}{3} = 53°8' = \theta.$$

Hence,

$$z = 3 + 4i = 5(\cos 53°8' + i\sin 53°8') = 5 \text{ cis } 53°8'$$

Example 2.91 Express the complex number 4 cis 120° in trigonometric and rectangular forms.

We have

$$\begin{aligned} 4 \text{ cis } 120° &= 4(\cos 120° + i\sin 120°) \\ &= 4(-0.5 + i0.866) \\ &= -2 + i3.464. \end{aligned}$$

The graphical representation is given in Fig. 2.62.

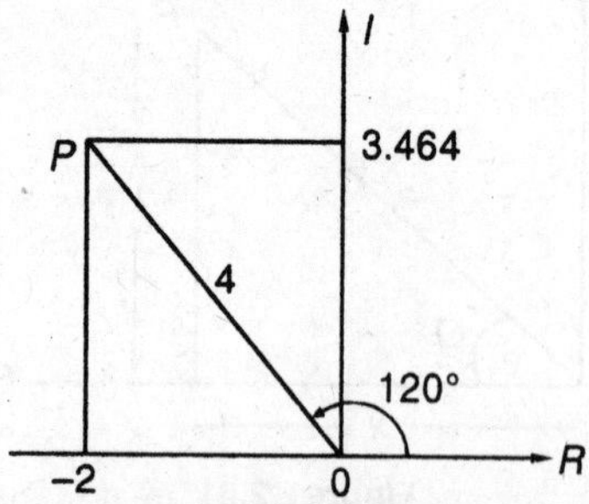

Figure 2.62

2.4.2 Multiplication and Division in Polar Forms

Let $z_1 = r_1(\cos\theta_1 + i\sin\theta_1)$ and $z_2 = r_2(\cos\theta_2 + i\sin\theta_2)$ be two complex numbers in polar form. Then,

$$\begin{aligned} z_1 z_2 &= r_1(\cos\theta_1 + i\sin\theta_1)\, r_2(\cos\theta_2 + i\sin\theta_2) \\ &= r_1 r_2[\cos\theta_1 \cos\theta_2 + i^2 \sin\theta_1 \sin\theta_2 + i(\cos\theta_1 \sin\theta_2 + \sin\theta_1 \cos\theta_2] \\ &= r_1 r_2[\cos\theta_1 \cos\theta_2 - \sin\theta_1 \sin\theta_2 + i(\cos\theta_1 + \sin\theta_2 + \sin\theta_1 \cos\theta_2] \\ &= r_1 r_2[\cos(\theta_1 + \theta_2) + i\sin(\theta_1 + \theta_2)], \end{aligned}$$

which shows that the product is a complex number whose modulus is the product of their moduli and whose argument is the sum of their arguments. Similarly,

$$\begin{aligned} \frac{z_1}{z_2} &= \frac{r_1(\cos\theta_1 + i\sin\theta_1)}{r_2(\cos\theta_2 + i\sin\theta_2)} \\ &= \frac{r_1}{r_2}\,\frac{(\cos\theta_1 + i\sin\theta_1)(\cos\theta_2 - i\sin\theta_2)}{(\cos\theta_2 + i\sin\theta_2)(\cos\theta_2 - i\sin\theta_2)} \\ &= \frac{r_1}{r_2}[\cos\theta_1 + \cos\theta_2 + \sin\theta_1 \sin\theta_2 + i(\sin\theta_1 \cos\theta_2 - \cos\theta_1 \sin\theta_2)] \\ &= \frac{r_1}{r_2}[\cos(\theta_1 - \theta_2) + i\sin(\theta_1 - \theta_2)]. \end{aligned}$$

This shows that the quotient is a complex number whose modulus is the quotient of their moduli and whose argument is the difference of their arguments.

2.4.3 De Moivre's Theorem

The law of multiplication of two complex numbers derived in the preceding section can easily be extended to n factors; thus we have

$$\begin{aligned} z_1 z_2 z_3 \ldots z_n &= (r_1 \operatorname{cis}\theta_1)(r_2 \operatorname{cis}\theta_2)(r_3 \operatorname{cis}\theta_3)\ldots(r_n \operatorname{cis}\theta_n) \\ &= r_1 r_2 \ldots r_n \operatorname{cis}(\theta_1 + \theta_2 + \cdots + \theta_n), \end{aligned}$$

where n is a positive integer. By setting $r_1 = r_2 = \cdots = r_n = r$, and $\theta_1 = \theta_2 = \cdots = \theta_n = \theta$, we obtain

$$z^n = r^n(\cos\theta + i\sin\theta)^n = r^n(\cos n\theta + i\sin n\theta).$$

If $r = 1$, we have the important result

$$(\cos\theta + i\sin\theta)^n = \cos n\theta + i\sin n\theta,$$

which is known as *De Moivre's theorem* and is true for all integral and rational values of n.

Proof

Case (i): n is a positive integer: We have

$$(\cos\theta + i\sin\theta)^2 = (\cos\theta + i\sin\theta)(\cos\theta + i\sin\theta)$$
$$= \cos^2\theta + i^2\sin^2\theta + i(\sin\theta\cos\theta + \sin\theta\cos\theta)$$
$$= \cos^2\theta - \sin^2\theta + i(2\sin\theta\cos\theta)$$
$$= \cos 2\theta + i\sin 2\theta.$$

Similarly,

$$(\cos\theta + \sin\theta)^3 = (\cos\theta + i\sin\theta)^2(\cos\theta + i\sin\theta)$$
$$= (\cos 2\theta + i\sin 2\theta)(\cos\theta + i\sin\theta)$$
$$= \cos 2\theta\cos\theta - \sin 2\theta\sin\theta + i(\sin 2\theta\cos\theta + \cos 2\theta\sin\theta)$$
$$= \cos 3\theta + i\sin 3\theta.$$

The procedure may be continued for any number of positive terms and hence we have

$$(\cos\theta + i\sin\theta)^n = \cos n\theta + i\sin n\theta,$$

when n is any positive integer.

Case (ii): n is a negative integer: In this case, let $n = -m$ where m is a positive integer. Then,

$$(\cos\theta + i\sin\theta)^n = (\cos\theta + i\sin\theta)^{-m}$$
$$= \frac{1}{(\cos\theta + i\sin\theta)^m}$$
$$= \frac{1}{\cos m\theta + i\sin m\theta}, \quad \text{(since } m \text{ is a positive integer)}$$
$$= \frac{\cos m\theta - i\sin m\theta}{(\cos m\theta + i\sin m\theta)(\cos m\theta - i\sin m\theta)}$$
$$= \cos m\theta - i\sin m\theta, \quad \text{(since the denominator equals unity)}$$
$$= \cos(-m)\theta + i\sin(-m)\theta$$

Then

$$(\cos\theta + i\sin\theta)^n = \cos n\theta + i\sin n\theta,$$

which proves the result when n is a negative integer.

Case (iii): n is real and rational: Let $n = p/q$, where p and q are integers. Since p and q are integers, we can write

$$(\cos\theta + i\sin\theta)^p = \cos p\theta + i\sin p\theta$$

and

$$\left(\cos\frac{p}{q}\theta + i\sin\frac{p}{q}\theta\right)^q = \cos p\theta + i\sin p\theta.$$

It follows that

$$[(\cos\theta + i\sin\theta)^p]^{1/q} = \cos\frac{p}{q}\theta + i\sin\frac{p}{q}\theta = \cos n\theta + i\sin n\theta.$$

This proves that

$$(\cos\theta + i\sin\theta)^n = \cos n\theta + i\sin n\theta,$$

for all integral and rational values of n.

Example 2.92 If $z = 3 + 4i$, find the values of z^{-3}.

We have

$$z = 3 + 4i = 5\,(\cos 53°8' + i\sin 53°8')$$

Then,

$$\begin{aligned} z^{-3} &= 5^{-3}[\cos 3\,(-53°8') + i\sin 3\,(-53°8')] \\ &= 5^{-3}(\cos 159°24' - i\sin 159°24') \\ &= 0.008\,(\cos 159°24' - i\sin 159°24'). \end{aligned}$$

2.4.4 Roots of a Complex Number

We have seen that

$$(\cos\theta + i\sin\theta)^{1/n} = \cos\frac{\theta}{n} + i\sin\frac{\theta}{n}.$$

This result serves a useful purpose for finding the roots of complex numbers. From the meaning of an nth root, we see that $[\cos(\theta/n) + i\sin(\theta/n)]$ is an nth root of $(\cos\theta + i\sin\theta)$. This is so because of the fact that θ/n is only one value of argument. To obtain all the values of the argument, we take its general value as $(\theta + 2k\pi)/n$, where k takes all integral values from zero to $(n-1)$. If we take $k \ge n$, we merely repeat one of the earlier values. Then, for the general solution, we have

$$r^{1/n}(\cos\theta + i\sin\theta)^{1/n} = r^{1/n}\left(\cos\frac{\theta + 2k\pi}{n} + i\sin\frac{\theta + 2k\pi}{n}\right),$$

where $k = 0, 1, 2, \ldots, (n-1)$.

The following examples illustrate the method of procedure.

Example 2.93 Find the cube roots of unity.

We have

$$1 = \cos 0° + i \sin 0°.$$

Hence,

$$(\cos 0° + i \sin 0°)^{1/3} = \cos \frac{0 + 2k\pi}{3} + i \sin \frac{0 + 2k\pi}{3}$$

for $k = 0, 1$ and 2.

Let z_1, z_2, z_3 be the three roots obtained with $k = 0$, 1, 2, respectively. Then,

$$z_1 = \cos 0 + i \sin 0 = 1,$$

$$z_2 = \cos \frac{2\pi}{3} + i \sin \frac{2\pi}{3} = -\frac{1}{2} + \frac{\sqrt{3}}{2} i,$$

$$z_3 = \cos \frac{4\pi}{3} + i \sin \frac{4\pi}{3} = -\frac{1}{2} - \frac{\sqrt{3}}{2} i.$$

It can be seen that $k = 3$ gives the root as that obtained with $k = 0$ and that the roots are merely repeated for higher values of k.

Example 2.94 Find the three cube roots of 27 and represent them on an Argand diagram.

We have

$$27 = 27 (\cos 0° + i \sin 0°).$$

Therefore,

$$27^{1/3} = 27^{1/3} \left(\cos \frac{0 + 2k\pi}{3} + i \sin \frac{0 + 2k\pi}{3} \right), \qquad k = 0, 1, 2.$$

Let z_1, z_2 and z_3 be the three roots corresponding to $k = 0$, 1, 2 respectively. Then,

$$z_1 = 3 (\cos 0° + i \sin 0°) = 3$$

$$z_2 = 3 \left(\cos \frac{2\pi}{3} + i \sin \frac{2\pi}{3} \right) = 3 (\cos 120° + i \sin 120°)$$

$$z_3 = 3 \left(\cos \frac{4\pi}{3} + i \sin \frac{4\pi}{3} \right) = 3 \cos (240° + i \sin 240°).$$

The three roots are shown plotted on an Argand diagram (Fig. 2.63).

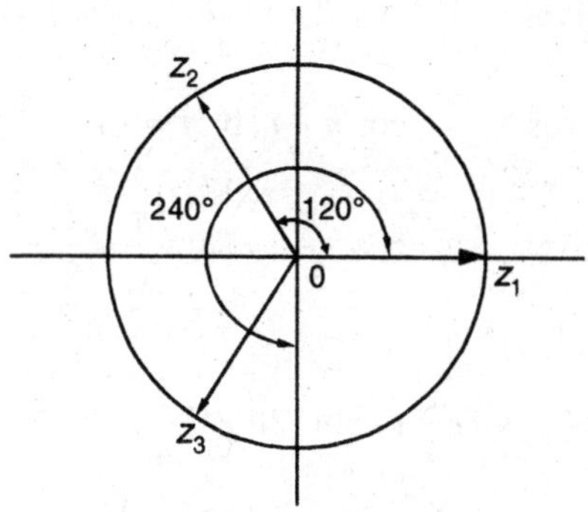

Figure 2.63

Example 2.95 If $(p+qi)$ is a root of

$$a_0 z^n + a_1 z^{n-1} + \cdots + a_n = 0 \qquad \text{(i)}$$

where $a_0 \neq 0$ and $a_1, a_2, \ldots, a_n$ are real, prove that $(p-qi)$ is also a root.

Let $z = p + qi = re^{i\theta}$. Substituting this in the given equation, we obtain:

$$a_0 r^n e^{in\theta} + a_1 r^{n-1} e^{i(n-1)\theta} + \cdots + a_{n-1} re^{i\theta} + a_n = 0$$

Taking the conjugates of both sides, we get:

$$a_0 r^n e^{-in\theta} + a_n r^{n-1} e^{-i(n-1)\theta} + \cdots + a_{n-1} re^{-i\theta} + a_n = 0,$$

which shows that $re^{-i\theta}$ is also a root of (i), i.e. $(p-qi)$ is also a root of Eq. (i).

This result is usually expressed as the statement: the roots of a polynomial equation with real coefficients occur in conjugate pairs.

Example 2.96 Solve: $x^7 + x^4 + x^3 + 1 = 0$.

The given equation can be written as

$$(x^3 + 1)(x^4 + 1) = 0$$

which gives $x = (-1)^{1/3}$ or $x = (-1)^{1/4}$.

Now

$$(-1)^{1/3} = (\cos\pi + i\sin\pi)^{(1/3)}$$

$$= \cos(2r+1)\frac{\pi}{3} + i\sin(2r+1)\frac{\pi}{3}, \qquad r = 0, 1, 2.$$

Here

$$r = 0 \quad \text{gives} \qquad \cos\frac{\pi}{3} + i\sin\frac{\pi}{3} = \frac{1}{2} + i\frac{\sqrt{3}}{2}.$$

$$r = 1 \quad \text{gives} \qquad \cos\pi + i\sin\pi = -1.$$

$$r = 2 \quad \text{gives} \qquad \cos\frac{5\pi}{3} + i\sin\frac{5\pi}{3} = \frac{1}{2} - i\frac{\sqrt{3}}{2}.$$

Also

$$(-1)^{1/4} = \cos(2r+1)\frac{\pi}{4} + i\sin(2r+1)\frac{\pi}{4}, \qquad r = 0, 1, 2, 3$$

$$r = 0 \quad \text{gives:} \qquad \cos\frac{\pi}{4} + i\sin\frac{\pi}{4} = \frac{1}{\sqrt{2}} + i\frac{1}{\sqrt{2}}.$$

$$r = 1 \quad \text{gives:} \qquad \cos\frac{3\pi}{4} + i\sin\frac{3\pi}{4} = -\frac{1}{\sqrt{2}} + i\frac{1}{\sqrt{2}}.$$

$$r = 2 \quad \text{gives:} \qquad \cos\frac{5\pi}{4} + i\sin\frac{5\pi}{4} = -\frac{1}{\sqrt{2}} - i\frac{1}{\sqrt{2}}.$$

$$r = 3 \quad \text{gives:} \qquad \cos\frac{7\pi}{4} + i\sin\frac{7\pi}{4} = \frac{1}{\sqrt{2}} - i\frac{1}{\sqrt{2}}.$$

Thus all the seven roots have been found.

2.4.5 Expansions for Powers of sine and cosine

An important application of De Moivre's theorem is to obtain expansions for powers of sines or cosines. Let

$$z = \cos\theta + i\sin\theta,$$

so that

$$z^n = \cos n\theta + i\sin n\theta.$$

Again

$$\frac{1}{z} = \cos\theta - i\sin\theta$$

and

$$\frac{1}{z^n} = \cos n\theta - i\sin n\theta.$$

It therefore follows that

$$\cos\theta = \frac{1}{2}\left(z + \frac{1}{z}\right), \qquad \cos n\theta = \frac{1}{2^n}\left(z^n + \frac{1}{z^n}\right);$$

and

$$\sin\theta = \frac{1}{2i}\left(z - \frac{1}{z}\right), \qquad \sin n\theta = \frac{1}{2^n i}\left(z^n - \frac{1}{z^n}\right)$$

Hence

$$\cos^n\theta = \frac{1}{2^n}\left(z + \frac{1}{z}\right)^n \quad \text{and} \quad \sin^n\theta = \frac{1}{(2i)^n}\left(z - \frac{1}{z}\right)^n.$$

The right-hand sides can now be expanded by Binomial theorem and powers of z are grouped together so that they can be expressed in terms of sines or cosines of multiple angles. The following examples illustrate the procedure.

Example 2.97 Express $\cos^4\theta$ in terms of cosines of multiple angles.

From the results obtained in the preceding section, we have

$$\begin{aligned}\cos^4\theta &= \frac{1}{2^4}\left(z + \frac{1}{z}\right)^4 \\ &= \frac{1}{16}\left(z^4 + 4z^3\frac{1}{z} + 6z^2\frac{1}{z^2} + 4z\frac{1}{z^3} + \frac{1}{z^4}\right) \\ &= \frac{1}{16}\left[\left(z^4 + \frac{1}{z^4}\right) + 4\left(z^2 + \frac{1}{z^2}\right) + 6\right].\end{aligned}$$

But

$$z^2 + \frac{1}{z^2} = 2\cos 2\theta \quad \text{and} \quad z^4 + \frac{1}{z^4} = 2\cos 4\theta.$$

Hence

$$\begin{aligned}\cos^4\theta &= \frac{1}{16}\,[2\cos 4\theta + 4\,(2\cos 2\theta) + 6] \\ &= \frac{1}{16}\,[2\cos 4\theta + 8\cos 2\theta + 6] \\ &= \frac{1}{8}\,(\cos 4\theta + 4\cos 2\theta + 3).\end{aligned}$$

We have thus expressed $\cos^4\theta$ in terms of $\cos 4\theta$ and $\cos 2\theta$. In a similar manner, a power of sine can be expressed in terms of sines of multiple angles.

Example 2.98 Expand $\sin^7\theta$ in a series of sines of multiples of θ.

Let $x = \cos\theta + i\sin\theta$. Then $1/x = \cos\theta - i\sin\theta$. Hence

$$x^n + \frac{1}{x^n} = 2\cos n\theta \quad \text{and} \quad x^n - \frac{1}{x^n} = 2i\sin n\theta.$$

Therefore,

$$(2i\sin\theta)^7 = \left(x - \frac{1}{x}\right)^7 = x^7 - 7x^5 + 21x^3 - 35x + \frac{35}{x} - \frac{21}{x^3} + \frac{7}{x^5} - \frac{1}{x^7}$$

or

$$2^7 i^7 \sin^7\theta = \left(x^7 - \frac{1}{x^7}\right) - 7\left(x^5 - \frac{1}{x^5}\right) + 21\left(x^3 - \frac{1}{x^3}\right) - 35\left(x - \frac{1}{x}\right)$$

or

$$-i2^7 \sin^7\theta = 2i\sin 7\theta - 7\,(2i\sin 5\theta) + 21\,(2i\sin 3\theta) - 35\,(i\sin\theta).$$

or

$$\sin^7\theta = -\frac{1}{64}(\sin 7\theta - 7\sin 5\theta + 21\sin 3\theta - 35\sin\theta).$$

Example 2.99 Prove that

$$\cos^8\theta = \frac{1}{2^7}(\cos 8\theta + 8\cos 6\theta + 28\cos 4\theta + 56\cos 2\theta + 35).$$

We have

$$(2\cos\theta)^8 = \left(x + \frac{1}{x}\right)^8$$

$$= x^8 + 8x^6 + 28x^4 + 56x^2 + 70 + \frac{56}{x^2} + \frac{28}{x^4} + \frac{8}{x^6} + \frac{1}{x^8}.$$

$$= \left(x^8 + \frac{1}{x^8}\right) + 8\left(x^6 + \frac{1}{x^6}\right) + 28\left(x^4 + \frac{1}{x^4}\right) + 56\left(x^2 + \frac{1}{x^2}\right) + 70$$

$$= 2\cos 8\theta + 16\cos 6\theta + 56\cos 4\theta + 112\cos 2\theta + 70$$

Hence

$$\cos^8\theta = \frac{1}{2^7}(\cos 8\theta + 8\cos 6\theta + 28\cos 4\theta + 56\cos 2\theta + 35).$$

Example 2.100 Prove that

$$\cos^5\theta \sin^4\theta = \frac{1}{2^8}(\cos 9\theta + \cos 7\theta - 4\cos 5\theta - 4\cos 3\theta + 6\cos\theta)$$

We have

$$x + \frac{1}{x} = 2\cos x \quad \text{and} \quad x - \frac{1}{x} = 2i\sin\theta.$$

Hence

$$\cos^5\theta \sin^4\theta = \frac{1}{2^5}\left(x + \frac{1}{x}\right)^5 \frac{1}{2^4}\left(x - \frac{1}{x}\right)^4, \qquad (\text{since } i^4 = 1).$$

$$= \frac{1}{2^9}\left(x^2 - \frac{1}{x^2}\right)^4 \left(x + \frac{1}{x}\right). \tag{i}$$

Now,

$$\left(x^2 - \frac{1}{x^2}\right)^4 = x^8 + 4(x^2)^3\left(-\frac{1}{x^2}\right) + 6(x^2)^2\left(-\frac{1}{x^2}\right)^2 + 4x^2\left(-\frac{1}{x^2}\right)^3 + \left(\frac{1}{x^2}\right)^4.$$

$$= x^8 - 4x^4 + 6 - \frac{4}{x^4} + \frac{1}{x^8}.$$

Hence (i) gives:

$$\cos^5\theta \sin^4\theta = \frac{1}{2^9}\left(x^8 - 4x^4 + 6 - \frac{4}{x^4} + \frac{1}{x^8}\right)\left(x + \frac{1}{x}\right)$$

$$= \frac{1}{2^9}\left(x^9 - 4x^5 + 6x - \frac{4}{x^3} + \frac{1}{x^7} + x^7 - 4x^3 + \frac{6}{x} - \frac{4}{x^5} + \frac{1}{x^9}\right)$$

$$= \frac{1}{2^9}\left[\left(x^9 + \frac{1}{x^9}\right) + \left(x^7 + \frac{1}{x^7}\right) - 4\left(x^5 + \frac{1}{x^5}\right) - 4\left(x^3 + \frac{1}{x^3}\right)\right.$$

$$\left. + 6\left(x + \frac{1}{x}\right)\right]$$

Hence

$$\cos^5\theta \sin^4\theta = \frac{1}{2^9}(2\cos 9\theta + 2\cos 7\theta - 8\cos 5\theta - 8\cos 3\theta + 12\cos\theta)$$

$$= \frac{1}{2^8}(\cos 9\theta + \cos 7\theta - 4\cos 5\theta - 4\cos 3\theta + 6\cos\theta).$$

2.4.6 Expansion of cos $n\theta$ and sin $n\theta$ in Powers of cos θ and sin θ

By De Moivre's theorem, we have

$$(\cos\theta + i\sin\theta)^n = \cos n\theta + i\sin n\theta.$$

Since n is a positive integer we can expand $(\cos\theta + i\sin\theta)^n$ by binomial theorem. Hence we have

$$\cos n\theta + i\sin n\theta = \cos^n\theta + n\cos^{n-1}\theta\,(i\sin\theta) + \frac{n(n-1)}{2}\cos^{n-2}\theta\,(i\sin\theta)^2$$

$$+\frac{n(n-1)(n-2)}{3!}\cos^{n-3}\theta\,(i\sin\theta)^3 + \cdots + i^n\sin^n\theta.$$

Equating the real and imaginary parts in the above, we obtain

$$\cos n\theta = \cos^n\theta - \frac{n(n-1)}{2}\cos^{n-2}\theta\sin^2\theta + \cdots$$

and

$$\sin n\theta = n\cos^{n-1}\theta\sin\theta - \frac{n(n-1)(n-2)}{3!}\cos^{n-3}\theta\sin^3\theta + \cdots$$

We can also obtain

$$\tan n\theta = \frac{\sin n\theta}{\cos n\theta}$$

$$= \frac{n\cos^{n-1}\theta\sin\theta - \dfrac{n(n-1)(n-2)}{3!}\cos^{n-3}\theta\sin^3\theta + \cdots}{\cos^n\theta - \dfrac{n(n-1)}{2}\cos^{n-2}\theta\cdot\sin^2\theta + \cdots}$$

$$= \frac{n\tan\theta - \dfrac{n(n^2-1)(n-2)}{3!}\tan^3\theta + \cdots}{1 - \dfrac{n(n-1)}{2}\tan^2\theta + \cdots}$$

Example 2.101 Express $\cos 6\theta$ in terms of powers of $\cos\theta$.

We have

$$(\cos\theta + i\sin\theta)^6 = \cos^6\theta + 6\cos^5\theta(i\sin\theta) + 15\cos^4\theta\,(i\sin\theta)^2$$

$$+ 20\cos^3\theta\cdot(i\sin\theta)^3 + 15\cos^2\theta\,(i\sin\theta)^4$$

$$+ 6\cos\theta\,(i\sin\theta)^5 + (i\sin\theta)^6.$$

$$= \cos^6\theta + i\,(6\cos^5\theta\sin\theta) - 15\cos^4\theta\sin^2\theta$$

$$-20\,i\cos^3\theta\sin^3\theta + 15\cos^2\theta\sin^4\theta$$

$$+ 6i\cos\theta\sin^5\theta - \sin^6\theta.$$

Since $(\cos\theta + i\sin\theta)^6 = \cos 6\theta + i\sin 6\theta$, we obtain on equating the real parts:

$$\cos 6\theta = \cos^6\theta - 15\cos^4\theta\sin^2\theta + 15\cos^2\theta\sin^4\theta - \sin^6\theta$$

$$= \cos^6\theta - 15\cos^4\theta(1-\cos^2\theta) + 15\cos^2\theta(1-2\cos^2\theta+\cos^4\theta)$$

$$-(1-\cos^6\theta - 3\cos^2\theta + 3\cos^4\theta)$$

$$= 32\cos^6\theta - 48\cos^4\theta + 18\cos^2\theta - 1.$$

Example 2.102 Prove that

$$\frac{\sin 6\theta}{\sin\theta} = 32\cos^5\theta - 32\cos^3\theta + 6\cos\theta.$$

We have,

$$\cos 6\theta + i\sin 6\theta = (\cos\theta + i\sin\theta)^6$$

$$= \cos^6\theta + 6i\cos^5\theta\sin\theta + 15\cos^4\theta\sin^2\theta$$

$$-20i\cos^3\theta\sin^3\theta + 15\cos^2\theta\sin^4\theta$$

$$+6i\cos\theta\sin^5\theta - \sin^6\theta.$$

Hence,

$$\sin 6\theta = 6\cos^5\theta\sin\theta - 20\cos^3\theta\sin^3\theta + 6\cos\theta\sin^5\theta$$

Therefore,

$$\frac{\sin 6\theta}{\sin\theta} = 6\cos^5\theta - 20\cos^3\theta\sin^2\theta + 6\cos\theta\sin^4\theta$$

$$= 6\cos^5\theta - 20\cos^3\theta(1-\cos^2\theta) + 6\cos\theta(1+\cos^4\theta - 2\cos^2\theta)$$

$$= 32\cos^5\theta - 32\cos^3\theta + 6\cos\theta.$$

Example 2.103 Expand $\tan 7\theta$ in terms of $\tan\theta$.

We have,

$$\tan 7\theta = \frac{{}^7C_1\tan\theta - {}^7C_3\tan^3\theta + {}^7C_5\tan^5\theta - \tan^7\theta}{1 - {}^7C_2\tan^2\theta + {}^7C_4\tan^4\theta - {}^7C_6\tan^6\theta}$$

$$= \frac{7\tan\theta - 35\tan^3\theta + 21\tan^5\theta - \tan^7\theta}{1 - 21\tan^2\theta + 35\tan^4\theta - 7\tan^6\theta}$$

2.4.7 Relation to Exponential Series

We have seen that the exponential series is given by

$$e^x = 1 + x + \frac{x^2}{2!} + \frac{x^3}{3!} + \cdots + \frac{x^n}{n!} + \cdots$$

Writing ix for x, we obtain

$$e^{ix} = 1 + ix + \frac{i^2x^2}{2!} + \frac{i^3x^3}{3!} + \cdots + \frac{i^nx^n}{n!} + \cdots$$

$$= 1 + ix - \frac{x^2}{2!} - i\frac{x^3}{3!} + \frac{x^4}{4!} + i\frac{x^5}{5!} \cdots$$

$$= \left(1 - \frac{x^2}{2!} + \frac{x^4}{4!} - \cdots\right) + i\left(x - \frac{x^3}{3!} + \frac{x^5}{5!} - \cdots\right)$$

It will be seen later that

$$1 - \frac{x^2}{2!} + \frac{x^4}{4!} - \cdots = \cos x$$

and

$$x - \frac{x^3}{3!} + \frac{x^5}{5!} - \cdots = \sin x.$$

Hence,

$$e^{ix} = \cos x + i \sin x$$

Similarly,

$$e^{-ix} = \cos x - i \sin x$$

From these relations, we obtain

$$\cos x = \frac{e^{ix} + e^{-ix}}{2} \quad \text{and} \quad \sin x = \frac{e^{ix} - e^{-ix}}{2i}$$

These expressions define the sine and cosine in terms of the exponential and enable us to express a complex number in terms of exponentials. The above relations bear a clear analogy to the hyperbolic functions which are defined as

$$\sinh x = \frac{e^x - e^{-x}}{2!} \quad \text{and} \quad \cosh x = \frac{e^x + e^{-x}}{2}$$

To deduce the relation between circular and hyperbolic functions, we observe that

$$\sin ix = \frac{e^{-x} - e^x}{2i} = i\frac{e^{-x} - e^x}{-2} = i\frac{e^x - e^{-x}}{2} = i \sinh x,$$

and

$$\cos ix = \frac{e^{-x} + e^{x}}{2} = \cosh x.$$

These relations enable us to deduce properties of hyperbolic functions from those of circular functions and vice versa. We give below two examples.

(i) We have

$$\cos^2 z + \sin^2 z = \left(\frac{e^{iz} + e^{-iz}}{2}\right)^2 + \left(\frac{e^{iz} - e^{-iz}}{2i}\right)^2$$

$$= \frac{e^{2iz} + 2 + e^{-2iz}}{4} + \frac{e^{2iz} - 2 + e^{-2iz}}{4i^2}$$

$$= \frac{e^{2iz} + 2 + e^{-2iz}}{4} - \frac{e^{2iz} - 2 + e^{-2iz}}{4}$$

$$= 1.$$

We can deduce a corresponding relation for hyperbolic functions by setting $z = ix$ in the above. We then obtain

$$\cos^2 (ix) + \sin^2 (ix) = 1.$$

But

$$\cos ix = \cosh x \text{ and } \sin ix = i \sinh x.$$

We therefore obtain

$$\cosh^2 x - \sinh^2 x = 1.$$

(ii) Suppose that we wish to evaluate $\cos (x + iy)$. From the addition formula, we have

$$\cos (x + iy) = \cos x \cos iy - \sin x \sin iy$$

$$= \cos x \cosh y - i \sin x \sinh y.$$

Thus, $\cos (x + iy)$ has been expressed in the form $u + iv$ where,

$$u = \cos x \cosh y \quad \text{and} \quad v = -\sin x \sinh y.$$

The following formulae are easily derived from the corresponding formulae for circular functions:

(i) $\cosh 2x = \cosh^2 x + \sinh^2 x$

(ii) $\sinh 3x = 3 \sinh x + 4 \sinh^3 x$

(iii) $\sinh (x + iy) = \sinh x \cosh y + \cosh x \sinh y$

(iv) $\tanh (x + iy) = \dfrac{\tanh x + \tanh y}{1 + \tanh x \tanh y}$

Example 2.104 Find all the nth roots of unity.

Let

$$z^n = 1 = \cos 2k\pi + i \sin 2k\pi = e^{2k\pi i}, \qquad k = 0, \pm 1, \pm 2, \ldots$$

Hence

$$z = \cos \frac{2k\pi}{n} + i \sin \frac{2k\pi}{n}, \qquad k = 0, 1, 2, \ldots, (n-1).$$

Thus, the required roots can be written as: 1, $e^{2\pi i/n}$, $e^{4\pi i/n}$, $e^{6\pi i/n}$, ..., $e^{2(n-1)\pi i/n}$.

Example 2.105 Prove that

(i) $\cos \frac{2\pi}{n} + \cos \frac{4\pi}{n} + \cdots + \cos \frac{2(n-1)\pi}{n} = -1.$

(ii) $\sin \frac{2\pi}{n} + \sin \frac{4\pi}{n} + \cdots + \sin \frac{2(n-1)\pi}{n} = 0.$

From Example 2.99, the roots of $z^n - 1 = 0$ are: 1, $e^{2\pi i/n}, e^{4\pi i/n}, e^{6\pi i/n}, \ldots, e^{2(n-1)\pi i/n}$. Since the sum of the roots is zero, we have

$$e^{2\pi i/n} + e^{4\pi i/n} + \cdots + e^{2(n-1)\pi i/n} = -1.$$

i.e.

$$\cos \frac{2\pi}{n} + i \sin \frac{2\pi}{n} + \cos \frac{4\pi}{n} + i \sin \frac{4\pi}{n} + \cdots$$

$$+ \cos \frac{2(n-1)\pi}{n} + i \sin \frac{2(n-1)\pi}{n} = -1.$$

Equating the real and imaginary parts of both sides, we get the required results.

Example 2.106 Find the periods of $\sinh x$, $\cosh x$ and $\tanh x$.

We have

$$\sinh (x + 2\pi i) = \frac{e^{x+2\pi i} - e^{-(x+2\pi i)}}{2} = \frac{1}{2}(e^x e^{2\pi i} - e^{-x} e^{-2\pi i}) \qquad \text{(i)}$$

But

$$e^{2\pi i} = \cos 2\pi + i \sin 2\pi = 1 \quad \text{and} \quad e^{-2\pi i} = 1.$$

Therefore,

$$\sinh (x + 2\pi i) = \frac{1}{2}(e^x - e^{-x}) = \sinh x.$$

Hence the period of $\sinh x$ is $2\pi i$.

Again,

$$\cosh(x+2\pi i)=\frac{1}{2}\left[e^{x+2\pi i}+e^{-(x+2\pi i)}\right]$$

$$=\frac{1}{2}(e^x e^{2\pi i}+e^{-x}e^{-2\pi i})$$

$$=\frac{1}{2}(e^x+e^{-x})$$

$$=\cosh x.$$

Hence the period of $\cosh x$ is $2\pi i$.

Finally,

$$\tanh(x+\pi i)=\frac{\sinh(x+\pi i)}{\cosh(x+\pi i)}$$

$$=\frac{e^{x+\pi i}-e^{-x}e^{-\pi i}}{e^{x+\pi i}+e^{-x}e^{-\pi i}}$$

$$=\frac{e^x-e^{-x}}{e^x+e^{-x}}, \quad \text{since } e^{\pi i}=e^{-\pi i}=-1$$

$$=\tanh x$$

Hence the period of $\tanh x$ is πi.

2.4.8 Inverse Hyperbolic Functions

The inverse hyperbolic functions are defined in the same way as inverse circular functions. If $y=\sinh x$, x is called the *inverse hyperbolic sine of* y and is written as $x=\sinh^{-1}y$. Similarly, we can define $\cosh^{-1}y$, $\tanh^{-1}y$, etc. The inverse hyperbolic functions can be expressed as logarithmic functions.

(i) Let $y=\sinh^{-1}x$. Then

$$x=\sinh y=\frac{e^y-e^{-y}}{2}=\frac{e^{2y}-1}{2e^y} \quad \text{or} \quad e^{2y}-2xe^y-1=0$$

Hence,

$$e^y=\frac{2x\pm\sqrt{4x^2+4}}{2}=x\pm\sqrt{x^2+1}$$

Since e^y is positive, it follows that

$$e^y=x+\sqrt{x^2+1} \quad \text{and} \quad y=\log_e(x+\sqrt{x^2+1})$$

(ii) Let $y = \cosh^{-1} x$. Then

$$x = \cosh y = \frac{e^{2y}+1}{2e^{y}} \quad \text{or} \quad e^{2y} - 2xe^{y} + 1 = 0$$

Hence,

$$e^{y} = \frac{2x \pm \sqrt{4x^2 - 4}}{2} = x \pm \sqrt{x^2 - 1}.$$

Therefore,

$$y = \log_e (x + \sqrt{x^2 - 1}) \quad \text{or} \quad y = -\log (x + \sqrt{x^2 - 1}).$$

We take the positive value called the *principal value*.

(iii) Let $y = \tanh^{-1} x$. Then

$$x = \tanh y = \frac{e^{2y}-1}{e^{2y}+1} \quad \text{or} \quad (e^{2y}+1)x = e^{2y} - 1$$

or

$$e^{2y}(x-1) = -1 - x$$

or

$$e^{2y} = \frac{1+x}{1-x}$$

or

$$y = \frac{1}{2} \log_e \frac{1+x}{1-x}.$$

The hyperbolic and inverse hyperbolic functions can be put into the form $a + ib$ (where a and b are real quantities), i.e. can be separated into real and imaginary parts. This is illustrated in the examples given below.

Example 2.107 Express $\tan(\alpha - i\beta)$ in the form $a + ib$.

We have,

$$\tan(\alpha - i\beta) = \frac{\sin(\alpha - i\beta)}{\cos(\alpha - i\beta)}.$$

$$= \frac{2\sin(\alpha - i\beta)\cos(\alpha + i\beta)}{2\cos(\alpha - i\beta)\cos(\alpha + i\beta)}$$

$$= \frac{\sin 2\alpha - \sin 2\beta i}{\cos 2\alpha + \cos 2\beta i}$$

$$= \frac{\sin 2\alpha - i \sinh 2\beta}{\cos 2\alpha + \cosh 2\beta}$$

Then,

$$a = \frac{\sin 2\alpha}{\cos 2\alpha + \cosh 2\beta} \quad \text{and} \quad b = \frac{-\sinh 2\beta}{\cos 2\alpha + \cosh 2\beta}$$

Example 2.108 Show that

$$\cosh (x + iy) = \cosh x \cos y + i \sinh x \sin y.$$

We know that $\cos (i\theta) = \cos\theta$. Therefore,

$$\begin{aligned}\cosh (x + iy) &= \cos [i (x + iy)] \\ &= \cos (ix - y) \\ &= \cos (ix) \cos y + \sin (ix) \sin y \\ &= \cosh x \cos y + i \sinh x \sin y.\end{aligned}$$

Example 2.109 Separate into real and imaginary parts The expression $\tan^{-1}(x + iy)$.

Let $\tan^{-1}(x + iy) = a + ib$. Then $\tan (a + ib) = x + iy$. Therefore, $\tan (a - ib) = (x - y)$. Now,

$$\begin{aligned}\tan 2a &= \tan (a + ib + a - ib) \\ &= \frac{\tan (a + ib) + \tan (a - ib)}{1 - \tan (a + ib) \tan (a - ib)} \\ &= \frac{x + iy + x - iy}{1 - (x^2 + y^2)} \\ &= \frac{2x}{1 - x^2 - y^2}.\end{aligned}$$

Hence,

$$a = \frac{1}{2} \tan^{-1} \frac{2x}{1 - x^2 - y^2}$$

Similarly,

$$b = \frac{1}{2} \tanh^{-1} \frac{2y}{1 + x^2 + y^2}.$$

Example 2.110 If $\sin (A + i\beta) = \tan x + i \sec x$, prove that

$$\cosh 2\beta = 3 \sec 2A.$$

We have $\sin(A+iB) = \tan x + i\sec x$.

or

$$\sin A\cos iB + \cos A\sin iB = \tan x + i\sec x$$

or

$$\sin A\cosh B + i\cos A\sinh B = \tan x + i\sec x \qquad \text{(i)}$$

Equating the real and imaginary parts in (i), we get

$$\sin A\cosh B = \tan x \quad \text{and} \quad \cos A\sinh B = \sec x$$

or

$$\sin^2 A\cosh^2 B = \tan^2 x \qquad \text{(ii)}$$

and

$$\cos^2 A\sinh^2 B = \sec^2 x \qquad \text{(iii)}$$

Subtracting (iii) from (ii), we have

$$\sin^2 A\cosh^2 B - \cos^2 A\sinh^2 B = \tan^2 x - \sec^2 x = -1$$

or

$$\cos^2 A\sinh^2 B - \sin^2 A\cosh^2 B = 1$$

or

$$\frac{1+\cos 2A}{2}\sinh^2 B - \frac{1-\cos^2 A}{2}\cosh^2 B = 1$$

or

$$\cos 2A\cosh 2B = 3$$

or

$$\cosh 2B = 3\sec 2A.$$

Example 2.111 If $a+ib = \tan^{-1}(x+iy)$, show that

$$b = \frac{1}{2}\log\frac{x^2+(1+y)^2}{x^2+(1-y)^2}.$$

We have

$$\tan(a+ib) = x+iy, \qquad \tan(a-ib) = x-iy.$$

Now

$$\tan(2ib) = \tan[a+ib-(a-ib)] = \frac{\tan(a+ib)-\tan(a-ib)}{1+\tan(a+ib)\tan(a-ib)}$$

or

$$i\tanh 2b = \frac{x+iy-(x-iy)}{1+(x^2+y^2)} = \frac{2iy}{1+x^2+y^2}$$

or

$$\tanh 2b = \frac{2y}{1+x^2+y^2}$$

or

$$b = \frac{1}{2}\tanh^{-1}\frac{2y}{1+x^2+y^2} = \frac{1}{4}\log\frac{x^2+(1+y)^2}{x^2+(1-y)^2},$$

since

$$\tanh^{-1}u = \frac{1}{2}\log\frac{1+u}{1-u}.$$

Example 2.112 Prove that

$$\cosh 5x = 16\cosh^5 x - 20\cosh^3 x + 5\cosh x.$$

Now,

$$16\cosh^5 x - 20\cosh^3 x + 5\cosh x = 16\left(\frac{e^x+e^{-x}}{2}\right)^5 - 20\left(\frac{e^x+e^{-x}}{2}\right)^3 + 5\,\frac{e^x+e^{-x}}{2}$$

$$= \frac{1}{2}(e^{5x} + 5e^{3x} + 10e^x + 10e^{-x} + 5e^{-3x} + e^{-5x})$$

$$-\frac{5}{2}(e^{3x} + 3e^x + 3e^{-x} + e^{-3x}) + \frac{5}{2}(e^x + e^{-x})$$

$$= \frac{1}{2}(e^{5x} + e^{-5x})$$

$$= \cosh 5x.$$

Example 2.113 If $a + ib = \cos(x + iy)$, show that $\cos^2 x$ and $\cosh^2 y$ are the roots of the equation:

$$u^2 - u(1 + a^2 + b^2) + a^2 = 0.$$

We have

$$a + ib = \cos(x + iy) = \cos x\,\cosh y = -i\sin x\,\sinh y.$$

Hence

$$a = \cos x\cosh y \quad \text{and} \quad b = -\sin x\sinh y$$

Therefore,

$$1 + a^2 + b^2 = 1 + \cos^2 x\,\cosh^2 y + \sin^2 x\,\sinh^2 y$$

$$= 1 + \cos^2 x\,(1 + \sinh^2 y) + (1 - \cos^2 x)\sinh^2 y$$

$$= \cos^2 x + \cosh^2 y$$

$$= \text{sum of the roots.}$$

Also,

$$\cos^2 x \cosh^2 y = \text{product of the roots} = a^2.$$

Hence the equation whose roots are $\cos^2 x$ and $\cosh^2 y$ is given by

$$u^2 - u(1 + a + a^2) + a^2 = 0.$$

2.4.9 The Logarithm of a Negative Quantity

It is known that the logarithm of a negative quantity, for example, $\log_e(-3)$, has no real value. We are now in a position to evaluate this.

We have

$$-1 = 1(\cos \pi + i \sin \pi) = e^{i\pi}$$

Using the definition of logarithm, we obtain from the above

$$\log_e(-1) = i\pi,$$

and this result will be useful in evaluating the logarithm of any negative quantity.

Now,

$$\log_e(-3) = \log_e 3(-1) = \log_e 3 + \log_e(-1)$$

$$= \log_e 3 + i\pi$$

$$= 1.0986 + i\,(3.1416).$$

2.4.10 Complex Numbers and Matrices

We have so far represented complex numbers as polynomials in i. We shall now see an interesting relationship between complex numbers and matrices, viz., the algebra of square matrices is identical with the algebra of complex numbers. We show this by considering the square matrix

$$A = \begin{bmatrix} p & -q \\ q & p \end{bmatrix}$$

This can be rewritten as

$$A = \begin{bmatrix} p & 0 \\ 0 & p \end{bmatrix} + \begin{bmatrix} 0 & -q \\ q & 0 \end{bmatrix}$$

$$= p\begin{bmatrix} 1 & 0 \\ 0 & 1 \end{bmatrix} + q\begin{bmatrix} 0 & -1 \\ 1 & 0 \end{bmatrix} \tag{2.109}$$

The first matrix on the right-hand side is I, the unit matrix of second order which has the property $I^2 = I$, and the second matrix corresponds to i since

$$\begin{bmatrix} 0 & -1 \\ 1 & 0 \end{bmatrix}\begin{bmatrix} 0 & -1 \\ 1 & 0 \end{bmatrix} = \begin{bmatrix} -1 & 0 \\ 0 & -1 \end{bmatrix} = -1\begin{bmatrix} 1 & 0 \\ 0 & 1 \end{bmatrix}$$

Hence the expression on the right side of (2.109) behaves like a complex number of the form $p + iq$. Thus, the complex number $3 + 4i$ is represented by the square matrix

$$\begin{bmatrix} 3 & -4 \\ 4 & 3 \end{bmatrix}.$$

Similarly, instead of the complex number $1+5i$, we write the matrix

$$\begin{bmatrix} 1 & -5 \\ 5 & 1 \end{bmatrix}.$$

Example 2.114 Write a 2×2 matrix corresponding to the complex number $(5+2i)(2+3i)$.

We have

$$(5+2i)(2+3i) = 10 - 6 + 19i = 4 + 19i \tag{i}$$

But the complex number $5+2i$ is

$$\begin{bmatrix} 5 & -2 \\ 2 & 5 \end{bmatrix},$$

and the complex number $2+3i$ is

$$\begin{bmatrix} 2 & -3 \\ 3 & 2 \end{bmatrix}.$$

Their product is

$$\begin{bmatrix} 5 & -2 \\ 2 & 5 \end{bmatrix}\begin{bmatrix} 2 & -3 \\ 3 & 2 \end{bmatrix} = \begin{bmatrix} 4 & -19 \\ 19 & 4 \end{bmatrix},$$

which corresponds to the complex number $4+19i$, as in (i).

EXERCISES 2.4

Express the following complex numbers in trigonometric and polar forms (Problems 1–10):

1. $2+4i$

2. $-6+3i$

3. $-10i$

4. $2-\sqrt{3}+i$

5. $4+3i$

6. $\frac{1}{i}$

7. $\frac{-3+2i}{1+i}$

8. $\frac{2+i\sqrt{3}}{\sqrt{3}+2i}$

9. $1+\cos\alpha+i\sin\alpha$

10. $1-\cos\alpha-i\sin\alpha$.

Express the following in the form $x+iy$, where x and y are real (Problems 11–22):

11. $\frac{1}{1+\cos\theta+i\sin\theta}$

12. $\frac{2}{3+2i}$

13. $\frac{(1+i)(2-i)}{3+i}$

14. $\frac{2+5i}{4-3i}$

15. $\frac{1+3i}{3-4i}$

16. $i(2-i)(1+3i)$

17. $\frac{2+i}{i(1+i)(1-i)}$

18. $(1-\sqrt{3}i)^3$

19. $\sqrt{i}$

20. $i^3+i^4+i^6$

21. $\frac{4-3i}{1+2i+i(3+i)}$

22. $(1+i)^7$.

Solve the following equations (Problems 23–28):

23. $x^2-6x+11=0$

24. $36x^2-36x+13=0$

25. $x^6+7x^3-8=0$

26. $x^2-3x+5=0$

27. $x^7+1=0$

28. $x^8+x^5+x^3+1=0$.

Find complex numbers z that satisfy the following equations (Problems 29–31):

29. $z^3-1=0$

30. $z^2-8=6i$

31. $z^4+1=0$.

Use De Moivre's theorem to find values of the following (Problems 32–37):

32. $(1+2i)^3$

33. $\frac{2}{(1-i)^4}$

34. $\frac{5}{(i-1)^3}$

35. $[3(\cos 75°+i\sin 75°)]^3$

36. $\left(\frac{1}{2}+\frac{\sqrt{3}}{2}i\right)^{15}$

37. $(1+i)^{10}$.

38. Find, (i) the cube roots of $-i$, (ii) the fourth roots of -16, (iii) the sixth roots of 8, (iv) fourth roots of $-2 -2\sqrt{3}i$.

Find all the roots and represent them on an Argand diagram (Problems 39–44):

39. $(4+3i)^{1/4}$ **40.** $(2-i)^{1/3}$

41. $(1+i)^{1/3}$ **42.** $i^{1/4}$

43. $\left(\dfrac{2+3i}{1+i}\right)^{1/2}$ **44.** $\dfrac{1}{(1+i)^{1/3}}$

If $z = x+iy$, write the following in the form $u+iv$ (Problems 45–53):

45. $\dfrac{1}{z}$ **46.** $\dfrac{1}{z-1}$

47. $\dfrac{1}{z-i}$ **48.** $z^2 - z + 1$

49. $\dfrac{1}{z^2+i}$ **50.** $|z|$

51. $\dfrac{z+i}{z-i}$ **52.** $\left|\dfrac{z-1}{z+1}\right|$

53. $\dfrac{1}{\bar{z}}$.

54. Express.

$$\frac{3}{2+\cos\theta + i\sin\theta}$$

in the form $x+iy$ and show that $x^2+y^2 = 4x-3$.

Verify the triangle inequality in the following cases (Problems 55–56):

55. $z_1 = -1-i,\ z_2 = -2+3i$ **56.** $z_1 = 5+i,\ z_2 = 1-2i$

57. In an Argand diagram, P represents the complex number z. Find the locus of P if

(i) $|z| = 1$, (ii) $|z-1| = 2$,

(iii) $|z+2| = 5$, (iv) $|z-i| = 3$,

(v) $|2z-3| = 4$, (iv) $|z-2-2i| = 0$.

58. Four points A, B, C, D are represented by the complex numbers $4+i$, $1+6i$, $-4+3i$, $-1-2i$, respectively. Show that $ABCD$ is a square.

59. A point z moves so that

$$\frac{z-1}{z+1} = 2.$$

Show that the locus of z is a circle and find its centre and radius.

60. Show that

$$\left(\frac{1+\cos\theta+i\sin\theta}{1+\cos\theta-i\sin\theta}\right)^n = \cos n\theta + i\sin n\theta.$$

61. If $x+1/x = 2\cos\theta$, prove that $x^n + 1/x^n = 2\cos n\theta$.

62. If $(1+ix_1)(1+ix_2)(1+ix_3) = A+iB$, show that

$$\tan(\tan^{-1}x_1 + \tan^{-1}x_2 + \tan^{-1}x_3) = \frac{B}{A}.$$

63. If $\cos x_1 + \cos x_2 + \cos x_3 = 0 = \sin x_1 + \sin x_2 + \sin x_3$, show that

$$\cos 3x_1 + \cos 3x_2 + \cos 3x_3 = 3\cos(x_1+x_2+x_3).$$

64. If n is a positive integer, show that

$$(1+i)^n + (1-i)^n = (\sqrt{2})^{n+2}\cos\frac{n\pi}{4}.$$

65. Prove that

$$\left(\frac{1+\sin\theta+i\cos\theta}{1+\sin\theta-i\cos\theta}\right)^n = \cos\left(\frac{n\pi}{2}-n\theta\right) + i\sin\left(\frac{n\pi}{2}-n\theta\right)$$

66. If $Z_n = \cos(\pi/2^n) + i\sin(\pi/2^n)$, prove that $\lim_{n\to\infty}[z_1z_2z_3 \ldots z_n] = -1$.

67. If $x+\frac{1}{x} = 2\cos\theta$, $y+\frac{1}{y} = 2\cos\phi$, and $z+\frac{1}{z} = 2\cos\psi$, show that

$$xyz + \frac{1}{xyz} = 2\cos(\theta+\phi+\psi).$$

68. If $\omega = \cos(2\pi/n) + i\sin(2\pi/n)$, so that all the nth roots of unity are $1, \omega, \omega^2, \ldots, \omega^{n-1}$. Deduce that $1+\omega+\omega^2+\cdots+\omega^{n-1} = 0$.

69. Equations like $\omega = f(z) = u+iv$, where $z = x+iy$, occur in fluid dynamics problems. Derive the Cauchy–Riemann equations:

$$\frac{\partial u}{\partial x} = \frac{\partial v}{\partial y} \quad \text{and} \quad \frac{\partial u}{\partial y} = -\frac{\partial v}{\partial x}$$

and deduce Laplace's equation in two dimensions.

Write a 2 × 2 matrix corresponding to the following complex numbers (Problems 70–73):

70. $3+5i$

71. $(3+2i)(4-i)$

72. $\dfrac{2+i}{3-i}$

73. $\dfrac{-3+2i}{1+i}$

Write the following matrices as complex numbers (Problems 74–77):

74. $\begin{bmatrix} 2 & -1 \\ 1 & 2 \end{bmatrix}$

75. $\begin{bmatrix} 5 & 3 \\ -3 & 5 \end{bmatrix}$

76. $\begin{bmatrix} -1 & 0 \\ 0 & -1 \end{bmatrix}$

77. $\begin{bmatrix} 1 & \sqrt{2} \\ -\sqrt{2} & 1 \end{bmatrix}.$

78. Represent the complex numbers $2 + 3i$. and $-1 - i$ as matrices A and B respectively and show that $AB = BA$.

Prove the following (Problems 79–85):

79. $\cos^8\theta = \dfrac{1}{128}(\cos 8\theta + 8\cos 6\theta + 28\cos 4\theta + 56\cos 2\theta + 35).$

80. $\dfrac{\sin 6\theta}{\sin\theta} = 32\cos^5\theta - 32\cos^3\theta + 6\cos\theta.$

81. $128\sin^2\theta\cos^6\theta = 5 + 4\cos 2\theta - 4\cos 4\theta - 4\cos 6\theta - \cos 8\theta.$

82. $y = \log\tan\left(\dfrac{\pi}{4} + \dfrac{\theta}{2}\right)$, where $\tan\dfrac{\theta}{2} = \tanh\dfrac{y}{2}.$

83. $\log(a + ib) = \dfrac{1}{2}\log(a^2 + b^2) + i\tan^{-1}\dfrac{b}{a}.$

84. $\log\dfrac{\sin(x + iy)}{\sin(x - iy)} = 2i\tan^{-1}(\cot x\tanh y).$

85. $\dfrac{\sinh 6x}{\cosh x} = 32\sinh^5 x + 32\sinh^3 x + 6\sinh x.$

3

Differentiation

3.1 INTRODUCTION

The reader is assumed to be familiar with the concept of the differentiation of a function of a single variable. We therefore present, in this chapter, a *review* of the techniques of differentiation and discuss a few other topics of differential calculus which are not usually covered in a first course in calculus. Points of inflexion and concavity, curvature, asymptotes, envelopes and evolutes are some of the topics discussed in the sequel. To aid the average reader, the concept development is supplemented throughout with a large number of illustrative examples and exercises.

3.1.1 The Derivative

Let $y = f(x)$ be a single-valued and real-valued function of a real variable defined in a neighbourhood of the concerned value of x.

The function $y = f(x)$ is said to be *differentiable* (or have a *derivative*) at a point x if and only if

$$\lim_{h \to 0} \frac{f(x+h) - f(x)}{h}$$

'exists' and is 'finite'. The derivative is denoted by the symbol $\frac{dy}{dx}$ *or* $f'(x)$ so that

$$f'(x) = \frac{dy}{dx} = \lim_{h \to 0} \frac{f(x+h) - f(x)}{h} \tag{3.1}$$

If $f'(x)$ is continuous, $f(x)$ is said to be *continuously differentiable.*

It is frequently important to consider the left-hand and right-hand derivatives. These are defined by

$$f'(x-) = \lim_{h \to 0} \frac{f(x) - f(x-h)}{h} \tag{3.2}$$

and

$$f'(x+) = \lim_{h \to 0} \frac{f(x+h) - f(x)}{h} \tag{3.3}$$

A function $y = f(x)$ is differentiable if both $f'(x-)$ and $f'(x+)$ are finite and equal.

If a function is differentiable at a point, it is necessarily continuous there. This can easily be proved by considering the equation

$$f(x+h) - f(x) = h \frac{f(x+h) - f(x)}{h},$$

and taking limits, as $h \to 0$, of both sides. We then obtain:

$$\lim_{h \to 0} [f(x+h) - f(x)] = \lim_{h \to 0} [hf'(x)] = 0,$$

which shows that $f(x)$ is continuous for all values of x.

It is important to note the *finiteness* of the derivative. For example, the signum function defined by

$$f(x) = \text{sgn}\,(x) = \begin{cases} 1, & x > 0 \\ 0, & x = 0 \\ -1, & x < 0 \end{cases} \tag{3.4}$$

has an infinite derivative at $x = 0$ but is not continuous there,

The converse, however, is not true, i.e. a continuous function need not have a derivative at every point. For example, the absolute value function defined by

$$f(x) = |x| = \begin{cases} x, & x \geq 0 \\ -x, & x < 0 \end{cases} \tag{3.5}$$

is continuous everywhere, but has no derivative at $x = 0$, since

$$f'(0-) = -1 \qquad \text{and} \qquad f'(0+) = 1.$$

Denoting $\Delta y = f(x + \Delta x) - f(x)$, the definition of a derivative becomes:

$$f'(x) = \frac{dy}{dx} = \lim_{\Delta x \to 0} \frac{\Delta y}{\Delta x} \tag{3.6}$$

Equation (3.6) is equivalent to saying that the expression

$$\frac{\Delta y}{\Delta x} - \frac{dy}{dx}$$

is an *infinitesimal* (i.e. its limit approaches 0 as $\Delta x \to 0$). We therefore write

$$\frac{\Delta y}{\Delta x} = \frac{dy}{dx} + \varepsilon, \quad \text{where} \quad \varepsilon \to 0 \text{ as } \Delta x \to 0$$

or as,

$$\Delta y = \frac{dy}{dx}\Delta x + \varepsilon \Delta x \tag{3.7}$$

In Eq. (3.7), the difference Δy is closely approximated by an expression involving the derivative.

The derivations of the formulae for differentiation are omitted here, since these are covered in a first course. We, however, list them below for ready reference by the reader.

Rules of differentiation

(i) $\dfrac{d}{dx}(u+v) = \dfrac{du}{dx} \pm \dfrac{dv}{dx}$

(ii) $\dfrac{d}{dx}(uv) = u\dfrac{dv}{dx} + v\dfrac{du}{dx}$

(iii) $\dfrac{d}{dx}\left(\dfrac{u}{v}\right) = \dfrac{v\dfrac{du}{dx} - u\dfrac{dv}{dx}}{v^2}$

(iv) $\dfrac{d}{dx} f[u(x)] = f'(u(x))\dfrac{du}{dx}$,

where u and v are functions of x.

Derivatives of standard functions

1. $\dfrac{d}{dx}(c) = 0$

2. $\dfrac{d}{dx}x^n = nx^{n-1}$, n is a constant

3. $\dfrac{d}{dx}(\sin x) = \cos x$

4. $\dfrac{d}{dx}(\cos x) = -\sin x$

5. $\dfrac{d}{dx}(\tan x) = \sec^2 x$

6. $\dfrac{d}{dx}(\cot x) = -\operatorname{cosec}^2 x$

7. $\dfrac{d}{dx}(\sec x) = \sec x \tan x$

8. $\dfrac{d}{dx}(\operatorname{cosec} x) = -\operatorname{cosec} x \cot x$

9. $\dfrac{d}{dx}(e^x) = e^x$

10. $\dfrac{d}{dx}(a^x) = a^x \log_e a$

11. $\dfrac{d}{dx}(\log_e x) = \dfrac{1}{x}$

12. $\dfrac{d}{dx}(\log_a x) = \dfrac{1}{x}\log_a e$

13. $\dfrac{d}{dx}(\sin^{-1} x) = \dfrac{1}{\sqrt{1-x^2}}$

14. $\dfrac{d}{dx}(\cos^{-1} x) = -\dfrac{1}{\sqrt{1-x^2}}$

15. $\dfrac{d}{dx}(\tan^{-1}x) = \dfrac{1}{1+x^2}$

16. $\dfrac{d}{dx}(\cot^{-1}x) = -\dfrac{1}{1+x^2}$

17. $\dfrac{d}{dx}(\sec^{-1}x) = \dfrac{1}{x\sqrt{x^2-1}}$

18. $\dfrac{d}{dx}(\text{cosec}^{-1}x) = -\dfrac{1}{x\sqrt{x^2-1}}$

19. $\dfrac{d}{dx}(\sinh x) = \cosh x$

20. $\dfrac{d}{dx}(\cosh x) = \sinh x$

21. $\dfrac{d}{dx}(\tanh x) = \text{sech}^2 x$

22. $\dfrac{d}{dx}(\text{sech}\, x) = -\text{sech}\, x \tanh x$

23. $\dfrac{d}{dx}(\coth x) = -\text{cosech}^2 x$

24. $\dfrac{d}{dx}(\text{cosech}\, x) = -\text{cosech}\, x \coth x$

25. $\dfrac{d}{dx}(\sinh^{-1}x) = \dfrac{1}{\sqrt{x^2+1}}$

26. $\dfrac{d}{dx}(\cosh^{-1}x) = \dfrac{1}{\sqrt{x^2-1}}$

27. $\dfrac{d}{dx}(\tanh^{-1}x) = \dfrac{1}{1-x^2}$

28. $\dfrac{d}{dx}(\coth^{-1}x) = -\dfrac{1}{x^2-1}$

29. $\dfrac{d}{dx}(\text{sech}^{-1}x) = -\dfrac{1}{x\sqrt{1-x^2}}$

30. $\dfrac{d}{dx}(\text{cosech}^{-1}x) = -\dfrac{1}{x\sqrt{x^2+1}}$.

The following examples illustrate the techniques of differentiation of different functions.

Example 3.1 Find the differential coefficient of $x^2 \sin x$.

Let $y = x^2 \sin x$. Here

$$\frac{d}{dx}x^2 = 2x \quad \text{and} \quad \frac{d}{dx}\sin x = \cos x$$

Hence,

$$\frac{dy}{dx} = 2x \sin x + x^2 \cos x$$

Example 3.2 Find the differential coefficient of $f(x) = (x^2 + x + 3)^{1/2}$.

Let $u = x^2 + x + 3$ so that $f(x) = u^{1/2}$. Therefore,

$$\frac{du}{dx} = 2x + 1$$

and

$$\frac{df}{du} = \frac{1}{2}u^{1/2-1} = \frac{1}{2u^{1/2}} = \frac{1}{2\,(x^2 + x + 3)^{1/2}}$$

Hence

$$\frac{df}{dx} = \frac{df}{du}\frac{du}{dx} = \frac{1}{2\,(x^2 + x + 3)^{1/2}}\,(2x + 1).$$

Example 3.3 Find the differential coefficients of

(i) $\sin(x^2 + x + 1)$ (ii) $\sqrt{\cos(1 + x^2)}$ (iii) $\sin\sqrt{\cos x}$

(i) Let $y = \sin u$ and $u = x^2 + x + 1$. We have

$$\frac{dy}{du} = \cos u = \cos(x^2 + x + 1) \quad \text{and} \quad \frac{du}{dx} = 2x + 1.$$

Hence,

$$\frac{dy}{dx} = \frac{dy}{du}\frac{du}{dx} = \cos(x^2 + x + 1)\,(2x + 1) = (2x + 1)\cos(x^2 + x + 1).$$

(ii) Let $y = \sqrt{u}$, $u = \cos(1 + x^2)$ and $v = 1 + x^2$. We require dy/dx. Here,

$$\frac{dy}{dx} = \frac{dy}{du}\frac{du}{dv}\frac{dv}{dx}.$$

Now, since $y = \sqrt{u}$, we have

$$\frac{dy}{du} = \frac{1}{2}u^{-1/2} = \frac{1}{2\sqrt{u}} = \frac{1}{2\sqrt{\cos(1 + x^2)}}$$

Similarly,

$$\frac{du}{dv} = -\sin v = -\sin(1 + x^2), \quad \text{and} \quad \frac{dv}{dx} = 2x.$$

Hence,

$$\frac{dy}{dx} = \frac{1}{2\sqrt{\cos(1 + x^2)}}\,[-\sin(1 + x^2)]\,(2x) = -\frac{x\sin(1 + x^2)}{\sqrt{\cos(1 + x^2)}}$$

(iii) Let

$$y = \sin\sqrt{\cos x}, \quad u = \sqrt{\cos x}, \quad v = \cos x.$$

or

$$y = \sin u, \quad u = \sqrt{v}, \quad v = \cos x.$$

Then,

$$\frac{dy}{du} = \cos u, \quad \frac{du}{dv} = \frac{1}{2}v^{-1/2} = \frac{1}{2\sqrt{v}}, \quad \frac{dv}{dx} = -\sin x$$

Hence,

$$\frac{dy}{dx} = \frac{dy}{du}\frac{du}{dv}\frac{dv}{dx} = \cos u \frac{1}{2\sqrt{v}}(-\sin x)$$

$$= \cos\sqrt{\cos x}\,\frac{1}{2\sqrt{\cos x}}(-\sin x)$$

$$= -\frac{\sin x}{2\sqrt{\cos x}}\cos\sqrt{\cos x}.$$

Example 3.4 Differentiate $\frac{1-x^2}{1+x^2}$ with respect to x.

Let

$$f(x) = \frac{1-x^2}{1+x^2}$$

Then,

$$f'(x) = \frac{(1+x^2)\frac{d}{dx}(1-x^2) - (1-x^2)\frac{d}{dx}(1+x^2)}{(1+x^2)^2}$$

$$= \frac{(1+x^2)(-2x) - (1-x^2)2x}{(1+x^2)^2}$$

$$= \frac{-4x}{(1+x^2)^2}.$$

Example 3.5 Differentiate $\frac{x\cos x}{1+x^2}$ with respect to x.

Let

$$f(x) = \frac{x\cos x}{1+x^2}.$$

Then, using the quotient rule,

$$f'(x) = \frac{df}{dx} = \frac{(1+x^2)\frac{d}{dx}(x\cos x) - x\cos x\frac{d}{dx}(1+x^2)}{(1+x^2)^2}$$

But, by the product rule

$$\frac{d}{dx}(x\cos x) = x(-\sin x) + \cos x = -x\sin x + \cos x$$

and

$$\frac{d}{dx}(1+x^2) = 2x.$$

Hence,

$$\frac{df}{dx} = \frac{d}{dx}\left(\frac{x\cos x}{1+x^2}\right) = \frac{(1+x^2)(-x\sin x + \cos x) - x\cos x\,(2x)}{(1+x^2)^2}$$

$$= \frac{-x\sin x\,(1+x^2) + (1-x^2)\cos x}{(1+x^2)^2}$$

Example 3.6 Find the derivative of

$$f(x) = \frac{x-2}{2x^2-3x-2}, \qquad x \neq 2$$

$$= \frac{1}{5}, \qquad x = 2.$$

at $x = 2$.

Now, using the quotient rule,

$$f'(x) = \frac{d}{dx}\left(\frac{x-2}{2x^2-3x-2}\right)$$

$$= \frac{(2x^2-3x-2)\frac{d}{dx}(x-2) - (x-2)\frac{d}{dx}(2x^2-3x-2)}{(2x^2-3x-2)^2},$$

$$= \frac{2x^2-3x-2-(x-2)(4x-3)}{(2x^2-3x-2)^2}$$

$$= \frac{-2x^2+8x-8}{(2x^2-3x-2)^2}$$

$$= \frac{-2(x-2)^2}{(2x^2-3x-2)^2}$$

$$= -2\left(\frac{x-2}{2x^2-3x-2}\right)^2$$

Hence,

$$[f'(x)]_{x=2} = -2\left[\left(\frac{x-2}{2x^2-3x-2}\right)^2\right]_{x=2} = -2\left(\frac{1}{5}\right)^2 = -\frac{2}{25}.$$

Example 3.7 Differentiate the following functions:

(a) $f(x) = 2x^2 + 3e^{4x}$

(b) $g(x) = xe^{2x}$

(c) $y(x) = \dfrac{e^{2x}}{1+e^x}$.

(a) We have $f(x) = 2x^2 + 3e^{4x}$. Hence,

$$f'(x) = \frac{d}{dx}(2x^2) + \frac{d}{dx}(3e^{4x})$$

$$= 2(2x) + 3(4e^{4x})$$

$$= 4x + 12e^{4x}.$$

(b) We have $g(x) = xe^{2x}$. Hence,

$$g'(x) = \frac{d}{dx}(xe^{2x})$$

$$= x\frac{d}{dx}(e^{2x}) + e^{2x}\frac{d}{dx}(x)$$

$$= x(2e^{2x}) + e^{2x}$$

$$= e^{2x}(2x+1).$$

(c) Here,

$$y(x) = \frac{e^{2x}}{1+e^x}.$$

Hence,

$$y'(x) = \frac{d}{dx}\left(\frac{e^{2x}}{1+e^x}\right)$$

$$= \frac{(1+e^x)\frac{d}{dx}(e^{2x}) - e^{2x}\frac{d}{dx}(1+e^x)}{(1+e^x)^2}$$

$$= \frac{2e^{2x}(1+e^x) - e^{2x}e^x}{(1+e^x)^2}$$

$$= \frac{2e^{2x} + 2e^{3x} - e^{3x}}{(1+e^x)^2}$$

$$= \frac{2e^{2x} + e^{3x}}{(1+e^x)^2}.$$

Example 3.8 In differentiating functions of the form $y = u^v$, where both u and v are functions of x, it is found useful to take logarithms of both the sides and then differentiate. This process is known as *logarithmic differentiation* and is also useful when y is a product of several factors. The present example illustrates logarithmic differentiation.

Find dy/dx when

(i) $y = (\cos x)^{\log_e x}$, (ii) $y = \dfrac{xe^x}{\cos x}$, (iii) $y = \sqrt{\dfrac{x^2+x+1}{x^2-x+1}}$

(i) We have $y = (\cos x)^{\log_e x}$. Taking logarithms of both sides, we obtain

$$\log_e y = \log_e x \log_e \cos x \qquad \text{(i)}$$

We now differentiate both sides of (i) with respect to x. We obtain

$$\frac{d}{dx}(\log_e y) = \frac{d}{dx}(\log_e x \log_e \cos x)$$

$$= \frac{d}{dx}(\log_e x)\log_e \cos x + \log_e x \frac{d}{dx}(\log_e \cos x) \qquad \text{(ii)}$$

Now,

$$\frac{d}{dx}(\log_e y) = \frac{d}{dy}(\log_e y)\frac{dy}{dx} = \frac{1}{y}\frac{dy}{dx},$$

$$\frac{d}{dx}(\log_e x) = \frac{1}{x},$$

$$\frac{d}{dx}(\log_e \cos x) = \frac{1}{\cos x}(-\sin x) = -\tan x.$$

Hence, (ii) gives

$$\frac{1}{y}\frac{dy}{dx} = \frac{1}{x}\log_e \cos x + \log_e x(-\tan x)$$

or

$$\frac{dy}{dx} = \frac{d}{dx}\left[(\cos x)^{\log x}\right] = y\left(\frac{1}{x}\log_e \cos x - \tan x \log_e x\right)$$

$$= (\cos x)^{\log_e x}\left(\frac{1}{x}\log_e \cos x - \tan x \log_e x\right)$$

(ii) We have,

$$y = \frac{xe^x}{\cos x}.$$

Taking logarithms of both sides, we get

$$\log_e y = \log_e\left(\frac{xe^x}{\cos x}\right) = \log_e x + \log_e e^x - \log_e \cos x$$

$$= \log_e x + x - \log_e \cos x \qquad \text{(i)}$$

Differentiating both sides of (i) with respect to x, we get

$$\frac{1}{y}\frac{dy}{dx} = \frac{1}{x} + 1 - \frac{1}{\cos x}(-\sin x) = \frac{1}{x} + 1 + \tan x.$$

Hence,

$$\frac{dy}{dx} = \frac{d}{dx}\left(\frac{xe^x}{\cos x}\right) = y\left(\frac{1}{x} + 1 + \tan x\right) = \frac{xe^x}{\cos x}\left(\frac{1}{x} + 1 + \tan x\right).$$

(iii) Given that

$$y = \sqrt{\frac{x^2 + x + 1}{x^2 - x + 1}}$$

Taking logarithms of both sides, we get

$$\log_e y = \log_e\left(\frac{x^2 + x + 1}{x^2 - x + 1}\right)^{1/2}$$

$$= \frac{1}{2}\log_e \frac{x^2 + x + 1}{x^2 - x + 1}$$

$$= \frac{1}{2}[\log_e (x^2 + x + 1) - \log_e (x^2 - x + 1)]$$

Hence,

$$\frac{d}{dx}\log_e y = \frac{1}{2}\frac{d}{dx}[\log_e (x^2 + x + 1) - \log_e (x^2 - x + 1)]$$

i.e.

$$\frac{1}{y}\frac{dy}{dx} = \frac{1}{2}\left[\frac{1}{x^2 + x + 1}(2x + 1) - \frac{1}{x^2 - x + 1}(2x - 1)\right]$$

$$\frac{dy}{dx} = \frac{1}{2}\sqrt{\frac{x^2 + x + 1}{x^2 - x + 1}}\left(\frac{2x + 1}{x^2 + x + 1} - \frac{2x - 1}{x^2 - x + 1}\right)$$

$$= \frac{1}{2}\sqrt{\frac{x^2+x+1}{x^2-x+1}} \frac{2(1-x^2)}{(x^2+x+1)(x^2-x+1)}$$

$$= \sqrt{\frac{x^2+x+1}{x^2-x+1}} \frac{1-x^2}{(x^2+x+1)(x^2-x+1)}$$

$$= \frac{1-x^2}{\sqrt{x^2+x+1}\,(x^2-x+1)^{3/2}}$$

Example 3.9 Find the derivatives of:

(i) $y = \sin^{-1}\sqrt{x}$, (ii) $y = \tan^{-1}(\cos\sqrt{x})$, (iii) $y = \cos^{-1}\left(\dfrac{a+b\cos x}{b+a\cos x}\right)$.

(i) We have $y = \sin^{-1}\sqrt{x}$. Let $\sqrt{x} = z$ that $y = \sin^{-1}z$, where $z = \sqrt{x}$.

Hence,

$$\frac{dy}{dz} = \frac{d}{dz}\sin^{-1}z = \frac{1}{\sqrt{1-z^2}} = \frac{1}{\sqrt{1-x}}$$

and

$$\frac{dz}{dx} = \frac{1}{2\sqrt{x}}.$$

Therefore,

$$\frac{d}{dx}\sin^{-1}\sqrt{x} = \frac{dy}{dx} = \frac{dy}{dz}\frac{dz}{dx} = \frac{1}{\sqrt{1-x}}\frac{1}{2\sqrt{x}} = \frac{1}{2\sqrt{x(1-x)}}.$$

(ii) Given that $y = \tan^{-1}(\cos\sqrt{x})$. Let $u = \cos\sqrt{x}$ and $v = \sqrt{x}$. Then

$$y = \tan^{-1}u, \qquad u = \cos v,\ v = \sqrt{x}.$$

Hence,

$$\frac{dy}{dx} = \frac{dy}{du}\frac{du}{dv}\frac{dv}{dx}.$$

Here

$$\frac{dy}{dx} = \frac{d}{du}\tan^{-1}u = \frac{1}{1+u^2} = \frac{1}{1+\cos^2\sqrt{x}}$$

$$\frac{du}{dv} = -\sin v = -\sin\sqrt{x},$$

$$\frac{dv}{dx} = \frac{1}{2\sqrt{x}}.$$

Hence,

$$\frac{dy}{dx} = \frac{d}{dx}\tan^{-1}(\cos\sqrt{x}) = \frac{1}{1+\cos^2\sqrt{x}}(-\sin\sqrt{x})\frac{1}{2\sqrt{x}}$$

$$= -\frac{\sin\sqrt{x}}{2\sqrt{x}\,(1+\cos^2\sqrt{x})}.$$

(iii) We have,

$$y = \cos^{-1}\left(\frac{a+b\cos x}{b+a\cos x}\right).$$

Let

$$z = \frac{a+b\cos x}{b+a\cos x}.$$

Then,

$$y = \cos^{-1} z \qquad \text{where} \qquad z = \frac{a+b\cos x}{b+a\cos x}$$

Hence,

$$\frac{dy}{dx} = \frac{dy}{dz}\frac{dz}{dx}.$$

Here,

$$\frac{dy}{dz} = \frac{d}{dz}(\cos^{-1} z) = -\frac{1}{\sqrt{1-z^2}}$$

$$= -\frac{1}{\sqrt{1-\left(\frac{a+b\cos x}{b+a\cos x}\right)^2}}$$

$$= -\frac{b+a\cos x}{\sqrt{(b+a\cos x)^2-(a+b\cos x)^2}}$$

$$= -\frac{b+a\cos x}{\sqrt{(b^2-a^2)\sin^2 x}}$$

$$= -\frac{(b+a\cos x)}{\sqrt{b^2-a^2}\sin x}$$

and

$$\frac{dz}{dx} = \frac{d}{dx}\left(\frac{a + b\cos x}{b + a\cos x}\right)$$

$$= \frac{(b + a\cos x)\dfrac{d}{dx}(a + b\cos x) - (a + b\cos x)\dfrac{d}{dx}(b + a\cos x)}{(b + a\cos x)^2}$$

$$= \frac{-b\sin x\,(b + a\cos x) - (a + b\cos x)(-a\sin x)}{(b + a\cos x)^2}$$

$$= \frac{(a^2 - b^2)\sin x}{(b + a\cos x)^2}$$

Therefore,

$$\frac{dy}{dx} = -\frac{b + a\cos x}{\sqrt{b^2 - a^2}\sin x}\,\frac{(a^2 - b^2)\sin x}{(b + a\cos x)^2}$$

$$= \frac{b^2 - a^2}{\sqrt{b^2 - a^2}\,(b + a\cos x)}$$

$$= \frac{\sqrt{b^2 - a^2}}{b + a\cos x}$$

Example 3.10 Find the derivatives of

(i) $y = x - \tanh x$, (ii) $y = 2\sqrt{\cosh x - 1}$,

(iii) $y = \log_e(\cosh x)$, (iv) $y = \sin^{-1}(\tanh x)$.

(i) Given that $y = x - \tanh x$. Therefore,

$$\frac{dy}{dx} = \frac{d}{dx}(x - \tanh x)$$

$$= \frac{d}{dx}x - \frac{d}{dx}\tanh x$$

$$= 1 - \text{sech}^2 x$$

$$= \tanh^2 x, \qquad \text{since} \quad 1 - \tanh^2 x = \text{sech}^2 x$$

(ii) $y = 2\sqrt{\cosh x - 1}$,

$$\frac{dy}{dx} = \frac{2}{2\sqrt{\cosh x - 1}} \sinh x$$

$$= \frac{\sinh x}{\sqrt{\cosh x - 1}}$$

$$= \frac{\sinh x}{\sqrt{\cosh x - 1}} \frac{\sqrt{\cosh x + 1}}{\sqrt{\cosh x + 1}}$$

$$= \frac{\sinh x \sqrt{\cosh x + 1}}{\sqrt{\cosh^2 x - 1}}$$

$$= \frac{\sinh x \sqrt{\cosh x + 1}}{\sinh x}$$

$$= \sqrt{\cosh x + 1}.$$

(iii) If $y = \log_e(\cosh x) = \log_e z$ where $z = \cosh x$, then.

$$\frac{dy}{dx} = \frac{dy}{dz} \frac{dz}{dx}.$$

Here

$$\frac{dy}{dz} = \frac{1}{z} = \frac{1}{\cosh x} \quad \text{and} \quad \frac{dz}{dx} = \sinh x.$$

Therefore,

$$\frac{dy}{dx} = \frac{1}{\cosh x} \sinh x = \tanh x.$$

(iv) $y = \sin^{-1}(\tanh x) = \sin^{-1} z$, where $z = \tanh x$, then,

$$\frac{dy}{dx} = \frac{dy}{dz} \frac{dz}{dx}.$$

Here,

$$\frac{dy}{dz} = \frac{d}{dz}(\sin^{-1} z) = \frac{1}{\sqrt{1 - z^2}}$$

$$= \frac{1}{\sqrt{1 - \tanh^2 x}}$$

$$= \frac{1}{\operatorname{sech} x}$$

$$= \cosh x$$

and

$$\frac{dz}{dx} = \frac{d}{dx}(\tanh x) = \operatorname{sech}^2 x.$$

Hence

$$\frac{dy}{dx} = \frac{d}{dx}[\sin^{-1}(\tanh x)]$$

$$= \cosh x \operatorname{sech}^2 x$$

$$= \operatorname{sech} x.$$

3.1.2 Higher Derivatives and Leibnitz's Theorem

We have seen how to obtain $f'(x)$, the derivative of $f(x)$ with respect to x, when we are given a differentiable function $f(x)$. If the function $f'(x)$ is itself defferentiable, then a repetition of the process of differentiation on $f'(x)$, yields $f''(x)$, the *second derivative of* $f(x)$ with respect to x. Similarly, the *third derivative* of $f(x)$ is denoted by $f'''(x)$, the *fourth derivative* by $f^{iv}(x)$, and so on. Clearly, this notation would be cumbersome in practice and better notation is given by

$$\frac{dy}{dx}, \frac{d^2y}{dx^2}, \frac{d^3y}{dx^3}, \ldots, \frac{d^ny}{dx^n}$$

or, alternatively, the suffix notation

$$y_1, y_2, y_3, \ldots, y_n,$$

where $y = f(x)$ and n is a positive integer giving the order of the derivative. The operator notation

$$\mathrm{D}y, \mathrm{D}^2y, \mathrm{D}^3y, \ldots, \mathrm{D}^ny,$$

where D symbolizes the operation of differentiation, is also in wide use, especially in the theory of differential equations.

Example 3.11 Determine $\frac{dy}{dx}, \frac{d^2y}{dx^2}$ and $\frac{d^3y}{dx^3}$, given that

$$y = f(x) = \frac{1}{1+x}.$$

We have

$$\frac{dy}{dx} = -\frac{1}{(1+x)^2}$$

$$\frac{d^2y}{dx^2} = \frac{(-1)(-2)}{(1+x)^3} = \frac{(-1)^2 2!}{(1+x)^3}$$

$$\frac{d^3y}{dx^3} = \frac{(-1)(-2)(-3)}{(1+x)^4} = (-1)^3 \frac{3!}{(1+x)^4}.$$

The result is easily generalized and we see that

$$\frac{d^ny}{dx^n} = (-1)^n \frac{n!}{(1+x)^{n+1}}.$$

Example 3.12 (i) Given that $y = x^3$, find $\frac{dy}{dx}$, $\frac{d^2y}{dx^2}$, $\frac{dx}{dy}$ and $\frac{d^2x}{dy^2}$

(ii) If $y = \sin x$, find y_1, y_2, y_3 ...

(i) Here $y = x^3$. We easily obtain

$$\frac{dy}{dx} = 3x^2 \quad \text{and} \quad \frac{d^2y}{dx^2} = 6x.$$

Since

$$\frac{dx}{dy} = \frac{1}{dy/dx} = \frac{1}{3x^2},$$

we have

$$\frac{d^2x}{dy^2} = \frac{d}{dy}\left(\frac{1}{3x^2}\right)$$

$$= \frac{1}{3}\frac{d}{dy}\left(\frac{1}{x^2}\right)$$

$$= \frac{1}{3}\frac{d}{dx}\left(\frac{1}{x^2}\right)\frac{dx}{dy}$$

$$= \frac{1}{3}\left(-\frac{2}{x^3}\right)\frac{1}{3x^2}$$

$$= -\frac{2}{3x^3}\frac{1}{3x^2}$$

$$= -\frac{2}{9x^5}$$

We thus see that d^2y/dx^2 is different from d^2x/dy^2. This result demonstrates the fact that although

$$\frac{dx}{dy} = \frac{1}{dx/dy},$$

it is not true that

$$\frac{d^2x}{dy^2} = \frac{1}{d^2y/dx^2}.$$

(ii) Here $y = \sin x$. We have

$$\frac{dy}{dx} = y_1 = \cos x = \sin\left(x + \frac{\pi}{2}\right)$$

$$y_2 = \cos\left(x + \frac{\pi}{2}\right) = \sin\left(x + \frac{\pi}{2} + \frac{\pi}{2}\right) = \sin\left(x + 2\frac{\pi}{2}\right)$$

$$y_3 = \cos\left(x + 2\frac{\pi}{2}\right) = \sin\left(x + 3\frac{\pi}{2}\right)$$

$$\vdots$$

Hence, it follows that

$$y_n = \cos\left(x + \overline{n-1}\frac{\pi}{2}\right) = \sin\left(x + \frac{n\pi}{2}\right).$$

Example 3.13 A function $H(x)$ is defined as follows:

$$H(x) = [f(x)]^2 + [g(x)]^2,$$

where $f(x)$ is a twice differentiable function such that $f''(x) = -f(x)$ and $f'(x) = g(x)$. Determine $H(8)$ given that $H(5) = 5$. **(IIT/JEE)**

We have,

$$H(x) = [f(x)]^2 + [g(x)]^2.$$

Hence,

$$H'(x) = 2f(x)\,f'(x) + 2g(x)\,g'(x) \tag{i}$$

Since $g(x) = f'(x)$, we have, $g'(x) = f''(x) = -f(x)$. Substituting in (i), we obtain

$$H'(x) = 2f(x)\,g(x) + 2g(x)[-f(x)] = 0,$$

whence $H(x) = \text{constant}$. Since $H(5) = 5$, it follows that $H(8) = 5$.

It is seen that for computing the higher derivatives, we have to repeat the process of differentiation of the function. However, in the particular case when the function is of the form of a product of two functions, e.g. $f = uv$, a formula exists for writing down directly the nth derivative of y in terms of the derivatives of u and v. This formula, due to Leibnitz, is derived here using an intuitive approach and also the suffix notation for derivatives.

We have

$$y = u(x)\,v(x).$$

By using the product formula, we obtain

$$\frac{dy}{dx} = \frac{du}{dx} v + u \frac{dv}{dx},$$

or in the suffix notation,

$$y_1 = u_1 v + u v_1 \tag{3.8}$$

Differentiating (3.8) with respect to x, we obtain

$$\begin{aligned} y_2 = (uv)_2 &= u_2 v + u_1 v_1 + u_1 v_1 + u v_2 \\ &= u_2 v + 2u_1 v_1 + u v_2 \end{aligned} \tag{3.9}$$

Differentiating (3.9) with respect to x, we obtain

$$\begin{aligned} y_3 (uv)_3 &= u_3 v + u_2 v_1 + 2u_2 v_1 + 2u_1 v_2 + u_1 v_2 + u v_3 \\ &= u_3 v + 3u_2 v_1 + 3u_1 v_2 + u v_3 \end{aligned} \tag{3.10}$$

It is seen that the coefficients of the terms on the right-hand side of (3.8), (3.9) and (3.10) are the binomial coefficients. Hence we write down

$$y_4 = (uv)_4 = u_4 v + 4u_3 v_1 + 6u_2 v_2 + 4u_1 v_3 + u v_4 \tag{3.11}$$

and, in general,

$$y_n = (uv)_n = u_n v + n u_{n-1} v_1 + \frac{n(n-1)}{2!} u_{n-2} v_2 + \frac{n(n-1)(n-2)}{3!} u_{n-3} v_3 + \cdots + u v_n, \tag{3.12}$$

which is *Leibnitz's theorem.*

The practical application of this theorem is best shown through examples.

Example 3.14 If $y = x^2 \cos x$, find y_n.

Let $u = \cos x$ and $v = x^2$. Then,

$$u_1 = -\sin x = \cos\left(x + \frac{\pi}{2}\right)$$

$$u_2 = -\sin\left(x + \frac{\pi}{2}\right) = \cos\left(x + 2\frac{\pi}{2}\right)$$

$$u_3 = \cos\left(x + 3\frac{\pi}{2}\right);$$

$$\vdots$$

$$u_n = \cos\left(x + n\frac{\pi}{2}\right);$$

and

$$v_1 = 2x, \quad v_2 = 2, \quad v_3 = v_4 = \cdots = 0.$$

Hence, by Leibnitz's theorem,

$$y_n = u_n v + n u_{n-1} v_1 + \frac{n(n-1)}{2!} u_{n-2} v_2 + \cdots + u v_n$$

$$= \cos\left(x + \frac{n\pi}{2}\right) x^2 + n \cos\left(x + \overline{n-1}\frac{\pi}{2}\right) 2x$$

$$+ \frac{n(n-1)}{2} \cos\left(x + \overline{n-2}\frac{\pi}{2}\right)(2)$$

$$= x^2 \cos\left(x + \frac{n\pi}{2}\right) + 2nx \cos\left(x + \overline{n-1}\frac{\pi}{2}\right)$$

$$+ n(n-1)\cos\left(x + \overline{n-2}\frac{\pi}{2}\right).$$

Example 3.15 If $y = (x^2 - 1)^n$, prove that

$$(x^2 - 1)\, y_{n+2} + 2xy_{n+1} - n(n+1) y_n = 0.$$

We have

$$y_1 = n(x^2 - 1)^{n-1}(2x)$$

$$= 2nx(x^2 - 1)^{n-1}$$

$$= \frac{2nx}{(x^2 - 1)^{n-1}}$$

$$= \frac{2nx\,(x^2 - 1)^n}{x^2 - 1}$$

$$= \frac{2nxy}{x^2 - 1}$$

i.e.

$$y_1(x^2 - 1) = 2nxy \tag{i}$$

Differentiating Eq. (i) with respect to x, we get

$$y_2(x^2 - 1) + y_1(2x) = 2n\,(xy_1 + y)$$

i.e.

$$y_2(x^2 - 1) + 2\,(1 - n)\, xy_1 = 2ny \tag{ii}$$

Now, by Leibnitz's theorem:

$$\frac{d^n}{dx^n}[y_2(x^2 - 1)] = y_{n+2}(x^2 - 1) + ny_{n+1}(2x) + \frac{n(n-1)}{2} y_n(2)$$

$$= y_{n+2}\,(x^2 - 1) + 2nxy_{n+1} + n(n-1) y_n$$

and

$$\frac{d^n}{dx^n}(xy_1) = y_{n+1}x + ny_n(1) = xy_{n+1} + ny_n.$$

Hence, applying Leibnitz's theorem to Eq. (ii) above, we obtain:

$$(x^2 - 1)\, y_{n+2} + 2nxy_{n+1} + n(n-1)y_n + 2(1-n)\,(xy_{n+1} + ny_n) = 2ny_n,$$

i.e.

$$(x^2 - 1)y_{n+2} + 2xy_{n+1} - n(n+1)y_n = 0,$$

which is the required equation.

3.1.3 Implicit Functions and Functions in Parametric Form

If the functional relationship between two variables x and y is given in the form $y = f(x)$, y is said to be an *explicit function* of x. But such a form is not always possible. For example, in the following equations:

$$x^2 + y^2 + 2gx + 2fy + c = 0 \tag{3.13}$$

and,

$$\log(x^2 + y^2) = 2\tan^{-1}\frac{y}{x}, \tag{3.14}$$

the relationship between x and y is implied, and so y is said to be an *implicit function* of x. In such cases, we obtain dx/dy by differentiating the equation with respect to x. For example, differentiating both sides of (3.13) with respect to x, we obtain

$$\frac{d}{dx}(x^2 + y^2 + 2gx + 2fy + c) = \frac{d}{dx}(0) = 0 \tag{3.15}$$

Now,

$$\frac{d}{dx}(x^2) = 2x, \qquad \frac{d}{dx}(2gx) = 2g(1) = 2g, \qquad \frac{d}{dx}(c) = 0,$$

$$\frac{d}{dx}(y^2) = 2y\frac{dy}{dx}, \qquad \frac{d}{dx}(2fy) = 2f\frac{dy}{dx}.$$

Hence, Eq (3.15) becomes:

$$2x + 2y\frac{dy}{dx} + 2g + 2f\frac{dy}{dx} = 0,$$

from which we obtain:

$$\frac{dy}{dx} = -\frac{g+x}{f+y} \tag{3.16}$$

Computation of dy/dx in the case of (3.14) is left as an exercise to the reader.

Finally, we consider functions given in the parametric form, viz., $x = f(t), y = \phi(t)$, where t is a parameter. From these relations, we can obtain dx/dt and dy/dt. From the relation $x = f(t)$, we see that t depends on x. Hence, we can say that y is a function of t, where t is itself a function of x, i.e. y is a function of a function. We therefore have

$$\frac{dy}{dx} = \frac{dy}{dt}\frac{dt}{dx} = \frac{dy/dt}{dx/dt} \tag{3.17}$$

As an example, let $x = at^2$, $y = 2at$. It is easily seen that

$$\frac{dy}{dt} = 2a \quad \text{and} \quad \frac{dx}{dt} = 2at.$$

Hence, we have

$$\frac{dy}{dx} = \frac{dy/dt}{dx/dt} = \frac{2a}{2at} = \frac{1}{t}.$$

Higher derivatives can also be obtained easily. Thus,

$$\begin{aligned}\frac{d^2y}{dx^2} &= \frac{d}{dx}\left(\frac{dy}{dx}\right) = \frac{d}{dx}\left(\frac{1}{t}\right)\\ &= \frac{d}{dt}\left(\frac{1}{t}\right)\frac{dt}{dx}\\ &= -\frac{1}{t^2}\frac{1}{2at}\\ &= -\frac{1}{2at^3}.\end{aligned}$$

We conclude this section with some illustrative examples.

Example 3.16 (a) If $y = \tan^{-1}(\sinh x)$ prove that

$$\frac{d^2y}{dx^2} + \tan y\left(\frac{dy}{dx}\right)^2 = 0$$

(b) If $y = (\sin^{-1}x)^2$, prove that

$$(1-x^2)y_{n+2} - (2n+1)\,xy_{n+1} - n^2y_n = 0.$$

(c) Differentiate $\tan^{-1}\frac{2x}{1-x^2}$ with respect to $\sin^{-1}\frac{2x}{1+x^2}$.

(d) Verify that $y = \cos(k\cos^{-1}x)$ satisfies that differential equation

$$(1-x^2)\frac{d^2y}{dx^2}-x\frac{dy}{dx}+k^2y=2 \quad (k \text{ a constant}).$$

(a) To prove (a), we observe that

$$\tan y=\sinh x \tag{i}$$

Differentiating both sides with respect to x, we obtain

$$\sec^2 y\,\frac{dy}{dx}=\cosh x.$$

Hence, using (i)

$$\frac{dy}{dx}=\frac{\cosh x}{\sec^2 y}$$

$$=\frac{\cosh x}{1+\tan^2 y}$$

$$=\frac{\cosh x}{1+\sinh^2 x}$$

$$=\frac{\cosh x}{\cosh^2 x}$$

$$=\frac{1}{\cosh x}$$

$$=\operatorname{sech} x \tag{ii}$$

Now,

$$\frac{d^2y}{dx^2}=\frac{d}{dx}\left(\frac{dy}{dx}\right)=\frac{d}{dx}\operatorname{sech} x$$

$$=-\operatorname{sech} x\tanh x$$

$$=-\sinh x\operatorname{sech}^2 x$$

$$=-\tan y\left(\frac{dy}{dx}\right)^2.$$

Hence it follows that

$$\frac{d^2y}{dx^2}+\tan y\left(\frac{dy}{dx}\right)^2=0.$$

(b) We have, $y=(\sin^{-1}x)^2$, and so using the formula for the differentiation of a function of a function.

$$\frac{dy}{dx} = y_1 = 2\sin^{-1}x \frac{1}{\sqrt{1-x^2}},$$

i.e.

$$y_1\sqrt{1-x^2} = 2\sin^{-1}x \tag{i}$$

Differentiating the above equation with respect to x, we obtain

$$y_2\sqrt{1-x^2} + y_1\frac{1}{2\sqrt{1-x^2}}(-2x) = \frac{2}{\sqrt{1-x^2}},$$

which gives on simplification:

$$y_2(1-x^2) - xy_1 = 2. \tag{ii}$$

We now apply Leibnitz's theorem to both the sides of (ii)

$$\frac{d^n}{dx^n}[y_2(1-x^2)] = y_{n+2}(1-x^2) + ny_{n+1}(-2x) + \frac{n(n-1)}{2}y_n(-2)$$

$$= (1-x^2)y_{n+2} - 2nxy_{n+1} - n(n-1)y_n,$$

and

$$\frac{d^n}{dx^n}(xy_1) = y_{n+1}x + ny_n(1) = xy_{n+1} + ny_n.$$

Application of Leibnitz's theorem to (ii) therefore yields:

$$(1-x^2)y_{n+2} - 2nxy_{n+1} - n(n-1)y_n - xy_{n+1} - ny_n = 0.$$

i.e.

$$(1-x^2)y_{n+2} - (2n+1)xy_{n+1} - n^2y_n = 0,$$

which is the required result.

(c) Let

$$y = \tan^{-1}\frac{2x}{1-x^2} \quad \text{and} \quad u = \sin^{-1}\frac{2x}{1+x^2}$$

We are now required to find dy/du.

Assuming $x = \tan\theta$, it is easily seen that

$$y = \tan^{-1}\frac{2\tan\theta}{1-\tan^2\theta} = \tan^{-1}(\tan 2\theta) = 2\theta.$$

so that $dy/d\theta = 2$.

Again, we see that

$$u = \sin^{-1}\frac{2x}{1+x^2} = \sin^{-1}\frac{2\tan\theta}{1+\tan^2\theta} = \sin^{-1}(\sin 2\theta) = 2\theta,$$

so that $du/d\theta = 2$. Since

$$\frac{dy}{du} = \frac{dy}{d\theta}\frac{d\theta}{du} = \frac{dy/d\theta}{du/d\theta},$$

we immediately obtain $du/d\theta = 1$.

(d) We have, $y = \cos(k\cos^{-1}x) = \cos z$, where $z = k\cos^{-1}x$. Then,

$$\frac{dy}{dx} = \frac{dy}{dz}\frac{dz}{dx} = -\sin z\left(-\frac{k}{\sqrt{1-x^2}}\right) = k\sin(k\cos^{-1}x)\frac{1}{\sqrt{1-x^2}}.$$

or,

$$\sqrt{1-x^2}\,\frac{dy}{dx} = k\sin(k\cos^{-1}x).$$

Differentiating throughout with respect to x, we get

$$\frac{d}{dx}\left(\sqrt{1-x^2}\,\frac{dy}{dx}\right) = \frac{d}{dx}[k\sin(k\cos^{-1}x]$$

i.e.

$$\frac{1}{2\sqrt{1-x^2}}(-2x)\frac{dy}{dx} + \sqrt{1-x^2}\,\frac{d^2y}{dx^2} = k\cos(k\cos^{-1}x)\left(-\frac{k}{\sqrt{1-x^2}}\right).$$

or,

$$-\frac{x}{\sqrt{1-x^2}}\frac{dy}{dx} + \sqrt{1-x^2}\,\frac{d^2y}{dx^2} = -\frac{k^2}{\sqrt{1-x^2}}\cos(k\cos^{-1}x) = -\frac{k^2}{\sqrt{1-x^2}}y.$$

Multiplying throughout by $\sqrt{1-x^2}$, we obtain

$$-x\frac{dy}{dx} + (1-x^2)\frac{d^2y}{dx^2} = -k^2y$$

or,

$$(1-x^2)\frac{d^2y}{dx^2} - x\frac{dy}{dx} + k^2y = 0,$$

which is the required result.

Example 3.17 For the infinite continued fraction $y = x/(1+y)$, express dy/dx as a continued fraction.

Putting $x = 1$, express $(\sqrt{5}-1)/2$ as a continued fraction. We have

$$y = \frac{x}{1+y} \qquad \text{(i)}$$

or

$$y^2 + y - x = 0.$$

Differentiating the above with respect to x, we get

$$2y\frac{dy}{dx} + \frac{dy}{dx} - 1 = 0$$

or

$$\frac{dy}{dx} = \frac{1}{2y+1} = \frac{1}{1+2x/(1+y)}, \qquad \text{[using (i)]}$$

Substituting for y repeatedly in the above, we obtain an infinite continued fraction for dy/dx.

Now, putting $x = 1$ in (i), we obtain

$$y^2 + y - 1 = 0.$$

Then,

$$y = \frac{-1 \pm \sqrt{1+4}}{2} = \frac{\pm\sqrt{5}-1}{2}$$

We choose the positive roots and obtain

$$y = \frac{\sqrt{5}-1}{2} = \cfrac{1}{1+\cfrac{1}{1+\cfrac{1}{1+}}}$$

EXERCISES 3.1

Find dy/dx from first principles (Problems 1–8):

1. $y = x^2$.

2. $y = \sqrt{x}$.

3. $y = \dfrac{1}{x}$.

4. $y = \dfrac{1}{\sqrt{x}}$.

5. $y = \sqrt{1+3x}$.

6. $y = \dfrac{1}{3x+5}$.

7. $y = \sin(x^2+1)$.

8. $y = \tan(2x+1)$.

Differentiate the following functions with respect to x (Problems 9–33):

9. $x^3(1-x)^2$.

10. $x^2 + \dfrac{1}{x^2}$.

11. $\dfrac{1-x^2}{1+x^2}$.

12. $\sin^2 x$.

13. $\tan^2(2x+3)$.

14. $\dfrac{\cos x - \sin x}{\cos x + \sin x}$.

15. $\dfrac{\cos x}{x}$.

16. $x \cos \dfrac{1}{x}$.

17. $\sqrt{\dfrac{1+x}{1-x}} \cdot \sqrt{\dfrac{2+x}{2-x}}$.

18. $e^{2x} \cos 3x$.

19. $e^{\sin x}$.

20. $\cos (e^{x \sin x+2})$.

21. $\log \dfrac{1-x}{1+x}$.

22. $x^{x \log x}$.

23. $\log (\tan^{-1} x)$.

24. $\tan^{-1}(e^x \cos x)$.

25. $\tan^{-1} \dfrac{2x}{1-x^2}$.

26. $\sinh (2x) \cosh^2 x$.

27. $e^{1+\cosh 3x}$.

28. $\cosh (\sin 2x)$.

29. $\sinh^{-1}(\cos x)$.

30. $\text{sech}^{-1} \dfrac{x}{a}$.

31. $xe^x \sin^2 x \tan^{-1} x$.

32. $\tanh^{-1}(x^2+1)$.

33. $\sin^{-1} \dfrac{1}{\sqrt{x}}$.

Find *dy*/*dx* in the following cases (Problems 34–53):

34. $y = 6\sqrt[3]{x} - 5\sqrt[4]{x}$.

35. $y = \left(1 - \dfrac{1}{\sqrt[3]{x}}\right)^2$.

36. $y = \dfrac{x^2}{x^2+1}$.

37. $y = \dfrac{\tan x}{\sqrt{x}}$.

38. $y = \dfrac{\cos x}{1+2 \sin x}$.

39. $y = \dfrac{1}{(1-x^2)^5}$.

40. $y = (4+3x)^{1/2}$.

41. $y = \sqrt{a^2+b^2-2ab \cos x}$

42. $y = \log \dfrac{a^2+x^2}{a^2-x^2}$.

43. $y = \sqrt{1+\sin 2x} - \sqrt{1-\sin 2x}$.

44. $y = 2\sqrt{x} - 4 \log (2+\sqrt{x})$.

45. $y = x^2 e^{-2x}$.

46. $y = \log \dfrac{x^2}{\sqrt{1-2x^4}}$.

47. $y = \sqrt{x}\, e^{\sqrt{x}}$.

48. $y = \dfrac{1+e^x}{1-e^x}$.

49. $y = \log (\sqrt{x} - \sqrt{x-1})$.

50. $y = \log (1+\sec x)$.

51. $y = \log \dfrac{2+\tan x}{2-\tan x}$.

52. $y = \tan^{-1} x + \log \sqrt{\dfrac{1+x}{1-x}}$.

53. $y = 4 \log (\sqrt{x-4} + \sqrt{x}) + \sqrt{x^2 - 5x}$.

54. Find d^2y/dx^2 when (i) $y = e^{-x^2}$, (ii) $y = \sin^{-1}(x/3)$.

55. If $x = \cos t$ and $y = \cos 2nt$, prove that

$$(1-x^2)\frac{d^2y}{dx^2} - x\frac{dy}{dx} + 4n^2 y = 0.$$

56. If $x = \sin t - \operatorname{cosec} t$, $y = \sin^n t - \operatorname{cosec}^n t$, prove that

$$(x^2+4)\frac{d^2y}{dx^2} + x\frac{dy}{dx} - n^2 y = 0.$$

57. Find dy/dx and d^2y/dx^2 when

(i) $x = a \cos \theta$, $y = b \sin \theta$.

(ii) $x = a(\theta + \cos\theta)$, $y = a(1 - \sin\theta)$.

(iii) $x = 3\cos\theta - \cos^3\theta$, $y = 3\sin\theta - \sin^3\theta$.

58. The radius of curvature (r) at a point on a curve is given by the formula

$$\rho = \frac{[1 + (dy/dx)^2]^{3/2}}{d^2y/dx^2}.$$

Determine the radius of curvature at the point $(0, a)$ on the curve defined by $x = a \sinh t$ and $y = a \cosh t$.

59. If $x = a \cos nt + b \sin nt$ show that

$$\frac{d^2x}{dt^2} + n^2 x = 0.$$

60. If $x = Ae^{-kt} \cos(nt + \alpha) + [a/(n^2 - p^2)] \cos pt$, show that

$$\frac{d^2x}{dt^2} + n^2 x = a \cos pt.$$

61. Show that the differential equation

$$\frac{d^4y}{dx^4} + 4y = 0$$

is satisfied by $y = e^x \cos x$.

62. If $f(x) = x^5$, show that

$$f(1)+\frac{f'(1)}{1!}+\frac{f''(1)}{2!}+\cdots+\frac{f^{(5)}(1)}{5!}=2^5.$$

63. Using Leibnitz's theorem, find the second derivatives of:

(a) $y=e^x \cos x$ (b) $y=x^2 \cos x$.

64. Using Leibnitz's theorem, find the derivatives of:

(a) $y=x^3e^{2x}$ (b) $y=x^2 \sin \dfrac{x}{3}$

(b) $y=x^2 \cos x$ (c) $y=a^x x^3$.

65. Find the *n*th derivatives of:

(i) $y=x \log_e x$ (ii) $y=e^x \sin^2 x$

(iii) $y=\sin 2x \cos 3x$ (iv) $y=\dfrac{x+2}{x+1}$

(v) $y=\dfrac{1}{(x+1)(x+2)}$.

66. If $y=a \cos(\log x)+b \sin(\log x)$ prove that

$$x^2 y_{n+2}+(2n+1)\, xy_{n+1}+(n^2+1)\, y_n=0.$$

67. Find dy/dx and dx/dy for the following implicit functions:

(i) $xy+x^2y^2=x^4$ (ii) $\sqrt{x^2+xy+y^2}=5$

(iii) $(\sqrt{x}+\sqrt{y})(\sqrt[3]{x}-\sqrt[3]{y})=2$ (iv) $(3x^2y^2)^{1/5}=2$.

68. Find d^2y/dx^2 from the following equations:

(i) $x=a\cos\theta,\ y=a\sin\theta$.

(ii) $x=t^2,\ y=t^3/3-t$.

(iii) $x=a(\theta-\sin\theta),\ y=a(1-\cos\theta)$

(iv) $x=\tan\theta,\ y=\cos^2\theta$

(v) $x=t^2,\ y=t+t^3$

(vi) $x=e^{2t},\ y=e^{3t}$.

69. Find dy/dx in the following cases:

(i) $x^2+y^2=4$ (ii) $y^2=4ax$

(iii) $(x^2/4)+(y^2/9)=1$ (iv) $x^2+xy+y^2=1$

(v) $x^2-xy+y^2=8$ (vi) $x^{2/3}+y^{2/3}=a^{2/3}$

(vii) $x=y+\cot^{-1} y$ (viii) $x^3+y^3=3axy$

(ix) $e^x \sin y-e^{-y}\cos x=0$ (x) $e^{xy}-x^2+y^3=0$.

70. If $f(x) = x^2 e^{-x/\alpha}$, show that

$$f^n(0) = \frac{n(n-1)(-1)^n}{a^{n-2}}.$$

3.2 APPLICATIONS OF DIFFERENTIATION

This section will be devoted to a brief description of some important applications of derivatives. These will include the geometrical application of derivatives, related rates, maxima and minima, and expansion of functions.

3.2.1 Geometrical Applications

Tangents and normals

If the equation of a curve is given, those of the tangent and normal to the curve at any point can be conveniently determined by using the concept of the derivative.

Let $y = f(x)$ be the equation of the curve and let $P(x_0, f(x_0))$ be any point on it. Let $Q(x_1, f(x_1))$ be another point on the curve and is close to P(Fig. 3.1). Join QP and produce it to meet the x-axis at S and let $PSX = \theta$.

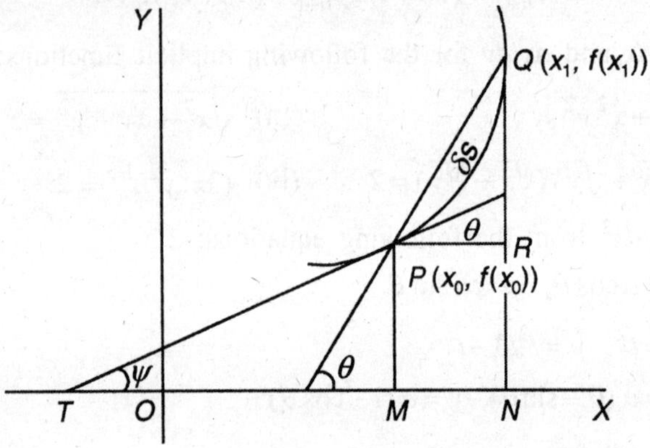

Figure 3.1

Let PT, where $PTX = \psi$, be the tangent to the curve at P. Let QN be perpendicular to OX, PR perpendicular to NQ, and PM perpendicular to OX. Now,

$$PR = ON - OM = x_1 - x_0,$$

and

$$RQ = NQ - MP = f(x_1) - f(x_0).$$

This line PQ is called the *secant* line and its equation is given by

$$\frac{y - f(x_0)}{x - x_0} = \frac{f(x_1) - f(x_0)}{x_1 - x_0}$$

or,

$$y - f(x_0) = \frac{f(x_1) - f(x_0)}{x_1 - x_0}(x - x_0), \tag{3.18}$$

where

$$\frac{f(x_1) - f(x_0)}{(x_1 - x_0)}$$

is the *slope* of the secant line

Now, the equation of the tangent line may be written as

$$y - f(x_0) = k\,(x - x_0) \tag{3.19}$$

where k, its slope, is to be determined. As Q approaches P and ultimately coincides with P, the secant line becomes the tangent to the curve at P. In other words,

$$k = \lim_{x_1 \to x_0} \frac{f(x_1) - f(x_0)}{x_1 - x_0}$$

If we write $x_1 = x_0 + h$, then by the definition of the derivative,

$$k = \lim_{h \to 0} \frac{f(x_0 + h) - f(x_0)}{h} = f'(x_0) \text{ or } \left[\frac{dy}{dx}\right]_{x = x_0},$$

we therefore see that the *slope of a curve* $y = f(x)$, at a given point is the value of the derivative *dy/dx* at that point. This is the geometrical significance of the derivative. We now obtain the equation of the tangent at the point $(x_0, f(x_0))$:

$$y - f(x_0) = \left[\frac{dy}{dx}\right]_{x = x_0}(x - x_0) \tag{3.20}$$

We know that the *normal* to the curve at a point P is a line passing through P and is perpendicular to the tangent line at P. Hence the slope of the normal at P is given by $-\frac{1}{[dy/dx]_{x=x_0}}$, and therefore the equation of the normal becomes

$$y - f(x_0) = -\frac{1}{[dy/dx]_{x = x_0}}(x - x_0)$$

or,s

$$(x - x_0) + [y - f(x_0)]\left[\frac{dy}{dx}\right]_{x = x_0} = 0 \tag{3.21}$$

Example 3.18 Find the equations of the tangent and normal at the point (x_1, y_1) of the ellipse $(x^2/a^2)/(y^2/b^2) = 1$.

To find dy/dx, we differentiate both sides of the equation with respect to x to get:

$$\frac{2x}{a^2}+\frac{2y}{b^2}\frac{dy}{dx}=0.$$

This gives

$$\frac{dy}{dx}=-\frac{x/a^2}{y/b^2}=-\frac{xb^2}{ya^2}$$

Hence

$$\left[\frac{dy}{dx}\right]_{(x_1,y_1)}=-\frac{x_1b^2}{y_1a^2}.$$

The equation of the tangent is given by

$$y-y_1=-\frac{x_1b^2}{y_1a^2}(x-x_1)$$

which becomes on simplification:

$$\frac{xx_1}{a^2}+\frac{yy_1}{b^2}=1.$$

Similarly, the equation of the normal is given by

$$x-x_1-\frac{x_1b^2}{y_1a^2}(y-y_1)=0,$$

or

$$\frac{(x-x_1)a^2}{x_1}-\frac{(y-y_1)b^2}{y_1}=0.$$

Example 3.19 Find the anlge of intersection of the curves

$$x^2-y^2=a^2 \tag{i}$$

and

$$x^2+y^2=\sqrt{2}a^2 \tag{ii}$$

Since the angle between the two curves is given by that between the tangents to the two curves at the point of intersection, we have from (Fig 3.2)

$$\theta=\psi_2-\psi_1,$$

where θ is the angle between the tangents, and ψ_1 and ψ_2 are the angles made by the tangents with the x-axis.

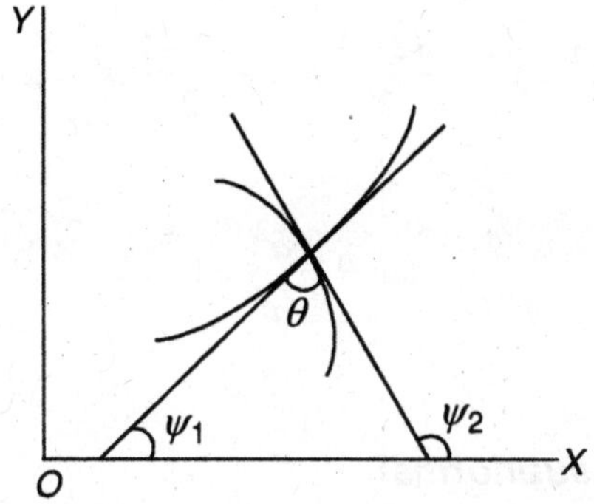

Figure 3.2

Solving (i) and (ii) and (ii) for the point of intersection, we obtain

$$x = \frac{(\sqrt{2}+1)^{1/2}}{\sqrt{2}}\, a, \qquad y = \frac{(\sqrt{2}-1)^{1/2}}{\sqrt{2}}\, a.$$

Now, differentiating (i) with respect to x, we get

$$2x - 2y\frac{dy}{dx} = 0$$

or

$$\frac{dy}{dx} = \frac{x}{y} = \tan\psi_1.$$

Again, differentiating (ii) with respect to x, we get

$$2x + 2y\frac{dy}{dx} = 0, \quad \text{or} \quad \frac{dy}{dx} = -\frac{x}{y} = \tan\psi_2.$$

Hence,

$$\tan\theta = \tan(\psi_2 - \psi_1) = \frac{\tan\psi_2 - \tan\psi_1}{1 + \tan\psi_2 \tan\psi_1} = \frac{(-x/y) - (x/y)}{1 - (x^2/y^2)}$$

$$= -\frac{(2x/y)}{(y^2 - x^2)/y^2}$$

$$= \frac{2xy}{x^2 - y^2}$$

But, we have

$$2xy = 2\frac{(\sqrt{2}+1)^{1/2}}{\sqrt{2}} \frac{a(\sqrt{2}-1)^{1/2}}{\sqrt{2}}\, a = a^2,$$

and

$$x^2 - y^2 = \frac{\sqrt{2}+1}{2}a^2 - \frac{\sqrt{2}-1}{2}a^2 = a^2.$$

Hence, we obtain

$$\tan\theta = \frac{a^2}{a^2} = 1,$$

which gives $\theta = \pi/4$.

Subtangent and subnormal

Let $y = f(x)$ be a curve and let $P(x, y)$ be any point on it. Let PT and PN respectively be the tangent and the normal at P and let PM be perpendicular from P on OX. Then TM and MN are defined to be the *subtangent* and *subnormal*, respectively (see Fig. 3.3). From the figure, it is clear that

$$\angle MPN = \angle PTM = \psi.$$

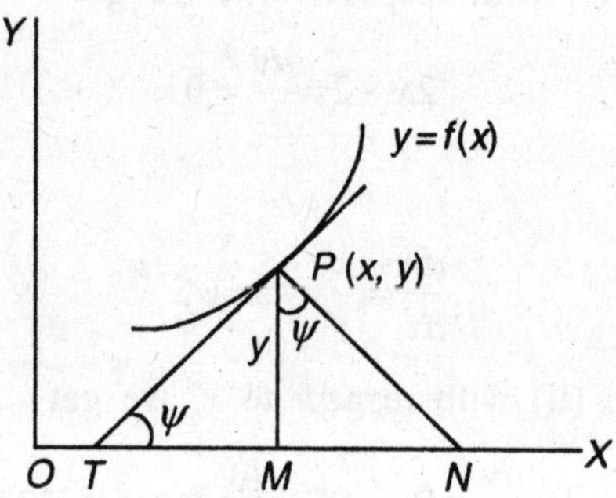

Figure 3.3

Since

$$\frac{dy}{dx} = \tan\psi,$$

we have

$$\frac{dy}{dx} = \tan\psi = \frac{y}{TM}$$

or,

$$TM = \frac{y}{dy/dx} \tag{3.22}$$

and

$$MN = y\tan\psi = y\frac{dy}{dx} \tag{3.23}$$

Also,

$$PT = \text{Length of the tangent} = \sqrt{TM^2 + MP^2}$$

$$= \sqrt{\frac{y^2}{(dy/dx)^2} + y^2}$$

$$= \frac{y\sqrt{1+(dy/dx)^2}}{dy/dx} \tag{3.24}$$

Finally,

$$PN = \text{Length of the normal} = \sqrt{y^2 + MN^2}$$

$$= \sqrt{y^2 + y^2\left(\frac{dy}{dx}\right)^2}$$

$$= y\sqrt{1+\left(\frac{dy}{dx}\right)^2}. \tag{3.25}$$

Example 3.20 Find the length of the subtangent and subnormal on the parabola $y^2 = 4ax$.

We find

$$2y\frac{dy}{dx} = 4a \quad \text{or} \quad \frac{dy}{dx} = \frac{2a}{y}.$$

Hence,

$$\text{Subnormal} = y\frac{dy}{dx} = y\frac{2a}{y} = 2a,$$

which is a constant. Similarly,

$$\text{Subtangent} = \frac{y}{dy/dx} = \frac{y^2}{2a} = \frac{4ax}{2a} = 2x,$$

which shows that the subtangent varies as the abscissa at the point of contact.

Example 3.21 Let the equation of the curve be given in the parametric form $x = a(\theta + \sin\theta)$, $y = a(1 - \cos\theta)$.

We have,

$$\frac{dy}{dx}=\frac{dy/d\theta}{dx/d\theta}=\frac{a\sin\theta}{a(1+\cos\theta)}=\frac{\sin\theta}{1+\cos\theta}=\frac{2\sin(\theta/2)\cos(\theta/2)}{2\cos^2(\theta/2)}=\tan\frac{\theta}{2}.$$

Hence,

(i) Length of the tangent $=\dfrac{y\sqrt{1+(dy/dx)^2}}{dy/dx}=\dfrac{a(1-\cos\theta)\sqrt{1+\tan^2(\theta/2)}}{\tan(\theta/2)}$

$$=\frac{2a\sin^2(\theta/2)\sec(\theta/2)}{\tan(\theta/2)}$$

$$=2a\sin\frac{\theta}{2}.$$

(ii) Length of the normal $=y\sqrt{1+\left(\dfrac{dy}{dx}\right)^2}=a(1-\cos\theta)\sec\dfrac{\theta}{2}$

$$=2a\sin\frac{\theta}{2}\tan\frac{\theta}{2}.$$

(iii) Subtangent $=\dfrac{y}{dy/dx}=\dfrac{a(1-\cos\theta)}{\tan(\theta/2)}=a\sin\theta.$

(iv) Subnormal $=y\dfrac{dy}{dx}=a(1-\cos\theta)\tan\dfrac{\theta}{2}=2a\sin^2\dfrac{\theta}{2}\tan\dfrac{\theta}{2}.$

Derivatives of arc-length

We shall first prove the limit

$$\lim_{Q\to P}\frac{\text{are } PQ}{\text{chord } PQ}=1, \tag{3.26}$$

which is useful in the following analysis.

In Fig. 3.4 let $\angle SPQ=\alpha$. It is clear that

$$\text{chord } PQ < \text{arc } PQ < PS + SQ.$$

Dividing throughout by PQ, we obtain

$$1<\frac{\text{arc } PQ}{PQ}<\frac{PS+SQ}{PQ}$$

i.e.

$$1<\frac{\text{arc } PQ}{PQ}<\cos\alpha+\sin\alpha.$$

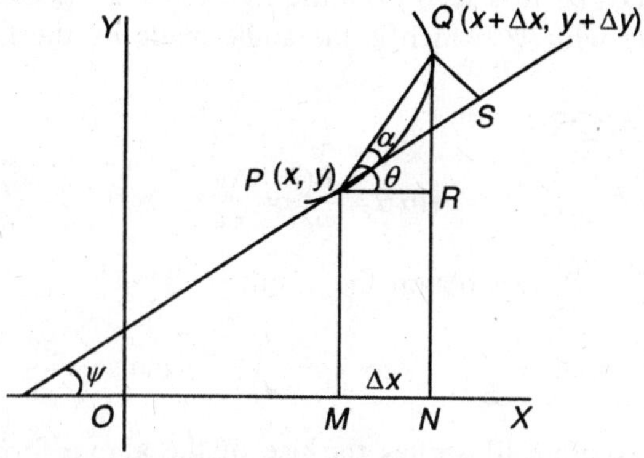

Figure 3.4

As $Q \to P$, the chord PQ tends to become the tangent to the curve at P, and $\alpha \to 0$. Then, $\cos\alpha + \sin\alpha \to 1$. Hence,

$$\lim_{Q\to P} \frac{\text{arc } PQ}{\text{chord } PQ} = 1,$$

from which it follows that

$$\lim_{Q\to P} \text{arc } PQ = \text{ chord } PQ.$$

From ΔPQR, we have

$$PQ^2 = (\Delta x)^2 + (\Delta y)^2$$

Since in the limiting position arc PQ = chord PQ, the above equation may be written as

$$(\Delta s)^2 = (\Delta y)^2 + (\Delta y)^2, \quad \text{where } \Delta s = \text{arc } PQ.$$

Dividing throughout by $(\Delta x)^2$ and proceeding to the limit, as $\Delta x \to 0$ (i.e. as $Q \to P$), we obtain the important relation:

$$\left(\frac{ds}{dx}\right)^2 = 1 + \left(\frac{dy}{dx}\right)^2 \tag{3.27}$$

In a similar way, the following relations can also be derived.

$$\left(\frac{ds}{dy}\right)^2 = 1 + \left(\frac{dx}{dy}\right)^2 \tag{3.28}$$

and

$$\left(\frac{ds}{dt}\right)^2 = \left(\frac{dx}{dt}\right)^2 + \left(\frac{dy}{dt}\right)^2 \tag{3.29}$$

Assuming $\angle RPQ = \theta$, it is seen from the figure that as $Q \to P$, the angle θ tends to coincide with ψ which is the angle made by the tangent with the x-axis.

Now from ΔPQR,

$$\sin\theta = \frac{RQ}{PQ} = \frac{\Delta y}{\Delta s}$$

and hence as $Q \to P$, we obtain the result

$$\sin\psi = \frac{dy}{dx}, \qquad \cos\psi = \frac{dx}{ds}, \qquad \tan\psi = \frac{dy}{dx} \tag{3.30}$$

The following example illustrates the use of the above formulae.

Example 3.22

(i) For the curve $y = a \cosh (x/a)$ find ds/dx.

(ii) For the curve $x = a(\theta + \sin\theta)$, $y = a(1 - \cos\theta)$, show that

$$\psi = \frac{\theta}{2}, \qquad \frac{ds}{dy} = \sqrt{\frac{2a}{y}} \quad \text{and} \quad \frac{ds}{d\theta} = 2a\cos\frac{\theta}{2}.$$

(i) We have, $y = a \cosh (x/a)$. Then,

$$\frac{dy}{dx} = a\left(\sinh\frac{x}{a}\right)\frac{1}{a} = \sinh\frac{x}{a}.$$

Hence,

$$\frac{ds}{dx} = \sqrt{1 + \left(\frac{dy}{dx}\right)^2} = \sqrt{1 + \sinh^2\frac{x}{a}} = \cosh\frac{x}{a} = \frac{y}{a}.$$

(ii) For the curve given by $x = a(\theta + \sin\theta)$, $y = a(1 - \cos\theta)$,

$$\frac{dy}{dx} = \frac{dy/d\theta}{dx/d\theta} = \frac{a\sin\theta}{a(1 + \cos\theta)} = \tan\frac{\theta}{2}.$$

Since $dy/dx = \tan\psi$, where ψ is the angle made by the tangent with the x-axis, we immediately deduce that $\psi = \theta/2$. Now,

$$\left(\frac{ds}{dy}\right)^2 = 1 + \left(\frac{dx}{dy}\right)^2 = 1 + \frac{1}{(dy/dx)^2} = 1 + \frac{1}{\tan^2(\theta/2)} = \frac{1}{\sin^2(\theta/2)}$$

$$= \frac{2a}{2a\sin^2(\theta/2)}$$

$$= \frac{2a}{a(1 - \cos\theta)}$$

$$= \frac{2a}{y},$$

and hence,

$$\frac{ds}{dy} = \sqrt{\frac{2a}{y}} = \frac{1}{\sin(\theta/2)}.$$

Finally,

$$\frac{ds}{d\theta} = \frac{ds}{dy}\frac{dy}{d\theta} = \frac{1}{\sin(\theta/2)}\, a\sin\theta = 2a\cos\frac{\theta}{2}.$$

Polar coordinates; pedal equation of a curve

Let $P(r, \theta)$ and $Q(r+\Delta r, \theta+\Delta\theta)$ be given on a curve with respect to the initial line OX. Refer to Fig. 3.5(a). Let PR be perpendicular to OQ. Then,

$$PR = r\sin\Delta\theta$$

and

$$RQ = OQ - OR = r + \Delta r - r\cos\Delta\theta = r(1-\cos\Delta\theta) + \Delta r$$

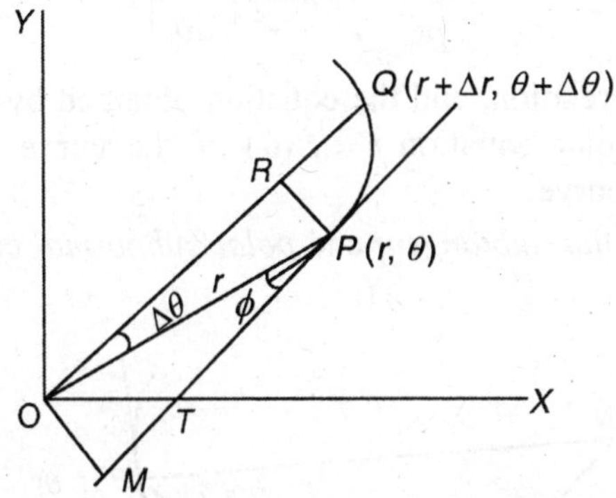

Figure 3.5(a)

In the limiting position as $Q \to P$, $\Delta\theta \to 0$, and $\sin\Delta\theta \to \Delta\theta$. Hence,

$$PR = r\Delta\theta \quad \text{and} \quad RQ = \Delta r.$$

If arc $PQ = \Delta s$, then from $\Delta\, PRQ$, we have the approximate relation

$$(\Delta s)^2 = (\Delta r)^2 + (r\Delta\theta)^2$$

This gives us the important relations:

$$\left(\frac{ds}{d\theta}\right)^2 = \sqrt{r^2 + \left(\frac{dr}{d\theta}\right)^2} \quad \text{and} \quad \frac{ds}{dr} = \sqrt{1 + r^2\left(\frac{d\theta}{dr}\right)^2} \qquad (3.31)$$

which will be useful in finding arcs of curves by integration.

If OM is drawn perpendicular to the tangent at P and if $\angle OPM = \phi$, then it is seen that in the limiting position as $Q \to P$, $\angle OQP$ tends to coincide with the angle ϕ. Hence,

$$\sin\phi = r\frac{d\theta}{ds}, \qquad \cos\phi = \frac{dr}{ds}, \qquad \tan\phi = r\frac{d\theta}{dr} \tag{3.32}$$

Further, if $OM = P$, then we have $p = r\sin\phi$, and hence

$$\frac{1}{p^2} = \frac{1}{r^2\sin^2\phi} = \frac{1}{r^2}\operatorname{cosec}^2\phi$$

$$= \frac{1}{r^2}\left[1 + \frac{1}{r^2(d\theta/dr)^2}\right]$$

$$= \frac{1}{r^2}\left[1 + \frac{(dr/d\theta)^2}{r^2}\right]$$

i.e.

$$\frac{1}{p^2} = \frac{1}{r^2} + \frac{1}{r^4}\left(\frac{dr}{d\theta}\right)^2, \tag{3.33}$$

which is a useful relation, and the equation obtained by eliminating θ from (3.33) and the polar equation $r = f(\theta)$ of the curve is called the *pedal equation* of the curve.

Finally, the *polar subtangent* and *polar subnormal* can be obtained from Fig. 3.5(b).

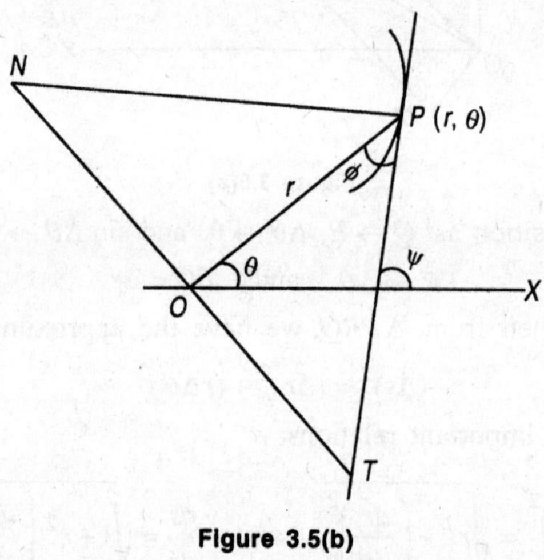

Figure 3.5(b)

Let PN be the normal and PT the tangent at P. Join OP and let the perpendicular to OP through O meet the normal and tangent at N and T, respectively. Then OT is the polar subtangent and ON the polar subnormal.

From ΔPOT, we have:

$$\text{Polar subtangent, } OT = OP \tan\phi = r\tan\phi$$

$$= r\left(r\,\frac{d\theta}{dr}\right)$$

$$= r^2\,\frac{d\theta}{dr}.$$

Similarly,

$$\text{Polar subnormal} = ON = \frac{dr}{d\theta}.$$

Example 3.23

(i) Find the polar subtangent to the curve $r = a\,(1-\cos\theta)$.

(ii) Show that the curves given by

$$r = \frac{a}{1+\cos\theta} \quad \text{and} \quad r = \frac{b}{1-\cos\theta}$$

intersect orthogonally.

(iii) find the pedal equation of the curve given in (i).

(i) The given curve is $r = a\,(1-\cos\theta)$. We obtain

$$\frac{dr}{d\theta} = a\sin\theta \quad \text{and} \quad \frac{d\theta}{dr} = \frac{1}{a\sin\theta}.$$

Here

$$\text{Polar subtangent} = r^2\,\frac{d\theta}{dr} = \frac{a^2\,(1-\cos\theta)^2}{a\sin\theta}$$

$$= a\,\frac{4\sin^4(\theta/2)}{2\sin(\theta/2)\,\cos(\theta/2)}$$

$$= 2a\,\frac{\sin^3(\theta/2)}{\cos(\theta/2)}$$

$$= 2a\sin^3\frac{\theta}{2}\,\sec\frac{\theta}{2}.$$

(ii) Let ϕ_1 and ϕ_2 be the angles made by the tangents to the two curves to the radius vector from 0. Then for the curve

$$r = \frac{a}{1+\cos\theta},$$

We have

$$\frac{dr}{d\theta} = \frac{d}{d\theta}\left[\frac{a}{2\cos^2(\theta/2)}\right] = \frac{a}{2}\frac{d}{d\theta}\left(\sec^2\frac{\theta}{2}\right)$$

$$= \frac{a}{2}\left(2\sec^2\frac{\theta}{2}\tan\frac{\theta}{2}\right)\frac{1}{2}$$

$$= \frac{a}{2}\sec^2\frac{\theta}{2}\tan\frac{\theta}{2}$$

i.e.

$$\tan\phi_1 = r\frac{d\phi}{dr} = \frac{\alpha}{2\cos^2(\theta/2)}\frac{1}{(a/2)\sec^2(\theta/2)\,\tan(\theta/2)} = \frac{1}{\tan(\theta/2)}$$

Similarly, for the curve

$$r = \frac{b}{1-\cos\theta} = \frac{b}{2\sin^2(\theta/2)} = \frac{b}{2}\operatorname{cosec}^2\frac{\theta}{2},$$

we have

$$\frac{dr}{d\theta} = \frac{b}{2}\left(2\operatorname{cosec}\frac{\theta}{2}\right)\left(-\operatorname{cosec}\frac{\theta}{2}\cot\frac{\theta}{2}\right)\frac{1}{2}$$

$$= -\frac{b}{2}\operatorname{cosec}^2\frac{\theta}{2}\cot\frac{\theta}{2}.$$

Hence,

$$\tan\phi_2 = r\frac{d\theta}{dr} = -\frac{b}{2\sin^2(\theta/2)}\frac{1}{(b/2)\operatorname{cosec}^2(\theta/2)\cot(\theta/2)} = -\frac{1}{\cot(\theta/2)}.$$

If α is the angle between the curves, then we have:

$$\tan\alpha = \tan(\phi_1 - \phi_2) = \frac{\tan\phi_1 - \tan\phi_2}{1+\tan\phi_1\tan\phi_2}.$$

Since $\tan\phi_1 \tan\phi_2 = -1$, the denominator vanishes and hence $\alpha = \pi/2$, which means that the curves intersect orthogonally.

(iii) For $r = a(1-\cos\theta)$, we have $dr/d\theta = a\sin\theta$. Hence,

$$\tan\phi = r\frac{d\theta}{dr} = \frac{a(1-\cos\theta)}{a\sin\theta} = \tan\frac{\theta}{2}.$$

It follows that $\phi = \theta/2$ and hence.

$$p = r\sin\phi = r\sin\frac{\theta}{2}.$$

But

$$r = a(1-\cos\theta) = 2a\sin^2\frac{\theta}{2}.$$

Eliminating θ between the two, we obtain:

$$r = 2a\frac{p^2}{r^2} \quad \text{or} \quad r^3 = 2ap^2,$$

which is the pedal equation of the curve.

3.2.2 Related Rates

The interpretation of derivative as a rate has important practical applications and will therefore be discussed now. Let a particle moves along the curve given by $s = f(t)$ and let $P(t, s)$ and $Q(t + \Delta t, s + \Delta s)$ be its positions at times t and $t + \Delta t$, respectively (see Fig. 3.6).

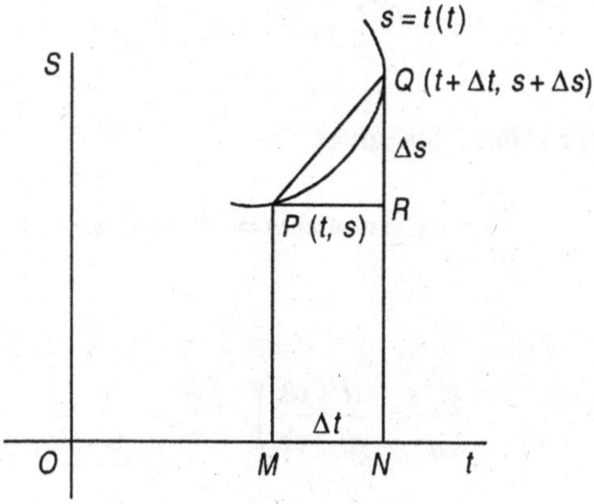

Figure 3.6

It is easily seen that $MN = \Delta t$ and $RQ = \Delta s$. Since $s = f(t)$, we have

$$s + \Delta s = f(t + \Delta t).$$

Hence,

$$\Delta s = \alpha f(t + \Delta) - f(t).$$

We now define the *average velocity* of the particle to be Δs divided by Δt:

$$v_{\text{av}} = \frac{\Delta s}{\Delta t} = \frac{f(t + \Delta t) - f(t)}{\Delta t}.$$

In order to obtain the *instantaneous velocity* at P, i.e. at time t, we take the average velocity over shorter and shorter intervals of time Δt. This means that the instantaneous velocity, or the *velocity* at any time t, is the limit of the average velocity as Δt approaches zero, i.e.

$$v = \lim_{\Delta t \to 0} \frac{\Delta s}{\Delta t} = \frac{ds}{dt}.$$

Similarly, the *instantaneous acceleration*, or the *acceleration* (a) is given by

$$a = \lim_{\Delta t \to 0} \frac{\Delta v}{\Delta t} = \frac{dv}{dt} = \frac{d}{dt}\left(\frac{ds}{dt}\right) = \frac{d^2 s}{dt^2}.$$

Thus, the second derivatives of s with respect to t has the physical significance of the acceleration of a particle at time t.

From the above discussion, we see that any derivative may be interpreted as the instantaneous rate of change of one variable per unit change in the other.

Example 3.24 A point in motion obeys the law $s = a \cos \omega t$. Determine its velocity and acceleration at time t and show that the acceleration $d^2 s/dt^2$ and the displacement s are related by the equation

$$\frac{d^2 s}{dt^2} + \omega^2 s = 0.$$

We have $s = a \cos \omega t$. This gives

$$\frac{ds}{dt} = a(-\sin \omega t)\, \omega = -a\omega \sin \omega t \tag{i}$$

and

$$\begin{aligned} \frac{d^2 s}{dt^2} &= \frac{d}{dt}\left(\frac{ds}{dt}\right) \\ &= \frac{d}{dt}(-a\omega \sin \omega t) \\ &= -a\omega\,(\cos \omega t)\, \omega \\ &= -a\omega^2 \cos \omega t \end{aligned} \tag{ii}$$

In (i) and (ii), we have the required expressions for the velocity and acceleration of the particle at time t.

Finally, from (ii)

$$\frac{d^2 s}{dt^2} = -a\omega^2 \cos \omega t = -\omega^2 (a \cos \omega t) = -\omega^2 s$$

or,

$$\frac{d^2 s}{dt^2} + \omega^2 s = 0,$$

which is the required equation.

Example 3.25 The relation $x = 3te^{-2t}$ defines the displacement of a certain mechanism at time t sec. Show that this function satisfies the differential equation

$$\frac{d^2x}{dt^2} + 4\frac{dx}{dt} + 4x = 0.$$

We have

$$x = 3te^{-2t} \tag{i}$$

Then,

$$\frac{dx}{dt} = 3e^{-2t} + 3t(-2e^{-2t}) = 3e^{-2t} - 6te^{-2t}, \tag{ii}$$

and

$$\frac{d^2x}{dt^2} = -6e^{-2t} - 6\,[e^{-2t} + te^{-2t}(-2)]$$

$$= -6e^{-2t} - 6e^{-2t} + 12te^{-2t}$$

$$= -12e^{-2t} + 12te^{-2t} \tag{iii}$$

Hence, substituting from (i), (ii) and (iii)

$$\frac{d^2x}{dt^2} + 4\frac{dx}{dt} + 4x = -12e^{-2t} + 12te^{-2t} + 4(3e^{-2t} - 6te^{-2t}) + 4\,(3te^{-2t}) = 0,$$

which shows that the function (i) satisfies the given differential equation.

Sometimes, the time rate of change in a variable may be related to some other rate. Such problems on related rates can be conveniently solved by using the formula for computing the derivative of a function of a function. The following examples illustrate the practical applications of the related rates.

Example 3.26 The radius of a sphere is increasing at a variable rate and is equal to 1 cm/sec, when the radius is 3 cm. Find the rate of change in volume at this time.

The volume V of the sphere is given by

$$V = \frac{4}{3}\pi r^3,$$

and

$$\frac{dV}{dt} = \frac{4}{3}\pi\,(3r^2)\frac{dr}{dt} = 4\pi r^2\,\frac{dr}{dt}.$$

Here we are given $dr/dt = 1$ when $r = 3$. Hence,

$$\frac{dV}{dt} = 36\pi \text{ cc/sec},$$

which means that the volume is changing at the rate of 36π cc per second.

Example 3.27 A vessel is in the form of an inverted cone of semivertical angle 45°. Water is poured into this vessel at the rate of 100 cc per second. Find the rate of rise in the water level when it is 2 cm deep.

At any instant, let y cm be the depth of the water and x cm the radius of the water level (see Fig. 3.7). Since the semivertical angle is 45°, we have $x = y$. Now, the volume V of the water inside the vessel is given by

$$V = \frac{1}{3}\pi x^2 y = \frac{1}{3}\pi y^3, \quad \text{since} \quad x = y.$$

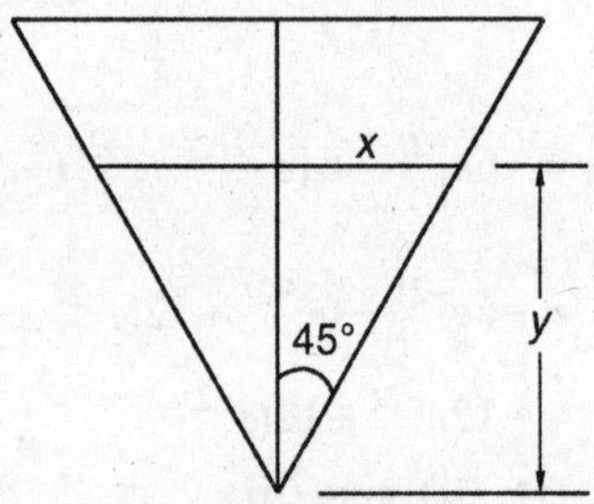

Figure 3.7

Hence,

$$\frac{dV}{dt} = \frac{1}{3}\pi(3y^2)\frac{dy}{dt} = \pi y^2\frac{dy}{dt}.$$

We are given that when $y = 2$ cm, $dV/dt = 100$. Therefore,

$$\frac{dy}{dt} = \frac{dV/dt}{\pi y^2} = \frac{100}{4\pi} = \frac{25}{\pi} \text{ cm/sec},$$

which means that the water level is rising at the rate of $(25/\pi)$ cm/sec, when the water is 2 cm deep.

3.2.3 Rolle's and Mean-value Theorems

From the fact that a function, continuous on a closed interval, is bounded and also attains its bounds, follow many results of importance. We discuss here two theorems which have useful applications in pure and applied mathematics.

Rolle's theorem

If $f(x)$ is continuous in a closed interval $a \le x \le b$, if $f'(x)$ exists in the open interval $a < x < b$, and if $f(a) = f(b) = 0$, then $f'(x)$ vanishes for at least one value of x in the interval $a < x < b$.

Figure 3.8 demonstrates the validity of the theorem. If the graph of $y = f(x)$ cuts the x-axis at A and B and if the curve is continuous and possesses a unique tangent at every point between A and B, then it is obvious that *the tangent line must be parallel to the x-axis at some intermediate point such* as P. Therefore the slope of the curve at P must be zero, i.e., $f'(x) = 0$.

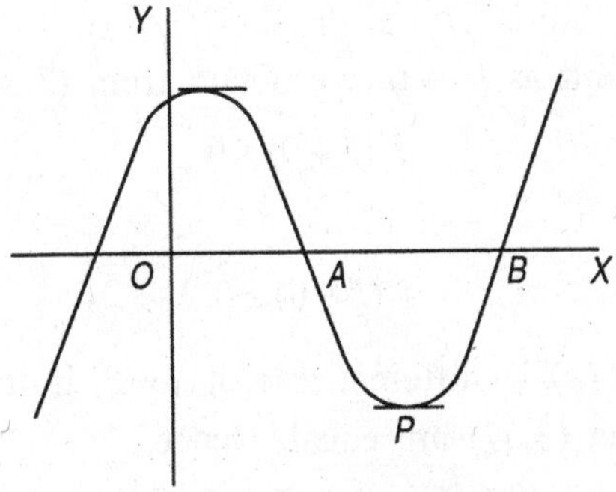

Figure 3.8(a)

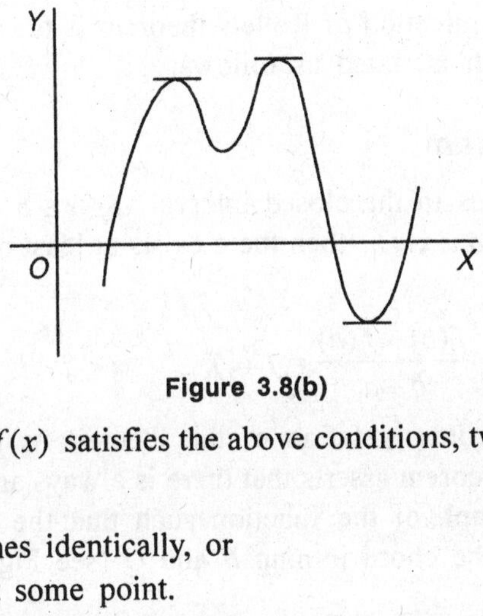

Figure 3.8(b)

Proof When $f(x)$ satisfies the above conditions, two possibilities arise. Either:

(i) $f(x)$ vanishes identically, or

(ii) $f(x) \neq 0$ at some point.

If $f(x) = 0$ throughout the interval $a \le x \le b$, then $f'(x) = 0$ at every point which shows that the theorem is true trivially. If $f(x)$ is somewhere positive, then it attains its maximum value at some interior point, say $x = \xi$ where $a < \xi < b$. Then,

$$f(\xi \pm h) \le f(\xi).$$

Hence,

$$\frac{f(\xi \pm h) - f(\xi)}{h} \le 0$$

which means that

$$\frac{f(\xi \pm h) - f(\xi)}{h} \le 0 \tag{3.34}$$

and

$$\frac{f(\xi - h) - f(\xi)}{-h} \ge 0 \tag{3.35}$$

Proceeding to the limit as $h \to 0$, we obtain from (3.34) and (3.35):

$$f'(\xi + 0) \le 0 \tag{3.36}$$

and

$$f'(\xi - 0) \ge 0, \tag{3.37}$$

respectively. Since $f(x)$ is differentiable at $x = \xi$, it follows that the two derivatives (3.36) and (3.37) are equal. Hence,

$$f'(\xi) = 0, \quad a < \xi < b,$$

which proves the theorem.

An important application of Rolle's theorem is the mean-value theorem for derivatives which is stated as follows:

Mean-value theorem

If $f(x)$ is continuous in the closed interval $a \le x \le b$ and if $f'(x)$ exists in the open interval $a < x < b$, then there exists at least one point, say $x = \xi$, such that

$$\frac{f(b) - f(a)}{b - a} = f'(\xi), \qquad a < \xi < b \tag{3.38}$$

The geometrical interpretation of the theorem is given in Fig. (3.9). Geometrically, the theorem asserts that there is always an intermediate point, say $x = \xi$, on the graph of the function such that the tangent line at this point is parallel to the chord joining P and Q (see Fig. 3.9).

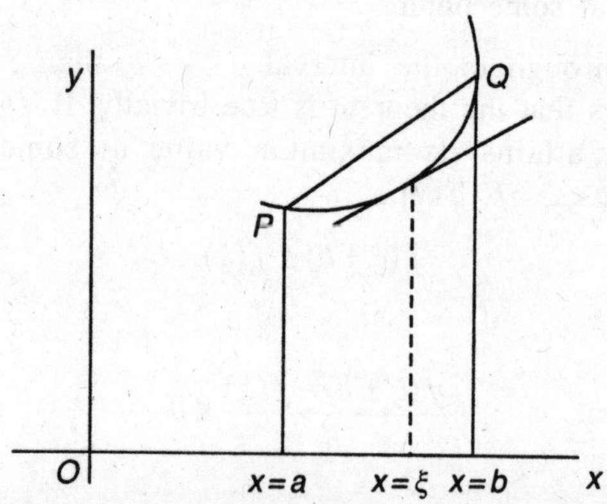

Figure 3.9

To prove the mean value theorem, we consider the function defined by

$$F(x) = A + Bx + f(x), \tag{3.39}$$

where $f(x)$ is continuous and differentiable. It therefore follows that $F(x)$ also satisfies these conditions, i.e. the first-two conditions of Rolle's theorem. The constants A and B in (3.39) are to be chosen by assuming that $F(a) = F(b) = 0$. We therefore have

$$A + Ba + f(a) = 0 \quad \text{and} \quad A + Ba + f(b) = 0,$$

from which it follows that

$$A = -f(a) + a\frac{f(b) - f(a)}{b - a} \quad \text{and} \quad B = \frac{f(b) - f(a)}{a - b}$$

The function $F(x)$ is therefore given by

$$F(x) = a\,\frac{f(b) - f(a)}{b - a} + \frac{f(b) - f(a)}{a - b}\,x + f(x) - f(a)$$

$$= f(x) - f(a) + \frac{f(b) - f(a)}{b - a}(a - x), \tag{3.40}$$

and this satisfies all the three conditions of Rolle's theorem. Hence, there exists at least one point, say $x = \xi$, such that

$$F'(\xi) = 0, \qquad a < \xi < b.$$

i.e.

$$f'(\xi) = \frac{f(b) - f(a)}{b - a}, \qquad a < \xi < b.$$

which proves the theorem.

Writing $b = a + h$, the above result can be expressed as

$$f(a + h) = f(a) + hf'(a + \theta h), \qquad 0 < \theta < 1 \tag{3.41}$$

a form that is often found more useful in practical applications.

Generalized mean-value theorem

Let $f(x)$ and its first $(n-1)$ derivatives are continuous in $a \le x \le b$ and its nth derivative exists in the open interval $a < x < b$. Then

$$f(b) = f(a) + (b - a)f'(a) + \frac{(b - a)^2}{2!}f''(a) + \cdots + \frac{(b - a)^{n-1}}{(n-1)!}f^{n-1}(a)$$

$$+ \frac{(b - a)^n}{n!}f^n\,[a + \theta\,(b - a)], \qquad 0 < \theta < 1 \tag{3.42}$$

Analogously with the mean-value theorem, we now assume that

$$F(x) = f(b) - f(x) - (b - x)f'(x) - \frac{(b - x)^2}{2!}f''(x) - \cdots$$

$$-\frac{(b-x)^{n-1}}{(n-1)!}f^{n-1}(x)-\frac{(b-x)^n}{n!}R, \tag{3.43}$$

where R is given by the equation $F(a)=0$, i.e.

$$f(b)-f(a)-(b-a)\,f'(a)-\frac{(b-a)^2}{2!}f''(a)-\cdots$$

$$-\frac{(b-a)^{n-1}}{(n-1)!}f^{n-1}(a)-\frac{(b-a)^n}{n!}R=0 \tag{3.44}$$

Again, $F(b)=0$ and therefore all the conditions of Rolle's theorem are satisfied. Hence $F'(x)$ vanishes for at least one value, say $x=\xi$, of x between a and b. Differentiation of $F(x)$ now gives

$$F'(x)=-f'(x)+f'(x)-(b-x)f''(x)+(b-x)f''(x)$$

$$-\frac{(b-x)^2}{2}f'''(x)-\cdots-\frac{(b-x)^{n-1}}{(n-1)!}f^n(a)$$

$$+\frac{(b-x)^{n-1}}{(n-1)!}R=\frac{(b-x)^{n-1}}{(n-1)!}[R-f^n(x)].$$

Setting $x=\xi$, the equation $F'(\xi)=0$ gives $R=f^n(\xi)$. Equation (3.44) therefore becomes

$$f(b)=f(a)+(b-a)f'(a)+\frac{(b-a)^2}{2!}f''(a)+\cdots$$

$$+\frac{b-a)^{n-1}}{(n-1)!}f^{n-1}(a)+\frac{(b-a)^n}{n!}f^n(\xi), \qquad a<\xi<b$$

Putting $b=a+h$, the above can be written as

$$f(a+h)=f(a)+hf'(a)+\frac{h^2}{2!}f''(a)+\cdots+\frac{h^{n-1}}{(n-1)!}f^{n-1}(a)$$

$$+\frac{h^n}{n!}f^n(a+\theta h), \qquad 0<\theta<1 \tag{3.45}$$

On the other hand, if we put $b=x$ in (3.44), we obtain

$$f(x)=f(a)+(x-a)f'(a)+\frac{(x-a)^2}{2!}f''(a)+\cdots$$

$$+\frac{(x-a)^n}{n!}f^n[a+\theta\,(x-a)], \qquad 0<\theta<1 \tag{3.46}$$

which is called *Taylor's theorem.*

Setting $x = 0$ in the above result, we obtain *Maclaurin's theorem*

$$f(x) = f(0) + xf'(0) + \frac{x^2}{2!} f''(0) + \cdots + \frac{x^n}{n!} f^n(\theta x), \qquad (3.47)$$

Taylor's and Maclaurin's theorems, deduced above, are important in the expansion of functions and will be discussed in detail in Section 3.2.7.

Example 3.28 If $f(x)$ and $g(x)$ are differentiable functions for $0 \le x \le 1$ such that $f(0) = 5,\ g(0) = 1,\ f(1) = 8,\ g(1) = 2$, then show that there exists c satisfying $0 < c < 1$ and $f'(c) = 3g'(c)$. **(IIT/JEE)**

Let the function $\phi(x)$ be defined by

$$\phi(x) = f(x) - 3g(x).$$

Since $f(x)$ and $g(x)$ are differentiable functions, $\phi(x)$ is also differentiable in $0 \le x \le 1$. Further, we have

$$\phi(0) = f(0) - 3g(0) = 5 - 3 = 2,$$

and

$$\phi(1) = f(1) - 3g(1) = 8 - 6 = 2.$$

It therefore follows that $\phi(x)$ satisfies the conditions of Rolle's theorem. Hence, there exists a number c, $0 < c < 1$, such that

$$\phi'(c) = 0 \quad \text{or} \quad f'(c) - 3g'(c) = 0 \quad \text{or} \quad f'(c) = 3g(c),$$

which is the required result.

Example 3.29 Use Rolle's theorem to prove that between any two real roots of $e^{-x} = \sin x$, there lies at least one real root of $e^{-x} = -\cos x$.

We consider the function defined by

$$f(x) = e^{-x} - \sin x.$$

Let two real roots of $f(x) = 0$ be given by $x = a$ and $x = b$ so that

$$f(a) = f(b) = 0.$$

Now $f'(x) = -e^{-x} \cos x$, which exists in $a < x < b$. Hence, $f(x)$ satisfies all conditions of Rolle's theorem. It follows that $f'(x) = 0$ for at least one value of x in $a < x < b$. That is,

$$-e^{-x} - \cos x = 0 \quad \text{or} \quad e^{-x} = -\cos x$$

for at least one value of x in $a < x < b$.

This proves that between any two real roots (in this case a and b) of $e^{-x} = \sin x$, there lies at least one real root of $e^{-x} = -\cos x$.

Example 3.30

(i) If $f(x) = x^3 + x,\ a = 0,\ b = 1$, determine ξ such that

$$\frac{f(b)-f(a)}{b-a}=f'(\xi), \quad a<\xi<b$$

(ii) Use the mean-value theorem to show that

$$3+\frac{1}{28}<\sqrt[3]{28}<3+\frac{1}{27}.$$

(i) Since $f(x)=x^3+x$, $a=0$, $b=1$, it follows that $f(a)=0$, $f(b)=2$, $b-a=1$, and $f'(\xi)=3\xi^2+1$. Hence,

$$3\xi^2+1=\frac{f(b)-f(a)}{b-a}=2,$$

which gives

$$3\xi^2=1 \quad \text{or} \quad \xi=\pm\frac{1}{\sqrt{3}}.$$

It is seen that the point $\xi=1/\sqrt{3}$ is an intermediate point of the interval [0, 1].

(ii) The mean-value theorem can be written as

$$f(x+h)=f(x)+hf'(x+\theta h), \quad 0<\theta<1$$

We take

$$f(x)=\sqrt[3]{x}=x^{1/3},$$

so that

$$f'(x)=\frac{1}{3x^{2/3}}.$$

Substituting $x=27$, $h=1$, the above gives:

$$\sqrt[3]{28}=\sqrt[3]{27}+\frac{1}{3(27+\theta)^{2/3}}, \quad 0<\theta<1$$

from which it is clear that $\sqrt[3]{28}$ lies between

$$3+\frac{1}{3(28)^{2/3}} \quad \text{and} \quad 3+\frac{1}{3(27)^{2/3}}$$

Now,

$$3+\frac{1}{3(27)^{2/3}}=3+\frac{1}{27}$$

and

$$3+\frac{1}{3(28)^{2/3}}=3+\frac{1}{(27)^{1/3}(28)^{2/3}}>3+\frac{1}{(28)^{1/3}(28)^{1/3}}=3+\frac{1}{28}$$

We therefore have the inequality

$$3+\frac{1}{28}<3+\frac{1}{3(28)^{2/3}}<\sqrt[3]{28}<3+\frac{1}{27},$$

i.e.,

$$3+\frac{1}{28}<\sqrt[3]{28}<3+\frac{1}{27},$$

which is the required inequality.

3.2.4 Indeterminate Forms

An important application of the mean-value theorem consists in the evaluation of limits of quotients of functions which tend to the forms 0/0 or ∞/∞, which are called *indeterminate forms.* Other indeterminate forms are

$$0\times\infty,\ \infty-\infty,\ 0^{\circ},\ \infty^{\circ},\ 1^{\infty},$$

All these are discussed below.

Form $\frac{0}{0}$ (L'Hospital's rule)

We consider

$$\lim_{x\to a}\frac{f(x)}{\phi(x)},\quad \text{where}\ \ f(a)=\phi(a)=0.$$

Let $x=a+h$ so that as $h\to 0$, $x\to a$. Then, by mean-value theorem for $0<\theta_1<1$, $0<\theta_2<1$ we have

$$\lim_{x\to a}\frac{f(x)}{\phi(x)}=\lim_{h\to 0}\frac{f(a+h)}{\phi(a+h)}=\lim_{h\to 0}\frac{f(a)+hf'(a+\theta_1 h)}{\phi(a)+h\phi'(a+\theta_2 h)}.$$

Since $f(a)=\phi(a)=0$, This becomes

$$\lim_{x\to a}\frac{f(x)}{\phi(x)}=\lim_{h\to 0}\frac{f'(a+\theta_1 h)}{\phi'(a+\theta_2 h)}=\frac{f'(a)}{\phi'(a)}=\lim_{x\to a}\frac{f'(x)}{\phi'(x)}$$

If, however, $f'(a)=\phi'(a)=0$, we repeat the process and obtain

$$\lim_{x\to a}\frac{f(x)}{\phi(x)}=\frac{f''(x)}{\phi''(x)}.$$

The process is repeated until the ratio ceases to be an indeterminate form.

Example 3.31 Evaluate the following limits:

(i) $\lim\limits_{x\to 0}\dfrac{\log(1-x^2)}{\log\cos x}$. (ii) $\lim\limits_{x\to 0}\dfrac{5\sin x-7\sin 2x+3\sin 3x}{\tan x-x}$.

(iii) $\lim\limits_{x\to 0}\dfrac{a^x-1}{x}$. (iv) $\lim\limits_{x\to 0}\dfrac{2^x-1}{(1+x)^{1/2}-1}$.

(i) As $x \to 0$, we observe that the ratio is of the form 0/0, and hence we can apply L'Hospital's rule. We therefore differentiate both numerator and denominator, and obtain

$$\lim_{x\to 0} \frac{\log(1-x^2)}{\log\cos x} = \lim_{x\to 0} \frac{-2x/(1-x^2)}{-\tan x}$$

$$= \lim_{x\to 0} 2\left(\frac{1}{1-x^2}\frac{x}{\tan x}\right)$$

$$= 2\lim_{x\to 0}\frac{x}{\tan x}, \quad \text{since } \lim_{x\to 0}\frac{1}{1-x^2} = 1$$

$$= 2\lim_{x\to 0}\left(\frac{x}{\sin x}\cos x\right)$$

$$= 2\lim_{x\to 0}\frac{x}{\sin x}\lim_{x\to 0}\cos x$$

$$= 2.$$

(ii) By L'Hospital's rule.

$$\lim_{x\to 0}\frac{5\sin x - 7\sin 2x + 3\sin 3x}{\tan x - x} - \lim_{x\to 0}\frac{5\cos x - 14\cos 2x + 9\cos 3x}{\sec^2 x - 1}, \left(\text{form } \frac{0}{0}\right)$$

This is also of the form 0/0. Hence, applying L'Hospital's rule again, the above limit becomes

$$\lim_{x\to 0}\frac{-5\sin x + 28\sin 2x - 27\sin 3x}{2\sec^2 x\tan x} = \lim_{x\to 0}\frac{-5\cos x + 56\cos 2x - 81\cos 3x}{2\sec^4 x + 4\sec^2 x\tan^2 x}$$

$$= \frac{-5+56-81}{2}$$

$$= \frac{-30}{2}$$

$$= -15.$$

(iii) We know that $a^x = e^{x\log_e a}$, and hence

$$\lim_{x\to 0}\frac{a^x - 1}{x} = \lim_{x\to 0}\frac{e^{x\log_e a} - 1}{x} \qquad \left(\text{form } \frac{0}{0}\right)$$

$$= \lim_{x\to 0}\frac{e^{x\log_e a}\log_e a}{1}$$

$$= \log_e a, \quad \text{since } \lim_{x\to 0} e^{x\log_e a} = e^0 = 1.$$

(iv) We first rationalize the denominator and obtain

$$\frac{2^x - 1}{(1+x)^{1/2} - 1} = \frac{(2^x - 1)[(1+x)^{1/2} + 1]}{[(1+x)^{1/2} - 1][(1+x)^{1/2} + 1]}$$

$$= \frac{(2^x - 1)[(1+x)^{1/2} + 1]}{1 + x - 1}$$

$$= \frac{(2^x - 1)[(1+x)^{1/2} + 1]}{x}$$

Hence

$$\lim_{x\to 0} \frac{2^x - 1}{(1+x)^{1/2} - 1} = \lim_{x\to 0} \frac{(2^x - 1)[(1+x)^{1/2} + 1]}{x} \qquad \left(\text{form } \frac{0}{0}\right)$$

$$= \lim_{x\to 0}\left(\frac{2^x - 1}{x}\right) \lim_{x\to 0} [(1+x)^{1/2} + 1]$$

But by (iii)

$$\lim_{x\to 0}\left(\frac{2^x - 1}{x}\right) = \log_e 2,$$

and

$$\lim_{x\to 0} [(1+x)^{1/2} + 1] = \lim_{x\to 0}\left(1 + \frac{1}{2}x - \frac{1}{8}x^2 + \cdots + 1\right) = 2$$

Hence,

$$\lim_{x\to 0} \frac{2^x - 1}{(1+x)^{1/2} - 1} = 2 \log_e 2.$$

Form $\frac{\infty}{\infty}$

To find

$$\lim_{x\to a} \frac{f(x)}{\phi(x)}, \quad \text{where } f(a) = \infty \quad \text{and} \quad \phi(a) = \infty$$

we write

$$\lim_{x\to a} \frac{f(x)}{\phi(x)} = \lim_{x\to a} \frac{1/\phi(x)}{1/f(x)} \qquad \left(\text{form } \frac{0}{0}\right)$$

$$= \lim_{x \to a} \frac{-\phi'(x)/[\phi(x)]^2}{f'(x)/[f(x)]^2}$$

$$= \lim_{x \to a} \frac{\phi'(x)}{f'(x)} \frac{[f(x)]^2}{[\phi(x)]^2}$$

i.e.

$$\lim_{x \to a} \frac{f(x)}{\phi(x)} = \lim_{x \to a} \frac{\phi'(x)}{f'(x)} \left[\lim_{x \to a} \frac{f(x)}{\phi(x)} \right]^2 \tag{i}$$

If we let

$$\lim_{x \to a} \frac{f(x)}{\phi(x)} = k,$$

three cases arise:

Case (i): Let $k \neq 0$ and $k \neq \infty$, then, we obtain from (i)

$$k = \lim_{x \to a} \frac{f(x)}{\phi(x)} = \lim_{x \to a} \frac{\phi'(x)}{f'(x)} k^2$$

or

$$\frac{1}{k} = \lim_{x \to a} \frac{\phi'(x)}{f'(x)}$$

from which we obtain

$$\lim_{x \to a} \frac{f'(x)}{\phi'(x)} = k$$

In other words,

$$\lim_{x \to a} \frac{f(x)}{\phi(x)} = \lim_{x \to a} \frac{f'(x)}{\phi'(x)} = k.$$

Case (ii): Let $k = 0$ Since

$$k = \lim_{x \to a} \frac{f(x)}{\phi(x)},$$

by case (i), we have

$$k + 1 = \lim_{x \to a} \frac{f(x) + \phi(x)}{\phi(x)} = \lim_{x \to a} \frac{f'(x) + \phi'(x)}{\phi'(x)} = \lim_{x \to a} \frac{f'(x)}{\phi'(x)} + 1$$

Therefore,

$$k = \lim_{x \to a} \frac{f'(x)}{\phi'(x)}.$$

Case (iii): Let $k=\infty$. Now by case (ii)

$$\frac{1}{k}=\lim_{x\to a}\frac{1}{[f(x)/\phi(x)]}=\lim_{x\to a}\frac{\phi(x)}{f(x)}=\lim_{x\to a}\frac{\phi'(x)}{f'(x)},$$

Therefore,

$$\lim_{x\to a}\frac{f(x)}{\phi(x)}=\lim_{x\to a}\frac{f'(x)}{\phi'(x)}.$$

This completes the proof.

Example 3.32 Evaluate the following limits:

(i) $\lim\limits_{x\to\infty}\dfrac{x^n}{e^x}$ (ii) $\lim\limits_{x\to 0}\dfrac{\log x}{\operatorname{cosec} x}$.

(i) Let

$$\begin{aligned}
y&=\lim_{x\to\infty}\frac{x^n}{e^x} && \left(\text{form }\frac{\infty}{\infty}\right)\\
&=\lim_{x\to\infty}\frac{nx^{n-1}}{e^x} && \left(\text{form }\frac{\infty}{\infty}\right)\\
&=\lim_{x\to\infty}\frac{n(n-1)x^{n-2}}{e^x} && \left(\text{form }\frac{\infty}{\infty}\right)\\
&\;\;\vdots\\
&=\lim_{x\to\infty}\frac{n(n-1)(n-2)\ldots 3.2.1}{e^x}\\
&=\lim_{x\to\infty}\frac{n!}{e^x}\\
&=\frac{n!}{\infty}\\
&=0.
\end{aligned}$$

(ii) Let

$$\begin{aligned}
y&=\lim_{x\to 0}\frac{\log x}{\operatorname{cosec} x}, && \left(\text{form }\frac{\infty}{\infty}\right)\\
&=\lim_{x\to\infty}\frac{1/x}{-\operatorname{cosec} x\cot x}\\
&=-\lim_{x\to 0}\frac{\sin x}{x}\tan x
\end{aligned}$$

$$= \lim_{x\to 0} \frac{\sin x}{x} \lim_{x\to 0} \tan x$$

$$= 1 \times 0$$

$$= 0.$$

Form $0 \times \infty$

We consider $\lim_{x\to a} [f(x)\,\phi(x)]$, where $\lim_{x\to a} f(x) = 0$ and $\lim_{x\to a} \phi(x) = \infty$. This can be written as

$$\lim_{x\to a} \frac{f(x)}{1/\phi(x)},$$

which is of the form $0/0$ and can therefore be evaluated by using L'Hospital's rule.

Example 3.33 Evaluate the following limits:

(i) $\lim_{x\to 1} (1-x) \tan \dfrac{\pi x}{2}$ (ii) $\lim_{x\to 0} x \log x.$

(i) Let

$$y = \lim_{x\to 1} (1-x) \tan \frac{\pi x}{2}, \qquad \text{(form } 0\times\infty\text{)}$$

$$= \lim_{x\to 1} \frac{1-x}{\cot(\pi x/2)} \qquad \left(\text{form } \frac{0}{0}\right)$$

$$= \lim_{x\to 1} \frac{-1}{-[\text{cosec}^2(\pi x/2)]\,(\pi/2)} = \frac{2}{\pi}$$

(ii) Let

$$y = \lim_{x\to 0} x \log x, \qquad \text{(form } 0\times\infty\text{)}$$

$$= \lim_{x\to 0} \frac{\log x}{1/x} \qquad \left(\text{form } \frac{\infty}{\infty}\right)$$

$$= \lim_{x\to 0} \frac{1/x}{-1/x^2}$$

$$= \lim_{x\to 0} (-x)$$

$$= 0.$$

Form $\infty - \infty$

We consider $= \lim_{x \to a} [f(x) - \phi(x)]$, where $\lim_{x\to a} f(x) = \infty$ and $\lim_{x\to a} \phi(x) = \infty$. This can be written as

$$\lim_{x\to a} \frac{[1/\phi(x)] - [1/f(x)]}{1/[f(x)\,\phi(x)]} \qquad \left(\text{form } \frac{0}{0}\right)$$

and can therefore be evaluated by using L'Hospital's rule.

Example 3.34 Evaluate

$$\lim_{x\to 0}\left(\frac{1}{x^2} - \frac{1}{\sin^2 x}\right).$$

We have

$$\lim_{x\to 0}\left(\frac{1}{x^2} - \frac{1}{\sin^2 x}\right) = \lim_{x\to 0} \frac{\sin^2 x - x^2}{x^2 \sin^2 x} \qquad \left(\text{form } \frac{0}{0}\right)$$

$$= \lim_{x\to 0} \frac{2 \sin x \cos x - 2x}{2x \sin^2 x + x^2\,(2) \sin x \cos x}$$

$$= \lim_{x\to 0} \frac{\sin 2x - 2x}{2x \sin^2 x + x^2 \sin 2x} \qquad \left(\text{form } \frac{0}{0}\right)$$

$$= \lim_{x\to 0} \frac{2 \cos 2x - 2}{2 \sin^2 x + 2x \sin 2x + 2x \sin 2x + x^2\,(2) \cos 2x}$$

$$= \lim_{x\to 0} \frac{\cos 2x - 1}{\sin^2 x + 2x \sin 2x + x^2 \cos 2x} \qquad \left(\text{form } \frac{0}{0}\right)$$

$$= \lim_{x\to 0} \frac{-2 \sin 2x}{\sin 2x + 2 \sin 2x + (2x)\, 2 \cos 2x + 2x \cos 2x + x^2(-2 \sin 2x)}$$

$$= \lim_{x\to 0} \frac{-2 \sin 2x}{3 \sin 2x + 6x \cos 2x - 2x^2 \sin 2x} \qquad \left(\text{form } \frac{0}{0}\right)$$

$$= \lim_{x\to 0} \frac{-4 \cos 2x}{6 \cos 2x + 6 \cos 2x + 6x\,(-2 \sin 2x) - 4x \sin 2x - 4x^2 \cos 2x}$$

$$= \frac{-4}{6+6+0+0+0}$$

$$= -\frac{4}{12}$$

$$= -\frac{1}{3}.$$

Form 0^0, ∞^0 and 1^∞

We, finally, consider $\lim_{x\to a}[f(x)]^{\phi(x)}$, where

(i) $\lim\limits_{x\to a} f(x)=0$ and $\lim\limits_{x\to a}\phi(x)=0$,

(ii) $\lim\limits_{x\to a} f(x)=\infty$ and $\lim\limits_{x\to a}\phi(x)=0$,

(iii) $\lim\limits_{x\to a} f(x)=1$ and $\lim\limits_{x\to a}\phi(x)=\infty$.

Let $y=\lim_{x\to a}[f(x)]^{\phi(x)}$. Therefore,

$$\begin{aligned}\log_e y &= \log_e \lim_{x\to a}\{[f(x)]^{\phi(x)}\}\\ &= \lim_{x\to a}\log_e [f(x)]^{\phi(x)}\\ &= \lim_{x\to a}[\phi(x)\log_e f(x)], \qquad (\text{form } 0\times\infty)\end{aligned}$$

This can therefore be evaluated by the method outlined in case (c) above.

Example 3.35 Evaluate the following limits:

(i) $\lim\limits_{x\to 0} x^x$ (ii) $\lim\limits_{x\to 0}(\cot x)^{\sin 2x}$

(iii) $\lim\limits_{x\to\pi/2}(\sin x)^{\tan x}$, (iv) $\lim\limits_{x\to 1} x^{1/(1-x)}$

(i) Let $y=\lim_{x\to 0}x^x$ (form 0^0). Therefore,

$$\begin{aligned}\log_e y &= \log_e(\lim_{x\to 0}x^x)\\ &= \lim_{x\to 0}(\log_e x^x)\\ &= \lim_{x\to 0} x\log_e x, \qquad (\text{form } 0\times\infty)\\ &= \lim_{x\to 0}\frac{\log_e x}{1/x}, \qquad \left(\text{form } \frac{\infty}{\infty}\right)\\ &= \lim_{x\to 0}\frac{1/x}{-1/x^2}=\lim_{x\to 0}(-x)\\ &= 0\end{aligned}$$

Hence,

$$y=\lim_{x\to 0}x^x=e^0=1.$$

(ii) Let $y = \lim_{x \to 0} (\cot x)^{\sin 2x}$ (form ∞^0). Therefore

$$\log_e y = \lim_{x \to 0} \sin 2x \log_e \cot x, \qquad (\text{form } 0 \times \infty)$$

$$= \lim_{x \to 0} \frac{\log_e \cot x}{\operatorname{cosec} 2x}, \qquad \left(\text{form } \frac{\infty}{\infty}\right)$$

$$= \lim_{x \to 0} \frac{(1/\cot x)(-\operatorname{cosec}^2 x)}{-2 \operatorname{cosec} 2x \cot 2x}$$

On simplification, we get

$$\log_e y = \lim_{x \to 0} \frac{\sin 2x}{\cos 2x} = 0.$$

Hence,

$$y = \lim_{x \to 0} (\cot x)^{\sin 2x} = e^0 = 1.$$

(iii) Let $y = \lim_{x \to \pi/2} (\sin x)^{\tan x}$ (form 1^∞). Therefore,

$$\log_e y = \lim_{x \to \pi/2} \tan x \log_e \sin x, \qquad (\text{form } 0 \times \infty)$$

$$= \lim_{x \to \pi/2} \frac{\log_e \sin x}{\cot x}, \qquad \left(\text{form } \frac{0}{0}\right)$$

$$= \lim_{x \to \pi/2} \frac{\cot x}{-\operatorname{cosec}^2 x}, \qquad (\text{by L'Hospital's rule})$$

$$= \lim_{x \to \pi/2} (-\cos x \sin x)$$

$$= 0.$$

Hence,

$$y = \lim_{x \to \pi/2} (\sin x)^{\tan x} = e^0 = 1.$$

(iv) Let $y = \lim_{x \to 1} x^{1/(1-x)}$ (form 1^∞). Then,

$$\log_e y = \lim_{x \to 1} \left(\frac{1}{1-x} \log_e x\right) \qquad \left(\text{form } \frac{0}{0}\right)$$

$$= \lim_{x \to 1} \frac{1/x}{-1}, \qquad (\text{by L'Hospital's rule})$$

$$= -1.$$

Hence,

$$y = \lim_{x \to 1} x^{1/(1-x)} = e^{-1} = \frac{1}{e}.$$

3.2.5 Maxima and Minima

The determination of maxima and minima is of greatest importance in engineering problems, particularly those concerned with the stability of the equilibrium states of mechanical and physical systems. We shall, in this section, derive conditions under which a function of one variable has maxima or minima.

A function $f(x)$ is said to have a *maximum* at $x = a$ if

$$f(a) \geq f(a+h)$$

for all positive and negative values of h sufficiently near zero. Similarly, $f(b)$ is said to be a *minimum* if

$$f(b) \leq f(b+h),$$

for values of h close to zero. Figure 3.10 gives the graph of a typical function. This function has maximum at A and minimum at B. It may be seen that there are values of the function which are greater than the maximum and values which are less than the minimum.

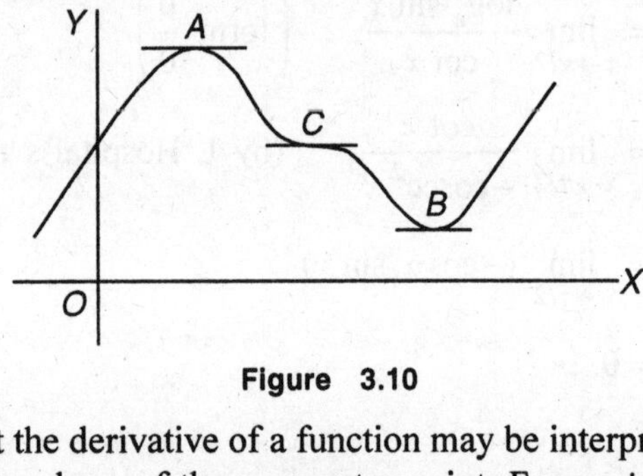

Figure 3.10

We know that the derivative of a function may be interpreted geometrically as representing the slope of the curve at a point. For an *increasing* function (i.e. a function for which y increases as x increases), the slope will be positive and so dy/dx would be a positive quantity. Similarly, if $y = f(x)$ is a *decreasing* function (i.e. a function for which *y decreases* as x increases), then dy/dx will be negative. From Fig. 3.10, it is easily seen that the tangents at the points A, B and C are all parallel to the x-axis so that if a, b, c are, respectively, the abscissae of these points, then each of $f'(a)$, $f'(b)$ and $f'(c)$ is zero. Such points are called *stationary* points (or *extreme* points).

We shall now prove the following theorem.

Theorem 3.1 If $f(x)$ is differentiable at $x = c$ and has a maximum or minimum there, then $f'(c) = 0$.

Proof We shall prove the theorem in the case of a maximum at $x = c$. From the definition of a maximum, we have

$$f(c) \geq f(c+h)$$

for all h close to zero, i.e. when $(c+h)$ is close to c. Further, we have

$$f'(c) = \lim_{h \to 0} \frac{f(c+h) - f(c)}{h},$$

since $f'(x)$ exists at $x = c$. We shall prove that this value is zero.

Here, h may be positive or negative. We therefore have

$$\frac{f(c+h) - f(c)}{h} \leq 0, \qquad \text{if } h > 0$$

and

$$\frac{f(c+h) - f(c)}{h} \geq 0, \qquad \text{if } h < 0,$$

since, in both the cases, the numerator is either zero or negative. It follows from the above that

$$\lim_{h \to 0+} \frac{f(c+h) - f(c)}{h} \leq 0$$

and

$$\lim_{h \to 0-} \frac{f(c+h) - f(c)}{h} \geq 0.$$

The above two relations will hold simultaneously if and only if

$$f'(c) = 0.$$

In a similar way, it can be shown that $f'(c) = 0$ if $f(c)$ is a minimum value of $f(x)$.

It should be understood that the vanishing of the first derivative is only a *necessary* condition for a stationary value. This means that not all points at which the derivative vanishes provide either a maximum or a minimum. For example, consider the function defined by

$$f(x) = x^3.$$

In this case, $f(x) = x^3$ and so $f'(0) = 0$; but $f(x) > 0$ for $x > 0$ and $f(x) < 0$ for $x < 0$. Thus, $f(x)$ has neither a maximum nor a minimum at $x = 0$ although $f'(0) = 0$.

Maximum and minimum values of a function are sometimes referred to as *extrema*. We shall now state *sufficient* conditions for extreme values.

Theorem 3.2 $f(c)$ is an extreme value of $f(x)$ if and only if $f'(x)$ changes sign as x passes through c.

We first consider point A in Fig. 3.10. It is clearly seen that before reaching A, the function is an increasing function and after passing through A, it is a decreasing function. Thus, dy/dx changes sign from positive to negative as x passes through A. In other words, d^2y/dx^2 is negative at A.

Hence, a function $y = f(x)$ is said to have a maximum (refer to Fig. 3.11) at $x = a$ if

(i) $dy/dx = 0$, at $x = a$ and

(ii) dy/dx changes sign from positive to negative as x passes through the value $x = a$, or d^2y/dx^2 is negative at $x = a$.

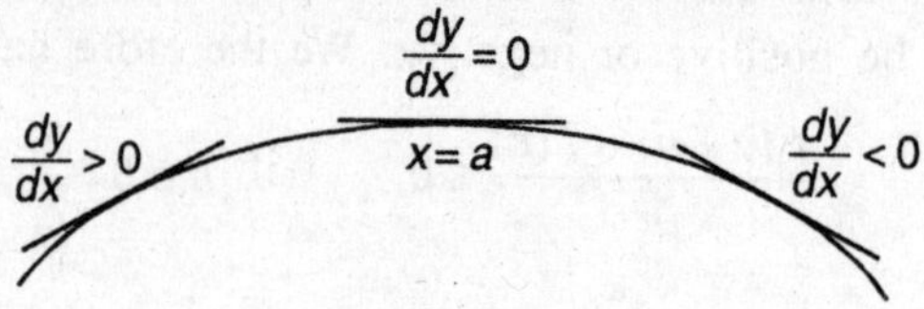

Figure 3.11

Similarly, at the point B (corresponding to $x = b$, say) which is a minimum (see Fig. 3.12), the conditions to be satisfied are:

(i) $dy/dx = 0$, at $x = b$ and

(ii) dy/dx changes sign from negative to positive as x passes through the value $x = b$, or d^2y/dx^2 is positive at $x = b$.

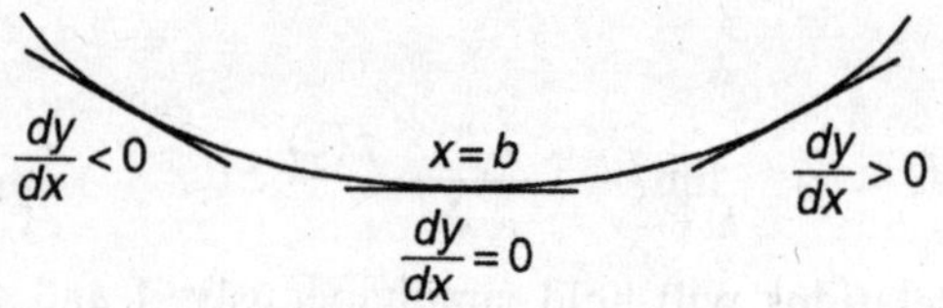

Figure 3.12

The conditions for maximum or minimum given above hold good only if $f'(c) = 0$ and $f''(c)$ is nonzero. However $f''(c)$, and possibly higher derivatives, may also be zero. In such a case, we have to examine the first nonzero higher derivative. It can be shown that if each of $f'(c), f''(c), \ldots, f^{(n-1)}(c)$ is zero and $f^n(c) \neq 0$, then

(i) $f(c)$ is neither a maximum nor a minimum if n is odd.

(ii) $f(c)$ is an extreme value if n is even. Also, $f(c)$ is a maximum or a minimum value according as

$$f^n(c) < 0 \quad \text{or} \quad f^n(c) > 0.$$

We shall now illustrate the determination of maxima or minima by several illustrative examples.

Example 3.36 Find the maximum and minimum values of the function defined by

$$y = (x + 2)^2 (x - 4).$$

We obtain

$$\frac{dy}{dx} = 2(x + 2)(x - 4) + (x + 2)^2 = 3(x + 2)(x - 2) = 3(x^2 - 4).$$

Hence, stationary points are given by $x=-2$ and $x=2$. We observe that dy/dx is positive when $x<-2$ and negative when $x>-2$. Thus, dy/dx changes sign from positive to negative as x passes through the value $x=-2$. It therefore follows that $f(x)$ is a maximum at $x=-2$. In a similar way, it can be shown that $x=2$ is a minimum point.

More simply, we can find the maximum or minimum by finding the second derivative. Thus, in the above example, we obtain

$$\frac{d^2y}{dx^2}=6x,$$

which is negative when $x=-2$ and positive for $x=2$. Hence, $x=2$ is a minimum point and $x=-2$ is a maximum point (see Fig. 3.13). The corresponding minimum and maximum values of y are -32 and 0, respectively.

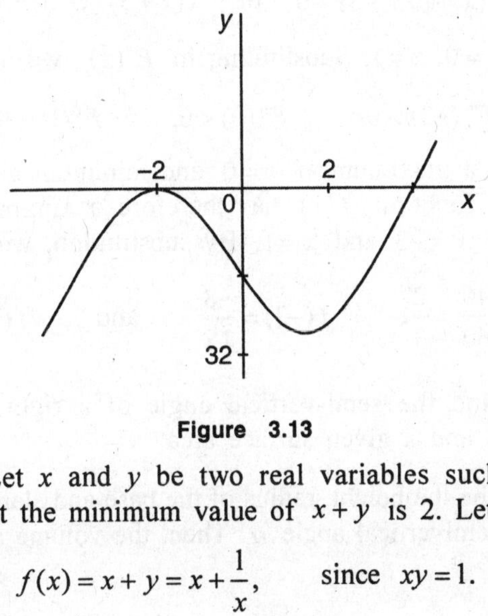

Figure 3.13

Example 3.37 Let x and y be two real variables such that $x>0$ and $xy=1$. Show that the minimum value of $x+y$ is 2. Let

$$f(x)=x+y=x+\frac{1}{x}, \qquad \text{since } xy=1.$$

Hence,

$$f'(x)=1-\frac{1}{x^2}.$$

For maxima or minima, $f'(x)=0$ and so

$$1-\frac{1}{x^2}=0, \qquad \text{which gives } x=\pm 1.$$

Since $x>0$, we have $x=1$. Further, $F''(x)=2/x^3$, which is positive for $x=1$. Hence, $f(x)$ has a minimum at $x=1$ and $f(1)=1+1=2$.

Example 3.38 Find the maxima and minima of the function defined by

$$f(x) = \frac{40}{3x^4 + 8x^3 - 18x^2 + 60}.$$

We first determine the maxima and minima of the function defined by

$$F(x) = 3x^4 + 8x^3 - 18x^2 + 60.$$

Clearly,

$$F'(x) = 12x^3 + 24x^2 - 36x = 12x(x^2 + 2x - 3)$$

and

$$F''(x) = 36x^2 + 48x - 36 = 12(3x^2 + 4x - 3)$$

For maxima or minima of $F(x)$, we must have $F'(x) = 0$, and hence we obtain

$$x(x^2 + 2x - 3) = 0 \quad \text{or} \quad x(x+3)(x-1) = 0,$$

whence $x = -3$, $x = 0$, $x = 1$. Substituting in $F''(x)$, we obtain

$$F''(-3) > 0, \qquad F''(0) < 0, \qquad F''(1) > 0.$$

Hence, $F(x)$ has a maximum at $x = 0$ and minimum at both $x = -3$ and $x = 1$. The given function $f(x)$ has therefore a minimum at $x = 0$ and maximum at both $x = -3$ and $x = 1$. By substitution, we obtain

$$f(0) = \frac{40}{60} = \frac{2}{3}, \qquad f(-3) = \frac{-8}{15} \qquad \text{and} \qquad f(1) = \frac{40}{53}.$$

Example 3.39 Find the semi-vertical angle of a right circular cone of maximum volume and a given surface area.

Let h, r and l be the height, radius of the base and slant height of a right circular cone of semivertical angle α. Then, the volume (V) of the cone is given by

$$V = \frac{1}{3}\pi r^2 h, \tag{i}$$

and the surface area S is

$$S = \pi r^2 + \pi r l \tag{ii}$$

Since S is given, we can rewrite (ii) as

$$l = \frac{S - \pi r^2}{\pi r} = \frac{S}{\pi r} - r.$$

We have

$$h^2 = l^2 - r^2$$

$$= \left(\frac{S}{\pi r} - r\right)^2 - r^2$$

$$= \frac{S^2}{\pi^2 r^2} + r^2 - \frac{2S}{\pi} - r^2$$

$$= \frac{S^2}{\pi^2 r^2} - \frac{2S}{\pi}.$$

Now from (i)

$$V^2 = \frac{1}{9}\pi^2 r^4 h^2 = \frac{1}{9}\pi^2 r^4 \left(\frac{S^2}{\pi^2 r^2} - \frac{2S}{\pi}\right) = \frac{1}{9} r^2 S^2 - \frac{2}{9}\pi S r^4. \qquad \text{(iii)}$$

We have thus expressed V in terms of a single variable r. Hence, differentiation with respect to r yields:

$$2V\frac{dV}{dr} = \frac{1}{9}S^2(2r) - \frac{2}{9}\pi S\,(4r^3)$$

or

$$V\frac{dV}{dr} = \frac{1}{9}S^2 r - \frac{4}{9}\pi S r^3. \qquad \text{(iv)}$$

Differentiating (iv) with respect to r, we obtain

$$V\frac{d^2V}{dr^2} + \left(\frac{dV}{dr}\right)^2 = \frac{1}{9}S^2 - \frac{4}{9}\pi S(3r^2) \qquad \text{(v)}$$

For maxima or minima, $dV/dr = 0$. Hence (iv) gives:

$$S^2 r = 4\pi S r^3 \quad \text{or} \quad r^2 = \frac{S}{4\pi} \quad \text{or} \quad r = \frac{1}{2}\sqrt{\frac{S}{\pi}}.$$

For this value of r, we obtain from (v):

$$V\frac{d^2V}{dr^2} = \frac{1}{9}S^2 - \frac{4}{3}\pi S\,\frac{S}{4\pi} = \frac{1}{9}S^2 - \frac{1}{3}S^2 = -\frac{2}{9}S^2 < 0.$$

Hence V is a maximum for $r^2 = S/(4\pi)$.

The slant height l is then given by

$$l^2 = \frac{S^2}{\pi^2 r^2} + r^2 - \frac{2S}{\pi}$$

$$= \frac{S^2}{\pi^2}\frac{4\pi}{S} + \frac{S}{4\pi} - \frac{2S}{\pi}$$

$$= \frac{4S}{\pi} + \frac{S}{4\pi} - \frac{2S}{\pi}$$

$$= \frac{9}{4}\frac{S}{\pi}.$$

It therefore follows that

$$\sin^2\alpha = \frac{r^2}{l^2} = \frac{S}{4\pi}\frac{4\pi}{9S} = \frac{1}{9},$$

Hence

$$\alpha = \sin^{-1}\frac{1}{3}.$$

Example 3.40 A window has the shape of a rectangle of width $2a$ and height b surmounted by a semicircle of radius a. For a given perimeter of the window, find the dimensions of the rectangle and semicircle so that the quantity of light admitted is a maximum.

If P is the perimeter of the window, then

$$P = \pi a + 2b + 2a \tag{i}$$

Area (A) of the window is given by

$$A = \frac{1}{2}\pi a^2 + 2ab$$

$$= \frac{1}{2}\pi a^2 + a(P - \pi a - 2a), \qquad \text{[using (i)]}$$

$$= \frac{1}{2}\pi a^2 + Pa - \pi a^2 - 2a^2$$

$$= Pa - 2a^2 - \frac{1}{2}\pi a^2. \tag{ii}$$

The quantity of light admitted through the window would be maximum provided that the area of the window is maximum, i.e. if A is maximum, the condition for which is $dA/da = 0$.

Hence, from (ii):

$$\frac{dA}{da} = 0 = P - 4a - \pi a$$

$$= \pi a + 2b + 2a - 4a - \pi a, \qquad \text{[using (i)]}$$

$$= 2b - 2a.$$

This gives $b = a$.

Further,

$$\frac{d^2A}{da^2} = -2,$$

which shows that the condition $b = a$ gives the maximum area, viz.,

$$A_{\text{max}} = a^2\left(2 + \frac{\pi}{2}\right).$$

Hence, in order that the quantity of light admitted is a maximum, the window should be designed so that its height is one-half its width.

Example 3.41 An open box is to be made from a rectangular piece of sheet metal, 12 by 18 cm, by cutting out equal squares from each corner and folding up the sides. Find the dimensions of the box of the largest volume that can be made in this manner.

Let x be the side of each square cut out. Then, the dimensions of the box are $(18 - 2x)$, $(12 - 2x)$ and x. Hence, the volume V of the box is given by

$$V = x(18 - 2x)(12 - 2x) = 4(54x - 15x^2 + x^3).$$

For maxima or minima, we have $dV/dx = 0$, and so

$$54 - 30x + 3x^2 = 0$$

which gives

$$x = \frac{10 \pm \sqrt{100 - 72}}{2} = 5 \pm \sqrt{7}$$

Further,

$$\begin{aligned}\frac{d^2V}{dx^2} &= 12(-10 + 2x)\\ &= 12[-10 + 2(5 - \sqrt{7})], \quad (\text{for } x = 5 - \sqrt{7})\\ &= -24\sqrt{7}.\end{aligned}$$

Hence, $x = 5 - \sqrt{7}$ gives the maximum volume. The required dimensions of the box are therefore

$$18 - 10 + 2\sqrt{7},\ 12 - 10 + 2\sqrt{7},\ 5 - \sqrt{7},$$

i.e.

$$8 + 2\sqrt{7},\ 2 + 2\sqrt{7},\ 5 - \sqrt{7}.$$

The reader should verify that the value $x = 5 + \sqrt{7}$ does not yield the maximum volume.

3.2.6 Concavity and Points of Inflexion

We have seen that in an interval where dy/dx is positive, the function is an increasing function of x, i.e. the curve is *rising* to the right while if dy/dx is negative, the curve is *falling* to the right. We have also found that at points where a curve changes its course (from rising to falling or vice versa), the first derivative vanishes and the tangent is horizontal. However, there may exist points at which $dy/dx = 0$ but these may neither be maxima nor minima.

At the point C, however, the curve crosses the tangent, i.e. there is a change in the direction of bending of the curve from *concavity upwards* to *concavity downwards*. Such a point is called a *point of inflexion* of the curve. At a point of inflexion, the curve changes from concavity upwards to concavity downwards or vice versa.

We shall now obtain the conditions for concavity upwards, concavity downwards and a point of inflexion. These conditions are given in terms of d^2y/dx^2 since the second derivative, being the rate of change of the gradient, is related to the direction of bending of the curve. We first consider point A where $x = a$. We observe that for values of x slightly less than a, the function is an increasing function of x and that dy/dx is positive. After passing through A, the curve bends downward to the right and is a decreasing function of x. Hence, dy/dx is negative for values of x slightly greater than a. We thus see that dy/dx changes sign from positive to negative as the curve passes through A. This means that dy/dx is itself a decreasing function so that d^2y/dx^2 is negative. Hence, at the point A, d^2y/dx^2 is negative. The direction of concavity is, therefore, *downward* if the second derivative is negative. Similarly, considering the point B (where $x = b$), we conclude that the direction of concavity is upward if the second derivative is *positive*.

A point where the direction of bending of a curve changes from concavity upwards to concavity downwards or vice versa is called a point of inflexion of the curve. This means that the value of d^2y/dx^2 changes sign as the curve passes through the point of inflexion and hence $d^2y/dx^2 = 0$ at the point of inflextion. In Fig. 3.10, the point C is a point of inflexion. The conditions to be satisfied by a point of inflextion may therefore be stated as follows:

The point $x = c$ is a point of inflexion on the curve defined by $y = f(x)$ if:

(i) $d^2y/dx^2 = 0$ at $x = c$ and

(ii) d^2y/dx^2 changes sign as the curve passes through $x = c$.

Example 3.42 Find the range of values of x for which the curve

$$y = 3x^5 - 40x^3 + 3x - 20$$

is concave upwards, or concave downwards. Also determine the points of inflexion.

We have

$$\frac{dy}{dx} = 15x^4 - 120x^2 + 3$$

and

$$\frac{d^2y}{dx^2} = 60x^3 - 240x = 60x(x^2 - 4) = 60x\,(x-2)\,(x+2).$$

Setting $d^2y/dx^2 = 0$, we obtain $x = -2$, $x = 0$ and $x = 2$. To test the point $x = -2$, we observe that

$$\text{For } x \text{ slightly less than } -2, \ \frac{d^2y}{dx^2} \text{ is negative}$$

and

$$\text{For } x \text{ slightly greater than } -2, \ \frac{d^2y}{dx^2} \text{ is positive.}$$

Therefore, d^2y/dx^2 changes sign from negative to positive as x passes through the value -2, i.e. the curve is concave downward in the interval $(-\infty, -2)$. Further, since $d^2y/dx^2 = 0$ at $x = -2$, we conclude that $x = -2$ is a point of inflexion. In a similar manner, the other points $x = 0$ and $x = 2$ may also be tested. In practice, however, it will be convenient to record the calculations in a tabular form as given below.

x	$y = f(x)$	y'	y''	Remarks
−3	+322	+128	−900	concave downwards
−2	+198	−237	0	point of inflexion
−1	+14	−102	+180	concave upwards
0	−20	+3	0	point of inflexion
1	−54	−102	−180	concave downwards
2	−238	−237	0	point of inflexion
3	−362	+128	+900	concave upwards.

From the table it is easily seen that the points (−2, 198), (0, −20) and (2, −238) are the points of inflexion and that the curve is concave downwards in [0, 2] and concave upwards in [−2, 0] and [2, ∞].

3.2.7 Taylor's Series and Maclaurin's Series

We are now in a position to derive formulae for the expansion of a function in ascending powers of the variable. These formulae are derived on the following assumptions:

(i) The expansion of the function in ascending powers of the variable is possible.

(ii) All the higher derivatives of the function exist and are finite, and

(iii) The infinite series is convergent.

Taylor's theorem

Let $f(x)$ be a function of x. If the function $f(x+h)$ can be expanded in a convergent series of positive integral powers of h, then this expansion is given by

$$f(x+h)=f(x)+hf'(x)+\frac{h^2}{2!}f''(x)+\cdots+\frac{h^n}{n!}f^{(n)}(x)+\cdots \tag{3.48}$$

We assume

$$f(x+h)=A_0+A_1h+A_2h^2+A_3h^3+\cdots+A_nh^n+\cdots \tag{3.49}$$

and differentiate successively with respect to h (i.e. regarding x as a constant) to get:

$$f'(x+h)=A_1+2A_2h+3A_3h^2+\cdots+nA_nh^{n-1}+\cdots$$

$$f''(x+h)=2A_2+3.2A_3h+\cdots+n(n-1)A_nh^{n-2}+\cdots$$

$$f'''(x+h)=3.2A_3+\cdots+n(n-1)(n-2)A_nh^{n-3}+\cdots$$

and, in general,

$$f^n(x+h)=n(n-1)(n-2)\cdots 3.2.1\,A_n+\text{powers of } h.$$

Putting $h=0$ successively in these equations, we obtain:

$$f(x)=A_0,$$

$$f'(x)=A_1,$$

$$f''(x)=2!\,A_2,$$

$$f^n(x)=n(n-1)(n-2)\cdots 3.2.1\,A_n=n!\,A_n.$$

Substituting for A_0, A_1, A_2,..., A_n,... in (3.49), we obtain the series given in (3.48).

If $f(x)$ is a polynomial, the series (3.48) terminates after a certain stage. If, on the other hand, $f(x)$ is not a polynomial, the series (3.48), terminated after $(n+1)$ terms, is only a *polynomial approximation* to the function. To find the remainder term, we write

$$f(b)=f(a)+(b-a)f'(a)+\frac{(b-a)^2}{2!}f''(a)+\cdots+\frac{(b-a)^n}{n!}f^{(n)}(a)+R,$$

where R is the remainder term and b is a particular point. Taylor's theorem relates the remainder to the $(n+1)$th derivative of the function. Following the procedure outlined in Section 3.2.3, we deduce Lagrange's form of the remainder:

$$R(a)=\frac{1}{(n+1)!}f^{n+1}(\xi)\,(b-a)^{n+1},$$

for some ξ in the range $a<\xi<b$. We therefore have the following theorem:

Theorem 3.3 (Taylor's theorem) If $f(x)$ is differentiable n times in the interval $a\le x\le b$ and $f^n(x)$ is

(i) differentiable in $a<x<b$,
(ii) continuous in $a\le x\le b$,

then there exists ξ in the interval $a<\xi<b$ such that

$$f(b)=f(a)+f(b-a)\,f'(a)+\frac{(b-a)^2}{2!}f''(a)+\cdots+\frac{(b-a)^n}{n!}f^{(n)}(a)$$

$$+\frac{1}{(n+1)!}(b-a)^{n+1}f^{n+1}(\xi),\qquad a<\xi<b. \tag{3.50}$$

Since b is any point we have taken, the above may be written as

$$f(x)=f(a)+(x-a)\,f'(a)+\frac{(x-a)^2}{2!}f''(a)+\cdots+\frac{(x-a)^n}{n!}f^n(a)$$

$$+\frac{(x-a)^{n+1}}{(n+1)!}f^{n+1}(\xi),\qquad a<\xi<x.$$

If we set $x=a+h$, we get

$$f(a+h)=f(a)+hf'(a)+\frac{h^2}{2!}f''(a)+\cdots+\frac{1}{n!}h^n f^n(a)$$

$$+\frac{1}{(n+1)!}h^{n+1}f^{n+1}(\xi),\qquad a<\xi<a+h.$$

which is another form of the above series.

An alternative form of the remainder term (Cauchy's form) can be obtained by applying the mean-value theorem directly to the function $F(x)$ in (3.43). This leads to

$$R=\frac{1}{n!}f^{n+1}(\xi)\,(b-a)\,(b-\xi)^n, \tag{3.51}$$

which is sometimes found more useful than Lagrange's form. The infinite series given in (3.48) is called *Taylor's series*.

Maclaurin's theorem

If the function $f(x)$ can be expanded in a convergent series of positive integral powers of x, then such an expansion is given by

$$f(x) = f(0) + xf'(0) + \frac{x^2}{2!} f''(0) + \frac{x^3}{3!} f'''(0) + \cdots + \frac{x^n}{n!} f^{(n)}(0) + \cdots \quad (3.52)$$

Let

$$f(x) = A_0 + A_1 x + A_2 x^2 + A_3 x^3 + \cdots + A_n x^n + \cdots \quad (3.53)$$

where $A_0, A_1, \ldots$ are to be determined. Differentiating (3.53) successively with respect to x, we obtain

$$f'(x) = A_1 + 2A_2 x + 3A_3 x^2 + \cdots + nA_n x^{n-1} + \cdots$$

$$f''(x) = 2A_2 + 3.2A_3 x + \cdots + n(n-1)\, A_n x^{n-2} + \cdots$$

$$f'''(x) = 3.2A_3 + \cdots + n(n-1)\,(n-2)\, A_n x^{n-3} + \cdots$$

$$\vdots$$

$$f^n(x) = n(n-1)\,(n-2) \ldots 3.2A_n + \text{powers of } x.$$

Putting $x = 0$ in the above equations, we obtain:

$$f(0) = A_0,$$

$$f'(0) = A_1,$$

$$f''(0) = 2.1 A_2,$$

$$\vdots$$

$$f^n(0) = n!A_n.$$

Substitution of $A_0, A_1, \ldots, A_n$ in (3.53) yields Maclaurin's series given by (3.52).

It is important to note that Taylor's and Maclaurin's series are not essentially different. In fact, each can be derived from the other. Thus, if we put $x = 0$ and $h = x$ in (3.48), we obtain

$$f(x) = f(0) + xf'(0) + \frac{x^2}{2!} f''(0) + \cdots + \frac{x^n}{n!} f^n(0) + \cdots$$

which is Maclaurin's series.

Similarly, in the Maclaurin's series for $g(x)$:

$$g(x) = g(0) + xg'(0) + \frac{x^2}{2!} g''(0) + \cdots + \frac{x^n}{n!} g^n(0) + \cdots, \quad (3.54)$$

put $g(x) = f(x+h)$. Then,

$$g'(x) = f'(x+h),\ g''(x) = f''(x+h), \ldots, g^n(x) = f^n(x+h)$$

Hence,

$$g(0) = f(h), \quad g'(0) = f'(h), \quad g''(0) = f''(h), \ldots, g^{(n)}(0) = f^{(n)}(h).$$

Substituting in (3.54), we obtain:

$$f(x+h) = f(h) + xf'(h) + \frac{x^2}{2!} f''(h) + \cdots + \frac{x^n}{n!} f^n(h) + \cdots$$

Interchanging x and h in the above, we obtain the Taylor's series.

Applications of Taylor's and Maclaurin's series are illustrated in the examples given below.

Example 3.43 Show that

$$\sin x = x - \frac{x^3}{3!} + \frac{x^5}{5!} - \cdots + \frac{x^n}{n!} \sin \frac{n\pi}{2} - \cdots$$

Let $f(x) = \sin x$, then $f(0) = 0$. Differentiating successively and putting $x = 0$, we obtain:

$$f'(x) = \cos x = \sin\left(x + \frac{\pi}{2}\right),$$

$$f''(x) = \cos\left(x + \frac{\pi}{2}\right) = \sin\left(x + \frac{2\pi}{2}\right),$$

$$f'''(x) = \cos\left(x + \frac{2\pi}{2}\right) = \sin\left(x + \frac{3\pi}{2}\right),$$

$$\vdots$$

$$f^n(x) = \sin\left(x + \frac{n\pi}{2}\right).$$

Then

$$f'(0) = 1, \quad f''(0) = 0, \quad f'''(0) = -1, \ldots, \qquad f^n(0) = \sin\frac{n\pi}{2}$$

Substituting in the Maclaurin's series (3.52), we obtain:

$$\sin x = x - \frac{x^3}{3!} + \frac{x^5}{5!} - \cdots + \frac{x^n}{n!} \sin \frac{n\pi}{2} + \cdots$$

In a similar manner, the following series can be derived:

(i) $e^x = 1 + x + \dfrac{x^2}{2!} + \dfrac{x^3}{3!} + \cdots + \dfrac{x^n}{n!} + \cdots$

(ii) $\log(1+x) = x - \dfrac{x^2}{2} + \dfrac{x^3}{3} - \cdots + (-1)^{n-1}\dfrac{x^n}{n} + \cdots$

(iii) $\cos x = 1 - \dfrac{x^2}{2!} + \dfrac{x^4}{4!} - \cdots + \dfrac{x^n}{n!}\cos\dfrac{n\pi}{2} + \cdots$

(iv) $\tan x = x + \dfrac{2}{3}x^3 + \dfrac{2}{15}x^5 + \cdots$

Example 3.44 Prove by Taylor's theorem that

$$\tan^{-1}(x+h) = \tan^{-1}x + h\sin\alpha\frac{\sin\alpha}{1} - (h\sin\alpha)^2\frac{\sin 2\alpha}{2} + (h\sin\alpha)^3\frac{\sin 3\alpha}{3} - \cdots,$$

where $x = \cot\alpha$.

Let $f(x) = \tan^{-1}x$. Then,

$$f'(x) = \frac{1}{1+x^2} = \frac{1}{(x+i)(x-i)} = \frac{1}{2i}\left(\frac{1}{x-i} + \frac{1}{x+i}\right),$$

$$f''(x) = \frac{1}{2i}\left[\frac{(-1)}{(x-i)^2} - \frac{(-1)}{(x+i)^2}\right],$$

$$f'''(x) = \frac{1}{2i}\left[\frac{(-1)(-2)}{(x-i)^3} - \frac{(-1)(-2)}{(x+i)^3}\right]$$

$$\vdots$$

$$f^n(x) = \frac{1}{2i}\left[\frac{(-1)(-2)\ldots(\overline{-n-1})}{(x-i)^n} - \frac{(-1)(-2)\ldots(\overline{-n-1})}{(x+i)^n}\right]$$

$$= \frac{(-1)^{n-1}(n-1)!}{2i}\left[\frac{1}{(x-i)^n} - \frac{1}{(x+i)^n}\right]$$

Let $x = r\cos\alpha$ and $1 = r\sin\alpha$. Then, $f^n(x)$ becomes:

$$f^n(x) = \frac{(-1)^{n-1}(n-1)!}{2i}\left[\frac{1}{r^n(\cos\alpha - i\sin\alpha)^n} - \frac{1}{r^n(\cos\alpha + i\sin\alpha)^n}\right]$$

$$= \frac{(-1)^{n-1}(n-1)!}{2ir^n}\left[\frac{1}{(\cos\alpha - i\sin\alpha)^n} - \frac{1}{(\cos\alpha + i\sin\alpha)^n}\right]$$

$$= \frac{(-1)^{n-1}(n-1)!}{2ir^n}\left[\frac{1}{\cos n\alpha - i\sin n\alpha} - \frac{1}{(\cos n\alpha + i\sin n\alpha)}\right],$$

$$= \frac{(-1)^{n-1}(n-1)!}{2ir^n}\frac{2i \sin n\alpha}{1}$$

$$= \frac{(-1)^{n-1}(n-1)! \sin n\alpha}{1/\sin^n \alpha}$$

i.e. $f^n(x) = (-1)^{n-1}(n-1)! \sin^n \alpha \sin n\alpha$, where $\cot \alpha = x$. Putting $n = 1, 2, 3, \ldots$, we get all the derivatives. Setting now $f(x) = \tan^{-1} x$ in the Taylor's series, we obtain:

$$\tan^{-1}(x+h) = \tan^{-1}x + h \sin \alpha \sin \alpha - \frac{h^2}{2!}\sin^2\alpha \sin 2\alpha + \frac{h^3}{3!} 2 \sin^3\alpha \sin 3\alpha - \cdots$$

$$= \tan^{-1}x + h \sin \alpha \frac{\sin \alpha}{1} - (h \sin \alpha)^2 \frac{\sin 2\alpha}{2} + (h \sin \alpha)^3 \frac{\sin 3\alpha}{3} - \ldots,$$

which is the required result.

Example 3.45 Expand $e^{\sin x}$ by Maclaurin's theorem as far as the term containing x^4.

Let $y = e^{\sin x} = f(x)$, then $f(0) = e^0 = 1$. Differentiating with respect to x, we obtain

$$y_1 = f'(x) = e^{\sin x} \cos x = y \cos x,$$

$$y_2 = f''(x) = y_1 \cos x - y \sin x,$$

$$y_3 = f'''(x) = y_2 \cos x - y_1 \sin x - y_1 \sin x - y \cos x$$

$$= y_2 \cos x - 2y_1 \sin x - y \cos x,$$

$$y_4 = f^{iv}(x) = y_3 \cos x - y_2 \sin x - 2y_2 \sin x - 2y_1 \cos x - y_1 \cos x + y \sin x$$

$$= y_3 \cos x - 3y_2 \sin x - 3y_1 \cos x + y \sin x,$$

Then

$$f'(0) = 1, \quad f''(0) = 1, \quad f'''(0) = 0, \quad f^{iv}(0) = -3$$

Substituting in the Maclaurin's series, we get:

$$e^{\sin x} = 1 + x + \frac{x^2}{2} - \frac{1}{8}x^4 + \cdots$$

Example 3.46 Show that

$$\tan\left(\frac{\pi}{4} + x\right) = 1 + 2x + 2x^2 + \frac{8}{3}x^3 + \frac{10}{3}x^4 + \cdots$$

and hence calculate the value of tan 46° correct to five decimal places, given that $\pi = 3.14159$.

Let

$$y = f(x) = \tan\left(\frac{\pi}{4} + x\right),$$

Successive differentiation gives:

$$y_1 = f'(x) = \sec^2\left(\frac{\pi}{4} + x\right) = 1 + y^2,$$

$$y_2 = f''(x) = 2yy_1,$$

$$y_3 = f'''(x) = 2yy_2 + 2y_1^2,$$

$$y_4 = f^{iv}(x) = 2yy_3 + 2y_1y_2 + 4y_1y_2 + 2yy_3 + 6y_1y_2,$$

Then

$$f(0) = 1, \quad f'(0) = 2, \quad f''(0) = 4, \quad f'''(0) = 16, \quad f^{iv}(0) = 80$$

Hence,

$$\tan\left(\frac{\pi}{4} + x\right) = 1 + 2x + \frac{x^2}{2}(4) + \frac{x^3}{6}(16) + \frac{x^4}{24}(80) + \ldots$$

$$= 1 + 2x + 2x^2 + \frac{8}{3}x^3 + \frac{10}{3}x^4 + \ldots$$

Now, let $x = 1^\circ = 0.017453$. Then,

$$x^2 0.000304, \; x^3 = 0.000005, \ldots$$

we therefore need the first-four terms only, and

$$\tan 46^\circ = 1 + 2\,(0.017453) + 2\,(0.000304) + \frac{8}{3}(0.000005)$$

$$= 1.035527$$

$$= 1.03553 \quad \text{(correct to five decimal places).}$$

Example 3.47 Maclaurin's series with Lagrange's remainder [see Eq. (3.47)] is given by

$$f(x) = f\,(0) + xf'(0) + \frac{x^2}{2!}f''(0) + \cdots + \frac{x^{n-1}}{(n-1)!}f^{n-1}(0) + \frac{x^n}{n!}f^n(\xi). \quad (3.47)$$

Find the remainders after n terms in the Maclaurin's expansions of sin x, cos x, and e^x and show that in each case the remainder terms tends to zero as $n \to \infty$.

Let $S_{nx}(x)$ denotes the sum of the first n terms and let $R_n(x)$ denotes Lagrange's remainder. Then, we have

$$f\,(x) = S_n(x) + R_n(x).$$

(i) Let $f(x) = \sin x$. We have

$$f^n(x) = \sin\left(x + n\frac{\pi}{2}\right) \quad \text{and} \quad R_n(x) = \frac{x^n}{n!}\sin\left(\theta x + n\frac{\pi}{2}\right)$$

Since

$$\left|\sin\left(\theta x + n\frac{\pi}{2}\right)\right| \le 1,$$

we must have

$$|R_n(x)| \le \left|\frac{x^n}{n!}\right|$$

and hence $R_n(x) \to 0$ *as* $n \to \infty$. It follows therefore that

$$\sin x = x - \frac{x^3}{3!} + \frac{x^5}{5!} - \cdots$$

(ii) The argument for $f(x) = \cos x$ is exactly the same as that for $\sin x$ and it follows that

$$\cos x = 1 - \frac{x^2}{2!} + \frac{x^4}{4!} - \cdots$$

(iii) Let $f(x) = e^x$. Then,

$$f'(x) = e^x, \qquad f''(x) = e^x, \qquad f^n(x) = e^x,$$

and

$$f^{(n)}(\theta x) = e^{\theta x}.$$

Putting $x = 0$ in $f(x), f'(x), \ldots,$ we obtain: $f(0) = 1,\ f'(0) = 1,\ f''(0) = 1, \ldots$ Hence, Maclaurin's series with the remainder term is

$$e^x = 1 + x + \frac{x^2}{2!} + \frac{x^3}{3!} + \cdots + \frac{x^{n-1}}{(n-1)!} + R_n(x)$$

where

$$R_n(x) = e^{\theta x}\frac{x^n}{n!}.$$

Now,

$$R_n(x) = e^{\theta x}\lim_{n\to\infty}\frac{x_n}{n!} = 0.$$

Hence,

$$e^x = 1 + x + \frac{x^2}{2!} + \frac{x^3}{3!} + \cdots + \frac{x^n}{n!} + \cdots$$

Examlpe 3.48 Obtain by Maclaurin's theorem the first-four terms of the expansion of $e^{x\cos x}$ in ascending powers of x.

Let $y = f(x) = e^{x\cos x}$, then $f(0) = 1$. Taking logarithms,we obtain

$$\log y = x\cos x.$$

Differentiating with respect to x, we obtain:

$$\frac{1}{y}\frac{dy}{dx} = \cos x - x\sin x.$$

or,

$$y_1 = f'(x) = y(\cos x - x\sin x).$$

Differentiating the above, we obtain:

$$y_2 = f''(x) = y_1(\cos x - x\sin x) + y(-\sin x - x\cos x - \sin x)$$
$$= y_1(\cos x - x\sin x) - y(2\sin x + x\cos x).$$

Again,

$$y_3 = f'''(x) = y_2(\cos x - x\sin x) - y(3\cos x - x\sin x).$$

Then

$$f'(0) = 1,\ f'''(0) = 1,\ f'''(0) = -2$$

Since

$$f(x) = f(0) + xf'(0) + \frac{x^2}{2}f''(0) + \frac{x^3}{3!}f'''(0) + \cdots,$$

we obtain

$$e^{x\cos x} = 1 + x + \frac{x^2}{2}(1) + \frac{x^3}{6}(-2) + \cdots = 1 + x + \frac{1}{2}x^2 - \frac{1}{3}x^3 + \cdots$$

Alternatively, the result may be obtained, more simply, in the following way:

By Maclaurin's theorem, we have

$$e^y = 1 + y + \frac{y^2}{2!} + \frac{y^3}{3!} + \cdots \qquad \text{(i)}$$

and

$$\cos x = 1 - \frac{x^2}{2!} + \frac{x^4}{4!} - \cdots \qquad \text{(ii)}$$

Putting $y = x\cos x$ in (*i*), we get:

$$e^{x\cos x} = 1 + x\cos x + \frac{x^2}{2}\cos^2 x + \frac{x^3}{6}\cos^3 x + \cdots$$

Using (ii), we get

$$e^{x\cos x} = 1 + x\left(1 - \frac{x^2}{2!} + \frac{x^4}{4!} - \cdots\right) + \frac{x^2}{2}\left(1 - \frac{x^2}{2!} + \frac{x^4}{4!} - \cdots\right)^2$$

$$+ \frac{x^3}{6}\left(1 - \frac{x^2}{2!} + \frac{x^4}{4} - \cdots\right)^3 + \cdots$$

$$= 1 + x\left(1 - \frac{x^2}{2!} + \cdots\right) + \frac{x^2}{2}(1 - \cdots) + \frac{x^3}{6}(1 - \cdots) + \cdots$$

$$= 1 + x - \frac{1}{2}x^3 + \frac{1}{2}x^2 + \frac{1}{6}x^3 + \cdots$$

$$= 1 + x + \frac{1}{2}x^2 - \left(\frac{1}{2} - \frac{1}{6}\right)x^3 + \cdots$$

$$= 1 + x + \frac{1}{2}x^2 - \frac{1}{3}x^3 + \cdots,$$

3.2.8 Curvature

Let AP $(=s)$ be an arc of a curve C (Fig. 3.14) and let the tangent at P be inclined at an angle ψ with the x-axis. Th direction of the curve at a point is determined by the slope at that point, but a measure of the rapidity with which the curve is turning (or bending) at the point is given by the rate of change in ψ with respect to s, the arc length. This rate of change in called the *curvature* of the curve at the particular point and is denoted by κ so that

$$\kappa = \frac{d\psi}{ds} \tag{3.55}$$

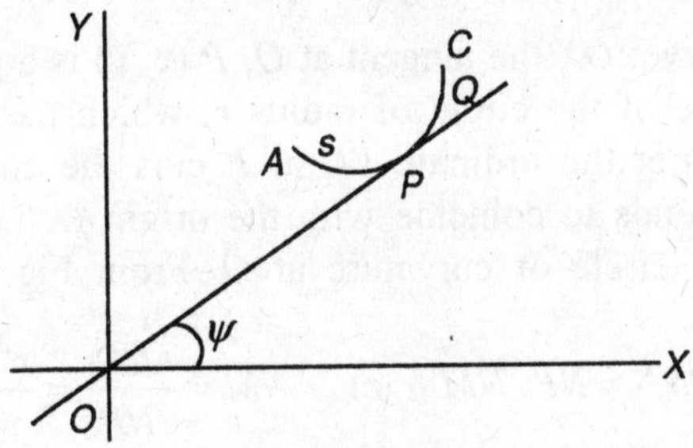

Figure 3.14

For example, if we consider a circle of radius a with centre at $(0, a)$, then $s = a\psi$, and therefore

$$\kappa = \frac{d\psi}{ds} = \frac{1}{a},$$

which shows that the curvature of a circle is the same at every point of its circumference. Due to this reason, the reciprocal of the curvature of a curve at a point is defined to be its *radius of curvature* and is denoted by ρ. Hence, we have

$$\rho = \frac{1}{\kappa} = \frac{ds}{d\psi} \tag{3.56}$$

This radius of curvature at any point P of a curve is the radius of the circle having the same curvature at P as the curve itself. If we draw the normal at the point P of the curve on its concave side and cut off a length $PC = \rho$ on this normal, the point C so obtained is called the *centre of curvature* of the curve at P.

We can now derive formulae for finding the radius of curvature ρ and the coordinates of the centre of curvature C.

Newton's method

This is a fundamental method but has a limited application since it assumes that the curve passes through the origin and has the x-axix as the tangent there (Fig. 3.15).

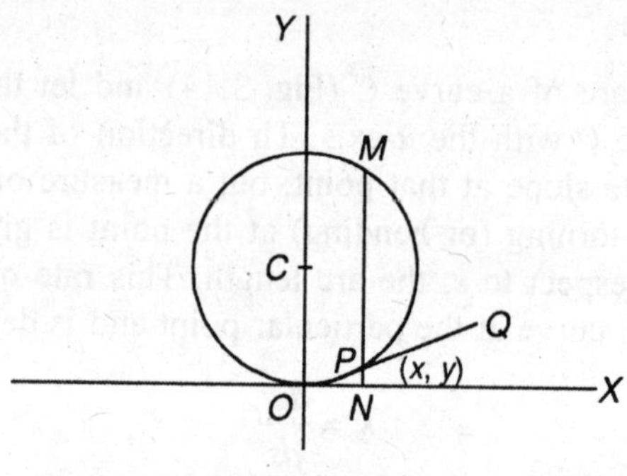

Figure 3.15

Let OPQ be the curve, OX the tangent at O, P (x, y) is a point on the curve. Let C be the centre of the circle of radius r, which passes through P and touches OX at O. Let the ordinate PN at P cuts the circle again at M. It is clear that as P tends to coincide with the origin O, i.e. as $x, y \to 0$, the circle becomes the circle of curvature at O. From Fig. 3.15,

$$ON^2 = NP \cdot NM \quad \text{or} \quad NM = \frac{ON^2}{NP} = \frac{x^2}{y}$$

As $x, y \to 0$, $NM \to 2\rho$, where ρ is the radius of curvature and hence

$$2\rho = \lim \frac{x^2}{y}$$

or,

$$\rho = \frac{1}{2}\lim \frac{x^2}{y} = \lim_{x,y \to 0} \frac{x^2}{2y} \tag{3.57a}$$

Instead of OX if OY is the tangent at the origin, the formula then takes the form:

$$\rho = \lim_{x,y \to 0} \frac{y^2}{2x} \tag{3.57b}$$

Cartesian coordinates

Let ψ be the inclination of the tangent at $P(x, y)$ with the x-axis. Then we know that

$$\frac{dy}{dx} = \tan\psi \quad \text{and} \quad \frac{dx}{ds} = \cos\psi$$

Differentiating the first relation with respect to s, we get

$$\frac{d}{ds}\left(\frac{dy}{dx}\right) = \frac{d\tan\psi}{ds}$$

or,

$$\frac{d^2y}{dx^2}\frac{dx}{ds} = \sec^2\psi\,\frac{d\psi}{ds}.$$

Since $d\psi/ds = 1/\rho$, we obtain from the above

$$\rho = \frac{\sec^2\psi}{\dfrac{d^2y}{dx^2}\dfrac{dx}{ds}} = \frac{\sec^2\psi}{\dfrac{d^2y}{dx^2}\cos\psi} = \frac{\sec^3\psi}{\dfrac{d^2y}{dx^2}} = \frac{(\sec^2\psi)^{3/2}}{\dfrac{d^2y}{dx^2}}.$$

But

$$\sec^2\psi = 1 + \tan^2\psi = 1 + \left(\frac{dy}{dx}\right)^2.$$

Hence, we have the important formula

$$\rho = \frac{\left[1 + (dy/dx)^2\right]^{3/2}}{d^2y/dx^2}, \tag{3.58a}$$

where $d^2y/dx^2 \neq 0$. In case dy/dx becomes infinite at any point (i.e. if the tangent at the point is parallel to the y-axis), a similar formula can be derived by starting with

$$\frac{dx}{dy} = \cot\psi.$$

Differentiating the above with respect to s, we get:

$$\frac{d}{dy}\left(\frac{dx}{dy}\right)\frac{dy}{ds} = -\text{cosec}^2\,\psi\,\frac{d\psi}{ds}.$$

Hence,

$$\rho = -\frac{\text{cosec}^3\,\psi}{d^2x/dy^2}, \qquad \text{since} \quad \frac{dy}{ds} = \sin\psi$$

$$= -\frac{\left[1+(dx/dy)^2\right]^{3/2}}{d^2x/dy^2}, \tag{3.58b}$$

where $d^2x/dy^2 \neq 0$.

Parametric equation

Let the curve be given by $x = f(t)$ and $y = g(t)$. Then

$$\frac{dy}{dx} = \frac{dy/dt}{dx/dt} = \frac{g'(t)}{f'(t)} = \frac{\dot{y}}{\dot{x}},$$

where a dot above the variable denotes differentiation with respect to t. Also,

$$\frac{d^2y}{dx^2} = \frac{d}{dx}\left(\frac{\dot{y}}{\dot{x}}\right) = \frac{d}{dt}\left(\frac{\dot{y}}{\dot{x}}\right)\frac{dt}{dx} = \frac{1}{\dot{x}}\,\frac{\dot{x}\ddot{y}-\dot{y}\ddot{x}}{\dot{x}^2} = \frac{\dot{x}\ddot{y}-\dot{y}\ddot{x}}{\dot{x}^3}$$

Hence the formula (3.58a) gives:

$$\rho = \frac{(1+\dot{y}^2/\dot{x}^2)^{3/2}}{\dot{x}\ddot{y}-\dot{y}\ddot{x}}\dot{x}^3 = \frac{(\dot{x}^2+\dot{y}^2)^{3/2}}{\dot{x}\ddot{y}-\dot{y}\ddot{x}} \tag{3.59}$$

Polar formula

Let $P\ (r, \theta)$ be a point on the curve whose equation is given in the form $r = f(\theta)$. Let PT be the tangent to the curve at P (Fig. 3.16).

Clearly,

$$\psi = \theta + \phi,$$

and hence,

$$\frac{d\psi}{d\theta} = 1 + \frac{d\phi}{d\theta}. \tag{3.60}$$

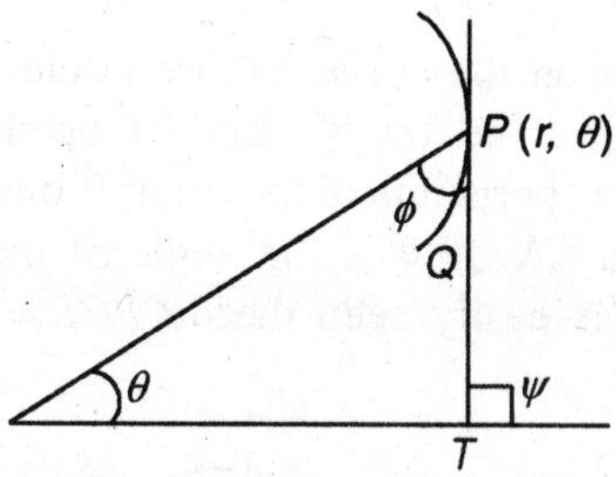

Figure 3.16

Now, we have from (3.31)

$$\left(\frac{ds}{d\theta}\right)^2 = r^2 + \left(\frac{dr}{d\theta}\right)^2 = r^2 + r_1^2, \qquad \text{where } r_1 = \frac{dr}{d\theta},$$

and so

$$\frac{ds}{d\theta} = \sqrt{r^2 + r_1^2} \tag{3.61}$$

we also know that

$$\tan\phi = r\frac{d\theta}{dr} = \frac{r}{r_1}.$$

Differentiating the above with respect to θ, we get:

$$\sec^2\phi\frac{d\phi}{d\theta} = \frac{r_1^2 - rr_2}{r_1^2}, \qquad \text{where } r_2 = \frac{d^2r}{d\theta^2}$$

But

$$\sec^2\phi = 1 + \tan^2\phi = 1 + \frac{r}{r_1^2} = \frac{r^2 + r_1^2}{r_1^2}.$$

Hence, the above equation becomes on simplification:

$$\frac{d\phi}{d\theta} = \frac{r_1^2 - rr_2}{r_1^2}\,\frac{r_1^2}{r_1^2 + r^2} = \frac{r_1^2 - rr_2}{r_1^2 + r^2}.$$

Substituting this in (3.60), we get

$$\frac{d\psi}{d\theta} = 1 + \frac{r_1^2 - rr_2}{r_1^2 + r^2} = \frac{2r_1^2 + r^2 - rr_2}{r_1^2 + r^2}.$$

Hence

$$\rho = \frac{ds}{d\psi} = \frac{ds/d\theta}{d\psi/d\theta} = \frac{\sqrt{r^2 + r_1^2}}{2r_1^2 + r^2 - rr_2}(r_1^2 + r^2) = \frac{(r^2 + r_1^2)^{3/2}}{2r_1^2 + r^2 - rr_2} \tag{3.62}$$

Centre of curvature

Let $P(x, y)$ be any point on the curve, whose circle of curvature has centre C. Let the corrdinates of C be (α, β). Let PT be the tangent to the curve at P. Let PM and CN be perpendiculars from P and C on OX. Let PR be perpendicular from P on CN. If ψ is the angle of inclination of the tangent with the x-axis, then it is easily seen that $\angle PCR = \psi$. We also know that $CP = \rho$ (see Fig. 3.17).

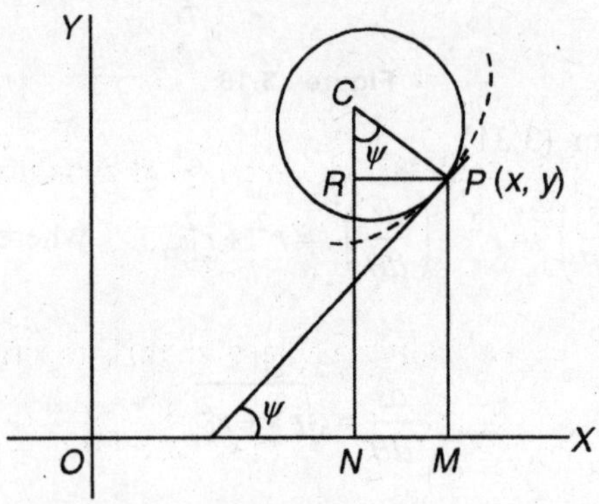

Figure 3.17

Now,

$$\alpha = ON = OM - NM = OM - RP = x - \rho \sin\psi \tag{3.63}$$

Also,

$$\beta = NC = NR + RC = MP + RC = y + \rho \cos\psi \tag{3.64}$$

Since $\tan\psi = dy/dx = y_1$, we have

$$\sin\psi = \frac{y_1}{\sqrt{1+y_1^2}} \quad \text{and} \quad \cos\psi = \frac{1}{\sqrt{1+y_1^2}}$$

We also have

$$\rho = \frac{(1+y_1^2)^{3/2}}{y_2}, \quad \text{where } y_2 = \frac{d^2y}{dx^2}.$$

Substituting these values in (3.63) and (3.64), we obtain

$$\alpha = x - \frac{(1+y_1^2)^{3/2}}{y_2}\frac{y_1}{\sqrt{1+y_1^2}} = x - \frac{y_1(1+y_1^2)}{y_2} \tag{3.65a}$$

and

$$\beta = y + \frac{(1+y_1^2)^{3/2}}{y_2}\frac{1}{\sqrt{1+y_1^2}} = y + \frac{1+y_1^2}{y_2} \tag{3.65b}$$

It is easily seen that the circle of curvature has the eqution

$$(x-\alpha)^2+(y-\beta)^2=\rho^2.$$

The locus of the centre of curvature of a curve is called its *evolute* and can be obtained by eliminating x and y from the formula for the centre of curvature and the equation of the given curve.

We conclude this chapter with some illustrative examples on radius of curvature.

Example 3.49 Find the radius of curvature at any point of the parabola $y^2=4ax$ and hence show that the radius of curvature at its vertex is equal to its semi latus rectum.

We have $y^2=4ax$. By differentiation,

$$2yy_1=4a \quad \text{or} \quad y_1=\frac{2a}{y},$$

and

$$y_2=-\frac{2a}{y^2}\frac{dy}{dx}=-\frac{2a}{y^2}\frac{2a}{y}=-\frac{4a^2}{y^3}.$$

Hence

$$\rho=\frac{(1+y_1^2)^{3/2}}{y_2}=\frac{(1+4a^2/y^2)^{3/2}}{-4a^2/y^3}=-\frac{(4a^2)+(4ax)^{3/2}}{4a^2}=-2a^{-1/2}(a+x)^{3/2}$$

or

$$|\rho|=2a^{-1/2}(a+x)^{3/2}.$$

At the vertex, i.e. at (0, 0), we have

$$(\rho)_{(0,0)}=2a^{-1/2}a^{3/2}=2a, \qquad \text{(in magnitude)}$$

$$=\frac{1}{2} \qquad \text{(latus rectum).}$$

Example 3.50 Find the radius of curvature at any point of the curve

$$x=a\,(\theta+\sin\theta),\; y=a\,(1-\cos\theta).$$

We have

$$\frac{dx}{d\theta}=a\,(1+\cos\theta)=x', \qquad \frac{dy}{d\theta}=y'=a\sin\theta$$

$$\frac{d^2x}{d\theta^2}=x''=-a\sin\theta, \qquad \frac{d^2y}{d\theta^2}=y''=a\cos\theta.$$

From (3.59) we have

$$\rho = \frac{(x'^2 + y'^2)^{3/2}}{x'y'' - y'x''}$$

$$= \frac{\left[a^2 (1+\cos\theta)^2 + a^2\sin^2\theta\right]^{3/2}}{a^2\cos\theta\,(1+\cos\theta) + a^2\sin^2\theta}$$

$$= \frac{\left[a^2(1+2\cos\theta+\cos^2\theta) + a^2\sin^2\theta\right]^{3/2}}{a^2\cos\theta + a^2\cos^2\theta + a^2\sin^2\theta}$$

$$= \frac{\left[2a^2(1+\cos\theta)\right]^{3/2}}{a^2(1+\cos\theta)}$$

$$= a(2^{3/2})\,(1+\cos\theta)^{1/2}$$

$$= a(2^{3/2})\left(2\cos^2\frac{\theta}{2}\right)^{1/2}$$

$$= 4a\cos\frac{\theta}{2}.$$

Example 3.51 If ρ and ρ' be the radii of curvature at the extremities of two conjugate diameters of an ellipse $(x^2/a^2) + (y^2/b^2) = 1$, prove that

$$(\rho^{2/3} + \rho'^{2/3})(ab)^{2/3} = a^2 + b^2.$$

Let CP and CD be two conjugate diameters of the given ellipse. If the coordinates of P are given by 't', i.e. $(a\cos t, b\sin t)$, then we know that those of D are given by $\pi/2 + t$ i.e.

$$\left[a\cos\left(\frac{\pi}{2}+t\right), b\sin\left(\frac{\pi}{2}+t\right)\right].$$

To find the radius of curvature at the point 't' of the ellipse, we observe that

$$\frac{dx}{dt} = \dot{x} = -a\sin t, \qquad \frac{dy}{dt} = b\cos t = \dot{y},$$

$$\frac{d^2x}{dt^2} = \ddot{x} = -a\cos t, \qquad \frac{d^2y}{dt^2} = \ddot{y} = -b\sin t.$$

Hence, substitution in (3.59) gives:

$$\rho = \frac{(a^2\sin^2 t + b^2\cos^2 t)^{3/2}}{ab\sin^2 t + ab\cos^2 t} = \frac{(a^2\sin^2 t + b^2\cos^2 t)^{3/2}}{ab},$$

or

$$(ab\rho)^{2/3} = a^2 \sin^2 t + b^2 \cos^2 t. \tag{i}$$

Putting $t = \pi/2 + t$ in the above, we deduce that

$$\rho' = \frac{(a^2 \cos^2 t + b^2 \sin^2 t)^{3/2}}{ab}$$

or

$$(ab\rho')^{2/3} = a^2 \cos^2 t + b^2 \sin^2 t \tag{ii}$$

Adding (i) and (ii), we obtain

$$(ab\rho)^{2/3} + (ab\rho')^{2/3} = a^2 + b^2,$$

i.e.

$$(\rho^{2/3} + \rho'^{2/3})(ab)^{2/3} = a^2 + b^2, \tag{ii}$$

which is the required result.

Example 3.52 Show that the radii of curvature at the origin for the curve $x^3 + y^3 = 3axy$ are each equal to $(3/2)a$.

The equation of the curve is $x^3 + y^3 = 3axy$. Since there is no constant term in the equation, it follows that the curve passes through the origin. Next, to determine the tangent at the origin, we equate to zero the lowest degree term in the equation. We therefore set

$$3axy = 0,$$

from which follows $x = 0$ or $y = 0$. This means that both the x-axis and y-axis are tangents to the curve at the origin. Thus, either of the formulae in (3.57) can be used to solve this problem. To apply (3.57a), we rewrite the equation of the curve as

$$\frac{x^3}{2xy} + \frac{y^3}{2xy} = \frac{3axy}{2xy},$$

i.e.

$$\frac{x^2}{2y} + \frac{y^2}{2x} = \frac{3a}{2} \tag{i}$$

We know that

$$\lim_{x,y \to 0} \frac{x^2}{2y} = \rho,$$

Also,

$$\lim_{x,y\to 0} \frac{y^2}{2x} = \lim_{x,y\to 0} \frac{2y}{x^2}\left(\frac{1}{4}xy\right)$$

$$= \lim_{x,y\to 0} \frac{2y}{x^2} \lim_{x,y\to 0} \frac{1}{4}xy$$

$$= \frac{1}{\rho}(0)$$

$$= 0.$$

Hence taking limits of both sides of (i), we obtain

$$\rho = \frac{3a}{2}.$$

In a similar manner, formula (3.57b) can also be used. This is left as an exercise to the reader.

Example 3.53 Find the coordinates of the centre of curvature for any point (x, y) on the parabola $y^2 = 4ax$. Also, find the equation of the evolute of the parabola.

For the parabola $y^2 = 4ax$, we have

$$2yy_1 = 4a \quad \text{or} \quad y_1 = \frac{4a}{2y} = \frac{2a}{y}$$

Also,

$$\frac{d^2y}{dx^2} = y_2 = -\frac{2a}{y^2}y_1 = -\frac{2a}{y^2}\frac{2a}{y} = -\frac{4a^2}{y^3}.$$

From formuale (3.65), we obtain

$$\alpha = x - \frac{y_1(1+y_1^2)}{y_2}$$

$$= x - \frac{(2a/y)(1+4a^2/y^2)}{-4a^2/y^3}$$

$$= x + \frac{y^2}{2a} + 2a$$

$$= x + \frac{4ax}{2a} + 2a$$

$$= 3x + 2a, \tag{i}$$

and

$$\beta = y + \frac{1+y_1^2}{y_2}$$

$$= y + \frac{1+4a^2/y^2}{-4a^2/y^3}$$

$$= y\left(1 - \frac{x+a}{a}\right)$$

$$= y\left(-\frac{x}{a}\right)$$

$$= -\frac{x}{a}\sqrt{4ax}$$

$$= -\frac{2}{\sqrt{a}}x^{3/2}$$

or

$$\beta^2 a = 4x^3. \qquad \text{(ii)}$$

We have thus obtained the coordinates (α, β) of the centre of curvature at any point of the parabola.

To obtain the evolute of the parabola, i.e. the locus of (α, β), we eliminate x between (i) and (ii). From (i), we have

$$\alpha - 2a = 3x \quad \text{or} \quad x = \frac{1}{3}(\alpha - 2a).$$

Substituting this in (ii), we obtain:

$$\beta^2 a = \frac{4}{27}(\alpha - 2a)^3 \quad \text{or} \quad 4(\alpha - 2a)^3 = 27\beta^2 a.$$

Hence the locus of (α, β) is $4(x-2a)^3 = 27ay^2$, which is the equation of the evoulte of the parabola.

3.2.9 Asymptotes

The reader is already familiar with the curves, parabola and hyperbola. These are the curves which extend to infinity. The concept of asymptote arises with these curves when the point at which a tangent to the curve is drawn moves further and further away from the orgin. It is possible that the tangent may tend to a definite straight line, or it might be possible that it will oscillate, or still it might go further away from the origin. In the first case, it is called an *asymptote.* We then define an asymptote of a curve to be a straight line which touches the curve at infinity but does not lie altogether at infinity.

In determining the asympotes of a curve, one of the follwing definitions will be useful:

1. An asymptote of a curve of infinite extent is a line which is the limit of a tangent to the curve at a point which recedes further away along the curve.
2. An asymptote of a curve of infinite extent is a line which is the limit of a secant to the curve as the two points in which the secant intersects the curve recede indefinitely in the same direction along the curve.

The above definitions are mathematically equivalent.

Let

$$f(x, y) = 0 \tag{3.66}$$

be the equation of any algebraic curve of the nth degree and let

$$y = mx + c \tag{3.67}$$

be the equation of any straight line. The abscissae of the points of intersection of the curve and the straight line are given by

$$f(x, mx + c) = 0 \tag{3.68}$$

Equation (3.68) can be expressed in the following form:

$$Ax^n + Bx^{n-1} + Cx^{n-2} + \cdots + K = 0, \tag{3.69}$$

where $A, B, C, \ldots, K$ are functions of m and c. If the two points of intersection of (3.66) and (3.67) recede indefinitely along the curve, then their abscissae become infinite unless the line (3.67) is approaching a vertical asymptote of the curve (3.66). We can therfore obtain the horizontal and oblique asymptotes of (3.66) by applying the condition that the Eq. (3.69) possesses a pair of infinite roots. To obtain the pairs of infinite roots, we first write down the equation, whose roots are the reciprocals of the roots of Eq. (3.69). For this, we put $x = 1/z$ in (3.69) and obtain equation:

$$A + Bz + Cz^2 + \cdots + Kz^n = 0 \tag{3.70}$$

It is now clear that two roots of Eq. (3.70) will be zero if $A = B = 0$. In such a case, two roots of Eq. (3.69) will become infinite. Thus, we calculate m and c such that

$$A = 0 \quad \text{and} \quad B = 0, \tag{3.71}$$

and then $y = mx + c$ will be an asymptote of the given curve.

It can easily be ascertained that a curve of nth degree does not posses more than n asympotes. For the hyperbola, in particular, its equation differs from that of the asympotes only by a constant.

Parallel asymptotes. It may sometimes happen that one or more values of m, obtained by putting $A = 0$, will make $B = 0$ vanish identically. Therfore, any value of c will give a line intersecting the curve in two points at infinity. In such a case, c is so chosen as to make the coefficient c vanish. Then each value of m will give rise to two values of c. This is the case of parallel asympotes in which case the two lines cut the curve at three points at infinity.

The following examples illustrate the calculation of asymptotes.

Example 3.54 Find the asymptotes of the curve

$$y^3 - x^2y + 2y^2 + 4y + x = 0. \tag{i}$$

Putting $y = mx + c$ in (i), we obtain

$$(mx + c)^3 - x^2(mx + c) + 2\,(mx + c)^2 + 4\,(mx + c) + x = 0$$

i.e.

$$(m^3 - m)x^3 + (3m^2c - c + 2m^2)x^2 + (3mc^2 + 4mc + 4m + 1)x$$
$$+ c^3 + 2c^2 + 4c = 0. \tag{ii}$$

Equating to zero the coefficients of the two highest powers of x in (ii), we get:

$$m^3 - m = 0 \tag{iii}$$

and

$$3m^2c - c + 2m^2 = 0 \tag{iv}$$

From (iii), we obtain $m = 0$, $m = 1$ and $m = -1$. Putting these values successively in (iv), we get $c = 0$, $c = -1$ or $c = -1$. Hence the three asymptotes are given by $y = 0$, $y = x - 1$ and $y = -x - 1$.

Example 3.55 Find the asymptotes of the curve

$$x^3 + 2x^2y - xy^2 - 2y^3 + 4y^2 + 2xy + y - 1 = 0. \tag{i}$$

Putting $y = mx + c$ in (i), we obtain

$$x^3 + 2x^2(mx + c) - x(mx + c)^2 - 2(mx + c)^3 + 4(mx + c)^2$$
$$+ 2x\,(mx + c) + mx + c - 1 = 0.$$

i.e.

$$x^3(1 + 2m - m^2 - 2m^3) + x^2\,(2c - 2mc - 6m^2c + 4m^2 + 2m)$$
$$+ x(-c^2 - 6mc^2 + 8mc + 2c + m) + (-2c^3 + 4c^2 + c - 1) = 0. \tag{ii}$$

For asymptotes, we therefore have:

$$1 + 2m - m^2 - 2m^3 = 0 \tag{iii}$$

and

$$2c - 2mc - 6m^2c + 4m^2 + 2m = 0 \tag{iv}$$

From (iii), we obtain: $m=-1/2$, $m=-1$ and $m=1$. We then obtain from (iv): $c=0$, $c=1$ and $c=1$, respectively. Hence the three asymptotes are given by

$$y=-\frac{1}{2}x, \quad y=-x+1, \quad y=x+1.$$

Example 3.56 Find the asymptotes of the curve

$$4x^3-3xy^2-y^3+2x^2-xy-y^2-1=0 \tag{i}$$

Putting $y=mx+c$ in (i), we obtain:

$$4x^3-3x(mx+c)^2-(mx+c)^3+2x^2-x(mx+c)-(mx+c)^2-1=0$$

i.e.

$$x^3(4-3m^2-m^3)+x^2(-6mc-3m^2c+2-m-m^2)$$
$$+x(-3c^2-3mc^2-c-2mc)+(-c^3-c^2-1)=0.$$

We therefore have

$$4-3m^2-m^3=0 \tag{ii}$$

and

$$-6mc-3m^2c+2-m-m^2=0 \tag{iii}$$

From (ii), we obtain: $m=1$ and $m=-2$ (twice). Now, corresponding to $m=1$, Eq. (iii) gives: $c=0$, and corresponding to $m=-2$, the left-hand side of (iii) becomes identically zero. Hence there are parallel asymptotes and therefore we set

$$-3c^2-3mc^2-c-2mc=0. \tag{iv}$$

Putting $m=-2$ in (iv), we obtain $c=0$ or $c=-1$. Hence the three asymptotes are given by $y=x$, $y=-2x$ and $y=-2x-1$.

3.2.10 Envelopes

To explain the concept of the envelope of a curve, we consider the following curves:

$$x\cos\alpha+y\sin\alpha=p \tag{3.72}$$

$$y=mx+\frac{a}{m} \tag{3.73}$$

$$y=mx\pm\sqrt{a^2m^2+b^2} \tag{3.74}$$

Equations (3.72)–(3.74) have the following characteristics:

(a) Each of them contains a parameter and represents a family of straight lines.
(b) Each of these straight lines touches a curve.

Thus Eq. (3.72) represents a family of straight lines for different values of the parameter α and touches the circle $x^2 + y^2 = p^2$; Eq. (3.73) represents a family of straight lines for different values of the parameter m and touches the parabola $y^2 = 4ax$; and Eq. (3.74) represents a family of straight lines for different values of the parameter m and touches the ellipse $x^2/a^2 + y^2/b^2 = 1$. The parameter in each of the above cases is called the *parameter of the family* and the curves are called *one-parametric family of curves*. In general case,

$$f(x, y, \theta) = 0 \tag{3.75}$$

let a consecutive member of the family of curves be given by

$$f(x, y, \theta + \Delta\theta) = 0 \tag{3.76}$$

The points of intersection of the curves represented by (3.75) and (3.76) are given by

$$f(x, y, \theta + \Delta\theta) - f(x, y, \theta) = 0,$$

or

$$\frac{f(x, y, \theta + \Delta\theta) - f(x, y, \theta)}{\Delta\theta} = 0 \tag{3.77}$$

Proceeding to the limit as $\Delta\theta \to 0$, the left side of (3.77) gives the derivative of $f(x, y, \theta)$ with respect to θ, treating the other variables x and y as constants. Such derivatives are called *partial derivatives* and are denoted by separate symbols — to distinguish them from the ordinary derivatives. For example, the partial derivative of $f(x, y, \theta)$ with respect to θ is denoted by $\partial f/\partial q$. In this book, partial derivatives are discussed in detail in chapter 5. With this notation, Eq. (3.77) gives (on proceeding to the limit as $\Delta\theta \to 0$):

$$\frac{\partial f}{\partial \theta} = 0 \tag{3.78}$$

Eliminating the parameter θ from the equations (3.75) and (3.78), we get the *envelope* of the family defined by Eq. (3.75). It follows, therefore, that the envelope of a one-parameter family of curves defined by $f(x, y, \theta)$ is the locus of the limiting position of the points of intersection of any two consecutive curves of the family when one of them tends to coincide with the other which is fixed.

Similarly, if the family of curves is given by the polar equation $f(r, \theta, \alpha) = 0$, where α is the parameter, then the envelope of the family is obtained by eliminating α between the equations.

$$f(r, \theta, \alpha) = 0 \tag{3.79}$$

and

$$\frac{\partial f}{\partial \alpha} = 0 \tag{3.80}$$

Now, we consider Eq. (3.72). In this case, the parameter is α and differentiating with respect to α, we obtain

$$-x \sin \alpha + y \cos \alpha = 0, \tag{3.81}$$

on treating x, y, p as constants.

To eliminate α between (3.72) and (3.81), we square both of them and then add to get

$$x^2 + y^2 = p^2,$$

which is a circle and is the envelope of the family given by (3.72).

On the other hand, if the family of curves is given by the quadratic in α:

$$\alpha^2 f_1(x) + \alpha f_2(x) + f_3(x) = 0 \tag{3.82}$$

where α is the parameter and $f_1(x)$, $f_2(x)$ and $f_3(x)$ are functions of x, then we differentiate (3.82) partially with respect to α, and get

$$2\alpha f_1(x) + f_2(x) = 0, \tag{3.83}$$

which gives:

$$\alpha = -\frac{1}{2}\frac{f_2(x)}{f_1(x)}.$$

Substituting this value of α in (3.82), we obtain:

$$\frac{1}{4}\left[\frac{f_2(x)}{f_1(x)}\right]^2 f_1(x) - \frac{1}{2}\frac{f_2(x)}{f_1(x)} f_2(x) + f_3(x) = 0$$

or

$$-\frac{1}{4}\frac{[f_2(x)]^2}{f_1(x)} + f_3(x) = 0$$

or

$$[f_2(x)]^2 - 4 f_1(x) f_3(x) = 0, \tag{3.84}$$

which is the envelope of the family given by (3.82).

Example 3.57 Find the envelope of the family of curves defined by $y = mx + a/m$, where m is the parameter of the family.

Let

$$y = mx + \frac{a}{m} \tag{i}$$

Differentiating (i) partially with respect to m, we obtain:

$$x-\frac{a}{m^2}=0,$$

which gives

$$m=\pm\sqrt{\frac{a}{x}} \tag{ii}$$

Putting this value of m in (i), we obtain:

$$y=\pm x\sqrt{\frac{a}{x}}\pm\frac{a}{\sqrt{a/x}}=\pm\sqrt{ax}\pm\sqrt{ax}=\pm 2\sqrt{ax}$$

This gives $y^2=4ax$, which is a parabola and is the required envelope.

Alternatively, Eq. (i) can be written as a quadratic in m, viz.,

$$m^2x-my+a=0 \tag{iii}$$

Hence, using (3.84), we find that the envelope is $y^2=4ax$, as before.

Example 3.58 Find the envelope of the family of curves given by

$$y=mx\pm\sqrt{(a^2m^2+b^2)},$$

where m is the parameter.

The given equation can be written as

$$y-mx=\pm\sqrt{a^2m^2+b^2} \tag{i}$$

Squaring (i), we obtain,

$$(y-mx)^2=a^2m^2+b^2$$

or

$$y^2+m^2(x^2-a^2)-2myx-b^2=0, \tag{iii}$$

which is of the form of (3.82). Hence the envelope given by Eq. (3.84), is

$$4y^2x^2=4(x^2-a^2)(y^2-b^2)$$

or

$$\frac{x^2}{a^2}+\frac{y^2}{b^2}=1,$$

which is an ellipse.

Example 3.59 Find the envelope of the family of circles $(x-\alpha)^2+y^2=2\alpha$, where α is a parameter.

Let

$$(x-\alpha)^2 + y^2 = 2\alpha \qquad \text{(i)}$$

Differentiating (i) partially with respect to α, we obtain

$$-2(x-\alpha) - 2 = 0,$$

which gives $\alpha = x+1$. Substituting this value of α in (i), we get

$$1 + y^2 = 2(x+1) \quad \text{or} \quad y^2 = 2x+1,$$

which is the required envelope.

Alternatively, Eq. (i) may be written as the quadratic:

$$\alpha^2 - 2\alpha\,(x+1) + (x^2 + y^2) = 0,$$

which is of the form of Eq. (3.82). Hence, Eq. (3.84) gives the envelope:

$$4(x+1)^2 = 4\,(x^2 + y^2)$$

Therefore, $y^2 = 2x+1$, as before.

Example 3.60 Find the envelope of $(x/a) + (y/b) = 1$, where a and b are connected by the relation $a^2 + b^2 = p^2$.

In this case, we have two parameters a and b and the method of procedure is as follows:

Let

$$\frac{x}{a} + \frac{y}{b} = 1 \qquad \text{(i)}$$

and

$$a^2 + b^2 = p^2 \qquad \text{(ii)}$$

Let a and b be both functions of t. Then differentiating (i) and (ii) with respect to t, we obtain

$$-\frac{x}{a^2}\frac{da}{dt} - \frac{y}{b^2}\frac{db}{dt} = 0 \qquad \text{(iii)}$$

and

$$2a\frac{da}{dt} + 2b\frac{db}{dt} = 0 \qquad \text{(iv)}$$

From (iv), we obtain

$$\frac{da}{db} = -\frac{b}{a} \qquad \text{(v)}$$

and (iii) gives:

$$\frac{da}{db} = \frac{y/b^2}{x/a^2} \tag{vi}$$

From (v) and (vi), we get

$$\frac{y/b^2}{x/a^2} = \frac{b}{a}$$

or

$$\frac{x/a}{a^2} = \frac{y/b}{b^2} = \frac{(x/a)+(y/b)}{a^2+b^2}$$

$$= \frac{1}{a^2+b^2}, \quad \left(\text{since } \frac{x}{a}+\frac{y}{b}=1\right)$$

$$= \frac{1}{p^2}, \quad \text{[using (ii)]}$$

we get

$$a^3 = xp^2 \quad \text{and} \quad b^3 = yp^2.$$

From the above equations, we obtain

$$a = x^{1/3}p^{2/3} \quad \text{and} \quad b = y^{1/3}p^{2/3}.$$

Hence

$$a^2 + b^2 = x^{2/3}p^{4/3} + y^{2/3}p^{4/3}$$

or

$$p^2 = p^{4/3}(x^{2/3} + y^{2/3}), \quad (\text{since } a^2 + b^2 = p^2)$$

We then obtain the equation $x^{2/3} + y^{2/3} = p^{2/3}$, which is the required equation of the envelope.

Example 3.61 Find the envelope of the normals to the parabola $y^2 = 4ax$.

Let $P(at^2, 2at)$ be any point on the parabola

$$y^2 = 4ax \tag{i}$$

Equation of the normal at P is given by

$$y + xt = 2at + at^3 \tag{ii}$$

We require the envelope of the family given by (ii). Differentiating (ii) with respect to t, we obtain.

$$x = 2a + 3at^2,$$

which gives

$$t = \sqrt{\frac{x-2a}{3a}} \tag{iii}$$

Substituting this value of t in (ii), we obtain

$$y + x\sqrt{\frac{x-2a}{3a}} = 2a\sqrt{\frac{x-2a}{3a}} + a\left(\frac{x-2a}{3a}\right)^{3/2},$$

which gives on simplification:

$$y = \sqrt{\frac{x-2a}{3a}}\left(\frac{4a}{3} - \frac{2x}{3}\right)$$

Squaring both sides and rearranging the terms, we get

$$27ay^2 = 4(x-2a)^3,$$

which is the required envelope.

Example 3.62 Circles are described on the double ordinates of $y^2 = 4\,ax$ as diameters. Show that the envelope is $y^2 = 4a\,(x+a)$.

Let $P(at^2, 2at)$ be a point on the parabola $y^2 = 4ax$. Then $M\,(at^2, 0)$ is the midpoint of the double ordinate at P. Hence the equation of the circle with centre at M and radius $2at$ is given by

$$(x - at^2)^2 + y^2 = 4a^2t^2 \tag{i}$$

Differentiating (i) with respect to t, we get

$$2\,(x - at^2)\,(-2at) = 8a^2t,$$

Which gives on simplification

$$t^2 = \frac{x+2a}{a} \tag{ii}$$

Putting the value of t^2 from (ii) in (i), we obtain

$$\left(x - a\frac{x+2a}{a}\right)^2 + y^2 = 4a^2\,\frac{x+2a}{a},$$

which simplifies to $y^2 = 4a\,(x+a)$, the required envelope.

Example 3.63 If the envelope of the straight line $(x/a) + (y/b) = 1$ is $x^r y^s = k^{r+s}$, show that

$$r^r s^s a^r b^s = k^{r+s}(r+s)^{r+s}.$$

Let

$$\frac{x}{a}+\frac{y}{b}=1 \tag{i}$$

and

$$r^r y^s = k^{r+s} \tag{ii}$$

From (i) we obtain

$$\frac{dy}{dx}=-\frac{b}{a} \tag{iii}$$

and from (ii), we get

$$\frac{dy}{dx}=-\frac{r/x}{s/y} \tag{iv}$$

From (iii) and (iv), we obtain

$$\frac{b}{a}=-\frac{r/x}{s/y} \quad \text{or} \quad a\frac{r}{x}=b\frac{s}{y}=\frac{1}{\lambda},$$

from which, we get

$$\frac{x}{a}=\lambda r \quad \text{and} \quad \frac{y}{b}=\lambda s.$$

Then (i) gives:

$$\lambda(r+s)=1 \quad \text{or} \quad \lambda=\frac{1}{r+s}.$$

Hence

$$\frac{x}{a}=\frac{r}{r+s} \quad \text{and} \quad \frac{y}{b}=\frac{s}{r+s}.$$

or

$$x=\frac{ar}{r+s} \quad \text{and} \quad y=\frac{bs}{r+s}.$$

Then $x^r y^s = k^{r+s}$ becomes

$$\left(\frac{ar}{r+s}\right)^r\left(\frac{bs}{r+s}\right)^s = k^{r+s}$$

which simplifies to

$$a^r b^s r^r s^s = k^{r+s}(r+s)^{r+s},$$

the required expression.

3.2.11 Evolutes and Involutes

The notion of the evolute of a curve has already been introduced (see Section 3.2.8) as the *locus* of the centre of curvature of a curve. It is obtained by eliminating x and y between the equation of the curve and the formulae for the centre of curvature.

Let the equation of the curve be given by

$$y = f(x) \tag{3.85}$$

and let $P(x, y)$ be a point on the curve. If (α, β) is the centre of curvature of the curve, then we have [see Eqs. (3.65a) and (3.65b)]:

$$\alpha = x - y_1 \frac{1+y_1^2}{y_2} \quad \text{and} \quad \beta = y + \frac{1+y_1^2}{y_2} \tag{3.86}$$

where $y_1 = dy/dx$ and $y_2 = d^2y/dx^2$. The evolute of (3.85) is then obtained by eliminating x and y from Eqs. (3.85) and (3.86).

Let

$$y = F(x) \tag{3.87}$$

be the evolute of (3.85). Then the curve which has $y = F(x)$ as its evolute is called an *involute*. It should be noted that the evolute is unique but the involutes are infinite in number. We give below two important properties of the evolute.

(a) The normals to a given curve are tangents to its evolute at its centre of curvature.

We consider again curve given by (3.85), viz., $y = f(x)$ and let $P(x, y)$ be a point on it and $Q(\alpha, \beta)$, the centre of curvature for the point P. Then from Eqs. (3.63) and (3.64), we have

$$\alpha = x - \rho \sin \psi \tag{3.88}$$

and

$$\beta = y + \rho \cos \psi, \tag{3.89}$$

where

$$\tan \psi = \frac{dy}{dx} \quad \text{and} \quad \rho = \frac{(1+y_1^2)^{3/2}}{y_2}$$

Differentiating (3.88) with respect to x, we obtain

$$\frac{d\alpha}{dx} = 1 - \rho \cos \psi \frac{d\psi}{dx} - \sin \psi \frac{d\rho}{dx}$$

$$= 1 - \frac{ds}{d\psi} \cos \psi \frac{d\psi}{dx} - \sin \psi \frac{d\rho}{dx}, \qquad \left(\text{since } \rho = \frac{ds}{d\psi}\right)$$

$$= 1 - \frac{ds}{d\psi}\frac{dx}{ds}\frac{d\psi}{dx} - \sin\psi\frac{d\rho}{dx}, \qquad \left(\text{using } \cos\psi = \frac{dx}{ds}\right)$$

$$= -\sin\psi\frac{d\rho}{dx} \tag{3.90}$$

Again, differentiating (3.89) with respect to x, we obtain

$$\frac{d\beta}{dx} = \frac{dy}{dx} + \rho(-\sin\psi)\frac{d\psi}{dx} + \cos\psi\frac{d\rho}{dx}$$

$$= \frac{dy}{dx} - \frac{ds}{d\psi}\frac{dy}{ds}\frac{d\psi}{dx} + \cos\psi\frac{d\rho}{dx}$$

$$= \cos\psi\frac{d\rho}{dx} \tag{3.91}$$

From (3.90) and (3.91), we obtain

$$\frac{d\alpha}{d\beta} = -\tan\psi \quad \text{or} \quad \frac{d\beta}{d\alpha} = -\cot\psi = -\frac{1}{dy/dx} \tag{3.92}$$

But $d\beta/d\alpha$ is the slope of the tangent at Q to the evolute. Hence it follows from (3.92) that the tangent at Q to the evolute and the normal at P to the given curve are the same. We have therefore proved that the normal at P to the given curve touches its evolute at the centre of curvature.

(b) The difference between the radii of curvatures at two points of a curve is equal to the length of the arc of the evolute between the two corresponding points.

Let P, Q be two points on the given curve and let L, M be the corresponding points on the evolute. Let ρ_1, ρ_2 be the radii of curvature at P, Q, respectively. Now, we square and add the Eqs. (3.90) and (3.91) and obtain:

$$\left(\frac{d\alpha}{dx}\right)^2 + \left(\frac{d\beta}{dx}\right)^2 = \left(\frac{d\rho}{dx}\right)^2 \tag{3.93}$$

Let S be the length of the arc of the evolute from a fixed point on it up to (α, β). Then from (3.93), we have

$$\left(\frac{ds}{dx}\right)^2 = \left(\frac{d\alpha}{dx}\right)^2 + \left(\frac{d\beta}{dx}\right)^2 = \left(\frac{d\rho}{dx}\right)^2 \tag{3.94}$$

Hence,

$$\frac{ds}{dx} = \pm\frac{d\rho}{dx} \quad \text{or} \quad S = \rho + C, \tag{3.95}$$

Now, let S_1 and S_2 be the values of S at L and M, respectively. Then (3.95) gives:

$$S_1 = \rho_1 + C \quad \text{and} \quad S_2 = \rho_2 + C,$$

so that

$$S_1 - S_2 = \rho_1 - \rho_2, \tag{3.96}$$

which is the required result.

Example 3.64 Find the equation of the evolute of the parabola, $y^2 = 4ax$.

Let $P(x, y)$ be a point on the given parabola. Then

$$y_1 = \frac{dy}{dx} = \frac{2a}{y} = \frac{2a}{\sqrt{4ax}} = \sqrt{\frac{a}{x}},$$

and,

$$y_2 = -\frac{2a}{y^2}\frac{dy}{dx} = -\frac{4a^2}{y^3} = -\frac{4a^2}{(4ax)^{3/2}} = \frac{-1}{2}\frac{\sqrt{a}}{x\sqrt{x}}$$

Hence,

$$\alpha = x - \frac{y_1(1+y_1^2)}{y_2} = x - \frac{\sqrt{a/x}\,(1+a/x)}{-(1/2)\,(\sqrt{a}/x^{3/2})}$$

$$= x + 2x\left(1+\frac{a}{x}\right)$$

$$= 3x + 2a. \tag{i}$$

and

$$\beta = y + \frac{1+y_1^2}{y_2} = y + \frac{1+a/x}{-(1/2)\,(\sqrt{a}/x^{3/2})}$$

$$= 2\sqrt{ax} - \frac{2x^{3/2}}{\sqrt{a}}\left(1+\frac{a}{x}\right)$$

$$= -\frac{2x^{3/2}}{\sqrt{a}}$$

$$= -\frac{2}{\sqrt{a}}\left(\frac{\alpha-2a}{3}\right)^{3/2}, \qquad \text{[using (i)]}$$

Squaring, we get

$$\beta^2 = \frac{4}{a}\left(\frac{\alpha - 2a}{3}\right)^3 \quad \text{or} \quad 27\,a\beta^2 = 4\,(\alpha - 2a)^3,$$

which gives the required equation of the evolute as

$$27ay^2 = 4\,(x - 2a)^3.$$

Alternative method. Let $P\,(at^2, 2\,at)$ be any point on the parabola $y^2 = 4ax$. Then

$$y_1 = \frac{2a}{y} = \frac{2a}{2\,at} = \frac{1}{t},$$

and

$$y_2 = -\frac{2a}{y^2}\frac{dy}{dx} = -\frac{2a}{4ax}\cdot\frac{1}{t} = -\frac{1}{2\,at^3}$$

Hence

$$\alpha = x - \frac{y_1(1 + y_1^2)}{y_2} = at^2 - \frac{(1/t)\,(1 + 1/t^2)}{-1/(2at^3)}$$

$$= 3at^2 + 2a \qquad \text{(i)}$$

and

$$\beta = y + \frac{1 + y_1^2}{y_2} = 2at + \frac{1 + 1/t^2}{-1/(2at^3)}$$

$$= 2at - 2at^3\left(1 + \frac{1}{t^2}\right)$$

$$= -2at^3 \qquad \text{(ii)}$$

Eliminating t between (i) and (ii), we obtain

$$\beta^2 = 4a^2t^6 = 4a^2\left(\frac{\alpha - 2a}{3a}\right)^3,$$

which gives: $27\,a\beta^2 = 4(\alpha - 2a)^3$, as before. This gives the required equation of the evolute as $27ay^2 = 4(x - 2a)^3$.

Example 3.65 Show that $(x + y)^{2/3} - (x - y)^{2/3} = (4c)^{2/3}$, is the evolute of the hyperbola $xy = c^2$.

Let $(ct, c/t)$ be any point on the hyperbola $xy = c^2$. Then,

$$y_1 = \frac{dy}{dx} = -\frac{c^2}{x^2} = -\frac{c^2}{c^2t^2} = -\frac{1}{t^2},$$

and

$$y_2 = \frac{d^2y}{dx^2} = \frac{2c^2}{x^3} = \frac{2c^2}{c^3t^3} = \frac{2}{ct^3}.$$

Hence

$$\alpha = x + \frac{(1/t^2)(1+1/t^4)}{2/(ct^3)}$$

$$= ct + \frac{1}{2t^2}ct^3\left(1+\frac{1}{t^4}\right)$$

$$= ct + \frac{ct}{2} + \frac{c}{2t^3}$$

$$= \frac{c}{2t^3}(1+3t^4) \tag{i}$$

and

$$\beta = y + \frac{1+1/t^4}{2/ct^3}$$

$$= \frac{c}{t} + \frac{ct^3}{2}\left(1+\frac{1}{t^4}\right)$$

$$= \frac{c}{t} + \frac{c}{2t} + \frac{ct^3}{2}$$

$$= \frac{3c}{2t} + \frac{ct^3}{2}$$

$$= \frac{c}{2t^3}(t^6 + 3t^2) \tag{ii}$$

From (i) and (ii), we obtain

$$\alpha + \beta = \frac{c}{2t^3}(1+3t^4+3t^2+t^6) = \frac{c}{2t^3}(1+t^2)^3 \tag{iii}$$

and

$$\alpha - \beta = \frac{c}{2t^3}(1+3t^4-3t^2-t^6) = \frac{c}{2t^3}(1-t^2)^3 \tag{iv}$$

Hence, we have:

$$(\alpha+\beta)^{2/3}+(\alpha-\beta)^{2/3}=\left(\frac{C}{2t^3}\right)^{2/3}\left[(1+t^2)^2-(1-t^2)^2\right]=(4c)^{2/3}$$

Hence the locus of (α, β) is

$$(x+y)^{2/3}+(x-y)^{2/3}=(4c)^{2/3}$$

which is the required equation of the evolute.

Example 3.66 Show that the evolute of the cycloid $x=a(\theta+\sin\theta)$, $y=a(1-\cos\theta)$ is given by $x=a(\theta-\sin\theta)$ and $y-2a=a(1+\cos\theta)$.

We have

$$y_1=\frac{dy}{dx}=\frac{a\sin\theta}{a(1+\cos\theta)}=\tan\frac{\theta}{2},$$

and

$$y_2=\frac{d^2y}{dx^2}=\frac{1}{2}\sec^2\frac{\theta}{2}\,\frac{1}{a(1+\cos\theta)}=\frac{1}{4a\cos^4(\theta/2)}$$

Hence

$$\alpha=x-\frac{\tan(\theta/2)\,[1+\tan^2(\theta/2)]}{1/[4a\cos^4(\theta/2)]}$$

$$=a(\theta+\sin\theta)-4a\cos^4\frac{\theta}{2}\tan\frac{\theta}{2}\sec^2\frac{\theta}{2}$$

$$=a(\theta+\sin\theta)-2a\sin\theta$$

$$=a(\theta-\sin\theta),$$

and

$$\beta=y+\frac{1+\tan^2(\theta/2)}{1/[4a\cos^4(\theta/2)]}=a(1-\cos\theta)+4a\cos^4\frac{\theta}{2}\sec^2\frac{\theta}{2}$$

$$=a(1-\cos\theta)+4a\cos^2\frac{\theta}{2}$$

or

$$\beta-2a=a(1-\cos\theta)+2a\cos\theta=a(1+\cos\theta).$$

Hence the evolute is given by

$$x=a(\theta-\sin\theta)\quad\text{and}\quad y-2a=a(1+\cos\theta).$$

Example 3.67 Find the evolute of the astroid $x^{2/3} + y^{2/3} = a^{2/3}$.

We have,

$$y_1 = \frac{dy}{dx} = -\frac{y^{1/3}}{x^{1/3}},$$

and

$$y_2 = \frac{d^2 y}{dx^2} = -\frac{(1/3)y^{-2/3}\, y_1{}^{1/3} - y^{1/3}\,(1/3)\, x^{-2/3}}{x^{2/3}}$$

$$= -\frac{1}{3x^{2/3}}\left(-\frac{x^{1/3}\, y^{1/3}/x^{1/3}}{y^{2/3}} - \frac{y^{1/3}}{x^{2/3}}\right)$$

$$= \frac{1}{3x^{2/3}}\left(\frac{1}{y^{1/3}} + \frac{y^{1/3}}{x^{2/3}}\right)$$

$$= \frac{1}{3x^{4/3}\, y^{1/3}}\,(x^{2/3} + y^{2/3})$$

$$= \frac{1}{3x^{4/3}\, y^{1/3}}\, a^{2/3}, \qquad (\text{since } x^{2/3} + y^{2/3} = a^{2/3})$$

Hence

$$\alpha = x - \frac{-(y^{1/3}/x^{1/3})\,(1 + y^{2/3}/x^{2/3})}{a^{2/3}/(3x^{4/3}\, y^{1/3})}$$

$$= x + \frac{3x^{4/3}\, y^{1/3}\, y^{1/3}}{x^{1/3}\, a^{2/3}}\, \frac{a^{2/3}}{x^{2/3}}$$

$$= x + 3x^{1/3}\, y^{2/3}.$$

and

$$\beta = y + \frac{1 + y^{2/3}/x^{2/3}}{a^{2/3}/(3x^{4/3}\, y^{1/3})} = y + \frac{3x^{4/3} y^{1/3} a^{2/3}}{a^{2/3} x^{2/3}} = y + 3x^{2/3} y^{1/3}. \tag{ii}$$

Then

$$\alpha + \beta = x + y + 3x^{1/3}\, y^{2/3} + 3x^{2/3}\, y^{1/3} = (x^{1/3} + y^{1/3})^3$$

and hence

$$(\alpha + \beta)^{2/3} = (x^{1/3} + y^{1/3})^2. \tag{iii}$$

Similarly, we obtain

$$(\alpha - \beta)^{2/3} = (x^{1/3} - y^{1/3})^2 \tag{iv}$$

Adding (iii) and (iv), we finally get

$$(\alpha+\beta)^{2/3}+(\alpha-\beta)^{2/3}=2\,(x^{2/3}+y^{2/3})=2a^{2/3}.$$

Hence the evolute of the astroid is given by

$$(x+y)^{2/3}+(x-y)^{2/3}=2a^{2/3}.$$

EXERCISES 3.2

Find the equations of the tangents and normals at the point (x_1, y_1) on each of the following curves (Problems 1–4):

1. $y^2=4ax$. **2.** $y=c\cosh\dfrac{x}{c}$.

3. $xy=a^2$. **4.** $\dfrac{x^2}{a^2}+\dfrac{y^2}{b^2}=1$.

5. Show that the curves given by $(x/a)^n+(y/b)^n=1$ for different values of n, have a common tangent at the point (a, b).
Find the equation of the common tangent.

6. Find the equation of the tangent at 't' on the curve $x=a\sin^3 t$, $y=b\cos^3 t$.

7. Find the equation of the tangent at the point 'θ' of the curve $x=a(\theta+\sin\theta)$ and $y=a(1-\cos\theta)$.

8. Prove that the condition that the straight line $x\cos\alpha+y\sin\alpha=p$ is a tangent to the curve $x^m y^n=a^{m+n}$ is

$$p^{m+n}m^m n^n=(m+n)^{m+n}a^{m+n}\sin^n\alpha\cos^m\alpha.$$

9. If x_1, y_1 be the part of the axes of x and y intercepted by the tangent at any point on the curve

$$\left(\frac{x}{a}\right)^{2/3}+\left(\frac{y}{b}\right)^{2/3}=1,$$

show that

$$\frac{x_1^2}{a^2}+\frac{y_1^2}{b^2}=1.$$

10. Write down the equation of the tangent to the curve $y(x^2+a^2)=ax^2$ at the point where $y=a/4$.

11. Prove that $(x/a)+(y/b)=1$ touches the curve $y=be^{-x/a}$ at the point where the curve crosses the axis of y.

12. If the tangent at (x_1, y_1) to the curve $x^3+y^3=a^3$ meets the curve again in (x_2, y_2), show that

$$\frac{x_2}{x_1}+\frac{y_2}{y_1}=1.$$

13. If $px+qy=1$ touches the curve $(ax)^n+(by)^n=1$. show that

$$\left(\frac{P}{a}\right)^{n(n-1)} + \left(\frac{q}{b}\right)^{n(n-1)} = 1,$$

14. If $x\cos\alpha + y\sin\alpha = p$ touches the curve

$$\left(\frac{x}{a}\right)^{n(n-1)} + \left(\frac{y}{b}\right)^{n(n-1)} = 1,$$

prove that $p^n = (a\cos\alpha)^n + (b\sin\alpha)^n$.

15. Find the condition that $x\cos\alpha + y\sin\alpha = p$ is normal to the curve

$$\frac{x^2}{a^2} + \frac{y^2}{b^2} = 1.$$

16. Find where the normal to $\sqrt{xy} = a + x$ makes equal intercepts on the axes.

17. Find the equations of the tangent and normal to the curve $y(x-2)(x-3) - x + 7 = 0$ at the point where it cuts the axis of x.

18. If the normal to the curve $x^{2/3} + y^{2/3} = a^{2/3}$ makes an angle θ with the x-axis, show that its equation is

$$y\cos\theta - x\sin\theta = a\cos 2\theta.$$

19. Show that the normal at any point of the curve

$$x = a\cos\theta + a\theta\sin\theta, \qquad y = a\sin\theta - a\theta\cos\theta,$$

is at a constant distance from the origin.

20. Show that in the curve $y = be^{-a/x}$, the subtangent varies as the square of the abscissa.

21. Show that in the curve $by^2 = (a+x)^3$, the square of the subtangent varies as the subnormal.

22. Find the lengths of the subtangent, the subnormal, the tangent and the normal to the curve $y = 2/(1+x^2)$ at the point $x = 1$.

23. For any curve, prove that

(Length of tangent)^2 · (Subnormal) = (Lenght of normal)^2 · (Subtangent).

24. Show that in the curve $y = a\log(x^2 - a^2)$, the sum of the tangent and subtangent varies as the product of the coordinates of the point.

25. In the curve $x^{m+n} = a^{m+n}y^{2n}$, prove that the mth power of the subtangent varies as the nth power of the subnormal.

Find the angle of intersection of the following pairs of curves (Problems 26–31):

26. $y = x^2/2$ and $y = 4 - x^2/2$.

27. $y = \sin x$ and the axis of x (positive side).

28. $x^2 - y^2 = 2a^2$ and $x^2 + y^2 = 4a^2$.

29. $y = x^3$ and $6y = 7 - x^2$.

30. $y^2 = ax$ and $x^3 + y^3 = 3axy$.
31. $y^2 = 4ax$ and $x^2 = 4by$.
32. For the curve defined by $x = a(\theta + \sin\theta)$ *and* $y = a(1 - \cos\theta)$ show that

$$\frac{ds}{dy} = \sqrt{\frac{2a}{y}} \quad \text{and} \quad \frac{ds}{d\theta} = 2a\cos\frac{\theta}{2}.$$

33. Find the angle of intersection of the curves

$$r = a(1 + \cos\theta) \quad \text{and} \quad r = b(1 - \cos\theta).$$

34. Find the condition that the curves $ax^2 + by^2 = 1$, $a_1x^2 + b_1y^2 = 1$ shall cut orthogonally.
35. Find ds/dx for the curves (i) $y^2 = 4ax$ (ii) $x^{2/3} + y^{2/3} = a^{2/3}$.
36. For the curve $x = a(\theta + \sin\theta)$, $y = a(1 - \cos\theta)$ prove that (i) $\psi = \theta/2$ (ii) $ds/d\theta = 2a\cos(\theta/2)$.
37. In the curve $y = a\log\sec(x/a)$ find ds/dx and prove that $x = a\psi$.
38. Find the pedal equations of the following curves: (i) $y^2 = 4a(x + a)$ (ii) $r^2 = a^2\cos 2\theta$.
39. For the curve given by $c^2(x^2 + y^2) = x^2y^2$, show that the pedal equation is given by

$$\frac{1}{p^3} + \frac{3}{r^2} + \frac{1}{c^2}.$$

40. Show that the pedal equation of the curve $r^m = a^m\cos m\theta$ is given by $pa^m = r^{m+1}$.
41. A body moves along a straight line OX according to the law

$$s = \frac{t^3}{3} - 2t^2 + 3t.$$

Determine the velocity and acceleration of its motion.

42. A sphercial balloon is being filled with air. Find the rate of change of its volume with respect to its radius. Find the rate of change of volume when the radius is 5 cm.
43. A body is projected vertically upwards with an intial velocity of Vm/s. At what height x will it be in t sec? Determine the velocity and acceleration of its motion. When will the body reach the highest point, and what distance from the earth's surface will this point be?
44. The volume V of a certain gas varies with pressure p according to the equation $p = 150/V$. Find the rate of change of p with respect to V when $V = 5$.
45. The volume V of a sphere is $V = 1/6\pi d^3$, where d is the diameter. Find the rate of change of V with respect to d when $d = 25$ cm.
46. The electric potential V at a distance r from a small object having a

certain electric charge is given by $V = 100/r$. Find the rate of change of V with respect to r.

47. A man 180 cm tall walks directly away from a lamplight 600 cm high so that his distance from the foot of the light at time t sec is 150 cm. (a) Find the rate at which the length of his shadow is increasing (b) Find the rate at which the tip of his shadow is moving.

48. A sphercial balloon is being filled with helium so that its volume increases at a constant rate of 8 cubic ft per minute. How fast is the radius increasing when the radius is 3 ft?

49. Water is withdrawn from a conical reservoir 2.5 m in diameter and 3 m deep (vertex down) at the constant rate of $0.15\,m^3/\text{min}$. How fast is the water level falling when the depth of the water in the reservoir is 2 m?

50. For a thin lens, the equation

$$\frac{1}{u}+\frac{1}{v}=\frac{1}{f},$$

where u, v and f respectively denote the object distance, image distance and the focal lenght, holds. An object is moving towards a lens with a constant focal lenght of 5 cm. At what rate is the image distance changing at the instant the object is 15 cm from the lens?

51. The total resistance R of a circuit containing two resistors, R_1 and R_2 in parallel is given by

$$r = \frac{1}{(1/R_1)+(1/R_2)}$$

If R_I is increasing at the rare of 6 ohm/min and R_2 is decreasing at the rate of 3 ohm min, what is the rate of change of the circuit resistance when $R_1 = R_2 = 10$ ohm?

52. Sand is falling into a conical pile at the rate of 150 cc/min. It is observed that the height of the pile is always twice the radius of the base. How fast is the height increasing when the pile is 100-cm high?

53. Explain why Rolle's theorem is inapplicable to the function $y = 1-(x-1)^{2/3}$ in $0 \le x \le 2$. What is the curve represented by this function?

54. Apply Rolle's theorem to the function $f(x) = x^3 - 12x$ in $0 \le x \le 2\sqrt{3}$.

55. Verify whether Rolle's theorem is applicable to the function $f(x) = 1 - x^{2/3}$ on the closed interval $[-1, 1]$.

56. Verify that $f(x) = x^2 + x + 4$ satisfies the conditions of Rolle's theorem on the interval $[-3, 2]$, and find the number c such that $f'(c) = 0$.

57. What point on the parabola $y = x^2 + x + 1$ satisfies the mean-value theorem in the range $-1 \le x \le 3$.

58. Use the mean-value theorem to show that

$$1-\frac{a}{b}<\log\frac{b}{a}<\frac{b}{a}-1.$$

Hence obtain $1/2<\log 2<1$.

59. Using the mean-value theorem, deduce the inequality

$$\left(\frac{101}{100}\right)^{100}<e<\left(\frac{100}{99}\right)^{100}$$

60. If x is positive, show that

$$x>\log(1+x)>x-\frac{1}{2}x^2.$$

61. Verify whether the mean-value theorem is applicalbe to the functions $4/x$ and $1-x^{2/3}$ on the interval $[-1, 2]$.

62. Derive the mean-value theorem by applying Rolle's theorem to the function $\phi(x)$ defined by

$$\phi(x)=\begin{vmatrix} x & f(x) & 1 \\ b & f(b) & 1 \\ a & f(a) & 1 \end{vmatrix}.$$

where $f(x)$ is continuous on the closed interval $[a, b]$ and has a derivative in the open interval (a, b).

63. If $f(x)=x^4+x^3-5x-7$, show that there is a point ξ in the interval $[0, 1]$ such that $f'(\xi)=0$.

64. If $f(x)=4x^3+3x^2+6x-27$, use the mean-value theorem to show that the equation $f(x)=0$ has a solution in the interval $(\pi/6, \pi/2)$.

65. Using the mean-value theorem, show that

$$3\sin 3x+2\sin 2x+\sin x=\frac{8}{\pi}$$

for at least one x in the interval $0<x<\pi/2$.

Evaluate the following limits (Problems 66–95):

66. $\lim\limits_{x\to 0}\dfrac{a^x-1}{x}$.

67. $\lim\limits_{x\to 0}\dfrac{\log(1-x^2)}{\log\cos x}$.

68. $\lim\limits_{x\to 0}\dfrac{\log(1+kx^2)}{1-\cos x}$.

69. $\lim\limits_{x\to\infty}\left(2^x\tan\dfrac{a}{2^x}\right)$.

70. $\lim\limits_{x\to 0}\dfrac{xe^x-\log(1+x)}{x^2}$.

71. $\lim\limits_{x\to 0}\left(\dfrac{1}{x}-\cot x\right)$.

72. $\lim\limits_{x\to 0}\left(\dfrac{\tan x}{x}\right)^{1/x}$.

73. $\lim\limits_{x\to 0}\left(\dfrac{\sin x}{x}\right)1/x^2$.

74. $\lim_{x\to 0} (\sin x)^{\tan x}$.

75. $\lim_{x\to 0} \frac{(1+x)^{1/x} - e}{x}$.

76. $\lim_{x\to 0} \frac{\tan x - x}{x^2 \tan x}$.

77. $\lim_{x\to 0} \frac{x\cos x - \sin x}{x^3}$.

78. $\lim_{x\to 0} \frac{x\sin x - \sin^2 x}{x^4}$.

79. $\lim_{x\to 0} \frac{e^x - e^{\sin x}}{x - \sin x}$.

80. $\lim_{x\to b} \frac{x^b - b^x}{x^x - b^b}$.

81. $\lim_{x\to \infty} \frac{\log n}{n}$.

82. $\lim_{x\to 0} \frac{\log x^2}{\log \cot^2 x}$.

83. $\lim_{x\to 0} \log_x \tan x$.

84. $\lim_{x\to 1} x^{1/(x-1)}$.

85. $\lim_{x\to \pi} (\pi - x) \tan \frac{x}{2}$.

86. $\lim_{x\to 0} x \log_e x$.

87. $\lim_{x\to \infty} x^n e^{-x}$.

88. $\lim_{x\to \infty} \left(1+\frac{3}{x}\right)^x$.

89. $\lim_{x\to 0} \frac{x - \tan^{-1} x}{x^3}$.

90. $\lim_{x\to a} \left(2-\frac{x}{a}\right) \tan\left(\frac{\pi x}{2a}\right)$.

91. $\lim_{x\to 0} (\cos \alpha x)^{1/x}$.

92. $\lim_{x\to \pi/2} (\sin x)^{\sec^2 x}$.

93. $\lim_{x\to 0} \left(\frac{a^x + b^x}{2}\right)^{1/x}$

94. $\lim_{x\to \pi/2} (\sec x)^{\cot x}$.

95. $\lim_{x\to 0} \left(\frac{1}{x}\right)^{\tan x}$

96. What is the maximum value of $x^2 y^3$, when $x + y = 10\ (x > 0, y > 0)$?

97. Find the maximum value of $x/(1 + x \tan x)$..

Find the maximum and minimum values of the following functions (Problems 98–104):

98. $2x^3 - 21x^2 + 36x - 20$.

99. $x^5 - 5x^4 + 5x^3 - 10$.

100. $\sin x\,(1 + \cos x)$.

101. $\sin^p x \cos^q x$.

102. $(\log x)/x$.

103. $a \operatorname{cosec}^2 x + b^2 \sec^2 x$.

104. $(\sin x)^{\sin x}$.

105. Show that the right circular cylinder of given surface and maximum volume is such that its height is equal to the diameter of the base.

106. Show that the volume of the greatest cylinder which can be inscribed in a cone of height h and semivertical angle α is $(4/27)\pi h^3 \tan^2 \alpha$.

107. A rectangle of the greatest area is inscribed in a triangle, whose base is a and height h. Find the area of the rectangle.

108. The electric current i in a circuit varies according to the equation

$$i = 2t + \frac{200}{t}$$

where i is in amperes and $t > 0$ is in second. Determine the minimum current.

109. A sector is cut out of a circular sheet of paper and the two straight edges of the remainder are folded up so as to form a cone. Find the angle of the sector removed so that the volume of the cone may be a maximum.

110. A tent is in the form of a circular cylindrical base surmounted by a concial portion of semivertical angle α . For a fixed base and a given volume, show that the amount of canvas used will be a minimum if $\alpha = \cos^{-1}(2/3)$.

111. In driving a motor boat, the petrol burnt per hour varies directly as the cube of its velocity. Find the most economical trip of the boat when going against a current of 2-metres per hour.

112. Two sources of light are situated 30-m apart. On the straight line connecting these sources find the least illuminated point if the ratio of candle powers of the light sources is 27:8.

113. The strength of a beam varies as the product of its breadth and the square of its depth. Find the dimensions of the strongest beam which can be cut from a circular log of radius C.

Find the range of the values of x in which the curves defined below (Problems 114–117) are concave upwards or concave downwards:

114. $y = \dfrac{x}{x+1}$.

115. $x^2 - xy + 4 = 0$.

116. $y = x^2 - 4x + 3$.

117. $y = 3x^5 - 40x^3 + 3x - 20$.

Find the points of inflexion of the following curves (Problems 118–125):

118. $y = \dfrac{x^3}{6} - x^2$.

119. $y = \dfrac{2x}{1+x^2}$.

120. $y = x^x$.

121. $y = x^{1/x}$.

122. $y = x^3 - 3x^2 + 2$.

123. $y = \dfrac{x}{x^2 + 2x + 2}$.

124. $y = \dfrac{(x+a)(x+b)}{(x-a)(x-b)}$ $(a > 0, b > 0)$.

125. $y = 2\sin^2 x - 3(1 + \cos x)^2$.

126. Show that

$$\sin\beta = \sin\alpha + (\beta-\alpha)\cos\alpha - \frac{(\beta-\alpha)^2}{2!}\sin\alpha - \frac{(\beta-\alpha)^2}{3!}\cos\xi,$$

where $\alpha < \xi < \beta$. Use this result to evaluate sin 51° given that sin 45° = $1/\sqrt{2}$ and show that your answer is accurate to three decimal places.

127. Expand $\sin(\pi/4+\theta)$ in a series of powers of θ.

128. Expand $\sin x$ in powers of $(x-\pi/2)$.

Using Maclaurin's theorem, expand the following functions (Problems 129–136):

129. $\tan x$.

130. $e^x \cos x$.

131. $e^{\sin x}$.

132. $\log \dfrac{\sin x}{x}$.

133. $\log \sec x$.

134. $e^x \sec x$.

135. $\cos x \cosh x$.

136. $\sin x \sinh x$.

137. Expand (i) $x/(e^x-1)$ (ii) $\log_e(1+\sin x)$ in ascending powers of x up to the term involving x^4.

138. Show that

(i) $\tan^{-1}(1+x) = \dfrac{\pi}{4} + \dfrac{1}{2}x - \dfrac{1}{4}x^2 + \dfrac{1}{12}x^3 - \cdots$

(ii) $\tan\left(\dfrac{\pi}{4}+x\right) = 1 + 2x + 2x^2 + \dfrac{8}{3}x^3 + \dfrac{10}{3}x^4 + \cdots$

139. Expand the function $f(x) = x^4 - 4x^2$ in a series of powers of $(x+2)$.

140. Expand the function $f(x) = \sin(\pi x/3)$ in a series of powers of $(x-1)$.

Find the radius if curvature at the point (x, y) of the following curves (Problems 141–146):

141. $y^2 = 4ax$.

142. $xy = a^2$.

143. $y = c \cosh \dfrac{x}{c}$.

144. $x^{2/3} + y^{2/3} = a^{2/3}$.

145. $y = a \log \sec \dfrac{x}{a}$.

146. $ay^2 = x^3$.

147. Show that the radius of curvature at the end of the major axis of the ellipse $x^2/a^2 + y^2/b^2 = 1$, is equal to the semi-latus rectum of the ellipse.

Find the radius of curvature at the point (r, θ) of the following curves (Problems 148–153):

148. $r = a\theta$.

149. $r = a \sin\theta$.

150. $r = a\cos\theta$.

151. $r^2 = a^2 \cos 2\theta$.

152. $r = a/(1+\cos\theta)$.

153. $r = a(1-\cos\theta)$.

Find the radius of curvature at the origin of the following curves (Problems 154–157):

154. $y = -x^3 + 3x^2 + x.$ **155.** $(y-x)(y-2x) = x^3 + y^3.$

156. $x^3 + y^3 - 2x^2 + 6y = 0.$ **157.** $r = a \sin n\theta.$

158. If ρ_1 and ρ_2 are the radii of curvature at the extremities of any chord of the cardioide $r = a(1 + \cos\theta)$, which pass through the pole, prove that $\rho_1^2 + \rho_2^2 = (16/9)a^2$.

159. For the curve defined by $x = at^2$, $y = 2at$, find the coordinates of the centre of curvature and hence find its evolute.

160. Find the evolute of the hyperbola $(x^2/a^2) - (y^2/b^2) = 1$.

161. Find the coordinates of the centre of curvature at a point (x, y) of the parabola $y^2 = 4ax$. Also find its evolute.

162. Find the evolute of the curve defined by $x = a\cos^3\theta$, $y = a\sin^3\theta$.

163. Show that the equation to the evolute of the curve is $x^{2/3} + y^{2/3} = a^{2/3}$ is $(x+y)^{2/3} + (x-y)^{2/3} = 2a^{2/3}$.

164. Find the radius of curvature and the centre of curvature for the curve $y = \tan x$ at the point where $x = \pi/4$.

165. Show that the radius of curvature for the curve defined by $x = 3/t^2$, $y = 1/t$ at the point $x = 0$, $y = 0$ is (1/6) unit.

Find the asymptotes of the following curves (Problems 166–169):

166. $y^3 - x^2y + 2y^2 + 4y + x = 0.$

167. $y^3 - 6xy^2 + 11x^2y - 6x^3 + x + y = 0.$

168. $y^3 + 2xy^2 + x^2y - y + 1 = 0.$

169. $x^3 - 4y^3 + 3x^2 + y - x + 3 = 0.$

170. Show that the evolute of the ellipse $(x^2/a^2) + (y^2/b^2) = 1$ is $(ax)^{2/3} + (by)^{2/3} = (a^2 - b^2)^{2/3}$.

171. Show that $(x+y)^{2/3} + (x-y)^{2/3} = 2a^{2/3}$ is the evolute of the hypocycloid $x^{2/3} + y^{2/3} = a^{2/3}$.

172. Show that the envelope of the line $(x/a) + (y/b) = 1$ where $(a/h) + (b/k) = 1$ is the parabola

$$\sqrt{\frac{x}{h}} + \sqrt{\frac{y}{k}} = 1.$$

173. Prove that the envelope of the lines $x\cos\alpha + y\sin\alpha = p\sin\alpha\cos\alpha$ (where α is the parameter) is the astroid, $x^{2/3} + y^{2/3} = p^{2/3}$.

174. Show that $r = 2a\cos\theta$ is the envelope of the straight lines drawn at right angles to the radii vectors of the cardioid, $r = a(1 + \cos\theta)$ through their extremities.

175. Prove that the evolute of the hyperbola $2xy = c^2$ is $(x+y)^{2/3} - (x-y)^{2/3} = 2c^{2/3}$.

4

Integration

4.1 INTRODUCTION

The basic problem of integral calculus may be stated as follows: Given that $y = f(x)$ where $a \le x \le b$, is a segment of a plane curve, it is required to find the area bounded by this curve, the x-axis and the ordinates $x = a$ and $x = b$ (Fig. 4.1).

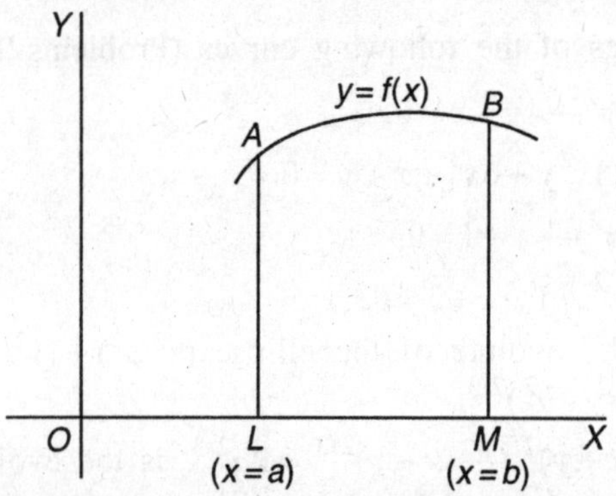

Figure 4.1

If $y = f(x)$ is a straight line, the computation of the area presents no difficulty since in that case the area can be divided into a number of rectangles whose areas can be obtained from geometrical considerations. It is only when the boundary or part of the boundary is curved, it would be necessary to introduce the concept of a limiting process.

4.1.1 Theory of the Integral

We start with the simple example of finding the area bounded by the straight line $y = 2x + 3$, the x-axis and the two ordinates $x = 1$ and $x = 3$. This area can easily be computed using the formula for the area of a trapezium, but our objective here is to describe a method which is applicable even to areas involving curved boundaries. The method consists in dividing *LM* into a number of subintervals and then adding up the areas of the rectangles. In each of the Figs. 4.2(a) and 4.2(b), the interval *LM* is divided into two equal

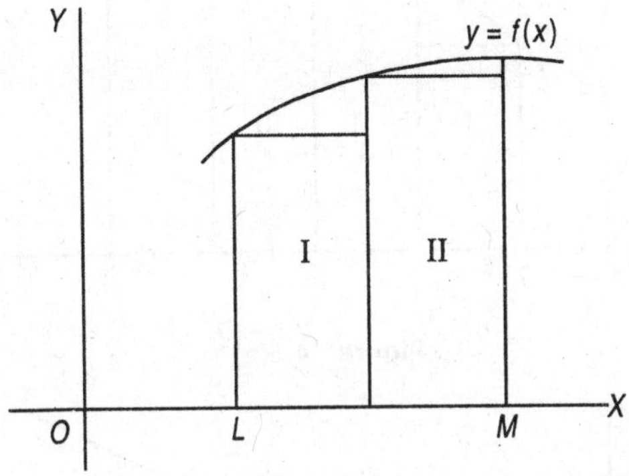

Figure 4.2(a)

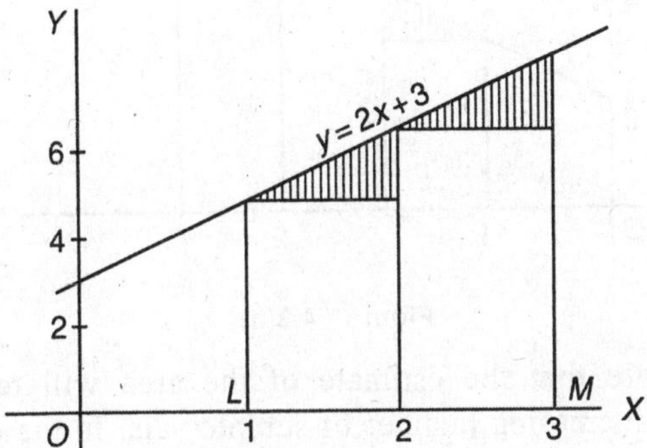

Figure 4.2(b)

subintervals and the required area is therefore equal to the sum of the areas I and II. It follows that the sum of the areas I and II in Fig. 4.2(b) is $5 + 7 = 12$. This is, obviously, inaccurate since the areas of the triangles (shaded portions) have not been included (the accurate value is 14 units, being given by the area of the trapezium *ABML*). Similarly, there is an error in the general case described in Fig. 4.2(a).

Now, suppose that *LM* is divided into four equal parts and complete the rectangles, as shown in Figs. 4.3(a) and 4.3(b). From Fig. 4.3(b), the required area is the sum of the area of the rectangles I, II, III and IV, and this is equal to

$$5\times\frac{1}{2}+6\times\frac{1}{2}+7\times\frac{1}{2}+8\times\frac{1}{2}=\frac{5}{2}+3+\frac{7}{2}+4=13,$$

which is an improvement over the previous result.

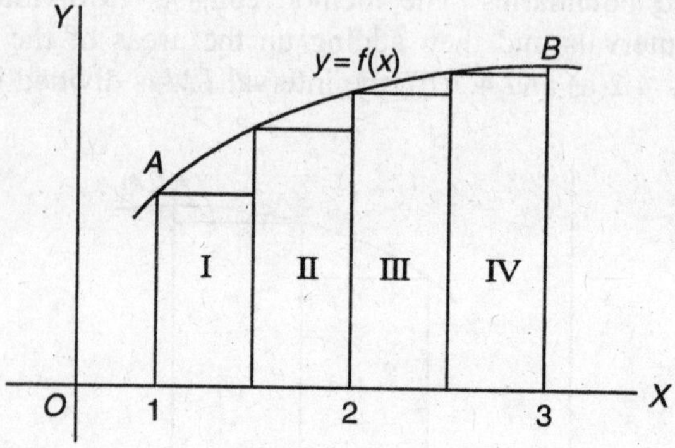

Figure 4.3(a)

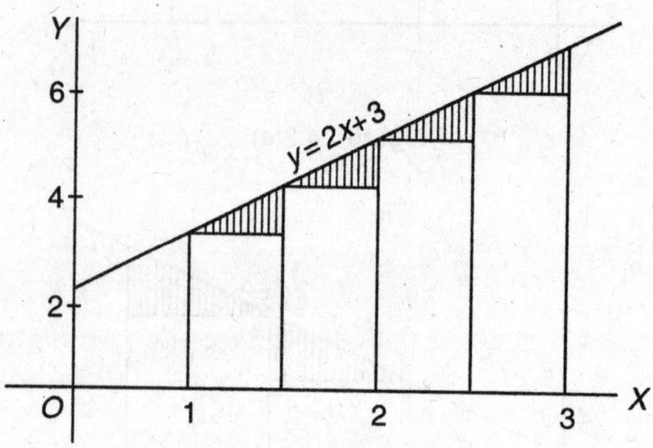

Figure 4.3(b)

It follows therefore that the estimate of the area will tend to be more accurate with the increasing number of subintervals. In the general case, let LM be divided into n subintervals of length δx_r. Then the estimate of the area becomes

$$y_1\,\delta x_1 + y_2\,\delta x_2 + y_3\delta x_3 + \cdots + y_n\,\delta x_n = \sum_{r=1}^{n} y_r\,\delta x_r,$$

where $y_1, y_2, \ldots, y_n$, are the heights of the rectangles. Since the actual area is the limit of the sums of the areas of all rectangles as the number of subintervals is increased indefinitely (i.e. as each $\delta x \to 0$), we have

$$\text{Area} = \lim_{n\to\infty} \sum_{r=1}^{n} y_r\ \delta x_r$$

This is defined as the *integral of* $y = f(x)$ with respect to x from $x = a$ to $x = b$ and is written as

$$I = \int_a^b f(x)\ dx \tag{4.1}$$

The function $f(x)$ is called the *integrand* and the interval $a \le x \le b$ is referred to as the *range of integration*. In (4.1), the variable of integration x is a *dummy variable* since the value of the integral depends on the lower and upper limits a and b. Due to this reason, the integral in (4.1) is usually referred to as a *definite integral* and we can write

$$I = \int_a^b f(x)\, dx = \int_a^b f(t)\, dt = \int_a^b f(u)\, du \tag{4.2}$$

We thus see that the integral is the value of a limit and when the limit exists we say that the function $f(x)$ *is integrable* (or *Riemann integrable*) over the interval $a \le x \le b$. Necessary and sufficient conditions exist for determining the integrability of a function but it would be sufficient to remark that the theory of the integral outlined here is concerned with the integration of *bounded functions* over *finite intervals*. The following example demonstrates the evaluation of a definite integral from the first principles.

Example 4.1 Evaluate

$$I = \int_0^\pi \sin x\, dx$$

from the first principles.

Let the range $[0, \pi]$ be divided into n subintervals so that $\delta x = \pi/n$. Then the integral I is defined as

$$I = \lim_{n\to\infty} \frac{\pi}{n} \sum_{r=1}^{n} \sin \frac{r\pi}{n} \tag{i}$$

From elementary trigonometry, we have

$$\cos \frac{2r-1}{2n}\pi - \cos \frac{2r+1}{2n}\pi = 2 \sin \frac{r\pi}{n} \sin \frac{\pi}{2n},$$

for any integer r. Hence, (i) becomes:

$$I = \lim_{n\to\infty} \frac{\pi}{n} \sum_{r=1}^{n} \frac{\cos \dfrac{(2r-1)\pi}{2n} - \cos \dfrac{(2r+1)\pi}{2n}}{2 \sin \dfrac{\pi}{2n}}$$

$$= \lim_{n\to\infty} \frac{\pi/(2n)}{\sin [\pi/(2n)]} \lim_{n\to\infty} \left[\cos \frac{\pi}{2n} - \cos \frac{3\pi}{2n} + \cos \frac{3\pi}{2n} - \cos \frac{5\pi}{2n} + \cdots \right.$$

$$\left. + \cos \frac{(2n-1)\pi}{2n} - \cos \frac{(2n+1)\pi}{2n} \right]$$

Now,

$$\lim_{n\to\infty} \frac{\pi/(2n)}{\sin[\pi/(2n)]} = 1$$

and

$$\lim_{n\to\infty} \left(\cos\frac{\pi}{2n} - \cos\frac{2n+1}{2n}\pi \right) = \cos 0 - \cos\pi = 2$$

Hence,

$$I = \int_0^{\pi} \sin x \, dx = 2.$$

Example 4.2 Evaluate

$$I = \int_1^3 (x^2 + 2x) \, dx$$

from the first principles.

Let the range [1, 3] be divided into n subintervals so that $\delta x = 2/n$. We therefore write

$$I = \lim_{n\to\infty} \frac{2}{n} \sum_{r=1}^{n} \left[\left(1 + \frac{2r}{n}\right)^2 + 2\left(1 + \frac{2r}{n}\right) \right]$$

$$= \lim_{n\to\infty} \frac{2}{n} \sum_{r=1}^{n} \left(3 + \frac{8r}{n} + \frac{4r^2}{n^2} \right)$$

$$= \lim_{n\to\infty} \frac{2}{n} \left[3n + \frac{8}{n}(1 + 2 + 3 + \cdots + n) + \frac{4}{n^2}(1^2 + 2^2 + 3^2 + \cdots + n^2) \right] \quad \text{(i)}$$

But we know that

$$1 + 2 + 3 + \cdots + n = \frac{n(n+1)}{2}$$

and

$$1^2 + 2^2 + 3^2 + \cdots + n^2 = \frac{n(n+1)(2n+1)}{6}.$$

Using the above results in (i), we obtain.

$$I = \lim_{n\to\infty} \frac{2}{n} \left[3n + \frac{8}{n}\frac{n(n+1)}{2} + \frac{4}{n^2}\frac{n(n+1)(2n+1)}{6} \right]$$

$$= \lim_{n\to\infty}\left[6+8\left(1+\frac{1}{n}\right)+\frac{4}{3}\left(1+\frac{1}{n}\right)\left(2+\frac{1}{n}\right)\right]$$

$$= 6+8+\frac{8}{3}$$

$$= 14+\frac{8}{3}$$

$$= \frac{50}{3}.$$

4.1.2 Properties of the Definite Integral

We give below some important properties of the definite integral. Since these are merely simple consequences of the definition, the proofs are omitted.

Linearity

The process of integration is a *linear operation.* That is, if $f(x)$ and $\phi(x)$ are two bounded functions each of which is integrable over the interval $a \le x \le b$ and if m and n are any real numbers, then

$$\int_a^b [mf(x)+n\phi(x)]\,dx = m\int_a^b f(x)\,dx + n\int_a^b \phi(x)\,dx \tag{4.3}$$

Range of integration

If $f(x)$ is integrable over $a \le x \le b$ and if c is any point such that $a < c < b$, then $f(x)$ is integrable over $a \le x \le c$ and $c \le x \le b$. That is,

$$\int_a^b f(x)\,dx = \int_a^c f(x)\,dx + \int_c^b f(x)\,dx \tag{4.4}$$

From the above result, if follows easily that

$$\int_a^b f(x)\,dx = -\int_b^a f(x)\,dx \tag{4.5}$$

Invariance

If $f(x)$ is integrable on $[0, \pi]$, then

$$I = \int_0^\pi f(x)\,dx = \int_0^\pi f(\pi - x)\,dx \tag{4.6}$$

This result is a direct consequence of the definition of the definite integral and can be shown easily by a simple change in the variable of integration. If we let $x = \pi - y$, then $dx = -dy$ and the new limits of integration are from $y = \pi$ to $y = 0$. Hence the given integral takes the form

$$I = \int_{\pi}^{0} f(\pi - y)\,(-dy)$$

$$= \int_{0}^{\pi} f(\pi - y)\,dy, \qquad \text{[using result (4.5)]}$$

$$= \int_{0}^{\pi} f(\pi - x)\,dx.$$

Example 4.3 Show that

$$\int_{0}^{\pi} x f(\sin x)\,dx = \frac{\pi}{2}\int_{0}^{\pi} f(\sin x)\,dx$$

Let

$$I = \int_{0}^{\pi} x f(\sin x)\,dx$$

$$= \int_{0}^{\pi} (\pi - x)\, f[\sin(\pi - x)]\,dx,$$

$$= \int_{0}^{\pi} (\pi - x)\, f(\sin x)\,dx, \quad \text{since } \sin(\pi - x) = \sin x$$

$$= \pi\int_{0}^{\pi} f(\sin x)\,dx - \int_{0}^{\pi} x f(\sin x)\,dx$$

$$= \pi\int_{0}^{\pi} f(\sin x)\,dx - I$$

i.e.

$$2I = \pi\int_{0}^{\pi} f(\sin x)\,dx$$

so that

$$I = \frac{\pi}{2}\int_{0}^{\pi} f(\sin x)\,dx.$$

The mean-value theorem

Let m and M be the least and greatest values of $f(x)$ in $a \le x \le b$ and let the interval (a, b) be divided into n subintervals $\delta x_r, r = 1, 2, ..., n$. Then we have

$$m\delta x_r \le f(x_r)\delta x_r \le M\delta x_r$$

and hence

$$m\sum_{r=1}^{n} \delta x_r \le \sum_{r=1}^{n} f(x_r)\, \delta x_r \le M\sum_{r=1}^{n} \delta x_r,$$

which means that

$$m(b-a) \le \sum_{r=1}^{n} f(x_r)\, \delta x_r \le M(b-a).$$

Since $f(x)$ is integrable over $a \le x \le b$, we obtain from the above

$$m(b-a) \le \int_a^b f(x)\, dx \le M(b-a)$$

or,

$$m \le \frac{1}{b-a}\int_a^b f(x)\, dx \le M \tag{4.7}$$

If we assume that $f(x)$ is continuous in $a \le x \le b$, then $f(x)$ assumes every value between m and M. Hence, there exists a number ξ such that $a < \xi < b$, for which.

$$f(\xi) = \frac{1}{b-a}\int_a^b f(x)\, dx \tag{4.8}$$

This result is known as the *first mean-value theorem* of the integral calculus and has considerable importance. Geometrically, it means that the area bounded by the curve $y = f(x)$, the x-axis and the ordinates $x = a$ and $x = b$, is equal to the area of the rectangle with length $(b - a)$ and height (ξ).

Example 4.4 Estimate the integral

$$I = \int_2^5 \sqrt{x^2 + 5}\, dx.$$

The function $f(x) = \sqrt{x^2 + 5}$ is an increasing function on the interval [2, 5]. Hence, we have

$$m = \sqrt{4+5} = 3, \quad \text{and} \quad M = \sqrt{25+5} = \sqrt{30} = 5.477.$$

Since $b-a=3$, it follows from (4.7) that

$$9 \le \int_2^5 \sqrt{x^2+5}\, dx \le 3\,(5.477).$$

Hence the given integral lies between 9 and 16.43.

The fundamental theorem of integral calculus

This theorem is a consequence of the mean-value theorem given above.

Let $f(x)$ be integrable over $a \le x \le b$ and define

$$F(x) = \int_a^x f(x)\, dx \tag{4.9}$$

Here $f(x)$ is integrable over $a \le x \le b$, so also is $|f(x)|$. Let M' be the maximum value attained by $|f(x)|$ in this interval. Then, for any two points x and $(x+h)$ in the interval $a \le x \le b$, we have,

$$|F(x+h)-F(x)| = \left| \int_a^{x+h} f(x)\, dx - \int_a^x f(x)\, dx \right|$$

$$= \left| \int_x^{x+h} f(x)\, dx \right|$$

$$\le \int_x^{x+h} |f(x)|\, dx$$

$$\le M' |h|.$$

Thus, by taking $|h|$ sufficiently small, we can make $|F(x+h)-F(x)|$ as small as we please. This means that even if the integral $f(x)$ possesses discontinuities, the function $F(x)$, defined by (4.9), is always continuous on $a \le x \le b$. But if $f(x)$ is continuous on $a \le x \le b$, then we have

$$F(x+h)-F(x) = hf(\xi), \quad x < \xi < x+h \tag{4.10}$$

or

$$\frac{F(x+h)-F(x)}{h} = f(\xi).$$

Hence

$$\lim_{h \to 0} \frac{F(x+h)-F(x)}{h} = \lim_{h \to 0} f(\xi),$$

which means that

$$F'(x) = f(x) \tag{4.11}$$

Thus, for a continuous integrand $f(x)$, the function $F(x)$ defined by (4.9), is not only continuous but is also differentiable on $a \le x \le b$, and its derivatives is equal to the original integrand. We thus have the theorem:

If $f(x)$ is continuous on $a \le x \le b$, then

$$\frac{d}{dx}\int_a^x f(x)\,dx$$

exists and is equal to $f(x)$. That is

$$\frac{d}{dx}\int_a^x f(x)\,dx = f(x) \tag{4.12}$$

This result is called *the fundamental theorem of the integral calculus* and provides a justification for regarding integration as the inverse of differentiation.

Let $f(x)$ be continuous on $a \le x \le b$ and let

$$F(x) = \int_a^x f(x)\,dx$$

so that

$$F'(x) = f(x).$$

From the fundamental theorem of the integral calculus, we have

$$\frac{d}{dx}\left[\int_a^x f(x)\,dx - F(x)\right] = f(x) - F'(x) = 0.$$

But we know that if the derivative of a function is zero throughout any interval, then the function must be a constant in that interval. Hence, we write

$$F(x) - \int_a^x f(x)\,dx = c, \tag{4.13}$$

where c is a constant. Putting $x = a$ in (4.13), we obtain:

$$F(a) - \int_a^a f(x)\,dx = c \quad \text{or} \quad c = F(a).$$

Putting again $x = b$ in (4.13), we obtain:

$$F(b) - \int_a^b f(x)\,dx = c = F(a),$$

i.e.

$$\int_a^b f(x)\,dx = F(b) - F(a). \tag{4.14}$$

This result forms the basis of the elementary methods of integration. If $f(x)$ is given, then we try to find a function $F(x)$ whose derivative is known to be $f(x)$ and then evaluate the integral $\int_a^b f(x)\,dx$ by using (4.14). Such a function $F(x)$ is called a *primitive* or an *indefinite integral* (or an *antiderivative*) and we write:

$$F(x) = \int f(x)\,dx. \tag{4.15}$$

Using known results from differential calculus, Table 4.1 of indefinite integrals can be prepared (the constants of integration having been omitted).

Example 4.5 Evaluate

$$\int_0^3 F(x)\,dx$$

where $F(x)$ is defined as follows:

$$\begin{aligned} F(x) &= x^2, \quad & 0 \le x \le 1 \\ &= \sqrt{x}, \quad & 1 \le x \le 3. \end{aligned}$$

We have

$$\begin{aligned} I &= \int_0^3 F(x)\,dx \\ &= \int_0^1 F(x)\,dx + \int_1^3 F(x)\,dx, \\ &= \int_0^1 x^2\,dx + \int_1^3 \sqrt{x}\,dx, \quad \text{[using the definition of } F(x)\text{]}. \\ &= \left[\frac{x^3}{3}\right]_0^1 + \frac{2}{3}\left[x^{3/2}\right]_1^3, \quad \text{[using the formula from Table 4.1]}. \\ &= \frac{1}{3} + \frac{2}{3}\left(\sqrt{27} - 1\right) \end{aligned}$$

$$= \frac{1}{3} + \frac{2}{3}\left(3\sqrt{3} - 1\right)$$

$$= 2\sqrt{3} - \frac{1}{3}.$$

Table 4.1 Indefinite Integrals of Some Functions

$f(x)$	$\int f(x)\,dx$
x^n	$\frac{x^{n+1}}{n+1}, n \neq -1$
$\frac{1}{x}$	$\log_e x$
e^x	e^x
$\sin x$	$-\cos x$
$\cos x$	$\sin x$
$\sec^2 x$	$\tan x$
$\text{cosec}^2 x$	$-\cot x$
$\sec x \tan x$	$\sec x$
$\text{cosec}\, x \cot x$	$-\text{cosec}\, x$
$\sinh x$	$\cosh x$
$\cosh x$	$\sinh x$
$\text{sech}^2 x$	$\tanh x$
$\text{cosech}^2 x$	$-\coth x$
$\frac{1}{\sqrt{a^2 - x^2}}$	$\sin^{-1}\frac{x}{a}$
$\frac{1}{\sqrt{x^2 - a^2}}$	$\cosh^{-1}\frac{x}{a}$
$\frac{1}{\sqrt{x^2 + a^2}}$	$\sinh^{-1}\frac{x}{a}$
$\frac{1}{x^2 + a^2}$	$\frac{1}{a}\tan^{-1}\frac{x}{a}$
$\frac{1}{a^2 - x^2}$	$\frac{1}{2a}\log_e\frac{a+x}{a-x}$
$\frac{1}{x^2 - a^2}$	$\frac{1}{2a}\log_e\frac{x-a}{x+a}$
$\int\sqrt{a^2 - x^2}\,dx$	$\frac{a^2}{2}\sin^{-1}\frac{x}{a} + \frac{x}{2}\sqrt{a^2 - x^2}$
$\int\sqrt{a^2 + x^2}\,dx$	$\frac{a^2}{2}\log(x + \sqrt{a^2 + x^2}) + \frac{x}{2}\sqrt{a^2 + x^2}$
$\int\sqrt{x^2 - a^2}\,dx$	$-\frac{a^2}{2}\log(x + \sqrt{x^2 - a^2}) + \frac{x}{2}\sqrt{x^2 - a^2}.$

4.1.3 Numerical Integration

From the fundamental theorem of the integral calculus, we have seen that the evaluation of a definite integral requires the determination of the indefinite integral of the integrand. Several examples occur where such a determination is not possible, e.g. the integral

$$I = \int_1^2 e^{-x^2} dx$$

cannot be expressed in terms of elementary functions. In situations like this, numerical evaluation of definite integrals will be of great importance. We derive here two simple rules for the numerical integration of definite integrals and demonstrate their use through illustratives examples, For more powerful methods of numerical integration, the reader is referred to Chapter 7 of this text.

Trapezoidal rule

Suppose that we wish to estimate the value of the definite integral

$$I = \int_a^b f(x)dx \tag{4.16}$$

Let the interval (a, b) be divided into n subintervals $(x_{i-1}, x_i), i = 1, 2, ..., n$ where $x_0 = a$ and $x_n = b$. We know that the integral represents the area bounded by the curve $y = f(x)$, the x-axis and the ordinates $x = a$ and $x = b$ (see Fig. 4.4).

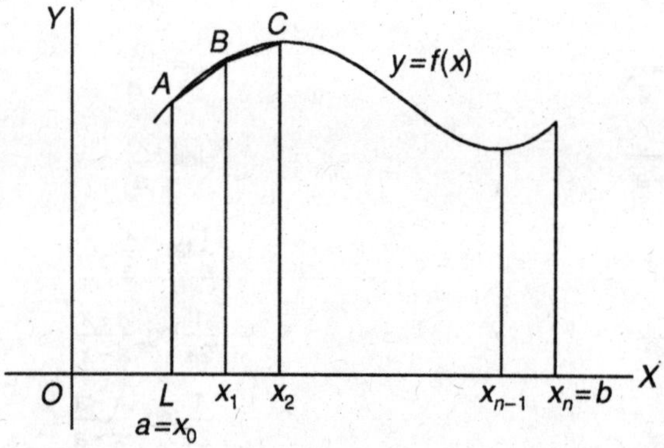

Figure 4.4

Consider the first subinterval, *viz.* (x_0, x_1) and let A and B be the points on the curve with coordinates $\{x_0, f(x)\}$ and $\{x_1, f(x)\}$ respectively. In the trapezoidal rule, the true area above $[x_0, x_1]$ is approximated by the trapezium obtained by replacing the arc of the curve through the points A and B by the chord joining these two points. then the area of the trapezium on the subinterval $[x_0, x_1]$ is

$$\frac{1}{2}[f(x_0)+f(x_1)]\,\Delta x_0, \quad \text{where } \Delta x_0 = x_1 - x_0.$$

Similarly, the area of the trapezium on the subinterval $[x_1, x_2]$ is

$$\frac{1}{2}[f(x_1)+f(x_2)]\,\Delta x_1 \quad \text{where } \Delta x_1 = x_2 - x_1,$$

and so on for the remaining subintervals. Adding the n contributions of this type, we obtain the *general trapezoidal rule*:

$$\int_a^b f(x)\,dx \approx \frac{1}{2}\{[f(x_0)+f(x_1)]\,\Delta x_0 + [f(x_1)+f(x_2)]\,\Delta x_1 + \cdots + [f(x_{n-1})+f(x_n)]\,\Delta x_{n-1}\}. \tag{4.17}$$

If the subintervals are all of equal length, say h, we have:

$$\Delta x_0 = \Delta x_1 = \cdots = \Delta x_{n-1} = h = \frac{b-a}{n}.$$

Hence the trapezoidal rule for equal intervals is

$$\int_a^b f(x)\,dx \approx \frac{h}{2}[f(x_0)+2f(x_1)+2f(x_2)+\cdots+2f(x_{n-1})+f(x_n)] \tag{4.18}$$

The error in the formula (4.18) can be obtained in the following way: Let $y = f(x)$ be continuous, well-behaved and possess continuous derivatives in $[x_0, x_n]$. Considering first the subinterval $[x_0, x_1]$, we have

$$\int_{x_0}^{x_1} y\,dx \approx \frac{h}{2}(y_0 + y_1), \quad \text{by the trapezoidal rule.}$$

Expanding y in a Taylor's series around $x = x_0$, we obtain:

$$\int_{x_0}^{x_1} y\,dx \approx \int_{x_0}^{x_1}\left[y_0 + (x-x_0)y'_0 + \frac{(x-x_0)^2}{2!}y''_0 + \cdots\right]dx$$

$$= hy_0 + \frac{h^2}{2}y'_0 + \frac{h^3}{6}y''_0 + \cdots \tag{4.19}$$

Similarly,

$$\frac{h}{2}(y_0 + y_1) = \frac{h}{2}\left(y_0 + y_0 + hy'_0 + \frac{h^2}{2!}y''_0 + \frac{h^3}{3!}y'''_0 + \cdots\right)$$

$$= hy_0 + \frac{h^2}{2} y_0' + \frac{h^3}{4} y_0'' + \cdots \tag{4.20}$$

From (4.19) and (4.20), we obtain:

$$\int_{x_0}^{x_1} y\,dx - \frac{h}{2}(y_0 + y_1) = -\frac{1}{12} h^3 y_0'' + \cdots, \tag{4.21}$$

which is the error in the subinterval $[x_0, x_1]$. Proceeding in a similar manner, we obtain the errors in the remaining subintervals, viz, $[x_1, x_2]$, $[x_2, x_3]$, ..., $[x_{n-1}, x_n]$. If the error in the formula (4.18) is E_T, we have

$$E_T = -\frac{1}{12} h^3 y_0'' - \frac{1}{12} h^3 y_1'' - \cdots - \frac{1}{12} h^3 y_{n-1}''$$

$$= -\frac{1}{12} h^3 (y_0'' + y_1'' + \cdots + y_{n-1}'') \tag{4.22}$$

Assuming that $y''(\xi)$ is the largest value of the n quantities on the right side of (4.22), we obtain

$$E_T \approx -\frac{1}{12} h^3 n y''(\xi) = -\frac{(b-a)}{12} h^2 y''(\xi), \tag{4.23}$$

since

$$nh = (b-a).$$

Example 4.6 The following table of values of x and $f(x)$ are given: Find the area bounded by the curve $y = f(x)$, the x-axis and the ordinates $x = 7.47$ and $x = 7.52$.

x:	7.47	7.48	7.49	7.50	7.51	7.52
$f(x)$:	1.93	1.95	1.98	2.01	2.03	2.06

We know that,

$$\text{Area} = \int_{7.47}^{7.52} f(x)\,dx$$

With $h = 0.01$, the trapezoidal rule (4.18) gives:

$$\text{Area} \approx 0.01\left(\frac{1.93}{2} + 1.95 + 1.98 + 2.01 + 2.03 + \frac{2.06}{2}\right) = 0.0996.$$

Example 4.7 Find the approximate value of

$$\int_{1.2}^{1.6} e^{-x^2}\,dx$$

from the following table:

x	1.2	1.3	1.4	1.5	1.6
e^{-x^2}	0.237	0.185	0.141	0.106	0.077

Using the trapezoidal rule, we have

$$\int_{1.2}^{1.6} e^{-x^2}\,dx \approx 0.1\left(\frac{0.237}{2}+0.185+0.141+0.106+\frac{0.077}{2}\right)=0.0589.$$

Simpson's rule

This rule requires the division of the interval $[a, b]$ into an *even* number of subintervals and then approximating the function over consecutive pairs of subintervals by a quadratic polynomial. Considering, for instance, the first two subintervals $[a, a+h]$ and $[a+h, a+2h]$, we have

$$\int_a^{a+2h} y\,dx \approx \int_a^{a+2h} (p_0 + p_1x + p_2x^2)\,dx$$

$$= \left[p_0x + p_1\frac{x^2}{2} + p_2\frac{x^3}{3}\right]_a^{a+2h}, \tag{4.24}$$

using the rule

$$\int x^n\,dx = \frac{x^{n+1}}{n+1}, \qquad n \neq -1,$$

given in Table 4.1. Now

$$\int_a^{a+2h} y\,dx = p_0(2h) + p_1(2h^2 + 2ah) + \frac{p_2}{3}(8h^3 + 12ah^2 + 6a^2h) \tag{4.25}$$

Let $y(a), y(a+h)$ and $y(a+2h)$ be the values of y corresponding to $x=a,\ a+h$ and $a+2h$, respectively. Substituting these pairs of values of x and y in the relation

$$y = p_0 + p_1x + p_2x^2,$$

we obtain the simultaneous equations

$$\left.\begin{aligned} y(a) &= p_0 + p_1a + p_2a^2 \\ y(a+h) &= p_0 + p_1(a+h) + p_2(a+h)^2 \\ y(a+2h) &= p_0 + p_1(a+2h) + p_2(a+2h)^2 \end{aligned}\right\} \tag{4.26}$$

Solving these equations and substituting for p_0, p_1 and p_2 in (4.25), we obtain:

$$\int_a^{a+2h} y\,dx \approx \frac{h}{3}[y(a)+4y(a+h)+y(a+2h)] \tag{4.27}$$

Similarly, for the next pair of subintervals, viz., $[a+2h,\ a+3h]$ and $[a+3h,\ a+4h]$, we obtain

$$\int_{a+2h}^{a+4h} y\,dx \approx \frac{h}{3}[y(a+2h)+4y(a+3h)+y(a+4h)],$$

and so on for other pairs. Adding up all these contributions, we obtain *Simpson's rule*:

$$\begin{aligned}\int_a^a y\,dx &\approx \frac{h}{3}[y(a)+4y(a+h)+y(a+2h)]\\ &\quad +y(a+2h)+4y(a+3h)+y(a+4h)+\cdots\\ &\quad +y(a+\overline{n-2}h)+4y(a+\overline{n-1}h)+y(a+nh)]\\ &=\frac{h}{3}\{y(a)+4[y(a+h)+y(a+3h)+\cdots+y(a+\overline{n-1}h)]\\ &\quad +2[y(a+2h)+y(a+4h)+\cdots+y(a+\overline{n+2}h)]+y(a+nh)\}\end{aligned} \tag{4.28}$$

The error E_s in (4.28) is given by

$$E_s = -\frac{b-a}{180}h^4 y^{iv}(\xi), \tag{4.29}$$

where $y^{iv}(\xi)$ is the largest value of the fourth derivatives. The method of derivation of (4.29) is the same as that of (4.23) and is left as an exercise to the reader.

Example 4.8 Evaluate,

$$I = \int_0^1 \frac{1}{1+x}\,dx,$$

correct to three decimal places.

We demonstrate, in this example, the relative accuracies of the trapezoidal and Simpson's rules. We choose $h = 0.5, 0.25$ and 0.125. Let $y = 1/(1+x)$ and also let I_T and I_s be the values obtained by the trapezoidal and Simpson's rules.

(i) When $h = 0.5$, the values of x and y are tabulated below:

x	0	0.5	1.0
y	1.0000	0.6667	0.5

$$I_T \approx \frac{1}{2}\left(\frac{1.000}{2} + 0.6667 + \frac{0.5}{2}\right) = 0.708$$

$$I_S \approx \frac{1}{6}[1.0000 + 4(0.6667) + 0.5] = 0.694.$$

(ii) When $h = 0.25$, the tabulated values of x and y are:

x	0	0.25	0.5	0.75	1.0
y	1.0000	0.8000	0.6667	0.5714	0.5

We obtain

$$I_T \approx \frac{1}{4}\left(\frac{1.0}{2} + 0.8000 + 0.6667 + 0.5714 + \frac{0.5}{2}\right) = 0.697$$

and

$$I_s \approx \frac{1}{12}[(1.0 + 4(0.8000 + 0.5714) + 2(0.6667) + 0.5] = 0.693.$$

(iii) When $h = 0.125$, the tabulated values of x and y are:

x	0	0.125	0.250	0.375	0.5	0.625	0.750	0.875	1.0
y	1.0	0.8889	0.8000	0.7273	0.6667	0.6154	0.5714	0.5333	0.5

We obtain

$$I_T \approx \frac{1}{8}\left(\frac{1.0}{2} + 0.8889 + 0.8000 + 0.7273 + 0.6667 + 0.6154 + 0.5714 + 0.5333 + \frac{0.5}{2}\right)$$

$$= 0.694,$$

and

$$I_S \approx \frac{1}{24}[1.0 + 4\,(0.8889 + 0.7273 + 0.6154 + 0.5333)$$
$$+ 2(0.8000 + 0.6667 + 0.5714) + 0.5]$$

$$= 0.693.$$

The value of I is therefore taken as 0.693 correct to three decimal places. The exact value of I is $\log_e 2$, which is equal to 0.693147. This example demonstrates that, in general, Simpson's rule yields more accurate results than the trapezoidal rule.

Example 4.9 Estimate the errors in the results obtained by evaluating the integral

$$I = \int_0^1 \frac{1}{1+x}\,dx$$

by trapezoidal and Simpson's rules.

To estimate the errors, we choose the approximate values obtained with $h = 0.25$, i.e. with $N = 4$. These were already computed in Example 4.3.

Using the trapezoidal rule, we obtained $I_T = 0.6970$. The correct answer is 0.6931 to four decimal places and the magnitude of the error in I_T is therefore 0.0039. From the estimate given in (4.23), we obtain

$$|E_T| = \left| \frac{(b-a)^3}{12n^2} y''(\xi) \right|$$

Here,

$$y''(x) = \frac{2}{(1+x)^3}, \qquad n = 4$$

Hence,

$$|E_T| < \frac{1}{12 \times 16} \times 2 = 0.0104.$$

Again, from Simpson's rule, we obtained $I_S = 0.6932$ and hence the magnitude of the error in the above result is 0.0001. From the estimate given in (4.29), we obtain

$$|E_S| = \left| \frac{(b-a)^5}{180n^4} y^{iv}(\xi) \right|$$

But

$$y^{iv}(x) = \frac{24}{(1+x)^5} \quad \text{and} \quad n = 4.$$

Hence

$$|E_S| < \frac{1}{180 \times 256} \times 24 = 0.0005.$$

The above results demonstrate that our bounds for the errors are correct.

EXERCISES 4.1

Evaluate the following definite integrals from first principles (Problems 1–6):

1. $\int_0^a x\,dx.$ **2.** $\int_0^a x^2\,dx.$

3. $\int_0^a e^x \, dx,$

4. $\int_a^b e^{kx} \, dx.$

5. $\int_0^{\pi/2} \cos x \, dx.$

6. $\int_0^{\pi} \sin x \, dx.$

7. The mean-value theorem cf integral calculus states that

$$\int_a^b f(x)\,dx = (b-a)\,f(\xi), \quad a < \xi < b.$$

Find ξ when (a) $f(x) = x$ (b) $f(x) = x^2$ and (c) $f(x) = 1/x^2$.

Evaluate the following integrals using the properties of the definite integral (Problems 8–10):

8. Find $\int_0^2 f(x)\,dx$ where

$$f(x) = x, \qquad 0 \le x \le 1$$
$$= \sqrt{x}, \qquad 1 \le x \le 2$$

9. $\int_0^2 |1-x|\,dx$

10. $\int_0^{\pi} \sqrt{\dfrac{1+\cos 2x}{2}}\,dx$

11. Using the trapezoidal rule, compute

$$\log_e 2 = \int_1^2 \frac{1}{x}\,dx$$

and estimate the error in the solution.

12. Use the trapezoidal rule to compute

$$I = \int_0^1 e^{-x^2}\,dx$$

accurate to three decimal places.

13. Use the trapezoidal rule with $h = 4, 2$ and 1 to evaluate the integral

$$\int_0^{12} \frac{1}{1+x}\,dx,$$

correct to 4 decimal places.

14. Evaluate

$$\int_1^3 \frac{1}{x}\,dx$$

by Simpson's rule with 4 subintervals and 8 subintervals respectively.

15. Evaluate the integral

$$\int_0^{12} \frac{1}{1+x}\, dx$$

using Simpson's rule with $h = 2$ and 1.

16. Evaluate the integral

$$I = \int_1^3 \frac{e^{-x}\, dx}{x}$$

using Simpson's rule with $h = 1$ and $h = 0.5$.

17. Compute

$$\pi/4 = \int_0^1 \frac{dx}{1+x^2}$$

by means of Simpson's rule and estimate the error in your result.

Compute the following integrals by Simpson's rule (Problems 18–20):

18. $\int_0^2 \sqrt{1+x^3}\, dx, \left(h = \frac{1}{2}\right)$

19. $\int_0^{\pi/2} (3 - \cos 2x)^{1/2}\, dx, \left(h = \frac{\pi}{12}\right)$

20. $\int_0^4 \frac{1}{1+x^4}\, dx, \ (h = 1).$

21. Find the value of

$$\int_3^7 x^2 \log_e x\, dx$$

by taking 4 subintervals.

22. Evaluate

$$\int_0^{\pi/2} \sqrt{\cos\theta}\, d\theta$$

by dividing the interval into six parts.

23. The following table gives the coordinates of seven points on a quadrant of a circle. Use the trapezoidal rule to calculate the area of the quadrant and, hence, find the approximate value for π.

x	0	1	2	3	4	5	6
y	6	5.92	5.66	5.20	4.47	3.32	0

24. Upon take-off, the velocity readings of a light vehicle have been noted every second for 10 seconds. These are: 0, 90, 125, 360, 495, 630, 765, 900, 1035, 1125, 1260 in metres per second, respectively. Using the trapezoidal rule, estimate the distance travelled by the flight vehicle during the first-ten seconds.

25. The speed, v metres per second, of a car, t seconds, after it starts, is shown in the following table:

t	0	12	24	36	48	60	72	84	96	108	120
v	0	3.60	10.08	18.90	21.60	18.54	10.26	5.40	4.50	5.40	9.00

Using Simpson's rule, find the distance travelled by the car in 2 minutes.

4.2 METHODS OF INTEGRATION

In the previous section, we have seen that an integral can be regarded either as the limit of a sum or as the reverse of differentiation. The former approach has been used to introduce the idea of a definite integral, but in solving problems the latter approach will be more convenient.

The integral

$$\int_a^b f(x)\,dx$$

is a definite integral, but an integral without limits is called an *indefinite integral.* For example,

$$\int f(x)\,dx$$

is an indefinite integral which is also often called an *anti-derivative.*

The reader is assumed to be familiar with the techniques of indefinite integration and therefore only a brief review of the different methods of integration is given below.

4.2.1 Integration by Substitution

We list below a few types of integrals which can be solved by using an appropriate substitution.

(i) Type $\int \dfrac{f'(x)}{f(x)}\,dx.$

Putting $f(x) = t$ so that $f'(x)\,dx = dt$, the given integral reduces to

$$\int \frac{dt}{t} = \log t + c = \log f(x) + c.$$

Example 4.10 Integrate

$$\int \operatorname{cosec} x \, dx.$$

$$\int \operatorname{cosec} x \, dx = \int \frac{\operatorname{cosec} x (\operatorname{cosec} x + \cot x)}{\operatorname{cosec} x + \cot x} dx$$

$$= -\int \frac{-\operatorname{cosec} x \cot x - \operatorname{cosec}^2 x}{\operatorname{cosec} x + \cot x} dx$$

$$= -\log (\operatorname{cosec} x + \cot x) + c.$$

Example 4.11 Integrate

$$\int \frac{\cos \sqrt{x}}{\sqrt{x}} dx.$$

Let $\sqrt{x} = t$, so that $(1/\sqrt{x})\, dx = 2\, dt$. Hence

$$\int \frac{\cos \sqrt{x}}{\sqrt{x}} dx = \int 2 \cos t \, dt = 2 \sin t + c = 2 \sin \sqrt{x} + c$$

(ii) Type $\int e^x [f(x) + f'(x)]\, dx.$

Put $e^x f(x) = t$, so that

$$[e^x f(x) + e^x f^1(x)]\, dx = dt.$$

Hence

$$\int e^x [f(x) + f'(x)]\, dx = \int dt = t + c = e^x f(x) + c.$$

Example 4.12 Integrate

$$\int e^x \frac{1 - \sin x}{1 - \cos x} dx.$$

The given integral

$$I = \int e^x \, \frac{1 - 2 \sin (x/2) \cos (x/2)}{2 \sin^2 (x/2)} dx = \int e^x \left(\frac{1}{2} \operatorname{cosec}^2 \frac{x}{2} - \cot \frac{x}{2} \right) dx$$

Now, put $e^x \cot (x/2) = t$, so that

$$\left(e^x \cot \frac{x}{2} - \frac{1}{2} e^x \operatorname{cosec}^2 \frac{x}{2} \right) dx = dt.$$

Hence

$$\int e^x \frac{1-\sin x}{1-\cos x}\,dx = \int -dt = -t + c = -e^x \cot\frac{x}{2} + c.$$

(iii) Type $\int \dfrac{a \sin x + b \cos x}{A \sin x + B \cos x}\,dx.$

Put

$$\begin{aligned} a \sin x + b \cos x &= l\,(A \sin x + B \cos x) + m\,\frac{d}{dx}(A \sin x + B \cos x) \\ &= l\,(A \sin x + B \cos x) + m\,(A \cos x - B \sin x) \end{aligned}$$

Equating the coefficients of $\sin x$ and $\cos x$ on both sides, we obtain

$$a = lA - mB \quad \text{and} \quad b = lB + mA,$$

from which l and m can be calculated. Then the given integral becomes

$$\int \frac{l\,(A \sin x + B \cos x) + m\,(A \cos x - B \sin x)}{A \sin x + B \cos x}\,dx$$

$$= lx + m \log (A \sin x + B \cos x) + c,$$

where l and m are known.

Example 4.13 Evaluate

$$\int \frac{\sin x + 8 \cos x}{2 \sin x + 3 \cos x}\,dx.$$

Put $\sin x + 8 \cos x = l\,(2 \sin x + 3 \cos x) + m\,(2 \cos x - 3 \sin x)$. Equating the coefficients of $\sin x$ and $\cos x$, we get

$$2l - 3m = 1 \quad \text{and} \quad 3l + 2m = 8$$

These equations give $l = 2$ and $m = 1$. Hence

$$\int \frac{\sin x + 8 \cos x}{2 \sin x + 3 \cos x}\,dx = 2x + \log (2 \sin x + 3 \cos x) + c.$$

(iv) Type $\int \dfrac{dx}{ax^2 + bx + c}.$

In this case, we complete the square in the denominator and reduce the integral to one of the standard forms.

Example 4.14 Integrate

$$\int \frac{1}{x^2 + x + 1}\,dx.$$

Here,

$$\int \frac{dx}{x^2+x+1} = \int \frac{1}{(x+1/2)^2+3/4}\,dx$$

$$= \frac{2}{\sqrt{3}} \tan^{-1} \frac{x+1/2}{\sqrt{3}/2} + c$$

$$= \frac{2}{\sqrt{3}} \tan^{-1} \frac{2x+1}{\sqrt{3}} + c.$$

Example 4.15 Evaluate

$$\int \frac{dx}{x^2+x-1}.$$

Here

$$\int \frac{dx}{x^2+x-1} = \int \frac{dx}{(x+1/2)^2 - \frac{5}{4}} = \frac{1}{\sqrt{5}} \log \frac{2x+1-\sqrt{5}}{2x+1+\sqrt{5}} + c$$

(v) Type $\int \dfrac{px+q}{ax^2+bx+c}\,dx.$

In this case, express the numerator as the derivative of the denominator plus a constant and use the method of type (iv).

Example 4.16 Evaluate

$$\int \frac{x+3}{x^2+4x+5}\,dx.$$

Here

$$\int \frac{x+3}{x^2+4x+5}\,dx = \frac{1}{2}\int \frac{2x+6}{x^2+4x+5}\,dx$$

$$= \frac{1}{2}\int \frac{2x+4+2}{x^2+4x+5}\,dx$$

$$= \frac{1}{2}\int \frac{2x+4}{x^2+4x+5}\,dx + \int \frac{1}{x^2+4x+5}\,dx$$

$$= \frac{1}{2}\log(x^2+4x+5) + \int \frac{1}{(x+2)^2+1}\,dx$$

$$= \frac{1}{2}\log(x^2+4x+5) + \tan^{-1}(x+2) + c.$$

(vi) Type $\int \sqrt{ax^2 + bx + c}\, dx$

An integral of this type is reduced to a standard form by completing the square.

Example 4.17 Evaluate

$$\int \sqrt{1 + x + x^2}\, dx.$$

We let

$$\int \sqrt{1+x+x^2}\, dx = \int \sqrt{\left(x + \frac{1}{2}\right)^2 + \frac{3}{4}}\, dx$$

$$= \frac{3}{8} \log\left(x + \sqrt{1+x+x^2}\right) + \frac{x}{2}\sqrt{1+x+x^2} + c.$$

(vii) Type $\int \dfrac{1}{\sqrt{ax^2 + bx + c}}\, dx.$

Example 4.18 Integrate

$$\int \frac{1}{\sqrt{1+x+x^2}}\, dx.$$

Here

$$\int \frac{dx}{\sqrt{1+x+x^2}} = \int \frac{1}{\sqrt{(x+1/2)^2 + 3/4}}\, dx$$

$$= \sinh^{-1} \frac{x + 1/2}{\sqrt{3}/2} + c$$

$$= \sinh^{-1} \frac{2x+1}{\sqrt{3}} + c.$$

(viii) Type $\int \dfrac{px + q}{\sqrt{ax^2 + bx + c}}\, dx.$

Example 4.19 Evaluate the integral

$$\int \frac{5 + 3x}{\sqrt{2 - 3x - 4x^2}}\, dx.$$

Let

$$\int \frac{5+3x}{\sqrt{2-3x-4x^2}}\, dx = \frac{1}{3} \int \frac{x + 5/3}{\sqrt{2-3x-4x^2}}\, dx = -\frac{1}{24} \int \frac{-8x - 40/3}{\sqrt{2-3x-4x^2}}\, dx$$

$$= -\frac{1}{24}\int \frac{-8x-3-31/3}{\sqrt{2-3x-4x^2}}\,dx$$

$$= -\frac{1}{24}\int \frac{-8x-3}{\sqrt{2-3x-4x^2}}\,dx + \frac{31}{72}\int \frac{1}{\sqrt{2-3x-4x^2}}\,dx$$

$$= -\frac{1}{24}(2\sqrt{2-3x-4x^2}) + \frac{31}{72}\int \frac{1}{\sqrt{41/16-(2x+3/4)^2}}\,dx$$

$$= -\frac{1}{12}\sqrt{2-3x-4x^2} + \frac{31}{72}\frac{1}{2}\sin^{-1}\frac{2x+3/4}{\sqrt{41}/4} + c$$

$$= -\frac{1}{12}\sqrt{2-3x-4x^2} + \frac{31}{144}\sin^{-1}\frac{8x+3}{\sqrt{41}} + c$$

(ix) Type $\int (px+q)\sqrt{ax^2+bx+c}\,dx$.

Example 4.20 Evaluate

$$\int (x+1)\sqrt{2x^2+3}\,dx.$$

We have

$$\int (x+1)\sqrt{2x^2+3}\,dx = \frac{1}{4}\int (4x+4)\sqrt{2x^2+3}\,dx$$

$$= \frac{1}{4}\int 4x\sqrt{2x^2+3}\,dx + \int \sqrt{2x^2+3}\,dx$$

$$= \frac{1}{4}\frac{(2x^2+3)^{3/2}}{3/2} + \sqrt{2}\int \sqrt{x^2+\frac{3}{2}}\,dx$$

$$= \frac{1}{6}(2x^2+3)^{3/2} + \sqrt{2}\,\frac{3}{4}\log\left(x+\sqrt{x^2+\frac{3}{2}}\right) + \frac{x}{2}\sqrt{x^2+\frac{3}{2}} + c$$

$$= \frac{1}{6}(2x^2+3)^{3/2} + \frac{3\sqrt{2}}{4}\log\left(x+\sqrt{x^2+\frac{3}{2}}\right) + \frac{x}{2\sqrt{2}}\sqrt{2x^2+3} + c.$$

(x) Type $\int \frac{dx}{(px+q)\sqrt{ax^2+bx+c}}$

Put $1/(px+q)=t$ and proceed.

Example 4.21 Evaluate

$$\int \frac{dx}{(x+1)\sqrt{x^2-1}}.$$

Putting $x+1=1/t$, we get $dx=(-1/t^2)\,dt$. Then

$$\int \frac{1}{(x+1)\sqrt{x^2-1}}\,dx = \int \frac{1}{(1/t)\sqrt{(1/t-1)^2-1}}\left(-\frac{1}{t^2}\right)dt$$

$$= \int \frac{1}{(1/t)\sqrt{(1/t^2)-(2/t)}}\left(-\frac{1}{t^2}\right)dt$$

$$= -\int \frac{dt}{\sqrt{1-2t}}$$

$$= (1-2t)^{1/2}+c$$

$$= \sqrt{1-\frac{2}{x+1}}+c$$

$$= \sqrt{\frac{x-1}{x+1}}+c$$

(xi) Type $\int \frac{dx}{(px+q)\sqrt{ax+b}}$

Put $ax+b=t^2$ and proceed.

Example 4.22 Integrate

$$\int \frac{dx}{(2x+3)\sqrt{x+5}}.$$

Putting $x+5=t^2$, we get $dx=2t\,dt$. Then

$$\int \frac{dx}{(2x+3)\sqrt{x+5}} = \int \frac{2t\,dt}{[2(t^2-5)+3]t}$$

$$= \int \frac{2\,dt}{2t^2-7}$$

$$= \int \frac{dt}{t^2 - (7/2)}$$

$$= \frac{1}{\sqrt{14}} \log \frac{\sqrt{2x+10} - \sqrt{7}}{\sqrt{2x+10} + \sqrt{7}} + C$$

(xii) Form $\int \frac{dx}{(px^2 + q)\sqrt{ax^2 + b}}$

Put $x = 1/t$ and proceed. Alternatively, the substitution $(ax^2 + b)/(px^2 + q) = t^2$ may also be used. If the factor $px^2 + q$ can be put into linear factors, then such an integral can be reduced into one of type (xi).

Example 4.23 Evaluate

$$\int \frac{dx}{x^2\sqrt{4 - x^2}}$$

Let $x = 1/t$ so that $dx = (-1/t^2)\,dt$. Then

$$\int \frac{dx}{x^2\sqrt{4 - x^2}} = \int \frac{(-1/t^2)dt}{(1/t^2)\sqrt{4 - 1/t^2}}$$

$$= -\int \frac{t\,dt}{\sqrt{4t^2 - 1}}$$

$$= -\frac{1}{4}\sqrt{4t^2 - 1} + c$$

$$= -\frac{1}{4}\sqrt{\frac{4}{x^2} - 1} + c$$

$$= -\frac{1}{4x}\sqrt{4 - x^2} + c.$$

Alternatively, putting $(4 - x^2)/x^2 = t^2$, we get

$$x = \frac{2}{\sqrt{t^2 + 1}} \quad \text{and} \quad dx = -\frac{2t\,dt}{(t^2 + 1)^{3/2}}$$

Also

$$\sqrt{4 - x^2} = \sqrt{4 - \frac{4}{t^2 + 1}} = \frac{2t}{\sqrt{t^2 + 1}}$$

Hence,

$$\int \frac{dx}{x^2\sqrt{4-x^2}} = \int \frac{-2t\,dt}{(t^2+1)^{3/2}} \frac{1}{4/(t^2+1)} \frac{1}{2t/\sqrt{t^2+1}}$$

$$= -\frac{1}{4}\int dt$$

$$= -\frac{1}{4}t + c$$

$$= -\frac{1}{4}\sqrt{\frac{4-x^2}{x^2}} + c$$

$$= -\frac{1}{4x}\sqrt{4-x^2} + c, \text{ as before.}$$

4.2.2 Integration by Parts

We know that if u and v are functions of x, then

$$\frac{d}{dx}(uv) = u\frac{dv}{dx} + v\frac{du}{dx}$$

or

$$u\frac{dv}{dx} = \frac{d}{dx}(uv) - v\frac{du}{dx}.$$

Integrating both sides with respect to x, we obtain:

$$\int u\frac{dv}{dx}\,dx = uv - \int v\frac{du}{dx}\,dx + C$$

or

$$\int u\,dv = uv - \int v\,du + C \tag{4.30}$$

We see that formula (4.30) expresses one integral, $\int u\,dv$, in terms of a second integral, $\int v\,du$. If we choose so that the second integral is simpler than the first, the evaluation can be done quite simply to arrive at the required answer. The process involved in formula (4.30) is known as *integration by parts*. The following examples illustrate the procedure.

Example 4.24 Evaluate the following integrals:

(a) $\int xe^x dx$ (b) $\int \log_e x\,dx$

(c) $\int_0^{\pi/2} x^2 \sin 2x\,dx$ (d) $\int e^{-2x}\cos 3x\,dx$

(e) $\int_{-1}^{3/2} |x\sin \pi x|\,dx.$

(a) We observe that

$$\frac{d}{dx}(e^x) = e^x$$

and hence we choose here $e^x = v$ and $x = u$.
Formula (4.30) now gives

$$\int xe^x dx = xe^x - \int e^x\, dx = xe^x - e^x + C = e^x(x-1) + C$$

(b) We choose $\log_e x = u$ and $x = v$. Formula (4.30) now gives:

$$\int \log_e x\, dx = x \log_e x - \int x \cdot \frac{1}{x}\, dx = x \log_e x - x + C.$$

In the same way as above, the integrals

$$\int \tan^{-1} x\, dx, \qquad \int x \sin^{-1} x\, dx,$$

etc., can also be evaluated.

(c) Here, we choose $dv = \sin 2x\, dx$ and $u = x^2$. Using the formula, we obtain

$$\int_0^{\pi/2} x^2 \sin 2x\, dx = \left[x^2 \left(-\frac{\cos 2x}{2} \right) \right]_0^{\pi/2} - \int_0^{\pi/2} -\frac{\cos 2x}{2} (2x)\, dx$$

$$= -\left[\frac{1}{2} x^2 \cos 2x \right]_0^{\pi/2} + \int_0^{\pi/2} x \cos 2x\, dx$$

Now,

$$\left[x^2 \cos 2x \right]_0^{\pi/2} = \left(\frac{\pi}{2} \right)^2 \cos \frac{2\pi}{2} - 0 = -\frac{\pi^2}{4},$$

and

$$\int_0^{\pi/2} x \cos 2x\, dx = \left[x \frac{\sin 2x}{2} \right]_0^{\pi/2} - \int_0^{\pi/2} \frac{\sin 2x}{2}\, dx$$

$$= \frac{1}{2} \left(\frac{\pi}{2} \sin \frac{2\pi}{2} - 0 \right) - \frac{1}{2} \left[-\frac{\cos 2x}{2} \right]_0^{\pi/2}$$

$$= 0 + \frac{1}{4} \left(\cos \frac{2\pi}{2} - \cos 0 \right)$$

$$= \frac{1}{4}(-1-1)$$

$$= -\frac{1}{2}.$$

Hence,

$$\int_0^{\pi/2} x^2 \sin 2x\, dx = -\frac{1}{2}\left(-\frac{\pi^2}{4}\right) + \left(-\frac{1}{2}\right) = \frac{\pi^2}{8} - \frac{1}{2}.$$

(d) We shall evaluate the more general integral given by

$$I = \int e^{ax} \cos bx\, dx$$

We choose $dv = \cos bx\, dx$ and $u = e^{ax}$. Formula (4.30) gives

$$I = e^{ax}\frac{\sin bx}{b} - \int \frac{\sin bx}{b} a e^{ax} dx$$

$$= \frac{1}{b} e^{ax} \sin bx - \frac{a}{b}\int e^{ax} \sin bx\, dx.$$

To evaluate the integral on the right side, we choose $\sin bx\, dx = dv$, $u = e^{ax}$ and integrate again using (4.30). We then obtain:

$$I = \frac{1}{b} e^{ax} \sin bx - \frac{a}{b}\left[e^{ax}\left(-\frac{\cos bx}{b}\right) - \int -\frac{\cos bx}{b} a e^{ax}\, dx\right]$$

$$= \frac{1}{b} e^{ax} \sin bx - \frac{a}{b}\left(-\frac{1}{b} e^{ax} \cos bx + \frac{a}{b}\int e^{ax} \cos bx\, dx\right)$$

$$= \frac{1}{b} e^{ax} \sin bx + \frac{a}{b^2} e^{ax} \cos bx - \frac{a^2}{b^2} I + c$$

so that

$$\left(1 + \frac{a^2}{b^2}\right) I = \frac{1}{b} e^{ax} \sin bx + \frac{a}{b^2} e^{ax} \cos bx + c$$

$$= \frac{e^{ax}}{b^2}(b \sin bx + a \cos bx) + c$$

Hence

$$I = \frac{e^{ax}}{a^2 + b^2}(b \sin bx + a \cos bx) + c'.$$

We therefore write down

$$\int e^{-2x} \cos 3x\, dx = \frac{e^{-2x}}{13}(3 \sin 3x - 2 \cos 3x) + c.$$

(e) We observe that as x varies from -1 to $3/2$, the product $x \sin \pi x$ changes sign at the point $x = 1$; it is *positive* if $-1 < x < 1$ and *negative* if $1 < x < 3/2$. We therefore split the range of integration into two parts and write the integral as

$$I = \int_{-1}^{1} |x \sin \pi x| \, dx + \int_{1}^{3/2} |x \sin \pi x| \, dx$$

$$= \int_{-1}^{1} x \sin \pi x \, dx - \int_{1}^{3/2} x \sin \pi x \, dx$$

$$= \left[\frac{x \cos \pi x}{-\pi} + \frac{\sin \pi x}{\pi^2} \right]_{-1}^{1} - \left[\frac{x \cos \pi x}{-\pi} + \frac{\sin \pi x}{\pi^2} \right]_{1}^{3/2}$$

$$= \left(\frac{\cos \pi}{-\pi} + \frac{\sin \pi}{\pi^2} \right) - \left[\frac{(-1) \cos (-\pi)}{-\pi} + \frac{\sin (-\pi)}{\pi^2} \right]$$

$$- \left(\frac{3/2 \cos (3\pi/2)}{-\pi} + \frac{\sin (3\pi/2)}{\pi^2} \right) + \left(\frac{\cos \pi}{-\pi} + \frac{\sin \pi}{\pi^2} \right)$$

$$= -\frac{2}{\pi} \cos \pi - \frac{1}{\pi} \cos \pi + \frac{1}{\pi^2}$$

$$= \frac{3}{\pi} + \frac{1}{\pi^2}.$$

4.2.3 Integrationby Partial Fractions

We have already discussed about partial fractions in Chapter 1. The use of partial fractions enables us to find certain integrals quite easily. We consider integrals whose integrands have linear or quadratic factors in denominators.

We first consider examples of fractions whose denominators contain only linear factors.

Example 4.25 Integrate

$$I = \int \frac{dx}{1 - x^2}.$$

By partial fractions

$$\frac{1}{1 - x^2} = \frac{A}{1 - x} + \frac{B}{1 + x}.$$

We then have

$$1 = A(1 + x) + B(1 - x) = A + B + (A - B)x$$

which gives $A + B = 1$ and $A - B = 0$. The solution of this system is given by $A = B = 1/2$. Hence,

$$I = \int \frac{1}{1-x^2}\, dx = \int \left(\frac{1/2}{1-x} + \frac{1/2}{1+x} \right) dx$$

$$= -\frac{1}{2} \log_e |1-x| + \frac{1}{2} \log_e |1+x| + C$$

$$= \frac{1}{2} \log_e \left| \frac{1+x}{1-x} \right| + C.$$

Example 4.26 Evaluate

$$I = \int \frac{7(x-3)}{2x^2 + 5x - 3}\, dx$$

We have

$$\frac{7(x-3)}{2x^2 + 5x - 3} = \frac{7(x-3)}{(2x-1)(x+3)} = \frac{A}{2x-1} + \frac{B}{x+3}$$

It follows

$$A(x+3) + B(2x-1) = 7(x-3)$$

Clearly, $A = -5$ and $B = 6$. Hence

$$I = \int \left(\frac{-5}{2x-1} + \frac{6}{x+3} \right) dx = -\frac{5}{2} \log_e (2x-1) + 6 \log_e (x+3) + C.$$

Our next example will be concerned with repeated linear factors in the denominator.

Example 4.27 Evaluate

$$I = \int \frac{x\, dx}{(x+2)^2 (x-1)}$$

We write

$$\frac{x}{(x+2)^2 (x-1)} = \frac{A}{x-1} + \frac{B}{x+2} + \frac{C}{(x+2)^2}.$$

From the theory of partial fractions, we obtain $A = 1/9$, $B = -1/9$ and $C = 2/3$. Hence

$$I = \int \left[\frac{1}{9} \frac{1}{x-1} - \frac{1}{9} \frac{1}{x+2} + \frac{2}{3} \frac{1}{(x+2)^2} \right] dx$$

$$= \frac{1}{9} \log_e (x-1) - \frac{1}{9} \log_e (x+2) + \frac{2}{3} \left(-\frac{1}{x+2} \right) + C$$

$$= \frac{1}{9} \log_e \frac{x-1}{x+2} - \frac{2}{3} \frac{1}{(x+2)} + C.$$

We, finally, consider the case where the denominator has quadratic factors. This is illustrated in the next two examples.

Example 4.28 Integrate

$$I = \int \frac{x}{(x+1)(x^2+4x+13)}\,dx.$$

In this case, we write

$$\frac{x}{(x+1)(x^2+4x+13)} = \frac{A}{x+1} + \frac{Bx+C}{x^2+4x+13}.$$

This gives

$$A(x^2+4x+13) + (x+1)(Bx+C) = x.$$

Comparing the coefficients of x^2 and x and equating the constant terms, we get:

$$A+B=0, \qquad 4A+B+C=1, \qquad 13A+C=0,$$

the solution of which is: $A=-1/10$, $B=1/10$ and $C=13/10$. Hence, the given integral can be written as:

$$I = \int \left(-\frac{1}{10(x+1)} + \frac{1}{10}\frac{x+13}{x^2+4x+13} \right) dx$$

$$= -\frac{1}{10}\log_e(x+1) + \frac{1}{10}\int \frac{x+13}{x^2+4x+13}\,dx.$$

The integral on the rightside can be written as:

$$\int \frac{x+13}{x^2+4x+13}\,dx = \frac{1}{2}\int \frac{2x+4}{x^2+4x+13}\,dx + \int \frac{11}{x^2+4x+13}\,dx$$

$$= \frac{1}{2}\log_e(x^2+4x+13) + 11\int \frac{1}{(x+2)^2+9}\,dx$$

$$= \frac{1}{2}\log_e(x^2+4x+13) + \frac{11}{3}\tan^{-1}\frac{x+2}{3} + C.$$

Hence,

$$I = -\frac{1}{10}\log_e(x+1) + \frac{1}{20}\log_e(x^2+4x+13) + \frac{11}{30}\tan^{-1}\frac{x+2}{3} + C.$$

Example 4.29 Evaluate the integral

$$I = \int \frac{dx}{(x+1)(x^2+1)}$$

Let

$$\frac{1}{(x+1)(x^2+1)} = \frac{A}{x+1} + \frac{Bx+C}{x^2+1}.$$

Then,

$$1 = A(x^2+1) + (Bx+C)(x+1),$$

from which we obtain:

$$A+B=0, \qquad B+C=0, \qquad A+C=1.$$

The solution of the above system is easily seen to be $A=C=1/2$ and $B=-1/2$. Hence

$$I = \int \frac{1}{(x+1)(x^2+1)}\,dx = \int \left(\frac{1/2}{x+1} - \frac{1}{2}\frac{x-1}{x^2+1}\right) dx$$

$$= \frac{1}{2}\log_e(x+1) - \frac{1}{4}\int \frac{2x\,dx}{x^2+1} + \frac{1}{2}\int \frac{1}{x^2+1}\,dx$$

$$= \frac{1}{2}\log_e(1+x) - \frac{1}{4}\log_e(x^2+1) + \frac{1}{2}\tan^{-1}x + C$$

$$= \frac{1}{4}\log_e \frac{(1+x)^2}{x^2+1} + \frac{1}{2}\tan^{-1}x + C.$$

4.2.4 Reduction Formulae

Suppose we wish to integrate

$$I_n = \int x^n e^{ax}\,dx \tag{4.31}$$

by parts. Taking $u = x^n$ and $dv = e^{ax}\,dx$, we obtain from formula (4.30)

$$I_n = x^n \frac{e^{ax}}{a} - \int \frac{e^{ax}}{a} n x^{n-1}\,dx$$

$$= \frac{x^n e^{ax}}{a} - \frac{n}{a}\int x^{n-1}e^{ax}\,dx$$

$$= \frac{x^n e^{ax}}{a} - \frac{n}{a} I_{n-1} \tag{4.32}$$

Formulae like the above are called *reduction formulae* since they can be used continuously to reduce the integral. Take, for example, $n=4$ and $a=2$, i.e. the integral

$$I_4 = \int x^4 e^{2x}\,dx.$$

By repeated application of (4.32), we obtain:

$$I_4 = x^4 \frac{e^{2x}}{2} - \frac{4}{2} I_3$$

$$= \frac{1}{2} x^4 e^{2x} - 2I_3$$

$$= \frac{1}{2} x^4 e^{2x} - 2\left(x^3 \frac{e^{2x}}{2} - \int 3x^2 \frac{e^{2x}}{2} dx \right)$$

$$= \frac{1}{2} x^4 e^{2x} - x^3 e^{2x} + 3I_2$$

$$= \frac{1}{2} x^4 e^{2x} - x^3 e^{2x} + 3\left(x^2 \frac{e^{2x}}{2} - \int \frac{e^{2x}}{2} (2x)\, dx \right)$$

$$= \frac{1}{2} x^4 e^{2x} - x^3 e^{2x} + \frac{3}{2} x^2 e^{2x} - 3I_1$$

$$= \frac{1}{2} x^4 e^{2x} - x^3 e^{2x} + \frac{3}{2} x^2 e^{2x} - 3\left(x \frac{e^{2x}}{2} - \int \frac{e^{2x}}{2} dx \right)$$

$$= \frac{1}{2} x^4 e^{2x} - x^3 e^{2x} + \frac{3}{2} x^2 e^{2x} - \frac{3}{2} x e^{2x} + \frac{3}{4} e^{2x} + C.$$

A special case is obtained when $a = -1$. If we assume that

$$I_n = \int x^n e^{-x} dx,$$

then the reduction formula is:

$$I_n = -x^n e^{-x} - \int \frac{e^{-x}}{-1} (nx^{n-1})\, dx$$

$$= -x^n e^{-x} + n \int x^{n-1} e^{-x}\, dx$$

$$= -x^n e^{-x} + nI_{n-1} \tag{4.33}$$

Now, if we consider the definite integral

$$I_n = \int_0^\infty x^n e^{-x} dx, \tag{4.34}$$

the formula (4.33) gives:

$$I_n = [-x^n e^{-x}]_0^\infty + nI_{n-1} = nI_{n-1} \tag{4.35}$$

The function defined by the integral in (4.34) is known as the *gamma function* and is denoted by $\Gamma(n+1)$ so that

$$\Gamma(n+1) = \int_0^\infty x^n e^{-x}\,dx \tag{4.36}$$

The reduction formula (4.35) now gives:

$$\begin{aligned}\Gamma(n+1) &= n\Gamma(n) = n(n-1)\Gamma(n-1)\ldots\\ &= n(n-1)(n-2)\ldots 1\cdot\Gamma(1)\\ &= n!\int_0^\infty e^{-x}\,dx\\ &= n!\left[\frac{e^{-x}}{-1}\right]_0^\infty\\ &= n!\end{aligned}$$

For a detailed discussion of gamma functions, the reader is referred to Chapter 6.

We now consider a reduction formula for the integral

$$\begin{aligned}I_n &= \int \cos^n\theta\,d\theta\\ &= \int \cos^{n-1}\theta\cos\theta\,d\theta\\ &= \cos^{n-1}\theta\sin\theta - \int(n-1)\cos^{n-2}\theta(-\sin\theta)\sin\theta\,d\theta\\ &= \cos^{n-1}\theta\sin\theta + (n-1)\int\cos^{n-2}\theta(1-\cos^2\theta)d\theta\\ &= \cos^{n-1}\theta\sin\theta + (n-1)\int\cos^{n-2}\theta\,d\theta - (n-1)I_n \qquad (4.37)\end{aligned}$$

i.e.

$$nI_n = \cos^{n-1}\theta\sin\theta + (n-1)I_{n-2}$$

or

$$I_n = \frac{1}{n}[\cos^{n-1}\theta\sin\theta + (n-1)I_{n-2}] \tag{4.38}$$

where $I_0 = \theta$ and $I_1 = \sin\theta$.

Similarly, the reduction formula

$$I_n = \int \sin^n\theta\,d\theta = \frac{1}{n}[-\sin^{n-1}\theta\cos\theta + (n-1)I_{n-2}] \tag{4.39}$$

can also be derived.

Finally, we derive a formula for the integral

$$I_{p,q} = \int \sin^p x \cos^q x\, dx \tag{4.40}$$

This can be written as

$$\begin{aligned} I_{p,q} &= \int \sin^p x \cos^{q-1} x \cos x\, dx \\ &= \int \cos^{q-1} x \sin^p x \cos x\, dx \\ &= \cos^{q-1} x \frac{\sin^{p+1} x}{p+1} - \int \frac{\sin^{p+1} x}{p+1} + (q-1)\cos^{q-2} x(-\sin x)\, dx \\ &= \frac{\cos^{q-1} x \sin^{p+1} x}{p+1} + \frac{q-1}{p+1} \int \sin^{p+2} x \cos^{q-2} x\, dx \\ &= \frac{\cos^{q-1} x \sin^{p+1} x}{p+1} + \frac{q-1}{p+1} \int \sin^p x \cos^{q-2} x(1-\cos^2 x)\, dx \\ &= \frac{\cos^{q-1} x \sin^{p+1} x}{p+1} + \frac{q-1}{p+1} \int \sin^p x \cos^{q-2} x\, dx - \frac{q-1}{p+1} I_{p,q} \end{aligned}$$

or

$$\frac{p+q}{p+1} I_{p,q} = \frac{\cos^{q-1} x \sin^{p+1} x}{p+1} + \frac{q-1}{p+1} I_{p,q-2}.$$

The reduction formula is therefore given by

$$I_{p,q} = \frac{\cos^{q-1} x \sin^{p+1} x}{p+q} + \frac{q-1}{p+q} I_{p,q-2} \tag{4.41a}$$

In a similar way, the reduction formula

$$I_{p,q} = -\frac{\cos^{q+1} x \sin^{p-1} x}{p+q} + \frac{p-1}{p+q} I_{p-2,q} \tag{4.41b}$$

can also be derived.

In particular, the definite integrals

$$I_n = \int_0^{\pi/2} \sin^n \theta\, d\theta = \int_0^{\pi/2} \cos^n \theta\, d\theta = \frac{n-1}{n} I_{n-2},$$

and

$$I_{p,q} = \int_0^{\pi/2} \sin^p x \cos^q x\, dx$$

$$= \frac{q-1}{p+q} I_{p,q-2}$$

$$= \frac{p-1}{p+q} I_{p-2,q}$$

are quite useful in practical applications.

Example 4.30 Evaluate

$$\int_0^{2a} \frac{x^{3/2}}{\sqrt{2a-x}}\, dx$$

Let $x = 2a\sin^2\theta$. Then $dx = 4a\sin\theta\cos\theta\, d\theta$. Then

$$\int_0^{2a} \frac{x^{3/2}}{\sqrt{2a-x}}\, dx = \int_0^{\pi/2} \frac{(2a)^{3/2}\sin^3\theta\,(4a)\sin\theta\cos\theta\, d\theta}{\sqrt{2a}\cos\theta}$$

$$= 8a^2 \int_0^{\pi/2} \sin^4\theta\, d\theta$$

$$= 8a^2\, \frac{3\cdot 1}{4\cdot 2}\, \frac{\pi}{2}$$

$$= \frac{3}{2}\pi a^2$$

Example 4.31 Evaluate

$$\int_0^{\pi/2} \sin^9\theta\, d\theta.$$

Using the formula directly, we obtain

$$\int_0^{\pi/2} \sin^9\theta\, d\theta = \frac{8\cdot 6\cdot 4\cdot 2}{9\cdot 7\cdot 5\cdot 3} \int_0^{\pi/2} \sin\theta\, d\theta$$

$$= \frac{8\cdot 6\cdot 4\cdot 2}{9\cdot 7\cdot 5\cdot 3}$$

$$= \frac{128}{315}$$

Example 4.32 Evaluate

$$\int_0^{\pi/2} \sin^6\theta\cos^4\theta\, d\theta.$$

Here $p=6,\ q=4$. We therefore use the formula

$$I_{p,q}=\frac{q-1}{p+q}I_{p,q-2}$$

Hence

$$\int_0^{\pi/2}\sin^6\theta\cos^4\theta\,d\theta=\frac{3\cdot 1}{10\cdot 8}\,\frac{5\cdot 3\cdot 1}{6\cdot 4\cdot 2}\,\frac{\pi}{2}=\frac{3\pi}{512}.$$

Example 4.33 Evaluate

$$I_{3,4}=\int_0^{\pi/2}\sin^3 x\cos^4 x\,dx.$$

Using formula (4.41), we obtain

$$I_{3,4}=\left[\frac{\sin^4 x\cos^3 x}{7}\right]_0^{\pi/2}+\frac{3}{7}\int_0^{\pi/2}\sin^3 x\cos^2 x\,dx.$$

$$=0+\frac{3}{7}\int_0^{\pi/2}\sin^3 x\,dx-\frac{3}{7}\int_0^{\pi/2}\sin^5 x\,dx$$

From formula (4.39), we have

$$\int_0^{\pi/2}\sin^3 x\,dx=-\frac{1}{3}\left[\sin^2 x\cos x\right]_0^{\pi/2}+\frac{2}{3}\int_0^{\pi/2}\sin x\,dx$$

$$=0+\frac{2}{3}\left[-\cos x\right]_0^{\pi/2}$$

$$=\frac{2}{3},$$

and

$$\int_0^{\pi/2}\sin^5 x\,dx=-\frac{1}{5}\left[\sin^4 x\cos x\right]_0^{\pi/2}+\frac{4}{5}\int_0^{\pi/2}\sin^3 x\,dx=\frac{4}{5}\,\frac{2}{3}=\frac{8}{15}.$$

Hence,

$$I_{3,4}=\frac{3}{7}\,\frac{2}{3}-\frac{3}{7}\,\frac{8}{15}=\frac{2}{7}-\frac{8}{35}=\frac{2}{35}.$$

4.2.5 Some Special Techniques

A large number of methods exist for evaluating particular types of integrals and we demonstrate here a few of them with illustrative examples.

The substitution $t = \tan(x/2)$

This substitution will be useful in integrating functions of the form

$$\frac{1}{a\sin x + b\cos x + c}.$$

Example 4.34 Evaluate

$$I = \int_0^{\pi/2} \frac{1}{2+\cos x}\,dx.$$

We write

$$I = \int_0^{\pi/2} \frac{1}{2+\dfrac{1-\tan^2(x/2)}{1+\tan^2(x/2)}}\,dx = \int_0^{\pi/2} \frac{\sec^2(x/2)\,dx}{3+\tan^2(x/2)}.$$

Setting $t = \tan(x/2)$, we obtain

$$dt = \frac{1}{2}\sec^2\frac{x}{2}\,dx, \quad \text{or} \quad \sec^2\frac{x}{2}\,dx = 2\,dt.$$

Further, when $x = 0$, $t = 0$ and also when $x = \pi/2$, $t = 1$. Hence, the above integral becomes:

$$I = \int_0^1 \frac{2\,dt}{3+t^2} = \int_0^1 \frac{2\,dt}{(\sqrt{3})^2+t^2}$$

$$= \frac{2}{\sqrt{3}}\left[\tan^{-1}\frac{t}{\sqrt{3}}\right]_0^1$$

$$= \frac{2}{\sqrt{3}}\tan^{-1}\frac{1}{\sqrt{3}}$$

$$= \frac{2}{\sqrt{3}}\frac{\pi}{6}$$

$$= \frac{\pi}{3\sqrt{3}}.$$

Functions involving x and $\sqrt{ax^2+bx+c}$

If $b = 0$, the integrand can be reduced to one involving only trigonometric or hyperbolic functions by using an appropriate transformation. If it is of trigonometric type, the substitution $t = \tan(x/2)$ can be used. On the other hand, for integrands of the hyperbolic type, the substitution $x = \sinh t$ will be helpful.

If $b \neq 0$, then completing the square under the square-root sign gives:

$$\sqrt{ax^2 + bx + c} = \sqrt{a\left[\left(x + \frac{b}{2a}\right)^2 + \left(\frac{c}{a} - \frac{b^2}{4a^2}\right)\right]}.$$

The substitution $u = x + b/(2a)$ will then reduce the problem to one of the known types. The following examples illustrate the method of procedure.

Example 4.35 Evaluate

$$I = \int \sqrt{x^2 + 1}\, dx.$$

Here $b = 0$ and the integrand is of the hyperbolic type. Hence, we put

$$x = \sinh t, \qquad \text{so that } dx = \cosh t\, dt.$$

Also,

$$\sqrt{x^2 + 1} = \sqrt{1 + \sinh^2 t} = \cosh t.$$

Hence,

$$I = \int \cosh t \cosh t\, dt$$

$$= \int \cosh^2 t\, dt = \int \frac{1 + \cosh 2t}{2}\, dt$$

$$= \frac{1}{2}t + \frac{1}{4}\sinh 2t + C$$

$$= \frac{1}{2}\sinh^{-1} x + \frac{1}{4}(2 \sinh t \cosh t) + C$$

$$= \frac{1}{2}\sinh^{-1} x + \frac{1}{2}x\sqrt{x^2 + 1} + C.$$

The substitution $x = \tan t$ can also be used here but the resulting integration will be more complicated than that given above.

Example 4.36 Evaluatc

$$I = \int \frac{x\, dx}{\sqrt{x^2 - 2x + 3}} = \int \frac{x\, dx}{\sqrt{(x-1)^2 + 2}}, \qquad \text{since } b \neq 0.$$

Put $x - 1 = t$ so that $dx = dt$. Hence,

$$I = \int \frac{t + 1}{\sqrt{t^2 + 2}}\, dt$$

$$= \frac{1}{2}\int \frac{2t\,dt}{\sqrt{t^2+2}} + \int \frac{1}{\sqrt{t^2+2}}\,dt$$

$$= \sqrt{t^2+2} + \sinh^{-1}\frac{t}{\sqrt{2}} + C$$

$$= \sqrt{(x-1)^2+2} + \sinh^{-1}\frac{x-1}{\sqrt{2}} + C.$$

Use of Maclaurin's series

The Maclaurin expansion of a function can sometimes be used to evaluate a definite integral as accurately as possible. Consider, for instance, the integral.

$$I = \int_0^c \frac{\sin x}{x}\,dx.$$

We know that the Maclaurin expansion of sin x is given by

$$\sin x = x - \frac{x^3}{3!} + \frac{x^5}{5!} - \cdots$$

and therefore

$$\frac{\sin x}{x} = 1 - \frac{x^2}{3!} + \frac{x^4}{5!} - \cdots$$

Hence,

$$\int_0^c \frac{\sin x}{x}\,dx = \int_0^c \left(1 - \frac{x^2}{3!} + \frac{x^4}{5!} - \cdots\right) dx$$

$$= \left[x - \frac{x^3}{3(3!)} + \frac{x^5}{5(5!)} - \cdots\right]_0^c$$

$$= c - \frac{c^3}{3(3!)} + \frac{c^5}{5(5!)} - \cdots,$$

which is convergent for all values of c. If we consider, for instance, $c = 0.5$, we obtain

$$\int_0^{0.5} \frac{\sin x}{x}\,dx = 0.5 - \frac{(0.5)^3}{3(3!)} + \frac{(0.5)^5}{5(5!)} - \cdots$$

$$= 0.5 - 0.00694 + \frac{0.03125}{600} - \cdots$$

$$= 0.5 - 0.00694 + 0.000052 - \cdots$$

$$= 0.4931, \qquad \text{correct to four decimal places.}$$

We conclude this section with some illustrative examples on definite integrals.

Example 4.37 Evaluate

$$\int_0^{\pi/4} \log(1 + \tan\theta)\, d\theta.$$

Let

$$I = \int_0^{\pi/4} \log(1 + \tan\theta)\, d\theta$$

$$= \int_0^{\pi/4} \log\left[1 + \tan\left(\frac{\pi}{4} - \theta\right) d\theta\right]$$

$$= \int_0^{\pi/4} \log\left[1 + \frac{\tan(\pi/4) - \tan\theta}{1 + \tan(\pi/4)\tan\theta}\right] d\theta$$

$$= \int_0^{\pi/4} \log\left(1 + \frac{1 - \tan\theta}{1 + \tan\theta}\right) d\theta$$

$$= \int_0^{\pi/4} \log\frac{2}{1 + \tan\theta}\, d\theta$$

$$= \int_0^{\pi/4} [\log 2 - \log(1 + \tan\theta)]\, d\theta$$

$$= \int_0^{\pi/4} (\log 2)\, d\theta - I$$

i.e.

$$2I = \frac{\pi}{4}\log_e 2$$

Hence,

$$I = \int_0^{\pi/4} \log(1 + \tan\theta)\, d\theta = \frac{\pi}{8}\log_e 2.$$

Example 4.38 Evaluate

$$\int_0^1 (tx + 1 - x)^n\, dx$$

where n is a positive integer and t is a parameter independent of x. Hence, show that

$$\int_0^1 x^k (1-x)^{n-k} dx = {}^nC_k (n+1)^{-1} [{}^nC_k (n+1)]^{-1}, \quad k = 0, 1, 2, ..., n.$$

Let

$$I = \int_0^1 (tx + 1 - x)^n dx$$

$$= \int_0^1 [(t-1)\,x + 1]^n \, dx$$

$$= \left\{ \frac{[(t-1)\,x+1]^{n+1}}{(t-1)\,(n+1)} \right\}_0^1$$

$$= \frac{1}{(t-1)\,(n+1)} (t^{n+1} - 1)$$

$$= \frac{1}{n+1} \frac{t^{n+1} - 1}{t-1}$$

$$= \frac{1}{n+1} (1 + t + t^2 + \cdots + t^k + \cdots + t^n) \qquad \text{(i)}$$

Again,

$$I = \int_0^1 [tx + (1-x)^n] \, dx$$

Using Binomial theorem for a positive index, we get

$$I = \int_0^1 \sum_{k=0}^n {}^nC_k (1-x)^{n-k} t^k x^k \, dx = \sum_{k=0}^n {}^nC_k \, t^k \int_0^1 x^k \, (1-x)^{n-k} \, dx \qquad \text{(ii)}$$

From (i) and (ii), we therefore have

$$\frac{1}{n+1} (1 + t + t^2 + \cdots + t^k + \cdots + t^n) = \sum_{k=0}^n {}^nC_k \, t^k \int_0^1 x^k \, (1-x)^{n-k} \, dx \qquad \text{(iii)}$$

Comparing the coefficients of t^k on both sides of (iii), we get:

$$\frac{1}{n+1} = {}^nC_k \int_0^1 x^k (1-x)^{n-k} \, dx,$$

or

$$\int_0^1 x^k (1-x)^{n-k}\,dx = {}^nC_k\,(n+1)^{-1}[{}^nC_k(n+1)]^{-1},$$

which is the required result.

Example 4.39 Given that

$$\int_0^\infty e^{-x^2}\,dx = \frac{\sqrt{\pi}}{2},$$

show that $\int_0^\infty x^{-1/2}e^{-x}\,dx = \sqrt{\pi}$, and hence evaluate the integral

$$\int_0^a \left(\log_e \frac{a}{x}\right)^{-1/2} dx.$$

Let

$$I = \int_0^\infty e^{-x^2}\,dx = \frac{\sqrt{\pi}}{2}$$

Put $x^2 = y$ so that $2x\,dx = dy$. Then

$$dx = \frac{dy}{2x} = \frac{dy}{2\sqrt{y}} = \frac{1}{2}\,y^{-1/2}\,dy.$$

Hence,

$$I = \int_0^\infty e^{-y}\,\frac{1}{2}\,y^{-1/2}\,dy = \frac{1}{2}\int_0^\infty y^{-1/2}\,e^{-y}\,dy$$

By the property of definite integrals, we get

$$I = \frac{1}{2}\int_0^\infty x^{-1/2}\,e^{-x}\,dx,$$

Since $I = \sqrt{\pi}/2$, it follows that

$$\int_0^\infty x^{-1/2}e^{-x}\,dx = \sqrt{\pi}$$

Now, put $x = \log_e(a/t)$ in the above integral. Then

$$dx = \frac{t}{a}\left(-\frac{a}{t^2}\right)dt = -\frac{1}{t}\,dt.$$

When $x=0$, $t=a$ and when $x=\infty$, $t=0$. Hence the above integral becomes:

$$\sqrt{\pi}=\int_a^0\left(\log_e\frac{a}{t}\right)^{-1/2}\frac{t}{a}\left(-\frac{1}{t}\right)dt, \qquad \text{since } e^{-x}=\frac{t}{a}$$

$$=\int_0^a\frac{1}{a}\left(\log_e\frac{a}{t}\right)^{-1/2}dt$$

$$=\int_0^a\frac{1}{a}\left(\log_e\frac{a}{x}\right)^{-1/2}dx$$

It follows that

$$\int_0^a\left(\log_e\frac{a}{x}\right)^{-1/2}dx=a\sqrt{\pi}.$$

EXERCISES 4.2

Evaluate the following integrals (Problems 1–45):

1. $\int\frac{dx}{\sin x-\cos x}$.

2. $\int\frac{dx}{x^3+8}$.

3. $\int\frac{\sin x}{\sqrt{1+\sin x}}\,dx$.

4. $\int e^x\frac{1+\sin x}{1+\cos x}\,dx$.

5. $\int\frac{dx}{x^4-1}$.

6. $\int\frac{dx}{(x^2-6x+10)^3}$

7. $\int\frac{3x^3}{\sqrt{x^8-5}}\,dx$.

8. $\int e^{-\sqrt{x}}dx$.

9. $\int\frac{dx}{x^4+4}$.

10. $\int\sin^{-1}\sqrt{\frac{t}{a+t}}\,dt$.

11. $\int\cos\left(2\cos^{-1}\sqrt{\frac{1-t}{1+t}}\right)dt$.

12. $\int\sec^{-1}x\,dx$.

13. $\int\frac{1-x^2}{(1+x^2)\sqrt{1+x^4}}\,dx$.

14. $\int\sqrt{\tan x}\,dx$.

15. $\int\cos^5x\,dx$.

16. $\int\frac{dx}{(x^2+a^2)^{5/2}}$.

17. $\int\frac{dx}{20+29\cos x}$.

18. $\int\frac{4x^4-15x^2+11x+13}{2x^3+3x^2-3x-2}\,dx$.

19. $\int \frac{(x^2+1)(x^2+2)}{(x^2+3)(x^2+4)}\,dx.$

20. $\int \frac{x^2\,dx}{(x^2+a^2)(x^2+b^2)}.$

21. $\int \frac{x^2+2}{(x-1)^3}\,dx.$

22. $\int \frac{2x-1}{x^4+x^2+1}\,dx.$

23. $\int \frac{\log x}{(x^2+1)^2}\,dx.$

24. $\int_0^1 \cot^{-1}(1-x+x^2)\,dx.$

25. $\int \sqrt{\frac{\sin(x-\alpha)}{\sin(x+\alpha)}}\,dx.$

26. $\int \sqrt{\frac{1-\sqrt{x}}{1+\sqrt{x}}}\,dx.$

27. $\int \frac{1}{\sqrt{3-2x-3x^2}}\,dx.$

28. $\int \frac{2x-1}{\sqrt{x^2+7x+3}}\,dx.$

29. $\int_{1/2}^{1} \frac{(x-x^4)^{1/4}}{x^5}\,dx.$

30. $\int_{\alpha}^{\beta} \frac{dx}{\sqrt{(x-\alpha)(\beta-x)}}.$

31. $\int \frac{3\sin x+12\cos x}{5\sin x+14\cos x}\,dx.$

32. $\int \frac{dx}{(1+x)\sqrt{1+x-x^2}}.$

33. $\int \frac{dx}{(x^2-1)\sqrt{x^2+1}}.$

34. $\int \frac{dx}{(x-1)\sqrt{5x-2-2x^2}}.$

35. $\int \frac{x^2}{(\sin x+\cos x)^2}\,dx.$

36. $\int \frac{8\cos x+\sin x}{3\cos x+2\sin x}\,dx.$

37. $\int \frac{dx}{\sqrt{\sin^3 x\sin(x+\alpha)}}.$

38. $\int \left(\sqrt{\tan x}+\sqrt{\cot x}\right)dx.$

39. $\int \frac{x^5+4}{x^5-x}\,dx.$

40. $\int_8^{15} \frac{dx}{(x-3)\sqrt{x+1}}.$

41. $\int \sqrt{\frac{3-2x}{3x+4}}\,dx.$

42. $\int \sqrt{\frac{5-x}{x-2}}\,dx.$

43. $\int_0^1 x^{3/2}(1-x)^{3/2}\,dx.$

44. $\int_0^{\pi} \frac{d\theta}{5+3\cos\theta}.$

45. $\int_0^1 x^6(1-x)^9\,dx.$

46. Show that

$$\int x^4\cos 2x\,dx = \frac{\sin 2x}{4}(2x^4-6x^2+3)+\frac{x\cos 2x}{2}(2x^2-3)+C$$

47. If $I_n = \int \tan^n\theta \, d\theta$, show that

$$I_n = \frac{\tan^{n-1}\theta}{n-1} - I_{n-2},$$

and hence, or otherwise, show that

$$\int_0^{\pi/4} \tan^5\theta = \frac{1}{4}(2\log 2 - 1).$$

48. Show that

$$\int \cos^p x \cos qx \, dx = \frac{1}{p+q}\cos^p x \sin qx + \frac{p}{p+q}\int \cos^{p-1} x \cos(q-1)x \, dx.$$

49. By using the series expansion of e^{-x^2}, show that $\int_0^{0.4} e^{-x^2} dx = 0.380$, correct to three decimal places.

50. Show that

$$\int_0^{\pi} f(\sin\theta)\, d\theta = 2\int_0^{\pi/2} f(\sin\theta)\, d\theta.$$

Prove the following results (Problems 51–63):

51. $\displaystyle\int_0^{\infty} e^{-ax}\cos mx \, dx = \frac{a}{a^2+m^2}.$

52. $\displaystyle\int_0^{\infty} e^{-ax}\sin mx \, dx = \frac{m}{a^2+m^2}.$

53. $\displaystyle\int_0^{\pi/2} \cos^5 x \, dx = \frac{8}{15}.$

54. $\displaystyle\int_0^{\pi/2} \sin^7 x \cos^5 x \, dx = \frac{1}{120}.$

55. $\displaystyle\int_{-a}^{a} x^3\sqrt{2a^2 - x^2}\, dx = 0$

56. $\displaystyle\int_1^{\infty} \frac{dx}{(x+1)^2(x^2+1)} = 0.077.$

57. $\displaystyle\int_0^{\pi/2} \frac{\sqrt{\sin x}}{\sqrt{\sin x} + \sqrt{\cos x}}\, dx = \frac{\pi}{4}.$

58. $\displaystyle\int_{\pi/8}^{\pi/4} x\cos 2x \, dx = \frac{\pi}{8} - \frac{\pi}{16\sqrt{2}} - \frac{1}{4\sqrt{2}}.$

59. $\displaystyle\int_0^{\pi/2} \sin 2x \log\tan x \, dx = 0$

60. $\displaystyle\int_0^{\pi/2} \frac{\sin\theta\cos\theta\, d\theta}{\cos^2\theta + 3\cos\theta + 2} = \log_e\frac{9}{8}.$

61. $\displaystyle\int_0^{\pi/2} x \sin x \cos^4 x\,dx = \frac{\pi}{5}$

62. $\displaystyle\int_0^{\pi/2} \sin^6\theta\,d\theta = \frac{5\pi}{32}$

63. $\displaystyle\int_0^{\pi} x \cos^4 x\,dx = \frac{3\pi^2}{16}.$

64. Obtain a reduction formula for $\int_0^{\pi/2} \sec^2 x\,dx$ and hence evaluate $\int_0^{\pi/4} \sec^5 x\,dx$.

65. If $I_n = \int_0^{\pi/2} x^n \cos x\,dx$, prove that

$$I_n + n(n-1)\,I_{n-2} = \left(\frac{\pi}{2}\right)^n.$$

4.3 APPLICATIONS OF THE DEFINITE INTEGRAL

In Section 4.1, we have seen how the concept of the limit of a sum is used to define a definite integral, i.e. to obtain the area bounded by a curve, the x-axis, and two given ordinates. We can use a similar approach to develop formulae for determining important physical quantities like volumes and surface areas of solids of revolution, length of a curve, moments of inertia, centre of gravity, etc. Before proceeding to do this, we shall first illustrate the use of definite integrals in calculating the areas bounded by curves.

Example 4.40 Find the area bounded by the curve $x^2 = 4y$ and the straight line $x = 4y - 2$.

The bounded area is shown in Fig. 4.5. The points of intersection of the parabola and the straight line are given by

$$x^2 = 4\,\frac{x+2}{4} = x + 2$$

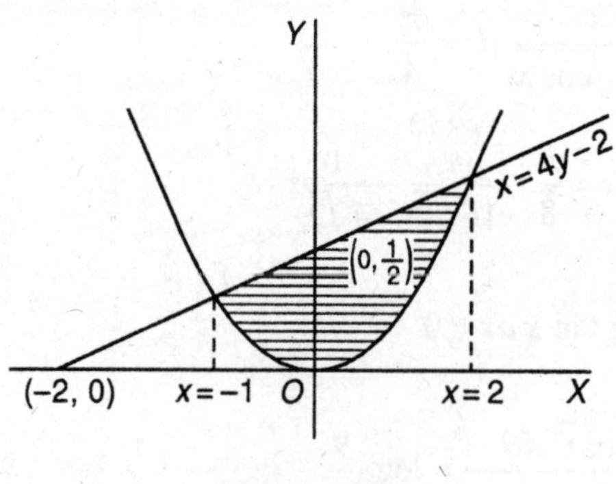

Figure 4.5

Solving, we get $x = -1, 2$. Hence, the area bounded is given by

$$A = \int_{-1}^{2} \left(\frac{x+2}{4} - \frac{x^2}{4} \right) dx = \left[\frac{(x+2)^2}{8} - \frac{x^3}{12} \right]_{-1}^{2} = \frac{9}{8}.$$

Example 4.41 For any real t,

$$x = \frac{e^t + e^{-t}}{2}, \qquad y = \frac{e^t - e^{-t}}{2}$$

is a point on the hyperbola $x^2 - y^2 = 1$. Show that the area bounded by this hyperbola and the lines joining its centre to the points corresponding to t_1 and $-t_1$ is t_1.

Let $B(t_1)$ and $C(-t_1)$ be two points on the hyperbola. The area bounded by the hyperbola and the two lines OB and OC is shown in Fig. 4.6. The required area is obtained by subtracting the portion $BACPB$ from the area of the triangle OBC. Now,

$$\text{The area of the } \Delta OBC = \frac{1}{2} CB \cdot OP$$

$$= PB \cdot OP$$

$$= \frac{e^{t_1} - e^{-t_1}}{2} \, \frac{e^{t_1} + e^{-t_1}}{2}$$

$$= \frac{1}{4}(e^{2t_1} - e^{-2t_1}).$$

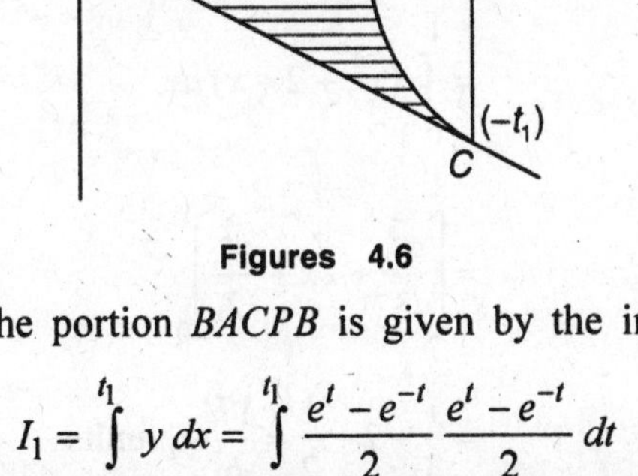

Figures 4.6

Also, the area of the portion $BACPB$ is given by the integral

$$I_1 = \int_{-t_1}^{t_1} y \, dx = \int_{-t_1}^{t_1} \frac{e^t - e^{-t}}{2} \, \frac{e^t - e^{-t}}{2} \, dt$$

$$= \int_{-t_1}^{t_1} \frac{1}{4}(e^t - e^{-t})^2 \, dt$$

$$= \int_{-t_1}^{t_1} \frac{1}{4}(e^{2t} + e^{-2t} - 2) \, dt$$

$$= \left[\frac{e^{2t}}{8} - \frac{e^{-2t}}{8} - \frac{1}{2} t \right]_{-t_1}^{t_1}$$

$$= \frac{e^{2t_1}}{8} - \frac{e^{-2t_1}}{8} - \frac{1}{2} t_1 - \left(\frac{e^{-2t_1}}{8} - \frac{e^{2t_1}}{8} + \frac{1}{2} t_1 \right)$$

i.e.

$$I_1 = \frac{1}{4} e^{2t_1} - \frac{1}{4} e^{-2t_1} - t_1.$$

Hence,

$$\text{The required area} = \textit{Area of the } \Delta OBC - I_1$$

$$= \frac{1}{4}(e^{2t_1} - e^{-2t_1}) - \left(\frac{1}{4} e^{2t_1} - \frac{1}{4} e^{-2t_1} - t_1 \right)$$

$$= t_1.$$

Example 4.42 Sketch the region bounded by the parabola $y = x^2 + 2$ and the straight lines $x = 0$, $x = 1$ and $y + x = 0$. Find the area of this region.

The area is shown in Fig. 4.7. The area is given by

$$A = \int_0^1 [x^2 + 2 - (-x)] \, dx$$

$$= \int_0^1 (x^2 + 2 + x) \, dx$$

$$= \left[\frac{x^3}{3} + 2x + \frac{x^2}{2} \right]_0^1$$

$$= \frac{1}{3} + 2 + \frac{1}{2} = \frac{17}{6} \text{ sq. units}$$

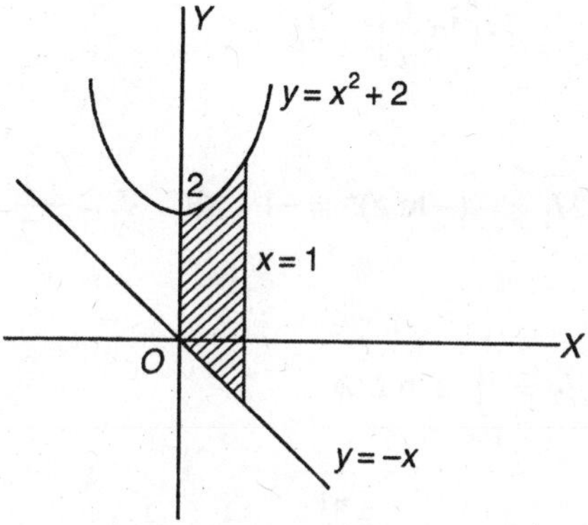

Figure 4.7

Example 4.43 Find the area bounded by the curves

$$y = ex \ln x \quad \text{and} \quad y = \frac{\ln x}{ex}, \qquad \text{where} \quad \ln e = 1.$$

The x-coordinates of the points of intersection of the two curves are given by

$$ex \ln x = \frac{\ln x}{ex}.$$

Hence,

$$\ln x = 0 \quad \text{or} \quad e^2x^2 = 1.$$

i.e.

$$x = 1 \quad \text{or} \quad x = \pm\frac{1}{e} = \frac{1}{e},$$

since the negative sign is inadmissible. Therefore,

$$\text{The required area} = \int_{1/e}^{1} \left(\frac{\ln x}{ex} - ex \ln x\right) dx$$

$$= \frac{1}{e} I_1 - eI_2, \qquad \text{say.} \tag{i}$$

Now,

$$I_1 = \int_{1/e}^{1} \frac{\ln x}{x}\, dx$$

$$= [(\ln x)^2]_{1/e}^{1} - \int_{1/e}^{1} (\ln x)\frac{1}{x}\, dx$$

$$= -\left(\ln\frac{1}{e}\right)^2 - I_1$$

Then

$$2I_1 = -(-\ln e)^2 = -1 \quad \text{or} \quad I_1 = -\frac{1}{2}.$$

Also,

$$I_2 = \int_{1/e}^{1} x \ln x \, dx$$

$$= \left[(\ln x)\frac{x^2}{2}\right]_{1/e}^{1} - \int_{1/e}^{1} \frac{x^2}{2}\frac{1}{x}\,dx.$$

$$= -\frac{1}{2e^2}\left(\ln\frac{1}{e}\right) - \left[\frac{x^2}{4}\right]_{1/e}^{1}$$

$$= \frac{1}{2e^2} - \left(\frac{1}{4} - \frac{1}{4e^2}\right)$$

$$= \frac{3}{4e^2} - \frac{1}{4}.$$

Hence from (i),

$$\text{The required area} = \frac{1}{e}\left(-\frac{1}{2}\right) - e\left(\frac{3}{4e^2} - \frac{1}{4}\right)$$

$$= -\frac{1}{2e} - \frac{3}{4e} + \frac{e}{4}$$

$$= \frac{e}{4} - \frac{5}{4e}.$$

4.3.1 Volumes and Surfaces of Solids of Revolution

Let $AA'BB'$ (Fig. 4.8) be the area bounded by the curve $y = f(x)$, the x-axis and the ordinates at $A'(x = a)$ and $B'(x = b)$. Suppose that this area is rotated about OX through four right angles (i.e. a complete revolution). Then it sweeps out a volume which is a solid of revolution, such as the one shown in Fig. 4.9. It is our objective now to find a formula for this volume. If $A'B'$ is divided into n intervals of width $P'Q'(= \Delta x)$, then the volume ΔV of the disc formed by rotating the strip $PQP'O'$ about OX is estimated as that of the circular slice formed by rotating the rectangle $PRP'Q'$ about

OX. If the coordinates of P are (x_i, y_i), then the volume of the circular disc is $\pi y_i^2 \,\Delta x$, and hence

$$\Delta V = \pi y_i^2 \,\Delta x.$$

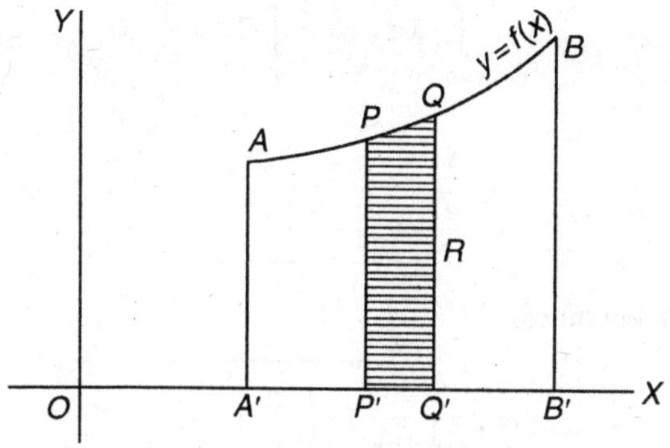

Figure 4.8

An estimate of the total volume is therefore given by

$$\sum_{i=1}^{n} \pi y_i^2 \,\Delta x \qquad \text{(since there are } n \text{ strips).}$$

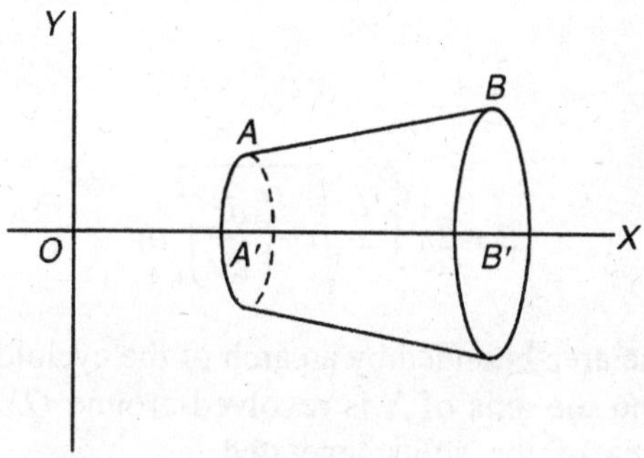

Figure 4.9

If the number of intervals increases indefinitely, i.e. as $\Delta x \to 0$, the above estimate tends to coincide with the actual volume. Hence, in the limit,

$$V = \lim_{\Delta x \to 0,\, n \to \infty} \sum_{i=1}^{n} \pi y_i^2 \Delta x = \int_a^b \pi y^2 \, dx, \tag{4.42}$$

where V is the required volume of the solid of revolution.

Further, an estimate of the curved surface area of the slice may be found by assuming that it is formed by rotating the chord PQ, instead of the arc PQ, about OX. Since the solid formed is a slice of a cone, the surface area is approximately equal to the circumference multiplied by PQ, i.e.

$\Delta S = 2\pi y, \Delta s$, where ΔS is the surface element and Δs is the arc PQ. Hence, the curved surface area S of the complete solid of revolution is given by

$$S = \int_{x=a}^{x=b} 2\pi y\, dx = \int_a^b 2\pi y \frac{ds}{dx}\, dx \tag{4.43}$$

Since

$$\frac{ds}{dx} = \sqrt{1 + \left(\frac{dy}{dx}\right)^2},$$

Equation (4.43) becomes

$$S = 2\pi \int_a^b y \sqrt{1 + \left(\frac{dy}{dx}\right)^2}\, dx, \tag{4.44}$$

in which form it is more convenient to use in practice.

If the area is rotated about OY instead of the X-axis, then the corresponding formulae for the volume and surface of the solid of revolution are

$$V = \int_{y=c}^{y=d} \pi x^2\, dy \tag{4.45}$$

and

$$S = 2\pi \int_c^d x \sqrt{1 + \left(\frac{dx}{dy}\right)^2}\, dy \tag{4.46}$$

Example 4.44 The area bounded by an arch of the cycloid $x = a(\theta - \sin\theta)$, $y = a(1 - \cos\theta)$ and the axis of X is revolved around OX. Find the volume and the surface area of the solid generated.

The volume of revolution is given by

$$V = \pi \int_{\theta=0}^{\theta=2\pi} y^2\, dx$$

But $y = a(1 - \cos\theta)$ and $dx = a(1 - \cos\theta)\, d\theta$. Hence,

$$V = \pi \int_0^{2\pi} a^3 (1 - \cos\theta)^3\, d\theta$$

$$= \pi a^3 \int_0^{2\pi} (1 - 3\cos\theta + 3\cos^2\theta - \cos^3\theta)\, d\theta$$

$$= \pi a^3 \int_0^{2\pi} \left(\frac{5}{2} - \frac{15}{4} \cos \theta + \frac{3}{2} \cos 2\theta - \frac{1}{4} \cos 3\theta \right) d\theta$$

$$= \pi a^3 \left[\frac{5}{2} \theta - \frac{15}{4} \sin \theta + \frac{3}{4} \sin 2\theta - \frac{1}{12} \sin 3\theta \right]_0^{2\pi}$$

$$= \pi a^3 \left[\frac{5}{2} (2\pi) \right]$$

$$= 5\pi^2 a^3.$$

The surface area is given by

$$S = 2\pi \int_{\theta=0}^{\theta=2\pi} y \sqrt{1 + \left(\frac{dy}{dx} \right)^2} \, dx$$

But,

$$\frac{dy}{dx} = \frac{dy/d\theta}{dx/d\theta} = \frac{a \sin \theta}{a (1 - \cos \theta)} = \frac{2a \sin (\theta/2) \cos (\theta/2)}{2a \sin^2 (\theta/2)} = \cot \frac{\theta}{2},$$

$$y = a (1 - \cos \theta) = 2a \sin^2 \frac{\theta}{2},$$

and

$$dx = a (1 - \cos \theta) \, d\theta = 2a \sin^2 \frac{\theta}{2} \, d\theta.$$

Hence,

$$S = 2\pi \int_0^{2\pi} 2a \sin^2 \frac{\theta}{2} \sqrt{1 + \cot^2 \frac{\theta}{2}} \, (2a) \sin^2 \frac{\theta}{2} \, d\theta$$

$$= 8\pi a^2 \int_0^{2\pi} \sin^4 \frac{\theta}{2} \operatorname{cosec} \frac{\theta}{2} \, d\theta$$

$$= 8\pi a^2 \int_0^{2\pi} \sin^3 \frac{\theta}{2} \, d\theta$$

$$= 8\pi a^2 \int_0^{2\pi} \frac{1}{4} \left(3 \sin \frac{\theta}{2} - \sin \frac{3\theta}{2} \right) d\theta$$

$$= 2\pi a^2 \left[-6 \cos \frac{\theta}{2} + \frac{2}{3} \cos \frac{3\theta}{2} \right]_0^{2\pi}$$

$$= -12\pi a^2(-2) + \frac{4\pi a^2}{3}(-2)$$

$$= 24\pi a^2 - \frac{8}{3}\pi a^2$$

$$= \frac{64}{3}\pi a^2.$$

Example 4.45 Find the volume of the solid obtained by revolving about the X-axis the area bounded by the parabola $y = (x^2/4) + 2$ and the straight line $8y = 5x + 14$.

The abscissae of the points of the intersection of $y = (x^2/4) + 2$ and $8y = 5x + 14$ are given by $2x^2 - 5x + 2 = 0$, which gives $x_1 = 1/2$ and $x_2 = 2$. Here,

$$V_1 = \pi \int_{1/2}^{2} \left(\frac{5x+14}{8}\right)^2 dx \quad \text{and} \quad V_2 = \pi \int_{1/2}^{2} \left(\frac{1}{4}x^2 + 2\right)^2 dx.$$

Hence the required volume is given by

$$V = \pi \int_{1/2}^{2} \left[\left(\frac{5x+14}{8}\right)^2 - \left(\frac{1}{4}x^2 + 2\right)^2\right] dx$$

$$= \pi \int_{1/2}^{2} \left(-\frac{1}{16}x^4 - \frac{39}{64}x^2 + \frac{35}{16}x - \frac{15}{16}\right) dx$$

$$= \pi \left[-\frac{1}{16}\frac{x^5}{5} - \frac{39}{64}\frac{x^3}{3} + \frac{35}{16}\frac{x^2}{2} - \frac{15}{16}x\right]_{1/2}^{2}$$

$$= \frac{891}{1280}\pi.$$

Example 4.46 The area bounded by the parabola $y^2 = 4x$ and the straight line $4x - 3y + 2 = 0$ is rotated about the Y-axis. Show that the volume of the solid formed is given by $\pi/20$.

The area is shown in Fig. 4.10. Eliminating x between the equations of the parabola and the straight line, we obtain

$$y^2 - 3y + 2 = 0,$$

which gives $y = 1$ and $y = 2$. If V_1 is the volume of the solid formed by rotating the area bounded by $4x - 3y + 2 = 0$ and the part of the Y-axis between $y = 1$ and $y = 2$, then we have

$$V_1 = \pi \int_1^2 \left(\frac{3y-2}{4}\right)^2 dy.$$

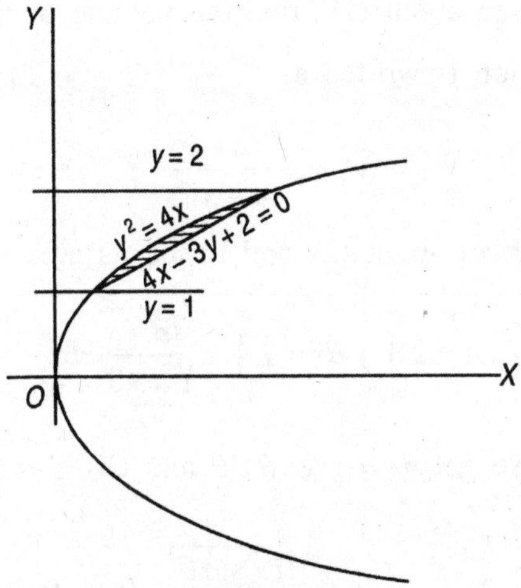

Figure 4.10

Similarly, the volume V_2 of the solid formed by rotating the area bounded by the parabola $y^2 = 4x$ and the part of the Y-axis between $y = 1$ and $y = 2$ given by

$$V_2 = \pi \int_1^2 \left(\frac{y^2}{4}\right)^2 dy$$

The required volume V is the difference of these volumes and we therefore have

$$V = V_1 - V_2 = \frac{\pi}{16} \int_1^2 \left[(3y-2)^2 - y^4\right] dy$$

$$= \frac{\pi}{16} \int_1^2 (9y^2 + 4 - 12y - y^4)\, dy$$

$$= \frac{\pi}{16} \left[3y^3 + 4y - 6y^2 - \frac{y^5}{5}\right]_1^2$$

$$= \frac{\pi}{16} \left[3(7) + 4\,(1) - 6(3) - \frac{1}{5}\,(31)\right]$$

$$= \frac{\pi}{20}.$$

Example 4.47 Find the area of the loop of the curve

$$x(x^2+y^2)=a(x^2-y^2).$$

If this loop is revolved about OX, find the volume of the solid generated.

The given equation is written as

$$y^2=\frac{x^2(a-x)}{a+x} \tag{i}$$

The loop is symmetrical about OX and hence its area is given by

$$A=2\int_0^a y\,dx=2\int_0^a x\sqrt{\frac{a-x}{a+x}}\,dx \tag{ii}$$

Putting $x=a\sin\theta$, we get $dx=a\cos\theta\,d\theta$ and (ii) gives:

$$A=2\int_0^{\pi/2} a\sin\theta\sqrt{\frac{1-\sin\theta}{1+\sin\theta}}\,(a\cos\theta)\,d\theta$$

$$=2a^2\int_0^{\pi/2}\sin\theta\,\frac{(1-\sin\theta)}{\cos\theta}\cos\theta\,d\theta$$

$$=2a^2\int_0^{\pi/2}(\sin\theta-\sin^2\theta)\,d\theta$$

$$=2a^2-2a^2\left(\frac{\pi}{4}\right)$$

$$=2a^2\left(1-\frac{\pi}{4}\right).$$

If this loop is revolved about OX, the volume of the solid formed is given by

$$V=\int_0^a \pi y^2\,dx=\pi\int_0^a\frac{x^2(a-x)}{a+x}\,dx.$$

$$=-\pi\int_0^a\frac{x^3-ax^2}{a+x}\,dx$$

$$=-\pi\int_0^a\left(x^2-2ax+2a^2-\frac{2a^3}{a+x}\right)dx$$

$$= \pi \int_0^a \left(-x^2 + 2ax - 2a^2 + \frac{2a^3}{a+x} \right) dx$$

$$= \pi \left[-\frac{x^3}{3} + ax^2 - 2a^2 x + 2a^3 \log (a+x) \right]_0^a$$

$$= \pi \left(-\frac{a^3}{3} + a^3 - 2a^3 + 2a^3 \log 2a - 2a^3 \log a \right)$$

$$= \pi \left(-\frac{4a^3}{3} + 2a^3 \log 2 \right).$$

Example 4.48 Find the area bounded by the curve $x^{2/3} + y^{2/3} = a^{2/3}$. If the part of the area lying between $\theta = -\pi/2$ to $\theta = \pi/2$ is rotated about OX, find the volume of the solid generated.

It would be convenient to use parametric equations of the curve, viz., $x = a \cos^3\theta$, $y = a \sin^3\theta$. Since the curve is symmetrical about both the axes, the required area is given by

$$A = 4 \int_0^a y \, dx$$

$$= 4 \int_{\pi/2}^0 a \sin^3\theta \, (-3a \cos^2\theta \sin\theta) \, d\theta$$

$$= 12 \int_0^{\pi/2} a^2 \sin^4\theta \cos^2\theta \, d\theta$$

$$= 12 a^2 \frac{1}{6} \frac{3 \cdot 1}{4 \cdot 2} \frac{\pi}{2}$$

$$= \frac{3}{8} \pi a^2.$$

The required volume is given by

$$V = \pi \int_0^a y^2 \, dx$$

$$= \pi \int_{\pi/2}^0 a^2 \sin^6\theta \, (-3a \cos^2\theta \sin\theta) \, d\theta$$

$$= 3a^3 \pi \int_0^{\pi/2} \sin^7\theta \cos^2\theta \, d\theta.$$

$$= 3a^3\pi\left(\frac{1}{9}\,\frac{6\cdot 4\cdot 2}{7\cdot 5\cdot 3\cdot 1}\right)$$

$$= \frac{16\pi a^3}{105}.$$

4.3.2 Length of an Arc of a Curve

Let AB be an arc of a curve and let PQ $(=\delta s)$ be a small element of it [Fig. 4.11(a)]. For smaller values of δs, the arc length is approximately equal to the chord PQ. From Fig. 4.11(b), we have

$$(\delta s)^2 = (\delta x)^2 + (\delta y)^2$$

so that

$$\delta s = \sqrt{1+\left(\frac{\delta y}{\delta x}\right)^2}\,\delta x = \sqrt{1+\left(\frac{\delta y}{\delta x}\right)^2}\,\delta x$$

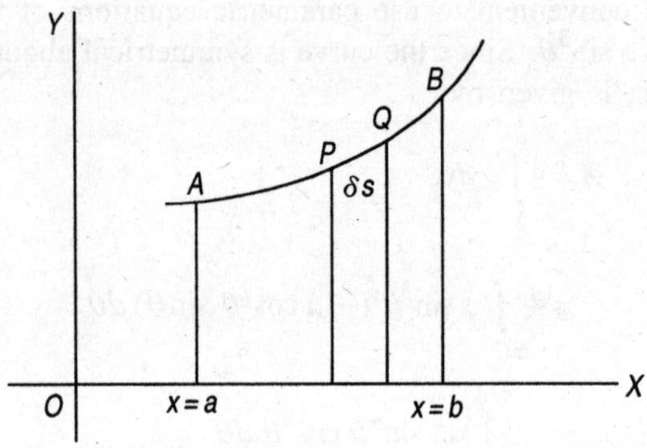

Figure 4.11(a)

If s is the total arc length, then an estimate of this length is given by the sum of all elements of the above type from $x = a$ to $x = b$, i.e.

$$s \approx \sum \sqrt{1+\left(\frac{\delta y}{\delta x}\right)^2}\,\delta x \qquad \text{from } x = a \text{ to } x = b.$$

Hence, the total arc length (s) is given by the limit of the above sum as $\delta x \to 0$, i.e.

$$s = \lim_{\delta x\to 0} \sum \sqrt{1+\left(\frac{\delta y}{\delta x}\right)^2}\,\delta x \qquad \text{from } x = a \text{ to } x = b,$$

i.e.

$$s = \int_a^b \sqrt{1+\left(\frac{dy}{dx}\right)^2}\,dx \tag{4.47}$$

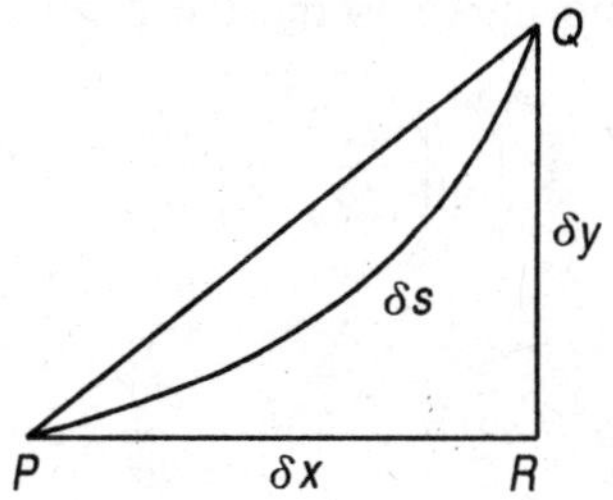

Figure 4.11(b)

If x and y are given as functions of a parameter θ, say $x = x(\theta)$ and $y = y(\theta)$, then

$$\left(\frac{\delta s}{\delta \theta}\right)^2 = \left(\frac{\delta x}{\delta \theta}\right)^2 + \left(\frac{\delta y}{\delta \theta}\right)^2$$

and

$$s = \int_{\theta_1}^{\theta_2} \sqrt{\left(\frac{dx}{d\theta}\right)^2 + \left(\frac{dy}{d\theta}\right)^2}\, d\theta, \tag{4.48}$$

where θ_1 and θ_2 are the values of θ corresponding to $x = a$ and $x = b$ respectively.

Example 4.49 Find the total length of the curve given by $x = a\cos^3\theta$, $y = a\sin^3\theta$.

As θ varies from 0 to $\pi/2$, we observe that the coordinates (x, y) change from $(a, 0)$ to $(0, a)$; from $\pi/2$ to π, they change from $(0, a)$ to $(-a, 0)$; from π *to* $3\pi/2$, they change from $(-a, 0)$ to $(0, -a)$; and finally from $3\pi/2$ to 2π, (x, y) change from $(0, -a)$ to $(a, 0)$. This shows that the curve is symmetrical about both the axes. This curve is called *hypocycloid* and its sketch is given in Fig. 4.12.

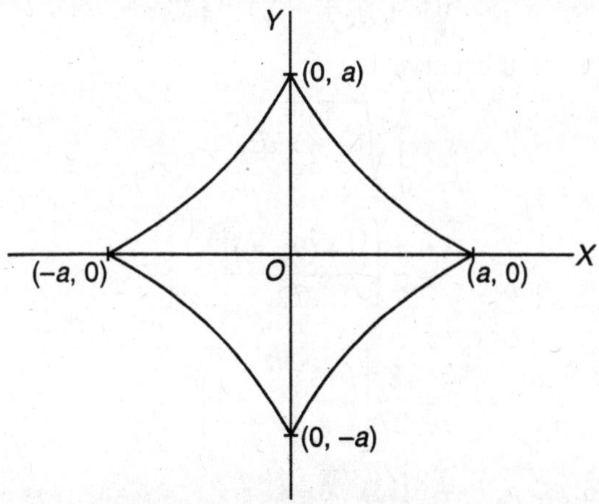

Figure 4.12

The total length (s) of the curve is therefore given by

$$s = 4\int_0^{\pi/2} \sqrt{\left(\frac{dx}{d\theta}\right)^2 + \left(\frac{dy}{d\theta}\right)^2}\, d\theta.$$

Here,

$$\frac{dx}{d\theta} = -3a\cos^2\theta \sin\theta, \quad \text{and} \quad \frac{dy}{d\theta} = 3a\sin^2\theta\cos\theta.$$

Hence,

$$s = 4\int_0^{\pi/2} \sqrt{9a^2\cos^4\theta\sin^2\theta + 9a^2\sin^4\theta\cos^2\theta}\, d\theta$$

$$= 4\int_0^{\pi/2} 3a\cos\theta\sin\theta\sqrt{\cos^2\theta + \sin^2\theta}\, d\theta$$

$$= 12a\int_0^{\pi/2} \sin\theta\cos\theta\, d\theta$$

$$= 12a\left[\frac{\sin^2\theta}{2}\right]_0^{\pi/2}$$

$$= 6a.$$

Example 4.50 Find the length of the curve $y = x^{3/2}$ for $0 \le x \le 1$.

Here $y = f(x) = x^{3/2}$. We therefore have

$$\frac{dy}{dx} = f'(x) = \frac{3}{2}x^{1/2},$$

and

$$\sqrt{1 + \left(\frac{dy}{dx}\right)^2} = \sqrt{1 + \frac{9}{4}x}.$$

Hence the length of the curve is

$$s = \int_0^1 \sqrt{1 + \frac{9}{4}x}\, dx.$$

$$= \frac{4}{9}\left\{\frac{[1 + (9x/4)]^{3/2}}{3/2}\right\}_0^1$$

$$= \frac{8}{27}\left[\left(\frac{13}{4}\right)^{3/2} - 1\right]$$

$$= \frac{(13)^{3/2} - 8}{27}.$$

4.3.3 Centroid of Area and Volume of Revolution

We consider the area bounded by the curve $y = f(x)$, the x-axis and the ordinates $x = a$ and $x = b$ (see Fig. 4.13). Let $PQP'Q'$ be a strip of width Δx with $P \equiv (x, y)$. Then the elemental area ΔA is given by

$$\Delta A = y \Delta x.$$

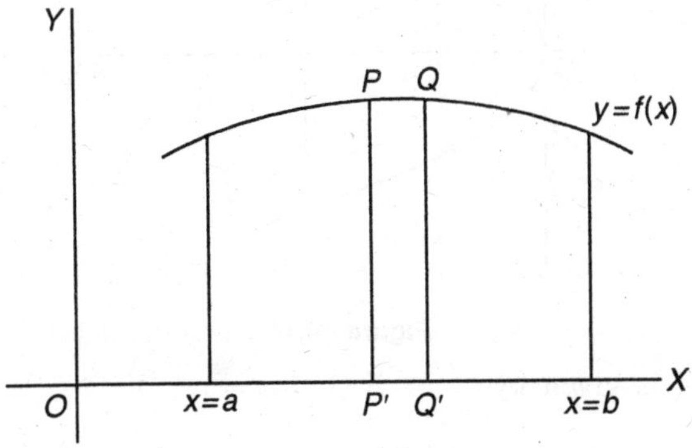

Figure 4.13

The sums $\Sigma(\Delta A)x$ and $\Sigma(\Delta A)y$ are called the *first moments* of the area about OY and OX, respectively. The point G whose coordinates $(\overline{x}, \overline{y})$ are given by

$$\overline{x} = \frac{\sum x(\Delta A)}{\sum (\Delta A)}, \qquad \overline{y} = \frac{\sum (y/2)\Delta A}{\sum \Delta A},$$

is called the *centroid* of the area. As $\Delta x \to 0$, i.e. as the number of such strips increases indefinitely, we have

$$\overline{x} = \frac{\int x\, dA}{\int dA} = \frac{\int_a^b xy\, dx}{\int_a^b y\, dx}, \qquad \overline{y} = \frac{(1/2)\int_a^b y^2\, dx}{\int_a^b y\, dx} \tag{4.49}$$

We have also seen that if the above area is rotated about OX, a solid of revolution will be formed. If the strip $PQP'Q'$ is rotated about OX, a thin circular disc of volume, $\Delta V = \pi y^2 \Delta x$, will be formed. Since the centroid of the volume of revolution lies on the X-axis, we have

$$\overline{x} = \frac{\int_a^b (\pi y^2\, dx)\, x}{\int_a^b \pi y^2\, dx} = \frac{\int_a^b \pi y^2 x\, dx}{\int_a^b \pi y^2\, dx} \tag{4.50}$$

Example 4.51 Let us consider the area bounded by the parabola $y^2 = 4ax$, the X-axis and the ordinate $x = c$ (Fig. 4.14).

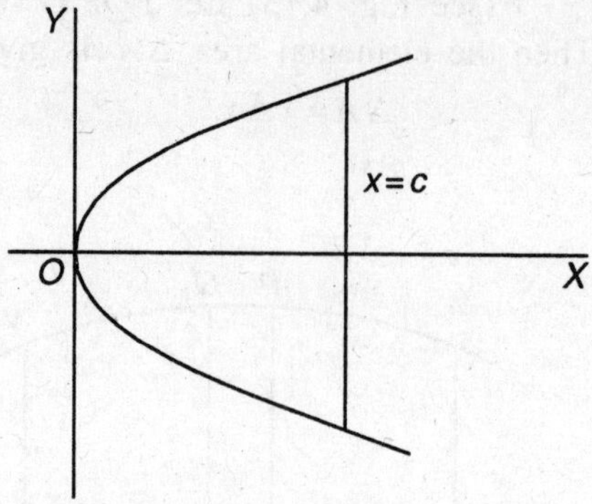

Figure 4.14

The area A is given by

$$A = \int_0^c y\,dx$$

$$= \int_0^c 2\sqrt{ax}\,dx$$

$$= 2\sqrt{a}\left[\frac{x^{3/2}}{3/2}\right]_0^c$$

$$= \frac{4}{3}\sqrt{a}\,c^{3/2}.$$

The coordinates of the centroid of A are given by

$$\bar{x} = \frac{\int_0^c xy\,dx}{A} \quad \text{or} \quad A\bar{x} = \int_0^c xy\,dx = \int_0^c 2x\sqrt{ax}\,dx$$

i.e.

$$A\bar{x} = 2\sqrt{a}\left[\frac{x^{5/2}}{5/2}\right]_0^c = \frac{4}{5}\sqrt{a}\,c^{5/2}$$

so that

$$\bar{x} = \frac{(4/5)\sqrt{a}\,c^{5/2}}{(4/3)\sqrt{a}\,c^{3/2}} = \frac{3}{5}c.$$

Similarly,

$$A\bar{y} = \int_0^c \frac{1}{2}y^2\,dx = \frac{1}{2}\int_0^c 4ax\,dx = 2a\frac{c^2}{2} = ac^2,$$

so that

$$\bar{y} = \frac{ac^2}{(4/3)\sqrt{a}\,c^{3/2}} = \frac{3}{4}\sqrt{a}\,c^{1/2} = \frac{3}{4}\sqrt{ac}.$$

For the solid of revolution, the centroid is given by

$$\bar{x} = \frac{\int_0^c \pi y^2 x\,dx}{\int_0^c \pi y^2\,dx} = \frac{\int_0^c 4ax^2\,dx}{\int_0^c 4ax\,dx} = \frac{c^3/3}{c^2/2} = \frac{2}{3}c \quad \text{and} \quad \bar{y} = 0.$$

4.3.4 Theorems of Pappus

Let the area A revolve about OX (which does not intersect it) through four right angles. This will result in a ring-shaped solid of volume V. To obtain an expression for V, let ΔA be a small element of A situated at a point $P(x, y)$ (Fig. 4.15). It is clear that the element ΔA will describe a circular filament of radius y and sectional area ΔA. Its volume ΔV is therefore given by

$$\Delta V = 2\pi y \Delta A$$

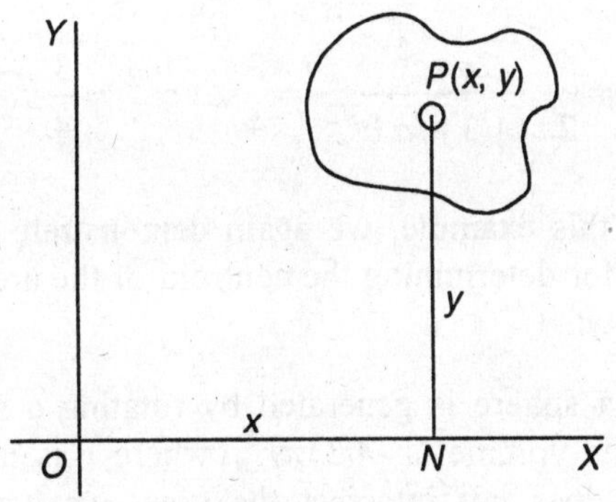

Figure 4.15

Hence the volume V of the solid is given by

$$V = \sum 2\pi y \Delta A \tag{4.51}$$

But

$$\frac{\sum y\Delta A}{\sum \Delta A} = \bar{y}$$

and hence

$$\sum \bar{y}A = \bar{y}\sum \Delta A = \bar{y}A.$$

Hence, (4.51) becomes:

$$V = 2\pi\bar{y}A, \tag{4.52}$$

where $2\pi\bar{y}$ is the total path described by the centroid of the area. The result given in (4.52) is called *Pappus first theorem* and is stated thus:

> The volume of the solid generated by revolving an area about an axis which does not intersect it is obtained by multiplying the area by the total path described by the centroid of the area.

Example 4.52 We consider again the area bounded by the parabola $y^2 = 4ax$, the x-axis and the ordinate $x = c$.

If V is the volume of the solid of revolution, then we know that

$$V = \int_0^c \pi y^2\, dx = \pi \int_0^c 4ax\, dx = 4a\pi \frac{c^2}{2} = 2a\pi c^2.$$

Hence, by Pappus first theorem, we have

$$2a\pi c^2 = 2\pi\bar{y}A,$$

where $\bar{y}$ is the y-coordinate of the centroid of the area. Further, we have

$$A = \frac{4}{3}\sqrt{a}\, c^{3/2}$$

Hence, we obtain

$$\bar{y} = \frac{2a\pi c^2}{2\pi(4/3)\sqrt{a}\, c^{3/2}} = \frac{3}{4}\sqrt{a}\, c^{1/2} = \frac{3}{4}\sqrt{ac}.$$

Example 4.53 In this example, we again demonstrate the use of the first theorem of Pappus for determining the centroid of the area when the volume of the solid is known.

We know that a sphere is generated by rotating a semicircle about its diameter and that its volume is $(4/3)\pi r^3$, where r is its radius. Since the axis of revolution does not intersect the area, we have by Pappus first theorem,

$$V = 2\pi\bar{y}A, \qquad \text{where } A = \frac{1}{2}\pi r^2$$

Hence,

$$\bar{y} = \frac{V}{2\pi A} = \frac{(4/3)\,\pi r^3}{2\pi\,(1/2)\,\pi r^2} = \frac{4}{3}\frac{r}{\pi}.$$

Again, suppose that an arc of a plane curve [Fig. 4.16(a)] is rotated about OX which does not intersect the arc. Let S be the surface area of the solid

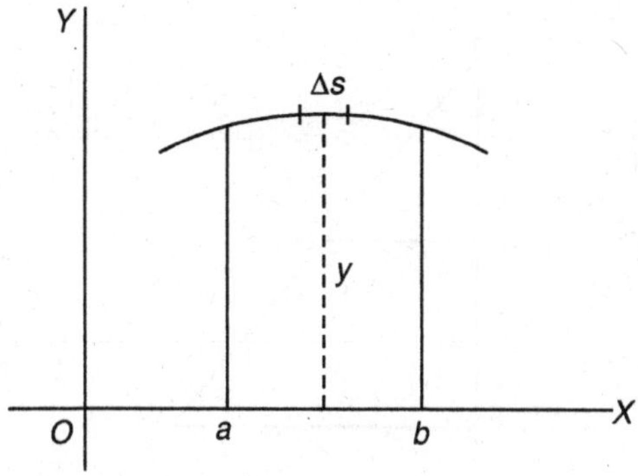

Figure 4.16(a)

generated and let Δs be an element of the arc. If the element Δs is revolved about OX, it describes a thin circular ribbon of area $\Delta S = 2\pi y \Delta s$. Hence, the total surface area S is given by

$$S = \sum 2\pi y \, \Delta s,$$

taken along the whole perimeter of the area. But

$$\frac{\sum y\Delta s}{\sum \Delta s} = \bar{y},$$

where $(\bar{x}, \bar{y})$ is now the centroid of the perimeter s. Hence, we obtain:

$$S = 2\pi\bar{y} \sum \Delta s = 2\pi\bar{y}s$$

This is the path described by the centroid of the perimeter multiplied by the perimeter. This is *Pappus second theorem* which states that

> The total surface area generated is equal to the perimeter multiplied by the path described by the centroid of the perimeter.

If V and S are known in a particular problem, Pappus theorems enable us to determine the Y-coordinate of the centroid of the area of perimeter.

Example 4.54 Using Pappus second theorem, find the area of the curved surface of a right circular cone.

We know that a right circular cone is formed by the revolution of a straight line AC about another line AB [Fig. 4.16(b)]. The centre of gravity of the straight line AC is at G, its midpoint. Let $BC = r$ and $AC = l$. Draw a straight line GD from G and perpendicular to AB. Since G is the mid-point of AC and GD parallel to BC, it follows that

$$DG = \frac{1}{2} BC = \frac{1}{2} r.$$

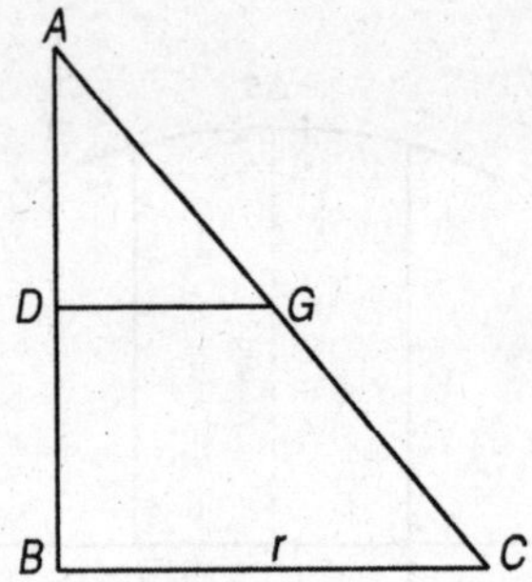

Figure 4.16(b)

Hence, by Pappus second theorem, the required

$$\text{Surface area} = AC\left(2\pi \frac{r}{2}\right) = \pi r l.$$

Example 4.55 The loop of the curve $2ay^2 = x(a-x)^2$ revolves about the line $y = a$. Using the first theorem of Pappus, find the volume of the solid formed.

The loop is between $x = 0$ to $x = a$ and is symmetrical about the x-axis. Therefore,

$$\text{Area of the loop} = 2\int_0^a y\,dx$$

$$= 2\int_0^a \frac{(a-x)\sqrt{x}}{\sqrt{2a}}\,dx$$

$$= \sqrt{\frac{2}{a}} \int_0^a (ax^{1/2} - x^{3/2})\,dx$$

$$= \sqrt{\frac{2}{a}} \left[ax^{3/2}\left(\frac{2}{3}\right) - x^{5/2}\left(\frac{2}{5}\right) \right]_0^a$$

$$= \frac{2\sqrt{2}}{\sqrt{a}} \left[\frac{1}{3} aa^{3/2} - \frac{1}{5} a^{5/2} \right]$$

$$= \frac{4\sqrt{2}}{15} a^2 \text{ sq. units.}$$

Because of symmetry, the centre of gravity of the loop lies on the X-axis and the perpendicular distance of the CG from the axis of revolution is a. Therefore,

$$\text{Length of the path traversed by the } CG = 2\pi a.$$

Hence, by pappus theorem,

$$\text{Volume of the solid generated} = \frac{4\sqrt{2}}{15}a^2(2\pi a) = \frac{8\sqrt{2}}{15}\pi a^3.$$

4.3.5 Areas and Volumes in Polar Coordinates

We consider the area OAB (Fig. 4.17), i.e. the area bounded by the curve $r = f(\theta)$ and the radii $\theta = \theta_1$ and $\theta = \theta_2$. Divide the area OAB into sectors of the type OPQ where $P(r, \theta)$ and $Q(r + \Delta r, \theta + \Delta\theta)$. If ΔA is the area of POQ, then we have,

$$\Delta A = \frac{1}{2}r^2\Delta\theta.$$

Hence,

$$A = \int_{\theta_1}^{\theta_2} \frac{1}{2}r^2\, d\theta = \frac{1}{2}\int_{\theta_1}^{\theta_2} [f(\theta)]^2\, d\theta \tag{4.53}$$

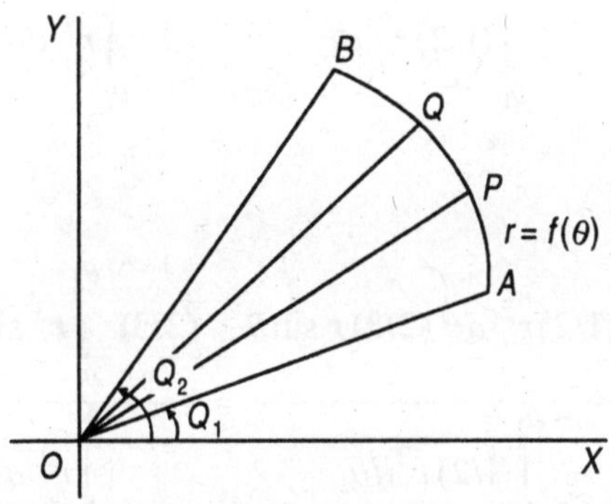

Figure 4.17

Let V be the volume formed by rotating the sector OAB about OX. To find V, we consider the element of volume generated when the elementary sector OPQ is rotated. In the limiting case the centroid G of this sector will lie on OP so that

$$OG = \frac{2}{3}OP = \frac{2}{3}r.$$

It follows that the coordinates of G are

$$\left(\frac{2}{3}r\cos\theta, \frac{2}{3}r\sin\theta\right).$$

Since the area of the elementary sector OPQ is $(1/2)r^2\Delta\theta$ and the total path described by G being $2\pi\,(2/3)\,r\sin\theta$, we have by Pappus first theorem,

$$\Delta V = \text{Elementary volume generated}$$

$$= \frac{1}{2}r^2(\Delta\theta)\,2\pi\left(\frac{2}{3}r\sin\theta\right)$$

$$= \frac{2}{3}\pi r^3 \sin\theta\,(\Delta\theta)$$

Hence,

$$V = \frac{2}{3}\pi\int_{\theta_1}^{\theta_2} r^3\sin\theta\,d\theta \tag{4.54}$$

If $(\bar{x}, \bar{y})$ are the coordinates of the centroid of the sector OAB, then we have

$$\bar{x} = \frac{\int_{\theta_1}^{\theta_2}\frac{2}{3}r\cos\theta\,(1/2)r^2\,d\theta}{\int_{\theta_1}^{\theta_2}(1/2)r^2\,d\theta} = \frac{\frac{2}{3}\int_{\theta_1}^{\theta_2} r^3\cos\theta\,d\theta}{\int_{\theta_1}^{\theta_2} r^2\,d\theta} \tag{4.55}$$

and

$$\bar{y} = \frac{\int_{\theta_1}^{\theta_2}(1/2)r^2\,d\theta\,(2/3)\,r\sin\theta}{\int_{\theta_1}^{\theta_2}(1/2)\,r^2 d\theta} = \frac{(2/3)\int_{\theta_1}^{\theta_2} r^3\sin\theta\,d\theta}{\int_{\theta_1}^{\theta_2} r^2\,d\theta} \tag{4.56}$$

Example 4.56 Compute the area bounded by a loop of the leminiscate $r^2 = a^2\cos 2\theta$ and find its centroid.

The complete curve is shown in Fig. 4.18. For the area of one loop, we have

$$A = 2\int_0^{\pi/4}\frac{1}{2}r^2\,d\theta$$

$$= \int_0^{\pi/4} a^2 \cos 2\theta \, d\theta$$

$$= a^2 \left[\frac{\sin 2\theta}{2} \right]_0^{\pi/4}$$

$$= \frac{a^2}{2}.$$

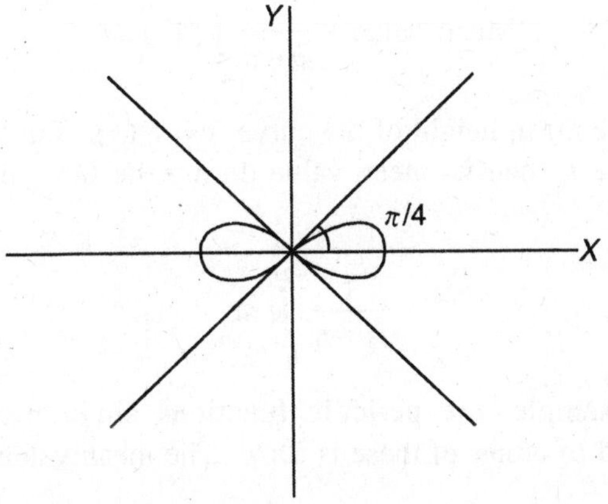

Figure 4.18

For its centroid, clearly $\bar{y} = 0$, and

$$\bar{x} = \frac{(2/3) \int_0^{\pi/4} r^3 \cos \theta \, d\theta}{\int_0^{\pi/4} r^2 \, d\theta} = \frac{2}{3} a \frac{\int_0^{\pi/4} (\cos 2\theta)^{3/2} \cos \theta \, d\theta}{\int_0^{\pi/4} \cos 2\theta \, d\theta}$$

$$= \frac{2}{3} a \frac{\int_0^{\pi/4} (1 - 2\sin^2 \theta)^{3/2} \cos \theta \, d\theta}{[(\sin 2\theta)/2]_0^{\pi/4}}$$

$$= \frac{4}{3} a \int_0^{\pi/4} (1 - 2\sin^2 \theta)^{3/2} \cos \theta \, d\theta$$

Put $\sqrt{2} \sin \theta = \sin \phi$, so that $\sqrt{2} \cos \theta \, d\theta = \cos \phi \, d\phi$. Hence,

$$\bar{x} = \frac{4}{3} a \int_0^{\pi/2} \frac{1}{\sqrt{2}} \cos^3 \phi \cos \phi \, d\phi = \frac{2\sqrt{2}}{3} a \int_0^{\pi/2} \cos^4 \phi \, d\phi$$

Using formula (4.38), we get

$$\bar{x} = \frac{2\sqrt{2}}{3} a \frac{1}{2} \frac{3}{4} \frac{\pi}{2} = \frac{1}{4\sqrt{2}} a\pi.$$

4.3.6 Mean-values and Root-mean-square Values

The mean value of a function $f(x)$ between the value $x = a$, $x = b$ of the variable is defined as:

$$\text{Mean value} = \frac{1}{b-a} \int_a^b f(x)\, dx \tag{4.57}$$

This is clearly the mean height of the curve $y = f(x)$. Similarly, if a current i varies with time t, then its mean value during the time interval t_1 to t_2 is given by

$$\frac{1}{t_2 - t_1} \int_{t_1}^{t_2} i\, dt$$

Take, for example, the periodic functions $\sin pt$ or $\cos pt$, or $i = I_M \sin pt$. The period of either of these is $2\pi/p$. The mean value of $I_M \sin pt$ is given by

$$\begin{aligned}
\text{Mean value} &= \frac{p}{2\pi} \int_0^{2\pi/p} I_M \sin pt\, dt \\
&= \frac{p}{2\pi} I_M \left[\frac{-\cos pt}{p}\right]_0^{2\pi/p} \\
&= -\frac{p}{2\pi} I_M \frac{1}{p} (\cos 2\pi - \cos 0) \\
&= 0,
\end{aligned}$$

which shows that the mean value of any periodic function over its period is zero. In such cases, the square root of the mean value of the square of the function is used. This is called the *root-mean-square value*, and is written as R.M.S. value. Thus, if $f(x)$ is defined between $x = a$ and $x = b$ $(b > a)$, then

$$\text{R.M.S. value} = \left\{\frac{1}{b-a} \int_a^b [f(x)]^2\, dx\right\}^{1/2}$$

Example 4.57 Find the R.M.S. value of $a \cos \omega t$ over a period.

Let $f(t) = a \cos \omega t$. The period of $f(t)$ is $2\pi/\omega$. Therefore,

$$\text{Mean value of } [f(t)]^2 = \frac{\omega}{2\pi} \int_0^{2\pi/\omega} [f(t)]^2 \, dt.$$

$$= \frac{\omega}{2\pi} \int_0^{2\pi/\omega} a^2 \cos^2(\omega t) \, dt$$

$$= \frac{\omega}{2\pi} \frac{a^2}{2} \int_0^{2\pi/\omega} (1 + \cos 2\,\omega t) \, dt$$

$$= \frac{\omega a^2}{4\pi} \left[t + \frac{\sin 2\omega t}{2\omega} \right]_0^{2\pi/\omega}$$

$$= \frac{\omega a^2}{4\pi} \frac{2\pi}{\omega}$$

$$= \frac{1}{2} a^2.$$

Hence,

$$\text{R.M.S. value} = \sqrt{\text{Mean value of } [f(t)]^2} = \frac{a}{\sqrt{2}},$$

which shows that the maximum value or peak value 'a' is given by $\sqrt{2}$ times the R.M.S. value. This is an important result, since it gives us the peak value when the R.M.S. value is known. For example, the statement that electricity is being supplied at 240 V really means that the R.M.S. value of the voltage is 240 and hence the peak value is $240\sqrt{2}$.

EXERCISES 4.3

1. Find the area bounded by the parabola $y^2 = 9x$ and the line $y = 3x$.
2. Find the area included between the parabolas $y^2 = 4x$ and $x^2 = 4y$.
3. Find the area bounded by the loop of the curve
$$4(y^2 - x^2) + x^3 = 0.$$
4. Find the area enclosed by the loop of the curve
$$y^2(2a - x) = x(x - a)^2.$$
5. Find the area bounded by the catenary
$$y = \frac{a}{2}(e^{x/a} + e^{-x/a})$$
and the lines $x = \pm a$ and the X-axis.
6. The curve defined by the equation $x^3 + y^3 = 3axy$ is called *Folium of Descartes*. Find the area of its loop.

7. Find the area of the smaller of the two parts into which the parabola $y^2 = 3ax$ divides the circle $x^2 + y^2 = 4a^2$.
8. Find the total area bounded by the cardioid $r = a(1 - \cos\theta)$.
9. Find the area bounded by the common portion of the ellipses
$$\frac{x^2}{a^2} + \frac{y^2}{b^2} = 1 \quad \text{and} \quad \frac{x^2}{b^2} + \frac{y^2}{a^2} = 1.$$
10. Find the area under one arc of the curve $y = \sin x$.
11. Find the area bounded by the curve $y = x^2 + x + 2$ and the lines $x + 2 = 0$, $x = 1$ and $y = 0$.
12. Find the area bounded by $y = \sin x$, the axis of x and the ordinates $x = 0$ and $x = 2\pi$.
13. Find the area bounded by $y = x^3$ and the lines $x + 2 = 0$, $x = 1$ and $y = 0$.
14. Find the area bounded by $y^2 = 2x$ and $y = x - 4$.
15. The area bounded by the parabola $y^2 = 2ax$ and the line $x = h$ is revolved about OX. Find the volume generated.
16. The area bounded by
$$y = a\cosh\frac{x}{a}, \qquad x = \pm a$$
and the X-axis is revolved about OX. Find the volume generated.
17. The area bounded by $y = x^2$ and $y = x$ is revolved about the X-axis. Find the volume generated.
18. The area bounded by $y = x^2$ and $x = 1$ and the X-axis is revolved about the Y-axis. Find the volume generated.
19. Find the centroid of the area bounded by an arc of the curve $y = b\sin(2\pi nx)$ and the axis of X. If this area is revolved about OX, find the volume generated.
20. The ellipse $(x^2/a^2) + (y^2/b^2) = 1$ is revolved about X-axis. Find the volume of the solid formed.
21. A cone is formed by revolving about the Y-axis the line joining the origin to the point (a, b). Find the volume of this cone.
22. Find the area bounded by an arc of the cycloid $x = a(\theta - \sin\theta)$, $y = a(1 - \cos\theta)$ and the axis of X. If this area is revolved about OX, find the volume of the solid formed.
23. The curve $y = f(x), 0 \le x \le b$ is rotated about the X-axis to form a solid. Given that the volume of this solid is $b^2 + b$ for all b, find the value of the function $f(x)$.
24. Find the volume that is common to the two cylinders
$$x^2 + y^2 = r^2 \quad \text{and} \quad y^2 + z^2 = r^2.$$

25. The area bounded by $y^2 = x - 1$, $x = 3$ and $y = 1$ is rotated about the Y-axis. Find the volume generated.

26. The astroid $x = a\sin^3\theta, y = a\cos^3\theta$ is revolved about the X-axis. Find the surface of the solid of revolution.

27. One arc of the cycloid $x = a(\theta - \sin\theta)$, $y = a(1 - \cos\theta)$ is revolved about the X-axis. Find the area of the surface formed.

28. Find the area of the surface generated by rotating the curve $x^{2/3} + y^{2/3} = a^{2/3}$ and the X-axis.

29. The area bounded by the parabola $y^2 = 4x$ and the line $y = 2x$ is revolved round that line. Find the volume of the solid generated.

30. The portion of the cardioid $r = a(1 + \cos\theta)$ between $\theta = -\pi/2$ to $\theta = \pi/2$ is rotated round the line $\cos\theta = 0$. Prove that the volume and the area of the surface so generated are

$$2\left(2 + \frac{5}{8}\pi\right)\pi a^3 \quad \text{and} \quad \frac{48}{5}\sqrt{2}\,(\pi a^2).$$

31. Find the length of the astroid defined by $x = a\cos^3\theta$ and $y = a\sin^3\theta$.

32. Find the length of the cardioid $r = a(1 + \cos\theta)$.

33. Find the entire length of each of the following curves:
    (a) The hypocycloid $x^{2/3} + y^{2/3} = a^{2/3}$
    (b) The curve given by $r = a\sin^3(\theta/3)$.

34. If s is the length of the arc of the cycloid $x = a(\theta + \sin\theta)$, $y = a(1 - \cos\theta)$ from the origin to the point (x, y), show that $s^2 = 8ay$.

35. Find the length of the curve

$$y = \frac{3}{4}x^{4/3} - \frac{1}{2}x^{1/2} \quad \text{from } x = 0 \text{ to } x = 3.$$

36. Find the total perimeter of the curve $x^{2/3} + y^{2/3} = a^{2/3}$.

37. The shape of a cable from an antenna tower is given by the equation $y = (4/3)x^{3/2}$ from $x = 0$ to $x = 20$ ft. Find the total length of the cable.

38. Find the centroid of the area bounded by $y = e^x$, $y = 0$, $x = 0$ and $x = 1$.

39. Find the centroid of the area bounded by $y = \log_e x$, $x = 1$ and $y = 1$.

40. A flat plate is in the shape of a triangle bounded by the lines $x = 0$, $y = 0$ and $y + 3x - 6 = 0$. Find the coordinates $(\bar{x}, \bar{y})$ of the centre of mass of the plate.

41. Find the area of the loop of the curve

$$y^2 = 4x^2\left(1 - \frac{x}{a}\right)$$

and also its centroid.

42. Determine the mean-value of the following functions on the given intervals:

(a) $y = x^2$ on $[a, b]$

(b) $y = \sin x$ on $[0, \pi]$

(c) $y = \dfrac{1}{1+x^2}$ on $[-1, 1]$

43. Find the mean-value of the function $y = 10 \sin 2t$ from $t = 0$ to $t = \pi/2$.

44. Find the R.M.S. value for one complete period of the function given by $f(t) = (1/2) + \cos t$.

45. Find the mean height of the portion of the parabola $y = (x-2)(3-x)$, which lies above the *X*-axis.

46. The equation of motion of a particle travelling in a straight line is $s = t^2 + t$. Find the average velocity on the interval from $t = 0$ to $t = 3$. Assume s in metres and t in seconds.

5

Functions of Several Variables

5.1 FUNDAMENTAL IDEAS, LIMITS AND CONTINUITY

We have so far considered functions of a single independent variable only. In practical situations, however, we come across functions which depend on two or more independent variables. For example,

(a) The area A of a rectangle with sides of lengths x and y is given by $A = xy$. Here, x and y are two independent variables and A is a function of both x and y.

(b) The pressure p of a given gas will depend on volume v and temperature T. Here, we say that p is a function of two independent variables v and T and express it explicitly as a functional relation of the form $p = f(v, T)$. It is also possible to express this implicitly as $\phi(p, v, T) = 0$, which is a function of three variables.

In this chapter, we extend our consideration to functions with more than one independent variable and it will be seen that most of the concepts, theorems and proofs are easy generalizations of those for functions of one variable.

For functions of two independent variables, say $z = f(x, y)$, *a* geometrical representation can be obtained by plotting the surface which provides a visual representation of the functional relationship. In Chapter 2, we have come across several examples of surfaces. For more than two independent variables, however, no such simple representation is possible.

The concept of a function of one independent variable can easily be extended to two or more independent variables. Thus, $z = f(x, y)$ is said to be a real-valued function of two independent real variables x and y if corresponding to every pair of values of x and y there exists a real number defined by $f(x, y)$. Here the domain of definition of the function is not specified, but is understood to be the largest possible domain that can be defined. This concept of the function can easily be generalized to functions with more than two independent variables.

5.1.1 Limits

Before considering the limit of a function of more than one independent variable, we shall first examine the notion of neighbourhood of a point and for this we shall confine our attention to functions $f(x, y)$ of two independent variables x and y, the generalization to functions of more than two independent variables following quite closely.

In the case of a function of one variable, a neighbourhood of a point, say a, is defined as an open interval of the form $(a-\varepsilon, a+\varepsilon)$. Equivalently, it is defined as the set of all points x such that

$$|x-a|<\varepsilon \tag{5.1}$$

For a function of two independent variables, two kinds of neighbourhoods are in common use—the *circular* and the *square* neighbourhoods. A circular neighbourhood of a point (a, b) is the set of all points (x, y) such that

$$\sqrt{(x-a)^2+(y-b)^2}<\varepsilon, \tag{5.2}$$

which consists of all points in a circle of radius ε and centre (a, b). On the other hand, a square neighbourhood of (a, b) is defined to be the set of all points (x, y) such that

$$|x-a|<\varepsilon \quad \text{and} \quad |y-b|<\varepsilon, \tag{5.3}$$

which consists of all points in a square with centre (a, b) and sides of length 2ε and parallel to the coordinate axes. By the neighbourhood of a point is normally meant either the circular neighbourhood or the square neighbourhood. The limit of a function can now be defined:

Definition 5.1 The function $f(x, y)$ is said to tend to the limit l as $x \to a$ and $y \to b$ if and only if the limit l is independent of the path followed by the point (x, y) as $x \to a$ and $y \to b$, and we write

$$\lim_{\substack{x\to a\\ y\to b}} f(x, y)=l \tag{5.4}$$

In terms of a circular neighbourhood, the above definition of the limit takes the form:

Definition 5.2 The function $f(x, y)$ is said to tend to the limit l as $x \to a$ and $y \to b$ if and only if corresponding to a positive number ε there exists another positive number $\delta=\delta(\varepsilon)$ such that

$$|f(x, y)-l|<\varepsilon, \quad \text{whenever } 0<\sqrt{(x-a)^2+(y-b)^2}<\delta. \tag{5.5}$$

A similar definition can be given for the square neighbourhoods.

Definition 5.3 If a function $f(x, y)$ has *distinct* limits as (x, y) approaches a point (a, b) along two distinct paths, the limit

$$\lim_{\substack{x\to a\\ y\to b}} f(x, y) \text{ does not exist.}$$

In a similar way, the limit to infinity can also be defined.

The following results are easy generalizations of the corresponding results for the function of a single variable:

If $f(x, y)$ and $g(x, y)$ are functions defined in some domain of the (x, y) plane, except possibly at (x_0, y_0), and if

$$\lim_{\substack{x\to x_0\\ y\to y_0}} f(x, y) = l_1, \quad \text{and} \quad \lim_{\substack{x\to x_0\\ y\to y_0}} g(x, y) = l_2,$$

then

(i) $\lim_{\substack{x\to x_0\\ y\to y_0}} f(x, y)$ is unique

(ii) $\lim_{\substack{x\to x_0\\ y\to y_0}} [f(x, y) \pm g(x, y)] = l_1 \pm l_2$

(iii) $\lim_{\substack{x\to x_0\\ y\to y_0}} [f(x, y) \cdot g(x, y) = l_1 \cdot l_2$

(iv) $\lim_{\substack{x\to x_0\\ y\to y_0}} \dfrac{f(x, y)}{g(x, y)} = \dfrac{l_1}{l_2}$ provided that $l_2 \neq 0$.

The following examples illustrate the method of evaluation of limits.

Example 5.1

(a) If $f(x, y) = \dfrac{2x}{x^2 + y^2 + 1}$,

(i) $\lim_{\substack{x\to 0\\ y\to 0}} f(x, y) = \dfrac{0}{1} = 0$

(ii) $\lim_{\substack{x\to 1\\ y\to 2}} f(x, y) = \dfrac{2}{6} = \dfrac{1}{3}$.

(b) If $f(x, y) = \dfrac{2xy + 1}{x^2 + 3y^2}$, $\lim_{\substack{x\to \infty\\ y\to 2}} f(x, y) = 0$.

(c) If $f(x, y)=\dfrac{3x^2 y}{2x^2+2y^2+1}$, $\lim\limits_{\substack{x\to 1\\ y\to 2}} f(x, y)=\dfrac{6}{2+8+1}=\dfrac{6}{11}$.

(d) If $f(x, y)=\dfrac{x+1}{4-y}$, $\lim\limits_{\substack{x\to 1\\ y\to 3}} f(x, y)=\dfrac{2}{1}=2$.

(e) If $f(x, y)=\sin\left(\dfrac{\pi^2}{4}-xy\right)$, $\lim\limits_{\substack{x\to \pi/2\\ y\to \pi/2}} f(x, y)=0$.

(f) If $f(x, y)=\sin\left(\dfrac{\pi x}{y+3}\right)$, $\lim\limits_{\substack{x\to 2\\ y\to -3}} f(x, y)$ does not exist.

(g) If $f(x, y)=\dfrac{x(y-1)}{y(x-1)}$, $\lim\limits_{\substack{x\to 1\\ y\to 1}} f(x, y)=\dfrac{0}{0}$.

If, however, $y=x$, then $\lim_{y=x,\ x\to 1} f(x, y)=1$. On the other hand, if $y=2-x$, then

$$\lim_{\substack{y\to 2-x\\ x\to 0}} f(x, y)=\lim_{x\to 1}\frac{x(1-x)}{(2-x)(x-1)}=\lim_{x\to 1}\frac{x}{x-2}=-1.$$

It follows therefore that $\lim_{y=x,\, x\to 1}$ does not exist.

(h) If $f(x, y)=\dfrac{xy}{x^2+y^2}$, $\lim\limits_{\substack{x\to 0\\ y\to 0}} f(x, y)=\dfrac{0}{0}$.

Let $y=mx$. Then

$$\lim_{x\to 0} f(x, y)=\lim_{x\to 0}\frac{mx^2}{x^2(1+m^2)}=\frac{m}{1+m^2},$$

which depends on the path of approach. Hence, the above limit does not exist.

5.1.2 Continuity

The concept of continuity of a function of several variables is, essentially, the same as that of a function of a single variable. The definition is as follows:

Definition 5.4 The function $f(x, y)$ is said to be *continuous* at the point (a, b) if

(i) $\lim_{x\to a,\, y\to b} f(x, y)$ exists and is equal to l, say, and

(ii) $f(a, b) = l$.

In terms of a circular neighbourhood, the above definition takes the form:

Definition 5.5 A function $f(x, y)$ is said to be *continuous* at (a, b) if and only if corresponding to an arbitrary positive number ε, there exists another positive number $\delta = \delta(\varepsilon)$ such that

$$|f(x, y) - f(a, b)| < \varepsilon, \quad \text{whenever } \sqrt{(x-a)^2 + (y-b)^2} < \delta.$$

A similar definition holds for a square neighbourhood.

If a function is continuous at all the points of a region, then we say that it is continuous in the region. As in the case of single variables, a function which is not continuous at a point is said to be *discontinuous* at that point or, alternatively, to have a *discontinuity* there.

Example 5.2

(a) $f(x, y) = \dfrac{2x}{x^2 + y^2 + 1}$ is continuous everywhere.

(b) $f(x, y) = \log_e(1 + x^2 + y^2) + e^x \sin y$ is continuous everywhere.

(c) $f(x, y) = \begin{cases} \dfrac{xy}{x^2 + y^2}, & (x, y) \neq (0, 0) \\ 0, & (x, y) = (0, 0) \end{cases}$

is discontinuous at (0, 0).

EXERCISES 5.1

Evaluate $f(1, 2)$, $f(-2, 1)$ and $f(0, 0)$ for each of the following functions (Problems 1–8):

1. $f(x, y) = x^2 + y^2 - x + 1.$

2. $f(x, y) = \dfrac{xy}{x^2 + y^2}\cdot$

3. $f(x, y) = \log(1 + x^2 + y^2).$

4. $f(x, y) = 2xe^y + ye^{2x}.$

5. $f(x, y) = \dfrac{x^3 y^3}{x^2 + y^2}\cdot$

6. $f(x, y) = \dfrac{x^2 - y^2}{x^2 + y^2}\cdot$

7. $f(x, y) = \dfrac{x^2 y^2}{x^4 + y^4}\cdot$

8. $f(x, y) = (2x - 1)^2 + (y - 2)^2\cdot$

Evaluate the following limits (Problems 9–16):

9. $\lim\limits_{\substack{x\to 1\\ y\to 1}} \dfrac{x^2}{x^2+y^2+1}$.

10. $\lim\limits_{\substack{x\to 1\\ y\to 2}} \dfrac{(x-1)^3}{(x-1)^2+(y-2)}$.

11. $\lim\limits_{\substack{y\to 2\\ x\to 1}} \dfrac{(x-1)^3}{(x-1)^2+(y-2)}$.

12. $\lim\limits_{\substack{x\to 0\\ y\to 0}} \dfrac{x^2+y^2}{xy}$.

13. $\lim\limits_{\substack{x\to 0\\ y\to 0}} \dfrac{2xy}{x^2+y^2}$.

14. $\lim\limits_{\substack{x\to 0\\ y\to 0}} \dfrac{x^2-y^2}{x^2+y^2}$.

15. $\lim\limits_{\substack{x\to 1\\ y\to 1}} \dfrac{x^2-2xy+y^2}{x^2y-y^3}$.

16. $\lim_{x\to 0,\ y\to 1} g(x, y)$, where

$$g(x, y) = xy, \quad xy \neq 0$$
$$= 1, \quad xy = 0$$

17. If $p(x) \to l_1$ and $q(x) \to l_2$ as $x \to 0$ show that

$$p(x)\,q(y) \to l_1 l_2 \quad \text{as} \quad (x, y) \to (0, 0).$$

Discuss the continuity of the following functions (Problems 18–25):

18. $\dfrac{x-y}{x+y}$.

19. $\dfrac{x+y}{x-y}$.

20. $\dfrac{xy}{x+y}$.

21. $\dfrac{x}{x^2+y^2+1}$.

22. $\dfrac{xy+1}{x^2-y}$.

23. $\dfrac{1}{1-(x^2+y^2)}$.

24. $\dfrac{x^2y^2}{x^4+y^4}$.

25. $\dfrac{x^4-y^4}{x^4+y^4}$.

26. Determine the set of points at which the function $f(x, y)$ defined by

$$f(x, y) = xy, \quad xy \neq 0$$
$$= 0, \quad xy = 0$$

is discontinuous.

27. Given that

$$g(x, y) = \frac{x^2-2xy+y^2}{y(x^2-y^2)}$$

$$g(1, 1) = 1,$$

discuss the continuity of $g(x, y)$ at the point (1, 1).

28. Given that

$$f(x, y) = \frac{x^2 - y^2}{x^2 + y^2}, \qquad \text{where } x^2 + y^2 \neq 0,$$

is it possible to assign a value for $f(0, 0)$ such that $f(x, y)$ is continuous at $(0, 0)$? Why?

29. Investigate the continuity of the function

$$\begin{aligned} f(x, y) &= \frac{2xy}{\sqrt{x^2 + y^2}}, \quad (x, y) \neq (0, 0) \\ &= 0, \quad (x, y) = (0, 0) \end{aligned}$$

at the origin.

30. Determine whether

$$\begin{aligned} f(x, y) &= x^2 + 2y, \quad (x, y) \neq (1, 2) \\ &= 0, \quad (x, y) = (1, 2) \end{aligned}$$

is continuous at $(1, 2)$.

5.2 PARTIAL DIFFERENTIATION

In the preceding section, we have extended the notions of limit and continuity to functions of several variables. We shall now extend the notion of a derivative in a similar fashion. For simplicity, we shall consider a function of two independent variables given in the explicit form $z = f(x, y)$, which represents a surface in three-dimensional space. To carry out the differentiation, we are now in a dilemma: Should we differentiate f with respect to x or with respect to y? The answer is simple: Treating y as a constant, differentiate f with respect to x and then treating x as constant differentiate it with respect to y. In this way, we define two different derivatives and call them *partial derivatives* to distinguish them from the (ordinary) derivative of a function of a single independent variable. These ideas will now be formalized.

5.2.1 Partial Derivatives

Definition 5.6 Let $z = f(x, y)$ be a function defined in a region R. Then the limit

$$\lim_{h \to 0} \frac{f(x+h, y) - f(x, y)}{h},$$

when it exists and is independent of the path of approach, is called the *first-order partial derivative* (or simply the *first partial derivative*) of f with respect to x and is denoted by $\partial z/\partial x$ (or $\partial f/\partial x$, or $f_x(x, y)$) so that

$$\frac{\partial z}{\partial x} = \frac{\partial f}{\partial x} = f_x(x, y) = \lim_{h \to 0} \frac{f(x+h, y) - f(x, y)}{h} \qquad (5.6)$$

Similarly, the limit

$$\lim_{k\to 0}\frac{f(x, y+k)-f(x, y)}{k},$$

when it exists and is independent of the path of approach, is called the *first-order partial derivative with respect to y*, and is denoted by $\partial z/\partial y$ [or $\partial f/\partial y$, or $f_y(x, y)$], so that

$$\frac{\partial z}{\partial y}=\frac{\partial f}{\partial y}=f_y(x, y)=\lim_{k\to 0}\frac{f(x, y+k)-f(x, y)}{k} \tag{5.7}$$

As in the case of ordinary differentiation, if $f(x, y)$ is differentiable partially with respect to x and y at all points of a certain region, then we say that $f(x, y)$ is differentiable in that region. Partial differentiation with respect to x and y is denoted by the operators $\partial/\partial x$ and $\partial/\partial y$, respectively. It follows easily that all the rules of differentiation developed for functions of a single variable are also rules of partial differentiation. Thus, when the operator $\partial/\partial x$ acts on a function of y only, say $g(y)$, then since this function is to be taken as a constant with respect to the operator $\partial/\partial x$, we have

$$\frac{\partial}{\partial x}[g(y)]=0.$$

Similarly,

$$\frac{\partial}{\partial y}[\phi(x)]=0.$$

Example 5.3 Find $\partial z/\partial x$ and $\partial z/\partial y$ when $z=x^3+y^3+3axy$.

We have $z=x^3+y^3+3axy$. Here

$$\begin{aligned}\frac{\partial z}{\partial x}&=\frac{\partial}{\partial x}x^3+\frac{\partial}{\partial x}y^3+\frac{\partial}{\partial x}(3axy)\\&=3x^2+0+3ay\,(1)\\&=3x^2+3ay\end{aligned}$$

and

$$\begin{aligned}\frac{\partial z}{\partial y}&=\frac{\partial}{\partial y}(x^3)+\frac{\partial}{\partial y}(y^3)+\frac{\partial}{\partial y}(3axy)\\&=0+3y^2+3ax\,(1)\\&=3y^2+3ax.\end{aligned}$$

Example 5.4 If $u=\sin^{-1}(x/y)$, prove that

$$x\frac{\partial u}{\partial x}+y\frac{\partial u}{\partial y}=0.$$

We have

$$u = \sin^{-1}\frac{x}{y}$$

Hence,

$$\frac{\partial u}{\partial x} = \frac{\partial}{\partial x}\left(\sin^{-1}\frac{x}{y}\right) = \frac{\partial}{\partial x}(\sin^{-1} v),$$

where $v = x/y$. Then

$$\frac{\partial u}{\partial x} = \frac{\partial}{\partial v}(\sin^{-1} v)\frac{\partial v}{\partial x}$$

$$= \frac{1}{\sqrt{1-v^2}}\frac{1}{y}$$

$$= \frac{1}{\sqrt{1-x^2/y^2}}\frac{1}{y}$$

$$= \frac{1}{\sqrt{y^2-x^2}}$$

Also,

$$\frac{\partial u}{\partial y} = \frac{\partial}{\partial y}\left(\sin^{-1}\frac{x}{y}\right) = \frac{1}{\sqrt{1-x^2/y^2}}\left(-\frac{x}{y^2}\right) = -\frac{x}{y\sqrt{y^2-x^2}}.$$

or,

$$y\frac{\partial u}{\partial y} = -\frac{x}{\sqrt{y^2-x^2}}.$$

It follows that

$$x\frac{\partial u}{dx} + y\frac{\partial u}{\partial y} = \frac{x}{\sqrt{y^2-x^2}} - \frac{x}{\sqrt{y^2-x^2}} = 0.$$

5.2.2 Geometrical Interpretation of $\partial z/\partial x$ and $\partial z/\partial y$

We have seen that the equation $z = f(x, y)$ represents a *surface* geometrically. We now ask the question: What is the geometrical interpretation of the first-order partial derivatives $\partial z/\partial x$ and $\partial z/\partial y$ taken at a given point? To see this, let $P[a, b, f(a, b)]$ be a point on the given surface $z = f(x, y)$. To find $\partial z/\partial x$, we keep y constant and then study the variation in the function with respect to x. In this case, the point P moves on the curve of intersection of the surface $z = f(x, y)$ and the plane $y = b$, i.e. the curve $z = f(x, b)$. Now, $[\partial z/\partial x]_{(a, b)}$ is clearly the ordinary derivative of $f(x, b)$ with respect to x at $x = a$. Since this is the slope of the tangent line to the curve at the point $x = a$, it follows that the first partial derivative $[\partial z/\partial x]_{(a, b)}$ denotes the tangent of the angle which the tangent at $P[a, b, f(a, b)]$ to the curve of intersection of the surface $z = f(x, y)$ and the plane $y = b$ makes with the x-axis.

Similarly, the first partial derivative $[\partial z/\partial y]_{(a, b)}$ denotes the tangent of the angle which the tangent at $P[a, b, f(a, b)]$ to the curve of intersection of the surface $z = f(x, y)$ and the plane $x = a$ makes with the y-axis.

5.2.3 Higher-order Partial Derivatives

If the function $f(x, y)$ is differentiable with continuous first derivatives f_x and f_y, then these partial derivatives may themselves be partially differentiable. We thus obtain:

$$\left.\begin{aligned}
&\text{(i)} && \frac{\partial}{\partial x}\left(\frac{\partial f}{\partial x}\right) = \frac{\partial^2 f}{\partial x^2} = f_{xx} \\
&\text{(ii)} && \frac{\partial}{\partial y}\left(\frac{\partial f}{\partial x}\right) = \frac{\partial^2 f}{\partial y \partial x} = f_{yx} \\
&\text{(iii)} && \frac{\partial}{\partial x}\left(\frac{\partial f}{\partial y}\right) = \frac{\partial^2 f}{\partial x \partial y} = f_{xy} \\
&\text{(iv)} && \frac{\partial}{\partial y}\left(\frac{\partial f}{\partial y}\right) = \frac{\partial^2 f}{\partial y^2} = f_{yy}
\end{aligned}\right\} \tag{5.8}$$

It should be noted that in cases (ii) and (iii), partial differentiation with respect to x and y is involved, but in different orders, since the limiting values are found in different orders. This does not necessarily mean that $f_{xy} = f_{yx}$. If, however, f_x, f_y and f_{xy} exist and are continuous in the neighbourhood of a point (a, b), then it can be shown that

$$\left[\frac{\partial^2 f}{\partial x \, \partial y}\right]_{(a, b)} = \left[\frac{\partial^2 f}{\partial y \, \partial x}\right]_{(a, b)} \tag{5.9}$$

In practical cases, the conditions are satisfied and we therefore assume that the order of differentiation is immaterial and that $f_{xy} = f_{yx}$.

Third and higher-order partial derivatives can be defined by an obvious extension. We can thus determine f_{xxx}, f_{yyy}, f_{xyx}, f_{x^2yx}, etc.

Example 5.5 Given that

$$z = x^2 \tan^{-1}\frac{y}{x} - y^2 \tan^{-1}\frac{x}{y},$$

prove that

$$\frac{\partial^2 z}{\partial y \partial x} = \frac{x^2 - y^2}{x^2 + y^2}.$$

We have,

$$z = x^2 \tan^{-1} \frac{y}{x} - y^2 \tan^{-1} \frac{x}{y}.$$

Then,

$$\frac{\partial z}{\partial x} = \frac{\partial}{\partial x}\left(x^2 \tan^{-1} \frac{y}{x} - y^2 \tan^{-1} \frac{x}{y}\right) = 2x \tan^{-1} \frac{y}{x} + x^2 \frac{1}{1 + y^2/x^2}\left(-\frac{y}{x^2}\right)$$

$$-y^2 \frac{1}{1 + x^2/y^2} \frac{1}{y}$$

$$= 2x \tan^{-1} \frac{y}{x} - \frac{yx^2}{x^2 + y^2} - \frac{y^3}{x^2 + y^2}$$

i.e.

$$\frac{\partial z}{\partial x} = 2x \tan^{-1} \frac{y}{x} - y.$$

Hence,

$$\frac{\partial^2 z}{\partial y \partial x} = \frac{\partial}{\partial y}\left(2x \tan^{-1} \frac{y}{x} - y\right).$$

$$= 2x \frac{1}{1 + y^2/x^2} \frac{1}{x} - 1$$

$$= \frac{2x^2}{x^2 + y^2} - 1$$

$$= \frac{x^2 - y^2}{x^2 + y^2}.$$

Example 5.6 If

$$u = \log \frac{x^2 + y^2}{xy},$$

verify that

$$\frac{\partial^2 u}{\partial x \partial y} = \frac{\partial^2 u}{\partial y \partial x}.$$

Since

$$u = \log \frac{x^2 + y^2}{xy}$$

$$= \log(x^2 + y^2) - \log xy,$$

$$= \log(x^2 + y^2) - \log x - \log y,$$

we obtain

$$\frac{\partial u}{\partial x} = \frac{1}{x^2 + y^2}(2x) - \frac{1}{x} - 0 = \frac{2x}{x^2 + y^2} - \frac{1}{x}$$

and

$$\frac{\partial u}{\partial y} = \frac{1}{x^2 + y^2}(2y) - 0 - \frac{1}{y} = \frac{2x}{x^2 + y^2} - \frac{1}{y}.$$

Also,

$$\frac{\partial^2 u}{\partial y \partial x} = \frac{\partial}{\partial y}\left(\frac{2x}{x^2 + y^2} - \frac{1}{x}\right) = 2x\left[-\frac{1}{(x^2 + y^2)^2}\right](2y) - 0 = -\frac{4xy}{(x^2 + y^2)^2},$$

and

$$\frac{\partial^2 u}{\partial x \partial y} = \frac{\partial}{\partial x}\left(\frac{2y}{x^2 + y^2} - \frac{1}{y}\right)$$

$$= 2y\left[-\frac{1}{(x^2 + y^2)^2}\right](2x) - 0$$

$$= -\frac{4xy}{(x^2 + y^2)^2}.$$

It follows that

$$\frac{\partial^2 u}{\partial y \partial x} = \frac{\partial^2 u}{\partial x \partial y}.$$

Example 5.7 Show that $z = f(x - ct) + g(x + ct)$ satisfies the partial differential equation

$$\frac{\partial^2 z}{\partial t^2} = c^2 \frac{\partial^2 z}{\partial x^2}.$$

We have

$$z = f(x - ct) + g(x + ct).$$

Then,

$$\frac{\partial z}{\partial x} = f'(x-ct)\frac{\partial}{\partial x}(x-ct) + g'(x+ct)\frac{\partial}{\partial x}(x+ct)$$

$$= f'(x-ct) + g'(x+ct)$$

and

$$\frac{\partial^2 z}{\partial x^2} = f''(x-ct) + g''(x+ct) \qquad \text{(i)}$$

Again,

$$\frac{\partial z}{dt} = f'(x-ct)\frac{\partial}{\partial t}(x-ct) + g'(x+ct)\frac{\partial}{\partial t}(x+ct)$$

$$= -cf'(x-ct) + cg'(x+ct),$$

and

$$\frac{\partial^2 z}{dt^2} = c^2 f''(x-ct) + c^2 g''(x+ct)$$

$$= c^2[f''(x-ct) + g''(x+ct)] \qquad \text{(ii)}$$

From (i) and (ii), follows that

$$\frac{\partial^2 z}{\partial t^2} = c^2\frac{\partial^2 z}{\partial x^2}.$$

This is an important partial differential equation occurring in wave motion and is called *wave equation*.

Example 5.8 If $u = (x^2 + y^2 + z^2)^{1/2}$, prove that

$$\frac{\partial^2 u}{\partial x^2} + \frac{\partial^2 u}{\partial y^2} + \frac{\partial^2 u}{\partial z^2} = \frac{2}{u}.$$

Here, $u = (x^2 + y^2 + z^2)^{1/2}$, and so

$$\frac{\partial u}{\partial x} = \frac{1}{2\sqrt{x^2+y^2+z^2}}(2x) = \frac{x}{u},$$

$$\frac{\partial u}{\partial y} = \frac{1}{2\sqrt{x^2+y^2+z^2}}(2y) = \frac{y}{u},$$

$$\frac{\partial u}{\partial z} = \frac{1}{2\sqrt{x^2+y^2+z^2}}(2z) = \frac{z}{u}.$$

Hence,

$$\frac{\partial^2 u}{\partial x^2} = \frac{\partial}{\partial x}\left(\frac{\partial u}{\partial x}\right) = \frac{\partial}{\partial x}\left(\frac{x}{u}\right) = \frac{u - x\,(\partial u/\partial x)}{u^2} = \frac{u - x\,(x/u)}{u^2} = \frac{1}{u} - \frac{x^2}{u^3}.$$

By symmetry, we write

$$\frac{\partial^2 u}{\partial z^2} = \frac{1}{u} - \frac{y^2}{u^3} \quad \text{and} \quad \frac{\partial^2 u}{\partial z^2} = \frac{1}{u} - \frac{z^2}{u^3}.$$

Hence,

$$\begin{aligned}\frac{\partial^2 u}{\partial x^2} + \frac{\partial^2 u}{\partial y^2} + \frac{\partial^2 u}{\partial z^2} &= \frac{3}{u} - \frac{x^2 + y^2 + z^2}{u^3} \\ &= \frac{3}{u} - \frac{u^2}{u^3} \\ &= \frac{3}{u} - \frac{1}{u} \\ &= \frac{2}{u}.\end{aligned}$$

Example 5.9 If (x, y) and (r, θ) are, respectively, the Cartesian and polar coordinates of a point P, find

$$\frac{\partial r}{\partial x}, \frac{\partial r}{\partial y}, \frac{\partial \theta}{\partial x}, \frac{\partial \theta}{\partial y}, \frac{\partial y}{\partial \theta} \text{ and } \frac{\partial x}{\partial r}$$

From the relations

$$x = r\cos\theta \quad \text{and} \quad y = r\sin\theta, \tag{i}$$

we obtain:

$$\frac{\partial x}{\partial r} = \cos\theta \quad \text{and} \quad \frac{\partial y}{\partial \theta} = r\cos\theta = x.$$

To obtain the other derivatives, we write from (i):

$$r^2 = x^2 + y^2 \quad \text{and} \quad \tan\theta = \frac{y}{x}. \tag{ii}$$

These relations express r and θ as functions of x and y. From the first equation in (ii), we obtain:

$$\frac{\partial r}{\partial x} = \frac{x}{r} = \cos\theta \quad \text{and} \quad \frac{\partial r}{\partial y} = \frac{y}{r} = \sin\theta.$$

Similarly, differentiating the second equation in (ii), we obtain:

$$\sec^2\theta \frac{\partial \theta}{\partial x} = -\frac{y}{x^2}$$

or

$$\frac{\partial \theta}{\partial x} = -\frac{y}{x^2}\cos^2\theta = -\frac{y}{x^2}\frac{x^2}{r^2} = -\frac{y}{r^2} = -\frac{\sin\theta}{r},$$

and

$$\frac{\partial \theta}{\partial y} = \frac{1}{x}\cos^2\theta = \frac{1}{x}\frac{x^2}{r^2} = \frac{x}{r^2} = \frac{\cos\theta}{r}.$$

5.2.4 Homogeneous Functions: Euler's Theorem

Definition 5.7 A function of two variables, say $f(x, y)$, is said to be *homogeneous* of degree n if, for any constant λ,

$$f(\lambda x, \lambda y) = \lambda^n f(x, y) \tag{5.10}$$

For example,

$x^3 + y^3 + 3x^2y$ is homogeneous of degree 3,

$\dfrac{2x+y}{x^2y^2}$ is homogeneous of degree -3, and

$\tan^{-1}\dfrac{y}{x}$ is homogeneous of degree 0.

It may be seen that expression $x^3 + y^3 + 3x^2y$ can also be written as

$$x^3\left(1 + \frac{y^3}{x^3} + 3\frac{y}{x}\right) = x^3 f\left(\frac{y}{x}\right)$$

Similarly,

$$\frac{2x+y}{x^2y^2} = x^{-3}\left(2\frac{x^2}{y^2} + \frac{x}{y}\right),$$

and

$$\tan^{-1}\frac{y}{x} = x^0 \tan^{-1}\frac{y}{x}.$$

Thus, we have an alternate definition of homogeneity which would be found useful.

A function $f(x, y)$ is said to be a homogeneous function of degree n in x and y if it can be expressed in the form

$$x^n \phi\left(\frac{y}{x}\right).$$

Euler's theorem

If $u = f(x, y)$ is a homogeneous function in x and y of degree n, then

$$x\frac{\partial u}{\partial x} + y\frac{\partial u}{\partial y} = nu \tag{5.11}$$

Proof We assume

$$u = f(x, y) = a_1x^{\alpha_1}y^{\beta_1} + a_2x^{\alpha_2}y^{\beta_2} + \cdots + a_nx^{\alpha_n}y^{\beta_n} \tag{5.12}$$

where

$$\alpha_1 + \beta_1 = \alpha_2 + \beta_2 = \cdots = \alpha_n + \beta_n = n.$$

Differentiating (5.12) with respect to x, we get:

$$\frac{\partial u}{\partial x} = a_1\alpha_1x^{\alpha_1-1}y^{\beta_1} + a_2\alpha_2x^{\alpha_2-1}y^{\beta_2} + \cdots + a_n\alpha_nx^{\alpha_n-1}y^{\beta_n}.$$

Differentiating (5.12) with respect to y, we get:

$$\frac{\partial u}{\partial y} = a_1x^{\alpha_1}\beta_1y^{\beta_1-1} + a_2x^{\alpha_2}\beta_2y^{\beta_2-1} + \cdots + a_nx^{\alpha_n}\beta_ny^{\beta_n-1}.$$

Hence, we obtain:

$$\begin{aligned}
x\frac{\partial u}{\partial x} + y\frac{\partial u}{\partial y} &= a_1\alpha_1x^{\alpha_1}y^{\beta_1} + a_2\alpha_2x^{\alpha_2}y^{\beta_2} + \cdots + a_n\alpha_nx^{\alpha_n}y^{\beta_n} \\
&\quad + a_1\beta_1x^{\alpha_1}y^{\beta_1} + a_2\beta_2x^{\alpha_2}y^{\beta_2} + \cdots + a_n\beta_nx^{\alpha_n}y^{\beta_n} \\
&= (\alpha_1 + \beta_1)a_1x^{\alpha_1}y^{\beta_1} + (\alpha_2 + \beta_2)a_2x^{\alpha_2}y^{\beta_2} + \cdots + (\alpha_n + \beta_n)a_nx^{\alpha_n}y^{\beta_n} \\
&= n\,[a_1x^{\alpha_1}y^{\beta_1} + a_2x^{\alpha_2}y^{\beta_2} + \cdots + a_nx^{\alpha_n}y^{\beta_n}]
\end{aligned}$$

Since $\alpha_1 + \beta_1 = \alpha_2 + \beta_2 = \cdots = \alpha_n + \beta_n = n$, we get

$$x\frac{\partial u}{\partial x} + y\frac{\partial u}{\partial y} = nu,$$

which proves the theorem.

Example 5.10 Verify Euler's theorem for the function $u = x^n \sin(y/x)$.

We obtain

$$\frac{\partial u}{\partial x} = nx^{n-1}\sin\frac{y}{x} + x^n\cos\frac{y}{x}\left(-\frac{y}{x^2}\right)$$

$$= nx^{n-1} \sin \frac{y}{x} - x^{n-2} y \cos \frac{y}{x},$$

whence,

$$x \frac{\partial u}{\partial x} = nx^n \sin \frac{y}{x} - x^{n-1} y \cos \frac{y}{x} = nu - x^{n-1} y \cos \frac{y}{x} \cdot$$

Again,

$$\frac{\partial u}{\partial y} = x^n \left(\cos \frac{y}{x} \right) \frac{1}{x} = x^{n-1} \cos \frac{y}{x},$$

and therefore

$$y \frac{\partial u}{\partial y} = x^{n-1} y \cos \frac{y}{x} \cdot$$

It is easy to see that

$$x \frac{\partial u}{\partial y} + y \frac{\partial u}{\partial y} = nu$$

which verifies Euler's theorem, since $u = x^n \sin (y/x)$ is a homogeneous function of degree n in x and y.

Example 5.11 If

$$u = \sin^{-1} \frac{x^2 + y^2}{x + y},$$

show that

$$x \frac{\partial u}{\partial x} + y \frac{\partial u}{\partial y} = \tan u.$$

Since

$$u = \sin^{-1} \frac{x^2 + y^2}{x + y},$$

we have

$$\sin u = \frac{x^2 + y^2}{x + y} = \frac{x^2 (1 + y^2/x^2)}{x (1 + y/x)} = x \frac{1 + y^2/x^2}{1 + y/x},$$

which shows that $\sin u$ is a homogeneous function of degree 1 in x and y. Applying Euler's theorem, we get:

$$x \frac{\partial}{\partial x} (\sin u) + y \frac{\partial}{\partial y} (\sin u) = \sin u$$

i.e.

$$x \cos u \frac{\partial u}{\partial x} + y \cos u \frac{\partial u}{\partial y} = \sin u$$

or,

$$x \frac{\partial u}{\partial x} + y \frac{\partial u}{\partial y} = \tan u,$$

which is the desired result.

Example 5.12 If

$$u = x^n f\left(\frac{y}{x}\right),$$

show that,

$$x^2 \frac{\partial^2 u}{\partial x^2} + 2xy \frac{\partial^2 u}{\partial x \partial y} + y^2 \frac{\partial^2 u}{\partial y^2} = n(n-1)\, u.$$

Now,

$$u = x^n f\left(\frac{y}{x}\right)$$

is a homogeneous function of degree n in x and y. Hence, by Euler's theorem:

$$x \frac{\partial u}{\partial x} + y \frac{\partial u}{\partial y} = nu \tag{i}$$

Differentiating (i) partially with respect to x, we get

$$x \frac{\partial^2 u}{\partial x^2} + \frac{\partial u}{\partial x} + y \frac{\partial^2 u}{\partial x \partial y} = n \frac{\partial u}{\partial x}$$

i.e.

$$x^2 \frac{\partial^2 u}{\partial x^2} + x \frac{\partial u}{\partial x} + xy \frac{\partial^2 u}{\partial x \partial y} = nx \frac{\partial u}{\partial x}$$

i.e.

$$x^2 \frac{\partial^2 u}{\partial x^2} + xy \frac{\partial^2 u}{\partial x \partial y} = (n-1)\, x \frac{\partial u}{\partial x} \tag{ii}$$

Again, differentiating (i) partially with respect to y, we get:

$$x \frac{\partial^2 u}{\partial y \partial x} + \frac{\partial u}{\partial y} + y \frac{\partial^2 u}{\partial y^2} = n \frac{\partial u}{\partial y}.$$

i.e.

$$xy\frac{\partial^2 u}{\partial y \partial x} + y\frac{\partial u}{\partial y} + y^2\frac{\partial^2 u}{\partial y^2} = ny\frac{\partial u}{\partial y}.$$

i.e.

$$xy\frac{\partial^2 u}{\partial y \partial x} + y^2\frac{\partial^2 u}{\partial y^2} = (n-1)\,y\frac{\partial u}{\partial y} \qquad \text{(iii)}$$

Adding (ii) and (iii), we obtain

$$x^2\frac{\partial^2 u}{\partial x^2} + 2xy\frac{\partial^2 u}{\partial x\,\partial y} + y^2\frac{\partial^2 u}{\partial y^2} = (n-1)\left(x\frac{\partial u}{\partial x} + y\frac{\partial u}{\partial y}\right) = (n-1)\,u,$$

since

$$\frac{\partial^2 u}{\partial x\,\partial y} = \frac{\partial^2 u}{\partial y\,\partial x}.$$

EXERCISES 5.2

Find f_x and f_y when $f(x, y)$ is given by the following functions (Problems 1–15):

1. $x^3 + 3x^2y - y^3$.

2. $\dfrac{xy}{y-x}$.

3. $\dfrac{x^2+3x+1}{y}$.

4. $\dfrac{x^2+y^2}{x^2+y}$.

5. $\dfrac{x^2+2xy}{y^3+x^2}$.

6. $\dfrac{y}{x}+\dfrac{x}{y}$.

7. $\log\left(\dfrac{1}{\sqrt[3]{x}} - \dfrac{1}{\sqrt[3]{y}}\right)$.

8. $ye^{-y/x}$.

9. $\sin^{-1}(y\sqrt{x})$.

10. $\log(x^2+y^2)$.

11. $\tan(x^2+y^2)$.

12. $e^{xy^2}\sec(x^2y)$.

13. $e^{xy}\cos x^2$.

14. $\log_e[\sin(xy)]$.

15. $\tan^{-1}(xy^2)$.

16. If $u = \log(\sqrt{x}+\sqrt{y})$, show that

$$x\frac{\partial u}{\partial x} + y\frac{\partial u}{\partial y} = \frac{1}{2}.$$

17. If $u = x^y$, prove that

$$\frac{x}{y}\frac{\partial u}{\partial x} + \frac{1}{\log x}\frac{\partial u}{\partial y} = 2u.$$

18. Given that $t = 2\pi\sqrt{l/g}$, show that

$$l\frac{\partial t}{\partial l} + g\frac{\partial t}{\partial g} = 0.$$

19. If $u = \sqrt{x^2 + y^2 + z^2}$, prove that

$$\left(\frac{\partial u}{\partial x}\right)^2 + \left(\frac{\partial u}{\partial y}\right)^2 + \left(\frac{\partial u}{\partial z}\right)^2 = 1.$$

20. Given that

$$z = \tan^{-1}\frac{xy}{\sqrt{1 + x^2 + y^2}}$$

show that

$$\frac{\partial^2 z}{\partial x\,\partial y} = (1 + x^2 + y^2)^{-3/2}.$$

21. If

$$\sin z = \frac{\sqrt{x} - \sqrt{y}}{\sqrt{x} + \sqrt{y}},$$

find

$$x\frac{\partial z}{\partial x} + y\frac{\partial z}{\partial y}.$$

22. Given that $u = f(x/y)$, obtain

$$x\frac{\partial u}{\partial x} + y\frac{\partial u}{\partial y}\cdot$$

23. If

$$z = \sin^{-1}\frac{x}{y} + \tan^{-1}\frac{y}{x},$$

show that

$$x\frac{\partial z}{\partial x} + y\frac{\partial z}{\partial y} = 0.$$

24. If $z = f(y + ax) + \phi(y - ax)$, show that

$$\frac{\partial^2 z}{\partial x^2} - a^2\frac{\partial^2 z}{\partial y^2} = 0.$$

25. If $u = f(r)$ where $r = \sqrt{x^2 + y^2}$, show that

$$u_{xx} + u_{yy} = f''(r) + \frac{1}{r} f'(r).$$

26. Given that $u = x(x^2 + y^2 + z^2)^{-3/2}$, show that $xu_x + yu_y + zu_z = -2u$.

Verify Euler's theorem for the following functions (Problems 27–32):

27. $x^3 \log \frac{y}{x}$.

28. $\frac{x(x^3 - y^3)}{x^3 + y^3}$.

29. $\frac{x^{1/4} + y^{1/4}}{x^{1/5} + y^{1/5}}$.

30. $x^3 - 3x^2 y + 5xy^2 - y^3$.

31. $(x^2 + xy + y^2)^{1/2}$

32. $e^{x/y}$.

33. Show that $V = (r^n + r^{-n}) \sin n\theta$ satisfies Laplace's equation in polar coordinates, viz.,

$$\frac{\partial^2 V}{\partial r^2} + \frac{1}{r}\frac{\partial V}{\partial r} + \frac{1}{r^2}\frac{\partial^2 V}{\partial \theta^2} = 0.$$

34. Given that

$$z = \frac{1}{x}[f(x - ct) + g(x + ct)],$$

show that

$$\frac{\partial}{\partial x}\left(x^2 \frac{\partial z}{\partial x}\right) = c^2 x^2 \frac{\partial^2 z}{\partial t^2}.$$

35. If

$$u = \tan^{-1} \frac{\cos x}{\sinh y} \quad \text{and} \quad v = \tanh^{-1} \frac{\sin x}{\cosh y},$$

show that

$$\frac{\partial u}{\partial x} = \frac{\partial v}{\partial y} \quad \text{and} \quad \frac{\partial u}{\partial y} = -\frac{\partial v}{\partial x}.$$

36. Show that $V = xyz(x^2 + y^2 + z^2)^{-1/2}$ satisfies

$$\frac{\partial^2 V}{\partial x^2} + \frac{\partial^2 V}{\partial y^2} + \frac{\partial^2 V}{\partial z^2} = 0.$$

37. If

$$u = \tan^{-1}\frac{x^3 + y^3}{x - y},$$

prove that

$$x\frac{\partial u}{\partial x} + y\frac{\partial u}{\partial y} = \sin 2u.$$

38. If $u = xyf(y/x)$, prove that

$$x\frac{\partial u}{\partial x} + y\frac{\partial u}{\partial y} = 2u.$$

39. If $x = r\cos\theta$ and $y = r\sin\theta$, transform the equation

$$\frac{\partial^2 u}{\partial x^2} + \frac{\partial^2 u}{\partial y^2} = 0$$

into polar coordinates.

40. If the curves $f(x, y) = 0$ and $\phi(x, y) = 0$ touch, show that at the point of contact $f_x\phi_y - f_y\phi_x = 0$.

5.3 DIFFERENTIALS

If y is a function of a single variable x, the differential coefficients dy/dx, when it exists, simply represents a number defined by a limit. We now ascribe a meaning to the symbols dx and dy which permit us to manipulate the derivative dy/dx as a quotient of two quantities dy and dx. These quantities, called the *differentials*, have important applications, e.g. in the solution of differential equations.

Let the function be given by $y = f(x)$. We define dy, the differential of y, by

$$dy = f'(x)\,\Delta x \tag{5.13}$$

Then, the differential dx of x is given by

$$dx = 1\,\Delta x, \tag{5.14}$$

since the derivative of x with respect to itself is unity. This shows that we can write Eq. (5.13) as

$$dy = f'(x)\,dx \tag{5.15}$$

Let Δy be the change (or *increment*) in y corresponding to a change Δx in x. Although Δx and dx are the same, according to Eq. (5.14), it should be noted that Δy and dy are two distinct quantities and the relation between them is shown in Fig. 5.1.

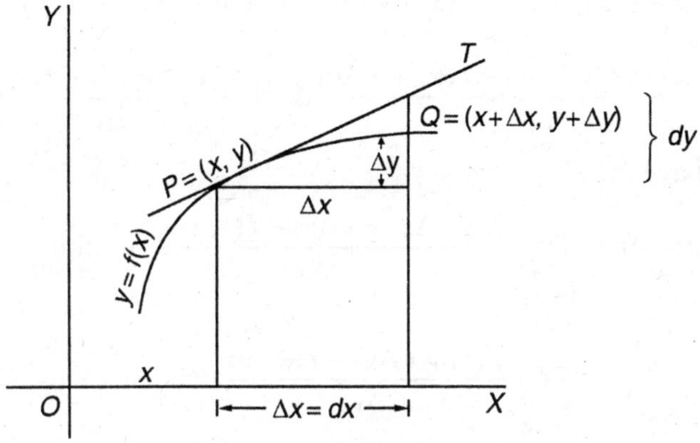

Figure 5.1

Thus, if $y = x^2 + \sin x$, then $dy/dx = 2x + \cos x$, and hence $dy = (2x + \cos x)\,dx$.

From Eq. (5.15) we immediately deduce:

(i) if $y = af(x) + bg(x)$, then $dy = af'(x)\,dx + bg'(x)\,dx$

(ii) if $y = f(x)\,g(x)$, then $dy = f'(x)\,g(x)\,dx + f(x)\,g'(x)\,dx$.

(iii) if $y = \dfrac{f(x)}{g(x)}$, then $dy = \dfrac{f'(x)\,g(x)\,dx - f(x)\,g'(x)dx}{[g(x)]^2}$, where $g(x) \neq 0$.

For example, if $f(x) = x^2$ and $g(x) = \sin x$, then

(i) $d(f+g) = 2x\,dx + \cos x\,dx$

(ii) $d(fg) = 2x \sin x\,dx + x^2 \cos x\,dx$

(iii) $d\left(\dfrac{f}{g}\right) = \dfrac{2x \sin x\,dx - x^2 \cos x\,dx}{\sin^2 x}$.

Now, let $u = f(x, y)$ be a function of two independent variables x and y. We have already seen that

$$\frac{\partial u}{\partial x} = \lim_{\Delta x \to 0} \frac{f(x+\Delta x, y) - f(x, y)}{\Delta x}$$

and

$$\frac{\partial u}{\partial x} = \lim_{\Delta y \to 0} \frac{f(x, y+\Delta y) - f(x, y)}{\Delta x}$$

Let Δu be the change in u when both x and y change by the amounts Δx and Δy respectively, so that $\lim_{\Delta x, \Delta y \to 0} \Delta u = du$. In this case, du is called the *total differential*.

Now,

$\Delta u = f(x+\Delta x, y+\Delta y) - f(x, y)$

$$= f(x+\Delta x, y+\Delta y) - f(x, y+\Delta y) + f(x, y+\Delta y) - f(x, y)$$

$$= \frac{f(x+\Delta x, y+\Delta y) - f(x, y+\Delta y)}{\Delta x}\Delta x + \frac{f(x, y+\Delta y) - f(x, y)}{\Delta y}$$

Hence we have

$$\lim_{\substack{\Delta x\to 0\\ \Delta y\to 0}} \Delta u = \lim_{\substack{\Delta x\to 0\\ \Delta y\to 0}} \frac{f(x+\Delta x, y+\Delta y) - f(x, y+\Delta y)}{\Delta x} \lim_{\Delta x\to 0} \Delta x$$

$$+ \lim_{\substack{\Delta x\to 0\\ \Delta y\to 0}} \frac{f(x, y+\Delta y) - f(x, y)}{\Delta y} \lim_{\Delta y\to 0} \Delta y$$

Now, on putting $\Delta y = 0$.

$$\lim_{\substack{\Delta x\to 0\\ \Delta y\to 0}} \frac{f(x+\Delta x, y+\Delta y) - f(x, y+\Delta y)}{\Delta x} = \lim_{\Delta x\to 0} \frac{f(x+\Delta x, y) - f(x, y)}{\Delta x} = \frac{\partial u}{\partial x},$$

and

$$\lim_{\substack{\Delta x\to 0\\ \Delta y\to 0}} \frac{f(x, y+\Delta y) - f(x, y)}{\Delta y} = \frac{\partial u}{\partial y},$$

where

$$\lim_{\Delta x\to 0} \Delta x = dx, \qquad \lim_{\Delta y\to 0} \Delta y = dy, \qquad \lim_{\substack{\Delta x\to 0\\ \Delta y\to 0}} \Delta u = du.$$

Hence, we obtain

$$du = \frac{\partial u}{\partial x} dx + \frac{\partial u}{\partial y} dy \tag{5.16}$$

Equation (5.16) is called the *total differential equation* and has great practical importance. This result can be generalized to functions of several independent variables; if for examples, $u = f(x, y, z)$ then,

$$du = \frac{\partial u}{\partial x} dx + \frac{\partial u}{\partial y} dy + \frac{\partial u}{\partial z} dz \tag{5.17}$$

If x, y, z are themselves functions of a single independent variable t, then formula (5.17) becomes:

$$\frac{du}{dt} = \frac{\partial u}{\partial x}\frac{dx}{dt} + \frac{\partial u}{\partial y}\frac{dy}{dt} + \frac{\partial u}{\partial z}\frac{dz}{dt}, \tag{5.18}$$

where du/dt is called the total differential coefficient.

On the other hand, if x, y, z are functions of two independent variables, say s and t, then in (5.18), instead of the total derivatives we write the partial derivatives. We thus have the two formulae instead:

$$\frac{\partial u}{\partial t} = \frac{\partial u}{dx}\frac{\partial x}{\partial t} + \frac{\partial u}{\partial y}\frac{\partial y}{\partial t} + \frac{\partial u}{\partial z}\frac{\partial z}{\partial t} \tag{5.19}$$

and

$$\frac{\partial u}{\partial s} = \frac{\partial u}{\partial x}\frac{\partial x}{\partial s} + \frac{\partial u}{\partial y}\frac{\partial y}{\partial s} + \frac{\partial u}{\partial z}\frac{\partial z}{\partial s} \tag{5.20}$$

Example 5.13 If $f(x, y) =$ constant, show that

$$\frac{dy}{dx} = -\frac{\partial f/\partial x}{\partial f/\partial y}.$$

Since $f(x, y) =$ constant, we have

$$df = \frac{\partial f}{\partial x}dx + \frac{\partial f}{\partial y}dy = 0$$

i.e.

$$\frac{df}{dx} = \frac{\partial f}{dx} + \frac{\partial f}{\partial y}\frac{dy}{\partial x} = 0.$$

Hence, it follows that

$$\frac{dy}{dx} = -\frac{\partial f/\partial x}{\partial f/\partial y}.$$

Example 5.14 If $x = r\cos\theta$, $y = r\sin\theta$ and x, y, r and θ are functions of t, prove that $\dot{x}\cos\theta + \dot{y}\sin\theta = \dot{r}$, where the dots denote total derivatives with respect to t.

We have $x = r\cos\theta$, $y = r\sin\theta$. Then

$$\dot{x} = \frac{dx}{dt} = \frac{\partial x}{\partial r}\frac{dr}{dt} + \frac{\partial x}{\partial\theta}\frac{d\theta}{dt} = (\cos\theta)\,\dot{r} - r\,(\sin\theta)\dot{\theta},$$

so that

$$\dot{x}\cos\theta = \dot{r}\cos^2\theta - r\,(\sin\theta\cos\theta)\,\dot{\theta}, \tag{i}$$

and

$$\dot{y} = \frac{dy}{dt} = \frac{\partial y}{\partial r}\frac{dr}{dt} + \frac{\partial y}{\partial\theta}\frac{d\theta}{dt} = \dot{r}\sin\theta + r\cos\theta\,.\,\dot{\theta}$$

so that

$$\dot{y}\sin\theta = \dot{r}\sin^2\theta + r\sin\theta\cos\theta\,.\,\dot{\theta} \tag{ii}$$

Adding (i) and (ii), we get $\dot{x}\cos\theta + \dot{y}\sin\theta = \dot{r}$, which is the required result.

Example 5.15 Given that $u = \sin(x^2 + y^2)$ where $x = 3t$ and $y = 1/(1+t^2)$, determine du/dt.

Since $u = \sin(x^2 + y^2)$, we have

$$\frac{\partial u}{\partial x} = 2x\cos(x^2 + y^2) \quad \text{and} \quad \frac{\partial u}{\partial y} = 2y\cos(x^2 + y^2).$$

Again, since $x = 3t$ and $y = 1/(1+t^2)$, we have,

$$\frac{dx}{dt} = 3 \quad \text{and} \quad \frac{dy}{dt} = -\frac{2t}{(1+t^2)^2}.$$

Substituting these values in the formula

$$\frac{du}{dt} = \frac{\partial u}{\partial x}\frac{dx}{dt} + \frac{\partial u}{\partial y}\frac{dy}{dt},$$

we obtain

$$\frac{du}{dt} = 2x\cos(x^2 + y^2)(3) + 2y\cos(x^2 + y^2)\left[-\frac{2t}{(1+t^2)^2}\right]$$

$$= 2\cos(x^2 + y^2)\left[3x - \frac{2yt}{(1+t^2)^2}\right].$$

Example 5.16 If $z = f(x, y)$ where $x = r\cos\theta$ and $y = r\sin\theta$, find $\partial z/\partial x$ and $\partial^2 z/\partial x^2$.

We have $r^2 = x^2 + y^2$ and $\theta = \tan^{-1}(y/x)$. Hence,

$$\frac{\partial r}{\partial x} = \frac{x}{r} = \cos\theta,$$

$$\frac{\partial r}{\partial y} = \frac{y}{r} = \sin\theta,$$

$$\frac{\partial \theta}{\partial x} = \frac{1}{1 + y^2/x^2}\left(-\frac{y}{x^2}\right) = -\frac{y}{x^2 + y^2} = -\frac{y}{r^2} = -\frac{\sin\theta}{r}$$

$$\frac{\partial \theta}{\partial y} = \frac{1}{1 + (y^2/x^2)}\frac{1}{x} = \frac{x}{r^2} = \frac{\cos\theta}{r}.$$

Substituting the above in the formula

$$\frac{\partial z}{\partial x} = \frac{\partial z}{\partial r}\frac{\partial r}{\partial x} + \frac{\partial z}{\partial \theta}\frac{\partial \theta}{\partial x},$$

we obtain:

$$\frac{\partial z}{\partial x}=\frac{\partial z}{\partial r}\cos\theta-\frac{\sin\theta}{r}\frac{\partial z}{\partial\theta}=\cos\theta\frac{\partial z}{\partial r}-\frac{\sin\theta}{r}\frac{\partial z}{\partial\theta},$$

from which follows:

$$\frac{\partial}{\partial x}\equiv\cos\theta\frac{\partial}{\partial r}-\frac{\sin\theta}{r}\frac{\partial}{\partial\theta}. \tag{i}$$

Hence,

$$\frac{\partial^2 z}{\partial x^2}=\frac{\partial}{\partial x}\left(\frac{\partial z}{\partial x}\right)$$

$$=\left(\cos\theta\frac{\partial}{\partial r}-\frac{\sin\theta}{r}\frac{\partial}{\partial\theta}\right)\left(\cos\theta\frac{\partial z}{\partial r}-\frac{\sin\theta}{r}\frac{\partial z}{\partial\theta}\right),$$

$$=\cos\theta\frac{\partial}{\partial r}\left(\cos\theta\frac{\partial z}{\partial r}-\frac{\sin\theta}{r}\frac{\partial z}{\partial\theta}\right)-\frac{\sin\theta}{r}\frac{\partial}{\partial\theta}\left(\cos\theta\frac{\partial z}{\partial r}-\frac{\sin\theta}{r}\frac{\partial z}{\partial\theta}\right)$$

$$=\cos\theta\left(\cos\theta\frac{\partial^2 z}{\partial r^2}+\frac{\sin\theta}{r^2}\frac{\partial z}{\partial\theta}-\frac{\sin\theta}{r}\frac{\partial^2 z}{\partial r\,\partial\theta}\right)$$

$$-\frac{\sin\theta}{r}\left(-\sin\theta\frac{\partial z}{\partial r}+\cos\theta\frac{\partial^2 z}{\partial r\,\partial\theta}-\frac{\cos\theta}{r}\frac{\partial z}{\partial\theta}-\frac{\sin\theta}{r}\frac{\partial^2 z}{\partial\theta^2}\right)$$

$$=\cos^2\theta\frac{\partial^2 z}{\partial r^2}+\frac{\cos\theta\sin\theta}{r^2}\frac{\partial z}{\partial\theta}-\frac{\cos\theta\sin\theta}{r}\frac{\partial^2 z}{\partial r\partial\theta}+\frac{\sin^2\theta}{r}\frac{\partial z}{\partial r}$$

$$-\frac{\sin\theta\cos\theta}{r}\frac{\partial^2 z}{\partial r\partial\theta}+\frac{\sin\theta\cos\theta}{r^2}\frac{\partial z}{\partial\theta}+\frac{\sin^2\theta}{r^2}\frac{\partial^2 z}{\partial\theta^2}$$

$$=\cos^2\theta\frac{\partial^2 z}{\partial r^2}+\frac{2\cos\theta\sin\theta}{r^2}\frac{\partial z}{\partial\theta}-\frac{2\cos\theta\sin\theta}{r}\frac{\partial^2 z}{\partial r\partial\theta}$$

$$+\frac{\sin^2\theta}{r}\frac{\partial z}{\partial r}+\frac{\sin^2\theta}{r^2}\frac{\partial^2 z}{\partial\theta^2}.$$

Example 5.17 Given that $F(x, y, z)=0$, prove that

$$\frac{\partial y}{\partial x}\frac{\partial z}{\partial y}\frac{\partial z}{\partial x}=-1.$$

Since $F(x, y, z)=0$, we have

$$dF=\frac{\partial F}{\partial x}dx+\frac{\partial F}{\partial y}dy+\frac{\partial F}{\partial z}dz=0.$$

If z is kept constant, then $dz = 0$ and we obtain:

$$\left[\frac{dy}{dx}\right]_{z=\text{constant}} = \frac{\partial y}{\partial x} = -\frac{\partial F/\partial x}{\partial F/\partial y}.$$

Similarly, we obtain

$$\frac{\partial z}{\partial x} = -\frac{\partial F/\partial z}{\partial F/\partial x} \quad \text{and} \quad \frac{\partial z}{\partial y} = -\frac{\partial F/\partial y}{\partial F/\partial z}.$$

Multiplying these relations together, we get:

$$\frac{\partial y}{\partial x}\frac{\partial z}{\partial x}\frac{\partial z}{\partial y} = -\frac{\partial F/\partial x}{\partial F/\partial y}\frac{\partial F/\partial z}{\partial F/\partial y}\frac{\partial F/\partial y}{\partial F/\partial z} = -1.$$

Jacobians

Let $u = f(x, y)$ and $v = \phi(x, y)$ be two continuous functions of the independent variables x and y such that

$$\frac{\partial u}{\partial x}, \quad \frac{\partial u}{\partial y}, \quad \frac{\partial v}{\partial x}, \quad \frac{\partial v}{\partial y}$$

are also continuous in x and y. Then, the determinant

$$\begin{vmatrix} \dfrac{\partial u}{\partial x} & \dfrac{\partial u}{\partial y} \\ \dfrac{\partial v}{\partial x} & \dfrac{\partial v}{\partial y} \end{vmatrix} \tag{5.21}$$

is called the *Jacobian* of u, v with respect to x and y and is denoted by

$$J\left(\frac{u, v}{x, y}\right) \quad \text{and} \quad \frac{\partial(u, v)}{\partial(x, y)}.$$

In the case of three variables u, v, w which are functions of x, y, z, the Jacobian of u, v, w, with respect to x, y, z is defined as

$$\begin{vmatrix} \dfrac{\partial u}{dx} & \dfrac{\partial u}{\partial y} & \dfrac{\partial u}{\partial z} \\ \dfrac{\partial v}{\partial x} & \dfrac{\partial v}{\partial y} & \dfrac{\partial v}{\partial z} \\ \dfrac{\partial w}{\partial x} & \dfrac{\partial w}{\partial y} & \dfrac{\partial w}{\partial z} \end{vmatrix} \tag{5.22}$$

and is denoted by

$$J\left(\frac{u, v, w}{x, y, z}\right) = \frac{\partial\,(u, v, w)}{\partial\,(x, y, z)}.$$

We now derive some important results concerning Jacobians.

(i) If u, v are functions of x, y and x, y are themselves functions of s, t, then

$$\frac{\partial(u, v)}{\partial(x, y)} \frac{\partial(x, y)}{\partial(s, t)} = \frac{\partial(u, v)}{\partial(s, t)}. \tag{5.23}$$

Now,

$$\frac{\partial(u, v)}{\partial(x, y)} \cdot \frac{\partial(x, y)}{\partial(s, t)} = \begin{vmatrix} \frac{\partial u}{\partial x} & \frac{\partial u}{\partial y} \\ \frac{\partial v}{\partial x} & \frac{\partial v}{\partial y} \end{vmatrix} \begin{vmatrix} \frac{\partial x}{\partial s} & \frac{\partial x}{\partial t} \\ \frac{\partial y}{\partial s} & \frac{\partial y}{\partial t} \end{vmatrix}$$

$$= \begin{vmatrix} \frac{\partial u}{\partial x}\frac{\partial x}{\partial s} + \frac{\partial u}{\partial y}\frac{\partial y}{\partial s} & \frac{\partial u}{\partial x}\frac{\partial x}{\partial t} + \frac{\partial u}{\partial y}\frac{\partial y}{\partial t} \\ \frac{\partial v}{\partial x}\frac{\partial x}{\partial s} + \frac{\partial v}{\partial y}\frac{\partial y}{\partial s} & \frac{\partial v}{\partial x}\frac{\partial x}{\partial t} + \frac{\partial v}{\partial y}\frac{\partial y}{\partial t} \end{vmatrix} \tag{5.24}$$

But since $u = f(x, y)$ $v = \phi(x, y)$, $x = f_1(s, t)$ and $y = f_2(s, t)$, we have

$$\frac{\partial u}{\partial s} = \frac{\partial u}{\partial x}\frac{\partial x}{\partial s} + \frac{\partial u}{\partial y}\frac{\partial y}{\partial s}$$

$$\frac{\partial u}{\partial t} = \frac{\partial u}{\partial x}\frac{\partial x}{\partial t} + \frac{\partial u}{\partial y}\frac{\partial y}{\partial t}$$

$$\frac{\partial v}{\partial s} = \frac{\partial v}{\partial x}\frac{\partial x}{\partial s} + \frac{\partial v}{\partial y}\frac{\partial y}{\partial s}$$

$$\frac{\partial v}{\partial t} = \frac{\partial v}{\partial x}\frac{\partial x}{\partial t} + \frac{\partial v}{\partial y}\frac{\partial y}{\partial t}.$$

Hence, (5.24) becomes:

$$\begin{vmatrix} \frac{\partial u}{\partial s} & \frac{\partial u}{\partial t} \\ \frac{\partial v}{\partial s} & \frac{\partial v}{\partial t} \end{vmatrix}$$

which is equal to $\partial(u, v)/\partial(s, t)$. This proves (5.23).

(ii) $$\frac{\partial(u, v)}{\partial(x, y)} \frac{\partial(x, y)}{\partial(u, v)} = 1. \tag{5.25}$$

In (i), we put $s = u$ and $t = v$. Then we have,

$$\frac{\partial(u, v)}{\partial(x, y)} \frac{\partial(x, y)}{\partial(u, v)} = \frac{\partial(u, v)}{\partial(u, v)}.$$

But,

$$\frac{\partial(u, v)}{\partial(u, v)} = \begin{vmatrix} \dfrac{\partial u}{\partial u} & \dfrac{\partial u}{\partial v} \\ \dfrac{\partial v}{\partial u} & \dfrac{\partial v}{\partial v} \end{vmatrix}$$

Since u and v are independent variables,

$$\frac{\partial u}{\partial v} = \frac{\partial v}{\partial u} = 0.$$

Hence,

$$\frac{\partial(u, v)}{\partial(u, v)} = \begin{vmatrix} 1 & 0 \\ 0 & 1 \end{vmatrix} = 1,$$

i.e.

$$\frac{\partial(u, v)}{\partial(x, y)} \frac{\partial(x, y)}{\partial(u, v)} = 1,$$

which proves (5.25).

(iii) The Jacobian is useful in determining whether or not two functions are functionally dependent. Two functions, say $f(x, y)$ and $\phi(x, y)$, are said to be functionally dependent if they are functions of each other. The vanishing of the Jacobian is a necessary and sufficient condition for the functional dependence of two functions.

To prove that the condition is *necessary*, let us assume that $f(x, y)$ and $\phi(x, y)$ are functionally dependent. Then a new function can be formed such that

$$F(f, \phi) = 0. \tag{5.26}$$

Differentiating the above, we obtain

$$\frac{\partial F}{\partial f} \frac{\partial f}{\partial x} + \frac{\partial F}{\partial \phi} \frac{\partial \phi}{\partial x} = 0 \tag{5.27}$$

and

$$\frac{\partial F}{\partial f} \frac{\partial f}{\partial y} + \frac{\partial F}{\partial \phi} \frac{\partial \phi}{\partial y} = 0. \tag{5.28}$$

From (5.27) and (5.28), we obtain

$$\begin{vmatrix} \dfrac{\partial f}{\partial x} & \dfrac{\partial \phi}{\partial x} \\ \dfrac{\partial f}{\partial y} & \dfrac{\partial \phi}{\partial y} \end{vmatrix} = 0$$

i.e.

$$\begin{vmatrix} f_x & f_y \\ \phi_x & \phi_y \end{vmatrix} = 0, \tag{5.29}$$

which proves that the condition is necessary. The proof of sufficiency of the condition is too involved and will therefore be omitted here.

Example 5.18 If $u = f(y-z, z-x, x-y)$, prove that

$$\frac{\partial u}{\partial x} + \frac{\partial u}{\partial y} + \frac{\partial u}{\partial z} = 0.$$

Let $X = y - z$, $Y = z - x$, $Z = x - y$. Then,

$$\frac{\partial X}{\partial x} = 0, \qquad \frac{\partial X}{\partial y} = 1, \qquad \frac{\partial X}{\partial z} = -1,$$

$$\frac{\partial Y}{\partial x} = -1, \qquad \frac{\partial Y}{\partial y} = 0, \qquad \frac{\partial Y}{\partial z} = 1,$$

$$\frac{\partial Z}{\partial x} = 1, \qquad \frac{\partial Z}{\partial y} = -1 \qquad \frac{\partial Z}{\partial z} = 0.$$

We have

$$\frac{\partial u}{\partial x} = \frac{\partial u}{\partial X}\frac{\partial X}{\partial x} + \frac{\partial u}{\partial Y}\frac{\partial y}{\partial x} + \frac{\partial u}{\partial Z}\frac{\partial Z}{\partial x} = -\frac{\partial u}{\partial Y} + \frac{\partial u}{\partial Z}, \tag{i}$$

$$\frac{\partial u}{\partial y} = \frac{\partial u}{\partial X}\frac{\partial X}{\partial y} + \frac{\partial u}{\partial Y}\frac{\partial Y}{\partial y} + \frac{\partial u}{\partial Z}\frac{\partial Z}{\partial y} = \frac{\partial u}{\partial X} - \frac{\partial u}{\partial Z} \tag{ii}$$

$$\frac{\partial u}{\partial z} = \frac{\partial u}{\partial X}\frac{\partial X}{\partial z} + \frac{\partial u}{\partial Y}\frac{\partial Y}{\partial z} + \frac{\partial u}{\partial Z}\frac{\partial Z}{\partial z} = -\frac{\partial u}{\partial X} + \frac{\partial u}{\partial Y} \tag{iii}$$

Adding (i), (ii) and (iii), we obtain

$$\frac{\partial u}{\partial x} + \frac{\partial u}{\partial y} + \frac{\partial u}{\partial z} = 0.$$

Example 5.19 Show that the functions

$$f(x, y) = x^2 + xy + y^2 = 0$$

$$\phi(x, y) = x^4 + 2x^3y + 3x^2y^2 + 2xy^3 + y^4 = 0$$

are functionally dependent.

We have,

$$f_x = 2x + y, \qquad f_y = x + 2y,$$

$$\phi_x = 4x^3 + 6x^2y + 6xy^2 + 2y^3, \qquad \phi_y = 2x^3 + 6xy^2 + 4y^3 + 6x^2y,$$

Hence,

$$\begin{vmatrix} f_x & f_y \\ \phi_x & \phi_y \end{vmatrix} = \begin{vmatrix} 2x + y & x + 2y \\ 4x^3 + 6x^2y + 6xy^2 + 2y^3 & 2x^3 + 6x^2y + 6xy^2 + 4y^3 \end{vmatrix}$$

$$= (2x + y)(2x^3 + 6x^2y + 6xy^2 + 4y^3)$$
$$-(x + 2y)(4x^3 + 6x^2y + 6xy^2 + 2y^3)$$
$$= 0.$$

Hence the functions $f(x, y)$ and $\phi(x, y)$ are functionally dependent.

Example 5.20 If $z = f(x^2 + y^2)$, show that

$$x\frac{\partial z}{\partial y} - y\frac{\partial z}{\partial x} = 0.$$

Let $u = x^2 + y^2$. Then $z = f(u)$, where $u = x^2 + y^2$. Hence,

$$\frac{\partial z}{\partial x} = \frac{dz}{du}\frac{\partial u}{\partial x} = 2x\, f'(u), \quad \text{and} \quad \frac{\partial z}{\partial y} = 2y\, f'(u).$$

It follows that

$$x\frac{\partial z}{\partial y} - y\frac{\partial z}{\partial x} = 2xyf'(u) - 2xyf'(u) = 0.$$

Example 5.21 If $u = x + y$ and $v = xy^2$, find $\partial y/\partial u$ and $\partial y/\partial v$.

Eliminating x from the two given relations, we obtain:

$$u = \frac{v}{y^2} + y \quad \text{or} \quad uy^2 = y^3 + v \tag{i}$$

Differentiating (i) partially with respect to u, we get:

$$y^2 + 2uy\frac{\partial y}{\partial u} = 3y^2\frac{\partial y}{\partial u} + 0$$

or,

$$\frac{\partial y}{\partial u} = \frac{y^2}{3y^2 - 2uy} = \frac{y}{3y - 2u}.$$

In a similar manner, it can be shown that

$$\frac{\partial y}{\partial v} = \frac{1}{2uy - 3y^2}.$$

Example 5.22 If $x = r\cos\theta$, $y = r\sin\theta$, prove that

$$\frac{\partial(x, y)}{\partial(r, \theta)} = r \quad \text{and} \quad \frac{\partial(r, \theta)}{\partial(x, y)} = \frac{1}{r}.$$

We have

$$\frac{\partial x}{\partial r} = \cos\theta, \qquad \frac{\partial x}{\partial \theta} = -r\sin\theta,$$

$$\frac{\partial y}{\partial r} = \sin\theta, \qquad \frac{\partial y}{\partial \theta} = r\cos\theta$$

Therefore,

$$\frac{\partial(x, y)}{\partial(r, \theta)} = \begin{vmatrix} \dfrac{\partial x}{\partial r} & \dfrac{\partial x}{\partial \theta} \\[2ex] \dfrac{\partial y}{\partial r} & \dfrac{\partial y}{\partial \theta} \end{vmatrix} = \begin{vmatrix} \cos\theta & -r\sin\theta \\ \sin\theta & r\cos\theta \end{vmatrix} = r.$$

and

$$\frac{\partial(r, \theta)}{\partial(x, y)} = \begin{vmatrix} \dfrac{\partial r}{\partial x} & \dfrac{\partial r}{\partial y} \\[2ex] \dfrac{\partial \theta}{\partial x} & \dfrac{\partial \theta}{\partial y} \end{vmatrix} = \begin{vmatrix} \dfrac{x}{r} & \dfrac{y}{r} \\[2ex] \dfrac{-\sin\theta}{r} & \dfrac{\cos\theta}{r} \end{vmatrix}$$

$$= \frac{x\cos\theta}{r^2} + \frac{y\sin\theta}{r^2} = \frac{r\cos^2\theta}{r^2} + \frac{r\sin^2\theta}{r^2}$$

$$= \frac{1}{r}.$$

Example 5.23 If $x = r\sin\theta\cos\phi$, $y = r\sin\theta\sin\phi$, $z = r\cos\theta$, show that

$$\frac{\partial(x, y, z)}{\partial(r, \theta, \phi)} = r^2\sin\theta.$$

We have

$$\frac{\partial x}{\partial r} = \sin\theta\cos\phi, \quad \frac{\partial x}{\partial \theta} = r\cos\theta\cos\phi, \quad \frac{\partial x}{\partial \phi} = -r\sin\theta\sin\phi$$

$$\frac{\partial y}{\partial r} = \sin\theta\sin\phi, \quad \frac{\partial y}{\partial \theta} = r\cos\theta\sin\phi, \quad \frac{\partial y}{\partial \phi} = r\sin\theta\cos\phi.$$

$$\frac{\partial z}{\partial r} = \cos\theta, \qquad \frac{\partial z}{\partial \theta} = -r\sin\theta, \qquad \frac{\partial z}{\partial \phi} = 0.$$

Hence

$$\frac{\partial(x, y, z)}{\partial(r, \theta, \phi)} = \begin{vmatrix} \sin\theta\cos\phi & r\cos\theta\cos\phi & -r\sin\theta\sin\phi \\ \sin\theta\sin\phi & r\cos\theta\sin\phi & r\sin\theta\cos\phi \\ \cos\theta & -r\sin\theta & 0 \end{vmatrix}$$

$$= \cos\theta\,[r^2\cos\theta\sin\theta\cos^2\phi + r^2\sin\theta\cos\theta\sin^2\phi]$$

$$+ r\sin\theta\,[r\sin^2\theta\cos^2\phi + r\sin^2\theta\sin^2\phi]$$

$$= r^2\cos^2\theta\sin\theta + r^2\sin^3\theta.$$

$$= r^2\sin^2\theta\,(\cos^2\theta + \sin^2\theta)$$

$$= r^2\sin\theta.$$

EXERCISES 5.3

1. Find dv/dt when
   (i) $v = xy + 2,\ x = t^2,\ y = 1/t.$
   (ii) $v = x^2 + y^2,\ x = t/(1+t)\ \ y = t^2/(1+t).$
2. Find $\partial z/\partial x$ and $\partial z/\partial y$ when (i) $x + y + z = \log z$ (ii) $z = x^2 + 8xy + 2.$
3. If (x, y) and (r, θ) are, respectively, the cartesian and polar coordinates of a point P, show that

$$\frac{\partial(x, y)}{\partial(r, \theta)} = r.$$

4. If (x, y, z) and (r, θ, ϕ) are, respectively, the cartesian and spherical polar coordinates of a point P, show that

$$\frac{\partial(x, y, z)}{\partial(r, \theta, \phi)} = -r^2\sin\theta.$$

   Find also

$$\frac{\partial(r, \theta, \phi)}{\partial(x, y, z)}.$$

5. If $x = r + s + t,\ y = st + tr + rs,\ z = rst,$ show that

$$\frac{\partial(x, y, z)}{\partial(r, s, t)} = (r - s)(r - t)(s - t).$$

   Also, find the value of

$$\frac{\partial(r, s, t)}{\partial(x, y, z)}.$$

6. If $r = x + iy$, $s = x - iy$, and u is a function of x and y, show that the equation

$$\frac{\partial^2 u}{\partial x^2} + \frac{\partial^2 u}{\partial y^2} = 0$$

transforms into

$$\frac{\partial^2 u}{\partial r \partial s} = 0.$$

7. If $z = xy$ where $x = r\cos\theta$ and $y = r\sin\theta$, express z in terms of r and θ. Find $\partial z/\partial r$ and $\partial z/\partial\theta$.

8. The equation $pV = RT$, where p, V, T refer to pressure, volume and absolute temperature and R is a gas constant, is called *Clapeyron's equation* for a gas. Show that

$$\frac{\partial V}{\partial T}\frac{\partial T}{\partial p}\frac{\partial p}{\partial V} = -1.$$

9. If $z = \tan^{-1}(y/x)$, where $x = u + \sin v$ and $y = u - \cos v$, find $\partial z/\partial u$ and $\partial z/\partial v$.

10. If $z = x + f(u)$ and $u = xy$, show that

$$x\frac{\partial z}{\partial x} - y\frac{\partial z}{\partial y} = x.$$

11. If $z = f(x, y)$, $x = r\cos\theta$ and $y = r\sin\theta$, show that

$$\left(\frac{\partial z}{\partial r}\right)^2 + \frac{1}{r^2}\left(\frac{\partial z}{\partial\theta}\right)^2 = \left(\frac{\partial z}{\partial x}\right)^2 + \left(\frac{\partial z}{\partial y}\right)^2.$$

12. Given the relations $x + y = uv$ and $xy = u - v$, obtain functional relationships of the form $x = f(u, v)$ and $y = \phi(u, v)$. Also find $\partial f/\partial u$, $\partial f/\partial v$, $\partial\phi/\partial u$ and $\partial\phi/\partial v$.

13. Given that

$$f(x, y, u, v) = x^3 + 2xy + uy^3 - u^2 + v = 0$$

and

$$\phi(x, y, u, v) = \sin(2xu) + y^2 e^{2v} = 0,$$

find $\partial u/\partial x$ and $\partial v/\partial y$.

14. Given that

$$f(x, y, u, v) = x^2 + y^2 + u + v^2$$

and

$$\phi(x, y, u, v) = x^4 + y^3 + u^2 + v,$$

find $\partial u/\partial x$.

15. For the function $f(z/x, x/y) = 0$, show that

$$y\frac{\partial x}{\partial y} + z\frac{\partial x}{\partial z} = x.$$

5.4 TAYLOR'S THEOREM

Let $f(x, y)$ be a function of two variables with continuous partial derivatives up to those of nth order, and let

$$F(\lambda) = f(x_0 + \lambda h, y_0 + \lambda k) \tag{5.30}$$

where x_0, y_0, h and k are constants.

Then $F(\lambda) = f(x, y)$, where $x = x_0 + \lambda h$ and $y = y_0 + \lambda k$. Hence, we have

$$F'(\lambda) = \frac{dF}{d\lambda} = \frac{\partial f}{\partial x}\frac{dx}{d\lambda} + \frac{\partial f}{\partial y}\frac{dy}{d\lambda}$$

$$= h\frac{\partial f}{\partial x} + k\frac{\partial f}{\partial y}$$

$$= \left(h\frac{\partial}{\partial x} + k\frac{\partial}{\partial y}\right) f \tag{5.31}$$

Similarly,

$$F''(\lambda) = \frac{d^2F}{d\lambda^2} = \left(h\frac{\partial}{\partial x} + k\frac{\partial}{\partial y}\right)^2 f$$

$$= h^2\frac{d^2 f}{\partial x^2} + 2hk\frac{\partial^2 f}{\partial x\,\partial y} + k^2\frac{\partial^2 f}{\partial y^2}, \tag{5.32}$$

and in general,

$$F^{(n)}(\lambda) = \frac{d^n F}{d\lambda^n} = \left(h\frac{\partial}{\partial x} + k\frac{\partial}{\partial y}\right)^n f \tag{5.33}$$

Setting $\lambda = 0$ in (5.30) to (5.33), we obtain

$$F(0) = f(x_0, y_0),$$

$$F'(0) = \left(h \frac{\partial}{\partial x} + k \frac{\partial}{\partial y} \right) f(x_0, y_0),$$

$$F''(0) = \left(h \frac{\partial}{\partial x} + k \frac{\partial}{\partial y} \right)^2 f(x_0, y_0),$$

$$\vdots$$

where

$$\left(h \frac{\partial}{\partial x} + k \frac{\partial}{\partial y} \right) f(x_0, y_0)$$

is the value of

$$\left(h \frac{\partial}{\partial x} + k \frac{\partial}{\partial y} \right) f(x, y)$$

at (x_0, y_0) and similarly for other terms. Again putting $\lambda = 1$ in (5.30), we obtain:

$$F(1) = f(x_0 + h, y_0 + k) = F(0) + F'(0) + \frac{1}{2} F''(0) + \cdots + \frac{1}{n!} F^n(\xi),$$

which is the Taylor expansion of $F(1)$. Hence,

$$f(x_0 + h, y_0 + k) = f(x_0, y_0) + \left(h \frac{\partial}{\partial x} + k \frac{\partial}{\partial y} \right) f(x_0, y_0)$$

$$+ \frac{1}{2!} \left(h \frac{\partial}{\partial x} + k \frac{\partial}{\partial y} \right)^2 f(x_0, y_0) + \cdots$$

$$+ \frac{1}{n!} \left(h \frac{\partial}{\partial x} + k \frac{\partial}{\partial y} \right)^n f(x_0 + \xi h, y_0 + \xi k), \quad (5.34)$$

on substituting for $F(0)$, $F'(0), \ldots, F^n(\xi)$.

This result is Taylor's theorem for a function of two variables and its extension to several dimensions is quite straightforward.

An important application of Taylor's theorem consists in estimating the effects of small changes in the variables. Let

$$f(x, y) = \sqrt{x^2 + y^2}$$

so that we have

$$\frac{\partial f}{\partial x} = \frac{x}{\sqrt{x^2 + y^2}}, \qquad \frac{\partial f}{\partial y} = \frac{y}{\sqrt{x^2 + y^2}},$$

and

$$f(x_0+h, y_0+k) = f(x_0, y_0) + \left(h\frac{\partial}{\partial x} + k\frac{\partial}{\partial y}\right) f(x_0, y_0) + \cdots$$

$$\approx f(x_0, y_0) + \left(h\frac{\partial}{\partial x} + k\frac{\partial}{\partial y}\right) f(x_0, y_0).$$

This result can be used to find an approximate value of $f(x_0+h, y_0+k)$.

Example 5.24 Find an approximate value of $\sqrt{(5.01)^2 + (12.02)^2}$.

Here $x = 5, y = 12, h = 0.01$ and $k = 0.02$. Then

$$f(x, y) = \sqrt{x^2+y^2}, \qquad \frac{\partial f}{\partial x} = \frac{x}{\sqrt{x^2+y^2}}, \qquad \frac{\partial f}{\partial y} = \frac{y}{\sqrt{x^2+y^2}}.$$

Hence

$$\sqrt{(5.01)^2+(12.02)^2} = \sqrt{5^2+12^2} + (0.01)\frac{5}{\sqrt{5^2+12^2}} + (0.02)\frac{12}{\sqrt{5^2+12^2}}.$$

$$= 13 + 0.01\left(\frac{5}{13}\right) + 0.02\left(\frac{12}{13}\right)$$

$$= 13.022.$$

Example 5.25 Expand $x^2y + 3y - 2$ in powers of $(x+1)$ and $(y-2)$.

Let $f(x, y) = x^2y + 3y - 2$. Then,

$$f_x = 2xy, \qquad f_y = x^2+3, \qquad f_{xx} = 2y,$$

$$f_{xy} = 2x, \qquad f_{yy} = 0, \qquad f_{xxx} = 0,$$

$$f_{xxy} = 2, \qquad f_{xyy} = 0, \qquad f_{yyy} = 0.$$

At $x = -1, y = 2$, we obtain

$$f(-1, 2) = 6, \qquad f_x(-1, 2) = -4, \qquad f_y(-1, 2) = 4,$$

$$f_{xx}(-1, 2) = 4, \qquad f_{xy}(-1, 2) = -2, \qquad f_{yy}(-1, 2) = 0,$$

$$f_{xxx}(-1, 2) = 0, \qquad f_{xxy}(-1, 2) = 2, \qquad f_{xyy}(-1, 2) = 0.$$

$$f_{yyy}(-1, 2) = 0,$$

all higher derivatives being zero. By Taylor's theorem for a function of two variables, we have,

$$f(x, y) = f(-1, 2) + (x+1) f_x(-1, 2) + (y-2) f_y(-1, 2)$$

$$+\frac{1}{2!}\Big[(x+1)^2 f_{xx}(-1, 2) + 2(x+1)(y-2) f_{xy}(-1, 2)$$

$$+(y-2)^2 f_{yy}(-1, 2)\Big]$$

$$+\frac{1}{3!}[(x+1)^3 f_{xxx}(-1, 2) + 3(x+1)^2(y-2) f_{xxy}(-1, 2)$$

$$+3(x+1)(y-2)^2 f_{xyy}(-1, 2) + (y-2)^3 f_{yyy}(-1, 2)] + \cdots$$

Substituting the values of the partial derivatives, we therefore obtain

$$f(x, y) = 6 + (x+1)(-4) + (y-2)4 + \frac{1}{2}[4(x+1)^2 + 2(x+1)(y-2)(-2)]$$

$$+\frac{1}{6}[3(x+1)^2(y-2)(2)]$$

$$= 6 - 4(x+1) + 4(y-2) + 2(x+1)^2 - 2(x+1)(y-2) + (x+1)^2(y-2),$$

which can be verified by direct expansion.

Example 5.26 Expand the function

$$\frac{(x+h)(y+k)}{x+y+h+k}$$

in powers of h and k.

Let

$$f(x+h, y+k) = \frac{(x+h)(y+k)}{(x+h)+(y+k)} \tag{i}$$

Then we have,

$$f(x, y) = \frac{xy}{x+y} \tag{ii}$$

By Taylor's series we have

$$f(x+h, y+k) = f(x, y) + \left(h\frac{\partial f}{\partial x} + k\frac{\partial f}{\partial y}\right) + \frac{1}{2}\left(h^2\frac{\partial^2 f}{\partial x^2} + 2hk\frac{\partial^2 f}{\partial x \partial y} + k^2\frac{\partial^2 f}{\partial y^2}\right) \tag{iii}$$

up to the second degree terms. Now,

$$\frac{\partial f}{\partial x} = \frac{y(x+y) - xy}{(x+y)^2} = \frac{y^2}{(x+y)^2},$$

$$\frac{\partial f}{\partial y} = \frac{x(x+y) - xy}{(x+y)^2} = \frac{x^2}{(x+y)^2},$$

$$\frac{\partial^2 f}{\partial y \partial x} = \frac{\partial}{\partial y}\left[\frac{y^2}{(x+y)^2}\right]$$

$$= \frac{2y(x+y)^2 - y^2 \cdot 2(x+y)}{(x+y)^4}$$

$$= \frac{2y(x+y) - 2y^2}{(x+y)^3}$$

$$= \frac{2xy}{(x+y)^3}.$$

$$\frac{\partial^2 f}{\partial x^2} = -\frac{2y^2}{(x+y)^3}$$

$$\frac{\partial^2 f}{\partial y^2} = \frac{\partial}{\partial y}\left[\frac{x^2}{(x+y)^2}\right] = -\frac{2x^2}{(x+y)^3}.$$

Hence, (iii) gives:

$$f(x+h, y+k) = \frac{xy}{x+y} + \left[\frac{hy^2}{(x+y)^2} + \frac{kx^2}{(x+y)^2}\right]$$

$$+ \frac{1}{2}\left[\frac{-2h^2y^2}{(x+y)^3} + \frac{4\,hkxy}{(x+y)^3} - \frac{2k^2x^2}{(x+y)^3}\right]$$

$$= \frac{xy}{x+y} + \frac{hy^2 + kx^2}{(x+y)^2} - \frac{h^2y^2 - 2hkxy + k^2x^2}{(x+y)^3},$$

where the terms containing h^3, k^3 and higher powers are neglected.

Example 5.27 Expand $e^x \log(1+y)$ in a Taylor's series in the neighbourhood of the origin. Give the first-six terms.

Let $f(x, y) = e^x \log(1+y)$. We know that

$$f(x, y) = f(0, 0) + \left(x\frac{\partial f_0}{\partial x} + y\frac{\partial f_0}{\partial y}\right) + \frac{1}{2}\left(x^2\frac{\partial^2 f_0}{\partial x^2} + 2xy\frac{\partial^2 f_0}{\partial x \partial y} + y^2\frac{\partial^2 f_0}{\partial y^2}\right)$$

$$+\frac{1}{6}\left(x^3\frac{\partial^3 f_0}{\partial x^3}+3x^2y\frac{\partial^3 f_0}{\partial x^2\partial y}+3xy^2\frac{\partial^3 f_0}{\partial x\partial y^2}+y^3\frac{\partial^3 f_0}{\partial y^3}\right)+\cdots \qquad \text{(i)}$$

where

$$\frac{\partial f_0}{\partial x}=\left[\frac{\partial f}{\partial x}\right]_{(0,\,0)}, \text{ etc.}$$

Now $f(0,0)=0$ and

$$\frac{\partial f}{\partial x}=e^x\log(1+y); \qquad \left[\frac{\partial f}{\partial x}\right]_{(0,\,0)}=0,$$

$$\frac{\partial f}{\partial y}=\frac{e^x}{y+1}; \qquad \left[\frac{\partial f}{\partial y}\right]_{(0,\,0)}=1,$$

$$\frac{\partial^2 f}{\partial x^2}=e^x\log(1+y); \qquad \left[\frac{\partial^2 f}{\partial x^2}\right]_{(0,\,0)}=0,$$

$$\frac{\partial^2 f}{\partial y^2}=-\frac{e^x}{(y+1)^2}; \qquad \left[\frac{\partial^2 f}{\partial y^2}\right]_{(0,\,0)}=-1$$

$$\frac{\partial^2 f}{\partial x\partial y}=\frac{e^x}{y+1}; \qquad \left[\frac{\partial^2 f}{\partial x\partial y}\right]_{(0,\,0)}=1$$

$$\frac{\partial^3 f}{\partial x^3}=e^x\log(1+y); \qquad \left[\frac{\partial^3 f}{\partial x^3}\right]_{(0,\,0)}=0$$

$$\frac{\partial^3 f}{\partial x^2\partial y}=\frac{e^x}{y+1}; \qquad \left[\frac{\partial^3 f}{\partial x^2\partial y}\right]_{(0,\,0)}=1$$

$$\frac{\partial^3 f}{\partial x\partial y^2}=-\frac{e^x}{(y+1)^2}; \qquad \left[\frac{\partial^3 f}{\partial x\partial y^2}\right]_{(0,\,0)}=-1$$

$$\frac{\partial^3 f}{\partial y^3}=\frac{2e^x}{(y+1)}3j; \qquad \left[\frac{\partial^3 f}{\partial y^3}\right]=2.$$

Substituting these values in (i), we obtain

$$e^x\log(1+y)=y+xy-\frac{y^2}{2}+\frac{1}{2}x^2y-\frac{1}{2}xy^2+\frac{1}{3}y^3+\cdots$$

which is the required expansion up to the first-six terms.

5.4.1 Maxima, Minima and Saddle Points

Let $z = f(x, y)$ be a function of two independent variables x and y which is defined in some region D of the xy-plane. The definition of maximum or minimum of the function $f(x, y)$ is essentially the same as for the case of a function of a single variable. We say that $f(x, y)$ has a *maximum* value for a certain pair of values of x and y, if this value, say $f_{\max}$, is greater than its value for all values of x and y in the immediate neighbourhood of the particular pair of values, i.e. $f(x, y) > f(x+h, y+k)$, where h and k have small values both positive and negative. In a similar way, a *minimum* value of $f(x, y)$ is defined. Thus, the quantity $f(x, y) - f(x+h, y+k)$ must retain a constant sign for small variations in h and k, this constant sign being *positive* for a maximum and *negative* for a minimum value of $f(x, y)$. If the sign does not remain constant, there will be neither a maximum nor a minimum, and then the point is said to be a *saddle point*.

The criterion for maxima or minima can be obtained by using Taylor's theorem for $f(x+h, y+k)$. We have

$$f(x+h, y+k) = f(x, y) + \left(h\frac{\partial f}{\partial x} + k\frac{\partial f}{\partial y}\right)$$

$$+\frac{1}{2}\left(h^2\frac{\partial^2 f}{\partial x^2} + 2hk\frac{\partial^2 f}{\partial x \partial y} + k^2\frac{\partial^2 f}{\partial y^2}\right) + \cdots$$

or

$$f(x+h, y+k) - f(x, y) = (hf_x + kf_y) + \frac{1}{2}(h^2 f_{xx} + 2hkf_{xy} + k^2 f_{yy}) + \cdots \quad (5.35)$$

For small variations in h and k, the first term dominates the series on the right side of (5.35), and hence for a constant sign on the right, we must have

$$f_x = f_y = 0 \quad (5.36)$$

as preliminary conditions for a maximum or minimum. A point satisfying the conditions in (5.36), is called a *stationary* point or a *critical* point. Thus, Eq. (5.35) becomes:

$$f(x+h, y+k) - f(x, y) = \frac{1}{2}(h^2 f_{xx} + 2hkf_{xy} + k^2 f_{yy}) + \cdots \quad (5.37)$$

Now

$$h^2 f_{xx} + 2hkf_{xy} + k^2 f_{yy} = \frac{1}{f_{xx}}(h^2 f^2{}_{xx} + 2hkf_{xy}f_{xx} + k^2 f_{xx}f_{yy})$$

$$= \frac{1}{f_{xx}}\left[(hf_{xx} + kf_{xy})^2 + k^2(f_{xx}f_{yy} - f_{xy}^2)\right].$$

If $f_{xx}y_{yy} - f_{xy}^2 > 0$, then the expression in the square brackets is always positive and there will be maximum if f_{xx} is negative, and minimum if f_{xx} is positive. On the other hand, if $f_{xx}f_{yy} - f_{xy}^2 < 0$, the quantity in the square brackets can vary in sign and hence there will be neither maxima nor minima. Thus, provided that f_{xx}, f_{yy} and f_{xy} are not all zero, the criteria for maxima or minima are given as follows:

(a) $f_x = f_y = 0$, the solution of which gives pairs of values of x and y which are called stationary points.

(b) Calculate f_{xx}, f_{xy} and f_{yy} for each (x, y). If $f_{xx}f_{yy} - f_{xy}^2 > 0$, there will be a maximum if f_{xx} (or f_{yy}) < 0 and a minimum if f_{xx} (or f_{yy}) > 0.

(c) If $f_{xx}\, f_{yy} - f_{xy}^2 < 0$, there will be neither maxima nor minima and the points are saddle points.

It might happen, but quite rarely, that f_{xx}, f_{xy} and f_{yy} are all zero. In such a case, further terms involving the third-and fourth-order partial derivatives on the right side of (5.35) must be considered and since the algebra is quite involved, this case is not pursued here further.

Example 5.28 Examine the function $f(x, y) = x^3 - 3x^2 - 4y^2 + 1$ for maxima or minima.

For the given function, we have

$$f_x = 3x^2 - 6x = 3x\,(x - 2) \quad \text{and} \quad f_y = -8y.$$

Thus, the stationary points are given by (0, 0) and (2, 0). Further,

$$f_{xx} = 6x - 6 = 6\,(x - 1), \quad f_{xy} = 0, \quad f_{yy} = -8.$$

Hence,

$$f_{xx}f_{yy} - f_{xy}^2 = -48\,(x - 1).$$

For (0, 0): $f_{xx}f_{yy} - f_{xy}^2 = 48$ and since $f_{xx} = -6$, it follows that the point (0, 0) is a maximum. Again for (2, 0):

$$f_{xx}f_{yy} - f_{xy}^2 = -48,$$

and hence there will be neither maximum nor minimum at (2, 0).

Example 5.29 Find the dimensions of rectangular box (without top) with a given volume so that the material used is least.

Let x, y, z be the dimensions of the box and let its volume be V. Then,

$$V = xyz \tag{i}$$

The surface area S is given by

$$S = xy + 2xz + 2yz \tag{ii}$$

Eliminating z between (i) and (ii), we get:

$$S = xy + (2x+2y)\frac{V}{xy} = xy + \frac{2V}{y} + \frac{2V}{x}. \qquad \text{(iii)}$$

Hence, we obtain

$$\frac{\partial S}{\partial x} = y - \frac{2V}{x^2}; \qquad \frac{\partial S}{\partial y} = x - \frac{2V}{y^2}.$$

The stationary points are given by

$$yx^2 = xy^2 = 2V \quad \text{or} \quad xy(x-y) = 0,$$

whence $x = 0$, $y = 0$ or $x = y$. Further,

$$\frac{\partial^2 S}{\partial x \partial y} = 1, \qquad \frac{\partial^2 S}{\partial x^2} = \frac{4V}{x^3}, \qquad \frac{\partial^2 S}{\partial y^2} = \frac{4V}{y^3}.$$

Now,

$$\frac{\partial^2 S}{\partial x^2}\frac{\partial^2 S}{\partial y^2} - \left(\frac{\partial^2 S}{\partial x\, \partial y}\right)^2 = \frac{16V^2}{x^3 y^3} - 1$$

For $x = y$, we have $y^3 = 2V$ or $y = (2V)^{1/3}$. Hence,

$$\frac{\partial^2 S}{\partial x^2}\frac{\partial^2 S}{\partial y^2} - \left(\frac{\partial^2 S}{\partial x \partial y}\right)^2 = \frac{16V^2}{4V^2} - 1 = 3, \qquad \text{which is positive.}$$

Further,

$$\frac{\partial^2 S}{\partial x^2} = \frac{4V}{x^3} = \frac{4V}{2V} = 2, \qquad \text{which is also positive.}$$

Since

$$\frac{\partial^2 S}{\partial x^2}\frac{\partial^2 S}{\partial y^2} - \left(\frac{\partial^2 S}{\partial x \partial y}\right)^2$$

is positive and $f_{xx} > 0$, it follows that $x = y$ gives a minimum for S and the required dimensions are $x = y = (2V)^{1/3}$, and

$$z = \frac{V}{xy} = \frac{Vx}{x^3} = \frac{Vx}{2V} = \frac{x}{2}.$$

Hence, the box should have a square base and an altitude which is half the length of the base.

Example 5.30 Find the minimum distance between the origin and the plane $2x + y - z = 5$.

Let $P(x, y, z)$ be any point on the plane. Then the distance OP (where O is the orign) is given by $|\overline{OP}| = \sqrt{(x^2 + y^2 + z^2)}$. This will be minimum provided that the function $f(x, y, z)$ given by $f(x, y, z) = x^2 + y^2 + z^2$ is minimum. From the equation of the plane, we have $z = 2x + y - 5$. Then, we have

$$f(x, y, z) = F(x, y) = x^2 + y^2 + (2x + y - 5)^2. \quad \text{(i)}$$

Clearly,

$$\frac{\partial F}{\partial x} = 2x + 2(2x + y - 5)(2) = 10x + 4y - 20.$$

$$\frac{\partial F}{\partial y} = 2y + 2(2x + y - 5)(1) = 4x + 4y - 10,$$

$$\frac{\partial^2 F}{\partial x^2} = 10, \quad \frac{\partial^2 F}{\partial y^2} = 4, \quad \frac{\partial^2 F}{\partial x \partial y} = 4.$$

Now,

$$\frac{\partial F}{\partial x} = 0 = \frac{\partial F}{\partial y}$$

gives the relations:

$$10x + 4y - 20 = 0 \quad \text{and} \quad 4x + 4y - 10 = 0,$$

which give $x = 5/3$, $y = 5/6$ and $z = -5/6$. Further,

$$F_{xx}F_{yy} - (F_{xy})^2 = 40 - 16 = 24, \text{ which is positive.}$$

Hence, $(5/3, 5/6, -5/6)$ is the point on the plane whose distance from the origin O is minimum, and the minimum distance is given by

$$\sqrt{\frac{25}{9} + \frac{25}{36} + \frac{25}{36}} = \sqrt{\frac{150}{36}} = \frac{5}{\sqrt{6}}.$$

Example 5.31 Find the rectangular solid of maximum volume that can be inscribed in a given sphere.

Let the rectangular solid be of dimensions x, y, z and also let d be the diameter of the given sphere. Since a diagonal of the solid passes through the centre of the sphere $x^2 + y^2 + z^2 = d^2$, length of the diameter is equal to d units. Then,

If V is the volume of the solid, we have

$$V = xyz = xy\sqrt{d^2 - x^2 - y^2}, \qquad \text{since } x^2 + y^2 + z^2 = d^2.$$

We also have,

$$V^2 = x^2y^2(d^2 - x^2 - y^2) \tag{i}$$

For V^2 to be maximum, we have

$$\frac{\partial f}{\partial x} = \frac{\partial f}{\partial y} = 0,$$

and,

$$\frac{\partial^2 f}{\partial x^2}\frac{\partial^2 f}{\partial y^2} - \left(\frac{\partial^2 f}{\partial x\,\partial y}\right) > 0, \qquad \frac{\partial^2 f}{\partial x^2}\left(\text{or } \frac{\partial^2 f}{\partial y^2}\right) < 0$$

Here

$$f(x, y) = V^2 = d^2x^2y^2 - x^4y^2 - x^2y^4 \tag{ii}$$

Then

$$\frac{\partial f}{\partial x} = 2d^2xy^2 - 4x^3y^2 - 2xy^4 = 2xy^2(d^2 - 2x^2 - y^2)$$

At (0, 0),

$$\frac{\partial f}{\partial x} = 0 \quad \text{or} \quad d^2 = 2x^2 + y^2 \tag{iii}$$

Also

$$\frac{\partial f}{\partial y} = 2d^2x^2y - 2x^4y - 4x^2y^3 = 2x^2y(d^2 - x^2 - 2y^2)$$

At (0, 0)

$$\frac{\partial f}{\partial y} = 0 \quad \text{or} \quad d^2 = x^2 + 2y^2 \tag{iv}$$

From (iii) and (iv) we obtain $x = y = d/\sqrt{3}$. Also,

$$\frac{\partial^2 f}{\partial x^2} = 2\,d^2y^2 - 12x^2y^2 - 2y^4$$

$$= 2y^2(d^2 - 6x^2 - y^2)$$

$$= 2\frac{d^2}{3}\left(d^2 - 6\frac{d^2}{3} - \frac{d^2}{3}\right)$$

$$= -8\frac{d^4}{9},$$

$$\frac{\partial^2 f}{\partial y^2} = 2d^2x^2 - 2x^4 - 12x^2y^2$$

$$= 2d^2\frac{d^2}{3} - 2\frac{d^4}{9} - 12\frac{d^2}{3}\frac{d^2}{3}$$

$$= \frac{2}{3}d^4 - \frac{14}{9}d^4$$

$$= -\frac{8}{9}d^4,$$

$$\frac{\partial^2 f}{\partial x \partial y} = 4d^2xy - 8x^3y - 8xy^3$$

$$= 4d^2\frac{d}{\sqrt{3}}\frac{d}{\sqrt{3}} - 8\frac{d^3}{3\sqrt{3}}\frac{d}{\sqrt{3}} - 8\frac{d}{\sqrt{3}}\frac{d^3}{3\sqrt{3}}$$

$$= \frac{4}{3}d^4 - \frac{8}{9}d^4 - \frac{8}{9}d^4$$

$$= -\frac{4d^4}{9}$$

Therefore,

$$\frac{\partial^2 f}{\partial x^2}\frac{\partial^2 f}{\partial y^2} = \frac{64}{81}d^8 \quad \text{and} \quad \left(\frac{\partial^2 f}{\partial x \partial y}\right) = \frac{16}{81}d^8.$$

Hence

$$\frac{\partial^2 f}{\partial x^2}\frac{\partial^2 f}{\partial y^2} > \left(\frac{\partial^2 f}{\partial x \partial y}\right)^2.$$

Further $\partial^2 f/\partial x^2$ and $\partial^2 f/\partial y^2$ are both negative.

Hence, $f(x, y)$ has a maximum when $x = y = d/\sqrt{3}$. Since $x^2 + y^2 + z^2 = d^2$, it follows that $z = d/\sqrt{3}$, i.e. $x = y = z = d/\sqrt{3}$. Therefore the rectangular solid is a cube.

5.4.2 Constrained Maxima or Minima: Lagrange's Method

Sometimes it would be necessary to find maxima or minima of a function subject to a condition (or constraint) being satisfied. Suppose we are required to find the maxima or minima of the function $z = f(x, y)$ subject to the condition $g(x, y) = 0$. If we can solve the latter equation explicitly for y and substitute for y in the function $f(x, y)$, then the problem is reduced to that of finding maxima or minima of a function of a single variable x. However, this is usually not possible and, in the general case, the problem is best solved by using Lagrange's method of undetermined multipliers which is described below.

Since $z = f(x, y)$ we have

$$\frac{dz}{dx} = \frac{\partial f}{\partial x} + \frac{\partial f}{\partial y}\frac{dy}{dx} \tag{5.38}$$

Again, since $g(x, y) = 0$, we write

$$\frac{\partial g}{\partial x} + \frac{\partial g}{\partial y}\frac{dy}{dx} = 0. \tag{5.39}$$

From (5.39), we have

$$\frac{dy}{dx} = -\frac{g_x}{g_y} \quad \text{on the curve } g(x, y) = 0.$$

But on $g(x, y) = 0$, the function $z = f(x, y)$ is only a function of x and hence its stationary points are given by $dz/dx = 0$, i.e.

$$\frac{\partial f}{\partial x} + \frac{\partial f}{\partial y}\left(-\frac{g_x}{g_y}\right) = 0,$$

or,

$$f_x g_y - f_y g_x = 0, \tag{5.40a}$$

which is solved subject to the condition

$$g(x, y) = 0. \tag{5.40b}$$

Algebraically, this is equivalent to finding the stationary points of the function

$$F(x, y) = f(x, y) + \lambda g(x, y), \tag{5.41}$$

where λ is a parameter to be determined. The stationary points of $F(x, y)$ are given by the equations

$$f_x + \lambda g_x = 0 \quad \text{and} \quad f_y + \lambda g_y = 0,$$

which have a solution if the condition (5.40a) is satisfied. The maxima or minima can then be located by examining the conditions in the neighbourhood of the stationary points. The following examples illustrate the procedure.

Example 5.32 Find the minimum value of $x^2 + y^2$ subject to the condition $ax + by = c$.

We have the function

$$F(x, y) = x^2 + y^2 + \lambda(ax + by - c), \tag{i}$$

from which we obtain

$$F_x = 2x + \lambda a = 0, \qquad F_y = 2y + \lambda b = 0.$$

The above equations give:

$$2xb - 2ay = 0, \quad \text{or} \quad x = \frac{a}{b}y.$$

Substituting for x in $ax + by = c$, we obtain:

$$y = \frac{bc}{a^2 + b^2} \quad \text{and} \quad x = \frac{ac}{a^2 + b^2}.$$

The stationary point is therefore

$$\left(\frac{ac}{a^2 + b^2}, \frac{bc}{a^2 + b^2}\right).$$

The minimum value of $x^2 + y^2$ is therefore

$$\frac{a^2c^2}{(a^2 + b^2)^2} + \frac{b^2c^2}{(a^2 + b^2)^2} = \frac{c^2}{a^2 + b^2}.$$

Example 5.33 Find the minimum distance of the origin from the plane whose equation is given by $ax + by + cz + d = 0$.

Geometrically, it is clear that such a minimum exists. Since the distance of the origin from any point (x, y, z) on the plane is $\sqrt{(x^2 + y^2 + z^2)}$ our problem is to minimize $f(x, y, z) = x^2 + y^2 + z^2$ subject to the constraint

$$g(x, y, z) = ax + by + cz + d = 0.$$

We now consider the function defined by

$$F(x, y, z) = x^2 + y^2 + z^2 + \lambda(ax + by + cz + d).$$

The necessary conditions for the minimum are

$$\frac{\partial F}{\partial x} = 2x + \lambda a = 0$$

$$\frac{\partial F}{\partial y} = 2y + \lambda b = 0.$$

$$\frac{\partial F}{\partial z} = 2z + \lambda c = 0.$$

$$\frac{\partial F}{\partial \lambda} = ax + by + cz + d = 0.$$

Solving the above set for λ, we obtain:

$$\lambda = \frac{2d}{a^2 + b^2 + c^2}.$$

The minimum occurs at the point

$$x=-\frac{ad}{a^2+b^2+c^2},\qquad y=-\frac{bd}{a^2+b^2+c^2},\qquad z=-\frac{cd}{a^2+b^2+c^2},$$

and the minimum distance is equal to

$$\frac{|d|}{\sqrt{a^2+b^2+c^2}}.$$

Example 5.34 Find the dimensions of the rectangular parallelopiped of maximum volume that can be inscribed in the ellipsoid:

$$\frac{x^2}{a^2}+\frac{y^2}{b^2}+\frac{z^2}{c^2}=1.$$

Let the required dimenstions be $2x, 2y$ and $2z$. Then its volume V is given by

$$V=f(x, y, z)=(2x)(2y)(2z)=8xyz \tag{i}$$

Let

$$\phi(x, y, z)\equiv\frac{x^2}{a^2}+\frac{y^2}{b^2}+\frac{z^2}{c^2}-1 \tag{ii}$$

Now, let $F(x, y, z)=f(x, y, z)+\lambda\phi(x, y, z)$.

Therefore,

$$\frac{\partial F}{\partial x}=\frac{\partial f}{\partial x}+\lambda\frac{\partial \phi}{\partial x}=8yz+\lambda\frac{2x}{a^2}=0 \tag{iii}$$

$$\frac{\partial F}{\partial y}=\frac{\partial f}{\partial y}+\lambda\frac{\partial \phi}{\partial y}=8xz+\lambda\frac{2y}{b^2}=0 \tag{iv}$$

$$\frac{\partial F}{\partial z}=\frac{\partial f}{\partial z}+\lambda\frac{\partial \phi}{\partial z}=8xy+\lambda\frac{2z}{c^2}=0 \tag{v}$$

Multiplying (iii) by x, (iv) by y and (v) by z and then adding, we get

$$24xyz+2\lambda\left(\frac{x^2}{a^2}+\frac{y^2}{b^2}+\frac{z^2}{c^2}\right)=0$$

or,

$$24xyz+2\lambda=0,$$

which gives $\lambda=-12xyz$.

Then (iii) gives:

$$8yz - \frac{24x^2yz}{a^2} = 0 \quad \text{or} \quad x = \frac{a}{\sqrt{3}}.$$

Similarly, (iv) and (v) give:

$$y = \frac{b}{\sqrt{3}} \quad \text{and} \quad z = \frac{c}{\sqrt{3}}.$$

Hence the required dimensions are

$$x = \frac{2a}{\sqrt{3}}, \quad y = \frac{2b}{\sqrt{3}} \quad z = \frac{2c}{\sqrt{3}}.$$

5.4.3 Differentiation under the Integral Sign

Let $f(x)$ be a continuous function in the closed interval (a, b) and let $F(x)$ be such that $F(x) = f(x)$. Then, by the fundamental theorem of the integral calculus, we have

$$\int_a^b f(x)\,dx = F(b) - F(a). \tag{5.42}$$

If, however, a and b are differentiable functions of a variable α, so that

$$a = a(\alpha) \quad \text{and} \quad b = b(\alpha), \qquad \alpha_0 \le \alpha \le \alpha_1,$$

then the right-hand side of (5.42) is a function of α, and we have

$$\frac{d}{d\alpha} F(b) = F'(b)\frac{db}{d\alpha} = f(b)\frac{db}{d\alpha} \tag{5.43}$$

$$\frac{d}{d\alpha} F(a) = F'(a)\frac{da}{d\alpha} = f(a)\frac{da}{d\alpha} \tag{5.44}$$

Hence, we have the important formula:

$$\frac{d}{d\alpha}\int_{a(\alpha)}^{b(\alpha)} f(x)\,dx = f(b)\frac{db}{d\alpha} - f(a)\frac{da}{d\alpha} \tag{5.45}$$

Suppose that the variable α occurs under the integral sign so that we have

$$\phi(\alpha) = \int_a^b f(x, \alpha)\,dx \tag{5.46}$$

Assuming that $f(x, \alpha)$ is continuous and has a continuous partial derivative $\partial f/\partial \alpha$, the differential coefficient $\phi'(\alpha)$ can be calculated as the limit of a difference quotient. We have

$$\phi(\alpha+\delta\alpha)-\phi(\alpha)=\int_a^b [f(x,\alpha+\delta\alpha)-f(x,\alpha)]\,dx,$$

so that

$$\frac{\phi(\alpha+\delta\alpha)-\phi(\alpha)}{\delta\alpha}=\int_a^b \frac{f(x,\alpha+\delta\alpha)-f(x,\alpha)}{\delta\alpha}\,dx.$$

Proceeding to the limit as $\delta\alpha \to 0$, we obtain the formula:

$$\phi'(\alpha)=\int_a^b \frac{\partial f(x,\alpha)}{d\alpha}\,dx \qquad (5.47)$$

In the general case, both the limits a and b and also the integrand depend on the variable α so that the integral may be written as

$$g(a,b,\alpha)=\int_{a(\alpha)}^{b(\alpha)} f(x,\alpha)\,dx \qquad (5.48)$$

Hence $dg/d\alpha$ is now the total differential coefficient and we have

$$\frac{dg}{d\alpha}=\frac{\partial g}{\partial \alpha}+\frac{\partial g}{\partial a}\frac{da}{d\alpha}+\frac{\partial g}{\partial b}\frac{db}{d\alpha}. \qquad (5.49)$$

Using the fundamental theorem of the integral calculus, we have

$$\left.\begin{aligned}\frac{\partial g}{\partial a}&=\frac{\partial}{\partial a}\int_a^b f(x,\alpha)\,dx=-f(x,a)\\ \frac{\partial g}{\partial b}&=\frac{\partial}{\partial b}\int_a^b f(x,\alpha)\,dx=f(x,b)\end{aligned}\right\} \qquad (5.50)$$

Also from (5.47)

$$\frac{\partial g}{\partial \alpha}=\int_a^b \frac{\partial f}{\partial \alpha}\,dx, \qquad (5.51)$$

Using these results in (5.49), we obtain the formula:

$$\frac{dg}{d\alpha}=\int_{a(\alpha)}^{b(\alpha)} \frac{\partial f(x,\alpha)}{\partial \alpha}\,dx-\frac{da}{d\alpha}f(x,a)+\frac{db}{d\alpha}f(x,b), \qquad (5.52)$$

which is called *Leibnitz's rule.*

Example 5.35 Evaluate

$$\frac{d}{d\alpha}\int_{\alpha}^{\alpha^2} \cos x^2 \, dx.$$

Let

$$F(\alpha) = \int_{\alpha}^{\alpha^2} \cos x^2 \, dx.$$

Here the variable occurs in the limits only. Leibnitz's rule therefore gives:

$$\frac{dF}{d\alpha} = \cos \alpha^4 \frac{d}{d\alpha}(\alpha^2) - \cos \alpha^2 \frac{d}{d\alpha}(\alpha) = 2\alpha \cos \alpha^4 - \cos \alpha^2.$$

Example 5.36 Given that

$$\int_0^{\pi} \frac{dx}{\alpha - \cos x} = \frac{\pi}{\sqrt{\alpha^2 - 1}}, \qquad \alpha^2 > 1,$$

find the value of integral

$$\int_0^{\pi} \frac{dx}{(\alpha - \cos x)^2}.$$

Let

$$g(\alpha) = \int_0^{\pi} \frac{dx}{\alpha - \cos x} = \frac{\pi}{\sqrt{\alpha^2 - 1}}.$$

Leibnitz's rule gives:

$$g'(\alpha) = -\int_0^{\pi} \frac{dx}{(\alpha - \cos x)^2} = \frac{d}{dx}\left(\frac{\pi}{\sqrt{\alpha^2 - 1}}\right) = \pi\left(-\frac{1}{2}\right)(\alpha^2 - 1)^{-3/2}(2\alpha)$$

Hence

$$\int_0^{\pi} \frac{dx}{(\alpha - \cos x)^2} = \frac{\pi\alpha}{(\alpha^2 - 1)^{3/2}}.$$

Example 5.37 If

$$f(\alpha) = \int_{\alpha}^{\alpha^2} \frac{\sin \alpha x}{x} \, dx,$$

find $f'(\alpha)$ where $\alpha \neq 0$.

Using Leibnitz's rule, we obtain

$$f'(\alpha) = \int_{\alpha}^{\alpha^2} \frac{x \cos \alpha x}{x}\, dx - (1)\frac{\sin \alpha^2}{\alpha} + (2\alpha)\frac{\sin \alpha^3}{\alpha^2}\cdot$$

$$= \int_{\alpha}^{\alpha^2} \cos \alpha x\, dx - \frac{1}{\alpha} \sin \alpha^2 + \frac{2}{\alpha} \sin \alpha^3$$

$$= \left[\frac{\sin \alpha x}{\alpha}\right]_{\alpha}^{\alpha^2} - \frac{1}{\alpha} \sin \alpha^2 + \frac{2}{\alpha} \sin \alpha^3$$

$$= \frac{1}{\alpha} \sin \alpha^3 - \frac{1}{\alpha} \sin \alpha^2 - \frac{1}{\alpha} \sin \alpha^2 + \frac{2}{\alpha} \sin \alpha^3$$

$$= \frac{3}{\alpha} \sin \alpha^3 - \frac{2}{\alpha} \sin \alpha^2.$$

EXERCISES 5.4

Obtain the Taylor expansions of the following functions (Problems 1–6):

1. $x^2 y + x^3 y + 1$ about (0, 1).
2. $(1 - x^2 - y^2)^{1/2}$ about the origin up to the third-degree terms.
3. $\tan^{-1}(y/x)$ about (1, 1) up to the second-degree terms.
4. $e^y \log_e(x+1)$ about (1, 1) up to the first-three terms.
5. y^2/x^3 about $(1, -1)$ up to and including second-degree terms.
6. $\sin(xy)$ in powers of $(x-1)$ and $(y - \pi/2)$ up to and including second-degree terms.
7. Write down the Taylor series for a function $f(x, y, z)$ in powers of $(x-a), (y-b), (z-c)$ and the values of f and its partial derivatives at (a, b, c).

Find the stationary points of (Problems 8–13):

8. $x^2 + 2xy$.
9. $x^3 + y^3 - 3xy$.
10. $x^3 - 3x^2 - 4y^2 + 1$.
11. $x^3 + 8y^3 - 12xy$.
12. $(x^2 + y^2)^2 - 2a^2(x^2 - y^2)$.
13. $\sin x + \sin y + \sin(x + y)$.
14. Show that the function $f(x, y, z) = x^3 - 3x + y^3 - 9y + z^3 - 9z + 6yz$ has a minimum value at the point $(-1, -3, -3)$.
15. Show that the function $xy(3 - x - y)$ has a maximum value when $x = y = 1$.
16. Find the stationary points of $\cos^2 x + \cos^2 y$ subject to the condition $y = x + \pi/4$.

17. Prove that $z = x^2y^2(x^2 + y^2 - 1)$ has a maximum value of $-1/27$ for four sets of values of x and y and a minimum value at $x = 0,\ y = 0$.
18. Prove that the maximum value of $\sin x + \sin y + \sin (x + y)$ is $3\sqrt{3}/2$.
19. Find the volume of the largest rectangular parallelopiped that can be inscribed in the ellipsoid

$$\frac{x^2}{a^2} + \frac{y^2}{b^2} + \frac{z^2}{c^2} = 1.$$

20. Find the dimensions of the rectangular box, without top, of maximum capacity whose surface area is S sq. m.
21. The area A of a triangle is given by

$$A = \sqrt{s\,(s-a)\,(s-b)\,(s-c)}$$

where $2s = a + b + c$. For a given perimeter, show that the equilateral triangle has maximum area.
22. Find three minimum distance between the point $(0, -2, -4)$ and the plane $x + y - 4z = 5$.
23. Find three numbers whose sum is s and whose product is a maximum.
24. Determine constants a and b such that the integral

$$\int_0^1 (ax + b - x^2)^2\, dx$$

will be as small as possible.
25. Maximize the expression $a_1x_1 + a_2x_2 + \cdots + a_nx_n$ subject to the condition $x_1^2 + x_2^2 + \cdots + x_n^2 = 1$, and that $a_1, a_2 i, \ldots, a_n$ are n positive numbers.
26. Find the minimum distance from the origin to the surface $xyz = 1$.
27. Find the minimum value of $f(x,\ y,\ z) = x^2 + y^2 + z^2$ subject to the condition $x + y + z = 1$.
28. Find the points on the ellipse

$$\frac{x^2}{4} + \frac{y^2}{1} = 1$$

whose distances from the straight line $2x + 3y = 6$ are the greatest and the least.
29. Find the extreme value of $z = xy$ provided that $x^2 + y^2 = 2$.
30. Find the maximum value of $f(x, y, z) = xyz$ subject to the condition $x + y + z = 1$.
31. Use the method of Lagrange multipliers to find the greatest and least distances of a point on the ellipse $x^2 + 4y^2 = 4$ from the straight line $x + y = 4$.
32. Find the values of x, y, z that maximize the function

$$3x+5y+z-x^2-y^2-z^2$$

subject to the constraint $x+y+z=6$.

Use Leibnitz's formula to evaluate (Problems 33–35):

33. $\dfrac{d}{dx}\displaystyle\int_x^{x^2} \cos u^2\, du.$ **34.** $\dfrac{d}{d\alpha}\displaystyle\int_\alpha^{\alpha^2} \dfrac{\sin u}{u}\, du.$

35. $\dfrac{d}{d\alpha}\displaystyle\int_0^{\alpha/2} (\alpha^2-u^2)^{1/2}\, du.$

5.5 CALCULUS OF VARIATIONS*

The calculus of variations was founded in the eighteenth centry by L. Euler and J.L. Lagrange and extensively developed by other mathematicians in the nineteenth century. At present, it is one of the most important branches of applied mathematics and is a powerful technique for the solution of problems in dynamics and statics of rigid bodies, the theory of plates and shells, vibration problems, optimization of orbits, etc. The subject is primarily concerned with the study of certain maxima and minima problems which cannot be treated by the techniques of elementary calculus. Only some elementary ideas and derivation of some significant classical examples are discussed here with the sole object of giving the student a glimpse into a topic whose importance is steadily growing in science and engineering.

5.5.1 Classical Problems

The thoery of the calculus of variations had its beginning in the solution of the brachistochrome (i.e. shortest time) problem which was formulated by John Bernoulli in 1696.

Consider two points A and B at two different levels and not in the same vertical plane (Fig. 5.2). A smooth wire of any length has its extremities at A and B. A heavy bead slides down the wire starting from rest at A and reaching B. It is required to find the path from A to B so that the descent to B is quickest. This path of the quickest descent is called the *brachistochrome* (cycloid).

Figure 5.2

*For more advanced information and applications, the reader is referred to standard texts on the subject, e.g. Schechter, R.S., *The Variational Method in Engineering*, McGraw Hill, New York, 1967.

In Fig. 5.3, let AP, measured along the curve be s. Then,

$$\dot{s} = \frac{ds}{dt} = \text{Velocity at } P.$$

Let m be the mass of the bead and let PM $(= y)$ be the perpendicular from P on to the horizontal at A. Then we have

$$\text{Kinetic energy} = \frac{1}{2} m\dot{s}^2$$

and potential energy $= -mgy$.

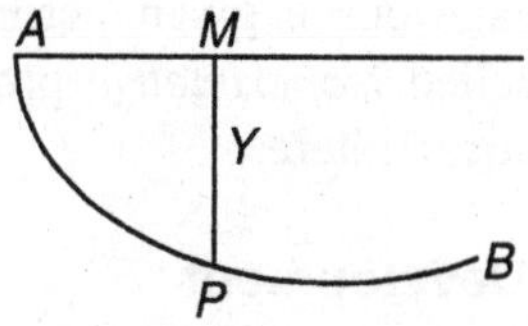

Figure 5.3

The principle of energy therefore gives

$$\frac{1}{2} m\dot{s}^2 - mgy = C \tag{5.53}$$

where C is a constant. But initially, i.e. at A, $\dot{s} = 0$ when $y = 0$. This gives $c = 0$ and hence (5.53) gives:

$$\dot{s}^2 = 2gy, \qquad \text{whence } \dot{s} = \frac{ds}{dt} = \sqrt{2gy}$$

Therefore,

$$\int_A^B dt = \int_A^B \frac{ds}{\sqrt{2gy}} = \int_A^B \frac{\sqrt{dx^2 + dy^2}}{\sqrt{2gy}} = \int_A^B \sqrt{\frac{1+y'^2}{2gy}}dx, \tag{5.54}$$

where $y' = dy/dx$. The general problem is to find $y = \phi(x)$ subject to the terminal conditions, namely that A and B lie on the curve so that

$$I = \int_A^B f(x, y, y')\, dx \tag{5.55a}$$

is stationary. In the particular problem considered above,

$$f(x, y, y') = \left(\frac{1+y'^2}{2gy}\right)^{1/2}$$

Another classical problem is to find the shortest smooth plane curve joining two points A and B. The length is given by

$$I = \int_A^B ds = \int_A^B \sqrt{dx^2 + dy^2} = \int_A^B \sqrt{1 + y'^2}\, dx$$

where I is to be minimized for AB to be the shortest path. Corresponding problems in higher dimensions may be formulated in a similar manner but these will not be attempted in this book.

Finally, mention must be made of problems in which it will be necessary to make an integral stationary while at the same time, one or more integrals involving the same variables have to be kept constant. A typical example is that of finding a closed plane curve given perimeter and maximum area. Problems of this type are called *isoperimetric* problems and are one of the earliest to engage mathematical interest.

5.5.2 Euler–Lagrange Development

In the study of different calculus, we have learnt the determination of maxima and minima of functions of one or more variables. The calculus of variations is primarily concerned with the determination of the extrema, i.e. the stationary values, for *functionals.* A *functional* is defined as a function which depends upon the entire course or path of one or more functions rather than on a number of discrete variables. In differential calculus, the domain of a function is a certain region in coordinate space, whereas the domain of a functional is a set of functions which belong to a function space. For example, in Eq. (5.55a), $f(x, y, y')$ is a functional.

Now we consider the basic problem of formulating Euler's equation for a class of problems, where the integrand is a functional of the form $f(x, y, y')$. Our object is to determine necessary conditions so that a function $y = f(x)$ makes the integral

$$I = \int_a^b f(x, y, y')\, dx \tag{5.55b}$$

stationary, where $y(a)$ and $y(b)$ are prescribed. Before proceeding to investigate this problem, we derive two useful lemmas.

Let PQ be a curve and $P'Q'$ be a neighbouring curve obtained by small δ-displacements, P goes to P' and Q to Q' (Fig. 5.4).

Let $MN = dx$, $MM' = \delta x$, $PM = y$, $P'M' = y + \delta y$, and $KP' = \delta y$. Let the tangent to the curve at P cuts $P'M'$ at L, so that

$$\frac{KL}{PK} = \frac{dy}{dx} = y' = \frac{KL}{MM'}.$$

Hence,

$$KL = y'\delta x \tag{5.56}$$

Now,

$$LP' = KP' - KL = \delta y - y'\delta x.$$

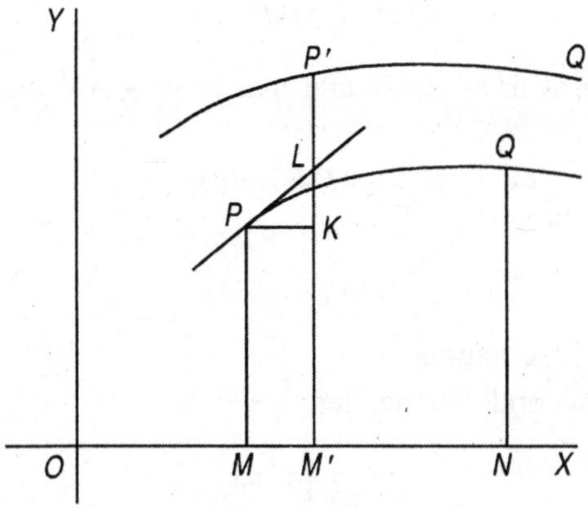

Figure 5.4

This combination occurs frequently in the study of calculus of variations and the term $y'\delta x$ is called a *transport term.* In this connection, we prove two lemmas:

Lemma 1 $\delta(dx) = d(\delta x)$.

Lemma 2 $\delta \int f(x)\, dx = \int \delta f(x)\, dx$.

To prove the lemmas, let $P'Q'$ be the neighbouring curve of PQ (Fig. 5.5.), and let $OM = x$, $MN = dx$ and $MM' = \delta x$, as before. Therefore,

$$\delta(dx) = \delta(MN).$$

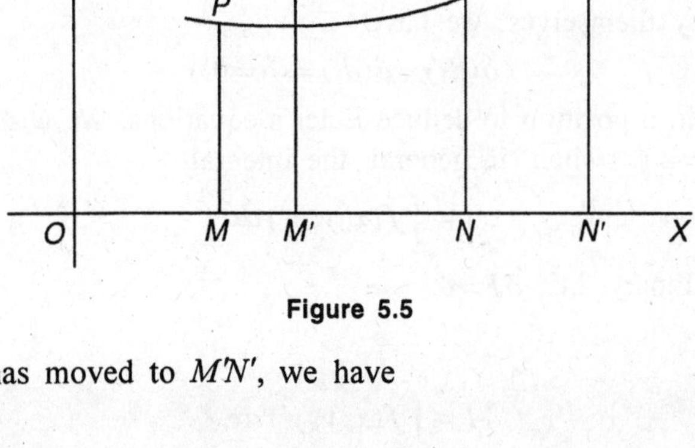

Figure 5.5

Since MN has moved to $M'N'$, we have

$$\begin{aligned}\delta(MN) &= M'N' - MN \\ &= M'N + NN' - MM' - M'N \\ &= NN' - MM'.\end{aligned}$$

Since the variation at M is MM' and that at N is NN' and since MN is dx, it follows that

$$NN' - MM' = d(\delta x).$$

Hence,

$$\delta(dx) = d(\delta x), \tag{5.57}$$

which proves the first lemma.

To prove the second lemma, let

$$z = \int V\,dx \tag{5.58}$$

so that

$$dz = V\,dx.$$

Hence,

$$\delta(dz) = \delta(V\,dx) = \delta V\,dx + V\delta\,(dx) \tag{5.59}$$

Since x has no variation in the present case, we have $\delta(dx) = 0$. Hence, Eq. (5.59) becomes:

$$\delta(dz) = \delta V\,dx$$

Using Lemma 1, we get

$$d(\delta z) = \delta V\,dx,$$

Therefore,

$$\delta z = \int \delta V\,dx$$

i.e.

$$\delta \int V\,dx = \int \delta V\,dx. \tag{5.60}$$

It should be noted that if $x, t, \ldots$ are independent variables, then since $x, t, \ldots$ do not change by themselves, we have

$$\delta(dx) = \delta(dt) = \cdots = 0.$$

We are now in a position to deduce Euler's equations. We wish to find the path, say $y = \phi\,(x)$ when, in general, the integral

$$I = \int f(x, y, y')\,dx$$

becomes stationary, i.e. $\delta I = 0$.

Let

$$I = \int_a^b f(x, y, y')\,dx, \tag{5.61}$$

where a and b are supposed to be fixed. Hence, by Lemma 2, we get

$$\delta I = \int_a^b \delta f(x, y, y')\, dx = \int_a^b \left(\frac{\partial f}{\partial y} \delta y + \frac{\partial f}{\partial y'} \delta y' \right) dx, \qquad \text{since } \delta x = 0 \qquad (5.62)$$

Now, we consider the term

$$\int_a^b \frac{\partial f}{\partial y'} \delta y'\, dx = \int_a^b \frac{\partial f}{\partial y'} \delta \left(\frac{dy}{dx} \right) dx, \qquad \text{since } y' = \frac{dy}{dx}.$$

$$= \int_a^b \frac{\partial f}{\partial y'} \delta (dy)$$

$$= \int_a^b \frac{\partial f}{\partial y'} d(\delta y), \qquad \text{using Lemma 1.}$$

Integrating by parts, we obtain

$$\int_a^b \frac{\partial f}{\partial y'} d(\delta y) = \left[\frac{\partial f}{\partial y'} \delta y \right]_a^b - \int_a^b \delta y \frac{d}{dx} \left(\frac{\partial f}{\partial y'} \right) dx.$$

We assume the terminal conditions to be $\delta y = 0$ at $x = a$ and $x = b$. Therefore,

$$\int_a^b \frac{\partial f}{\partial y'} d(\delta y) = -\int_a^b \delta y \frac{d}{dx} \left(\frac{\partial f}{\partial y'} \right) dx.$$

Hence, Eq. (5.62) becomes

$$\delta I = \int_a^b \left[\frac{\partial f}{\partial y} \delta y - \frac{d}{dx} \left(\frac{\partial f}{\partial y'} \right) \delta y \right] dx$$

$$= \int_a^b \left[\frac{\partial f}{\partial y} - \frac{d}{dx} \left(\frac{\partial f}{\partial y'} \right) \right] \delta y\, dx. \qquad (5.63)$$

If I is to be stationary, $\delta I = 0$. Hence, for an arbitrary variation δy, the right-hand side of (5.63) vanishes. It follows therefore that

$$\frac{\partial f}{\partial y} - \frac{d}{dx} \left(\frac{\partial f}{\partial y'} \right) = 0. \qquad (5.64)$$

Equation (5.64) is called *Euler's equation* and is frequently used in the study of calculus of variations. On carrying out the differentiation indicated in (5.64), we obtian

$$\frac{\partial f}{\partial y} - \frac{\partial^2 f}{\partial x \partial y'} - \frac{\partial^2 f}{\partial y \partial y'} y' - \frac{\partial^2 f}{\partial y'^2} y'' = 0,$$

which is a differential equation of the second order.

In the particular case when f does not contain x explicitly, it can be shown that

$$y' \frac{\partial f}{\partial y'} - f = \text{constant} \tag{5.65}$$

is an integral of (5.64). To prove this, we differentiate (5.65) and obtain

$$\frac{d}{dx}\left(y' \frac{\partial f}{\partial y'} - f \right) = 0$$

i.e.

$$y'' \frac{\partial f}{\partial y'} + y' \frac{d}{dx}\left(\frac{\partial f}{\partial y'} \right) - \frac{df}{dx} = 0. \tag{5.66}$$

Since $f = f(y, y')$, we have

$$\frac{df}{dx} = \frac{\partial f}{\partial y}\frac{dy}{dx} + \frac{\partial f}{\partial y'}\frac{dy'}{dx} = \frac{\partial f}{\partial y} y' + \frac{\partial f}{\partial y'} y''.$$

Hence, Eq. (5.66) becomes:

$$y'' \frac{\partial f}{\partial y'} + y' \frac{d}{dx}\left(\frac{\partial f}{\partial y'} \right) - \frac{\partial f}{\partial y} y' - \frac{\partial f}{\partial y'} y'' = 0$$

i.e.

$$\left[\frac{d}{dx}\left(\frac{\partial f}{\partial y'} \right) - \frac{\partial f}{\partial y} \right] y' = 0,$$

which is satisfied by virtue of (5.64). Hence, it follows that

$$y' \frac{\partial f}{\partial y'} - f = \text{constant}$$

is an integral of (5.64).

Any curve corresponding to a solution of Euler's equation (5.64) is called an *extremal*. Even in simple problems, it may be impossible to pass an extremal through A and B for all possible positions. In many cases, however, it will be physically evident that there can be no maximum.

If higher derivatives are present, for example, as in the following integral

$$I = \int_a^b f(x, y, y', y'')\, dx \tag{5.67}$$

then the necessary condition for I to be minimum is

$$\frac{\partial f}{\partial y} - \frac{d}{dx}\left(\frac{\partial f}{\partial y'} \right) + \frac{d^2}{dx^2}\left(\frac{\partial f}{\partial y''} \right) = 0. \tag{5.68}$$

Similarly, Euler's equation can be written down even when the integrand contains a number of dependent variables (see Section 5.5.4).

5.5.3 Applications of Euler Equations

We shall now apply Euler's equation (5.64) to solve classical problems of the calculus of variations (see Section 5.5.4).

To find the shortest path from A to B

Let A and B be given by (x_0, y_0) and (x_1, y_1) respectively. The arc length of AB is given by

$$\int_{x_0}^{x_1} ds = \int_{x_0}^{x_1} \sqrt{dx^2 + dy^2} = \int_{x_0}^{x_1} (1 + y'^2)^{1/2}\, dx.$$

For AB to be the shortest path, I is stationary and hence $\delta I = 0$. It follows from (5.64) then:

$$\frac{\partial f}{\partial y} - \frac{d}{dx}\left(\frac{\partial f}{\partial y'}\right) = 0, \qquad \text{where } f = (1 + y'^2)^{1/2}.$$

Here f does not contain y explicitly, and hence $\partial f/\partial y = 0$. Thus, Euler's equation reduces to:

$$-\frac{d}{dx}\left(\frac{y'}{\sqrt{1 + y'2}}\right) = 0.$$

Therefore,

$$\frac{y'}{\sqrt{1 + y'^2}} = \text{constant} = k, \text{ say.}$$

i.e.

$$y' = k\,(1 + y'^2)^{1/2}, \quad \text{or} \quad y'^2 = k^2\,(1 + y'^2),$$

whence

$$y' = \frac{k}{\sqrt{1 - k^2}} = \text{constant}. \tag{5.69}$$

The integral of (5.69) may be written as

$$y = mx + n, \tag{5.70}$$

where m and n are constants. Since the path passes through (x_0, y_0) and (x_1, y_1), it is easily written as

$$\begin{vmatrix} y & x & 1 \\ y_0 & x_0 & 1 \\ y_1 & x_1 & 1 \end{vmatrix} = 0 \tag{5.71}$$

which is a straight line.

To find the path of quickest descent

This is John Bernoulli's problem and has already been formulated in Section 5.5.1. We have

$$I = \int dt = \int_{x_0}^{x_1} \frac{(1+y'^2)^{1/2}}{\sqrt{2gy}}\, dx.$$

If t is to be minimum, $\delta I = 0$. Here,

$$f = \left(\frac{1+y'^2}{2gy}\right)^{1/2}, \tag{5.72}$$

and hence f does not contain x explicitly. It follows, from Section 5.5.2, that

$$y'\frac{\partial f}{\partial y'} - f = \text{constant} = k, \text{ say} \tag{5.73}$$

is an integral of (5.64) with f given by (5.72).

Now,

$$\frac{\partial f}{\partial y'} = \frac{y'}{1+y'^2}\frac{1}{\sqrt{2gy}}$$

Hence, Eq. (5.73) gives:

$$y'\frac{y'}{\sqrt{1+y'^2}}\frac{1}{\sqrt{2gy}} - \frac{(1+y'^2)^{1/2}}{\sqrt{2gy}} = k,$$

i.e.

$$\sqrt{1+y'^2}\,\sqrt{2gy} = -\frac{1}{k},$$

or,

$$\sqrt{1+y'^2}\,\sqrt{y} = -\frac{1}{k\sqrt{2g}}.$$

or,

$$y(1+y'^2)=\frac{1}{2gk^2}.$$

i.e.

$$y\sec^2\psi=\frac{1}{2gk^2},\qquad \text{since}\quad y'=\tan\psi.$$

Therefore,

$$y=\frac{1}{2gk^2}\cos^2\psi=2a\cos^2\psi,\qquad \text{where}\quad \frac{1}{2gk^2}=2a,$$

and hence

$$2a-y=2a\sin^2\psi, \tag{5.74}$$

which indicates an arc of a cycloid with cusps on $y=0$, i.e. on the line of zero velocity.

Example 5.38 Find the brachistochrone if a particle is moving under the influence of a force perpendicular to and proportional to the distance from the line of zero velocity (see Fig. 5.6).

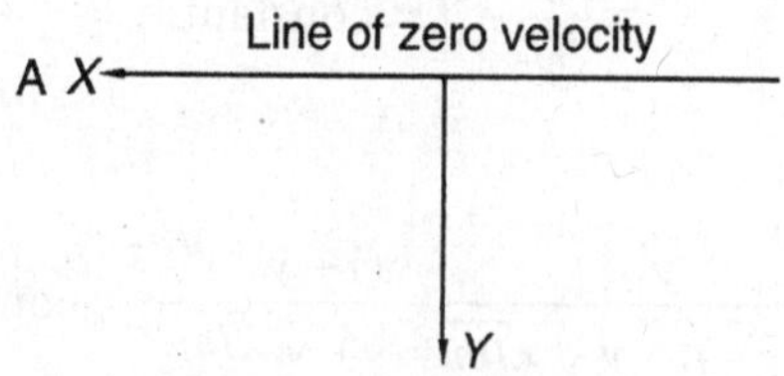

Figure 5.6

Since the force is perpendicular to the line of zero velocity and is also proportional to y, we have

$$F_y=\lambda y \qquad \text{and} \qquad F_x=0$$

If $\phi(x, y)$ is the potential energy, we have,

$$-\frac{\partial\phi}{\partial x}=F_x=0 \qquad \text{and} \qquad -\frac{\partial\phi}{\partial y}=F_y=\lambda y$$

Hence,

$$\phi=-\frac{1}{2}\lambda y^2+c.$$

If $v=ds/dt$ is the velocity at any time, the principle of energy gives:

$$\frac{1}{2}mv^2+c-\frac{1}{2}\lambda y^2=\text{constant}=k,\qquad \text{say.}$$

But at A (refer to Fig. 5.6), $v = 0$ and $y = 0$. Hence $k = c$.

The above equation therefore becomes:

$$\frac{1}{2}mv^2 = \frac{1}{2}\lambda y^2.$$

i.e.

$$\left(\frac{ds}{dt}\right)^2 = \frac{\lambda}{m}y^2, \quad \text{or} \quad \frac{ds}{dt} = \sqrt{\frac{\lambda}{m}}\,y,$$

the positive sign has been chosen since the velocity is increasing as y is increasing. Hence,

$$\int_A^B dt = \int_A^B \frac{ds}{y\sqrt{\lambda/m}} = \int_{x_0}^{x_1} \frac{(1+y'^2)^{1/2}}{y\sqrt{\lambda/m}}\,dx.$$

Here

$$f = \frac{(1+y'^2)^{1/2}}{y\sqrt{\lambda/m}}.$$

Since f does not contain x explicitly, we have

$$y'\frac{\partial f}{\partial y'} - f = \text{constant}.$$

i.e.

$$y'\frac{y'}{(1+y'^2)^{1/2}\,y\sqrt{\lambda/m}} - \frac{(1+y'^2)^{1/2}}{y\sqrt{\lambda/m}} = \text{constant}.$$

Hence,

$$y(1+y'^2)^{1/2} = \text{constant},$$

or,

$$y \sec\psi = \text{constant} = a, \quad \text{say}.$$

Therefore,

$$y = a\cos\psi \quad \text{or} \quad dy = -a\sin\psi\,d\psi.$$

Further, since $dy = \tan\psi\,dx$, it follows that

$$dx = \frac{dy}{\tan\psi} = -a\cos\psi\,d\psi.$$

Now,

$$ds^2 = dx^2 + dy^2.$$

Therefore,

$$ds = \sqrt{dx^2 + dy^2} = \pm a\,d\psi.$$

i.e.

$$\frac{ds}{d\psi} = \pm a, \quad \text{which is a constant.}$$

It therefore follows that the brachistochrone is a circle.

Example 5.39 An arc C of a curve is rotated about OX to form a surface of revolution. Find the arc C such that the surface area S is minimum.

Let the arc be bounded by the ordinates $x = a$ and $x = b$. The surface area S is given by

$$S = 2\pi \int_a^b y\, ds = 2\pi \int_a^b y\sqrt{1 + y'^2}\, dx, \tag{i}$$

where $y' = dy/dx$.

Let

$$I = \int_a^b F(x, y, y')\, dx \tag{ii}$$

where $F(x, y, y') = y\sqrt{(1 + y'^2)}$. It is evident that S will be least when I is least. Applying Euler's equation to the integral in (ii), we obtain:

$$\sqrt{1 + y'^2} - \frac{d}{dx}\left(\frac{yy'}{\sqrt{1 + y'^2}}\right) = 0$$

i.e.

$$1 + y'^2 - yy'' = 0. \tag{iii}$$

If we set $dy/dx = p$, then

$$y'' = \frac{dp}{dx} = \frac{dp}{dy}\frac{dy}{dx} = p\frac{dp}{dy}.$$

Hence (iii) becomes:

$$1 + p^2 - yp\frac{dp}{dy} = 0 \quad \text{or} \quad \frac{p\, dp}{1 + p^2} = \frac{dy}{y}.$$

Integrating both sides, we get:

$$\log_e y = \frac{1}{2}\log(1 + p^2) + \log c_1$$

i.e.

$$y = c_1\sqrt{1 + p^2} \qquad \text{or} \qquad p = \frac{dy}{dx} = \frac{1}{c_1}\sqrt{y^2 - c_1^2}$$

i.e.

$$dx = c\frac{dy}{\sqrt{y^2 - c_1^2}}.$$

Integration gives:

$$x = c_1 \cosh^{-1}\frac{y}{c_1} + c_2,$$

which shows that the extremals are catenaries with base, the axis of x.

5.5.4 Several Dependent Variables

It is easy to extend the theory to problems where instead of a single dependent variable y and its derivative y', there are several variables to be determined as functions of the same independent variable. That is, when

$$I = \int_{x_0}^{x_1} f(x, y_1, y_2, \ldots, y_n, y_1', y_2', \ldots, y_n')\, dx \tag{5.75}$$

and I is to be made a maximum or minimum. Here $y_1, y_2, \ldots, y_n$ are n independent function of x and $y_r' = dy_r/dx$. Now,

$$\delta f = \frac{\partial f}{\partial y_1}\delta y_1 + \frac{\partial f}{\partial y_2}\delta y_2 + \cdots + \frac{\partial f}{\partial y_n}\delta y_n$$

$$+ \frac{\partial f}{\partial y_1'}\delta y_1' + \frac{\partial f}{\partial y_2'}\delta y_2' + \cdots + \frac{\partial f}{\partial y_n'}\delta y_n'$$

Here,

$$\delta y_1' = \delta\left(\frac{dy_1}{dx}\right) = \frac{d}{dx}(\delta y_1).$$

From (5.75), we have

$$\delta I = \int_{x_0}^{x_1}\left(\frac{\partial f}{\partial y_1}\delta y_1 + \frac{\partial f}{\partial y_2}\delta y_2 + \cdots + \frac{\partial f}{\partial y_n}\delta y_n\right.$$

$$\left. + \frac{\partial f}{\partial y_1'}\delta y_1' + \frac{\partial f}{\partial y_2'}\delta y_2' + \cdots + \frac{\partial f}{\partial y_n'}\delta y_n'\right) dx. \tag{5.76}$$

Now,

$$\int_{x_0}^{x_1} \frac{\partial f}{\partial y_1'}\delta y_1'\, dx = \int_{x_0}^{x_1} \frac{\partial f}{\partial y_1'}\frac{d}{dx}(\delta y_1)\, dx$$

By integration by parts

$$\int_{x_0}^{x_1} \frac{\partial f}{\partial y_1'} \delta y_1' \, dx = \left[\frac{\partial f}{\partial y_1'} \delta y_1 \right]_{x_0}^{x_1} - \int_{x_0}^{x_1} \delta y_1 \frac{d}{dx}\left(\frac{\partial f}{\partial y_1'} \right) dx = - \int_{x_0}^{x_1} \delta y_1 \frac{d}{dx}\left(\frac{\partial f}{\partial y_1'} \right) dx,$$

since δy_1 vanishes at the terminal points which are fixed. Hence, Eq. (5.76) simplifies to:

$$\delta I = \int_{x_0}^{x_1} \left[\frac{\partial f}{\partial y_1} - \frac{d}{dx}\left(\frac{\partial f}{\partial y_1'} \right) \right] \delta y_1 \, dx + \int_{x_0}^{x_1} \left[\frac{\partial f}{\partial y_2} - \frac{d}{dx}\left(\frac{\partial f}{\partial y_2'} \right) \right] \delta y_2 \, dx + \cdots$$

$$+ \int_{x_0}^{x_n} \left[\frac{\partial f}{\partial y_n} - \frac{d}{dx}\left(\frac{\partial f}{\partial y_n'} \right) \right] \delta y_n \, dx \qquad (5.77)$$

But $\delta I = 0$ and $\delta y_1, \delta y_2, \ldots, \delta y_n$ are all arbitrary. Hence, we have

$$\frac{\partial f}{\partial y_1} - \frac{d}{dx}\left(\frac{\partial f}{\partial y_1'} \right) = 0,$$

$$\frac{\partial f}{\partial y_2} - \frac{d}{dx}\left(\frac{\partial f}{\partial y_2'} \right) = 0,$$

$$\vdots$$

$$\frac{\partial f}{\partial y_n} - \frac{d}{dx}\left(\frac{\partial f}{\partial y_n'} \right) = 0.$$

or, more simply,

$$\frac{\partial f}{\partial y_r} - \frac{d}{dx}\left(\frac{\partial f}{\partial y_r'} \right) = 0, \qquad r = 1, 2, \ldots, n. \qquad (5.78)$$

which are Euler's equations for the n variables. If, however, x does not appear explicitly in f, we have

$$\frac{\partial f}{\partial x} = 0,$$

and

$$\sum_{r=1}^{n} y_r' \frac{\partial f}{\partial y_r'} - f = \text{constant} \qquad (5.79)$$

Verification of (5.79) can be made as in the case of a single dependent variable and is left as an exercise to the reader.

5.5.5 Applications to Discrete Mechanics

An important concept of mathematical physics is known as *Hamilton's principle*, which states that for a dynamical system

$$\delta \int_{t_1}^{t_2} (T - V)\, dt = 0 \tag{5.80}$$

where T = kinetic energy; V = potential energy; and $T - V$ is called the *Lagrangian function.*

Starting from Hamilton's principle, it is possible to deduce directly Lagrange's equations, which in turn may be used to establish Newton's second law.

We consider discrete systems, i.e. systems in which the motion of one or more particles plays a dominant role. Let (x, y, z) be the coordinates of a particle of mass m at time t. Then, we have

$$T = \frac{1}{2} m (\dot{x}^2 + \dot{y}^2 + \dot{z}^2) \quad \text{and} \quad V = V(x, y, z).$$

Therefore,

$$I = \int (T - V)\, dt = \int \left[\frac{1}{2} m (\dot{x}^2 + \dot{y}^2 + \dot{z}^2) - V(x, y, z) \right] dt.$$

We have three dependent variables here and hence Euler's equations are:

$$\left.\begin{aligned} \frac{\partial}{\partial x}(T - V) - \frac{d}{dt}\frac{\partial}{\partial \dot{x}}(T - V) &= 0 \\ \frac{\partial}{\partial y}(T - V) - \frac{d}{dt}\frac{\partial}{\partial \dot{y}}(T - V) &= 0 \\ \frac{\partial}{\partial z}(T - V) - \frac{d}{dt}\frac{\partial}{\partial \dot{z}}(T - V) &= 0. \end{aligned}\right\} \tag{5.81}$$

Since

$$T - V = \frac{1}{2} m (\dot{x}^2 + \dot{y}^2 + \dot{z}^2) - V,$$

the above equations yield on simplification:

$$m\ddot{x} = -\frac{\partial V}{\partial x}, \qquad m\ddot{y} = -\frac{\partial V}{\partial y}, \qquad m\ddot{z} = -\frac{\partial V}{\partial z}, \tag{5.82}$$

which are the *equations of motion of the particle.*

We shall next derive Lagrange's and Hamilton's equations. For this, we start with D'Alembert's principle which states that a system of particles in motion under externally applied forces can be treated at each instant as a statical system of forces in equilibrium. There will be one set of forces called the *reversed effective forces* and another set consisting of *externally applied forces.*

The condition of statical equilibrium may be expressed as an equation of virtual work from which all forces of the nature of action and reaction drop out. We therefore have

$$\left[\sum(m\ddot{x}-X)\,\delta x+\sum(m\ddot{y}-Y)\,\delta y+\sum(m\ddot{z}-Z)\,\delta z\right]dt=0 \qquad (5.83)$$

where (t_2-t_1) is the duration of motion.

Now,

$$\int_{t_1}^{t_2} m\ddot{x}\delta x\,dt=\left[m\dot{x}\delta x\right]_{t_1}^{t_2}-\int_{t_1}^{t_2} m\dot{x}\frac{d}{dt}(\delta x)\,dt$$

Since δx vanishes at both the end points.

$$\int_{t_1}^{t_2} m\ddot{x}\,\delta x\,dt=-\int_{t_1}^{t_2} m\dot{x}\frac{d}{dt}(\delta x)\,dt,$$

$$=-\int_{t_1}^{t_2} m\dot{x}\delta\dot{x}\,dt, \qquad \text{since } \frac{d}{dt}(\delta x)=\delta\left(\frac{dx}{dt}\right)=\delta\dot{x}$$

$$=-\int_{t_1}^{t_2}\delta\left(\frac{1}{2}m\dot{x}^2\right)dt.$$

Hence, we have

$$\int_{t_1}^{t_2}\sum(m\ddot{x}\delta x+m\ddot{y}\delta y+m\ddot{z}\delta z)\,dt=-\int_{t_1}^{t_2}\delta\sum\left(\frac{1}{2}m\dot{x}^2+\frac{1}{2}m\dot{y}^2+\frac{1}{2}m\dot{z}^2\right)dt$$

$$=-\int_{t_1}^{t_2}\delta T\,dt, \qquad (5.84)$$

where

$$T=\sum\frac{1}{2}\left(m\dot{x}^2+m\dot{y}^2+m\dot{z}^2\right),$$

is the kinetic energy. In case there exists a potential energy function $V=V(x,y,z)$ such that

$$X=-\frac{\partial V}{\partial x},\quad Y=-\frac{\partial V}{\partial y},\quad Z=-\frac{\partial V}{\partial z},$$

then

$$\delta V=-\sum(X\,\delta x+Y\,\delta Y+Z\,\delta z),$$

and

$$\int_{t_1}^{t_2}-\sum(X\,\delta x+Y\,\delta y+Z\,\delta z)\,dt=\int_{t_1}^{t_2}\delta V\,dt. \qquad (5.85)$$

Hence, Eq. (5.83) simplifies to:

$$\int_{t_1}^{t_2} \delta (T - V)\, dt = 0$$

i.e.

$$\delta \int_{t_1}^{t_2} (T - V)\, dt = 0, \tag{5.86}$$

which is Hamilton's principle. We have thus shown that the equation of virtual work leads to Hamilton's principle.

In rigid dynamics, the coordinates (x_1, y_1, z_1), (x_2, y_2, z_2),... are normally expressed as

$$q_1 = f(x_1, y_1, z_1, \ldots), \quad q_2 = f(x_2, y_2, z_1, \ldots), \text{ etc.}$$

where $q_1, q_2, \cdots$ are called *generalized coordinates*. A system of particles may be given by a number of generalized coordinates $q_1, q_2, \ldots, q_n$. Such a system is said to be *holonomic* if the number of independent coordinates equals the number of degrees of freedom.

Let the Lagrangian L be expressed in generalized coordinates and velocities as

$$L = L(tq_1, q_2, \ldots, \dot{q}_n, \dot{q}_1, \dot{q}_2, \ldots, \dot{q}_n) \tag{5.87}$$

The equations of motion are given by

$$\delta \int L\, dt = 0. \tag{5.88}$$

The Euler equations that follow from the above are of the form:

$$\frac{\partial L}{\partial q_1} - \frac{d}{dt}\left(\frac{\partial L}{\partial \dot{q}_1}\right) = 0$$

$$\frac{\partial L}{\partial q_2} - \frac{d}{dt}\left(\frac{\partial L}{\partial \dot{q}_2}\right) = 0$$

$$\vdots$$

or, in the compact form

$$\frac{\partial L}{\partial q_r} - \frac{d}{dt}\left(\frac{\partial L}{\partial \dot{q}_r}\right) = 0, \tag{5.89}$$

which are usually referred to as *Lagrange's equations*. We now define *generalized momenta* p_r by the relations

$$p_r = \frac{\partial L}{\partial \dot{q}_r}, \qquad r = 1, 2, \ldots, n. \tag{5.90}$$

If we define the *Hamiltonian H* by the relation,

$$H = \sum p_r \dot{q}_r - L, \tag{5.91}$$

we have

$$\delta H = \sum \delta p_r \dot{q}_r + \sum p_r \delta \dot{q}_r - \sum \frac{\partial L}{\partial q_r} \delta q_r - \sum \frac{\partial L}{\partial \dot{q}_r} \delta \dot{q}_r, \quad \text{since } \delta t = 0$$

$$= \sum \delta p_r \dot{q}_r - \sum \frac{\partial L}{\partial q_r} \partial q_r + \sum \left(p_r - \frac{\partial L}{\partial \dot{q}_r} \right) \delta \dot{q}_r$$

$$= \sum \delta p_r \dot{q}_r - \sum \frac{\partial L}{\partial q_r} \delta q_r, \quad \text{using (5.90)} \tag{5.92}$$

But by Lagrange's equations,

$$\frac{\partial L}{\partial q_r} = \frac{d}{dt}\left(\frac{\partial L}{\partial \dot{q}_r} \right)$$

Hence, Eq. (5.92) becomes:

$$\delta H = \sum \delta p_r \dot{q}_r - \sum \frac{d}{dt}\left(\frac{\partial L}{\partial \dot{q}_r} \right) \delta q_r$$

$$= \sum \delta p_r \dot{q}_r - \sum \frac{d}{dt}(p_r)\, \delta q_r, \quad \text{by using (5.90).}$$

$$= \sum \delta p_r \dot{q}_r - \sum \dot{p}_r \delta q_r \tag{5.93}$$

But δH can also be written as

$$\delta H = \sum \frac{\partial H}{\partial p_r} \delta p_r + \sum \frac{\partial H}{\partial q_r} \delta q_r \tag{5.94}$$

Hence, comparing the coefficients of δp_r and δq_r in Eqs. (5.93) and (5.94), we obtain

$$\frac{\partial H}{\partial p_r} = \dot{q}_r \quad \text{and} \quad \frac{\partial H}{\partial q_r} = -\dot{p}_r \tag{5.95}$$

which are known as *Hamilton's equations.*

5.5.6 The Isoperimetric Problem

Problems sometimes arise in which it is necessary to make an integral stationary while at the same time, one or more integrals involving the same variables are to be kept constant. A typical example, discussed below, is that of finding the closed plane curve of given perimeter and maximum area. This example, being one of the earliest to engage mathematical interest, has given this class of problems the name "isoperimetric".

We desire to minimize the integral

$$I = \int_a^b f(x, y, y')\, dx \tag{5.96}$$

taken between fixed end points, while keeping a second integral J constant, viz.,

$$J = \int_b^b g\,(x, y, y')\, dx = K \tag{5.97}$$

In such a case, the desired curves satisfy the Euler equation for minimizing

$$\int_a^b [f(x, y, y') - \lambda g\,(x, y, y')]\, dx, \tag{5.98}$$

where λ is a constant.

The condition is given by

$$\frac{\partial f}{\partial y} - \lambda \frac{\partial g}{\partial y} - \frac{d}{dx}\left(\frac{\partial f}{\partial y'}\right) + \lambda \frac{d}{dx}\left(\frac{\partial g}{\partial y'}\right) = 0 \tag{5.99}$$

In this case, the values of the two constants of integration and the constant λ are determined by the three conditions, viz., that the curve passes through the fixed end points and that the second integral J has the given constant value K. The following examples illustrate the method of procedure.

Example 5.40 Find the arc C of given length L which minimizes S, the area of the surface of revolution generated by the rotation of C about OX.

We have to make

$$I = \frac{S}{2\pi} \int_a^b y\sqrt{1 + y'^2}\, dx \tag{i}$$

a minimum for

$$\int_a^b \sqrt{1 + y'^2}\, dx = L \tag{ii}$$

The extremals satisfy Euler's equation for

$$\int_a^b (y - \lambda)\sqrt{1 + y'^2}\, dx.$$

Hence,

$$\sqrt{1+y'^2} - \frac{d}{dx}\left[\frac{(y-\lambda)\,y'}{\sqrt{1+y'^2}}\right] = 0$$

or,

$$1+y'^2 - (y-\lambda)\,y'' = 0.$$

Following the same procedure as before the solution of this equation may be found to be

$$y = \lambda + c_1 \cosh \frac{x-c_2}{c_1},$$

which are catenaries with base lines parallel to the x-axis.

Example 5.41 Find the plane curve of fixed perimeter and maximum area.

Here

$$A = \frac{1}{2}\int_0^{2\pi} r^2\, d\theta,$$

and

$$S = \int_0^{2\pi} dS = \int_0^{2\pi} \sqrt{(dr)^2 + r^2(d\theta^2)} = \int_0^{2\pi} \sqrt{r^2 + r'^2}\, d\theta.$$

Now,

$$I = \int_0^{2\pi} \left[\frac{1}{2} r^2 + \lambda\,(r^2 + r'^2)^{1/2}\right] d\theta,$$

where λ is a constant. $\delta I = 0$ implies the Euler equation

$$\frac{\partial f}{\partial r} - \frac{d}{d\theta}\left(\frac{\partial f}{\partial r'}\right) = 0.$$

This gives:

$$r + \frac{\lambda r}{(r^2 + r'^2)^{1/2}} - \frac{d}{d\theta}\left[\frac{\lambda r'}{(r^2 + r'^2)^{1/2}}\right] = 0.$$

i.e.

$$r + \frac{\lambda r}{(r^2 + r'^2)^{1/2}} - \frac{\lambda r''}{(r^2 + r'^2)^{1/2}} + \frac{\lambda r'\,(rr' + r'\,r'')}{(r^2 + r'^2)^{3/2}} = 0$$

i.e.

$$\frac{r}{\lambda} + \frac{r\,(r^2 + r'^2) - r''\,(r^2 + r'^2) + r'\,(rr' + r'r'')}{(r^2 + r'^2)^{3/2}} = 0.$$

i.e.

$$\frac{1}{\lambda}+\frac{r^2+r'^2-rr''+r'^2}{(r^2+r'^2)^{3/2}}=0.$$

i.e.

$$\frac{1}{\lambda}+\frac{1}{\rho}=0$$

i.e.

$$\rho=\text{constant}.$$

Hence, the plane curve is a circle.

Example 5.42 Prove that the sphere is the solid figure of revolution which, for a given surface area, has maximum volume.

The surface area and volume of a sphere are

$$A=2\pi\int y\,ds=2\pi\int_0^a y\,(1+y'^2)^{1/2}\,dx \quad\text{and}\quad V=\pi\int_0^a y^2\,dx.$$

Here A is fixed and V is to be maximum, and

$$I=\int\left[y^2+\lambda y\,(1+y'^2)^{1/2}\right]dx,$$

dropping out the constant factors. Since $\delta I=0$, we have

$$f-y'\frac{\partial f}{\partial y'}=\text{constant},$$

and this gives

$$y^2+\lambda y\,(1+y'^2)^{1/2}-\lambda yy'^2\,(1+y'^2)^{-1/2}=c.$$

But $c=0$, since the curve has to pass through two terminal points, where $y=0$. Therefore, we obtain

$$y+\frac{2\lambda}{(1+y'^2)^{1/2}}=0.$$

or

$$1+y'^2=\frac{4\lambda^2}{y^2}$$

or

$$y'^2=\frac{4\lambda^2-y^2}{y^2}=\left(\frac{dy}{dx}\right)^2$$

Hence,

$$\frac{dy}{dx} = \sqrt{\frac{4\lambda^2 - y^2}{y}}$$

or

$$\frac{y\,dy}{\sqrt{4\lambda^2 - y^2}} = dx,$$

which on integration gives:

$$-\sqrt{4\lambda^2 - y^2} = x - x_0$$

i.e.

$$(x - x_0)^2 + y^2 = 4\lambda^2.$$

Thus, we note that the figure is a sphere with centre on the x-axis at x_0 and of radius 2λ.

EXERCISES 5.5

1. Find the curve of minimum length joining a given pair of points in a plane.
2. A bead slides freely on a smooth wire in a vertrical plane under the action of gravity. Find the form of the wire so that the time of descent from one point of the wire to another is the least possible.
3. For the integral

$$I = \int_{\alpha}^{\beta} \frac{1}{y}\left[1 + \left(\frac{dy}{dx}\right)^2\right]^{1/2} dx,$$

show that the general solution of the Euler equation is

$$(x - c_1)^2 + y^2 = c_2.$$

4. Using Euler's equations,
   (i) derive the equations of mechanics
   (ii) derive an expression for the radial acceleration in three-dimensional dynamics.
   (iii) obtain expressions for the acceleration in polar coordinates in three dimensions.
5. On any surface, the line of shortest length joining two points may be found by minimizing $\int ds$. The extremals obtained in such cases are called *geodesics*. Find the geodesics in the following cases:

   (a) on a plane taking $ds = [1 + (dy/dx)^2]^{1/2}\, dx$.

(b) on the surface of a sphere.
(c) on the surface of a cone.

6. Solve the Euler equations for the following cases (Problems):

(a) $\int_{\alpha}^{\beta} \frac{1}{y}\, ds$

(b) $\int_{\alpha}^{\beta} \sqrt{y}\, ds$

(c) $\int_{\alpha}^{\beta} y\left[1-\left(\frac{dy}{dx}\right)^2\right]^{1/2} dx.$

7. Find the closed convex curve of given circumference which encloses the greatest area.

8. Two rings, each of radius a, are placed parallel with their centres $2d$ apart and on a common normal. A soap film, which is assumed to be axially symmetric, is formed between the rings without covering their ends. Show that the film assumes the shape of a catenary.

9. If a uniform rope is suspended by its ends from two points at equal heights, show that it assumes the shape of a catenary.

10. Find the least value of the integral

$$\int_{A}^{B} \frac{1}{y}\left[1+\left(\frac{dy}{dx}\right)^2\right]^{1/2} dx$$

where A is $(-1, 1)$ and B is $(1, 1)$.

11. Minimize $\int ds$ subject to the condition $x^2 + y^2 + z^2 = a^2$.

12. Given two points a and b on the X-axis, and an arc length $L > b - a$, find the shape of the curve of length L joining the given points which, with the X-axis, encloses the largest area.

13. Between any two given points A and B, a light beam follows a path which makes the time of transit stationary. This is called Fermat's principle in optics. Using this principle, find the path followed by a light beam if the index of refraction is proportional to the functions $f(x)$ given below:

(a) $f(x) = 1$ (b) $f(x) = x$

(c) $f(x) = \frac{1}{x}$ (d) $f(x) = \sqrt{x}$

(e) $f(x) = \frac{1}{\sqrt{x}}$.

14. Find the equation of motion of a particle moving in a straight line with the potential energy $V = (1/2)kx^2$.

15. Using Lagrange's equations, find the equations of motion of a particle in terms of the polar coordinates r and θ.

16. Using Hamilton's principle, find the equations of motion of a single particle of mass m moving (near the earth) under gravity.

17. If the generalized coordinates q_i are transformed by the equations $\bar{q}_i = \bar{q}_i\,(q_1, q_2, \ldots, q_n)$, show that

(i) $\dot{\bar{q}}_i = \dfrac{\partial \bar{q}_i}{\partial q_j}\,\dot{q}_j$ and $\dot{q}_i = \dfrac{\partial q_i}{\partial \bar{q}_j}\,\dot{\bar{q}}_j$

(ii) $\dfrac{\partial \dot{q}_i}{\partial \bar{q}_j} = \dfrac{\partial^2 q_i}{\partial \bar{q}_j\,\partial \bar{q}_k}\,\dot{\bar{q}}_k$

Hence show that the form of Lagrange's equations is independent of the coordinate system.

6

Multiple Integrals

6.1 DOUBLE INTEGRALS

In Chapter 4, we have seen how to use the method of integration to calculate the area or centroid of a bounded region. We shall now show how to use the method of double integration to calculate these quantities. These ideas can be extended to multiple integrals, i.e. integrals of functions of more than two variables, but we shall not attempt to present all details. Triple integrals and their applications will, however, be discussed in Section 6.3. Before taking up these applications, we shall first define the double integral of a function $f(x, y)$ of two variables in x and y.

6.1.1 Definition and Evaluation of Double Integrals

The definite integral was defined in Section 4.1.1. The double integral can also be defined in the same way. Let $f(x, y)$ be a continuous and single-valued function of x and y within a region R bounded by a closed curve C and upon the boundary C. Let the region R be subdivided in any manner into n subregions of areas $\Delta A_1, \Delta A_2, \ldots, \Delta A_n$.

Let (ξ_t, η_i) be any point in the subregion of area ΔA_i and consider the sum

$$\sum_{i=1}^{n} f(\xi_i, \eta_i)\, \Delta A_i.$$

The limit of this sum as $n \to \infty$ and $\Delta A_i \to 0\ (i = 1, 2, \ldots)$ is defined as the *double integral of* $f(x, y)$ over the region R and is written as

$$\iint f(x, y)\, dA.$$

We therefore have

$$\iint_R f(x, y)\, dA = \lim_{n \to \infty} \sum_{i=1}^{n} f(\xi_i, \eta_i)\, \Delta A_i, \tag{6.1}$$

the region R being called the *region of integration* and this corresponds to the interval of integration (a, b) in the case of the definite integral.

In order to simplify the evaluation of the double integral, we often consider subregions R_i and one common choice is the rectangular subregions or rectangular grid, obtained by subdividing R by lines parallel to the coordinate axes as in Fig. 6.1. Since the area of a typical rectangle is $\Delta x_i\, \Delta y_j$, it follows from (6.1) that,

$$\iint_R f(x, y)\, dx\, dy = \lim_{n\to\infty} \sum_{i,j=1}^{n} f(x_i, y_j)\, \Delta x_i\, \Delta y_j \tag{6.2}$$

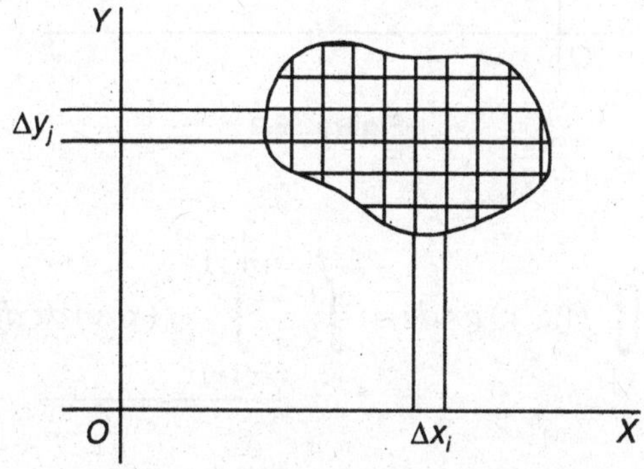

Figure 6.1

The definition (6.1) can obviously be generalized to higher dimensions. Before discussing methods of evaluating double integrals, we state below some properties of double integrals which follow from the definition.

If $f(x, y)$ and $g(x, y)$ are defined, bounded and continuous in a region R of the (x, y) plane and if the region R may be subdivided into two regions R_1 and R_2, then

(a) $\displaystyle\iint_R kf(x, y)\, dx\, dy = k \iint_R f(x, y)\, dx\, dy$, where k is a constant.

(b) $\displaystyle\iint_R [f(x, y) \pm g(x, y)]\, dx\, dy = \iint_R f(x, y)\, dx\, dy \pm \iint_R g(x, y)\, dx\, dy.$

(c) $\displaystyle\iint_R f(x, y)\, dx\, dy = \iint_{R_1} f(x, y)\, dx\, dy + \iint_{R_2} f(x, y)\, dx\, dy.$

The usefulness of the concept of the double integral in terms of the limit of sums is apparently limited in finding numerical answers to specific problems. It is, however, possible to evaluate a double integral over a region by two successive single integrations. This is described below.

Suppose that the region R (see Fig. 6.2) is described by the inequalities:

$$c \le y \le d \quad \text{and} \quad g(y) \le x \le h(y) \tag{6.3}$$

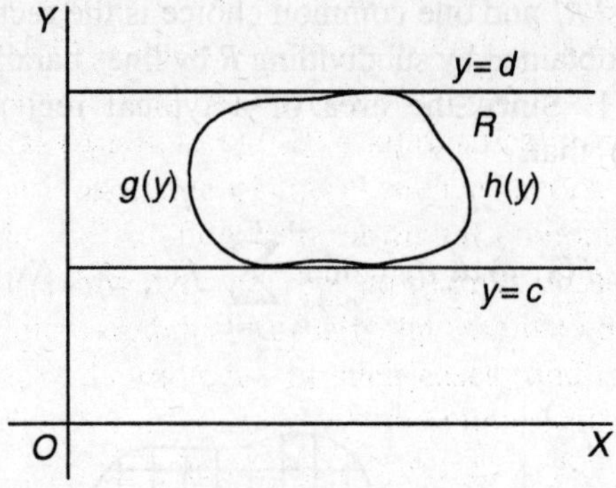

Figure 6.2

Then,

$$\iint_R f(x, y)\, dx\, dy = \int_{y=c}^{y=d} \underbrace{\int_{x=g(y)}^{x=h(y)} f(x, y)\, dx}\, dy \tag{6.4}$$

The inner integral

$$\int_{x=g(y)}^{x=h(y)} f(x, y)\, dx$$

is first evaluated with respect to x. Since the integrand and the limits are functions of y, the result would be a function of y and let this be $\phi(y)$. Hence, (6.4) becomes:

$$\iint f(x, y)\, dx\, dy = \int_c^d \phi(y)\, dy,$$

which is readily evaluated.

Alternatively, if the region R is described by the inequalities

$$a \le x \le b \quad \text{and} \quad r(x) \le y \le s(x) \tag{6.5}$$

as in Fig. 6.3, then

$$\iint_R f(x, y)\, dx\, dy = \iint_R f(x, y)\, dy\, dx = \int_{x=a}^{x=b} \underbrace{\int_{y=r(x)}^{y=s(x)} f(x, y)\, dy}\, dx. \tag{6.6}$$

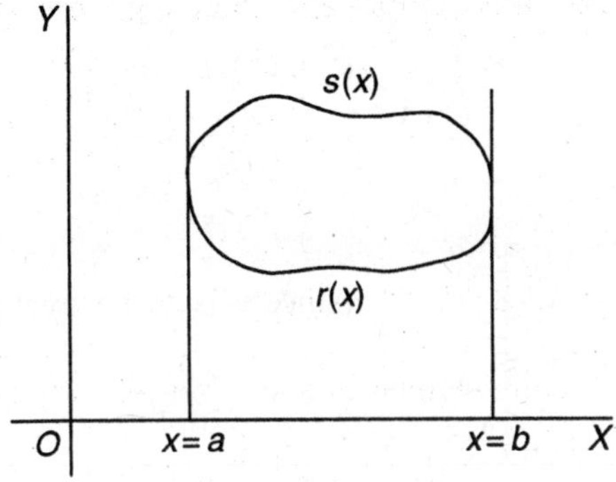

Figure 6.3

The inner integral

$$\int_{y=r(x)}^{y=s(x)} f(x, y)\, dy$$

is now a function of x, say $F(x)$. Hence, (6.6) becomes:

$$\iint_R f(x, y)\, dx\, dy = \iint_R f(x, y)\, dy\, dx = \int_a^b F(x)\, dx, \tag{6.7}$$

which can be easily evaluated.

If the region R is a rectangle bounded by the lines $x = a$, $x = b$, $y = c$, $y = d$, then

$$\iint_R f(x, y)\, dx\, dy = \int_a^b \int_c^d f(x, y)\, dy\, dx = \int_c^d \int_a^b f(x, y)\, dx\, dy, \tag{6.8}$$

from which it is clear that the *order of integration is immaterial for constant limits*.

We thus have two ways of evaluating double integrals and these corresponds to the two orders of integration. To change the order of integration, one should sketch the region R and from it determine the limits of x and y.

The evaluation of double integrals is illustrated by the following examples.

Example 6.1 Evaluate the double integral

$$\int_0^2 \int_1^2 (x^2 + y^2)\, dx\, dy.$$

Here the order of integration is first with respect to x and then with respect to y. We therefore write the double integral as

$$I = \int_0^2 \left[\int_1^2 (x^2 + y^2)\, dx \right] dy.$$

Now,

$$\int_1^2 (x^2 + y^2)\, dx = \left[\frac{x^3}{3} + xy^2 \right]_{x=1}^{x=2}, \qquad \text{(since } y \text{ is to be treated as a constant.)}$$

$$= \frac{1}{3}(8-1) + y^2(2-1)$$

$$= \frac{7}{3} + y^2.$$

Hence,

$$I = \int_0^2 \left(\frac{7}{3} + y^2 \right) dy$$

$$= \frac{7}{3}\left[y \right]_0^2 + \frac{1}{3}\left[y^3 \right]_0^2$$

$$= \frac{7}{3}(2) + \frac{1}{3}(8)$$

$$= \frac{22}{3}.$$

The reader may verify that the double integral

$$\int_1^2 \int_0^2 (x^2 + y^2)\, dy\, dx$$

also yields the same value as above, since the limits are constants in this case and therefore the order of integration is immaterial.

Example 6.2 Evaluate

$$\iint_R xy\, dx\, dy$$

where R is the positive quadrant of the circle $x^2 + y^2 = a^2$.

The order of integration is first with respect to x. Hence we consider an elementary strip parallel to the x-axis, as in Fig. 6.4. It is seen that such a strip extends from the Y-axis, i.e. $x = 0$, to the circle $x^2 + y^2 = a^2$ on which $x = \sqrt{(a^2 - y^2)}$. Hence the limits of x are $x = 0$ to $x = \sqrt{(a^2 - y^2)}$. To cover the entire region, such strips should start with the minimum value of y, i.e. the x-axis to the maximum value of y, i.e. the point where the curve

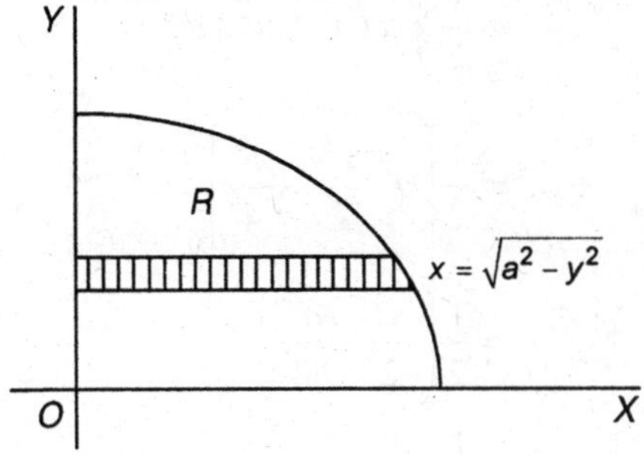

Figure 6.4

intersects the y-axis. The limits of y are therefore from $y = 0$ to $y = a$. The above double integral is therefore written as

$$I = \int_0^a \int_0^{\sqrt{a^2-y^2}} xy \, dx \, dy$$

$$= \int_0^a \left[\frac{x^2}{2} y \right]_0^{\sqrt{a^2-y^2}} dy$$

$$= \frac{1}{2} \int_0^a (a^2 - y^2) y \, dy$$

$$= \frac{1}{2} \left[a^2 \frac{y^2}{2} - \frac{y^4}{4} \right]_0^a$$

$$= \frac{1}{2} \left(\frac{a^4}{2} - \frac{a^4}{4} \right)$$

$$= \frac{1}{8} a^4.$$

Suppose that we wish to change the order to integration and evaluate the double integral

$$\iint_R xy \, dy \, dx.$$

Now, the order to integration is first with respect to y. Hence, we take an elementary strip parallel to the Y-axis, as in Fig. 6.5. The limits for y are, obviously, $y = 0$ to $y = \sqrt{(a^2 - x^2)}$. The minimum and maximum values of x are $x = 0$ to $x = a$ which are therefore the limits for x. Hence the double integral can be written as

$$\int_0^a \int_0^{\sqrt{a^2-x^2}} xy \, dy \, dx = \int_0^a \left[x \frac{y^2}{2} \right]_{y=0}^{y=\sqrt{a^2-x^2}} dx$$

$$= \frac{1}{2}\int_0^a x\,(a^2 - x^2)\,dx$$

$$= \frac{1}{2}\left[a^2\frac{x^2}{2} - \frac{x^4}{4}\right]_0^a$$

$$= \frac{1}{2}\left(\frac{a^4}{2} - \frac{a^4}{4}\right)$$

$$= \frac{1}{8}a^4.$$

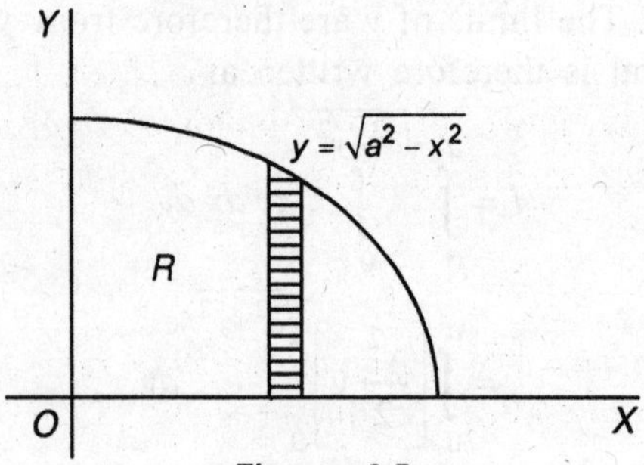

Figure 6.5

Example 6.3 Evaluate the double integral

$$\iint xy\,(x + y)\,dx\,dy$$

over the region bounded by the curves $y = x$ and $y = x^2$.

The region of integration is shown in Fig. 6.6.

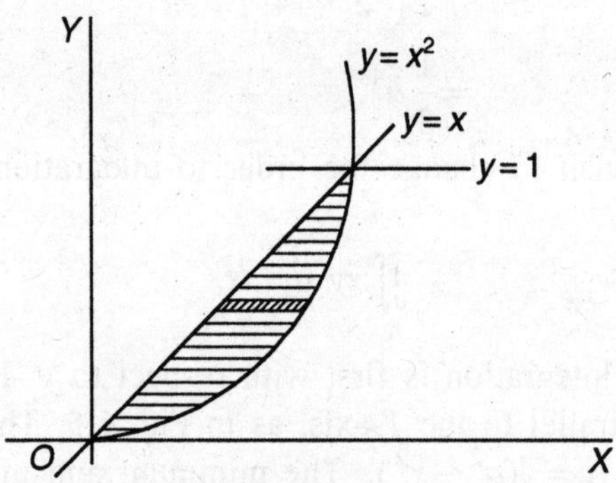

Figure 6.6

We have

$$I = \iint xy\,(x + y)\,dx\,dy.$$

Now the order of integration is first with respect to x and then with respect to y. Hence, we take an elementary strip parallel to the x-axis. The limits for x are easily seen to be from $x = y$ to $x = \sqrt{y}$. The limits for y are then from $y = 0$ to $y = 1$. Hence, we write

$$I = \int_0^1 \int_y^{\sqrt{y}} xy\,(x + y)\,dx\,dy$$

$$= \int_0^1 \left[\frac{x^3}{3} y + \frac{x^2}{2} y^2 \right]_{x=y}^{x=\sqrt{y}} dy$$

$$= \int_0^1 \left(\frac{y^{3/2} y}{3} + \frac{y^3}{2} - \frac{y^4}{3} - \frac{y^4}{2} \right) dy$$

$$= \int_0^1 \left(\frac{y^{5/2}}{3} + \frac{y^3}{2} - \frac{5}{6} y^4 \right) dy$$

$$= \left[\frac{2}{7} \frac{y^{7/2}}{3} + \frac{y^4}{8} - \frac{5}{6} \frac{y^5}{5} \right]_0^1$$

$$= \frac{2}{21} + \frac{1}{8} - \frac{1}{6}$$

$$= \frac{3}{56}.$$

Example 6.4 Change the order of integration and hence evaluate the double integral

$$\int_0^1 \int_x^{2-x} \frac{x}{y}\,dy\,dx.$$

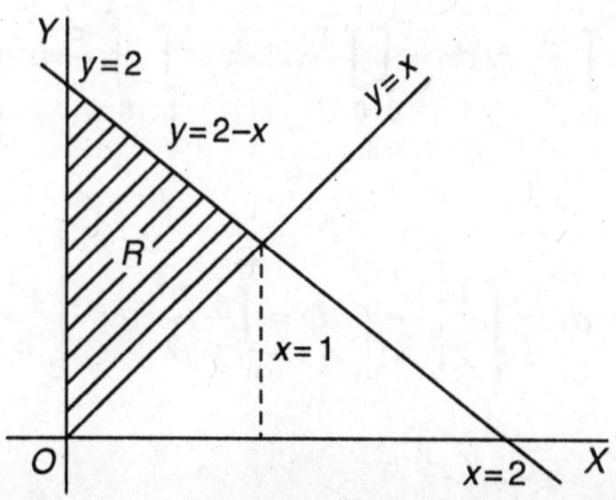

Figure 6.7

The region of integration is shown in Fig. 6.7. If the order of integration is changed, the above integral may be written as

$$\iint_R \frac{x}{y}\, dx\, dy,$$

where the limits are to be determined from the region R. Since the integration is first with respect to x, we take elementary strips parallel to the X-axis. Two such strips are shown in Fig. 6.8, and it is seen that they touch two different lines on the right, viz.

$$y = 2 - x \quad \text{and} \quad y = x.$$

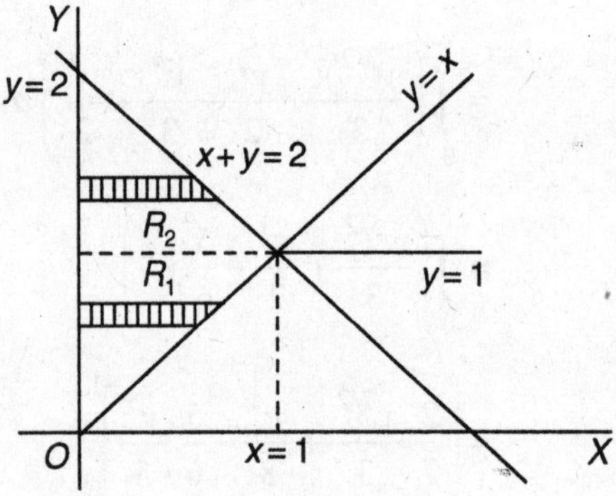

Figure 6.8

To change the order of integration, we therefore divide the integral into two distinct parts corresponding to the two regions R_1 and R_2. Hence we have

$$\int_0^1 \int_x^{2-x} \frac{x}{y}\, dy\, dx = \iint_{R_1} \frac{x}{y}\, dx\, dy + \iint_{R_2} \frac{x}{y}\, dx\, dy,$$

where the limits are obtained from the respective regions. Clearly, R_1 is bounded by $0 \le x \le y$, $0 \le y \le 1$ and R_2 is bounded by $0 \le x \le 2 - y$, $1 \le y \le 2$. Therefore,

$$\int_0^1 \int_x^{2-x} \frac{x}{y}\, dy\, dx = \int_0^1 \int_0^y \frac{x}{y}\, dx\, dy + \int_1^2 \int_0^{2-y} \frac{x}{y}\, dx\, dy. \qquad \text{(i)}$$

Now,

$$\int_0^1 \int_0^y \frac{x}{y}\, dx\, dy = \int_0^1 \frac{1}{y}\left[\frac{x^2}{2}\right]_0^y dy = \int_0^1 \frac{1}{y}\frac{y^2}{2}\, dy = \left[\frac{y^2}{4}\right]_0^1 = \frac{1}{4},$$

and

$$\int_1^2 \int_0^{2-y} \frac{x}{y}\, dx\, dy = \int_1^2 \frac{1}{y}\left[\frac{x^2}{2}\right]_0^{2-y} dy$$

$$= \int_1^2 \frac{1}{2y}(2-y)^2\, dy = \frac{1}{2}\int_1^2 \frac{1}{y}(4+y^2-4y)\, dy$$

$$= \frac{1}{2}\left[4 \log_e y + \frac{y^2}{2} - 4y\right]_1^2$$

$$= \frac{1}{2}\left(4 \log_e 2 + 2 - 8 - 0 - \frac{1}{2} + 4\right)$$

$$= \frac{1}{2}\left(4 \log_e 2 - \frac{5}{2}\right)$$

$$= 2 \log_e 2 - \frac{5}{4}.$$

Hence,

$$\int_0^1 \int_x^{2-x} \frac{x}{y}\, dy\, dx = \frac{1}{4} + 2 \log_e 2 - \frac{5}{4} = 2 \log_e 2 - 1.$$

This example demonstrates the fact that it is sometimes easier to evaluate a double integral by changing the order of integration. The reader may verify this by attempting to evaluate the double integral on the left-hand side of (i).

Example 6.5 Evaluate the double integral

$$I = \int_0^a \int_0^x \frac{f'(y)}{\sqrt{(a-x)(x-y)}}\, dy\, dx.$$

Here the order of integration is first with respect to y and then with respect to x. Clearly, the integration with respect to y is rather complicated. We, therefore, change the order of integration. The region of integration is shown in Fig. 6.9.

To change the order of integration, we first consider an elementary strip parallel to OX as in the figure. Its limits are seen to be from $x = y$ to $x = a$. The limits for y are then from $y = 0$ to $y = a$. Hence, we write

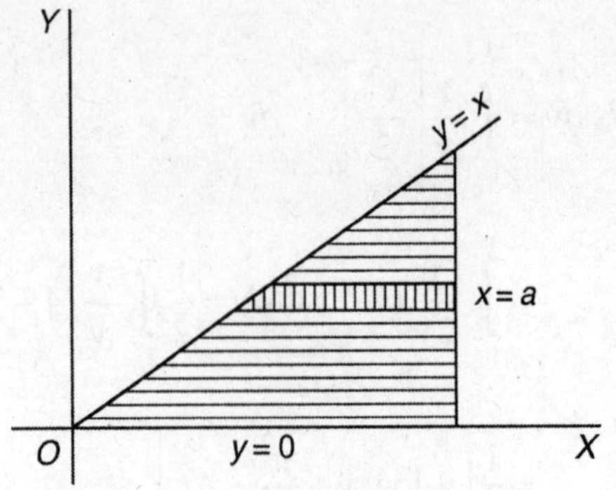

Figure 6.9

$$I = \int_0^a \int_y^a \frac{f'(y)}{\sqrt{(a-x)(x-y)}}\, dx\, dy = \int_0^a f'(y)\, dy \int_y^a \frac{dx}{\sqrt{(a-x)(x-y)}}$$

We first evaluate the integral

$$I_1 = \int_y^a \frac{1}{\sqrt{(a-x)(x-y)}}\, dx = \int_y^a \frac{1}{\sqrt{[(a-y)/2]^2 - [x-(a+y)/2]^2}}\, dx.$$

Using the substitution

$$x - \frac{a+y}{2} = \frac{a-y}{2} \sin\theta,$$

it will be found that the value of this integral is $-\pi$. Hence, we obtain

$$I = \int_0^a f'(y)\, dy\, (-\pi) = -\pi\, [f(a) - f(0)]$$

6.1.2 Double Integrals in Polar Coordinates

Certain types of double integrals may be evaluated more conveniently by changing to polar coordinates. In this case, we subdivide the region R in the following way. Let the region R be bounded by $\theta = \theta_1, \theta = \theta_2$ and the curves $r = a$ and $r = b$, as in Fig. 6.10.

Let the radial interval (a, b) be divided into m parts by concentric circular arcs at intervals of Δr, and, similarly, let the angular interval (θ_1, θ_2) be divided into n parts by drawing radial lines at intervals of $\Delta\theta$, so that

$$\Delta r = \frac{b-a}{m} \quad \text{and} \quad \Delta\theta = \frac{\beta - \alpha}{n}. \tag{6.9}$$

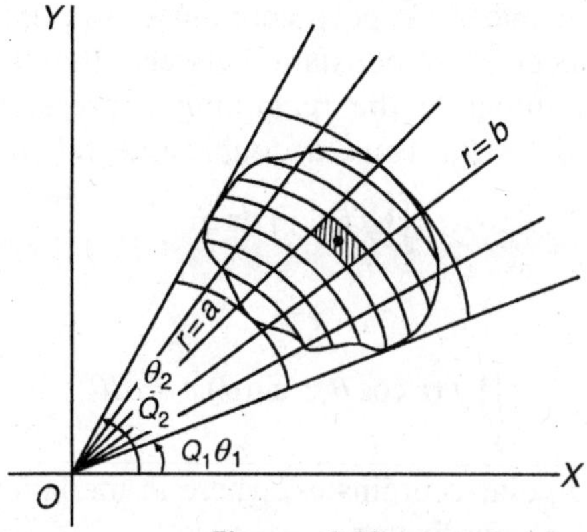

Figure 6.10

We have thus divided the range into infinitesimal curvilinear rectangles and partial rectangles. But the total area of the partial rectangles will be small provided that Δr and $\Delta\theta$ are sufficiently small.

Let (r_i, θ_i) be the coordinates of the centre of a typical element so that area ΔA_i is given by

$$\Delta A_i = \frac{1}{2}\left(r_i + \frac{1}{2}\Delta r\right)^2 \Delta\theta - \frac{1}{2}\left(r_i - \frac{1}{2}\Delta r\right)^2 \Delta\theta.$$

$$= \frac{1}{2}(2r_i \,\Delta r\, \Delta\theta)$$

$$= r_i \,\Delta r\, \Delta\theta. \tag{6.10}$$

We evaluate the function, $f(r, \theta)$ say, at the centre of each of these elements and from the sum of the products of these functional values and the corresponding values of ΔA_i. We then obtain

$$S = \sum_{i=1}^{N} f(r_i, \theta_i)\Delta A_i = \sum_{i=1}^{N} f(r_i, \theta_i)\, r_i \,\Delta r\, \Delta\theta, \tag{6.11}$$

where N is the total number of elements.

Then, $\lim_{N\to\infty} \Sigma_{i=1}^{N} f(r_i, \theta_i)\, r_i \,\Delta r\, \Delta\theta$ is defined as the *double integral* of f over R and we write

$$\lim_{N\to\infty} \sum_{i=1}^{N} f(r_i, \theta_i)\, r_i \,\Delta r\, \Delta\theta = \iint_R f(r, \theta)\, r\, dr\, d\theta$$

$$= \int_{\theta=\theta_1}^{\theta=\theta_2} \int_{r=f_1(\theta)}^{r=f_2(\theta)} f(r, \theta) r\, dr\, d\theta \tag{6.12}$$

To evaluate the double integral in polar coordinates, we first integrate $f(r, \theta)r$ with respect to r keeping θ constant between the limits $r = f_1(\theta)$ and $r = f_2(\theta)$, and then integrate the remaining expression with respect to θ between $\theta = \theta_1$, to $\theta = \theta_2$. Thus, a double integral in Cartesian form

$$\iint_R f(x, y)\, dx\, dy$$

becomes

$$\iint_R f(r\cos\theta, r\sin\theta)\, r\, dr\, d\theta$$

when transformed to polar coordinates, where in the latter integral R should be expressed in polar coordinates.

The general problem of change in variables will be considered later in this chapter. The following examples demonstrate the usefulness of the polar coordinates in the evaluation of double integrals.

Example 6.6 Evaluate

$$\iint_R xy\, dx\, dy$$

where R is the positive quadrant of the circle $x^2 + y^2 = a^2$ (Fig. 6.11).

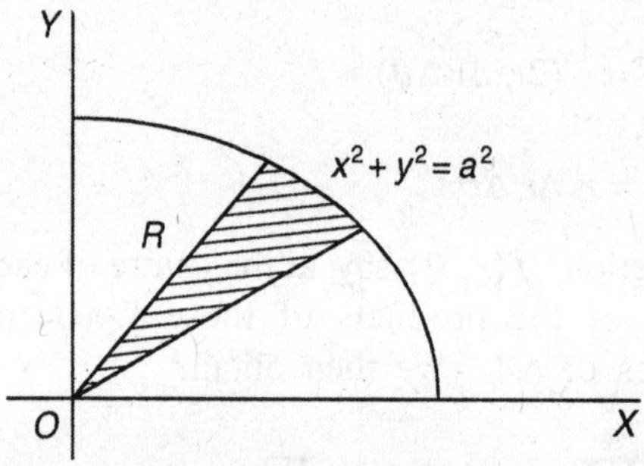

Figure 6.11

Changing to polars, the region R becomes $r = a$ from $\theta = 0$ to $\theta = \pi/2$. Hence, the given integral becomes:

$$\int_0^{\pi/2}\int_0^{a} (r\cos\theta)(r\sin\theta)\, r\, dr\, d\theta = \int_0^{\pi/2}\int_0^{a} r^3\, dr \cos\theta\sin\theta\, d\theta$$

$$= \left[\frac{r^4}{4}\right]_0^a \left[\frac{\sin^2\theta}{2}\right]_0^{\pi/2}$$

$$= \frac{1}{2}\,\frac{a^4}{4}$$

$$= \frac{a^4}{8}.$$

Example 6.7 Evaluate

$$\int_0^{2a} \int_0^{\sqrt{2ax-x^2}} (x^2 + y^2)\, dx\, dy$$

by changing into polar coordinates.

The region of integration is the upper half of the circle $x^2 + y^2 = 2ax$, (Fig. 6.12). If we change to polars, this region becomes $r = 2a \cos\theta$ from $\theta = 0$ to $\theta = \pi/2$.

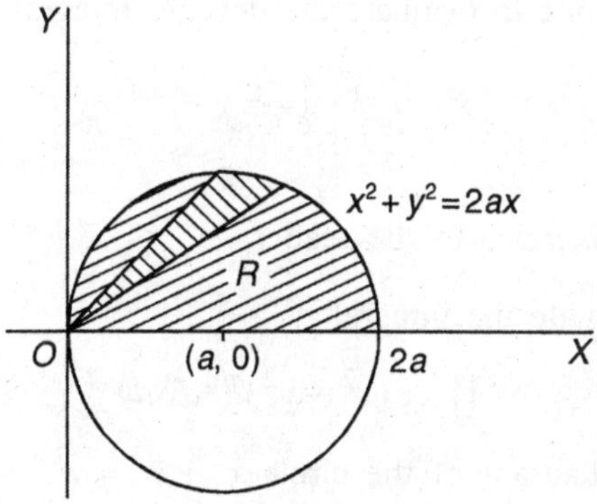

Figure 6.12

Hence the given integral becomes:

$$\int_0^{\pi/2} \int_0^{2a\cos\theta} r^2 (r\, dr)\, d\theta = \int_0^{\pi/2} \int_0^{2a\cos\theta} r^3\, dr\, d\theta$$

$$= \int_0^{\pi/2} \left[\frac{r^4}{4}\right]_0^{2a\cos\theta} dx$$

$$= \frac{1}{4} \int_0^{\pi/2} 16a^4 \cos^4\theta\, d\theta$$

$$= 4a^4 \int_0^{\pi/2} \cos^4\theta\, d\theta = 4a^4 \frac{3}{16} \pi$$

$$= \frac{3}{4} a^4 \pi.$$

Example 6.8 Evaluate

$$\int_0^{\infty}\int_0^{\infty} e^{-(x^2+y^2)}\,dx\,dy,$$

by changing into polar coordinates.

The region of integration here is clearly, the first quadrant of the *XY*-plane. If we use polar coordinates, the integral becomes:

$$\int_0^{\pi/2}\int_0^{\infty} e^{-r^2}(r\,dr)\,d\theta = \frac{\pi}{2}\left[\frac{e^{-R}}{-2}\right]_0^{\infty} = -\frac{\pi}{4}(0-1) = \frac{\pi}{4}.$$

This result enables one to evaluate the definite integral

$$I = \int_0^{\infty} e^{-x^2}\,dx.$$

This is left as an exercise to the reader.

Example 6.9 Evaluate the integral

$$\iint xy\,(x^2+y^2)^{n/2}\,dx\,dy$$

over the positive quadrant of the circle $x^2+y^2=a^2$, supposing $n+3>0$.

Refer to Fig. 6.13. Changing into polars, the region becomes $r=a$ from $\theta=0$ to $\theta=\pi/2$. Hence the given integral becomes:

$$\int_0^{\pi/2}\int_0^{a} r\cos\theta\,(r\sin\theta)\,r^n(r\,dr)\,d\theta = \int_0^{\pi/2}\int_0^{a} r^{n+3}\cos\theta\sin\theta\,d\theta.$$

$$= \int_0^{\pi/2} \sin\theta\cos\theta\,d\theta \int_0^{a} r^{n+3}\,dr$$

$$= \left[\frac{\sin^2\theta}{2}\right]_0^{\pi/2}\left[\frac{r^{n+4}}{n+4}\right]_0^{a}, \quad \text{since } n+3>0.$$

$$= \frac{1}{2}\frac{a^{n+4}}{n+4}$$

$$= \frac{a^{n+4}}{2(n+4)}.$$

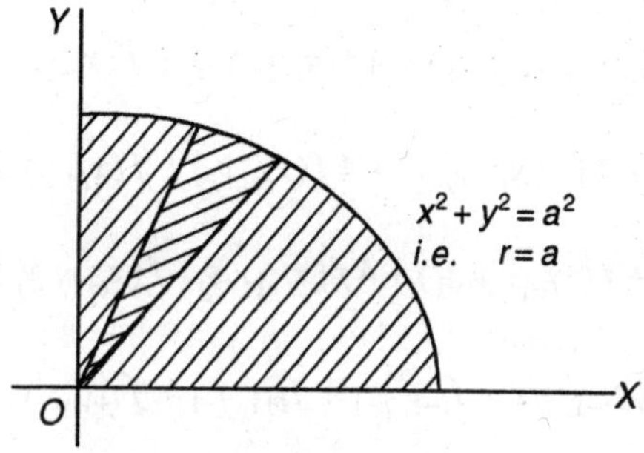

Figure 6.13

6.1.3 Numerical Double Integration

Formulae for the evaluation of a double integral can be obtained by repeatedly applying the trapezoidal and Simpson's rules derived in Section 4.1.3. We consider, as an example, the double integral defined by

$$I = \int_{y_j}^{y_{j+1}} \int_{x_i}^{x_{i+1}} f(x, y)\, dx\, dy \tag{6.13}$$

where $x_{i+1} = x_i + h$ and $y_{j+1} = y_j + k$. By the repeated application of the trapezoidal rule to (6.13), we get

$$I = \frac{h}{2} \int_{y_j}^{y_{j+1}} [f(x_i, y) + f(x_{i+1}, y)]\, dy$$

$$= \frac{hk}{4}\left[f(x_i, y_j) + f(x_{i+1}, y_j) + f(x_i, y_{j+1}) + f(x_{i+1}, y_{j+1})\right]$$

$$= \frac{hk}{4}(f_{i,j} + f_{i+1,j} + f_{i,j+1} + f_{i+1,j+1}) \tag{6.14}$$

where $f_{i,j} = f(x_i, y_j)$, etc.

Similarly, applying Simpson's rule to the integral

$$I = \int_{y_{j-1}}^{y_{j+1}} \int_{x_{i-1}}^{x_{i+1}} f(x, y)\, dx\, dy, \tag{6.15}$$

We obtain,

$$I = \frac{h}{3} \int_{y_{j-1}}^{y_{j+1}} [f(x_{i-1}, y) + 4f(x_i, y) + f(x_{i+1}, y)]\, dy$$

$$= \frac{hk}{9}\{f(x_{i-1}, y_{j-1}) + 4f(x_{i-1}, y_j) + f(x_{i-1}, y_{j+1})$$

$$+ 4[f(x_i, y_{j-1}) + 4f(x_i, y_j) + f(x_i, y_{j+1})]$$

$$+ f(x_{i+1}, y_{j-1}) + 4f(x_{i+1}, y_j) f(x_{i+1}, y_{j+1})\}$$

$$= \frac{hk}{9}[f_{i-1,\, j-1} + f_{i-1,\, j+1} + f_{i+1,\, j-1} + f_{i+1,\, j+1}$$

$$+ 4(f_{i-1,\, j} + f_{i,j-1} + f_{i,j+1} + f_{i+1,\, j}) + 16 f_{i,\, j}] \tag{6.16}$$

The following numerical example demonstrates the use of trapezoidal and Simpson's rules in evaluating a double integral.

Example 6.10 Evaluate

$$I = \int_0^1 \int_0^1 e^{x+y}\, dx\, dy,$$

using the trapezoidal and Simpson's rules.

With $h = k = 0.5$, the following table of values of e^{x+y} is prepared.

		x		
		0	0.5	1.0
	0	1	1.6487	2.7183
y	0.5	1.6487	2.7183	4.4817
	1.0	2.7183	4.4817	7.3891

(i) Using the trapezoidal rule (6.14) repeatedly, we obtain:

$$I = \frac{0.25}{4}[1.0 + 4(1.6487) + 6(2.7183) + 4(4.4817) + 7.3891]$$

$$= \frac{12.3050}{4}$$

$$= 3.0762.$$

(ii) If we use Simpson's rule (6.16) repeatedly, we have

$$I = \frac{0.25}{9}[1.0 + 2.7183 + 7.3891 + 2.7183$$

$$+4(1.6487+4.4817+4.4817+1.6487)+16(2.7183)]$$

$$=\frac{26.59042}{9}$$

$$=2.9545.$$

The exact value of the double integral is 2.9525 and therefore the result obtained by using Simpson's rule is about sixty times more accurate than that obtained by the trapezoidal rule.

EXERCISES 6.1

Evaluate the following double integrals (Problems 1–15):

1. $\int_0^1\int_1^2 (x^2+y^2)\,dx\,dy.$

2. $\int_0^3\int_1^2 xy\,(x+y)\,dx\,dy.$

3. $\int_0^a\int_0^b (x^2+y^2)\,dx\,dy.$

4. $\int_1^2\int_3^4 \frac{1}{(x+y)^2}\,dx\,dy.$

5. $\int_0^1\int_x^1 \frac{y^2\,dy}{\sqrt{x^2+y^2}}.$

6. $\int_1^4\int_0^{\sqrt{4-x}} xy\,dy\,dx.$

7. $\int_1^2\int_x^{x\sqrt{3}} xy\,dx\,dy.$

8. $\int_1^2\int_1^x xy^2\,dy\,dx.$

9. $\int_0^{\pi/4}\int_0^{\pi/2} \sin(x+y)\,dx\,dy.$

10. $\int_0^a\int_0^{\sqrt{a^2-x^2}} y^3\,dy\,dx$

11. $\int_0^1\int_{\sqrt{y}}^{2-y} x^2\,dx\,dy.$

12. $\int_0^2\int_{x^2}^{2x} (2x+3y)\,dy\,dx.$

13. $\int_0^a\int_0^{\sqrt{a^2-x^2}} xy\,dy\,dx.$

14. $\int_0^1\int_0^{1-x} (x^2+y^2)\,dy\,dx.$

15. $\int_0^a\int_{x^2/a}^{2a-x} xy\,dy\,dx.$

16. Sketch the region of integration when it is bounded by
    (a) the circle $x^2+y^2=a^2$
    (b) the cardioid $r=a(1+\cos\theta)$

(c) the X-axis and the parabola $y = 1 - x^2$
(d) $x = 2, x = 3, y = -1$ and $y = 5$
(e) the line $x = y$ and the parabola $y = x^2$
(f) $y = 0, y = x$ and $x = 1$.

17. Find the value of

$$\iint xy \, dx \, dy$$

taken over the positive quadrant of the ellipse

$$\frac{x^2}{a^2} + \frac{y^2}{b^2} = 1.$$

18. Evaluate

$$\iint (x^2 + y^2) \, dx \, dy$$

over the region bounded by $x = 0$, $y = 0$ and $x + y = 1$.

19. Find the value of

$$\iint xy\,(x + y) \, dx \, dy$$

taken over the region enclosed by the curves $y = x$ and $y = x^2$.

20. Evaluate

$$\iint (x - y) \, dx \, dy$$

over the region between the line $x = y$ and the parabola $y = x^2$.

21. Evaluate

$$\iint_R (x + y) \, dx \, dy$$

where R is the triangular region bounded by $y = 2x$, $y = x/2$ and $y = 3 - x$.

22. Evaluate

$$\iint_R xy \, dx \, dy$$

where R is the triangular region with vertices $A(-6, 2)$, $B(-1, 3)$ and $C(9, -7)$.

23. Find

$$\iint_R xy \, dx \, dy$$

where R is the region bounded by $x = 1$, $x = 2$, $y = 0$ and $xy = 1$.

Change to polar coordinates and then evaluate the following double integrals (Problems 24–30):

24. $\iint (x^2 + y^2)\, dx\, dy$ over the circle $x^2 + y^2 = a^2$.

25. $\int_0^a \int_0^{\sqrt{a^2-x^2}} \sqrt{a^2 - x^2 - y^2}\, dy\, dx.$

26. $\int_0^2 \int_0^{\sqrt{2x-x^2}} \dfrac{x}{x^2 + y^2}\, dy\, dx.$

27. $\int_0^\infty \int_0^\infty \dfrac{1}{(x^2 + y^2 + a^2)^2}\, dx\, dy.$

28. $\int_0^{2a} \int_0^{\sqrt{2ax-x^2}} (x^2 + y^2)\, dy\, dx.$

29. $\iint_R (4 - x^2 - y^2)\, dx\, dy$ over the circular region $x^2 + y^2 \le 4$.

30. $\int_0^a \int_y^a \dfrac{x\, dy\, dx}{x^2 + y^2}.$

Change the order of integration in the following integrals and hence, or otherwise, evaluate them (Problems 31–44):

31. $\int_1^2 \int_3^4 (x^2 + y^2)\, dx\, dy.$

32. $\int_0^1 \int_{x^3}^{\sqrt{x}} dy\, dx.$

33. $\int_0^a \int_{x^2/a}^{2a-x} xy\, dy\, dx.$

34. $\int_0^\infty \int_x^\infty \dfrac{e^{-y}}{y}\, dy\, dx.$

35. $\int_0^{4a} \int_{x^2/(4a)}^{2\sqrt{ax}} dy\, dx.$

36. $\int_0^a \int_0^x \dfrac{f'(y)\, dy\, dx}{\sqrt{(a-x)(x-y)}}.$

37. $\int_0^{2a} \int_{y^2/(4a)}^{3a-y} (x^2 + y^2)\, dx\, dy.$

38. $\int_0^c \int_x^c \dfrac{x}{\sqrt{x^2 + y^2}}\, dy\, dx.$

39. $\int_0^a \int_0^x \dfrac{1}{\sqrt{(a-x)(x-y)(a-y)(a+y)}}\, dy\, dx.$

40. $\int_0^1 \int_{-\sqrt{y}}^{\sqrt{y}} dx\, dy + \int_1^9 \int_{(y-3)/2}^{\sqrt{y}} dx\, dy.$

41. $\int_0^{\pi}\int_0^{\sin x} x^2 y\, dy\, dx.$ **42.** $\int_{-2}^{1}\int_{x^2+4x}^{3x+2} dy\, dx.$

43. $\int_0^{2}\int_0^{x^2} (\sin xy)\, dy\, dx.$

44. Evaluate the double integral

$$I = \int_4^{4.2}\int_2^{2.3} \frac{1}{xy}\, dx\, dy.$$

using the trapezoidal and Simpson's rules and compare the results with the exact value.

6.2 APPLICATIONS OF DOUBLE INTEGRALS

In this section, we shall discuss some important applications of double integrals which occur quite often in science and engineering. These include problems involving area, volume, mass, centroid, centre of mass, moment of inertia, etc. The simplest application is that of finding the area of a region of the xy-plane and this is considered first.

6.2.1 Area by Double Integration

We wish to find the area enclosed between $x = a$, $x = b$, $y = f(x)$ and $y = F(x)$. Let the area be divided into rectangular elements of the type $PQRS$ (Fig. 6.14) of area $\Delta x\, \Delta y$ where $P \equiv (x, y)$ and $R \equiv (x, +\Delta x, y + \Delta y)$.

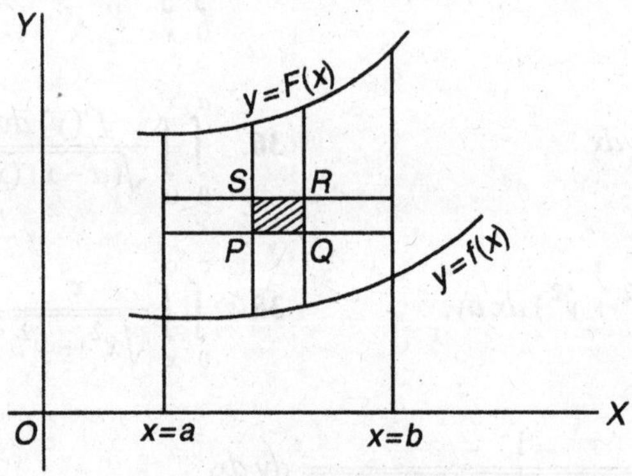

Figure 6.14

If this element is moved along the vertical strip from $y = f(x)$ to $y = F(x)$, we obtain

$$\Delta x \lim_{\Delta y \to 0} \sum_{y=f(x)}^{y=f(x)} \Delta y,$$

since Δx is a constant. This can be written as

$$\Delta x \int_{f(x)}^{F(x)} dy,$$

which represents the area of the vertical strip. Adding up all such strips between $x = a$ and $x = b$, we get the required area as

$$\lim_{\Delta x \to 0} \sum_{a}^{b} \Delta x \int_{f(x)}^{F(x)} dy = \int_{a}^{b} dx \int_{f(x)}^{F(x)} dy = \int_{a}^{b} \int_{f(x)}^{F(x)} dx\, dy. \tag{6.17}$$

In a similar way, the area A, in polar coordinates, becomes:

$$A = \iint r\, dr\, d\theta, \tag{6.18}$$

over the given region R.

Example 6.11 Find the area bounded by the ellipse $(x^2/a^2) + (y^2 + b^2) = 1$.

Because of symmetry, we consider the area in the first quadrant and write

$$A = 4 \int_{0}^{b} \int_{0}^{a\sqrt{1-y^2/b^2}} dx\, dy$$

$$= 4 \int_{0}^{b} a \sqrt{1 - \frac{y^2}{b^2}}\, dy$$

$$= 4a \int_{0}^{b} \sqrt{1 - \frac{y^2}{b^2}}\, dy.$$

Setting $y = b \sin\theta$, we obtain $dy = b\cos\theta\, d\theta$. Then,

$$A = 4a \int_{0}^{\pi/2} \cos\theta\, (b\cos\theta)\, d\theta$$

$$= 4ab \int_{0}^{\pi/2} \frac{1 + \cos 2\theta}{2}\, d\theta$$

$$= 2ab \times \frac{\pi}{2}$$

$$= \pi ab.$$

Example 6.12 Find by double integration, the area which lies inside the cardioid $r = a(1 + \cos\theta)$ and outside the circle $r = a$.

The area is shown in Fig. 6.15. Because of symmetry, we need consider only the portion of the area above the X-axis. We therefore write

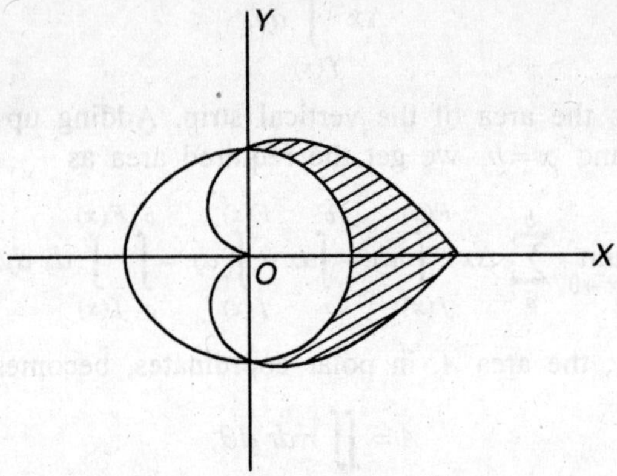

Figure 6.15

$$A = 2 \int_0^{\pi/2} \int_{r=a}^{a(1+\cos\theta)} r\, dr\, d\theta$$

$$= 2 \int_0^{\pi/2} \left[\frac{1}{2} r^2\right]_{r=a}^{r=a(1+\cos\theta)} d\theta$$

$$= a^2 \int_0^{\pi/2} \left[(1+\cos\theta)^2 - 1\right] d\theta$$

$$= a^2 \int_0^{\pi/2} (2\cos\theta + \cos^2\theta)\, d\theta$$

$$= a^2 \int_0^{\pi/2} \left(2\cos\theta + \frac{1+\cos 2\theta}{2}\right) d\theta$$

$$= a^2 \left[2\sin\theta + \frac{1}{2}\left(\theta + \frac{\sin 2\theta}{2}\right)\right]_0^{\pi/2}$$

$$= a^2 \left[2 + \frac{1}{2}\left(\frac{\pi}{2} + 0\right)\right]$$

$$= a^2 \left(2 + \frac{\pi}{4}\right).$$

Example 6.13 Find the area in the *XY*-plane bounded by the lemniscate $r^2 = a^2 \cos 2\theta$.

The area is shown in Fig. 6.16. By symmetry, the required area is

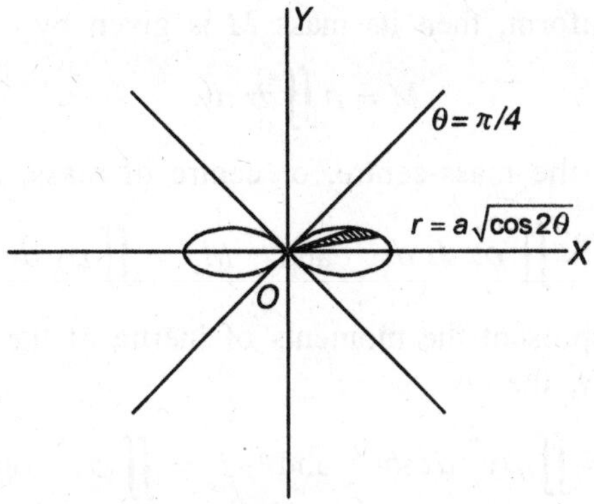

Figure 6.16

$$A = 4 \int_0^{\pi/4} \int_0^{a\sqrt{\cos 2\theta}} r\, dr\, d\theta$$

$$= 4 \int_0^{\pi/4} \left[\frac{r^2}{2}\right]_0^{a\sqrt{\cos 2\theta}}$$

$$= 2 \int_0^{\pi/4} a^2 \cos 2\theta\, d\theta$$

$$= 2a^2 \left[\frac{\sin 2\theta}{2}\right]_0^{\pi/4}$$

$$= a^2.$$

6.2.2 Moments and Centroid of Area

The coordinates of the centroid of the area are given by

$$\bar{x} = \frac{\iint_R x\, dx\, dy}{\iint_R dx\, dy}, \qquad \bar{y} = \frac{\iint_R y\, dx\, dy}{\iint_R dx\, dy}, \tag{6.19}$$

where R is the region of integration.

Instead of the plane area, if we have a lamina of surface density $\rho = f(x, y)$ then the mass of this lamina is given by

$$\iint \rho\, dx\, dy = \iint f(x, y)\, dx\, dy.$$

If the lamina is uniform, then its mass M is given by

$$M = \rho \iint dx\, dy.$$

The coordinates of the mass-centre, or centre of mass, are now given by

$$M\overline{x} = \iint \rho x\, dx\, dy \quad \text{and} \quad M\overline{y} = \iint \rho y\, dx\, dy. \tag{6.20}$$

If I_x and I_y represent the moments of inertia of the lamina about OX and OY respectively, then

$$I_x = \iint \rho y^2\, dx\, dy \quad \text{and} \quad I_y = \iint \rho x^2\, dx\, dy \tag{6.21}$$

If OZ is perpendicular to both OX and OY, then the moment of inertia I_0 of the lamina about OZ is given by

$$I_0 = \iint \rho r^2\, dx\, dy = \iint \rho (x^2 + y^2)\, dx\, dy = I_x + I_y, \tag{6.22}$$

which is called the *perpendicular axes* theorem.

Example 6.14 Find $\overline{x}, \overline{y}, I_x, I_y$ and I_0 for the area bounded by the circle $x^2 + y^2 = a^2$ in the first quadrant. Assume $\rho(x, y) = 1$.

We have

$$A = \iint dx\, dy = \int_0^{\pi/2} \int_0^a r\, dr\, d\theta = \frac{1}{4}\pi a^2.$$

Hence,

$$A\overline{x} = \iint x\, dA = \int_0^{\pi/2} \int_0^a (r \cos\theta)\, r\, dr\, d\theta$$

$$= \int_0^{\pi/2} \cos\theta\, d\theta \left(\frac{a^3}{3}\right)$$

$$= \frac{1}{3}a^3,$$

and so,

$$\overline{x} = \frac{(1/3)\, a^3}{(1/4)\, \pi a^2} = \frac{4a}{3\pi}.$$

By symmetry, it follows that

$$\bar{y} = \frac{4a}{3\pi}.$$

Also,

$$I_x = \iint y^2\,dx\,dy = \int_0^{\pi/2}\int_0^a (r^2 \sin^2\theta)\,r\,dr\,d\theta$$

$$= \frac{1}{4}a^4 \int_0^{\pi/2} \frac{1-\cos 2\theta}{2}\,d\theta$$

$$= \frac{1}{4}a^4\,\frac{1}{2}\,\frac{\pi}{2}$$

$$= \frac{1}{16}\pi a^4.$$

It follows, similarly, that

$$I_y = \frac{1}{16}\pi a^4.$$

More easily, we obtain:

$$I_x = I_y = \frac{1}{2}I_0 = \frac{1}{2}\iint_R (x^2 + y^2)\,dx\,dy$$

$$= \frac{1}{2}\int_0^{\pi/2}\int_0^a (r^2)\,r\,dr\,d\theta$$

$$= \frac{1}{2}\,\frac{a^4}{4}\,\frac{\pi}{2}$$

$$= \frac{1}{16}\pi a^4.$$

Example 6.15 A plane lamina of nonuniform density is in the form of a quadrant of the ellipse $(x^2/a^2) + (y^2/b^2) = 1$. If the density at any point (x, y) be kxy, where k is a constant, find the coordinates of the centroid of the lamina.

If $(\bar{x}, \bar{y})$ are the coordinates of the centroid, then

$$\bar{x} = \frac{\int_0^a \int_0^{b\sqrt{1-x^2/a^2}} k\, xy\, dx\, dy\,(x)}{\int_0^a \int_0^{b\sqrt{1-x^2/a^2}} k\, xy\, dx\, dy} \qquad \text{(i)}$$

and

$$\bar{y} = \frac{\int_0^a \int_0^{b\sqrt{1-x^2/a^2}} k\, xy\, dx\, dy\,(y)}{\int_0^a \int_0^{b\sqrt{1-x^2/a^2}} k\, xy\, dx\, dy} \qquad \text{(ii)}$$

In (i),

$$\text{The numerator} = \int_0^a \int_0^{b\sqrt{1-x^2/a^2}} k\, x^2 y\, dx\, dy$$

$$= k \int_0^a \int_0^{b\sqrt{1-x^2/a^2}} x^2 y\, dy\, dx$$

$$= k \int_0^a x^2 \frac{1}{2} b^2 \left(1 - \frac{x^2}{a^2}\right) dx$$

$$= \frac{k}{2} \frac{b^2}{a^2} \int_0^a x^2 (a^2 - x^2)\, dx$$

$$= \frac{kb^2}{2a^2} \left[a^2 \frac{x^3}{3} - \frac{x^5}{5} \right]_0^a$$

$$= \frac{kb^2}{2a^2} \left(\frac{a^5}{3} - \frac{a^5}{5} \right)$$

$$= \frac{k}{15} b^2 a^3,$$

and

$$\text{the denominator} = \int_0^a \int_0^{b\sqrt{1-x^2/a^2}} k\, xy\, dx\, dy$$

$$= k \int_0^a x \left[\frac{y^2}{2} \right]_0^{b\sqrt{1-x^2/a^2}} dx$$

$$= \frac{k}{2} \int_0^a xb^2 \left(1 - \frac{x^2}{a^2} \right) dx$$

$$= \frac{kb^2}{2} \left[\frac{x^2}{2} - \frac{1}{a^2} \frac{x^4}{4} \right]_0^a$$

$$= \frac{kb^2}{2} \left(\frac{a^2}{2} - \frac{a^2}{4} \right)$$

$$= \frac{k}{8} b^2 a^2.$$

Hence,

$$\bar{x} = \frac{8}{15} a,$$

and, by symmetry,

$$\bar{y} = \frac{8}{15} b.$$

6.2.3 Volumes of Solids

Volume of a solid of revolution

We consider the area bounded by $y = f(x)$, the ordinates $x = a$ and $x = b$, and the X-axis. Let this area revolves about the X-axis. To find the volume of the solid of revolution which is generated, we consider an elementary area $PQRS$ where $P \equiv (x, y)$ and $Q \equiv (x + \Delta x, y + \Delta y)$. The area of the rectangle is $\Delta x \Delta y$. When revolved about X-axis, the elementary area generates a solid whose volume is $\pi [(y + \Delta y)^2 - y^2] \Delta x$. Refer to Fig. 6.17. That is

$$\Delta V = \pi [(y + \Delta y)^2 - y^2] \Delta x$$

$$= \pi [2y\, \Delta y\, \Delta x + (\Delta y)^2 \Delta x]$$

$$= 2\pi y \Delta y\, \Delta x,$$

to the first order of infinitesimals. Hence the total volume is given by

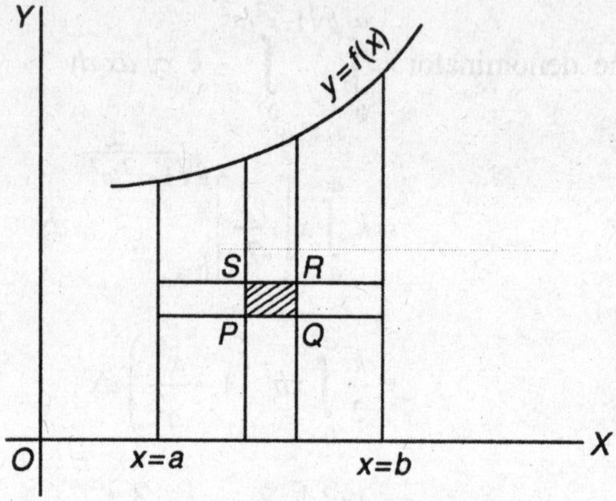

Figure 6.17

$$V = \lim_{\Delta x \to 0} \sum_{x=a}^{x=b} \sum_{y=0}^{y=f(x)} 2\pi y \,\Delta y \,\Delta x$$

$$= 2\pi \int_{x=a}^{x=b} \int_{y=0}^{y=f(x)} y \, dy \, dx. \qquad (6.23)$$

Volumes of solids as double integrals

Let S be the portion of the surface $z = f(x, y)$ and S' be its orthogonal projection on the xy-plane (Fig. 6.18). We wish to show that the volume of the solid bounded by $z = f(x, y)$ and the surrounding cylindrical surface can be expressed as a double integral.

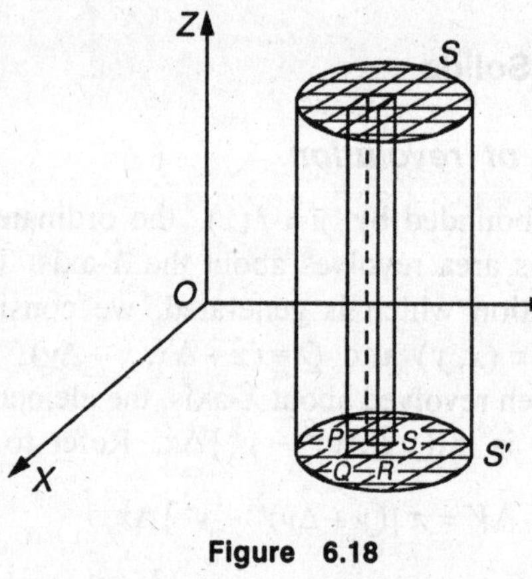

Figure 6.18

Let S' be divided into elementary rectangles by drawing lines parallel to OX and OY. Let $PQRS$ (of area $\Delta x\, \Delta y$) be a typical elementary area. On each of these rectangles as base, let a prism be constructed with its length parallel to the Z-axis. The volume of the typical prism shown in Fig. 6.18 is given by

$$\Delta V = z\Delta x\, \Delta y = f(x, y)\, \Delta x\, \Delta y.$$

Hence, the total volume is given by

$$V = \iint_{S'} f(x, y)\, dx\, dy, \tag{6.24}$$

where S' is the region of integration.

Example 6.16 Find the volume enclosed by the coordinate planes and that portion of the plane $x + y + z = 1$, which lies in the first quadrant.

The required volume is given by

$$V = \iint_R z\, dx\, dy$$

where $z = 1 - x - y$ and R is the region bounded by $x = 0$, $y = 0$, and $x + y = 1$ (see Fig. 6.19). Hence, we have:

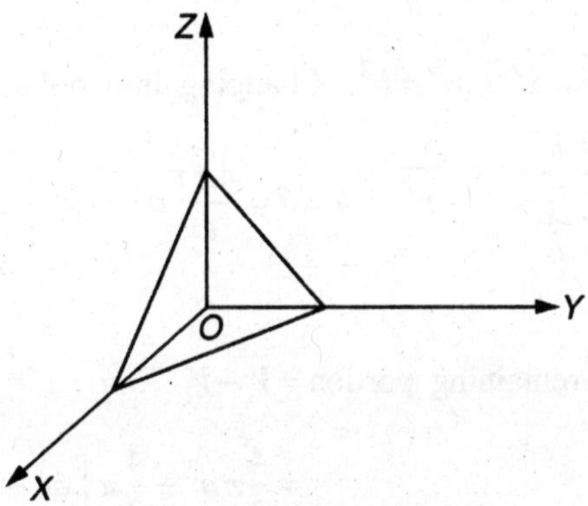

Figure 6.19

$$V = \int_0^1 \int_0^{1-y} (1 - x - y)\, dx\, dy$$

$$= \int_0^1 \left[(1 - y)\, x - \frac{x^2}{2} \right]_{x=0}^{x=1-y} dy$$

$$= \int_0^1 \left[(1 - y)^2 - \frac{(1 - y)^2}{2} \right] dy$$

$$= \frac{1}{2} \int_0^1 (1-y)^2 \, dy = \frac{1}{6}.$$

Example 6.17 **A circular hole of radius b is made centrally through a sphere of radius a. Find the volume of the remaining portion of the sphere.**

Let the centre of the sphere be at the origin and let the axis of the hole be along the z-axis. The volume V of the sphere is $(4/3)\pi a^3$ and that of the circular hole is obtained as follows:

$$\text{Volume of the upper half of the hole} = \iint_R f(x, y) \, dx \, dy$$

$$= \iint_R z \, dx \, dy$$

where z is obtained from the equation $x^2 + y^2 + z^2 = a^2$ and R is the circle in the XY-plane, i.e. $x^2 + y^2 = b^2$. Hence, the volume V_1 of the circular hole is

$$V_1 = 2 \iint_R \sqrt{a^2 - x^2 - y^2} \, dx \, dy,$$

where R is given by $x^2 + y^2 = b^2$. Changing into polars, we obtain,

$$V_1 = 2 \int_0^{2\pi} \int_0^b \sqrt{a^2 - r^2} \, r \, dr \, d\theta = \frac{4\pi}{3} \left[a^3 - (a^2 - b^2)^{3/2} \right].$$

Hence

$$\text{Volume of the remaining portion} = V - V_1$$

$$= \frac{4}{3}\pi a^3 - \frac{4}{3}\pi \left[a^3 - (a^2 - b^2)^{3/2} \right]$$

$$= \frac{4}{3}\pi (a^2 - b^2)^{3/2}.$$

EXERCISES 6.2

1. Find the total area inside the lemniscate $r^2 = a^2 \cos 2\theta$.
2. Find the area which is inside the circle $r = 3a \cos\theta$ and outside the cardioid $r = a(1 + \cos\theta)$.
3. Find the area bounded by the cardioid $r = a(1 + \cos\theta)$.
4. Find the area which is inside the circle $r = 2a \cos\theta$ and outside the circle $r = a$.

5. Find the area common to the cardioids $r = a(1 + \cos\theta)$ and $r = a(1 - \cos\theta)$.
6. A portion of the cone $x^2 + y^2 = z^2$, $z \geq 0$ lies inside the circular cylinder $x^2 + y^2 = 2y$. Find the area of this portion.
7. Find the coordinates of the centroid of the area bounded by $x + y = a$, $x = 0$, $y = 0$.
8. Find the coordinates of the centroid of the area bounded by $y^2 = -x$ and the line $y = x + 2$.
9. Find the centroid of the area of the circle $x^2 + y^2 = a^2$ in the first quadrant.
10. Find the centroid of the area bounded by the parabola $x = 2y - y^2$ and the line $x = -2y$.
11. The density of the material of a right circular cylinder of radius 'a' varies as the distance from the axis and as the distance from one end. Find the moment of inertia of the cylinder about the axis.
12. The density of a semicircular lamina of radius 'a' varies as the distance from the bounding diameter. Find the distance of its mass centre from this diameter.
13. Find the moment of inertia about the x-axis of the region bounded by the ellipse $(x^2/a^2) + (y^2/b^2) = 1$. Assume $\rho(x, y) = 1$.
14. A thin plate is in the form of a quadrant of the ellipse $(x^2/a^2) + (y^2/b^2) = 1$. The density at any point in the plane varies as the product of its distances from the axes. Find the mass.
15. The cardioid $r = a(1 + \cos\theta)$ is rotated about its axis. Find the volume generated.
16. Find the moment of inertia about the x-axis of the area of the triangle with vertices (1, 1), (2, 1) and (3, 3).

6.3 TRIPLE INTEGRALS—VOLUME

The triple integral of a function $f(x, y, z)$ over a closed rectangular parallelopiped is defined in the same manner as a double integral. Proceeding as in the case of the double integral, we obtain sums of the form

$$S = \sum_{i=1}^{N} f(x_i, y_i, z_i)\,\Delta x\,\Delta y\,\Delta z,$$

and the triple integral of $f(x, y, z)$ over the region V is the limiting value to which these sums tend, as $\Delta x, \Delta y$ and Δz tend to zero. Thus,

$$\iiint_V f(x, y, z)\,dV = \lim \sum_{i=1}^{N} f(x_i, y_i, z_i)\,\Delta x\,\Delta y\,\Delta z \qquad (6.25)$$

If $f(x, y, z) \equiv 1$, we immediately deduce that the triple integral is just the volume of V. If, on the other hand, $f(x, y, z)$ is the density at (x, y, z), then the integral is the mass in V. As in the case of the double integral, the definition (6.25) is seldom used for the evaluation of the triple integral. Instead, it is usually evaluated as an iterated integral. For example, suppose V is bounded below by a surface $z = z_1(x, y)$ and above by the surface $z = z_2(x, y)$, where $z_1(x_1, y)$ and $z_2(x, y)$ are functions defined in a region R of the XY-plane, then if $f(x, y, z)$ is continuous in V, we have that

$$\iiint_V f(x, y, z)\, dV = \iint_R \left[\int_{z=z_1(x, y)}^{z=z_2(x, y)} f(x, y, z)\, dz \right] dx\, dy \tag{6.26}$$

If (6.26), we first integrate with respect to z and then integrate the resulting function of x and y over the region R.

Example 6.18 Find the volume of the region bounded by the paraboloids

$$z = x^2 + y^2 \quad \text{and} \quad z = 6 - \frac{x^2 + y^2}{2}.$$

The curve of intersection of the two paraboloids is given by

$$x^2 + y^2 = 6 - \frac{x^2 + y^2}{2} \quad \text{or} \quad x^2 + y^2 = 4.$$

Hence, the volume V is given by

$$V = \iint_R \int_{x^2+y^2}^{6-(x^2+y^2)/2} dz\, dy\, dx,$$

where R is the region in the (x, y) plane bounded by $x^2 + y^2 = 4$. Therefore,

$$V = \int_{-2}^{2} \int_{-\sqrt{4-x^2}}^{\sqrt{4-x^2}} \left[6 - \frac{1}{2}(x^2 + y^2) - (x^2 + y^2) \right] dy\, dx$$

$$= 6 \int_0^2 \int_0^{\sqrt{4-x^2}} \left[4 - (x^2 + y^2) \right] dy\, dx$$

$$= 6 \int_0^{\pi/2} \int_0^2 (4 - r^2)\, r\, dr\, d\theta$$

$$= 12\pi.$$

Example 6.19 Find the volume of the region bounded by the paraboloid $z = x^2 + y^2$ and the plane $z = 4$.

The volume is given by

$$V = \iint_R \int_{x^2+y}^{4} dz\, dy\, dx,$$

where R is the region in the (x, y) plane bounded by the circle $x^2 + y^2 = 4$. We therefore have

$$V = \iint_R [4 - (x^2 + y^2)]\, dy\, dx$$

$$= 4 \int_0^{\pi/2} \int_0^{2} (4 - r^2)\, r\, dr\, d\theta$$

$$= 4 \frac{\pi}{2} \left[4 \frac{r^2}{2} - \frac{r^4}{4} \right]_0^2$$

$$= 2\pi \left[2r^2 - \frac{r^4}{4} \right]_0^2$$

$$= 2\pi\, (8 - 4)$$

$$= 8\pi.$$

Example 6.20 Find the volume of the solid enclosed by the ellipsoid

$$\frac{x^2}{a^2} + \frac{y^2}{b^2} + \frac{z^2}{c^2} = 1.$$

Because of symmetry, we need compute the volume in the first octant only. Hence, if V is the total volume, then

$$V = 8 \iint_R \int_0^{c\sqrt{1-(x^2/a^2)-(y^2/b^2)}} dz\, dy\, dx,$$

where R is the region bounded by the ellipse $(x^2/a^2) + (y^2/b^2) = 1$ in the first quadrant. Therefore,

$$V = 8 \int_0^{a} \int_0^{b\sqrt{1-x^2/a^2}} c\sqrt{1 - \frac{x^2}{a^2} - \frac{y^2}{b^2}}\, dy\, dx$$

$$= 8c \int_0^a \int_0^{b\sqrt{1-x^2/a^2}} \sqrt{1-\frac{x^2}{a^2}-\frac{y^2}{b^2}}\, dy\, dx$$

We shall first evaluate the inner integral. Let

$$I_1 = \int_0^{b\sqrt{p}} \sqrt{p-\frac{y^2}{b^2}}\, dy, \qquad \text{where} \quad p = 1-\frac{x^2}{a^2}.$$

Setting $y/b = \sqrt{p}\sin\theta$, we obtain $dy = b\sqrt{p}\cos\theta\, d\theta$. Then,

$$I_1 = \int_0^{\pi/2} (\sqrt{p}\cos\theta)\, b\sqrt{p}\cos\theta\, d\theta$$

$$= bp \int_0^{\pi/2} \cos^2\theta\, d\theta$$

$$= bp \int_0^{\pi/2} \frac{1+\cos 2\theta}{2}\, d\theta$$

$$= \frac{bp}{2}\left[\theta + \frac{\sin 2\theta}{2}\right]_0^{\pi/2}$$

$$= \frac{bp}{2}\frac{\pi}{2}$$

$$= \frac{bp\pi}{4}.$$

We therefore have

$$V = 8c \int_0^a \frac{bp\pi}{4}\, dx$$

$$= 2cb\pi \int_0^a \left(1-\frac{x^2}{a^2}\right) dx$$

$$= 2cb\pi \left[x - \frac{1}{a^2}\frac{x^3}{3}\right]_0^a$$

$$= 2cb\pi\left(a - \frac{1}{3}a\right)$$

$$= \frac{4}{3} cab\pi$$

$$= \frac{4}{3} \pi abc$$

Example 6.21 Find the volume of the region in space bounded above by the surface $z = 1 - (x^2 + y^2)$, on the sides by the planes $x = 0, y = 0, x + y = 1$, and below by the plane $z = 0$. The volume is given by

$$V = \int_0^1 \int_0^{1-x} \int_0^{1-x^2-y^2} dz\, dy\, dx$$

$$= \int_0^1 \int_0^{1-x} (1 - x^2 - y^2)\, dy\, dx$$

$$= \int_0^1 \left[y - x^2 y - \frac{1}{3} y^3 \right]_0^{1-x} dx$$

$$= \int_0^1 \left[1 - x - x^2(1-x) - \frac{1}{3}(1-x)^3 \right] dx$$

$$= \left[x - \frac{1}{2} x^2 - \frac{1}{3} x^3 + \frac{1}{4} x^4 - \frac{1}{3} \frac{(1-x)^4}{-4} \right]_0^1$$

$$= 1 - \frac{1}{2} - \frac{1}{3} + \frac{1}{4} + \frac{1}{12} [(1-x)^4]_0^1$$

$$= 1 - \frac{1}{2} - \frac{1}{3} + \frac{1}{4} - \frac{1}{12}$$

$$= \frac{1}{3}.$$

6.3.1 Cylindrical Coordinates

The use of cylindrical coordinates is particularly suited for problems in which there is an axis of symmetry of the solid. Instead of the volume element $dx\, dy\, dz$, we use, in cylindrical coordinates, the volume element $r\, dr\, d\theta\, dz$, i.e. an element with a cross-sectional area $r\, dr\, d\theta$ and altitude dz.

Example 6.22 If the radius of the base and altitude of a right circular cone are given by a and h respectively, express its volume as a triple integral and evaluate it using cylindrical coordinates.

Taking r, θ, and z as cylindrical coordinates, the equation of the cone may be written as

$$\frac{z}{r} = \frac{h}{a} \quad \text{or} \quad z = r\frac{h}{a}.$$

Hence, its volume is given by

$$V = \int_0^{2\pi} \int_0^a \int_{z=rh/a}^{z=h} r\,dr\,d\theta\,dz$$

$$= \int_0^{2\pi} \int_0^a \left(h - \frac{rh}{a}\right) r\,dr\,d\theta$$

$$= h \int_0^{2\pi} \left[\frac{r^2}{2} - \frac{r^3}{3a}\right]_0^a d\theta$$

$$= h(2\pi)\left(\frac{a^2}{2} - \frac{a^2}{3}\right)$$

$$= 2\pi h \frac{1}{6} a^2$$

$$= \frac{1}{3}\pi a^2 h.$$

Example 6.23 Find the volume bounded by the sphere $x^2 + y^2 + z^2 = a^2$.

In cylindrical coordinates, the equation of the sphere becomes

$$r^2 + z^2 = a^2$$

and the volume element is $r\,dr\,d\theta\,dz$. Then the volume is given by

$$V = \int_0^{2\pi} \int_0^a \int_{-\sqrt{a^2-r^2}}^{\sqrt{a^2-r^2}} r\,dr\,d\theta\,dz$$

$$= 2\int_0^{2\pi} \int_0^a \sqrt{a^2 - r^2}\, r\,dr\,d\theta$$

$$= \int_0^{2\pi} \int_0^{a^2} \sqrt{a^2 - R}\, dR\, d\theta$$

$$= \int_0^{2\pi} \left[\frac{(a^2 - R)^{3/2}}{-3/2} \right]_0^{a^2} d\theta$$

$$= -\frac{2}{3} (2\pi) (-a^3)$$

$$= \frac{4}{3} \pi a^3.$$

6.3.2 Spherical Polar Coordinates

If there is symmetry about a point in any problem, then it would be convenient to use spherical polar coordinates. These are given by the equations (refer to Fig. 6.20):

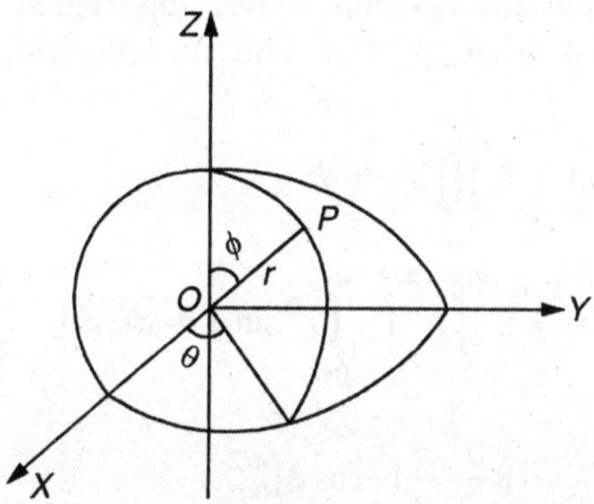

Figure 6.20

$$x = r \cos\theta \sin\phi, \qquad y = r \sin\theta \sin\phi, \qquad z = r \cos\phi. \tag{6.27}$$

The volume element in spherical polar coordinates is approximately

$$dV = dr\,(r\, d\phi)\,(r \sin\phi\, d\theta), \quad \text{as shown in Fig. 6.21.}$$

$$= r^2 \sin\phi\, dr\, d\theta\, d\phi \tag{6.28}$$

In the above, the values of ϕ are restricted to the range

$$0 \le \phi \le \pi. \tag{6.29}$$

The proof of (6.28) is omitted here.

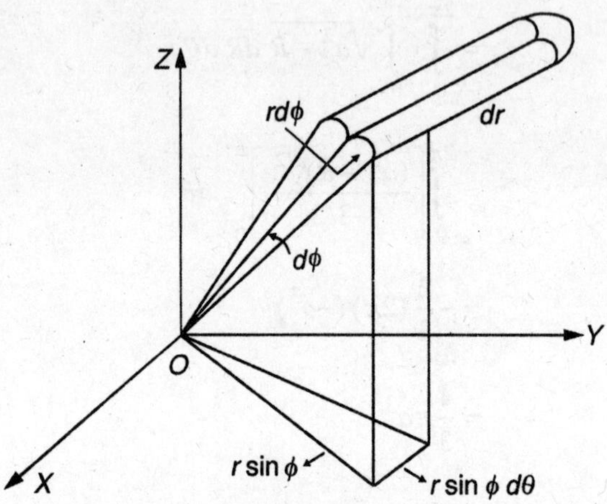

Figure 6.21

Example 6.24 Find the volume of the sphere $x^2 + y^2 + z^2 = a^2$.

The required volume V is eight times that enclosed in the first octant. We integrate the constant function 1 over this region using the volume element $dV = r^2 \sin\phi\, dr\, d\theta\, d\phi$, and limits in spherical polar coordinates. We then obtain

$$V = 8 \iiint dx\, dy\, dz$$

$$= 8 \int_0^{\pi/2} \int_0^{\pi/2} \int_0^{a} r^2 \sin\phi\, dr\, d\theta\, d\phi$$

$$= 8 \frac{a^3}{3} \frac{\pi}{2} [-\cos\phi]_0^{\pi/2}$$

$$= \frac{4}{3}(\pi a^3)\left(-\cos\frac{\pi}{2} + 1\right)$$

$$= \frac{4}{3}\pi a^3.$$

Example 6.25 Evaluate $\iiint xyz\, dx\, dy\, dz$ over the positive octant of the sphere $x^2 + y^2 + z^2 = a^2$.

Using (6.27) and (6.28), the given integral transforms into

$$I = \int_0^{\pi/2} \int_0^{\pi/2} \int_0^{a} (r\cos\theta\sin\phi)(r\sin\theta\sin\phi)(r\cos\phi)\, r^2 \sin\phi\, dr\, d\theta\, d\phi$$

$$= \int_0^{\pi/2} \int_0^{\pi/2} \int_0^a r^5 \cos\theta \sin\theta \sin^3\phi \cos\phi \, d\phi \, d\theta \, dr$$

$$= \left[\frac{r^6}{6}\right]_0^a \left[\frac{\sin^2\theta}{2}\right]_0^{\pi/2} \left[\frac{\sin^4\phi}{4}\right]_0^{\pi/2}$$

$$= \frac{1}{6}a^6 \frac{1}{2}\frac{1}{4}$$

$$= \frac{1}{48}a^6.$$

6.3.3 Applications of Triple Integrals

If $\rho = \rho(x, y, z)$ is the density at the point (x, y, z) of a region V of xyz-space, then the mass M is given by

$$M = \iiint \rho \, dV = \iiint \rho \, dx \, dy \, dz \tag{6.30}$$

The coordinates of the centre of gravity are given by

$$\bar{x} = \frac{\iiint x\rho \, dV}{\iiint \rho \, dV}, \quad \bar{y} = \frac{\iiint y\rho \, dV}{\iiint \rho \, dV} \quad \bar{z} = \frac{\iiint z\rho \, dV}{\iiint \rho \, dV} \tag{6.31}$$

Similarly, the moment of inertia about the X-axis is given by

$$I_x = \iiint \rho(y^2 + z^2) \, dV \tag{6.32}$$

with similar integrals for I_y and I_z.

The above integrals may be evaluated as triple integrals by taking $dV = dx\,dy\,dz$ or the corresponding volume element in other coordinate systems.

Example 6.26 Assuming $\rho(x, y, z) = 1$, find the centroid of the portion of the sphere $x^2 + y^2 + z^2 = a^2$ in the first octant.

Now,

$$V = \frac{1}{8}\frac{4}{3}\pi a^3 = \frac{1}{6}\pi a^3.$$

Here,

$$\bar{z} = \frac{\iiint z \, dV}{\iiint dV}$$

i.e.

$$\frac{1}{6}\pi a^3\bar{z} = \iiint z\, dV$$

$$= \int_0^{\pi/2}\int_0^{\pi/2}\int_0^{a} (r\cos\phi) r^2 \sin\phi\, dr\, d\theta\, d\phi$$

$$= \frac{a^4}{4}\left[\frac{\sin^2\phi}{2}\right]_0^{\pi/2} \frac{\pi}{2}$$

$$= \frac{1}{16}\pi a^4.$$

Therefore,

$$\bar{z} = \frac{(1/16)\,\pi a^4}{(1/6)\,\pi a^3} = \frac{3}{8}a.$$

By symmetry, we have

$$\bar{x} = \bar{y} = \bar{z} = \frac{3}{8}a.$$

Example 6.27 Find the mass and moment of inertia of a sphere of radius a with respect to a diameter if the density is proportional to the distance from the centre.

Let $\rho = kr$. Then,

$$M = \iiint \rho\, dV = 8\int_0^{\pi/2}\int_0^{\pi/2}\int_0^{a} (kr)(r^2\sin\phi)\, dr\, d\theta\, d\phi$$

$$= 8k\frac{1}{4}a^4\left[-\cos\phi\right]_0^{\pi/2} \frac{\pi}{2}$$

$$= k\pi a^4.$$

We have

$$I_x + I_y + I_z = 2I_0.$$

But, by symmetry,

$$I_x = I_y = I_z.$$

Hence,

$$I_x = I_y = I_z = \frac{2}{3}I_0.$$

Since,

$$I_0 = \iiint \rho(x^2 + y^2 + z^2)\, dV.$$

We have,

$$I_z = \frac{2}{3} I_0 = \frac{2}{3} \times 8 \int_0^{\pi/2} \int_0^{\pi/2} \int_0^{a} (kr)(r^2)\, r^2 \sin\phi \, dr\, d\theta\, d\phi$$

$$= \frac{16}{3} k \frac{1}{6} a^6 \left[-\cos\phi\right]_0^{\pi/2} \frac{\pi}{2}$$

$$= \frac{4}{9} k\pi a^6.$$

$$= \frac{4}{9} Ma^2.$$

6.3.4 Change of Variables

In Section 4.2.1, we have seen that the evaluation of an integral can be made easier by a suitable substitution. Similarly, the evaluation of a double integral of the form

$$\iint_R f(x, y)\, dx\, dy$$

can also be made easier by using a substitution of the form

$$x = x(u, v) \qquad \text{and} \qquad y = y(u, v).$$

We assume that the above equations can be solved for u and in terms of x and y so that the transformation from (x, y) coordinates to (u, v) coordinates is one-to-one.

In Fig. 6.22, let the coordinates of P be (x, y) and let u and $u + \Delta u$ are represented by the curves PS and QR respectively. Similarly, let v and $v + \Delta v$ are represented by the curves PQ and SR, respectively.

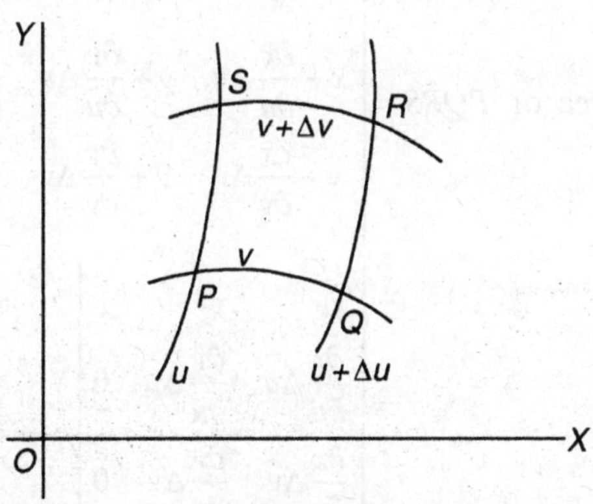

Figure 6.22

The coordinates of Q are $[x(u+\Delta u, v), y(u+\Delta u, v)]$. But by Taylor's theorem,

$$x(u+\Delta u, v) = x(u, v) + \frac{\partial x}{\partial u}\Delta u + \text{terms involving higher powers of } \Delta u.$$

$$= x + \frac{\partial x}{\partial u}\Delta u, \text{ approximately.}$$

Similarly,

$$y(u+\Delta u, v) = y + \frac{\partial y}{\partial u}\Delta u, \text{ approximately.}$$

Thus, the coordinates of Q are

$$\left(x + \frac{\partial x}{\partial u}\Delta u,\ y + \frac{\partial y}{\partial u}\Delta u\right).$$

Similarly, the coordinates of R are

$$\left(x + \frac{\partial x}{\partial u}\Delta u + \frac{\partial x}{\partial v}\Delta v,\ y + \frac{\partial y}{\partial u}\Delta u + \frac{\partial y}{\partial v}\Delta v\right),$$

and the coordinates of S are

$$\left(x + \frac{\partial x}{\partial v}\Delta v,\ y + \frac{\partial y}{\partial v}\Delta v\right).$$

Hence, the slopes of PQ and RS are

$$\frac{(\partial y/\partial u)\, u}{(\partial x/\partial u)\, u} \quad \text{and} \quad \frac{(\partial y/\partial u)\, u}{(\partial x/\partial u)\, u}$$

respectively, which shows that PQ is parallel to RS. Similarly, it can be shown that PS is parallel to QR. Thus, $PQRS$ is a parallelogram and hence its area is given by

$$\text{Area of } PQRS = \begin{vmatrix} x & y & 1 \\ x + \frac{\partial x}{\partial u}\Delta u & y + \frac{\partial y}{\partial u}\Delta u & 1 \\ x + \frac{\partial x}{\partial v}\Delta v & y + \frac{\partial y}{\partial v}\Delta v & 1 \end{vmatrix}$$

$$= \begin{vmatrix} x & y & 1 \\ \frac{\partial x}{\partial u}\Delta u & \frac{\partial y}{\partial u}\Delta u & 0 \\ \frac{\partial x}{\partial v}\Delta v & \frac{\partial y}{\partial v}\Delta v & 0 \end{vmatrix}$$

$$= \begin{vmatrix} \dfrac{\partial x}{\partial u}\Delta u & \dfrac{\partial y}{\partial u}\Delta u \\ \dfrac{\partial x}{\partial v}\Delta v & \dfrac{\partial y}{\partial v}\Delta v \end{vmatrix}$$

$$= \begin{vmatrix} \dfrac{\partial x}{\partial u} & \dfrac{\partial y}{\partial u} \\ \dfrac{\partial x}{\partial v} & \dfrac{\partial y}{\partial v} \end{vmatrix} \Delta u \, \Delta v \tag{6.33}$$

The quantity

$$\begin{vmatrix} \dfrac{\partial x}{\partial u} & \dfrac{\partial y}{\partial u} \\ \dfrac{\partial x}{\partial v} & \dfrac{\partial y}{\partial v} \end{vmatrix}$$

is often expressed symbolically as

$$\frac{\partial(x, y)}{\partial(u, v)}$$

and is called the *Jacobian* of the transformation $x = x(u, v)$ and $y = y(u, v)$. Hence,

$$\text{Area of } PQRS = \frac{\partial(x, y)}{\partial(u, v)} \Delta u \, \Delta v.$$

But since this is equal to $\Delta x \, \Delta y$, it follows that

$$\iint_R f(x, y)\, dx\, dy = \iint_R f[x(u, v), y(u, v)] \frac{\partial(x, y)}{\partial(u, v)}\, du\, dv. \tag{6.34}$$

An important property of the Jacobians is that

$$\frac{\partial(x, y)}{\partial(u, v)} \frac{\partial(u, v)}{\partial(x, y)} = 1.$$

The proof of this is left as an exercise to the reader. In a similar way, one can derive the corresponding result for triple integrals:

$$\iiint f(x, y, z)\, dx\, dy\, dz = \iiint F(u, v, w) \frac{\partial(x, y, z)}{\partial(u, v, w)}\, du\, dv\, dw, \tag{6.35}$$

where x, y, z are given in terms of u, v, *and* w.

Particular cases

(i) Let the transformation be from cartesian to polar coordinates, i.e. from (x, y) to (r, θ). Here, $x = r\cos\theta$, $y = r\sin\theta$. Therefore,

$$\frac{\partial(x, y)}{\partial(r, \theta)} = \begin{vmatrix} \cos\theta & \sin\theta \\ -r\sin\theta & r\cos\theta \end{vmatrix} = r\cos^2\theta + r\sin^2\theta = r.$$

Hence,

$$\Delta x\, \Delta y = r\, \Delta r\, \Delta\theta.$$

(ii) Let the transformations be from cartesian to spherical polar coordinates, i.e. from (x, y, z) to (r, ϕ, θ). Then,

$$x = r\cos\theta\sin\phi, \qquad y = r\sin\theta\sin\phi, \qquad z = r\cos\phi.$$

Therefore,

$$\frac{\partial(x, y, z)}{\partial(r, \phi, \theta)} = \begin{vmatrix} \dfrac{\partial x}{\partial r} & \dfrac{\partial y}{\partial r} & \dfrac{\partial z}{\partial r} \\ \dfrac{\partial x}{\partial \phi} & \dfrac{\partial y}{\partial \phi} & \dfrac{\partial z}{\partial \phi} \\ \dfrac{\partial x}{\partial \theta} & \dfrac{\partial y}{\partial \theta} & \dfrac{\partial z}{\partial \theta} \end{vmatrix}$$

$$= \begin{vmatrix} \cos\theta\sin\phi & \sin\theta\sin\phi & \cos\phi \\ r\cos\theta\cos\phi & r\sin\theta\cos\phi & -r\sin\phi \\ -r\sin\theta\sin\phi & r\cos\theta\sin\phi & 0 \end{vmatrix}$$

$$= r^2\sin^2\phi.$$

Hence,

$$\Delta x\, \Delta y\, \Delta z = r^2\sin^2\phi\, \Delta r\, \Delta\theta\, \Delta\phi.$$

Example 6.28 Evaluate the integral

$$\int_0^\infty \int_0^\infty e^{-(x+y)} \sin\frac{\pi y}{x+y}\, dx\, dy,$$

by means of transformation $u = x + y$ and $v = y$. We have

$$u = x + y \quad \text{and} \quad v = y \tag{i}$$

Form this, we obtain

$$u - v = x \quad \text{and} \quad v = y \tag{ii}$$

In the (x, y) plane, the region of integration is the first quadrant, i.e. $x \ge 0$ and $y \ge 0$. To obtain the region R' in the (u, v) plane, we use equations (ii) and obtain the region bounded by $u = v$ and $v = 0$. This is shown in Fig. 6.23. The Jacobian of the transformation is given by

$$\frac{\partial(x, y)}{\partial(u, v)} = \begin{vmatrix} \dfrac{\partial x}{\partial u} & \dfrac{\partial y}{\partial u} \\ \dfrac{\partial x}{\partial v} & \dfrac{\partial y}{\partial v} \end{vmatrix} = \begin{vmatrix} 1 & 0 \\ -1 & 1 \end{vmatrix} = 1.$$

Hence, $\Delta x\,\Delta y = \Delta u\,\Delta v$. That is, $dx\,dy = du\,dv$.

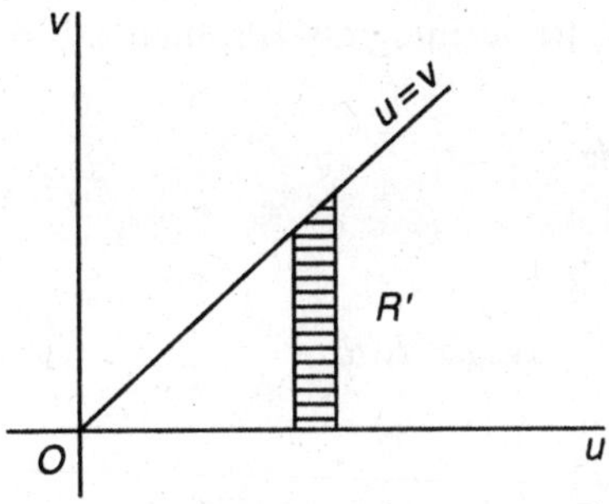

Figure 6.23

The given integral therefore transforms into

$$\iint_{R'} e^{-u}\sin\frac{\pi v}{u}\,du\,dv = \int_0^{\infty}\int_{v=0}^{v=u} e^{-u}\sin\frac{\pi v}{u}\,dv\,du$$

$$= \int_0^{\infty} e^{-u}\left[\frac{-\cos(\pi v/u)}{\pi/u}\right]_{v=0}^{v=u} du = -\frac{1}{\pi}\int_0^{\infty} ue^{-u}(-2)\,du$$

$$= \frac{2}{\pi}\int_0^{\infty} ue^{-u}\,du = \frac{2}{\pi}\cdot$$

Example 6.29 Evaluate the integral

$$\iiint \frac{1}{x^2+y^2+z^2}\,dx\,dy\,dz,$$

taken throughout the positive octant of the sphere $x^2+y^2+z^2=a^2$.

We transform the integral by means of the spherical polar coordinates (r,ϕ,θ). We therefore have

$$\frac{\partial(x,y,z)}{\partial(r,\phi,\theta)} = r^2\sin\phi.$$

Hence, the given integral becomes:

$$\int_0^{\pi/2}\int_0^{\pi/2}\int_0^{a}\frac{1}{r^2}r^2\sin\phi\,dr\,d\theta\,d\phi. = a\frac{\pi}{2}[-\cos\phi]_0^{\pi/2}$$

$$= \frac{a\pi}{2}\cdot$$

EXERCISES 6.3

Evaluate the following triple integrals (Problems 1–6):

1. $\int_0^a \int_0^x \int_0^y xyz \, dx \, dy \, dz.$

2. $\int_0^a \int_0^{\sqrt{a^2-x^2}} \int_0^{\sqrt{a^2-x^2-y^2}} xyz \, dx \, dy \, dz.$

3. $\int_0^1 \int_0^{1-x} \int_0^{1-x-y} (x^2 + y^2 + z^2) \, dz \, dy \, dx.$

4. $\int_0^1 \int_{y^2}^{\sqrt{y}} \int_0^{9-x^2-y^2} dz \, dx \, dy.$

5. $\int_1^e \int_x^{x^2} \int_0^{\log_e x} dz \, dy \, dx.$

6. $\int_0^1 \int_0^{1-y} \int_0^{1-x^2-y^2} dz \, dy \, dx.$

7. Find the volume common to the two cylinders $x^2 + y^2 = a^2$ and $x^2 + z^2 = a^2$.

8. Find the volume within the cylinder $(x-2)^2 + y^2 = 4$ cut off by the cylinder $z^2 = 4x$.

9. Find the volume cut off the sphere $x^2 + y^2 + z^2 = a^2$ by the cone $x^2 + y^2 = z^2$.

10. Find the volume of the ellipsoid

$$\frac{x^2}{a^2} + \frac{y^2}{b^2} + \frac{z^2}{c^2} = 0.$$

11. Evaluate $\iiint (x - y + z^2) \, dx \, dy \, dz$ over the region bounded above by $z = 1 + x^2 + y^2$, below by $z = 0$, and the sides by $x^2 + y^2 = 4$.

12. Evaluate

$$\iiint \frac{1}{(x^2 + y^2 + z^2)^{3/2}} \, dx \, dy \, dz$$

over the region bounded by the spheres $x^2 + y^2 + z^2 = 9$ and $x^2 + y^2 + z^2 = 1$.

13. Find the volume of the solid bounded by $x^2 + y^2 + z^2 = 4$ and the paraboloid $x^2 + y^2 = 3z$.

14. Find the moment of inertia of a sphere of constant density with respect to a diameter.
15. Find the moment of inertia of a homogeneous solid right circular cylinder with radius of base a, height h and density k with respect to its axis.
16. Find the moment of inertia about OZ of a solid bounded by the surfaces $z^2 = 2ax,\ z = 0,\ x^2 + y^2 - ax = 0$.
17. Find the mass of a solid bounded by the surfaces $2x + z = 2a,\ x + z = a,\ y^2 = ax,\ y = 0$ if the density at each point is equal to the ordinate y of this point.
18. If $x = r\cos\theta,\ y = r\sin\theta$, show that

$$\frac{\partial(x, y)}{\partial(r,\theta)} = r.$$

By changing to polar coordinates, evaluate

$$\iint xy(x^2 + y^2)^{n/2}\,dx\,dy$$

over the positive quadrant of the circle $x^2 + y^2 = a^2$, supposing $n + 3 > 0$.

19. If $x + y = u$ and $y = uv$, show that

$$\frac{\partial(x, y)}{\partial(u, v)} = u.$$

20. Prove that

$$\frac{\partial(x, y)}{\partial(u, v)}\frac{\partial(u, v)}{\partial(x, y)} = 1.$$

21. Using the transformation $x + y = u$ and $y = uv$, evaluate the integral $\iint xy\sqrt{1 - x - y}\,dx\,dy$ taken over the area of the triangle with sides $x = 0,\ y = 0$ and $x + y = 1$.
22. Evaluate $\int_0^1\int_0^x \sqrt{x^2 + y^2}$ using the transformation $x = u,\ y = uv$.
23. Using the transformation $u = x + y,\ v = x/y$, evaluate the integral

$$\iint_R \frac{x + y}{x^2}\,e^{x+y}\,dx\,dy,$$

where R is the triangle with the vertices (0, 0), (2, 0) and (1, 1).

24. If $x = u/(1+v)^2$ and $y = uv/(1+v^2)$, find $\partial(x, y)/\partial(u, v)$.
25. Use the transformation $x + y = u,\ y = uv$ to calculate the integral $\int_0^1\int_x^{1-x} e^{y/(x+y)}\,dy\,dx$.

26. Use the transformation $x-y=u$, $v=y$ to evaluate the integral $\int_0^2 \int_0^2 \left[4+(x-y)^2\right]^{-1/2} dx\,dy$.

6.4 IMPROPER INTEGRALS

The definite integral

$$I = \int_a^b f(x)\,dx$$

defined in Section 4.1.1, has meaning only when the limits a and b are finite and the integrand $f(x)$ is bounded on the interval $[a, b]$. In this section, we shall extend the definition when the range of integration is infinite or when the integrand is unbounded. Such integrals are called *improper integrals.*

6.4.1 Infinite Integrals

Let $f(x)$ be bounded and integrable in $a \le t \le x$, where a is some fixed number and x is any number greater than a. If

$$\int f(t)\,dt = F(t),$$

then the *improper* or *infinite integral* of $f(t)$ *is* defined by the relation

$$\int_a^\infty f(t)\,dt = \lim_{x\to\infty} \int_a^x f(t)\,dt = \lim_{x\to\infty} F(x) - F(a) \tag{6.36}$$

provided that the limit exists. In such a case, we say that the integral converges. If, on the other hand,

$$\int_a^x f(t)\,dt \to \infty, \qquad \text{as } x \to \infty,$$

we say that the infinite integral diverges to $+\infty$, or that it does not exist.

Again, if the infinite integral is of the type

$$\int_{-\infty}^b f(t)\,dt,$$

then we define

$$\lim_{x\to-\infty} \int_x^b f(t)\,dt = F(b) - \lim_{x\to-\infty} F(x) \tag{6.37}$$

provided that the limit exists. In such a case, the infinite integral is said to converge.

Finally, for integrals over the range $(-\infty,\infty)$, we write

$$\int_{-\infty}^{\infty} f(t)\,dt = \lim_{x\to-\infty}\int_{x}^{a} f(t)\,dt + \lim_{x\to\infty}\int_{a}^{x} f(t)\,dt \tag{6.38}$$

The infinite integral on the left is said to converge if both the integrals on the right also converge. Its value is then equal to the sum of the two infinite integrals on the right. Obviously, a may be any finite number we please, i.e. the value of the integral is independent of the number 'a' used.

Example 6.30 We consider the integral

$$\int_{a}^{\infty} \frac{1}{t^p}\,dt, \qquad \text{where } a>0 \text{ and } p\neq 1.$$

We write

$$\int_{a}^{\infty} \frac{1}{t^P}\,dt = \lim_{x\to\infty}\int_{a}^{x} \frac{1}{t^P}\,dt$$

$$= \lim_{x\to\infty}\left[\frac{t^{-p+1}}{-p+1}\right]_a^x$$

$$= \lim_{x\to\infty}\left[-\frac{x^{-(p-1)}}{p-1}\right] + \frac{a^{1-p}}{p-1}.$$

But

$$\lim_{x\to\infty} -\frac{x^{-(p-1)}}{p-1} \to 0, \qquad \text{if } p>1,$$

and diverges to $+\infty$ if $p<1$. Hence,

$$\int_{a}^{\infty} \frac{1}{t^p}\,dt = \frac{a^{1-p}}{p-1}, \qquad \text{if } p>1,$$

Also, if $p=1$,

$$\int_{a}^{\infty} \frac{1}{t}\,dt = \lim_{x\to\infty}\int_{a}^{\infty} \frac{1}{t}\,dt = \lim_{x\to\infty} \log_e x - \log_e a,$$

where $\lim_{x\to\infty} \log_e x$ tends to $+\infty$ as $x\to\infty$. Hence the above integral diverges.

Thus, the integral

$$\int_a^{\infty} \frac{1}{t^p}\, dt$$

converges for $p > 1$ and diverges for $p \le 1$.

In particular,

$$\int_2^{\infty} \frac{1}{t^2}\, dt = \lim_{x\to\infty} \int_1^{x} \frac{1}{t^2}\, dt = \lim_{x\to\infty}\left(-\frac{1}{x}\right) + 1 = 1,$$

and

$$\int_1^{\infty} \frac{dt}{\sqrt{t}} = \lim_{x\to\infty} \int_1^{x} \frac{dt}{\sqrt{t}} = \lim_{x\to\infty} 2\sqrt{x} - 2, = \infty$$

Hence the integral $\int_1^{\infty} (1/\sqrt{t}\,)\, dt$ diverges.

Example 6.31 We consider the integral

$$\int_{-\infty}^{0} t \sin t\, dt.$$

We write

$$\int_{-\infty}^{0} t \sin t\, dt = \lim_{x\to-\infty} \int_{x}^{0} t \sin t\, dt$$

Since

$$I = \int t \sin t\, dt = t(-\cos t) + \sin t,$$

we obtain:

$$I = \int_{-\infty}^{0} t \sin t\, dt = \lim_{x\to-\infty} [-t \cos t + \sin t]_x^0$$

$$= 0 - \lim_{x\to-\infty} [-x \cos x + \sin x]$$

$$= \lim_{x\to-\infty} (x \cos x - \sin x)$$

Since $\cos x$ and $\sin x$ oscillate between ± 1, it follows that the above integral diverges.

Example 6.32 Discuss the convergence of

$$\int_{-\infty}^{\infty} \frac{1}{1+x^2}\,dx.$$

We write

$$\int_{-\infty}^{\infty} \frac{1}{1+x^2}\,dx = \lim_{t\to-\infty} \int_{t}^{0} \frac{1}{1+x^2}\,dx + \lim_{t\to\infty} \int_{t}^{0} \frac{1}{1+x^2}\,dx$$

$$= \tan^{-1}0 - \lim_{t\to-\infty} \tan^{-1}t + \lim_{t\to\infty} \tan^{-1}t - \tan^{-1}0$$

$$= \frac{\pi}{2} + \frac{\pi}{2}$$

$$= \pi.$$

6.4.2 Integrals with Unbounded Integrands: Cauchy Principal Value

If $f(t)$ becomes unbounded as t tends to 'a' in $a \le t \le b$, then we write

$$\int_{a}^{b} f(t)\,dt = \lim_{\varepsilon\to0} \int_{a+\varepsilon}^{b} f(t)\,dt, \qquad (\varepsilon > 0) \tag{6.39}$$

provided that the limit exists.

Further, if $f(t)$ becomes unbounded as t tends to b in $a \le t \le b$, then we write

$$\int_{a}^{b} f(t)\,dt = \lim_{\varepsilon\to0} \int_{a}^{b-\varepsilon} f(t)\,dt, \qquad (\varepsilon > 0) \tag{6.40}$$

whenever the limit exists.

On the other hand, if both a and b are points of discontinuity, then $\int_a^b f(t)\,dt$ is defined as the sum of the two infinite integrals

$$\lim_{\varepsilon\to0} \int_{a+\varepsilon}^{c} f(t)\,dt + \lim_{\varepsilon\to0} \int_{c}^{b-\varepsilon} f(t)\,dt, \qquad a < c < b. \tag{6.41}$$

Finally, if $f(t)$ is unbounded at an interior point c such that $a < c < b$, then we write

$$\int_a^b f(t)\,dt = \lim_{\varepsilon_1 \to 0} \int_a^{c-\varepsilon_1} f(t)\,dt + \lim_{\varepsilon_2 \to 0} \int_{c+\varepsilon_2}^{b} f(t)\,dt \tag{6.42}$$

where ε_1 and ε_2 are two arbitrary positive quantities tending to zero independently.

It may, however, happen that although the limits on the right-hand side of (6.42) do not exist independently, we may set $\varepsilon_1 = \varepsilon_2 = \varepsilon$ and then find that we obtain a finite answer, since the unbounded parts of the two integrals cancel out. The value so obtained is called the *Cauchy principal value* of the integral and is written as

$$P\int_a^b f(t)\,dt = \lim_{\varepsilon \to 0}\left[\int_a^{c-\varepsilon} f(t)dt + \int_{c+\varepsilon}^{b} f(t)\,dt\right] \tag{6.43}$$

If there are a finite number of points of infinite discontinuity of $f(t)$ at $t_1, t_2, \ldots, t_n$ in (a, b), where

$$a \le t_1 \le t_2 \le \cdots \le t_n \le b,$$

then we write

$$\int_a^b f(t)\,dt = \int_a^{t_1} f(t)\,dt + \int_{t_1}^{t_2} f(t)\,dt + \cdots + \int_{t_n}^{b} f(t)\,dt,$$

where the integrals on the right side exist according to the definitions given above.

Example 6.33 Discuss the convergence of

$$\int_a^b \frac{dx}{(a-x)^2}.$$

Here, the integrand $1/(a-x)^2$ has an infinity at the lower limit $x = a$. By definition,

$$\int_a^b \frac{1}{(a-x)^2}\,dx = \lim_{\varepsilon \to 0} \int_{a+\varepsilon}^{b} \frac{1}{(a-x)^2}\,dx, \qquad \text{if this limit exists.}$$

$$= \lim_{\varepsilon \to \infty}\left[\frac{1}{a-x}\right]_{a+\varepsilon}^{b}$$

$$= \frac{1}{a-b} - \lim_{\varepsilon \to 0} \frac{1}{-\varepsilon}, \qquad \text{which does not exist.}$$

Hence, the integral diverges.

Example 6.34 Discuss the convergence of

$$\int_0^1 \frac{dx}{\sqrt{1-x^2}}.$$

In this case, the integrand $1/\sqrt{1-x^2}$ becomes infinite at the upper limit $x=1$. We have, by definition,

$$\int_0^1 \frac{dx}{\sqrt{1-x^2}} = \lim_{\varepsilon\to 0}\int_0^{1-\varepsilon} \frac{1}{\sqrt{1-x^2}}\,dx.$$

$$= \lim_{\varepsilon\to 0}\Big[\sin^{-1}x\Big]_0^{1-\varepsilon}$$

$$= \lim_{\varepsilon\to 0}\sin^{-1}(1-\varepsilon) - 0$$

$$= \sin^{-1}1$$

$$= \frac{\pi}{2}.$$

Hence, the integral exists and is equal to $\pi/2$.

Example 6.35 Evaluate the principal value of the integral

$$\int_1^4 \frac{dx}{(x-2)^3}.$$

By definition,

$$\int_1^4 \frac{dx}{(x-2)^3} = \lim_{\varepsilon_1\to 0}\int_1^{2-\varepsilon_1} \frac{dx}{(x-2)^3} + \lim_{\varepsilon_2\to 0}\int_{2+\varepsilon_2}^{4} \frac{dx}{(x-2)^3}$$

$$= \lim_{\varepsilon_1\to 0}\left[-\frac{1}{2(x-2)^2}\right]_1^{2-\varepsilon_1} + \lim_{\varepsilon_2\to 0}\left[-\frac{1}{2(x-2)^2}\right]_{2+\varepsilon_2}^{4}$$

$$= \lim_{\varepsilon_1\to 0}\left(\frac{1}{2}-\frac{1}{2\varepsilon_1^2}\right) + \lim_{\varepsilon_2\to 0}\left(-\frac{1}{8}+\frac{1}{2\varepsilon_2^2}\right) \qquad \text{(i)}$$

In (i), it is seen that both the limits do not exist and hence the integral diverges. But, if we set $\varepsilon_1 = \varepsilon_2 = \varepsilon$, then (i) gives:

$$\int_1^4 \frac{dx}{(x-2)^3} = \lim_{\varepsilon\to 0}\left(\frac{1}{2}-\frac{1}{2\varepsilon^2}-\frac{1}{8}+\frac{1}{2\varepsilon^2}\right) = \lim_{\varepsilon\to 0}\left(\frac{1}{2}-\frac{1}{8}\right) = \frac{3}{8},$$

which gives the principal value. Hence,

$$P\int_1^4 \frac{dx}{(x-2)^3} = \frac{3}{8}.$$

6.4.3 Comparison Test for Integrals

It may not always be necessary to explicitly evaluate an improper integral. In several problems, however, it may be sufficient to know about the existence of the integral. For this purpose, the *comparison test for integrals* may be used and this is stated below:

Let $\int_a^b f(t)\,dt$ be an improper integral. Then

(a) If there exists a function g such that $|f(t)| \le g(t)$, for $a \le t \le b$, and $\int_a^b g(t)\,dt$ converges, then $\int_a^b f(t)\,dt$ converges.

(b) If there exists a function g such that $f(t) \ge |g(t)|$ for $a \le t \le b$, and $\int_a^b |g(t)|\,dt$ diverges, then $\int_a^b f(t)\,dt$ diverges.

The function $g(t)$ is called a *comparison function* and is often determined from the behaviour of the given function $f(t)$. Alternatively, one may apply the *limit comparison test* which is stated below:

Let $f(x) > 0$ and $g(x) > 0$ for all $x \ge a$. If

$$\lim_{x\to+\infty} \frac{f(x)}{g(x)} = c, \qquad \text{where } c \ne 0, \tag{6.44}$$

then both the integrals, $\int_a^\infty f(x)\,dx$ and $\int_a^\infty g(x)\,dx$ converge or both diverge.

If the limit in (6.44) is 0, we can conclude only that convergence of $\int_a^\infty g(x)\,dx$ implies that of $\int_a^\infty f(x)\,dx$.

The proofs of both the comparison tests stated above are similar to the corresponding result for series and are left as exercises. The following examples demonstrate the use of the comparison tests.

Example 6.36 Discuss the convergence of

$$I = \int_0^\infty e^{-t^2}\,dt.$$

For $t \ge 1$, $t^2 \ge t$ and so $e^{-t^2} \le e^{-t}$. Hence, we choose $g(t) = e^{-t}$ so that $f(t) \le g(t)$ for $t \ge 1$. Now, if we write

$$I = \int_0^1 e^{-t^2}\,dt + \int_1^\infty e^{-t^2}\,dt,$$

we find that the convergence of I depends on that of the second integral on the right, since the first integral, i.e. $\int_0^1 e^{-t^2}dt$ exists. But $\int_1^\infty e^{-t}dt$ converges and hence, by the comparison test, $\int_1^\infty e^{-t^2}dt$ also converges. It follows therefore that the given integral, viz., $I=\int_0^\infty e^{-t^2}dt$ converges.

Example 6.37 Discuss the convergence of

$$\int_1^\infty e^{-x}x^s\,dx$$

for each real s. We choose $g(x)=x^{-2}$. Then,

$$\lim_{x\to+\infty}\frac{f(x)}{g(x)}=\lim_{x\to+\infty}\frac{e^{-x}x^s}{x^{-2}}=\lim_{x\to+\infty}(e^{-x}x^{s+2})=0.$$

But we know that $\int_1^\infty x^{-2}dx$ converges.

Hence, by the limit comparison test, it follows that $\int_1^\infty e^{-x}x^s dx$ also converges.

Example 6.38 (*The gamma function*). Discuss the convergence of the integral

$$I=\int_0^\infty e^{-t}t^{s-1}\,dt \qquad (s>0).$$

We write

$$I=\int_0^1 e^{-t}t^{s-1}\,dt+\int_1^\infty e^{-t}t^{s-1}\,dt.$$

The second integral on the right converges, by the example given above.

To test the first integral, we put $t=1/u$ so that $dt=(-1/u^2)\,du$. Then,

$$\int_x^1 e^{-t}t^{s-1}\,dt=-\int_{1/x}^1 e^{-1/u}u^{-s-1}\,du=\int_1^{1/x} e^{-1/u}u^{-s-1}\,du$$

With $g(u)=u^{-s-1}\,(s>0)$, we find that

$$\int_1^\infty u^{-s-1}\,du$$

converges. Hence, by the comparison test, it follows that

$$\int_0^t e^{-1/u}\, u^{-s-1}\, du\, dt$$

converges. Therefore, the integral

$$\int_0^t e^{-t}\, t^{-s-1}\, dt$$

converges for $s > 0$. When $s > 0$, this integral is called the *gamma function* and is denoted by $\Gamma(s)$. It has the interesting property that $\Gamma(n+1) = n!$ when n is any integer ≥ 0.

6.4.4 Beta and Gamma Functions

The *beta function*, which is also called *Euler's integral of the first kind*, is defined as

$$\beta(m,n) = \int_0^1 x^{m-1}(1-x)^{n-1}\, dx, \qquad (m,n > 0). \tag{6.45}$$

Putting $x = 1 - y$, we obtain

$$\begin{aligned}\beta(m,n) &= \int_1^0 (1-y)^{m-1} y^{n-1}(-dy)\\ &= \int_0^1 y^{n-1}(1-y)^{m-1}\, dy = \beta(n,m),\end{aligned}$$

which shows that the beta function is symmetric in m and n. Again, put

$$x = \frac{y}{1+y} \qquad \text{so that } dx = \frac{1}{(1+y)^2}\, dy.$$

Then, Eq. (6.45) becomes:

$$\beta(m,n) = \int_0^\infty \frac{y^{m-1}}{(1+y)^{m-1}} \frac{1}{(1+y)^{n-1}} \frac{1}{(1+y)^2}\, dy = \int_0^\infty \frac{y^{m-1}}{(1+y)^{m+n}}\, dy \tag{6.46}$$

Finally, setting $x = \sin^2\theta$ in (6.45), we obtain:

$$\begin{aligned}\beta(m,n) &= \int_0^{\pi/2} \sin^{2m-2}\theta \cos^{2n-2}\theta\, 2\sin\theta\cos\theta\, d\theta\\ &= 2\int_0^{\pi/2} \sin^{2m-1}\theta \cos^{2n-1}\theta\, d\theta\end{aligned} \tag{6.47}$$

Euler's integral of the second kind, or *the gamma function*, is defined as

$$\Gamma(n) = \int_0^\infty x^{n-1} e^{-x}\, dx, \qquad (n > 0) \tag{6.48}$$

and we have seen that this integral converges for $n > 0$. Now, we have by definition,

$$\Gamma(n+1) = \int_0^\infty x^n e^{-x}\, dx = \left[-x^n e^{-x}\right]_0^\infty + n\int_0^\infty x^{n-1} e^{-x}\, dx.$$

In the above,

$$\lim_{x\to\infty} x^n e^{-x} = \lim_{x\to\infty} \frac{x^n}{e^x} = 0,$$

and

$$\lim_{x\to\infty} x^n e^{-x} = 0 \qquad \text{if } n > 0,$$

Hence,

$$\Gamma(n+1) = n\Gamma(n), \qquad \text{if } n > 0. \tag{6.49}$$

Formula (6.49) is the recurrence formula for gamma functions. From (6.49), we can write

$$\begin{aligned}\Gamma(n+1) &= n(n-1)\Gamma(n-1)\\ &= n(n-1)(n-2)\Gamma(n-2)\\ &= n(n-1)(n-2)\cdots 1\cdot\Gamma(1).\end{aligned}$$

But, from the definition

$$\Gamma(1) = \int_0^\infty e^{-x}\, dx = \left[\frac{e^{-x}}{-1}\right]_0^\infty = 1.$$

Hence, we obtain

$$\Gamma(n+1) = n(n-1)(n-2), \cdots 1 = n!$$

There exists a very useful relation between beta and gamma functions and this will be proved now. We start with

$$\Gamma(m) = \int_0^\infty x^{m-1} e^{-x}\, dx.$$

Putting $x = t^2$, we obtain $dx = 2t\, dt$. Hence,

$$\Gamma(m) = 2\int_0^\infty t^{2m-2} e^{-t^2}\, t\, dt = 2\int_0^\infty t^{2m-1} e^{-t^2}\, dt$$

Similarly, we can write

$$\Gamma(n) = 2\int_0^\infty s^{2n-1} e^{-s^2}\, ds \tag{6.50}$$

Therefore,

$$\Gamma(m)\Gamma(n) = 4\int_0^\infty\int_0^\infty t^{2m-1} s^{2n-1}\, e^{-(s^2+t^2)}\, dt\, ds.$$

Transforming to polar coordinates $s = r\cos\theta$ and $t = r\sin\theta$, we have

$$dt\, ds = r\, dr\, d\theta.$$

Hence,

$$\Gamma(m)\Gamma(n) = 4\int_0^{\pi/2}\int_0^\infty r^{2m+2n-1}\, \sin^{2m-1}\theta \cos^{2n-1}\theta\, e^{-r^2}\, dr\, d\theta$$

$$= \left[2\int_0^\infty r^{2m+2n-1} e^{-r^2}\, dr\right]\left[2\int_0^{\pi/2} \sin 2^{2m-1}\theta \cos^{2n-1}\theta\, d\theta\right]$$

$$= \Gamma(m+n)\,\beta(m+n), \quad \text{using (6.50) and (6.47).}$$

It therefore follows that

$$\beta(m+n) = \frac{\Gamma(m)\Gamma(n)}{\Gamma(m+n)}, \tag{6.51}$$

which is the useful relation between beta and gamma functions. Setting $m = n = 1/2$ in (6.51), we obtain:

$$\beta\left(\frac{1}{2}, \frac{1}{2}\right) = \frac{\Gamma\left(\frac{1}{2}\right)\Gamma\left(\frac{1}{2}\right)}{\Gamma(1)}.$$

But, from (6.47)

$$\beta\left(\frac{1}{2}, \frac{1}{2}\right) = 2\int_0^{\pi/2} d\theta = \pi,$$

'and $\Gamma(1) = 1$. Hence,

$$\left[\Gamma\left(\frac{1}{2}\right)\right]^2 = \pi, \quad \text{which gives } \Gamma\left(\frac{1}{2}\right) = \sqrt{\pi}.$$

Alternatively, this important result can be deduced as follows. Putting $n = 1/2$ in (6.50), we obtain

$$\Gamma\left(\frac{1}{2}\right)=2\int_0^\infty e^{-s^2}\,ds.$$

Similarly, we can write

$$\Gamma\left(\frac{1}{2}\right)=2\int_0^\infty e^{-t^2}\,dt.$$

Hence,

$$\left[\Gamma\left(\frac{1}{2}\right)\right]^2=4\int_0^\infty\int_0^\infty e^{-(s^2+t^2)}\,ds\,dt.$$

Transforming to polar coordinates, the above equation becomes:

$$\begin{aligned}\left[\Gamma\left(\frac{1}{2}\right)\right]^2&=4\int_0^{\pi/2}\int_0^\infty e^{-r^2}\,r\,dr\,d\theta\\&=4\frac{\pi}{2}\int_0^\infty e^{-r^2}\,r\,dr.\\&=2\pi\frac{1}{2}\int_0^\infty e^{-r^2}\,d(r^2)\\&=\pi\left[\frac{e^{-r^2}}{-1}\right]_0^\infty\\&=\pi,\end{aligned}$$

whence

$$\Gamma\left(\frac{1}{2}\right)=\sqrt{\pi}.$$

We are now in a position to evaluate $\beta(m,n)$ and $\Gamma(n)$, where m and n are of the form $p+\frac{1}{2}$ where p is a positive integer. Thus, for example,

$$\Gamma\left(\frac{7}{2}\right)=\frac{5}{2}\Gamma\left(\frac{5}{2}\right)=\frac{5}{2}\frac{3}{2}\Gamma\left(\frac{3}{2}\right)=\frac{5}{2}\frac{3}{2}\frac{1}{2}\Gamma\left(\frac{1}{2}\right)=\frac{15}{8}\sqrt{\pi},$$

and

$$\beta\left(\frac{7}{2},\frac{5}{2}\right)=\frac{\Gamma\left(\frac{7}{2}\right)\Gamma\left(\frac{5}{2}\right)}{\Gamma\left(\frac{7}{2}+\frac{5}{2}\right)}=\frac{\frac{15}{8}\sqrt{\pi}\,\frac{3}{4}\sqrt{\pi}}{5!}=\frac{3}{256}\pi.$$

We now deduce a useful general formula for the integral

$$I_{m,n} = \int_0^{\pi/2} \sin^m\theta \cos^n\theta \, d\theta$$

Putting $x = \sin^2\theta$, we obtain:

$$dx = 2\sin\theta\cos\theta \, d\theta$$

or,

$$d\theta = \frac{dx}{2\sqrt{x}\sqrt{1-x}}.$$

Therefore,

$$I_{m,n} = \int_0^1 x^{m/2}(1-x)^{n/2} \frac{dx}{2\sqrt{x}\sqrt{1-x}}$$

$$= \frac{1}{2}\int_0^1 x^{(m-1)/2}(1-x)^{(n-1)/2} dx$$

$$= \frac{1}{2}\beta\left(\frac{m+1}{2}, \frac{n+1}{2}\right), \quad \text{from (6.45)}$$

i.e

$$I_{m,n} = \frac{\Gamma\left(\frac{m+1}{2}\right)\Gamma\left(\frac{n+1}{2}\right)}{2\Gamma\left(\frac{m+n+2}{2}\right)}$$

It follows that

$$I_{m,n} = \int_0^{\pi/2} \sin^m\theta\cos^n\theta \, d\theta = \frac{\Gamma\left(\frac{m+1}{2}\right)\Gamma\left(\frac{n+1}{2}\right)}{2\Gamma\left(\frac{m+n+2}{2}\right)} \tag{6.52}$$

Integrals of this type occur quite frequently in applications and hence the above formula will be quite useful in practice. The following examples illustrate the use of beta anad gamma functions.

Example 6.39 Express the integral

$$\int_0^1 x^m(1-x^n)^p \, dx$$

in terms of gamma functions and evaluate

$$\int_0^1 x^2(1-x^2)^4 \, dx.$$

Let

$$I = \int_0^1 x^m (1-x^n)^p \, dx.$$

Putting $x^n = y$, we obtain $nx^{n-1}\,dx = dy$, and hence

$$dx = \frac{dy}{nx^{n-1}}.$$

Therefore,

$$I = \frac{1}{n}\int_0^1 x^m (1-y)^p \frac{dy}{x^{n-1}}$$

$$= \frac{1}{n}\int_0^1 x^{m-n+1}(1-y)^p \, dy$$

$$= \frac{1}{n}\int_0^1 y^{(m-n+1)/n} (1-y)^p \, dy$$

$$= \frac{1}{n}\int_0^1 y^{\{(m+1)/n\}-1} (1-y)^p \, dy$$

$$= \frac{1}{n}\beta\left(\frac{m+1}{n}, p+1\right)$$

$$= \frac{1}{n}\frac{\Gamma\left(\frac{m+1}{n}\right)\Gamma(p+1)}{\Gamma\left(\frac{m+1}{n}+p+1\right)}$$

Putting $m = 2$, $n = 2$ and $p = 4$, we obtain:

$$\int_0^1 x^2 (1-x^2)^4 \, dx = \frac{1}{2}\frac{\Gamma\left(\frac{3}{2}\right)\Gamma(5)}{\Gamma\left(\frac{13}{2}\right)}$$

$$= \frac{1}{2}\frac{\frac{1}{2}\sqrt{\pi}\,4!}{\frac{11}{2}\frac{9}{2}\frac{7}{2}\frac{5}{2}\frac{3}{2}\frac{1}{2}\sqrt{\pi}}$$

$$= \frac{128}{3465}.$$

Example 6.40 Evaluate

$$\int_0^{\pi/2} \sin^7\theta \cos^5\theta \, d\theta.$$

Putting $m = 7$ and $n = 5$ in (6.52), we obtain:

$$\int_0^{\pi/2} \sin^7\theta \cos^5\theta \, d\theta = \frac{1}{2} \frac{\Gamma\left(\frac{7+1}{2}\right)\Gamma\left(\frac{5+1}{2}\right)}{\Gamma(4+3)}$$

$$= \frac{1}{2} \frac{\Gamma(4)\,\Gamma(3)}{\Gamma(7)}$$

$$= \frac{1}{2} \frac{3!\,2!}{6!}$$

$$= \frac{1}{2} \frac{6\,(2)}{120\,(6)}$$

$$= \frac{1}{120}.$$

Example 6.41 Evaluate $\int_0^{\pi/2} \sin^{10}\theta \, d\theta$.

Putting $m = 10$ and $n = 0$ in (6.52), we obtain:

$$\int_0^{\pi/2} \sin^{10}\theta \, d\theta = \frac{\Gamma\left(\frac{11}{2}\right)\Gamma\left(\frac{1}{2}\right)}{2\Gamma(6)} = \frac{\frac{9}{2}\,\frac{7}{2}\,\frac{5}{2}\,\frac{3}{2}\,\frac{1}{2}\,\pi}{2(5!)} = \frac{63}{512}\pi.$$

Example 6.42 Evaluate the integral

$$\iint_R x^m y^n \, dx \, dy$$

where R represents the region bounded by the positive quadrant of the circle $x^2 + y^2 = a^2$. Determine also the coordinates of its centroid.

Let

$$I = \iint_R x^m y^n \, dx \, dy,$$

Putting

$$\frac{x}{a} = X^{1/2} \quad \text{and} \quad \frac{y}{a} = Y^{1/2},$$

we obtain

$$dx = \frac{a}{2\sqrt{X}}\, dX \quad \text{and} \quad dy = \frac{a}{2\sqrt{Y}}\, dY.$$

Thus, the region now becomes the one bounded by $X = 0$, $Y = 0$ and $X + Y = 1$. The given integral becomes:

$$I = \iint\limits_{R'} a^m X^{m/2}\, a^n Y^{n/2}\, \frac{a}{2\sqrt{X}}\, dX\, \frac{a}{2\sqrt{Y}}\, dy$$

$$= \frac{1}{4} \iint\limits_{R'} a^{m+n+2} X^{(m-1)/2} Y^{(n-1)/2}\, dX\, dY$$

where R' is the region bounded by the triangle $X = 0$, $Y = 0$ and $X + Y = 1$ (see Fig. 6.24). From the figure, it is easily seen that the limits for X are from $X = 0$ to $X = 1 - Y$ and the limits for Y are from $Y = 0$ to $Y = 1$. We therefore have

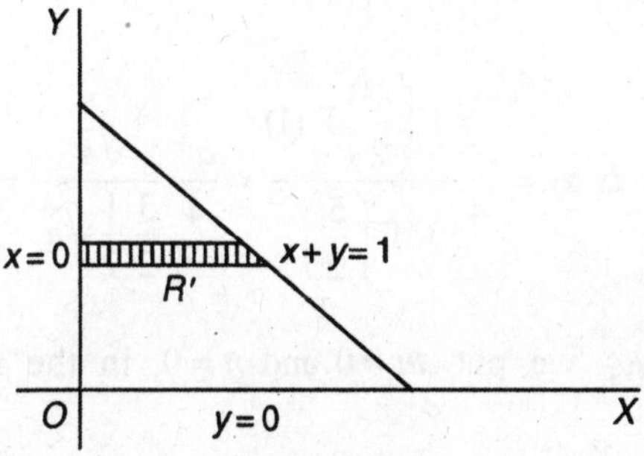

Figure 6.24

$$I = \frac{1}{4} \int_0^1 \int_0^{1-Y} a^{m+n+2} X^{(m-1)/2}\, Y^{(n-1)/2}\, dX\, dY$$

$$= \frac{a^{m+n+2}}{4} \int_0^1 Y^{(n-1)/2} \left[\frac{X^{(m+1)/2}}{(m+1)/2} \right]_0^{1-Y} dY$$

$$= \frac{a^{m+n+2}}{2\,(m+1)} \int_0^1 Y^{(n-1)/2}\, (1 - Y)^{(m+1)/2}\, dY$$

$$= \frac{a^{m+n+2}}{2(m+1)} \beta\left(\frac{n+1}{2}, \frac{m+3}{2}\right)$$

$$= \frac{a^{m+n+2}}{2(m+1)} \frac{\Gamma\left(\frac{n+1}{2}\right)\Gamma\left(\frac{m+3}{2}\right)}{\Gamma\left(\frac{m+n+4}{2}\right)}$$

$$= \frac{a^{m+n+2}}{4} \frac{\Gamma\left(\frac{n+1}{2}\right)\Gamma\left(\frac{m+1}{2}\right)}{\Gamma\left(\frac{m+n}{2}+2\right)}$$

If $(\bar{x}, \bar{y})$ are the coordinates of the centroid, then

$$\bar{x} = \frac{\iint x\, dx\, dy}{\iint dx\, dy} \quad \text{and} \quad \bar{y} = \frac{\iint y\, dx\, dy}{\iint dx\, dy}.$$

To determine $\int\int_R x\, dx\, dy$, we put $m = 1$, $n = 0$ in the above integral and obtain:

$$\iint_R x\, dx\, dy = \frac{a^3}{4} \frac{\Gamma\left(\frac{1}{2}\right)\Gamma(1)}{\Gamma\left(\frac{5}{2}\right)} = \frac{a^3}{4} \frac{\sqrt{\pi}}{\frac{3}{2}\frac{1}{2}\sqrt{\pi}} = \frac{a^3}{3}.$$

To obtain $\int\int_R dx\, dy$, we put $m = 0$ and $n = 0$ in the above integral and obtain:

$$\iint_R dx\, dy = \frac{a^2}{4} \frac{\Gamma\left(\frac{1}{2}\right)\Gamma\left(\frac{1}{2}\right)}{\Gamma(2)} = \frac{a^2}{4}\frac{\pi}{1} = \frac{a^2}{4}\pi.$$

Hence,

$$\bar{x} = \frac{a^3/3}{(1/4)\pi a^2} = \frac{4a}{3\pi}.$$

Similarly,

$$\bar{y} = \frac{4a}{3\pi}.$$

EXERCISES 6.4

Discuss the convergence of the following improper integrals (Problems 1–25):

1. $\int_0^\infty e^{-x}\,dx.$

2. $\int_0^\infty \frac{dx}{a^2+x^2}.$

3. $\int_0^\infty \frac{x}{\sqrt{x^4+1}}\,dx.$

4. $\int_1^\infty \frac{1}{\sqrt{x}}\,dx.$

5. $\int_1^\infty \log x\,dx.$

6. $\int_0^\infty \frac{1}{\sqrt{x^3+1}}\,dx.$

7. $\int_{-\infty}^0 \cosh x\,dx.$

8. $\int_{-\infty}^0 \sinh x\,dx.$

9. $\int_{-\infty}^0 \frac{dx}{(1-3x)^2}.$

10. $\int_0^\infty \frac{e^{-\sqrt{x}}}{x^{1/2}}\,dx.$

11. $\int_0^\infty \frac{\sin x}{x}\,dx.$

12. $\int_{-\infty}^\infty \frac{dx}{1+x^2}.$

13. $\int_a^b \frac{dx}{(a-x)^2}.$

14. $\int_{-1}^1 x^{-1/3}\,dx.$

15. $\int_{-1}^1 \frac{1}{x^2}\,dx.$

16. $\int_{-1}^1 \frac{dx}{\sqrt{x}}.$

17. $\int_{-1}^1 \frac{1}{x}\,dx.$

18. $\int_0^1 \frac{dx}{x}.$

19. $\int_0^\infty \frac{dx}{1+e^x}.$

20. $\int_0^{\pi/2} \tan x\,dx.$

21. $\int_0^1 \frac{1+x^2}{\sqrt{2x-x^2}}\,dx.$

22. $\int_1^\infty \frac{\log_e x}{1+x^2}\,dx.$

23. $\int_1^\infty \frac{e^{-x}}{x}\,dx.$

24. $\int_1^\infty \frac{\sin x}{x^2}\,dx.$

25. $\displaystyle\int_0^\infty \frac{dx}{(1+x)\sqrt{x}}.$

26. Determine a and b such that

$$\lim_{n\to+\infty}\int_{-n}^{+n}\frac{x^3+ax^2+bx}{x^2+x+1}\,dx,$$

exists and is equal to unity.

Show that the following improper integrals converge and evaluate them (Problems 27–36):

27. $\displaystyle\int_0^\infty \frac{1}{x^2+1}\,dx.$

28. $\displaystyle\int_{-1}^{+1}\sqrt{\frac{1+x}{1-x}}\,dx.$

29. $\displaystyle\int_0^3 \frac{dx}{(x-1)^{2/3}}.$

30. $\displaystyle\int_{-\infty}^{\infty}\frac{x}{1+x^2}\,dx.$

31. $\displaystyle\int_1^\infty \frac{dx}{x^2\sqrt{x^2-1}}.$

32. $\displaystyle\int_1^\infty \frac{1}{x^2+x}\,dx.$

33. $\displaystyle\int_{-1}^{1}\frac{1}{x}\,dx.$

34. $\displaystyle\int_{-1}^{1}x^{-1/3}\,dx.$

35. $\displaystyle\int_{-\pi/4}^{\pi/2}\cot x\,dx.$

36. $\displaystyle\int_0^2 \frac{x^2}{1-x^2}\,dx.$

Evaluate the following integrals using beta and gamma functions (Problems 37–54):

37. $\displaystyle\int_0^\infty e^{-x^2}\,dx.$

38. $\displaystyle\int_0^1 x^m\left(\log\frac{1}{x}\right)^n dx.$

39. $\displaystyle\int_0^{\pi/2}\sqrt{\tan x}\,dx.$

40. $\displaystyle\int_0^\infty x^4 e^{-x^4}\,dx.$

41. $\displaystyle\int_0^\infty e^{-x^3}\,dx.$

42. $\displaystyle\int_0^{\pi/2}\frac{dx}{\sqrt{\cos x}}.$

43. $\displaystyle\int_0^1 \frac{dx}{\sqrt{1-x^3}}.$

44. $\displaystyle\int_0^{\pi/2}\sin^6 x\cos^7 x\,dx.$

45. $\int_0^\infty \frac{x^2}{1+x^4}\,dx\cdot$

46. $\int_0^1 \frac{dx}{\sqrt{1-x^n}}\cdot$

47. $\int_0^1 x^{5/2}(1-x^3)^4\,dx.$

48. $\int_0^1 \frac{x^4}{(1-x^2)^{1/2}}\,dx.$

49. $\int_0^1 \left(x\log\frac{1}{x}\right)^{1/3} dx.$

50. $\int_0^1 \frac{1}{(1-x^4)^{1/2}}\,dx.$

51. $\int_0^2 \frac{x^2}{\sqrt{2-x}}\,dx.$

52. $\int_0^{\pi/2} \sin^4\theta\cos^6\theta\,d\theta.$

53. $\int_0^{\pi/2} \frac{d\theta}{\sqrt{\sin\theta}}\cdot$

54. $\int_0^a x^3(a^3-x^3)^5\,dx.$

55. Show that

(a) $\beta\left(n,\frac{1}{2}\right)=\frac{2^{2n}(n!)^2}{n\,(2n)!}$

(b) $\beta\left(n+\frac{1}{2},\frac{1}{2}\right)=\frac{(2n)!}{2^{2n-1}(n!)^2}\frac{\pi}{2}$

(c) $\Gamma\left(n+\frac{1}{2}\right)=\frac{1\cdot3\cdot5\cdots(2n-1)}{2^n}\sqrt{\pi}$

(d) $\frac{\beta(m,n+1)}{n}=\frac{\beta(m+1,n)}{m}=\frac{\beta(m,n)}{m+n}.$

56. Using the relation

$$\beta(m,n)=\int_0^\infty \frac{x^{m-1}}{(1+x)^{m+n}}\,dx,$$

deduce that

$$\Gamma(p)\,\Gamma(1-p)=\int_0^\infty \frac{x^{p-1}}{1+x}\,dx,\qquad 0<p<1.$$

57. Show that

$$\int_0^{\pi/2}\sqrt{\sin\theta}\,d\theta\int_0^{\pi/2}\frac{1}{\sqrt{\sin\theta}}\,d\theta=\pi.$$

58. Find the value, in terms of gamma functions, of the integral

$$\iint x^{m-1} y^{n-1}\, dx\, dy$$

over the positive quadrant of the ellipse $(x^2/a^2)+(y^2/b^2)=1$. Determine also the coordinates of its centroid.

59. Using beta functions, find the area bounded by the curve $x^3+y^3=8$ in the first quadrant.

60. Find an expression in terms of gamma functions for the area bounded by the curve $(x/a)^{2m}+(y/a)^{2n}=1$. Hence evaluate the area enclosed by the astroid $x^{2/3}+y^{2/3}=a^{2/3}$.

61. Evaluate

$$\iint \sqrt{xy\,(1-x-y)}\, dx\, dy$$

over the region bounded by $x=0$, $y=0$ and $x+y=1$.

62. Find the volume of the ellipsoid

$$\frac{x^2}{a^2}+\frac{y^2}{b^2}+\frac{z^2}{c^2}=1.$$

63. Show that

$$\iiint \frac{du\, dv\, dw}{\sqrt{1-u^2-v^2-w^2}}=\frac{\pi^2}{8}.$$

64. Evaluate the integral

$$\iiint_V x^{p-1} y^{q-1} z^{r-1} dx\, dy\, dz$$

where V is the volume of the tetrahedron bounded by

$$x \ge 0,\ y \ge 0,\ z \ge 0 \quad \text{and} \quad x+y+z \le 1.$$

7

Numerical Methods

7.1 ERRORS IN NUMERICAL CALCULATIONS

Scientific and engineering research problems may sometimes be very complicated, and therefore cannot be solved analytically. In such cases, recourse will have to be taken to the numerical methods of solution. This chapter presents an introduction to some basic numerical methods which include interpolation formulae, curve fitting, numerical differentiation and integration, and solution of nonlinear equations. Numerical methods for the solutions of ordinary and partial differential equations will be discussed in the next volume.

Since the solutions obtained by numerical methods are only approximations to the true solutions, the study of errors in the numerical solutions is a central concern of numerical analysis. We therefore start with a brief introduction to various kinds of errors which may occur in a problem and derive some elementary results on error propagation in numerical calculations.

7.1.1 Representation of Numbers in the Computer

Since digital computers are the principal means of calculation in numerical analysis, we first consider methods for the representation of numbers in computers and discuss some of their arithmetic operations in the later sections.

There are two main modes of operation which are commonly used on automatic computers—the *fixed point* and *floating-point* computations. The fixed-point mode is used only to represent integers, whereas the floating-point form is used to represent real numbers of different magnitudes.

Most digital computers use the binary number system or some simple variant of it, e.g. the octal system (base 8) or the hexadecimal system (base 16). Usually, the decimal numbers are converted into binary for internal use and, for output, the binary numbers are reconverted to decimal base. As an example of a binary representation and its conversion to a decimal base, we have

$$(10111.101)_2 = 1\times2^4 + 0\times2^3 + 1\times2^2 + 1\times2^1 + 1\times2^0 + 1\times2^{-1} + 0\times2^{-2} + 1\times2^{-3}$$
$$= 16 + 0 + 4 + 2 + 1 + 0.5 + 0 + 0.125$$
$$= 23.625$$

i.e.

$$(10111.101)_2 = (23.625)_{10}$$

where the left-hand side is the binary representation and the right-hand side is its decimal equivalent.

A floating-point number is usually of the form

$$x = \pm\,(0.a_1a_2 \ldots a_n)\times b^m \tag{7.1}$$

where b is the base of the number system used, the characters $a_i\,(i = 1, 2, \ldots, n)$ are all digits in the base b system with $a_1 \neq 0$, and the number m, called the *exponent* or the *characteristic*, is such that $M_1 \le m \le M_2$, where the numbers M_1 and M_2 vary with the computer. The number $a_1\, a_2 \ldots a_n$, containing n digits, is called the *mantissa* and therefore, restricts the length of a number. If in a computer the floating-point number has a fixed mantissa of n digits, we shall say that *word length* of the computer is n.

Following are some examples of floating-point numbers:

(a) *IBM 360 Series:* $b = 16$, $-64 \le m \le 63$. There are single-precision and double-precision floating-point numbers and the length of the mantissa therefore varies accordingly. For single-precision numbers, $n = 6$ in the hexadecimal system and this gives about 7 decimal digits. For double-precision numbers, $n = 14$ which is equivalent to about 16 decimal digits.

(b) *CDC 1600 Series:* For single-precision numbers, $b = 2$ and $-975 \le m \le 1071$, and $n = 48$ which gives about 14 significant digits.

(c) *CDC cyber 170/730:* For single-precision numbers $b = 2$ and $-975 \le m \le 1071$, and $n = 48$ which gives about 14 decimal digits, whereas for double-precision numbers we obtain about 29 decimal digits.

Thus, floating-point numbers in a computer have a finite word length and therefore a limited precision. This means that the results of all arithmetic calculations must be rounded to the number n of digits, i.e. the length of the mantissa. Such an error is called *round-off* error and is discussed in the next section.

7.1.2 Sources of Errors

We start with some simple definitions about errors. If x_T and x are, respectively, the true and computed values of a number, the error ε_x is given by

$$\varepsilon_x = x_T - x \tag{7.2}$$

Usually, we require the relative error (r_x) given by

$$r_x = \frac{x_T - x}{x} \tag{7.3}$$

We now discuss two sources of error, the *round off* and the *chopping* (or truncation) errors. If we are working with decimal computer which has a fixed word length 4 digits, round off of a number like 18.569 produces the number 18.57 whereas chopping gives the number 18.56.

To see the error of chopping, we assume a decimal computer of word length 4 and consider the following examples:

(a) Given $a_T = 0.3456789 \times 10^n$. Here

$$a_T = 0.3457 \times 10^n.$$

Hence,

$$\begin{aligned}\varepsilon_a &= a_T - a = (0.3457 - 0.3456) \times 10^n \\ &= 0.0001 \times 10^n = 10^{-4} \times 10^n \\ &= 10 \times 10^{-5} \times 10^n.\end{aligned}$$

We therefore have

$$|\varepsilon_a| = 10 \times 10^{-5} \times 10^n,$$

and

$$r_a = \frac{10^{-4}}{0.3456} = 2.89 \times 10^{-4}$$

Hence,

$$|r_a| = 2.89 \times 10^{-4}.$$

(b) Let $a_T = -0.3456789 \times 10^n$. Proceeding as in the above case, it can be shown that

$$|r_a| = 2.89 \times 10^{-4}.$$

To derive an expression for absolute relative error in chopping, we consider a floating-point number

$$a_T = [c + d \times b^{-1}] \times b^n, \qquad 0.1 \le |c| < 1 \quad \text{and} \quad 0 \le |d| < 1 \tag{7.4}$$

where b is the base of the system. For example, the floating-point number

$$a_T = 0.3456789 \times 10^n$$

can be written in the form

$$a_T = (0.3456 + 0.789 \times 10^{-4}) \times 10^n.$$

Let $a = c \times b^n$ Hence,

$$|a_T - a| = d \times b^{-t} \times b^n$$

and

$$r_a = \frac{d \times b^{-t} \times b_n}{c \times b^n} = \frac{d}{c} \times b^{-t}$$

Hence

$$|r_a| = \frac{|d|}{|c|} \times b^{-t} \tag{7.5}$$

Now, the maximum value of $|d|/|c|$ is 10, since the maximum value of $|d|$ is 1 and the minimum value of $|c|$ is 0.1. If the system used is binary the maximum value of $|d|/|c|$ can be shown to be 2, and, in general, the maximum value of the ratio is b, where b is the base of the system used.

Hence, formula (7.5) becomes

$$|r_a| \le b \times b^{-t} \tag{7.6}$$

where t is the word length of the computer.

Example 7.1 (i) For IBM 360 series computers, b = 16 and hence

$$|r_a| \le 16 \times 16^{-6} = 16^{-5}, \quad \text{for} \quad t = 6$$

$$= 9.54 \times 10^{-7}.$$

(ii) For CDC cyber, $b = 2$ and $t = 48$.

Hence,

$$|r_a| \le 2 \times 2^{-48} = 2^{-47} = 7.10 \times 10^{-15}.$$

To derive an expression for the rounding error, we proceed thus. Let the floating-point number be written as

$$a_T = \pm(.a_1 a_2 a_3 \ldots a_n a_{n+1} a_{n+2} \ldots)_b \times b^n \tag{7.7}$$

where $a \ne 0$. Assuming that b is even, the rounded version of a_T is given by

$$\left.\begin{aligned} a &= \pm(.a_1 a_2 \ldots a_n) \times b^n & 0 \le a_{n+1} < \frac{b}{2} \\ &= \pm[(.a_1 a_2 \ldots a_n) + (.000\ldots01)_b] \times b^n, & \frac{b}{2} \le a_{n+1} < b \end{aligned}\right\} \tag{7.8}$$

Let $0 \le a_{n+1} < b/2$. In this case, we have

$$a_T - a = \pm(0.00\ldots0a_{n+1}a_{n+2}\ldots)_b \times b^n = \pm(.a_{n+1}\, a_{n+2}\ldots)_b \times b^{-t} \times b^n.$$

Hence

$$|a_T - a| \le \frac{b}{2} \times b^{-t} \times b^n$$

and

$$|r_a| \leq \frac{1}{2} \times b^{1-t} \tag{7.9}$$

The absolute relative error in the second case can also be obtained in the same way. It follows that for a binary computer with t digits in the mantissa.

$$|r_a| \leq 2^{-t}$$

and for a decimal computer, $|r_a| \leq 5 \times 10^{-t}$.

7.1.3 Errors in Addition and Subtraction

Suppose that we wish to find the sum

$$S = a + b, \tag{7.10}$$

where a and b have errors due to rounding, etc. Then, we have:

$$dS = da + db, \tag{7.11}$$

Assuming that $dS = \Delta S$, the error, the above relation can be written as

$$\varepsilon_S = \varepsilon_a + \varepsilon_b \tag{7.12}$$

and the relative error r_s is given by

$$r_s = \frac{\varepsilon_a}{S} + \frac{\varepsilon_b}{S}.$$

But

$$\frac{\varepsilon_a}{a} = r_a \quad \text{and} \quad \frac{\varepsilon_b}{b} = r_b.$$

i.e.

$$r_s = \frac{a}{S} r_a + \frac{b}{S} r_b, \tag{7.13}$$

where r_a and r_b are the relative errors in a and b respectively. Hence we obtain:

$$|r_S| \leq \left|\frac{a}{S}\right| |r_a| + \left|\frac{b}{S}\right| |r_b| \tag{7.14}$$

Let the maximum of $|r_a|$ and $|r_b|$ be r. Then, (7.16) gives:

$$|r_s| \leq \frac{1}{|S|} \left(|a| + |b|\right) r \tag{7.15}$$

For most computers, the value of r is known. Thus, for example, for the IBM 360 computers,

$$r \le 9.54 \times 10^{-7}.$$

The results for subtraction are precisely analogous with those for addition.

Example 7.2 Find the sum of the following approximate numbers, each correct to the indicated significant digits:

$$136.426, \quad 28.3 \quad 321, \quad 68.243, \quad 17.482.$$

Since the number 321 is known to only three significant digits, we round off the other numbers to 0.1 and get

$$136.4, \quad 28.3, \quad 321, \quad 68.2 \quad \text{and} \quad 17.5.$$

The total of these numbers = 571.4 = 571.

To determine the total error of the result, we determine the individual errors:

(a) the sum of the errors of the original data:

$$\begin{aligned} \varepsilon_1 &= 10^{-3} + 10^{-1} + 10^{0} + 10^{-3} + 10^{-3} \\ &= .001 + .1 + 1 + .001 + .001 \\ &= 1.103, \end{aligned}$$

(b) the absolute value of the sum of the rounding errors in the given numbers:

$$\varepsilon_2 = |.026 + 0 + 0 + .043 + .018| = .051$$

(c) and finally, the round-off error in the sum: $\varepsilon_3 = .5$. Hence, the total error (ε) is given by

$$\varepsilon = \varepsilon_1 + \varepsilon_2 + \varepsilon_3 = 1.103 + .051 + .5 = 1.654 < 2$$

The required sum is therefore 571 ± 2.

Example 7.3 Find the difference $D = \sqrt{10} - \pi$ correct to five significant digits.

Since

$$\sqrt{10} = 3.1622776... \quad \text{and} \quad \pi = 3.1415926...,$$

the required result is

$$D = 3.1622776..., -3.1415926... = 0.0206850.$$

Example 7.4 Evaluate $D = \sqrt{3.01} - \sqrt{3}$ correct to 3 significant digits.

We have

$$D = \sqrt{3.01} - \sqrt{3} = \frac{(\sqrt{3.01} - \sqrt{3})(\sqrt{3.01} + \sqrt{3})}{\sqrt{3.01} + \sqrt{3}}$$

$$= \frac{0.01}{\sqrt{3.01} + \sqrt{3}}$$

$$= \frac{0.01}{1.73 + 1.73}$$

$$= \frac{0.01}{3.46}$$

$$= 10^{-2} \times 0.2890$$

$$= 2.89 \times 10^{-3}.$$

Example 7.5 Add the numbers 83.72 and 1.529 in a decimal computer with a fixed word length of 4.

Assuming that the machine does chopping, the sum S is given by

$$S = 83.72 + 1\ 52 = 85.24$$

The absolute and relative errors are given by

$$\varepsilon_S = 0.009 \quad \text{and} \quad r_s = \frac{0.009}{85.24} = 0.0001055 = 0.106 \times 10^{-4}.$$

7.1.4 Floating-point Multiplication and Division

Let

$$P = ab \tag{7.16}$$

Then

$$P - \Delta P = (a - \Delta a)\,(b - \Delta b)$$

or

$$P - \varepsilon_p = (a - \varepsilon_a)\,(b - \varepsilon_b)$$

or

$$\varepsilon_p = ab - (a - \varepsilon_a)\,(b - \varepsilon_b) = a \cdot \varepsilon_b + b \cdot \varepsilon_a - \varepsilon_a \cdot \varepsilon_b$$

Hence

$$\frac{\varepsilon_p}{P} = r_p = \frac{a}{P}\varepsilon_b + \frac{b}{P}\varepsilon_a - \frac{\varepsilon_a \varepsilon_b}{P}$$

$$= \frac{a}{ab}\varepsilon_b + \frac{b}{ab}\varepsilon_a - \frac{\varepsilon_a \varepsilon_b}{ab}$$

$$= \frac{\varepsilon_b}{b} + \frac{\varepsilon_a}{a} - \frac{\varepsilon_a}{a}\frac{\varepsilon_b}{b},$$

i.e.

$$r_p = \frac{\varepsilon_a}{a} + \frac{\varepsilon_b}{b} - \frac{\varepsilon_a}{a} \cdot \frac{\varepsilon_b}{b},$$

where r_P is the relative error in P. Hence,

$$r_P = a_a + r_b - r_a \cdot r_b \tag{7.17}$$

If $|r_a|, |r_b| << 1$, we have

$$r_P \approx r_a + r_b \tag{7.18}$$

Thus, the relative error in the product of two numbers is approximately equal to the sum of the relative errors in the numbers. Proceeding in a similar way, it can be shown that the relative error in the quotient of two numbers is approximately equal to the difference of the relative errors in the numbers.

A simple formula for the absolute error of a quotient can be found directly.

Let

$$Q = \frac{a}{b} \tag{7.19}$$

Then

$$\varepsilon_Q = \frac{a + \varepsilon_a}{b + \varepsilon_b} - \frac{a}{b}$$

$$= \frac{b\varepsilon_a - a\varepsilon_b}{b(b + \varepsilon_b)}$$

$$= \frac{\varepsilon_a - (a/b)\varepsilon_b}{b + \varepsilon_b}$$

$$= \frac{a\,(\varepsilon_a/a) - a\,(\varepsilon_b/b)}{b + \varepsilon_b} \tag{7.20}$$

Let ε be the greatest absolute value of ε_a/a or ε_b/b. Then,

$$\varepsilon_Q \le \frac{a(\varepsilon + \varepsilon)}{b - b\varepsilon} = \frac{2a\varepsilon}{b(1 - \varepsilon)} \tag{7.21}$$

It ε is negligible in comparison with unity, the above relation simplifies to:

$$\varepsilon_Q \le \frac{2a}{b}\,\varepsilon, \tag{7.22}$$

where it is assumed that a and b have errors of the same order of magnitude. On the other hand, the formula is

$$\varepsilon_Q \leq \frac{a}{b}\varepsilon, \tag{7.23}$$

if only a or b is subject to error.

Example 7.6 Find the product of the approximate numbers $a = 2.5$ and $b = 72.398$ correct to the number of digits given.

We take $a = 2.5$ and $b = 72.4$ (after rounding). Hence

$$P = ab = 2.5 \times 72.4 = 181.0 = 1.8 \times 10^2.$$

Example 7.7 Find the product of the numbers

$$a = 512.7 \quad \text{and} \quad b = 0.11576$$

assuming that the computer is a decimal one chopping after 4 digits in the mantissa.

We have

$$\text{True value} = P = 512.7 \times 0.11576 = 59.350152.$$

Now

$$\begin{aligned} \text{approximate value of the product } &= 0.5127 \times 10^3 \times 0.1157 \\ &= 0.05932 \times 10^3 \\ &= 59.32 \\ &= 59. \end{aligned}$$

Hence,

$$\text{Absolute error} = 0.350152$$

and

$$\text{Relative error } = \frac{0.350152}{59.350152} = 5.93 \times 10^{-3}.$$

Example 7.8 Find the number of trustworthy figures in the quotient of 786.5/289.3, assuming that both the numbers are approximate and true only to the number of digits given.

The maximum relative error is given by

$$r_Q = \frac{0.05}{289.3} = 0.0001728.$$

and

$$\text{The quotient} = \frac{786.5}{289.3} = 2.7220.$$

Hence, formula (7.22) gives

$$\varepsilon_Q = 2(2.7220)\,(0.0001728) = 0.0009406.$$

Since the error affects the fourth decimal place only, we conclude that the quotient is correct to three decimal places, and that

$$Q = 2.722.$$

7.1.5 General Formula for Errors

Let

$$y = f(x) \tag{7.24}$$

where $f(x)$ may be any of the functions, say sine, cosine, logarithmic, exponential, etc. Let us suppose that the quantity x is subject to an error, say Δx. Then the error in y is given by

$$\Delta y = f(x + \Delta x) - f(x) \tag{7.25}$$

By mean-value theorem, we have

$$f(x + \Delta x) = f(x) + \Delta x\, f'(\theta) \tag{7.26}$$

Hence, Eq. (7.25) becomes:

$$\Delta y = \Delta x f'(\theta)$$

i.e.

$$\varepsilon_y = \varepsilon_x f'(\theta) \approx \varepsilon_x f'(x + \Delta x). \tag{7.27}$$

In case of a function of several variables, say, $y = f(x_1, x_2, x_3, .., x_n)$. we have

$$\Delta y = \frac{\partial y}{\partial x_1}\Delta x_1 + \frac{\partial y}{\partial x_2}\Delta x_2 + \frac{\partial y}{\partial x_3}\Delta x_3 + \cdots + \frac{\partial y}{\partial x_n}\Delta x_n \tag{7.28}$$

To find the errors $\Delta x_1, \Delta x_2, ..., \Delta x_n$, we assume that all the partial differentials

$$\frac{\partial y}{\partial x_1}\Delta x_1, \frac{\partial y}{\partial x_2}\Delta x_2, ...,$$

contribute the same amount to the total error Δy, so that all the terms in the right-hand side of (7.28) are equal to each other. Hence, we have

$$\Delta y = n\frac{\partial y}{\partial x_1}\Delta x_1 = n\frac{\partial y}{\partial x_2}\Delta x_2 = \cdots = n\frac{\partial y}{\partial x_n}\Delta x_n \tag{7.29}$$

$$\Delta x_1 = \frac{\Delta y}{n(\partial y/\partial x_1)}, \Delta x_2 = \frac{\Delta y}{n(\partial y/\partial x_2)}, ..., \Delta x_n \frac{\Delta y}{n(\partial y/\partial x_n)} \tag{7.30}$$

As an example, let $f(x)$ be the polynomial

$$f(x) = 2x^3 - 3x^2 + 2.$$

Let x be approximated by $x = 1.001$ with an error $\varepsilon = 0.001$. Now,

$$f^1(x) = 6x^2 - 6x = 6x(x-1).$$

We have

$$f^1(0.001) = 6\,(1.001)\,(0.001) \approx 0.6 \times 10^{-2}.$$

Hence,

$$\varepsilon_y = 0.001 \times 0.6 \times 10^{-2} = 0.6 \times 10^{-5}.$$

In the above analysis, the round-off errors which may be generated in the evaluation of the function f are not taken into account.

Example 7.9 In the formula

$$2hR = r^2 + h^2$$

the percentage error in R is not to exceed 0.2%. Find the permissible percentage errors in r and h when $r = 40$ mm and $h = 48$ mm.

We have

$$R = \frac{40 \times 40}{2 \times 48} + 24 = \frac{122}{3}$$

Then,

$$\frac{\partial R}{\partial r} = \frac{r}{h} = \frac{5}{6},$$

and

$$\frac{\partial R}{\partial h} = -\frac{r^2}{2h^2} + \frac{1}{2} = \frac{1}{2} - \frac{1600}{2 \times 48 \times 48} = \frac{1}{2} - \frac{25}{72} = \frac{11}{72}$$

Now,

$$\Delta r = \frac{\Delta R}{2\,(\partial R/\partial r)} = \frac{3}{5}\,\Delta R$$

Hence,

$$100 \times \frac{\Delta r}{r} = \frac{3}{5r} \times 100 \times \frac{\Delta R}{R} \times R = \frac{3}{5 \times 40} \times 0.2 \times \frac{122}{3} = 0.122\%$$

Again,

$$\Delta h = \frac{\Delta R}{2(\partial R/\partial h)} = \frac{72}{11 \times 2} \times \Delta R = \frac{36}{11} \times \Delta R$$

Hence,

$$100 \times \frac{\Delta h}{h} = 100 \times \frac{\Delta R}{R} \times R \times \frac{36}{11 \times 48}$$

$$= 0.2 \times \frac{122}{3} \times \frac{36}{11 \times 48}$$

$$= 0.56\%$$

EXERCISES 7.1

1. An approximate number is given by $x = 36.145$. If the error is ± 0.00074, find the percentage relative error.
2. Round-off the following numbers successively: 2.75864, 3.14159, 0.56452, 4.2925, 0.60652.
3. Round-off the number $\pi = 3.141592653$ to four significant digits and state the absolute error in it.
4. Find the sum of the approximate numbers 0.258, 0.1724, 325.4, 215.2, 12.75, 9.28, 0.0849, 0.0215 and 0.000354.
5. Find the difference $d = \sqrt{3.01} - \sqrt{3}$ to three correct digits.
6. Find the product of the approximate numbers $x = 2.6$ and $y = 65.398$ if both the numbers are correct to the number of digits given.
7. The Maclaurin's expansion of e^x is given by.

$$e^x = 1 + x + \frac{x^2}{2!} + \cdots + \frac{x^{n-1}}{(n-1)!} + \frac{x^n}{n!} e^{\xi}, \quad 0 < \xi < x.$$

   Find the number of terms so that their sum yields the value of e correct to 8 decimal places when $x = 1$.
8. Find the product of the numbers $x = 56.54$ and $y = 12.4$ if both the numbers are correct to the digits given.
9. Find the quotient

$$q = \frac{x}{y},$$

   where $x = 4.536$ and $y = 1.32$, both x and y being correct to the digits given. Find also the relative error in the result.
10. Use the series

$$\log_e \frac{1+x}{1-x} = 2\left(x + \frac{x^3}{3} + \frac{x^5}{5} + \cdots\right)$$

   to compute the value of $\log_e 1.2$ correct to seven decimal places. Determine the number of terms required if the series for $\log_e (1+x)$ were used instead.

7.2 INTERPOLATION AND APPROXIMATION

Let $f(x)$ be a single valued and continuous function: If $f(x)$ is known explicitly, then the values of $f(x)$ corresponding to certain given values of x, say $x_0, x_1, ..., x_n$, can easily be computed and tabulated.

The central problem of numerical analysis is the converse one: Given the set of tabular values (x_i, y_i), $i = 0, 1, ..., n$, satisfying the relation $y = f(x)$, where the explicit nature of $f(x)$ is not known, it is required to find a simpler function, say $\phi(x)$, such that $f(x)$ and $\phi(x)$ agree at the set of tabulated points. Such a process is called *interpolation.* If $\phi(x)$ is a polynomial, then the process is called *polynomial interpolation* and $\phi(x)$ is called the *interpolating polynomial.*

For 'equally' spaced values of x, the formulae of Newton, Stirling, Bessel and Everett will be stated without derivations—but demonstrated with illustrative examples. The formulae of Lagrange and Newton for arbitrary intervals will, however, be derived and demonstrated. In Section 7.2.5 we shall discuss the important problem of the least squares curve fitting.

7.2.1 Lagrange Interpolation

Let $y = f(x)$ be a continuously differentiable function and is defined by the set of $(n+1)$ data points (x_i, y_i), $i = 0, 1, 2, ..., n$. Lagrange's formula gives an approximating function $\phi_n(x)$ [i.e. $\phi_n(x) \approx f(x)$] which is a polynomial in x of degree at most n such that

$$\phi_n(x_i) = f(x_i), \tag{7.31}$$

This means that both the unknown function $f(x)$ and the approximating function $\phi_n(x)$ agree at the points $x = x_i$.

In terms of the data points, the function $\phi_n(x)$ is given by

$$\phi_n(x) = \sum_{i=0}^{n} \frac{\pi_{n+1}(x)}{(x - x_i)\,\pi'_{n+1}(x_i)}\, y_i, \tag{7.32}$$

where

$$\pi_{n+1}(x) = (x - x_0)(x - x_1)(x - x_2)\cdots, (x - x_n) \tag{7.33}$$

and the prime denotes differentiation with respect to x. To understand formula (7.32), we first consider two data points $P_0(x_0, y_0)$ and $P_1(x_1, y_1)$. The approximating function $\phi_1(x)$, in this case, is the straight line joining P_0 and P_1 and is unique. If the equation of P_0P_1 is $y = ax + b$, then a and b can be eliminated using the data points and we readily obtain:

$$\begin{vmatrix} y & x & 1 \\ y_0 & x_0 & 1 \\ y_1 & x_1 & 1 \end{vmatrix} = 0, \tag{7.34a}$$

where $y = f(x) \approx \phi_1(x)$. Equation (7.34a) can be expressed as

$$y = f(x) \approx \phi_1(x) = \frac{x - x_1}{x_0 - x_1} y_0 + \frac{x - x_0}{x_1 - x_0} y_1$$

$$= \frac{(x - x_0)(x - x_1)}{(x - x_0)(x_0 - x_1)} y_0 + \frac{(x - x_0)(x - x_1)}{(x - x_1)(x_1 - x_0)} y_1$$

$$= \frac{\pi_2(x)}{(x - x_0)\pi_2'(x_0)} y_0 + \frac{\pi_2(x)}{(x - x_1)\pi_2'(x_1)} y_1$$

$$= \sum_{i=0}^{1} \frac{\pi_2(x)}{(x - x_i)\pi_2'(x_i)} y_i, \tag{7.34b}$$

which is of the form of (7.32) for the two data points. It is also easily seen that

$$\sum_{i=0}^{1} l_i(x) = \sum_{i=0}^{1} \frac{\pi_2(x)}{(x - x_i)\pi_2'(x_i)} = 1, \tag{7.35}$$

which is an important property of $l_i(x)$, which are called the *Lagrangian coefficients*. Formula (7.34b) is called the *two-point Lagrange interpolation formula*.

From (7.35),

$$l_0(x) = \text{Coeffcient of } y_0 \text{ in } (7.34\text{b}) = \frac{x - x_1}{x_0 - x_1},$$

so that $l_0(x_0) = 1$ and $l_0(x_1) = 0$. Similarly $l_1(x_0) = 0$ and $l_1(x_1) = 1$.

Now, let $P_0(x_0, y_0)$, $P_1(x_1, y_1)$ and $P_2(x_2, y_2)$ be three data points satisfying $y = f(x)$ and $\phi_2(x)$, the second degree polynomial in x satisfying the above points. If

$$y = \phi_2(x) = a + bx + cx^2, \tag{7.36}$$

Then we have,

$$y_0 = a + bx_0 + cx_0^2, \tag{7.37}$$

$$y_1 = a + bx_1 + cx_1^2, \tag{7.38}$$

$$y_2 = a + bx_2 + cx_2^2 \tag{7.39}$$

Eliminating a, b, c from Eqs. (7.36) to (7.39), we obtain

$$\begin{vmatrix} \phi_2(x) & 1 & x & x^2 \\ y_0 & 1 & x_0 & x_0^2 \\ y_1 & 1 & x_1 & x_1^2 \\ y_2 & 1 & x_2 & x_2^2 \end{vmatrix} = 0 \tag{7.40}$$

Simplification of Eq. (7.40) gives the three point Lagrange interpolation formula:

$$\phi_2(x) = \frac{(x-x_1)(x-x_2)}{(x_0-x_1)(x_0-x_2)} y_0 + \frac{(x-x_0)(x-x_2)}{(x_1-x_0)(x_1-x_2)} y_1 + \frac{(x-x_0)(x-x_1)}{(x_2-x_0)(x_2-x_1)} y_2$$

$$= \sum_{i=0}^{2} \frac{\pi_3(x)}{(x-x_i)\,\pi_3'(x)} y_i, \tag{7.41}$$

where

$$\pi_3(x) = (x-x_0)(x-x_1)(x-x_2) \quad \text{and} \quad \sum_{i=0}^{2} \frac{\pi_3(x)}{(x-x_i)\,\pi_3'(x_i)} = 1.$$

The procedure, outlined above, can easily be generalized to obtain the $(n+1)$ point Lagrange interpolation polynomial given in (7.32).

Before presenting a formal proof of the $(n+1)$ point formula, we discuss some numerical examples.

Example 7.10 Given the points (0, 1), (1, 3) and (3, 19) satisfying the function $y = f(x)$, obtain the Lagrange interpolation polynomial $\phi_2(x)$ which approximates $f(x)$ at the set of data points.

We have,

$$\pi_3(x) = x(x-1)(x-3),$$

and

$$\pi_3'(x) = (x-1)(x-3) + x(x-1) + x(x-3).$$

Then, $\pi_3'(0) = 3$, $\pi_3'(1) = -2$ and $\pi_3'(3) = 6$. Hence,

$$\phi_2(x) = \frac{(x-1)(x-3)}{3}(1) + \frac{x(x-3)}{-2}(3) + \frac{x(x-1)}{6}(19).$$

$$= \frac{1}{3}(x^2 - 4x + 3) - \frac{3}{2}(x^2 - 3x) + \frac{19}{6}(x^2 - x)$$

$$= \frac{1}{6}(12x^2 + 6)$$

$$= 2x^2 + 1,$$

which is exactly the function from which we have calculated the given data points.

Example 7.11 Given the points (0, 0), ($\pi/2$, 1) and (π, 0) satisfying the relation $y = \sin x$, find the Lagrange interpolation polynomial which approximates the above data.

We have,

$$\pi_3(x) = x\left(x - \frac{\pi}{2}\right)(x - \pi),$$

and

$$\pi_3'(x) = \left(x - \frac{\pi}{2}\right)(x - \pi) + x\,(x - \pi) + x\left(x - \frac{\pi}{2}\right).$$

Then

$$\pi_3'(0) = \frac{\pi^2}{2}, \qquad \pi_3'\left(\frac{\pi}{2}\right) = -\frac{\pi^2}{4}, \qquad \pi_3'(\pi) = \frac{\pi^2}{2}.$$

Hence

$$\phi_2(x) = \frac{x\,(x - \pi)}{-\pi^2/4}(1) + \frac{x\,(x - \pi/2)}{\pi^2/2}(0)$$

$$= -\frac{4x}{\pi^2}(x - \pi)$$

$$= -\frac{4}{\pi^2}(x^2 - \pi x)$$

$$= \frac{4}{\pi^2}(\pi x - x^2).$$

Putting $x = \pi/4$ in the above function, we get

$$\sin\frac{\pi}{4} \approx \phi_2\left(\frac{\pi}{4}\right) = \frac{4}{\pi^2}\left(\frac{\pi^2}{4} - \frac{\pi^2}{16}\right) = \frac{3}{4} = 0.75,$$

in which the percentage error is about 6%.

Example 7.12 Given the points (0, –12), (1, 0), (3, 12) and (4, 24) satisfying the function $y = f(x)$, find the Lagrange interpolation polynomial which approximates $f(x)$.

Since $y = 0$ when $x = 1$, it follows that $(x-1)$ is a factor of $f(x)$. Let $f(x) = (x-1)\,R(x)$. Then

$$R(x) = \frac{f(x)}{x-1}$$

we now tabulate the values of x and $R(x)$. We have,

x	$R(x)$
0	12
3	6
4	8

Using Lagrange's formula, we obtain:

$$R(x) \approx \frac{(x-3)(x-4)}{12}(12) + \frac{x(x-4)}{-3}(6) + \frac{x(x-3)}{4}(8)$$

$$= x^2 - 7x + 12 - 2x^2 + 8x + 2x^2 - 6x$$

$$= x^2 - 5x + 12$$

Hence the Lagrange interpolation polynomial approximating $f(x)$ is given by

$$\phi_3(x) = (x-1)(x^2 - 5x + 12).$$

We now give a formal derivation of the Lagrange interpolation formula. Let

$$\phi_n(x) = a_0 + a_1 x + a_2 x^2 + \cdots + a_n x^n \tag{7.42}$$

be the polynomial of the nth degree such that the conditions (usually called the *interpolating conditions*)

$$\phi_n(x_i) = y_i, \qquad i = 0, 1, 2, \ldots, n \tag{7.43}$$

are satisfied. Substituting these conditions in (7.42), we obtain the system of equations:

$$\left.\begin{aligned} y_0 &= a_0 + a_1 x_0 + a_2 x_0^2 + \cdots + a_n x_0^n \\ y_1 &= a_0 + a_1 x_1 + a_2 x_1^2 + \cdots + a_n x_1^n \\ y_2 &= a_0 + a_1 x_2 + a_2 x_2^2 + \cdots + a_n x_2^n \\ &\vdots \\ y_n &= a_0 + a_1 x_n + a_2 x_n^2 + \cdots + a_n x_n^n. \end{aligned}\right\} \tag{7.44}$$

The set of equations (7.44) will have a solution if

$$\begin{vmatrix} 1 & x_0 & x_0^2 & \cdots & x_0^n \\ 1 & x_1 & x_1^2 & \cdots & x_1^n \\ 1 & x_2 & x_2^2 & \cdots & x_2^n \\ \vdots & \vdots & \vdots & & \vdots \\ 1 & x_n & x_n^2 & \cdots & x_n^n \end{vmatrix} \neq 0 \tag{7.45}$$

The value of this determinant, called Vandermonde's determinant, is

$$(x_0\ x_1)(x_0 - x_2)\cdots(x_0 - x_n)(x_1 - x_2)(x_1 - x_3)\cdots(x_1 - x_n)\cdots(x_{n-1} - x_n)$$

Eliminating $a_0, a_1, a_2, ..., a_n$ from Eqs. (7.42) and (7.44), we obtain:

$$\begin{vmatrix} \phi_n(x) & 1 & x & x^2 & \cdots & x^n \\ y_0 & 1 & x_0 & x_0^2 & \cdots & x_0^n \\ y_1 & 1 & x_1 & x_1^2 & \cdots & x_1^n \\ \vdots & \vdots & \vdots & \vdots & & \vdots \\ y_n & 1 & x_n & x_n^2 & \cdots & x_n^n \end{vmatrix} = 0, \tag{7.46}$$

which shows that $\phi_n(x)$ is a linear combination of $y_0, y_1, ..., y_n$. Hence we write

$$\phi_n(x) = \sum_{i=0}^{n} l_i(x)\, y_i, \tag{7.47}$$

where $l_i(x)$ are polynomials in x of degree n. But $\phi_n(x_j) = y_j,\ j = 0, 1, 2, ..., n$. Hence it follows that

$$\left.\begin{aligned} l_i(x_j) &= 0 \quad \text{if } i \neq j \text{ and} \\ li(x_i) &= 1 \quad \text{for all } i. \end{aligned}\right\} \tag{7.48}$$

Hence $l_i(x)$ may be written as

$$l_i(x) = \frac{(x - x_0)(x - x_1)\cdots(x - x_{i-1})(x - x_{i+1})\cdots(x - x_n)}{(x_i - x_0)(x_i - x_1)\cdots(x_i - x_{i-1})(x_i - x_{i+1})\cdots(x_i - x_n)}, \tag{7.49}$$

which obviously satisfies the conditions (7.48).

If we now set

$$\pi_{n+1}(x) = (x - x_0)(x - x_1)\ldots(x - x_{i-1})(x - x_i)(x - x_{i+1})\ldots,(x - x_n), \tag{7.50}$$

Then,

$$\pi'_{n+1}(x_i) = \frac{d}{dx}\pi_{n+1}(x)\bigg|_{x=x_i}$$

$$= (x_i - x_0)(x_i - x_1)(x_i - x_2)\cdots(x_i - x_n) \tag{7.51}$$

so that (7.49) becomes

$$l_i(x) = \frac{\pi_{n+1}(x)}{(x - x_i)\,\pi'_{n+1}(x_i)} \tag{7.52}$$

Hence (7.47) gives:

$$\phi_n(x) = \sum_{i=0}^{n} \frac{\pi_{n+1}(x)}{(x - x_i)\,\pi'_{n+1}(x_i)}\, y_i, \tag{7.53}$$

which is Lagrange's interpolation formula. Interchanging x and y in (7.53) we obtain the formula

$$\psi_n(y) = \sum_{i=0}^{n} \frac{\pi_{n+1}(y)}{(y - y_i)\,\pi'_{n+1}(y_i)}\, x_i, \tag{7.54}$$

which is useful *for inverse interpolation.*

A major advantage of this formula is that the coefficients in (7.54) are easily determined. Its disadvantage is that it is inconvenient to pass from one interpolation polynomial to another of degree one greater. In practice, the Lagrangian coefficients, $l_i(x)$, can conveniently be computed by the following scheme: we first compute the differences, row-wise, as given below:

$$\left.\begin{matrix} x - x_0 & x_0 - x_1 & x_0 - x_2 & \cdots & x_0 - x_n \\ x_1 - x_0 & x - x_1 & x_1 - x_2 & \cdots & x_1 - x_n \\ x_2 - x_0 & x_2 - x_1 & x - x_2 & \cdots & x_2 - x_n \\ \vdots & \vdots & \vdots & & \vdots \\ x_n - x_0 & x_n - x_1 & x_n - x_2 & \cdots & x - x_n \end{matrix}\right\} \tag{7.55}$$

It is easily seen that the product of the elements along the diagonal line is $\pi_{n+1}(x)$. Similarly, the product of the elements of the first row is $(x - x_0)$ $\pi'_{n+1}(x_0)$, of the second row is $(x - x_1)\,\pi'_{n+1}(x_1)$, etc. The Lagrangian coefficients can then be computed using formula (7.52).

We now derive a formula for finding the error associated with the Lagrange interpolation polynomial. Let $y = f(x)$ be defined by the $(n+1)$ points (x_i, y_i), $i = 0, 1, 2, \ldots, n$ and is continuous and differentiable $(n+1)$ times. Since $y(x)$ and $\phi_n(x)$ are equal at all $x = x_i$ we put

$$y(x) - \phi_n(x) = k\pi_{n+1}(x), \tag{7.56}$$

where k is to be determined such that Eq. (7.56) holds for any intermediate value of x, say $x = x'$, $x_0 < x' < x_n$. Clearly,

$$k = \frac{y(x') - \phi_n(x')}{\pi_{n+1}(x')} \tag{7.57}$$

We now construct a function $F(x)$ such that

$$F(x) = y(x) - \phi_n(x) - k\pi_{n+1}(x), \tag{7.58}$$

where k is given by (7.57).

It is easily seen that

$$F(x_0) = F(x_1) = F(x_2) = \cdots = F(x_n) = F(x') = 0,$$

i.e. $F(x)$ vanishes $(n+2)$ times in the interval $x_0 \le x \le x_n$. Hence, by the repeated application of Rolle's theorem, $F'(x)$ must vanish $(n+1)$ times, $F''(x)$ must vanish n times, etc., in the interval $x_0 \le x \le x_n$. In particular, $F^{n+1}(x)$ must vanish once in the above interval. Let this point be given by $x = \xi$, $x_0 < \xi < x_n$. On differentiating (7.58) $(n+1)$ times with respect to x and putting $x = \xi$, we obtain

$$0 = y^{n+1}(\xi) - k(n+1)!,$$

so that

$$k = \frac{y^{n+1}(\xi)}{(n+1)!} \tag{7.59}$$

Comparing (7.57) and (7.59), we obtain

$$y(x') - \phi_n(x') = \frac{y^{n+1}(\xi)}{(n+1)!}\pi_{n+1}(x').$$

The general error formula is therefore

$$y(x) - \phi_n(x) = \frac{\pi_{n+1}(x)}{(n+1)!} y^{n+1}(\xi), \qquad x_0 < \xi < x_n \tag{7.60}$$

Let

$$E_\phi = \max_{[x_0,\, x_n]} \left| y(x) - \phi_n(x) \right|,$$

and

$$\left| y^{n+1}(\xi) \right| \le M_{n+1}, \qquad x_0 < \xi < x_n.$$

Then (7.60) gives:

$$E_\phi \le \frac{M_{n+1}}{(n+1)!} \max_{[x_0,\, x_n]} \left| \pi_{n+1}(x) \right| \tag{7.61}$$

Example 7.13 Find the Lagrange interpolating polynomial of degree 2 approximating the function $y = \ln x$ defined by the following table of values. Hence determine the value of ln 2.7 and estimate the error in the value obtained.

x	$y = \ln x$
2.0	0.69315
2.5	0.91629
3.0	1.09861

We have

$$l_0(x) = \frac{(x-2.5)(x-3.0)}{(-0.5(-1.0)} = 2x^2 - 11x + 15,$$

$$l_1(x) = \frac{(x-2)(x-3)}{(0.5)(-0.5)} = -4(x^2 - 5x + 6)$$

$$l_2(x) = \frac{(x-2)(x-2.5)}{(1)(0.5)} = 2x^2 - 9x + 10.$$

Hence,

$$\phi_2(x) = (2x^2 - 11x + 15)(0.69315) - (4x^2 - 20x + 24)(0.91629) + (2x^2 - 9x + 10)(1.09861)$$

$$= -0.08164x^2 + 0.81366x - 0.60761,$$

which is the required Lagrange interpolating polynomial. Putting $x = 2.7$ in $\phi_2(x)$, we obtain

$$\ln 2.7 \approx \phi_2(2.7) = -0.08164(2.7)^2 + 0.81366(2.7) - 0.60761 = 0.9941164$$

Exact value of $\ln 2.7 = 0.9932518$, so that

$$|\text{Error}| = 0.0008646.$$

Now, we obtain an estimate of the error using formula (7.61). We have $y' = 1/x$, $y'' = -1/x^2$. and $y''' = 2/x^3$. Therefore, $y'''(\xi) = 2/\xi^3$.

Clearly, the continuity conditions on $y(x)$ and its derivatives are satisfied in [2, 3]. Hence,

$$y(x) - \phi_2(x) = \frac{(x-2)(x-2.5)(x-3.0)}{6}\frac{2}{\xi^3}, \qquad 2 < \xi < 3.$$

But

$$\left|\frac{1}{\xi^3}\right| \le \frac{1}{2^3} = \frac{1}{8}.$$

When $x = 2.7$, we then obtain

$$|\,\text{Error}\,| \le \left| \frac{(0.7)\,(0.2)\,(-0.3)}{6}\,\frac{2}{8} \right| = \frac{0.7 \times 0.2 \times 0.3}{24} = 0.00175,$$

Which agrees with the actual error obtained above.

Example 7.14 Given the set of values of x and y as (5, 12), (6, 13), (9, 14) and (11, 16), find the value of x corresponding to $y = 15$.

For inverse interpolation, the set of tabular values y and x are written as

y	x
12	5
13	6
14	9
16	11

Lagrange's formula gives:

$$x(y) = \frac{(y-13)\,(y-14)\,(y-16)}{(-1)\,(-2)\,(-4)} \times 5 + \frac{(y-12)\,(y-14)\,(y-16)}{(1)\,(-1)\,(-3)} \times 6$$

$$+ \frac{(y-12)\,(y-13)\,(y-16)}{(2)\,(1)\,(-2)} \times 9 + \frac{(y-12)\,(y-13)\,(y-14)}{(4)\,(3)\,(2)} \times 11$$

$$= -\frac{5}{8}(y-13)\,(y-14)\,(y-16) + 2\,(y-12)\,(y-14)\,(y-16)$$

$$-\frac{9}{4}(y-12)\,(y-13)\,(y-16) + \frac{11}{24}(y-12)\,(y-13)\,(y-14)$$

Hence

$$x(15) = -\frac{5}{8}(2)\,(1)\,(-1) + 2(3)\,(1)\,(-1) - \frac{9}{4}(3)\,(2)\,(-1) + \frac{11}{24}(3)\,(2)\,(1)$$

$$= \frac{5}{4} - 6 + \frac{27}{2} + \frac{11}{4}$$

$$= -2 + \frac{27}{2}$$

$$= \frac{23}{2}$$

$$= 11.5$$

7.2.2 Newton's Divided Difference Formula

Let $(x_i, y_i), i = 0, 1, 2, ..., n$ be the set of values of x and y satisfying the function $y = f(x)$, the values of x_i being not necessarily equally spaced. Then the *first divided difference* of $f(x)$ for the arguments x_0 and x_1 is defined as

$$\frac{f(x_1) - f(x_0)}{x_1 - x_0}$$

and is denoted by $[x_0, x_1]$ so that

$$[x_0, x_1] = \frac{f(x_1) - f(x_0)}{x_1 - x_0} \tag{7.62}$$

Similarly, we have

$$[x_1, x_2] = \frac{f(x_2) - f(x_1)}{x_2 - x_1}, \ldots, [x_{n-1}, x_n] = \frac{f(x_n) - f(x_{n-1})}{x_n - x_{n-1}}.$$

Again, the *second divided difference* of $f(x)$ for the arguments x_0, x_1 and x_2 is defined as

$$[x_0, x_1, x_2] = \frac{[x_1, x_2] - [x_0, x_1]}{x_2 - x_0}, \tag{7.63}$$

and

$$[x_1, x_2, x_3] = \frac{[x_2, x_3] - [x_1, x_2]}{x_3 - x_1}, \text{ etc.}$$

Proceeding in this way, we define

$$[x_0, x_1, ..., x_{n-1}] = \frac{[x_1, x_2, ..., x_{n-1}] - [x_0, x_1, ..., x_{n-2}]}{x_{n-1} - x_0} \tag{7.64}$$

which is the nth *order divided difference* of $f(x)$ for the arguments $x_0, x_1, ..., x_{n-1}$. We note the following properties of divided differences:

(a) From (7.62), we have

$$[x_0, x_1] = \frac{f(x_1) - f(x_0)}{x_1 - x_0} = \frac{f(x_0) - f(x_1)}{x_0 - x_1} = [x_1, x_0].$$

Again,

$$[x_0, x_1, x_2] = \frac{[x_1, x_2] - [x_0, x_1]}{[x_2 - x_0]}$$

$$= \frac{[x_2, x_1] - [x_1, x_0]}{x_2 - x_0}$$

$$= \frac{[x_1, x_0] - [x_2, x_1]}{x_0 - x_2}$$

$$= [x_2, x_1, x_0],$$

which shows that the divided difference is symmetric with respect to its arguments.

(b) The nth divided differences of a polynomial of the nth degree are constant.

Let $y = p(x)$ be a polynomial in x of degree n, and let $y_i = p(x_i)$, $i = 0, 1, 2, .., n-1$. Then

$$[x, x_0] = \frac{y - y_0}{x - x_0},$$

which is a polynomial in x of degree $(n-1)$. Also

$$[x, x_0, x_1] = \frac{[x, x_0] - [x_0, x_1]}{x - x_1},$$

which is a polynomial in x of degree $(n-2)$. By a similar reasoning, we see that $[x, x_0, x_1, ..., x_{n-1}]$, which is a divided difference of the nth order, is a constant.

To derive a formula involving a function $y = f(x)$ and its divided differences, we observe that

$$[x, x_0] = [x_0, x] = \frac{y - y_0}{x - x_0},$$

so that

$$y = y_0 + (x - x_0)[x, x_0] \tag{7.65}$$

Again,

$$[x, x_0, x_1] = \frac{[x_0, x_1] - [x, x_0]}{x_1 - x} = \frac{[x, x_0] - [x_0, x_1]}{x - x_1}$$

Therefore,

$$[x, x_0] = [x_0, x_1] + (x - x_1)[x, x_0, x_1] \tag{7.66}$$

Substituting (7.66) in (7.65), we obtain

$$y = y_0 + (x - x_0)\{[x_0, x_1] + (x - x_1)[x, x_0, x_1]\}$$

$$= y_0 + (x - x_0)[x_0, x_1] + (x - x_0)(x - x_1)[x, x_0, x_1] \tag{7.67}$$

Further, we have

$$[x, x_0, x_1, x_2] = \frac{[x, x_0, x_1] - [x_0, x_1, x_2]}{x - x_2}$$

Therefore,

$$[x, x_0, x_1] = [x_0, x_1, x_2] + (x - x_2)[x, x_0, x_1, x_2] \tag{7.68}$$

Putting this value of $[x, x_0, x_1]$ in (7.67), we obtain

$$y = y_0 + (x - x_0)[x_0, x_1] + (x - x_0)(x - x_1)\{[x_0, x_1, x_2] + (x - x_2)[x, x_0, x_1, x_2]\}$$

$$= y_0 + (x - x_0)[x_0, x_1] + (x - x_0)(x - x_1)[x_0, x_1, x_2] + (x - x_0)(x - x_1)(x - x_2)[x, x_0, x_1, x_2] \tag{7.69}$$

Proceeding in this way, we obtain the formula

$$y = y_0 + (x - x_0)[x_0, x_1] + (x - x_0)(x - x_1)[x_0, x_1, x_2] + (x - x_0)(x - x_1)(x - x_2)[x_0, x_1, x_2, x_3] + \cdots + (x - x_0)(x - x_1)(x - x_2)\cdots(x - x_n)[x, x_0, x_1, \ldots, x_n], \tag{7.70}$$

where the last term is the remainder term in the formula. Formula (7.70) is called *Newton's divided difference interpolation* formula for unequal intervals.

Example 7.15 Using Newton's divided difference formula, find the cubic from the following table of values:

x	$f(x)$
0	1
2	15
3	40
5	156

We form the table of divided differences:

x	$f(x)$			
0	1			
		7		
2	15		6	
		25		1
3	40		11	
		58		
5	156			

Hence

$$f(x) = 1 + x(7) + x(x-2)(6) + x(x-2)(x-3)(1)$$

$$= 1 + 7x + 6x^2 - 12x + x^3 - 5x^2 + 6x$$

$$= x^3 + x^2 + x + 1,$$

which is the required cubic.

Example 7.16 Find $f(4)$ from the following table:

x	$f(x)$
0	8
1	11
3	35
5	123

We form the difference table:

x	$f(x)$			
0	8			
		3		
1	11		3	
		12		1
3	35		8	
		44		
5	123			

We obtain

$$f(x) = 8 + x(3) + x(x-1)(3) + x(x-1)(x-3)(1)$$
$$= 8 + 3x + 3x(x-1) + x(x-1)(x-3)$$

Hence

$$f(4) = 8 + 12 + 36 + 12 = 68$$

7.2.3 Finite Differences: Newton's Formulae

Assume that we have a set of values (x_i, y_i), $i = 0, 1, ..., n$ of function $y = f(x)$, the values of x being equally spaced, i.e., $x_i = x_0 + ih, i = 0, 1, ..., n$. Suppose that we are required to recover the values of $f(x)$ for some intermediate values of x, or to obtain the derivative of $f(x)$ for some x in the range $x_0 \le x \le x_n$. The methods for the solution of these problems are based on the concept of the *differences* of a function which we now proceed to define.

When the values of x are equally spaced as above, interpolation formulae are based on three types of differences, viz., (a) forward (b) backward and (c) central differences. Newton's formulae use forward and backward differences whereas Stirling's, Bessel's and Everett's formulae employ central differences. In what follows we state these formulae, without proof, but demonstrate their use with numerical examples. Forward and backward differences were already introduced in Chapter 1.

Forward differences

If $y_0, y_1, y_2, ..., y_n$ denote a set of values of y, then $y_1 - y_0, y_2 - y_1, ...,$ $y_n - y_{n-1}$ are called the *differences* of y. Denoting these differences by $\Delta y_0, \Delta y_1, ..., \Delta y_{n-1}$ respectively, we have

$$\Delta y_0 = y_1 - y_0, \quad \Delta y_1 = y_2 - y_1, ..., \ \Delta y_{n-1} = y_n - y_{n-1},$$

where Δ is called the *forward difference operator* and $\Delta y_0, \Delta y_1, ...,$ are called *first forward differences.* The differences of the first forward differences are called *second forward differences* and are denoted by $\Delta^2 y_0, \Delta^2 y_1, ...$ Similarly, one can define *third forward differences, fourth forward difference,*etc. Thus,

$$\Delta^2 y_0 = \Delta y_1 - \Delta y_0 = y_2 - y_1 - (y_1 - y_0) = y_2 - 2y_1 + y_0,$$

$$\Delta^3 y_0 = \Delta^2 y_1 - \Delta^2 y_0 = y_3 - 2y_2 + y_1 - (y_2 - 2y_1 + y_0)$$
$$= y_3 - 3y_2 + 3y_1 - y_0,$$

and

$$\Delta^4 y_0 = \Delta^3 y_1 - \Delta^3 y_0 = y_4 - 3y_3 + 3y_2 - y_1 - (y_3 - 3y_2 + 3y_1 - y_0)$$
$$= y_4 - 4y_3 + 6y_2 - 4y_1 + y_0.$$

It is therefore clear that any higher-order difference can easily be expressed in terms of the ordinates, since the coefficients occurring on the right side are the binomial coefficients. The following table shows how the forward differences of all orders can be formed (Table 7.1).

Table 7.1 Forward Difference Table

x	y	Δ	Δ^2	Δ^3	Δ^4	Δ^5	Δ^6
x_0	y_0						
		Δy_0					
x_1	y_1		$\Delta^2 y_0$				
		Δy_1		$\Delta^3 y_0$			
x_2	y_2		$\Delta^2 y_1$		$\Delta^4 y_0$		
		Δy_2		$\Delta^3 y_1$		$\Delta^5 y_0$	
x_3	y_3		$\Delta^2 y_2$		$\Delta^4 y_1$		$\Delta^6 y_0$
		Δy_3		$\Delta^3 y_2$		$\Delta^5 y_1$	
x_4	y_4		$\Delta^2 y_3$		$\Delta^4 y_2$		
		Δy_4		$\Delta^3 y_3$			
x_5	y_5		$\Delta^4 y_4$				
		Δy_5					
x_6	y_6						

Let $(x_0, y_0), (x_1, y_1), \ldots, (x_n, y_n)$ be a set of $(n+1)$ values of x and y and let $y_n(x)$ be a polynomial of the nth degree such that y and $y_n(x)$ agree at the tabulated points given above, where the values of x are equally spaced, i.e.,

$$x_1 - x_0 = x_2 - x_1 = \cdots = x_n - x_{n-1} = h.$$

Then *Newton's forward difference interpolation formula* is given by

$$y_n(x) = y_0 + p\Delta y_0 + \frac{p(p-1)}{2!}\Delta^2 y_0 + \frac{p(p-1)(p-2)}{3!}\Delta^3 y_0 + \cdots$$
$$+ \frac{p(p-1)(p-2)\cdots(p-n+1)}{n!}\Delta^n y_0 \tag{7.71}$$

where $x = x_0 + ph$. This formula is useful for interpolation *near the beginning* of a set of tabular values. The error in this formula is given by

$$y(x) - y_n(x) = \frac{p(p-1)(p-2)\cdots(p-n)}{(n+1)!}\Delta^{n+1} y(\xi), \qquad (x_0 < \xi < x_n). \tag{7.72}$$

Backward differences

The differences $y_1 - y_0, y_2 - y_1, ..., y_n - y_{n-1}$ are called *first backward differences* if they are denoted by $\nabla y_1, \nabla y_2, ..., \nabla y_n$ respectively, so that

$$\nabla y_1 = y_1 - y_0,\ \nabla y_2 = y_2 - y_1, ...,\ \nabla y_n = y_n - y_{n-1},$$

where ∇ is called the *backward difference operator*. In a similar way, one can define backward differences of higher orders. Thus we obtain

$$\nabla^2 y_2 = \nabla y_2 - \nabla y_1 = y_2 - y_1 - (y_1 - y_0) = y_2 - 2y_1 + y_0,$$
$$\nabla^3 y_3 = \nabla^2 y_3 - \nabla^2 y_2 = y_3 - 3y_2 + 3y_1 - y_0, \text{ etc.}$$

The backward difference table is given as in Table 7.2

Table 7.2 Backward Difference Table

x	y	∇	∇^2	∇^3	∇^4	∇^5	∇^6
x_0	y_0						
x_1	y_1	∇y_1					
x_2	y_2	∇y_2	$\nabla^2 y_2$				
x_3	y_3	∇y_3	$\nabla^2 y_3$	$\nabla^3 y_3$			
x_4	y_4	∇y_4	$\nabla^2 y_4$	$\nabla^3 y_4$	$\nabla^4 y_4$		
x_5	y_5	∇y_5	$\nabla^2 y_5$	$\nabla^3 y_5$	$\nabla^4 y_5$	$\nabla^5 y_5$	
x_6	y_6	∇y_6	$\nabla^2 y_6$	$\nabla^3 y_6$	$\nabla^4 y_6$	$\nabla^5 y_6$	$\nabla^6 y_6$

Newton's backward difference interpolation formula is given by

$$y_n(x) = y_n + p\nabla y_n + \frac{p(p+1)}{2!}\nabla^2 y_n + \cdots + \frac{p(p+1)\cdots(p+n-1)}{n}\nabla^n y_n, \quad (7.73)$$

where $p = (x - x_n)/h$. This formula is useful for interpolation *near the end* of the tabular values. The error in the formula is given by

$$y(x) - y_n(x) = \frac{p(p+1)(p+2)\cdots(p+n)}{(n+1)!} h^{n+1} y^{n+1}(\xi), \quad \text{where } x_0 < \xi < x_n \quad (7.74)$$

Use of formulae (7.71) and (7.73) is demonstrated in the examples given below.

Example 7.17 Find the polynomial of degree three which has values equal to 1, 15, 85 and 259 corresponding to the values 0, 2, 4, and 6 of the argument. Let the variables be x and y. Then we have the following difference table:

x	y			
0	1			
		14		
2	15		56	
		70		48
4	85		104	
		174		
6	259			

Here $h = 2$. To use formula (7.71), we take $x_0 = 0$. Then $x = x_0 + ph$ gives $p = x/2$. With $\Delta y_0 = 14$, $\Delta^2 y_0 = 56$ and $\Delta^3 y_0 = 48$, formula (7.71) gives:

$$y_3(x) = 1 + \frac{x}{2}(14) + \frac{\frac{x}{2}\left(\frac{x}{2} - 1\right)}{2}(56) + \frac{\frac{x}{2}\left(\frac{x}{2} - 1\right)\left(\frac{x}{2} - 2\right)}{6}(48)$$

$$= 1 + 7x + 7x(x - 2) + x(x - 2)(x - 4)$$

$$= 1 + 7x^2 - 7x + x(x^2 - 6x + 8)$$

$$= 1 + 7x^2 - 7x + x^3 - 6x^2 + 8x$$

$$= x^3 + x^2 + x + 1,$$

which is the required cubic.

Again, to use formula (7.73), we take $x = x_n + ph = 6 + 2p$. Hence $p = (x - 6)/2$ and $y_n = 259$. With $\nabla y_n = 174$, $\nabla^2 y_n = 104$ and $\nabla^3 y_n = 48$, formula (7.73) gives:

$$y_3(x) = 259 + \frac{(x-6)}{2}(174) + \frac{\left(\frac{x-6}{2}\right)\left(\frac{x-6}{2} + 1\right)}{2}(104)$$

$$+ \frac{\left(\frac{x-6}{2}\right)\left(\frac{x-6}{2} + 1\right)\left(\frac{x-6}{2} + 2\right)}{6}(48)$$

$$= 259 + 87(x - 6) + 13(x - 6)(x - 4) + (x - 6)(x - 4)(x - 2)$$

$$= 259 + 87x - 522 + 13x^2 - 130x + 312 + x^3 - 12x^2 + 44x - 48$$

$$= x^3 + x^2 + x + 1.$$

Example 7.18 The wind pressure p corresponding to velocity v is given by the following table.

v	p
10	1.2
20	2.0
30	4.5
40	8.0

Estimate the pressure p when v is equal to 25 and 35.

The difference table is:

v	p			
10	1.2			
		0.8		
20	2.0		1.7	
		2.5		–0.7
30	4.5		1.0	
		3.5		
40	8.0			

Here $h = 10$ and $v = v_0 + nh = 10 + 10n$. When $v = 25$, we have $n = 15/10 = 1.5$. Hence formula (7.71) gives:

$$p(25) = 1.2 + (1.5)(0.8) + \frac{(1.5)(0.5)}{2}(1.7) + \frac{(1.5)(0.5)(-0.5)}{6}(-0.7)$$

$$= 1.2 + 1.20 + 0.6375 + 0.04375$$

$$= 3.08125.$$

To use formula (7.73), we take $v = v_n + nh = 40 + 10n$. Therefore, when $v = 35$, we obtain $n = -0.5$. Hence formula (7.73) gives:

$$p(35) = 8.0 - 0.5\,(3.5) - \frac{(0.5)(0.5)}{2}(1.0) + \frac{(-0.5)(0.5)(1.5)}{6}(-0.7)$$

$$= 8.0 - 1.75 - 0.125 + 0.04375$$

$$= 6.16875.$$

Example 7.19 Values of x and $\tan x$ are given as (0.10, 0.1003), (0.15, 0.1511), (0.20, 0.2027), (0.25, 0.2553) and (0.30, 0.3093). Using Newton's formulae, find approximate values of tan 0.12 and tan 0.26.

The table of differences is

x	$y = \tan x$	Δ	Δ^2	Δ^3	Δ^4
0.10	0.1003				
		0.0508			
0.15	0.1511		0.0008		
		0.0516		0.0002	
0.20	0.2027		0.0010		0.0002
		0.0526		0.0004	
0.25	0.2553		0.0014		
		0.0540			
0.30	0.3093				

To find tan 0.12, we have $x_0 = 0.10$, $h = 0.05$. Then

$$p = \frac{0.12 - 0.10}{0.05} = \frac{0.02}{0.05} = 0.4$$

Hence

$$\tan 0.12 \approx 0.1003 + 0.4\,(0.0508) + \frac{0.4\,(0.4-1)}{2}(0.0008)$$

$$+ \frac{0.4\,(0.4-1)\,(0.4-2)}{6}(0.0002)^2$$

$$+ \frac{0.4\,(0.4-1)\,(0.4-2)\,(0.4-3)}{24}(0.0002)$$

$$= 0.1003 + 0.0203 - 0.0001,$$

the last two terms having been neglected, since their magnitudes are each less than 0.0001. Therefore, $\tan 0.12 \approx 0.1205$.

To find $\tan 0.26$, we have $0.26 = 0.30 + p(0.05)$, so that $p = -0.8$. Hence formula (7.73) gives:

$$\tan 0.26 \approx 0.3093 - 0.8\,(0.0540) + \frac{-0.8\,(-0.8+1)}{2}(0.0014)$$

$$+ \frac{-0.8(-0.8+1)\,(-0.8+2)}{6}(0.0004)$$

$$+ \frac{-0.8\,(-0.8+1)\,(-0.8+2)\,(-0.8+3)}{24}(0.0002)$$

$$= 0.3093 - 0.0432 - 0.0001$$

$$= 0.2660.$$

7.2.4 Central Differences: Stirling's, Bessel's and Everett's Formulae

The central difference operator δ is defined by the relations

$$y_1 - y_0 = \delta y_{1/2}, \qquad y_2 - y_1 = \delta y_{3/2}, \ldots, y_n - y_{n-1} = \delta y_{n-1/2}.$$

Similarly, higher order central differences can be defined. With the values of x and y as in the earlier cases, a central difference table (Table 7.3) can be formed thus:

Table 7.3 Central Difference Table

x	y	δ	δ^2	δ^3	δ^4	δ^5	δ^6
x_0	y_0						
		$\delta y_{1/2}$					
x_1	y_1		$\delta^2 y_1$				
		$\delta y_{3/2}$		$\delta^3 y_{3/2}$			
x_2	y_2		$\delta^2 y_2$		$\delta^4 y_2$		
		$\delta y_{5/2}$		$\delta^3 y_{5/2}$		$\delta^5 y_{5/2}$	
x_3	y_3		$\delta^2 y_3$		$\delta^4 y_3$		$\delta^6 y_3.$
		$\delta y_{7/2}$		$\delta^3 y_{7/2}$		$\delta^5 y_{7/2}$	
x_4	y_4		$\delta^2 y_4$		$\delta^4 y_4$		
		$\delta y_{9/2}$		$\delta^3 y_{9/2}$			
x_5	y_5		$\delta^2 y_5$				
		$\delta y_{11/2}$					
x_6	y_6						

It is clear from the three tables that in a definite numerical case, the same numbers occur in the same positions whether we use forward, backward or central differences.

The central difference formulae are most suited for interpolation near the middle of a tabulated set. The most important central difference formulae are those due to Stirling, Bessel and Everett. These are stated and illustrated below. Proofs of these formulae are omitted.

We state the central difference formulae with reference to the following difference table in which the central ordinate is taken for convenience as y_0 corresponding to $x = x_0$.

x	y						
x_{-3}	y_{-3}						
		Δy_{-3}					
x_{-2}	y_{-2}		$\Delta^2 y_{-3}$				
		Δy_{-2}		$\Delta^3 y_{-3}$			
x_{-1}	y_{-1}		$\Delta^2 y_{-2}$		$\Delta^4 y_{-3}$		
		Δy_{-1}		$\Delta^3 y_{-2}$		$\Delta^5 y_{-3}$	
x_0	y_0		$\Delta^2 y_{-1}$		$\Delta^4 y_{-2}$		$\Delta^6 y_{-3}$
		Δy_0		$\Delta^3 y_{-1}$		$\Delta^5 y_{-2}$	
x_1	y_1		$\Delta^2 y_0$		$\Delta^4 y_{-1}$		
		Δy_1		$\Delta^3 y_0$			
x_2	y_2		$\Delta^2 y_1$				
		Δy_2					
x_3	y_3						

Stirling's formula

$$y_p = y_0 + p\,\frac{\Delta y_{-1} + \Delta y_0}{2} + \frac{p^2}{2}\Delta^2 y_{-1} + \frac{p\,(p^2-1)}{3!}\,\frac{\Delta^3 y_{-1} + \Delta^3 y_{-2}}{2}$$

$$+\frac{p^2(p^2-1)}{4!}\,\Delta^4 y_{-2} + \cdots \tag{7.75}$$

Bessel's formula

$$y_p = y_0 + p\,\Delta y_0 + \frac{p(p-1)}{2!}\,\frac{\Delta^2 y_{-1} + \Delta^2 y_0}{2} + \frac{p(p-1)\left(p-\frac{1}{2}\right)}{3!}\Delta^3 y_{-1}$$

$$+\frac{(p+1)\,p(p-1)\,(p-2)}{4!}\,\frac{\Delta^4 y_{-2} + \Delta^4 y - 1}{2} + \cdots \tag{7.76}$$

Everett's formula

$$y_p = q y_0 + \frac{q(q^2-1^2)}{3!}\Delta^2 y_{-1} + \frac{q\,(q^2-1^2)\,(q^2-2^2)}{5!}\,\Delta^4 y_{-2} + \cdots$$

$$+p y_1 + \frac{p(p^2-1^2)}{3!}\,\Delta^2 y_0 + \frac{p(p^2-1^2)\,(p^2-2^2)}{5!}\Delta^4 y_{-1} + \cdots \tag{7.77}$$

where $q = 1 - p$ and $x = x_0 + ph$.

Note: Stirling's formula gives the most accurate result for $-1/4 \le p \le 1/4$, and Bessel's formula is most efficient for $1/4 \le p \le 3/4$. In order to use these formulae, therefore, we have to choose x_0 so that p satisfies the appropriate inequality.

Example 7.20 The following table gives the values of e^x for certain equidistant values of x. Find the value of e^x when $x = 0.644$.

x	$y = e^x$
0.61	1.8404
0.62	1.8589
0.63	1.8776
0.64	1.8965
0.65	1.9155
0.66	1.9348
0.67	1.9542

We form the table of differences:

x	$y = e^x$	Δ	Δ^2	Δ^3
0.61	1.8404			
		0.0185		
0.62	1.8589		0.0002	
		0.0187		0
0.63	1.8776		0.0002	
		0.0189		– 0.0001
0.64	1.8965		0.0001	
		0.0190		0.0002
0.65	1.9155		0.0003	
		0.0193		– 0.0002
0.66	1.9348		0.0001	
		0.0194		
0.67	1.9542			

Now,

$$p = \frac{0.644 - 0.64}{0.01} = \frac{0.004}{0.01} = 0.4.$$

Stirling's formula (7.75) gives:

$$y(0.644) = 1.8965 + 0.4 \cdot \frac{0.0189 + 0.0190}{2} + \frac{0.16}{2}(0.0001) + \cdots$$

$$= 1.8965 + 0.0076 = 1.9041 \cdot$$

Bessel's formula (7.76) gives:

$$y(0.644) = 1.8965 + 0.4\,(0.0190) + \frac{0.4(0.4-1)}{2}\,\frac{0.0001+0.0003}{2} + \cdots = 1.8965$$

$$+0.0076 = 1.9041$$

Everett's formula (7.77) gives:

$$y(0.644) = 0.6\,(1.8965) + \frac{0.6(0.36-1)}{2}\,0.0001 + \cdots + 0.4\,(1.9155)$$

$$+\frac{0.4(0.16-1)}{2}\,0.0003 + \cdots$$

$$= 1.1379 + 0.7662$$

$$= 1.9041.$$

Example 7.21 From the table of Example 7.20, find the value of e^x when $x = 0.638$, using Stirling's and Bessel's formulae.

We know that Stirling's formula gives the most accurate result for $-1/4 \le p \le 1/4$ and Bessel's formula is most efficient for $1/4 \le p \le 3/4$. In order to use these formulae, therefore, we have to choose x_0 so that p satisfies the appropriate inequality.

To use Stirling's formula, we choose $x_0 = 0.64$ and $x_n = 0.638$ so that $p = -0.2$. Hence we have

$$y\,(0.638) = 1.8965 - 0.2 \cdot \frac{0.0189 + 0.0190}{2} + \frac{0.04}{2}\,(0.0001) + \cdots$$

$$= 1.8965 - 0.1\,(0.0379)$$

$$= 1.8927,$$

which is correct to the last decimal place.

For Bessel's formula, we choose $x_0 = 0.63$, $x_n = 0.638$, so that $p = 0.8$. Hence we obtain

$$y(0.638) = 1.8776 + 0.8(0.0189) + \frac{0.8(0.8-1)}{2}\,\frac{0.0002+0.0001}{2} + \cdots = 1.8927.$$

Example 7.22 Certain values of x and y are given as (20, 2854), (24, 3162), (28, 3544) and (32, 3992).

Using Bessel's formula find the value of y when $x = 25$. The table of differences is given below:

x	y	Δ	Δ^2	Δ^3
20	2854			
		308		
24	3162		74	
		382		– 8
28	3544		66	
		448		
32	3992			

Let $x_0 = 24$, $x_n = 25$ so that $p = 0.25$. Hence, Bessel's formula gives:

$$y(25) = 3162 + 0.25(382) + \frac{0.25\,(0.25-1)}{2}\,\frac{(74+66)}{2}$$

$$+\frac{0.25(0.25-1)\left(0.25-1/2\right)}{6}\,(-8)$$

$$= 3162 + 95.50 - 6.5625 - 0.0625$$

$$= 3162 + 89$$

$$= 3251.$$

Example 7.23 The following table gives the values of the probability integral

$$f(x) = \frac{2}{\sqrt{\pi}}\int_0^{\infty} e^{-x^2}\,dx$$

for certain equidistant values of x. Find the value of this integral when $x = 0.5437$ using Everett's formula.

x	$y(x)$	Δ	Δ^2	Δ^3	Δ^4
0.51	0.5292437				
		0.0086550			
0.52	0.5378987		–0.0000896		
		0.0085654		–0.0000007	
0.53	0.5464641		–0.0000903		0
		0.0084751		–0.0000007	
0.54	0.5549392		–0.0000910		0
		0.0083841		–0.0000007	
0.55	0.5633233		–0.0000917		0.0000001
		0.0082924		–0.0000006	
0.56	0.5716157		–0.0000923		
		0.0082001			
0.57	0.5798158				

Taking $x_0 = 0.54,\ x = 0.5437,\ h = 0.01$, we obtain

$$p = \frac{0.0037}{0.01} = 0.37, \qquad 1 - p = q = 0.63$$

Hence Everett's formula (7.77) gives:

$$f(0.5437) = py_0 + \frac{p(p^2-1)}{3!}\Delta^2 y_{-1} + \cdots + qy_1 + \frac{q(q^2-1)}{3!}\Delta^2 y_0 + \cdots,$$

since the fourth differences are negligible. Now,

$$f(0.5437) = 0.37(0.5549392) + \frac{0.37(0.1369-1)}{6}(-0.0000910)$$

$$+ 0.63\,(0.5633233) + \frac{0.63\,(0.3969-1)}{6}(-0.0000917)$$

$$= 0.205327504 + 0.00004843429 + 0.354893679 + 0.0000580694835$$

$$= 0.5603277.$$

7.2.5 Empirical Laws and Least Squares Curve Fitting

In the preceding section we considered polynomial interpolation for a given data which were assumed to be true, i.e., free of errors. For a set of $(n+1)$ data points, for example, we replaced the unknown function $f(x)$ with the polynomial $p(x)$, of degree utmost n, which is such that $|\,p(x) - f(x)|$ is very small in the interval $x_0 \le x \le x_n$. In such a case, the polynomial $p(x)$ provides a very satisfactory approximating function in that interval. On the other hand, if the data are obtained from a measurement, it would probably be subject to errors—round-off and systematic. In such a case, it would be pointless to use a polynomial approximation as it would also exactly reproduce all the errors of the data values. Further, since the data values are either measured or observed values, we would not know of any rule (or law) connecting the variables concerned. If we draw a graph of these points, the curve obtained is called the *approximating curve* and its equation (or law) is called an *empirical equation* (or empirical law). Since different curves can be obtained for the same data, we are often interested in finding the *best fit*, and this process is called *curve fitting.* The method of least squares, described below, is probably the most systematic procedure to obtain a best fit to given data points and is widely used in practical computations. It can also be easily implemented on a digital computer.

The method of least squares

Let $(x_i, y_i), i = 1, 2, ..., m$ be a set of m data points and let the curve $y = f(x)$ be fitted to this data. Let

$$e_i = y_i - f(x_i) \tag{7.78}$$

where e_i is the error of approximation at the point $x = x_i$. Then $e_1^2 + e_2^2 + \cdots + e_m^2$ is the sum of the squares of the errors at all data points.

Let

$$S = e_1^2 + e_2^2 + \cdots + e_m^2$$

$$= [y_1 - f(x_1)]^2 + [y_2 - f(x_2)]^2 + \cdots + [y_m - f(x_m)]^2. \quad (7.79)$$

In (7.79), S is the sum of the squares of the errors and the method of least squares consists in minimizing S. We discuss below both *linear* and *nonlinear* least squares curve fitting.

Fitting a straight line. Let $y = a_0 + a_1 x$ be the straight line to be fitted to the data (x_i, y_i), $i = 1, 2, \ldots, m$. Then, corresponding to Eq. (7.79), we have

$$S = [y_1 - (a_0 + a_1 x_1)]^2 + [y_2 - (a_0 + a_1 x_2)]^2 + \cdots + [y_m - (a_0 + a_1 x_m)]^2 \quad (7.80)$$

Now, the conditions for minimum S are:

$$\frac{\partial S}{\partial a_0} = 0, \quad \frac{\partial S}{\partial a_1} = 0, \quad \text{and} \quad \frac{\partial^2 S}{\partial a_0^2} > 0, \quad \frac{\partial^2 S}{\partial a_1^2} > 0, \quad (7.81)$$

From Eq. (7.80) we then obtain:

$$\frac{\partial S}{\partial a_0} = -2[y_1 - (a_0 + a_1 x_1)] - 2[y_2 - (a_0 + a_1 x_2)] - \cdots - 2[y_m - (a_0 + a_1 x_m)].$$

$$\frac{\partial S}{\partial a_1} = -2x_1[y_1 - (a_0 + a_1 x_1)] - 2x_2[y_2 - (a_0 + a_1 x_2)] - \cdots - 2[y_m - (a_0 + a_1 x_m)]x_m \quad (7.82)$$

and

$$\frac{\partial^2 S}{\partial a_0^2} = 2 + 2 + \cdots + m \text{ times} = 2m, \text{ a positive quantity.}$$

$$\frac{\partial^2 S}{\partial a_1^2} = 2x_1^2 + 2x_2^2 + \cdots + 2x_m^2, \text{ a positive quantity.}$$

Using the conditions in (7.81), we obtain

$$\left.\begin{aligned} m a_0 + a_1 \sum_{i=1}^{m} x_i &= \sum_{i=1}^{m} y_i \\ a_0 \sum_{i=1}^{m} x_i + a_1 \sum_{i=1}^{m} x_i^2 &= \sum_{i=1}^{m} x_i y_i \end{aligned}\right\} \quad (7.83)$$

Equations (7.83) are called *normal equations* and from these we obtain the values of a_0 and a_1. Since the conditions for a minimum are satisfied, the equation $y = a_0 + a_1 x$ is the least squares straight line fit to the given data.

From the first equation in (7.83), we see that

$$a_0 + a_1 \bar{x} = \bar{y},$$

where $(\bar{x}, \bar{y})$ is the centroid of the given data points. Thus, it follows that the fitted straight line passes through the centroid of the given points.

Example 7.24 Fit a straight line $y = a_0 + a_1 x$ to the data given by (0, 1), (1, 3), (3, 2), (6, 5) and (8, 4).

We have $m = 5$ and

$$\sum x_i = 0 + 1 + 3 + 6 + 8 = 18,$$

$$\sum y_i = 1 + 3 + 2 + 5 + 4 = 15,$$

$$\sum x_i^2 = 0 + 1 + 9 + 36 + 64 = 110,$$

$$\sum x_i y_i = 0 + 3 + 6 + 30 + 32 = 71.$$

The normal equations (7.83) give:

$$5a_0 + 18a_1 = 15 \tag{i}$$

$$18a_0 + 110a_1 = 71 \tag{ii}$$

Equations (i) and (ii) give the solution,

$$a_0 = 1.646 \quad \text{and} \quad a_1 = 0.3761.$$

Hence the straight line to be fitted to the given data is given by

$$y = 1.646 + 0.3761x.$$

Fitting a parabola. Let

$$y = a_0 + a_1 x + a_2 x^2 \tag{7.84}$$

be a parabola fitted to the given data $(x_i, y_i), i = 1, 2, \ldots, m$. Then we have

$$S = [y_1 - (a_0 + a_1 x_1 + a_2 x_1^2)]^2 + [y_2 - (a_0 + a_1 x_2 + a_2 x_2)]^2 + \cdots$$

$$+ [y_m - (a_0 + a_1 x_m + a_2 x_m^2)]^2 \tag{7.85}$$

Differentiating with respect to a_0, a_1 and a_2, we get

$$\frac{\partial S}{\partial a_0} = -2[y_1 - (a_0 + a_1 x_1 + a_2 x_1^2)] - 2[y_2 - (a_0 + a_1 x_2 = a_2 x_2^2)].$$

$$-2[y_m - (-a_0 + a_1 x_m + a_2 x_m^2)].$$

$$= -2\left[\sum_{i=1}^{m} y_i - \left(ma_0 + a_1\sum_{i=1}^{m} x_i + a_2\sum_{i=1}^{m} x_i\right)^2\right] \tag{7.86a}$$

$$\frac{\partial S}{\partial a_1} = -2x_1[y_1 - (a_0 + a_1x_1 + a_2x_1^2)] - 2x_2[y_2 - (a_0 + a_1x_2 + a_2x_2^2)]$$

$$- \cdots - 2x_m[y_m - (a_0 + a_1x_m + a_2x_m^2)]$$

$$= -2\left[\sum_{i=1}^{m} x_iy_i - \left(a_0\sum_{i=1}^{m} x_i + a_1\sum_{i=1}^{m} x_i^2 + a_2\sum_{i=1}^{m} x_i^3\right)\right] \tag{7.86b}$$

$$\frac{\partial S}{\partial a_2} = -2x_1^2[y_1 - (a_0 + a_1x_1 + a_2x_1^2)] - 2x_2^2[y_2 - (a_0 + a_1x_2 + a_2x_2^2)]$$

$$- \cdots - 2x_m^2[y_m - (a_0 + a_1x_m + a_2x_m^2)]$$

$$= -2\left[\sum_{i=1}^{m} x_i^2y_i - \left(a_0\sum_{i=1}^{m} x_i^2 + a_1\sum_{i=1}^{m} x_i^3 + a_2\sum_{i=1}^{m} x_i^4\right)\right] \tag{7.86c}$$

It is also easily seen that

$$\frac{\partial^2 S}{\partial a_0^2}, \quad \frac{\partial^2 S}{\partial a_1^2}, \quad \frac{\partial^2 S}{\partial a_2^2}$$

are all positive. Setting now

$$\frac{\partial S_1}{\partial a_0} = \frac{\partial S}{\partial a_1} = \frac{\partial S}{\partial a_2} = 0,$$

Equations (7.86a)–(7.86c) give:

$$\left.\begin{aligned} m\,a_0 + a_1\sum_{i=1}^{m} x_i + a_2\sum_{i=1}^{m} x_i^2 &= \sum_{i=1}^{m} y_i \\ a_0\sum_{i=1}^{m} x_i + a_1\sum_{i=1}^{m} x_i^2 + a_2\sum_{i=1}^{m} x_i^3 &= \sum_{i=1}^{m} x_iy_i \\ a_0\sum_{i=1}^{m} x_i^2 + a_1\sum_{i=1}^{m} x_i^3 + a_2\sum_{i=1}^{m} x_i^4 &= \sum_{i=1}^{m} x_i^2y_i. \end{aligned}\right\} \tag{7.87}$$

which are the normal equations for a_0, a_1 and a_2. These values when substituted in (7.84) give the desired parabola of the best fit.

If we fit a polynomial of the *n*th degree, viz.,

$$y = a_0 + a_1 x + a_2 x^2 + \cdots + a_n x^n, \tag{7.88}$$

to the given *m* data points and proceed as above, then the normal equation would be

$$\left.\begin{aligned}
ma_0 + a_1 \sum_{i=1}^{m} x_i + a_2 \sum_{i=1}^{m} x_i^2 + \cdots + a_n \sum_{i=1}^{m} x_i^n &= \sum_{i=1}^{m} y_i \\
a_0 \sum_{i=1}^{m} x_i + a_1 \sum_{i=1}^{m} x_i^2 + \cdots + a_n \sum_{i=1}^{m} x_i^{n+1} &= \sum_{i=1}^{m} x_i y_i \\
a_0 \sum_{i=1}^{m} x_i^2 + a_1 \sum_{i=1}^{m} x_i^3 + \cdots + a_n \sum_{i=1}^{m} x_i^{n+2} &= \sum_{i=1}^{m} x_i^2 y_i \\
&\vdots \\
a_0 \sum_{i=1}^{m} x_i^n + a_1 \sum_{i=1}^{m} x_i^{n+1} + \cdots + a_n \sum_{i=1}^{m} x_i^{2n} &= \sum_{i=1}^{m} x_i^n y_i
\end{aligned}\right\} \tag{7.89}$$

These are $(n+1)$ equations in $(n+1)$ unknowns $a_0, a_1, \ldots, a_n$, and hence can be solved. Equation (7.88) then gives the required *n*th degree polynomial in *x*.

Example 7.25 Fit a parabola of the form $y = a_0 + a_1 x + a_2 x^2$ to the data (0, 1), (2, 7), (4, 21), (6, 43), (8, 73), using the method of least squares.

We form the following table:

x	y	x^2	x^3	x^4	xy	$x^2 y$
0	1	0	0	0	0	0
2	7	4	8	16	14	28
4	21	16	64	256	84	336
6	43	36	216	1296	258	1548
8	73	64	512	4096	584	4672
Total 20	145	120	800	5664	940	6584

Hence the normal equations are

$$5a_0 + 20a_1 + 120a_2 = 145 \tag{i}$$

$$20a_0 + 120a_1 + 800a_2 = 940 \tag{ii}$$

$$120a_0 + 800a_1 + 5664a_2 = 6584 \tag{iii}$$

Simplifying and solving the above equations, we obtain $a_0 = 1$, $a_1 = 1$ and $a_2 = 1$. The required parabola is therefore given by $y = 1 + x + x^2$.

Fitting an exponential curve

Let

$$y = ae^{bx} \tag{7.90}$$

be the curve to be fitted to the data $(x_i, y_i), i = 1, 2, ..., m$. Taking logarithms of both sides of (7.90), we obtain

$$\ln y = \ln a + bx,$$

which is of the form

$$z = A + bx, \tag{7.91}$$

where

$$z = \ln y \quad \text{and} \quad A = \ln a. \tag{7.92}$$

The problem therefore reduces to finding a least squares straight line through (x_i, z_i), $i = 1, 2, ..., m$.

Example 7.26 Determine the constants a and b such that the curve $y = ae^{bx}$ fits the following data:

x	y
0	1.0
1	1.64872
2	2.71828
3	4.48169
4	7.38906

To fit a straight line of the form $z = A + bx$, we require the values of x and $\ln y = z$. We therefore form the following table of values of x and $z (= \ln y)$:

x	$z = \ln y$
0	0.0
1	0.5
2	1.0
3	1.5
4	2.0

Then we obtain:

$$\Sigma x_i = 10, \quad \Sigma z_i = 5.0, \quad \Sigma x_i^2 = 30 \quad \text{and} \quad \Sigma z_i x_i = 15$$

The normal equations are therefore:

$$5A + 10b = 5 \quad \text{and} \quad 10A + 30b = 15$$

which give $A = 0$ and $b = 1/2$. Since $A = \ln a$, it follows that $a = e^0 = 1$. Hence the required exponential fit is given by $y = e^{x/2}$. In a similar way, it is also possible to fit a curve of the type $y = a_1 e^{b_1 x} + a_2 e^{b_2 x}$ to a given data.*

EXERCISES 7.2

1. Show that

$$\sum_{i=0}^{n} \frac{\pi_{n+1}(x)}{(x - x_i)\,\pi'_{n+1}(x_i)} = 1.$$

2. A function $y = f(x)$ is defined by the set of values (5, 12), (6, 13), (9, 14) and (11, 16). Using Lagrange's formula find an approximate value of $f(10)$.
3. Certain corresponding values of x and $\log_{10} x$ are given by (321.0, 2.50651), (322.8, 2.50893), (324.2, 2.51081), and (325.0, 2.51188). Using Lagrange's formula, find an approximate value of $\log_{10} 323.5$.
4. Given $y_1 = 22$, $y_2 = 30$, $y_4 = 82$, $y_7 = 106$, $y_8 = 206$, find an approximate value of y_6, using Lagrange's formula.
5. Using Lagrange's formula, find an approximate value of x when $y = 1$ from the following table:

x	−1	−2	2	4
y	−1	−9	11	69

6. Find the Lagrange polynomial satisfying the set of points (0, 659), (2, 705), (3, 729) and (6, 804).
7. Obtain the estimates of the missing figures in the following table:

x	2.0	2.1	2.2	2.3	2.4	2.5	2.6
y	0.135	0.123	?	0.100	?	0.082	0.074

8. The following table of cube roots is given:

n	3	5	6	8	10
$\sqrt[3]{n}$	1.442	1.710	1.817	2.000	2.154

   Obtain an approximate value of $\sqrt[3]{7}$.
9. Use Newton's divided difference formula to find an approximate value of $f(21)$, given the set of data values: (11, 14646), (17, 83526), (19, 1301981), and (23, 279846).
10. Given $y_0 = -12$, $y_2 = 4$, $y_3 = 6$ and $y_4 = 12$, find y_1 using Newton's formula.

*For this, the reader is referred to Sastry [1992].

11. If $\ln(x)$ is the Lagrange interpolating polynomial approximating a function $f(x)$ defined on $[a, b]$, show that an estimate of the absolute error in the formula is given by

$$|f(x)-\ln(x)|\le\frac{M_{n+1}}{(n+1)!}|\pi_{n+1}(x)|,$$

where $M_{n+1}=\max_{a\le x\le b}|f^{(n+1)}(x)|$.

12. The probability integral

$$f(x)=\frac{2}{\sqrt{\pi}}\int_0^x e^{-x^2}\,dx$$

is defined by the following table:

x	0.51	0.52	0.53	0.54	0.55
$f(x)$	0.5292	0.5379	0.5465	0.5549	0.5633

Find an approximate value of $f(0.543)$.

13. The following data defines the function $y=f(x)=x^2/(1+x^2)$:

x	1	2	3	4	5	6
$f(x)$	0.5	0.8	0.9	0.9412	0.9615	0.9730

Estimate the value of x that corresponds to $f(x)=0.93$.

14. The following table of values of x and $f(x)=e^x$ are given:

x	2.0	2.2	2.3	2.5	2.6	2.8
$f(x)$	7.3891	9.0250	9.9742	12.182	13.464	16.445

Using inverse interpolation, find the value of x which corresponds to $f(x)=11.023$.

15. Given $y_1=22$, $y_2=30$, $y_4=82$, $y_7=106$, $y_{12}=206$, estimate the value of y_8 using divided differences.

16. The pressure p of wind corresponding to velocity v is given by the following table:

v	10	20	30	40
p	1.1	2.0	4.4	7.9

Using Newton's formulae, estimate the values of p when $v=15$ and $v=35$.

17. The probability integral

$$f(x) = \frac{2}{\sqrt{\pi}} \int_0^x e^{x^2}\, dx$$

is tabulated below from $x = 0.51$ to $0.57°$.

x	0.51	0.52	0.53	0.54	0.55	0.56	0.57
$f(x)$	0.5292	0.5379	0.5465	0.5549	0.5633	0.5716	0.5798

Find $f(0.5437)$ using Stirling's formula.

18. Given the data points (0, 143), (4, 158), (8, 177), (12, 199), satisfying the function $y = f(x)$, find $f(5)$ using Bessel's formula.

19. Given $y_{20} = 2854$, $y_{24} = 3162$, $y_{28} = 3544$ and $y_{32} = 3992$, use Everett's formula to estimate the value of y_{25}.

20. Prove the following relations:

(a) $\Delta^n = \delta^n E^{n/2}$ (b) $\dfrac{\Delta^2}{E^2} = \dfrac{1}{E^2} - \dfrac{2}{E} + 1$

(c) $\Delta + \nabla = \dfrac{\Delta}{\nabla} - \dfrac{\nabla}{\Delta}$ (d) $\Delta^2 = (1 + \Delta)\,\delta^2$.

Use the method of least squares in the following (Problems 21–32):

21. Fit a straight line of the form $y = a_1 x + a_2$ to the data given by (1, 6.9), (2, 8.0), (3, 9.1), (4, 10.2) and (5, 11.3).

22. Fit a straight line of the form $y = a_1 + a_2 x$ to the following data: (0, 0.20), (1, 1.50), (2, 2.70), (3, 4.0) and (4, 5.0).

23. Fit a straight line of the form $y = a + bx$ to the following data:

x	1	2	3	4
y	1.6	1.9	2.3	3.3

24. In the tensile testing of a metal bar, the observed values of the load (x in tons) and the elongation (y in 1/100 mm) are given below. If the two variables follow a linear law of the type $y = a + bx$, find best values of a and b.

x	1	2	3	4	5
y	13	28	40	54	70

25. Fit a formula of the type $y = a + bx^2$ to the data given by (1, 21.0), (2, 24.0), (3, 29.0), (4, 36.0) and (5, 45.0).

26. Fit a formula of the type $y = a + bx + cx^2$ to the data given by

x	0	1	2	3	4
y	1.0	1.8	1.3	2.4	6.5

27. Find the best values of a, b, c so that $y = a + bx + cx^2$ fits the following data:

x	10	20	30	40	50	60
y	15.7	17.9	21.0	25.2	30.2	36.1

28. Fit a parabolic curve of the type $y = a + bx + cx^2$ to the following data:

x	1.0	1.5	2.0	2.5	3.0	3.5	4.0
y	0.11	0.13	0.16	0.20	0.27	0.34	0.41

29. Find the values of a, b, c if $y = a + bx + cx^2$ fits the data given below:

x	1	2	3	4	5	6	7	8	9
y	1.0	3.0	3.5	4.0	5.0	5.5	5.5	5.0	4.5

30. Determine the values of a_1 and a_2 so that the curve $y = a_1 e^{a_2 x}$ fits the data given below:

x	0	2	4
y	0.5013	1.001	3.160

31. Fit a formula of the type $y = ae^{bx}$ to the following data:

x	0	1	2	3
y	10.5	21.0	38.5	83.0

32. Fit a formula of the form $y = ab^x$ to the following data:

x	1	2	3	4
y	0.4	1.1	3.5	10.0

7.3 SOLUTION OF NONLINEAR EQUATIONS

A frequently occurring problem in applied mathematics is to find one or more roots of the equation

$$f(x) = 0. \tag{7.93}$$

If $f(x)$ is a quadratic, cubic or a biquadratic expression, the roots may be evaluated by using the available algebraic formulae. On the other hand, when $f(x)$ is a polynomial of higher degree or an expression involving transcendental functions, e.g. $1+\cos x-5x$, $x\tan x-\cosh x$, etc., algebraic methods are not available, and recourse must be taken to find the roots by approximate methods.

In this section, we describe some numerical methods for the solution of equations of the form (7.93), where $f(x)$ may be algebraic, transcendental or combination of both. Solution of systems of such equations will be described in Section 7.3.6.

The methods discussed below usually require knowledge of one or more initial guesses x_0 for the desired root α of $f(x)$. Such a guess can generally be found from the formulation of the problem itself; otherwise, a simple graph of $y=f(x)$ will usually be sufficient to estimate x_0.

7.3.1 The Bisection Method

This method is based on an important property of continuous functions, viz., if a function is continuous in the closed interval (a, b) and if $f(a)$ and $f(b)$ are of opposite signs, then there exists at least one root between a and b. For definiteness, let $f(a)$ be negative and $f(b)$ be positive. Then the root lies between a and b and let its approximate value be given by

$$x_0=\frac{a+b}{2}.$$

If $f(x_0)=0$, we conclude that x_0 is a root of the equation $f(x)=0$. Otherwise, the root lies either between x_0 and b, or between x_0 and a depending on whether $f(x_0)$ is negative or positive. Then, as before, we bisect the interval and repeat the process until the root is known to the desired accuracy.

***Example** 7.27* Find a real root of the equation $f(x)=x^3-x-1=0$.

Since $f(1)$ is negative and $f(2)$ positive, a root lies between 1 and 2. Therefore, we take $x_0=3/2$. Then, $f(x_0)=7/8$, which is positive. Hence, the root lies between 1 and 1.5 and we obtain

$$x_1=\frac{1+1.5}{2}=1.25.$$

Now, $f(x_1)=-19/64$, which is negative. Hence, we conclude that the root lies between 1.25 and 1.5. It follows that

$$x_2=\frac{1.25+1.5}{2}=1.375.$$

Repetition of the process yields the successive approximations

$$x_3=1.3125, \qquad x_4=1.34375, \qquad x_5=1.328125, \text{ etc.}$$

7.3.2 The Regula-falsi Method

In this method, we choose two points x_0 and x_1 such that $f(x_0)$ and $f(x_1)$ are of *opposite* signs. Since the graph of $y = f(x)$ crosses the x-axis between these two points, a root must lie between these points. The equation of the chord joining the points $[x_0, f(x_0)]$ and $[x_1, f(x_1)]$ is:

$$\frac{y - f(x_0)}{x - x_0} = \frac{f(x_1) - f(x_0)}{x_1 - x_0}. \tag{7.94}$$

The method consists in replacing the part of the curve between the points $[x_0, f(x_0)]$ and $[x_1, f(x_1)]$ by means of the chord joining these points and taking the points of intersection of the chord with the x-axis as an approximation to the root. In the present case, the point of intersection is obtained by setting $y = 0$ in (7.94) This gives

$$x = x_0 - \frac{f(x_0)}{f(x_1) - f(x_0)}(x_1 - x_0).$$

Hence, a second approximation to the root of $f(x) = 0$ is given by

$$x_2 = x_0 - \frac{f(x_0)}{f(x_1) - f(x_0)}(x_1 - x_0). \tag{7.95}$$

If now $f(x_2)$ and $f(x_0)$ are of opposite signs, then the root lies between x_0 and x_2, we replace x_1 by x_2 in (7.95) and obtain the next approximation. Otherwise, we replace x_0 by x_2 and generate the next approximation. The procedure is repeated until the root is obtained to the desired accuracy.

The following examples illustrate the application of the method.

Example 7.28 Find a real root of Walls equation $f(x) = x^3 - 2x - 5 = 0$.

It is easily seen that $f(2) = -1$ and $f(3) = 16$. Therefore, a root lies between 2 and 3. Equation (7.95) then gives:

$$x_2 = 2 + \frac{1}{17} = 2.059.$$

Now, $f(x_2) = -0.386$, and hence the root lies between 2.059 and 3.0. Equation (7.95) now gives:

$$x_3 = 2.059 + \frac{0.386}{16.386}(3 - 2.059) = 2.0812.$$

Repeating the process we obtain: $x_4 = 2.0904$, $x_5 = 2.0934$, etc. The correct value is 2.0945..., so that x_5 is correct to only two decimal places. It is obvious that this method is very slow.

Example 7.29 A root of the equation $2x - \log_{10}x = 7$ lies between 3.5 and 4.0. Using the method of false position, find this root correct to five places of decimals.

Let $f(x) = 2x - \log_{10}x - 7$. We have

$$f(3.5) = 7.0 - \log_{10}3.5 - 7 = -0.544068$$

and

$$f(4.0) = 8.0 - \log_{10}4.0 - 7 = 0.397940$$

Taking $x_0 = 3.5$ and $x_1 = 4.0$, Eq. (7.95) gives:

$$x_2 = 3.5 + \frac{0.544068}{0.397940 + 0.544068}(0.5) = 3.788781.$$

Now

$$f(x_2) = 3.788781 \times 2 - \log_{10}3.788781 - 7 = -0.000937529.$$

But

$$f(3.790000) = 3.79 \times 2 - \log_{10}3.79 - 7.0 = 0.001360790$$

Hence, the root lies between 3.788781 and 3.790000. Taking $x_0 = 3.788781$ and $x_1 = 3.79$, we again apply (7.95) and have

$$f(x_0) = -0.000937529, \qquad f(x_1) = 0.001360790.$$

We then obtain:

$$x_2 = 3.788781 + \frac{0.000937529}{0.001360790 + 0.000937529}(0.001219) = 3.789278$$

Now $f(x_2) = -0.0000004$, and the value of x_2 obtained above is correct to five decimal places.

7.3.3 Newton–Raphson Method

The solution to an equation $f(x) = 0$ may often be found by a simple procedure known as the *Newton–Raphson* method. This method consists in replacing the part of the curve between a point, say, $[x_0, f(x_0)]$ and the X-axis by means of the tangent to the curve at the point, and is described graphically in Fig. 7.1. The x-intercept of the tangent is then used as the first approximation.

From the figure, we have

$$f'(x_0) = \tan\theta = \frac{f(x_0)}{x_0 - x_1}$$

and this gives

$$x_1 = x_0 - \frac{f(x_0)}{f'(x_0)}. \qquad (7.96)$$

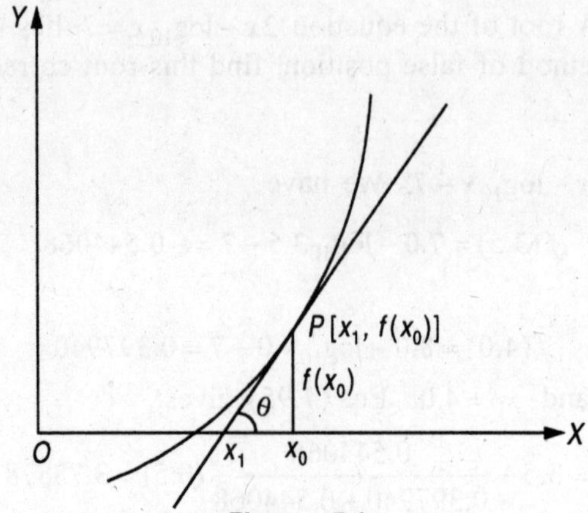

Figure 7.1

An iterative sequence can now be set up as follows:

$$x_{n+1} = x_n - \frac{f(x_n)}{f'(x_n)}. \tag{7.97}$$

This formula can be repeatedly used to find improved approximations to the real root. But it is obvious from formula (7.97) that this method fails if the slope of the curve becomes zero.

The recurrence relation (7.97) may also be derived from the Taylor's series expansion of $f(x)$ about x_0.

Let x_0 be an approximate root of $f(x)=0$ and let $x_1 = x_0 + h$ be the correct root so that $f(x_1)=0$, i.e. $f(x_0+h)=0$. Expanding $f(x_0+h)$ by Taylor's series, we obtain

$$f(x_0) + h f'(x_0) + \frac{h^2}{2!} f''(x_0) + \cdots = 0.$$

Neglecting the second and higher order derivatives, we have

$$f(x_0) + hf'(x_0) = 0,$$

which gives

$$h = -\frac{f(x_0)}{f'(x_0)}.$$

A better approximation than x_0 is therefore given by x_1 where

$$x_1 = x_0 - \frac{f(x_0)}{f'(x_0)}.$$

Successive approximations are given by $x_2, x_3, \ldots, x_{n+1}$ where

$$x_{n+1} = x_n - \frac{f(x_n)}{f'(x_n)}.$$

which is formula (7.97).

To obtain the rate of convergence of this method, we assume that ξ is a root of $f(x)=0$ so that $f(\xi)=0$. Taylor's expansion about x_n gives

$$f(x_n)+(\xi-x_n)f'(x_n)+\frac{1}{2}(\xi-x_n)^2 f''(x_n)+\cdots=0,$$

from which we obtain:

$$-\frac{f(x_n)}{f'(x_n)}=(\xi-x_n)+\frac{1}{2}(\xi-x_n)^2\frac{f''(x_n)}{f'(x_n)}, \tag{7.98}$$

after neglecting higher order terms.

From (7.97) and (7.98), we obtain:

$$x_{n+1}-\xi=\frac{1}{2}(x_n-\xi)^2\frac{f''(x_n)}{f'(x_n)}. \tag{7.99}$$

Setting

$$\varepsilon_n=x_n-\xi, \tag{7.100}$$

equation (7.99) becomes:

$$\varepsilon_{n+1}=\frac{1}{2}\varepsilon_n^2\frac{f''(x_n)}{f'(x_n)}. \tag{7.101}$$

so that the Newton–Raphson method has a second-order or quadratic convergence.

Example 7.30 Find a root of the equation $x^3-3x-5=0$ by using Newton –Raphson method.

Here $f(x)=x^3-3x-5$ and $f'(x)=3x^3-3$. Hence, formula (7.97) gives:

$$x_{x+1}=x_n-\frac{x_n^3-3x_n-5}{3x_n^2-3}. \tag{i}$$

Now, $f(2)=-3$ andt $f(3)=13$. Hence, a root lies between 2 and 3. We choose $x_0=3$ and obtain successively:

$$x_1=3-\frac{13}{24}=2.46, \quad x_2=2.295, \quad x_3=2.279, \quad x_4=2.279$$

Example 7.31 Using Newton–Raphson method, compute approximately the square root of 24.

Let

$$f(x)=x^2-24=0. \tag{i}$$

Now, $f'(x) = 2x$ and hence formula (7.97) gives:

$$x_{n+1} = x_n - \frac{x_n^2 - 24}{2x_n}.$$

i.e.

$$x_{n+1} = \frac{1}{2}\left(x_n + \frac{24}{x_n}\right) \tag{ii}$$

We choose arbitrarily $x_0 = 1$. Then, we obtain from (ii):

$$x_1 = \frac{1}{2}(25) = 12.5$$

$$x_2 = \frac{1}{2}\left(12.5 + \frac{24}{12.5}\right) = 7.21$$

$$x_3 = 5.2693,$$

$$x_4 = 4.9119,$$

$$x_5 = 4.8989, \text{ etc.}$$

Example 7.32 Find a root of the equation $x \sin x + \cos x = 0$.

Let $f(x) = x \sin x + \cos x$; then $f'(x) = x \cos x$. Formula (7.97) gives:

$$x_{n+1} = x_n - \frac{x_n \sin x_n + \cos x_n}{x_n \cos x_n}. \tag{i}$$

Choosing $x_0 = \pi = 3.1416$, we obtain:

$$f(x_0) = -1.0, \qquad f'(x_0) = -3.1416,$$

and

$$x_1 = 3.1416 + \frac{1.0}{-(3.1416)} = 2.8233.$$

Using (i) repeatedly, we obtain:

$$x_2 = 2.7986, \qquad x_3 = 2.7984, \qquad x_4 = 2.7984, \text{etc.}$$

7.3.4 Generalized Newton's Method

If ξ is a root of $f(x) = 0$ with multiplicity p, then the Newton–Raphson formula (7.97) may be written as:

$$x_{n+1} = x_n - p\frac{f(x_n)}{f'(x_n)} \tag{7.102}$$

which means that $(1/p)f'(x_n)$ is the slope of the straight line passing through (x_n, y_n) and intersecting the X-axis at the point $(x_{n+1}, 0)$.

Formula (7.102) is called the *generalized Newton's* formula and reduces to (7.97) when $p=1$. Since ξ is a root of $f(x)=0$ with multiplicity p, it follows that ξ is also a root of $f'(x)=0$ with multiplicity $(p-1)$, of $f''(x)=0$ with multiplicity $(p-2)$, and so on. Hence the expressions

$$x_0 - p\frac{f(x_0)}{f'(x_0)}, \qquad x_0-(p-1)\frac{f'(x_0)}{f''(x_0)}, \qquad x_0-(p-2)\frac{f''(x_0)}{f'''(x_0)},$$

must have the same value if there is a root with multiplicity p, provided that the initial approximation x_0 is chosen sufficiently close to the root.

Example 7.33 Find the double root of the equation $f(x)=x^3-x^2-x+1=0$, choosing $x_0=0.8$

We have

$$f'(x)=3x^2-2x-1 \quad \text{and} \quad f''(x)=6x-2.$$

With $x_0=0.8$, we obtain:

$$x_0-2\frac{f(x_0)}{f'(x_0)}=0.8-2\times\frac{0.072}{-0.68}=1.012,$$

and

$$x_0-\frac{f'(x_0)}{f''(x_0)}=0.8-\frac{-0.68}{2.8}=1.043.$$

The closeness of these values indicate that there is a double root nearer unity. For the next approximation, we choose $x_1=1.01$ and obtain:

$$x_1-2\frac{f(x_1)}{f'(x_1)}=1.01-0.0099=1.0001$$

and

$$x_1-\frac{f'(x_1)}{f''(x_1)}=1.01-0.0099=1.001.$$

We therefore conclude that there is a double root at $x=1.0001$ which is sufficiently close to the actual root, i.e. unity.

7.3.5 Muller's Method

This is a useful method for obtaining both real and complex roots of a polynomial and can also be applied to functions other than polynomials. The method is iterative and converges almost quadratically. In the equations, the function $f(x)$ is approximated by a second degree curve in the vicinity of a root and the roots of the quadratic are then assumed to be the approximations to the roots of the given equation.

Let x_{i-2}, x_{i-1}, x_i be three distinct approximations to a root of $f(x)=0$ and let y_{i-2}, y_{i-1} and y_i be the corresponding values of $y=f(x)$. Assuming that

$$y = Ax^2 + Bx + C \tag{7.103}$$

is the parabola passing through the three points $(x_{i-2}, y_{i-2}), (x_{i-1}, y_{i-1})$ and (x_i, y_i), we obtain immediately:

$$y_{i-2} = Ax_{i-2}^2 + Bx_{i-2} + C \tag{7.104}$$

$$y_{i-1} = Ax_{i-1}^2 + Bx_{i-1} + C \tag{7.105}$$

$$y_i = Ax_i^2 + Bx_i + C \tag{7.106}$$

Eliminating A, B, C from Eqs. (7.104), (7.105) and (7.106), we obtain:

$$\begin{vmatrix} y & x^2 & x & 1 \\ y_{i-2} & x_{i-2}^2 & x_{i-2} & 1 \\ y_{i-1} & x_{i-1}^2 & x_{i-1} & 1 \\ y_i & x_i^2 & x_i & 1 \end{vmatrix} = 0,$$

which can be written in the more convenient form

$$y = \frac{(x-x_{i-1})(x-x_i)}{(x_{i-2}-x_{i-1})(x_{i-2}-x_i)} y_{i-2} + \frac{(x-x_{i-2})(x-x_i)}{(x_{i-1}-x_{i-2})(x_{i-1}-x_i)} y_{i-1}$$

$$+ \frac{(x-x_{i-2})(x-x_{i-1})}{(x_i-x_{i-2})(x_i-x_{i-1})} y_i. \tag{7.107}$$

If we now define the quantities

$$\lambda = \frac{x-x_1}{x_i - x_{i-1}} \tag{7.108}$$

$$\lambda_i = \frac{x_i - x_{i-1}}{x_{i-1} - x_{i-2}} \tag{7.109}$$

and

$$\delta_i = \frac{x_i - x_{i-2}}{x_{i-1} - x_{i-2}} = 1 + \lambda_i, \tag{7.110}$$

equation (7.107) simplifies to

$$y = \frac{y_{i-2}\lambda_i^2 - y_{i-1}\lambda_i\delta_i + y_i\lambda_i}{\delta_i}\lambda^2 + \frac{y_{i-2}\lambda_i^2 - y_{i-1}\delta_i^2 + y_i(\lambda_i + \delta_i)}{\delta_i}\lambda + y_i \tag{7.111}$$

From (7.108), it follows that

$$x = x_i + \lambda(x_i - x_{i-1}), \tag{7.112}$$

and therefore λ is the unknown quantity which we need in order to determine a better approximation to the root. Further, since x is close to x_i, λ should be small in magnitude. To determine λ, we set $y = 0$ in Eq. (7.111) and obtain the quadratic equation

$$\lambda_i(y_{i-2}\lambda_i - y_{i-1}\delta_i + y_i)\lambda^2 + g_i\lambda + \delta_i y_i = 0 \tag{7.113}$$

where

$$g_i = y_{i-2}\lambda_i^2 - y_{i-1}\delta_i^2 + y_i(\lambda_i + \delta_i). \tag{7.114}$$

A direct solution of Eq. (7.113) usually leads to inaccurate results and therefore we rewrite it in the form

$$\frac{1}{\lambda^2} y_i\delta_i + \frac{1}{\lambda} g_i + \lambda_i(y_{i-2}\lambda_i - y_{i-1}\delta_i + y_i) = 0, \tag{7.115}$$

so that

$$\frac{1}{\lambda} = \frac{-g_i \pm [g_i^2 - 4y_i\delta_i\lambda_i(y_{i-2}\lambda_i - y_{i-1}\delta_i + y_i)]^{1/2}}{2y_i\delta_i}.$$

Hence,

$$\lambda = \frac{-2y_i\delta_i}{g_i \pm [g_i^2 - 4y_i\delta_i\lambda_i(y_{i-2}\lambda_i - y_{i-1}\delta_i + y_i)]^{1/2}} \tag{7.116}$$

in which the sign should be so chosen as to make the denominator largest in magnitude. Equation (7.112) then gives a better approximation to the root. The following example illustrates the application of Muller's method.

Example 7.34 Find the root of the equation $y(x) = x^3 - 3x - 5 = 0$ which lies between 2 and 3.

To start with, we choose $x_{i-2} = 1$, $x_{i-1} = 2$ and $x_i = 3$, and obtain: $y_{i-2} = -7$, $y_{i-1} = -3$ and $y_i = 13$, $\lambda = x - 3$, $y_i = 1$, $\delta_i = 2$, and $g_i = 44$. From Eq. (7.115), we obtain the quadratic equation for $1/\lambda$.

$$\frac{26}{\lambda^2} + \frac{44}{\lambda} + 12 = 0.$$

which gives

$$\frac{1}{\lambda} = \frac{-44 \pm \sqrt{688}}{52}$$

We choose the negative sign so that the numerator is largest in magnitude and obtain $\lambda = -0.7403$. The next approximation is therefore given by

$$x_{i+1} = 3 - 0.7403 = 2.26,$$

which is a better result than that obtained by the Newton–Raphson method in Example (7.30).

7.3.6 Solution of Systems of Nonlinear Equations

We now consider the solution of simultaneous nonlinear equations by the Newton–Raphson method. The first approximations are usually obtained from a graph. For simplicity, we consider the case of two equations in two unknowns.

Let the equations be

$$f(x, y) = 0 \quad \text{and} \quad g(x, y) = 0 \tag{7.117}$$

Let (x_0, y_0) be an initial approximation to the root of the system (7.117). If $(x_0 + h, y_0 + k)$ is the root of the system, we must have

$$f(x_0 + h, y_0 + k) = 0 \quad \text{and} \quad g(x_0 + h, y_0 + k) = 0 \tag{7.118}$$

Assuming that f and g are sufficiently differentiable, we expand (7.118) by Taylor's theorem to obtain:

$$\left.\begin{aligned} f_0 + h\frac{\partial f}{\partial x_0} + k\frac{\partial f}{\partial y_0} + \cdots = 0 \\ g_0 + h\frac{\partial g}{\partial x_0} + k\frac{\partial g}{\partial y_0} + \cdots = 0 \end{aligned}\right\} \tag{7.119}$$

where

$$\frac{\partial f}{\partial x_0} = \left[\frac{\partial f}{\partial x}\right]_{x=x_0}, \qquad f_0 = f(x_0, y_0), \text{ etc.}$$

Neglecting the second and higher order terms, we obtain the system of linear equations

$$h\frac{\partial f}{\partial x_0} + k\frac{\partial f}{\partial y_0} = -f_0, \qquad h\frac{\partial g}{\partial x_0} + k\frac{\partial g}{\partial y_0} = -g_0 \tag{7.120}$$

which can be readily solved for h and k. We have thus obtained a new approximation given by

$$x_1 = x_0 + h \quad \text{and} \quad y_1 = y_0 + k.$$

The process can be repeated till we obtain the value to the desired accuracy. We demonstrate the use of this procedure with a simple numerical example.

Example 7.35 Find a real root of the equations:

$$x^2 - y^2 = 4 \quad \text{and} \quad x^2 + y^2 = 16.$$

The solution of the system is trivial and we observe that it is

$$x = \sqrt{10} = 3.162 \quad \text{and} \quad y = \sqrt{6} = 2.449.$$

To apply the Newton–Raphson method, a starting solution is necessary. In the general case, this has to be obtained from a rough graph, but in this particular example we can obtain the first approximation (x_0, y_0) by replacing the hyperbola by means of its asymptote $y = x$. This gives

$$x_0 = y_0 = 2\sqrt{2} = 2.828.$$

Here we have

$$f = x^2 - y^2 - 4 \qquad \text{and} \qquad g = x^2 + y^2 - 16$$

so that

$$\frac{\partial f}{\partial x} = 2x, \qquad \frac{\partial f}{\partial y} = -2y$$

$$\frac{\partial g}{\partial x} = 2x, \qquad \frac{\partial g}{\partial y} = 2y,$$

Equations (7.120) will then give

$$h - k = 0.707 \quad \text{and} \quad h + k = 0,$$

so that $h = 0.354 = -k$. Hence, the second approximation to the root is given by

$$x_1 = x_0 + h = 3.182$$

and

$$y_1 = y_0 + k = 2.474.$$

Replacing (x_0, y_0) by (x_1, y_1) and repeating the above process, we obtain

$$h = -0.0204 \quad \text{and} \quad k = -0.0242,$$

and therefore

$$x_2 = 3.162 \quad \text{and} \quad y_2 = 2.450.$$

EXERCISES 7.3

Obtain a root of the following equations correct to three decimal places using the bisection method (Problems 1–4):

1. $x^3 - 18 = 0.$ **2.** $x^3 - x - 4 = 0.$

3. $x^3 - 4x - 9 = 0.$ **4.** $x^3 + x^2 + x + 7 = 0.$

5. Solve Problems 1–4 using the regula-falsi method.

6. Show that $x^3 - 2x - 17 = 0$ has only one real root and find its value correct to two decimal places.

7. Obtain a root of the equation $x \log_{10} x - 1.2 = 0$ by regula falsi. Give the result correct to two decimal places.

Use Newton–Raphson method to obtain a root of the following equations to the stated accuracy (Problems 8–16):

8. $xe^x - 3 = 0$ (2D). **9.** $x \log_{10} x - 1.2 = 0$ (3D).

10. $3x = \cos x + 1$ (4D). **11.** $\sqrt{12}$ (4D).

12. $\sin x - 1 = x^3$ (3D). **13.** $x + \log_{10} x = 3.375$ (2 significant figures)

14. $x^2 + 4 \sin x = 0$ (4D). **15.** $x^3 - 788 = 0$ (4D, take $x_0 = 14.0$).

16. $\tan x - x - 1/3 = 0$ (2 significant figures, root near 0.9)

17. Establish the formula

$$x_{n+1} = \frac{1}{2}\left(x_n + \frac{N}{x_n}\right)$$

and prove algebraically that

$$\frac{1}{2}\left(x + \frac{N}{x}\right)$$

is more nearly equal to $\sqrt{N}$ than is x.

18. Use Muller's method to find a root of the equations:

(a) $x^3 - x - 1 = 0$ (b) $x^3 - x^2 - x - 1 = 0$.

19. Solve the following systems:

(a) $x^2 + y = 11,\ y^2 + x = 7$. (b) $x^2 = 3xy - 7,\ y = 2(x+1)$.

7.4 NUMERICAL DIFFERENTIATION

In Section 7.2, we considered the general problem of interpolation, viz., given a set of values (x_i, y_i), $i = 0, 1, \ldots, n$ of two variables x and $y(x)$, we obtained a polynomial $\phi(x)$, called the *interpolating polynomial*, such that $\phi(x)$ agrees with $y(x)$ at all x_i. In the present section we shall be concerned with the problem of numerical differentiation whereas the next section will be devoted for a discussion of some numerical integration methods.

Given the set of values of x and y as above, we wish to derive formulae for computing the approximate values of

$$\frac{dy}{dx}, \frac{d^2y}{dx^2}, \ldots \text{ for any } x \in [x_0, x_n]$$

and

$$\int_{x_0}^{x_n} y\, dx.$$

Only an introductory treatment is given here. For more advanced treatment, the reader is referred to a standard text on numerical methods.*

*See, for example, Iyengar and Jain [2003].

7.4.1 Numerical Differentiation Formulae

The general method for deriving the numerical differentiation formulae is to differentiate the interpolating polynomial. Hence, corresponding to each of the formulae derived in Section 7.2, we may derive a formula for the derivative. We illustrate the derivation with Newton's forward difference formula only, the method of derivation being the same in each case.

Newton's forward difference formula is

$$y = y_0 + p\Delta y_0 + \frac{p(p-1)}{2}\Delta^2 y_0 + \frac{p(p-1)(p-2)}{6}\Delta^3 y_0 + \cdots \quad (7.121)$$

where

$$x = x_0 + ph. \quad (7.122)$$

Then

$$\frac{dy}{dx} = \frac{dy}{dp}\frac{dp}{dx} = \frac{1}{h}\left(\Delta y_0 + \frac{2p-1}{2}\Delta^2 y_0 + \frac{3p^2-6p+2}{6}\Delta^3 y_0 + \cdots\right) \quad (7.123)$$

Equation (7.123) can be used for computing the value of dy/dx for any non-tabular value of x in $[x_0, x_n]$.

For *tabular* values of x, the formula takes a simpler form. By setting $x = x_0$ we obtain $p = 0$ from (7.122), and hence (7.123) gives:

$$\left[\frac{dy}{dx}\right]_{x=x_0} = \frac{1}{h}\left(\Delta y_0 - \frac{1}{2}\Delta^2 y_0 + \frac{1}{3}\Delta^3 y_0 - \frac{1}{4}\Delta^4 y_0 + \cdots\right) \quad (7.124)$$

Differentiating (7.123) once again, we obtain

$$\frac{d^2y}{dx^2} = \frac{1}{h^2}\left(\Delta^2 y_0 + \frac{6p-6}{6}\Delta^3 y_0 + \frac{12p^2-36p+22}{24}\Delta^4 y_0 + \cdots\right) \quad (7.125)$$

from which we obtain

$$\left[\frac{d^2y}{dx^2}\right]_{x=x_0} = \frac{1}{h^2}\left(\Delta^2 y_0 - \Delta^3 y_0 + \frac{11}{12}\Delta^4 y_0 - \cdots\right) \quad (7.126)$$

Formulae for computing higher derivatives may be obtained by successive differentiation. In a similar way, different differentiation formulae can be derived by starting with other interpolation formulae. Thus,

(a) Newton's backward difference formula gives:

$$\left[\frac{dy}{dx}\right]_{x=x_n} = \frac{1}{h}\left(\nabla y_n + \frac{1}{2}\nabla^2 y_n + \frac{1}{3}\nabla^3 y_n + \cdots\right) \quad (7.127)$$

and

$$\left[\frac{d^2y}{dx^2}\right]_{x=x_n}=\frac{1}{h^2}\left(\nabla^2 y_n+\nabla^3 y_n+\frac{11}{12}\nabla^4 y_n+\frac{5}{6}\nabla^5 y_n+\cdots\right) \tag{7.128}$$

(b) Stirling's formula gives:

$$\left[\frac{dy}{dx}\right]_{x=x_0}=\frac{1}{h}\left(\frac{\Delta y_{-1}+\Delta y_0}{2}-\frac{1}{6}\frac{\Delta^3 y_{-2}+\Delta^3 y_{-1}}{2}+\frac{1}{30}\frac{\Delta^5 y_{-3}+\Delta^5 y_{-2}}{2}+\cdots\right) \tag{7.129}$$

and

$$\left[\frac{d^2y}{dx^2}\right]_{x=x_0}=\frac{1}{h^2}\left(\Delta^2 y_{-1}-\frac{1}{12}\Delta^4 y_{-2}+\frac{1}{90}\Delta^6 y_{-3}+\cdots\right) \tag{7.130}$$

Interpolation and allied tables [1956] gives a number of differentiation formulae. If a derivative is required *near the end* of a table, one of the following formulae may be used to obtain better accuracy:

$$hy_0'=\left(\Delta-\frac{1}{2}\Delta^2+\frac{1}{3}\Delta^3-\frac{1}{4}\Delta^4+\frac{1}{5}\Delta^5-\frac{1}{6}\Delta^6+\cdots\right)y_0 \tag{7.131}$$

$$=\left(\Delta+\frac{1}{2}\Delta^2-\frac{1}{6}\Delta^3+\frac{1}{12}\Delta^4-\frac{1}{20}\Delta^5+\frac{1}{30}\Delta^6-\cdots\right)y_{-1} \tag{7.132}$$

$$h^2y_0''=\left(\Delta^2-\Delta^3+\frac{11}{12}\Delta^4-\frac{5}{6}\Delta^5+\frac{137}{180}\Delta^6-\frac{7}{10}10^7+\cdots\right)y_0 \tag{7.133}$$

$$=\left(\Delta^2-\frac{1}{12}\Delta^4+\frac{1}{12}\Delta^5-\frac{13}{180}\Delta^6+\cdots\right)y_{-1} \tag{7.134}$$

$$hy_n'=\left(\nabla+\frac{1}{2}\nabla^2+\frac{1}{3}\nabla^3+\frac{1}{4}\nabla^4+\frac{1}{5}\nabla^5+\cdots\right)y_n \tag{7.135}$$

$$=\left(\nabla-\frac{1}{2}\nabla^2-\frac{1}{6}\nabla^3-\frac{1}{12}\nabla^4-\frac{1}{20}\nabla^5+\cdots\right)y_{n+1} \tag{7.136}$$

$$h^2y_n''=\left(\nabla^2+\nabla^3+\frac{11}{12}\nabla^4+\frac{5}{6}\nabla^5+\frac{137}{180}\nabla^6+\cdots\right)y_n \tag{7.137}$$

$$=\left(\nabla^2-\frac{1}{12}\nabla^4-\frac{1}{12}\nabla^5-\frac{13}{180}\nabla^6-\frac{11}{180}\nabla^7-\cdots\right)y_{n+1}. \tag{7.138}$$

Example 7.36 Find the values of dy/dx and d^2y/dx^2 at $x = 1.2$ from the following table:

x	1.0	1.2	1.4	1.6	1.8	2.0	2.2
y	2.7183	3.3201	4.0552	4.9530	6.0496	7.3891	9.0250

The difference table is:

x	y	Δ	Δ^2	Δ^3	Δ^4	Δ^5	Δ^6
1.0	2.7183						
		0.6018					
1.2	3.3201		0.1333				
		0.7351		0.0294			
1.4	4.0552		0.1627		0.0067		
		0.8978		0.0361		0.0013	
1.6	4.9530		0.1988		0.0080		0.0001
		1.0966		0.0441		0.0014	
1.8	6.0496		0.2429		0.0094		
		1.3395		0.0535			
2.0	7.3891		0.2964				
		1.6359					
2.2	9.0250						

Here $x_0 = 1.2$, $y_0 = 3.3201$ and $h = 0.2$. Hence (7.131) gives:

$$\left[\frac{dy}{dx}\right]_{x=1.2} = \frac{1}{0.2}\left[0.7351 - \frac{1}{2}(0.1625) + \frac{1}{3}(0.0361) - \frac{1}{4}(0.0080) + \frac{1}{5}(0.0014)\right]$$

$$= 3.3205.$$

On the other hand, (7.132) gives:

$$\left[\frac{dy}{dx}\right]_{x=1.2} = \frac{1}{0.2}\left[0.6018 + \frac{1}{2}(0.1333) - \frac{1}{6}(0.0294) + \frac{1}{12}(0.0067) - \frac{1}{20}(0.0013) + \frac{1}{30}(0.0001)\right]$$

$$= 3.3205.$$

Similarly, (7.133) gives:

$$\left[\frac{d^2y}{dx^2}\right]_{x=1.2} = \frac{1}{0.04}\left[0.1627 - 0.0361 + \frac{11}{12}(0.0080) - \frac{5}{6}(0.0014)\right] = 3.318$$

Finally, (7.134) gives:

$$\left[\frac{d^2y}{dx^2}\right]_{x=1.2} = \frac{1}{0.04}\left[0.1333 - \frac{1}{12}(0.0067) + \frac{1}{12}(0.0013)\right] = 3.32.$$

Example 7.37 A rod is rotating in a plane about one of its ends. If the following table gives the angle θ radians through which the rod has turned for different values of time t seconds, find its angular velocity and angular acceleration when $t = 0.7$ seconds.

t (secs)	0.0	0.2	0.4	0.6	0.8	1.0
θ (radians)	0.0	0.12	0.48	1.10	2.0	3.20

The difference table is:

t	θ	Δ	Δ^2	Δ^3	Δ^4
0	0.0				
		0.12			
0.2	0.12		0.24		
		0.36		0.02	
0.4	0.48		0.26		0
		0.62		0.02	
0.6	1.10		0.28		0
		0.90		0.02	
0.8	2.00		0.30		
		1.20			
1.0	3.20				

We require $d\theta/dt$ and $d^2\theta/dt^2$ at $t = 0.7$, i.e. at the end of the table. Hence we use Newton's backward difference formula. Here $h = 0.2$, $t_n = 1.0$, $\theta_n = 3.20$ and $t = 0.7$. We have, $1.0 = 0.7 - p(0.2)$, which gives $p = -3/2 = -1.5$.

Newton's backward difference formula is:

$$\theta_n(t) = \theta_n + p\nabla\theta_n + \frac{p(p+1)}{2}\nabla^2\theta_n + \frac{p(p+1)(p+2)}{6}\nabla^3\theta_n + \cdots,$$

where $p = (t - t_n)/h$. Differentiating the above equation successively twice, we get

$$\frac{d\theta}{dt} = \frac{1}{h}\left(\nabla\theta_n + \frac{2p+1}{2}\nabla^2\theta_n + \frac{3p^2+6p+2}{6}\nabla^3\theta_n + \frac{2p^3+9p^2+11p+3}{12}\nabla^4\theta_n + \cdots\right) \tag{i}$$

and

$$\frac{d^2\theta}{dt^2} = \frac{1}{h^2}\left(\nabla^2\theta_n + \frac{6p+6}{6}\nabla^3\theta_n + \cdots\right) \tag{ii}$$

Substituting the differences from the table, we obtain

$$\left[\frac{d\theta}{dt}\right]_{t=0.7} = 5\left[1.2 - \frac{2(1.5)-1}{2}(0.30) + \frac{3(1.5)^2-6(1.5)+2}{6}(0.02)\right]$$

$$= 5\,(1.2 - 0.30 - 0.0008$$

$$= 4.50 \text{ radians/sec.}$$

Also,

$$\left[\frac{d^2\theta}{dt^2}\right]_{t=0.7} = 25\,(0.3-0.01) = 7.25 \text{ radians/sec}^2.$$

Example 7.38 Find the first and second derivatives of the function, $y = 2e^x - x - 1$, tabulated below, at the point $x = 0.6$.

x	0.4	0.5	0.6	0.7	0.8
y	1.5836494	1.7974426	2.0442376	2.3275054	2.6510818

The difference table is:

x	y	Δ	Δ^2	Δ^3	Δ^4
0.4	1.5836494				
		0.2137932			
0.5	1.7974426		0.0330018		
		0.2467950		0.0034710	
0.6	2.0442376		0.0364728		0.0003648
		0.2832678		0.0038358	
0.7	2.3275054		0.0403086		
		0.3235764			
0.8	2.6510818				

Let $x_0 = 0.6$. Then (7.129) and (7.130) give, respectively:

$$\left[\frac{dy}{dx}\right]_{x=0.6} = \frac{1}{0.1}\left(\frac{0.2467950+0.2832678}{2} - \frac{1}{6}\,\frac{0.0034710+0.0038358}{2}\right)$$

$$= 2.6442, \text{ correct to four decimal places.}$$

and

$$\left[\frac{d^2y}{dx^2}\right]_{x=0.6} = 100\left[0.0364728 - \frac{1}{12}(0.0003648)\right]$$

$$= 3.6442, \text{ correct to four decimal places.}$$

Example 7.39 Find the first and second derivatives of y from the following table at $x = 1$.

x	0	1	2	4
y	–5	–6	5	111

We observe that the values of x are not equally spaced. Hence we first find Lagrange's interpolating polynomial for the above data. Let $\phi_3(x)$ be the Lagrange interpolating polynomial. Then we obtain:

$$\phi_3(x) \approx y = \frac{(x-1)(x-2)(x-4)}{(-1)(-2)(-4)}(-5) + \frac{x(x-2)(x-4)}{1(-1)(-3)}(-6)$$

$$+\frac{x(x-1)(x-4)}{2(1)(-2)}(5) + \frac{x(x-1)(x-2)}{4(3)(2)}(111)$$

$$= \frac{5}{8}(x^3 - 7x^2 + 14x - 8) - 2(x^3 - 6x^2 + 8x)$$

$$-\frac{5}{4}(x^3 - 5x^2 + 4x) + \frac{37}{8}(x^3 - 3x^2 + 2x)$$

$$= 2x^3 - 3x - 5, \text{ on simplification.}$$

Hence

$$\frac{dy}{dx} = 6x^2 - 3 \quad \text{and} \quad \left[\frac{dy}{dx}\right]_{x=1} = 3.$$

Also

$$\frac{d^2y}{dx^2} = 12x \quad \text{and} \quad \left[\frac{d^2y}{dx^2}\right]_{x=1} = 12.$$

7.4.2 Maximum and Minimum Values of a Tabulated Function

We know that the maximum and minimum values of a function can be found by equating the first derivative to zero and solving for the variable. The same procedure can be applied to determine the maxima and minima of a tabulated function.

We consider Newton's forward difference formula:

$$y = y_0 + p\Delta y_0 + \frac{p(p-1)}{2!}\Delta^2 y_0 + \frac{p(p-1)(p-2)}{6}\Delta^3 y_0 + \cdots$$

Differentiating the above with respect to p, we obtain:

$$\frac{dy}{dp} = \Delta y_0 + \frac{2p-1}{2}\Delta^2 y_0 + \frac{3p^2-6p+2}{6}\Delta^3 y_0 + \cdots \qquad (7.139)$$

For maxima or minima, $dy/dp = 0$. Terminating the right-hand side after the third difference and equating it to zero, we obtain:

$$\Delta y_0 + \frac{2p-1}{2}\Delta^2 y_0 + \frac{3p^2-6p+2}{6}\Delta^3 y_0 = 0,$$

which simplifies to:

$$\left(\Delta y_0 - \frac{1}{2}\Delta^2 y_0 + \frac{1}{3}\Delta^3 y_0\right) + (\Delta^2 y_0 - \Delta^3 y_0)p + \frac{1}{2}(\Delta^3 y_0)p^2 = 0 \qquad (7.140)$$

Equation (7.140) is a quadratic in p and can be solved.

The values of x can then be found from the relation $x = x_0 + ph$.

Example 7.40 Determine the maxima and minima of the function $y = f(x)$ tabulated below:

x	0	1	2	3
y	–5	–7	–3	13

The difference table is:

x	y	Δ	Δ^2	Δ^3
0	–5			
		–2		
1	–7		6	
		4		6
2	–3		12	
		16		
3	13			

Now,

$$\Delta y_0 - \frac{1}{2}\Delta^2 y_0 + \frac{1}{3}\Delta^3 y_0 = -2 - \frac{1}{2}(6) + \frac{1}{3}(6) = -3$$

and

$$\Delta^2 y_0 - \Delta^3 y_0 = 6 - 6 = 0.$$

Hence Eq. (7.140) gives: $3p^2 - 3 = 0$, and therefore $p = \pm 1$. Since $x_0 = 0$ and $h = 1$, we obtain $x = \pm 1$.

Again, putting $p = x$ in the Newton's forward difference formula, we obtain

$$y = -5 + x(-2) + \frac{x(x-1)}{2}(6) + \frac{x(x-1)(x-2)}{6}(6)$$

$$= -5 - 2x + 3x^2 - 3x + x^3 - 3x^2 + 2x$$

$$= x^3 - 3x - 5.$$

Therefore,

$$\frac{dy}{dx} = 3x^2 - 3 \quad \text{and} \quad \frac{d^2y}{dx^2} = 6x.$$

We therefore have a maximum for $x = -1$ and a minimum for $x = 1$.

EXERCISES 7.4

1. Find $\frac{d}{dx} J_0(x)$ at $x = 0.1$ from the following table:

x	0.0	0.1	0.2	0.3	0.4
$J_0(x)$	1.0000	0.9975	0.9900	0.9776	0.9604

2. Find dy/dx when $x = 0.04$ from the following table:

x	0.01	0.02	0.03	0.04	0.05	0.06
y	0.1023	0.1047	0.1071	0.1096	0.1122	0.1148

3. Values of x and $\sqrt{x}$ are given in the following table:

x	15	17	19	21	23	25
$\sqrt{x}$	3.873	4.123	4.359	4.583	4.796	5.000

Find the first and second derivatives of $\sqrt{x}$ at $x = 15$.

4. The following table gives the angular displacements θ (radians) at different values of t (seconds):

θ	0.052	0.105	0.168	0.242	0.327	0.408	0.489
t	0	0.02	0.04	0.06	0.08	0.10	0.12

Determine the angular velocity at $t = 0.06$.

5. Given that $y(0.96) = 0.7825$, $y(0.98) = 0.7739$, $y(1.0) = 0.7651$, $y(1.02) = 0.7563$, and $y(1.04) = 0.7473$, find an approximate value of d^2y/dx^2 at $x = 0.96$.

6. The values of x and y are as given below:

x	0	1	3	6
y	18	10	–18	90

Evaluate dy/dx and d^2y/dx^2 at $x = 2$.

7. Values of x and y are as given below:

x	0.35	0.40	0.45	0.50	0.55	0.60	0.65
y	1.520	1.506	1.486	1.468	1.445	1.420	1.385

Find the value of dy/dx at $x = 0.4$.

8. Certain values of x and y are tabulated below:

x	0.0	0.1	0.2	0.3	0.4	0.5	0.6
y	3.013	3.162	3.287	3.364	3.395	3.381	3.324

Find dy/dx and d^2y/dx^2 at $x = 0.3$.

9. For the set of points (0, 2), (2, –2) and (3, –1), satisfying a relation $y = f(x)$, evaluate dy/dx at $x = 2$.

10. Certain values of x and $f(x)$ are tabulated below:

x	1.2	1.3	1.4	1.5	1.6
$f(x)$	1.5095	1.6984	1.9043	2.1293	2.3756

Find $f'(1.4)$. Find also the value of x for which y is maximum and find this value of y.

7.5 NUMERICAL INTEGRATION

In numerical integration, we are required to compute the value of the definite integral

$$I = \int_a^b y\,dx, \tag{7.141}$$

where the function $y = f(x)$ is defined by the set of data points (x_i, y_i), $i = 0, 1, 2, \ldots, n$. To derive a general formula for numerical integration, we substitute Newton's forward difference formula for $y(x)$ in (7.141), where the interval $[a, b]$ is divided to n equal subintervals such that $a = x_0 < x_1 < x_2 < \cdots < x_n = b$, and $x_i = x_0 + ih$, $i = 0, 1, 2, \ldots, n$. We then obtain

$$I = \int_{x_0}^{x_n} \left[y_0 + p\Delta y_0 + \frac{p(p-1)}{2}\Delta^2 y_0 + \frac{p(p-1)(p-2)}{6}\Delta^3 y_0 + \cdots \right] dx,$$

where $x = x_0 + ph$. Since $dx = h\,dp$, the above integral becomes:

$$I = h\int_0^n \left[y_0 + p\Delta y_0 + \frac{p(p-1)}{2}\Delta^2 y_0 + \frac{p(p-1)(p-2)}{6}\Delta^3 y_0 + \cdots \right] dp$$

Integrating, we obtain:

$$I = \int_{x_0}^{x_n} y\,dx = nh\left[y_0 + \frac{n}{2}\Delta y_0 + \frac{n(2n-3)}{12}\Delta^2 y_0 + \frac{n(n-2)^2}{24}\Delta^3 y_0 + \cdots \right] \quad (7.142)$$

Formula (7.142) is the *general quadrature* formula and is called *Newton–Cote's* formula. Different numerical integration formulae can be derived by putting $n = 1, 2, 3, \ldots$ in (7.142). The trapezoidal and Simpson's rules, obtained by putting $n = 1$ and 2 respectively in (7.142), are found to give sufficient accuracy for use in practical problems.

7.5.1 Trapezoidal Rule

Setting $n = 1$ in (7.142) we find that differences of second and higher order vanish, and we obtain

$$\int_{x_0}^{x_1} y\,dx = h\left(y_0 + \frac{1}{2}\Delta y_0 \right) = \frac{h}{2}(2y_0 + y_1 - y_0) = \frac{h}{2}(y_0 + y_1)$$

For subsequent intervals we similarly obtain:

$$\int_{x_1}^{x_2} y\,dx = \frac{h}{2}(y_1 + y_2),$$

$$\int_{x_2}^{x_3} y\,dx = \frac{h}{2}(y_2 + y_3),$$

$$\vdots$$

$$\int_{x_{n-1}}^{x_n} y\,dx = \frac{h}{2}(y_{n-1} + y_n).$$

Combining the above formulae, we obtain the rule:

$$\int_{x_0}^{x_n} y\,dx = \frac{h}{2}\left[y_0 + 2(y_1 + y_2 + \cdots + y_{n-1}) + y_n \right], \quad (7.143)$$

which is known as the *trapezoidal* rule.

An expression for the error in the trapezoidal rule can be derived by using Taylor's series for y around $x = x_0$. Thus we have

$$\int_{x_0}^{x_1} y\,dx = \int_{x_0}^{x_1} \left[y_0 + (x - x_0) y_0' + \frac{(x - x_0)^2}{2} y_0'' + \cdots \right] dx,$$

where it is assumed that y and its higher derivatives are continuous. Integrating the right-hand side and simplifying, we get

$$\int_{x_0}^{x_1} y\,dx = hy_0 + \frac{h^2}{2} y_0' + \frac{h^3}{6} y_0'' + \cdots \tag{7.144}$$

Similarly,

$$\frac{h}{2}(y_0 + y_1) = \frac{h}{2}\left(y_0 + y_0 + hy_0' + \frac{h^2}{2} y_0'' + \frac{h^3}{6} y_0''' + \cdots \right)$$

$$= hy_0 + \frac{h^2}{2} y_0' + \frac{h^3}{4} y_0'' + \cdots \tag{7.145}$$

From (7.144) and (7.145), we see that

$$\int_{x_0}^{x_1} y\,dx - \frac{h}{2}(y_0 + y_1) = -\frac{1}{12} h^3 y_0'' + \cdots \tag{7.146}$$

In a similar manner, the errors in the remaining subintervals can be derived. We thus have

$$E = -\frac{1}{12} h^3 \left(y_0'' + y_1'' + \cdots + y_{n-1}'' \right), \tag{7.147}$$

where E is the total error. If $y''(\xi)$ is the largest value of the n quantities, viz., y_0'', y_1'', $\cdots$, y_{n-1}'', then we obtain:

$$E = -\frac{1}{12} h^3 n y''(\xi) = -\frac{1}{12}(b - a)\, h^2 y''(\xi), \qquad \text{since } nh = b - a. \tag{7.148}$$

7.5.2 Simpson's $\frac{1}{3}$-rule

Setting $n = 2$ in (7.142), we observe that differences of third and higher orders vanish, and we obtain:

$$\int_{x_0}^{x_2} y\,dx = 2h\left(y_0 + \Delta y_0 + \frac{1}{6}\Delta^2 y_0 \right) = \frac{h}{3}(y_0 + 4y_1 + y_2).$$

Similarly,

$$\int_{x_2}^{x_4} y\,dx = \frac{h}{3}(y_2 + 4y_3 + y_4),$$

$$\vdots$$

$$\int_{x_{n-2}}^{x_n} y\,dx = \frac{h}{3}(y_{n-2} + 4y_{n-1} + y_n).$$

Summing up the above results, we obtain

$$\int_{x_0}^{x_n} y\,dx = \frac{h}{3}[y_0 + 4(y_1 + y_3 + \cdots + y_{n-1}) + 2(y_2 + y_4 + \cdots + y_{n-2}) + y_n] \tag{7.149}$$

which is known as *Simpson's* $\frac{1}{3}$*-rule, and requires an even number of subintervals of width h.*

The total error in this rule is given by

$$E = -\frac{b-a}{180} h^4 y^{iv}(\xi), \tag{7.150}$$

where $y^{iv}(\xi)$ is the largest value of the fourth derivatives.

7.5.3 Simpson's $\frac{3}{8}$-rule

Setting $n = 3$ in (7.142) and proceeding as above, we obtain

$$\int_{x_0}^{x_3} y\,dx = 3h\left(y_0 + \frac{3}{2}\Delta y_0 + \frac{3}{4}\Delta^2 y_0 + \frac{1}{8}\Delta^3 y_0\right) = \frac{3h}{8}(y_0 + 3y_1 + 3y_2 + y_3),$$

$$\int_{x_3}^{x_6} y\,dx = \frac{3h}{8}(y_3 + 3y_4 + 3y_5 + y_6), \text{ etc.}$$

Adding up all these integrals, we obtain:

$$\int_{x_0}^{x_n} y\,dx = \frac{3h}{8}[y_0 + 3(y_1 + y_2 + y_4 + y_5 + \cdots + y_{n-2} + y_{n-1})$$

$$+ 2(y_3 + y_6 + \cdots + y_{n-3}) + y_n], \tag{7.151}$$

which is called *Simpson's* $\frac{3}{8}$*-rule.* The error in this rule can be shown to be

$$E = -\frac{3}{80} h^5 y^{iv}(\xi).$$

7.5.4 Romberg Integration

Let the definite integral

$$I = \int_a^b y\,dx$$

be evaluated numerically with two different subintervals of widths h_1 and h_2, and let the approximate values of the integral be I_1 and I_2, respectively. Let the errors in these values of I_1 and I_2 be E_1 and E_2, respectively. If trapezoidal rule is used, then we know that

$$E_1 = -\frac{1}{12}(b-a)\,h_1^2\,y''(\xi_1),$$

and

$$E_2 = -\frac{1}{12}(b-a)\,h_2^2\,y''(\xi_2).$$

Assuming that $y''(\xi_1) \approx y''(\xi_2)$, since both of them represent the largest value of $y''(x)$, we obtain

$$\frac{E_1}{E_2} \approx \frac{h_1^2}{h_2^2},$$

and the above implies:

$$\frac{E_2}{E_2 - E_1} = \frac{h_2^2}{h_2^2 - h_1^2}$$

or

$$E_2 = \frac{h_2^2}{h_2^2 - h_1^2}(I_2 - I_1) \qquad \text{since } E_2 - E_1 = I_2 - I_1.$$

A new approximation for I can therefore be written as

$$I_3 = I_2 - E_2 = \frac{I_1 h_2^2 - I_2 h_1^2}{h_2^2 - h_1^2} \tag{7.152}$$

If we now set

$$h_2 = \frac{1}{2}h_1 = \frac{1}{2}h,$$

Equation (7.152) can be written in the more convenient form

$$I\left(h, \frac{1}{2}h\right) = \frac{1}{3}\left[4\,I\left(\frac{1}{2}h\right) - I(h)\right], \tag{7.153}$$

where

$$I(h) = I_1, \quad I\left(\frac{1}{2}h\right) = I_2, \quad I\left(h, \frac{1}{2}h\right) = I_3.$$

The successive computations can be tabulated as below:

$$\begin{array}{llll}
I(h) & & & \\
 & I\left(h, \frac{1}{2}h\right) & & \\
I\left(\frac{1}{2}h\right) & & I\left(h, \frac{1}{2}h, \frac{1}{4}h\right) & \\
 & I\left(\frac{1}{2}h, \frac{1}{4}h\right) & & I\left(h, \frac{1}{2}h, \frac{1}{4}h, \frac{1}{8}h\right) \\
I\left(\frac{1}{4}h\right) & & I\left(\frac{1}{2}h, \frac{1}{4}h, \frac{1}{8}h\right) & \\
 & I\left(\frac{1}{4}h, \frac{1}{8}h\right) & & \\
I\left(\frac{1}{8}h\right) & & &
\end{array}$$

The computations can be stopped when two successive values are close to each other. Such a systematic tabulation of the approximate values of I is known as *Romberg integration.*

Example 7.41 Find, from the following table, the area bounded by the curve and the x-axis from $x = 7.47$ to $x = 7.52$.

x	7.47	7.48	7.49	7.50	7.51	7.52
$f(x)$	1.93	1.95	1.98	2.01	2.03	2.06

We know that

$$\text{Area} = \int_{7.47}^{7.52} f(x)\, dx.$$

With $h = 0.01$, the trapezoidal rule gives:

$$\text{Area} = \frac{0.01}{2}[1.93 + 2(1.95 + 1.98 + 2.01 + 2.03) + 2.06]$$

$$= \frac{0.01}{2}(19.93)$$

$$= 0.0996.$$

Example 7.42 Taking $h = \pi/6$, find by Simpson's rule the area under the curve $y = \sin x$, from $x = 0$ to $x = \pi$. Compare your result with the actual value.

We first tabulate the values of the function from $x = 0$ to $x = \pi$.

x	0	$\pi/6$	$2\pi/6$	$\pi/2$	$4\pi/6$	$5\pi/6$	π
$y = \sin x$	0	0.5	0.861	1.0	0.861	0.5	0

Hence, by Simpson's $\frac{1}{3}$- rule, we have

$$\text{Area} = \frac{1}{3}\frac{\pi}{6}[0 + 4(0.5 + 1.0 + 0.5) + 2(0.861 + 0.861) + 0]$$

$$= \frac{\pi}{18}(11.444)$$

$$= 1.9983.$$

The actual value is

$$I = \int_0^\pi \sin x\, dx = [-\cos x]_0^\pi = 2,$$

which shows that the computed value is very close to the true value.

Example 7.43 Certain values of x and y are given below:

x	0	0.25	0.5	0.75	1.0
y	1	0.9394	0.7788	0.5694	0.3678

Estimate the value of the integral $\int_0^1 y\, dx$, using both the trapezoidal and Simpson's rule.

We have $h = 0.25$. Trapezoidal rule gives

$$I \approx \frac{0.25}{2}[1.0 + 2(0.9394 + 0.7788 + 0.5694) + 0.3678]$$

$$= \frac{0.25}{2}(1.3678 + 4.5752) = \frac{0.25}{2}(5.9430)$$

$$= 0.7429.$$

Simpson's $\frac{1}{3}$-rule gives:

$$I \approx \frac{0.25}{3}[1.0 + 4(0.9394 + 0.5694) + 2(0.7788) + 0.3678]$$

$$= \frac{0.25}{3}(1.3678 + 6.0352 + 1.5576)$$

$$= 0.7467.$$

Example 7.44 The velocity v (metre/sec) of a particle at distance x (metre) is tabulated below:

x	0	10	20	30	40	50	60
v	47	58	65	64	58	55	35

Using Simpson's $\frac{1}{3}$-rule, find the time taken by the particle to travel 60 metre.

We know that $v = dx/dt$. Therefore,

$$t = \int \frac{1}{v}\, dx.$$

Hence the time taken to travel 60 metre is given by

$$T = \int_0^{60} \frac{1}{v}\, dx = \int_0^{60} y\, dx \qquad \text{where } y = \frac{1}{v}.$$

We therefore require a table of values of x and $y = 1/v$. We obtain

x	0	10	20	30	40	50	60
$y = \frac{1}{v}$	0.02128	0.01724	0.01538	0.01562	0.01724	0.01818	0.02857

Simpson's $\frac{1}{3}$-rule gives:

$$T = \frac{10}{3}[0.02128 + 4(0.01724 + 0.01562 + 0.01818$$

$$+ 2(0.01538 + 0.01724) + 0.02857]$$

$$= \frac{10}{3}[0.02128 + 0.20416 + 0.06524 + 0.02857]$$

$$= 1.064 \text{ sec.}$$

Hence the time taken to travel 60 metre is 1.064 sec.

Example 7.45 Compute the values of

$$I = \int_0^1 \frac{1}{1+x}\, dx,$$

by using the trapezoidal rule with $h = 0.5$, 0.25 and $h = 0.125$. Then obtain a better estimate by using the Romberg method. Compare your result with the true value.

(a) Let $h = 0.5$. The values of x and $y = 1/(1+x)$ are given below:

x	0	0.5	1.0
$y = 1/(1+x)$	1.0000	0.6667	0.5

Using the trapezoidal rule, we obtain

$$I(h) = \frac{0.5}{2}[1.0000 + 2(0.6667) + 0.5] = 0.7084$$

(b) Let $h = 0.25$. The tabulated values of x and y are now:

x	0	0.25	0.5	0.75	1.0
y	1.0000	0.8000	0.6667	0.5714	0.5

Trapezoidal rule gives:

$$I\left(\frac{h}{2}\right) = \frac{0.25}{2}[1.0000 + 2(0.8000 + 0.6667 + 0.5714) + 0.5]$$

$$= \frac{0.25}{2}(1.5 + 4.0762)$$

$$= 0.6970.$$

(c) Finally, let $h = 0.125$. Then the tabulated values are:

x	0	0.125	0.250	0.375	0.5	0.625	0.750	0.875	1.0
y	1.0	0.8889	0.8000	0.7273	0.6667	0.6154	0.5714	0.5333	0.5

Using trapezoidal rule, we obtain

$$I\left(\frac{h}{4}\right) = \frac{0.125}{2}[1.0 + 2(0.8889 + 0.8000 + 0.7273 + 0.6667$$

$$+ 0.6154 + 0.5714 + 0.5333) + 0.5]$$

$$= 0.6941.$$

Now, we use formula (7.153). We obtain:

$$I\left(h, \frac{1}{2}h\right) = \frac{1}{3}[4(0.6970) - 0.7084] = 0.6932,$$

$$I\left(\frac{1}{2}h, \frac{1}{4}h\right) = \frac{1}{3}[4(0.6941) - 0.6970] = 0.6931,$$

$$I\left(h, \frac{1}{2}h, \frac{1}{4}h\right) = 0.6931.$$

We therefore conclude that $I \approx 0.6931$. The exact value is $\log_e 2$, which is equal to 0.693147...

It follows that the computed value is correct to four decimal places.

EXERCISES 7.5

1. If $y = A + Bx + Cx^2$ and y_0, y_1, y_2 are the corresponding values of $x = a,\ a+h,\ a+2h$ respectively, prove that

$$\int_a^{a+2h} y\,dx = \frac{h}{3}(y_0 + 4y_1 + y_2).$$

2. Evaluate $\int_0^{\pi/2} \sqrt{\sin x}\,dx$, given that

x	0	$\pi/12$	$\pi/6$	$\pi/4$	$\pi/3$	$5\pi/12$	$\pi/2$
$\sqrt{\sin x}$	0.0	0.5087	0.7071	0.8409	0.9306	0.9828	1.0

3. Evaluate

$$\int_{0.2}^{1.4} (\sin x - \log_e x + e^x)\,dx$$

using (i) Trapezoidal rule (ii) Simpson's $\frac{1}{3}$-rule.

4. A curve is defined by the following table of values:

x	0	0.5	1.0	1.5	2.0	2.5	3.0	3.5	4.0
y	23	19	14	11	12.5	16	19	20	20

Calculate the area bounded by the curve, the x-axis and the ordinates $x = 0$ to $x = 4.0$.

5. Find the approximate value of $\int_0^{\pi/2} \sqrt{\cos\theta}\,d\theta$ by dividing the interval into six parts.

6. Evaluate the integral

$$I = \int_0^1 \frac{1}{1+x}\,dx$$

using Simpson's $\frac{1}{3}$-rule with $h = 0.5$, 0.25 and 0.125.

7. A river is 80-feet wide. The depth 'd' in feet at a distance x feet from one bank is given by the following table:

x	0	10	20	30	40	50	60	70	80
d	0	4	7	9	12	15	14	8	3

find approximately the area of cross-section of the river using Simpson's $\frac{1}{3}$-rule.

8. Evaluate

$$\int_0^1 \frac{dx}{\sqrt{1+x^3}}$$

by (i) trapezoidal rule (ii) Simpson's rule, with $h = 0.1$.

9. Evaluate $\int_0^{\pi/2} e^{\sin x}\,dx$ using Simpson's rule by dividing the interval $[0, \pi/2]$ into eight equal parts.

10. Find the value of $\int_3^7 x^2 \log x\,dx$ by taking four strips.

11. Find an approximate value of $\int_0^4 e^x\,dx$ given that $e^0 = 1$, $e^1 = 2.72$, $e^2 = 7.39$ $e^3 = 20.09$ and $e^4 = 54.60$. Use both the trapezoidal and Simpson's $\frac{1}{3}$-rule.

12. Determine the maximum error in evaluating the integral $\int_0^{\pi/2} \cos x\,dx$ by both the trapezoidal and Simpson's rules using four subintervals.

13. Find the value of the integral

$$\int_0^1 \frac{x^2 dx}{1+x^3}$$

using Simpson's rule with $h = 1/4$.

14. Evaluate

$$\int_0^2 \frac{dx}{x^3+x+1}$$

to three decimal places by Simpson's rule with $h = 1/4$.

15. Compute the values of

$$I = \int_0^1 \frac{dx}{1+x^2}$$

using the trapezoidal rule with $h = 0.5$, 0.25 and 0.125. Then obtain a better estimate by Romberg's method. Compare your result with the true value.

Bibliography

Ayre, H.G., R. Stephens and G.D. Mock, *Analytic Geometry* (Two and Three Dimensions), D. Van Nostrand, New York, 1967.

Bajpai, A.C., I.M. Calus and J.A. Fairley, *Mathematics for Engineers and Scientists*, Vol. 1, John Wiley & Sons, New York, 1973.

Benny, L.B., *Mathematics for Students of Engineering and Applied Science*, Oxford University Press, London, 1961.

Cairns, E.J., *Mathematics for Applied Engineering*, Prentice Hall, Englewood Cliffs, New Jersey, 1967.

Chambers, LL.G., *A Course in Vector Analysis* (Science Paperbacks), Chapman & Hall, London, 1969.

Dettman, J., *Mathematical Methods in Physics and Engineering*, McGraw Hill, New York, 1969.

Elsgole, L.E., *Calculus of Variations*, Pergamon Press, London, 1961.

Fox, C., *An Introduction to Calculus of Variations*, Oxford University Press, London, 1950.

Fraleigh, J.B., *Calculus with Analytic Geometry*, Addison-Wesley, Reading (MA), 1980.

Goldstein, L.J., D.C. Lay and D.I. Schneider, *Calculus and Its Applications*, Prentice Hall, Englewood Cliffs, New Jersey, 1977.

Goult, R.J., R.F. Hoskins, J.A. Milner and M.J. Pratt, *Applicable Mathematics*, Macmillan Press, London 1973.

Greenspan, H.P. and D.J. Benny, *Calculus—An Introduction to Applied Mathematics*, McGraw Hill, New York, 1973.

Hilderbrand, F.B., *Introduction to Numerical Analysis*, in Ch. 9, 'Continued Fractions', McGraw Hill, New York, 1956.

Hyslop, J.H., *Infinite Series*, Oliver & Boyd, London, 1965.

Jain, M.K., S.R.K. Iyengar and R.K. Jain, *Numerical Methods for Scientific and Engineering Computations*, New Age International, New Delhi, 2003.

Johnson, R.E., F.L. Kiokemeister and E.S. Wolk, *Calculus with Analytic Geometry*, Allyn & Bacon, Reading (MA.), 1974.

Noble, B and J.W. Daniel, *Applied Linear Algebra*, Prentice Hall, Englewood Cliffs, New Jersey, 1977.

Olmsted, J.M.H., *Advanced Calculus*, Eurasia Publishing House, New Delhi, 1970.

Peter Lax, S. Burstein and Lax Anneli, *Calculus with Applications and Counting*, Springer-Verlag, New York, 1976.

Sastry, S.S., *Introductory Methods of Numerical Analysis*, Prentice-Hall of India, New Delhi, 1992.

Schechter, R.S., *The Variational Method in Engineering*, McGraw Hill, New York, 1967.

Shere, W. and G. Love, *Applied Mathematics for Engineering and Science*, Prentice Hall, New York, 1969.

Sokolnikoff, I.S. and R.M. Redheffer, *Mathematics of Physics and Modern Engineering*, International Student Edition, McGraw Hill, Kogakusha (Japan), 1966.

Spain, D., *Vector Analysis*, D. Van Nostrand, London, 1967.

Spiegel, M.R., *Theory and Problems of Advanced Calculus*, Schaum Publishing, New York, 1963.

Srivastava, R.S.L., *Engineering Mathematics*, Vol. 1, Tata McGraw-Hill, New Delhi, 1980.

Thomas Jr., G.B. *Calculus with Analytic Geometry*, 3rd ed., Addison-Wesley, Reading (MA), 1960.

Wilf, H.S., *Mathematics for the Physical Sciences*, Dover Publications, New York, 1978.

Williams, I.P., *Matrices for Scientists*, Hutchinson University Library, London, 1972.

Answers to Exercises

Exercises 1.1

1. $\dfrac{3}{x-2}+\dfrac{1}{x-3}$.

3. $\dfrac{-10}{x-1}+\dfrac{13}{x-2}$.

5. $1-\dfrac{1}{x+1}+\dfrac{1}{(x+1)^2}$.

7. $\dfrac{1}{x-1}+\dfrac{2}{(x-1)^2}-\dfrac{1}{x+2}$.

9. $\dfrac{2}{x-1}-\dfrac{2x+1}{x^2+x+1}$.

10. $\dfrac{1}{5(x-2)}-\dfrac{x+2}{5(x^2+1)}$

11. $-\dfrac{2}{x^2}+\dfrac{3}{x^2+1}$.

14. $\dfrac{3}{x-1}+\dfrac{1}{(x-1)^2}-\dfrac{7}{(x-1)^3}+\dfrac{5}{(x-1)^4}$.

16. $\dfrac{1}{8x}+\dfrac{1/2-(1/8)\,x}{x^2+4x+8}$.

19. $\dfrac{1}{x-1}-\dfrac{1}{x+2}-\dfrac{3}{(x+2)^2}$.

21. $-\dfrac{1}{x+2}+\dfrac{2}{(x-1)^3}-\dfrac{3}{(x-1)^2}+\dfrac{1}{x-1}$.

22. $\dfrac{3}{x+1}-\dfrac{3}{2(x+2)}+\dfrac{1-(3/2)x}{x^2}$.

24. $\dfrac{1}{x-1}-\dfrac{3}{x^2+1}$.

25. $\dfrac{2}{x}+\dfrac{3x-1}{(x^2+1)^2}+\dfrac{x+2}{x^2+1}$.

27. $k_1=-3x-10,\ k_2=3x^2+10x+9$.

Exercises 1.2

1. Convergent.
2. Convergent.
3. Convergent.
5. Divergent.
7. Divergent.
9. Divergent.
10. Convergent.
12. Divergent.
13. Convergent for all x.
15. Convergent if $|x|<1$ and divergent if $|x|\geq 1$.
17. Divergent.
19. Convergent.
20. Convergent if $p>1/2$ and divergent if $p\leq 1/2$.
22. Convergent.
23. Divergent.
24. Divergent.
25. Divergent.
26. Convergent if $x<1$ and divergent if $x\geq 1$.
27. Convergent if $x>1$ or $x<1$ and divergent if $x=1$.

29. For $x<1$, convergent and for $x>1$, divergent for all values of p. For $x=1$, convergent if $p>1$ and divergent if $p<1$.

32. Absolutely convergent. **34.** Divergent. **36.** Convergent.

37. Divergent. **38.** Convergent. **40.** $|x|<1$.

41. $|x-2|\le 1$.

42. $|x+3|<1$; Convergent for $x=-4$ and divergent for $x=-2$.

Exercises 1.3

1. (a) 11.7473. (b) 9.0205 (c) 5.0100.

2. 0.01804. **3.** 4.242; $2\left(1-\frac{1}{2}x\right)$. **8.** -14.

13. 923/925. **18.** -7.875; 26.056. **21.** 3.708; 3.893.

23. $(-1)^n(3n^2-5n+1)/n!$ **28.** $3/n$.

Exercises 1.4

1. (a) $\frac{1}{4+}\,\frac{1}{1+}\,\frac{1}{4+}\,\frac{1}{1+}\,\frac{1}{4}$, $2+\frac{1}{6+}\,\frac{1}{1+}\,\frac{1}{1+}\,\frac{1}{11+}\,\frac{1}{2}$,

$\frac{1}{2+}\,\frac{1}{2+}\,\frac{1}{3+}\,\frac{1}{1+}\,\frac{1}{4+}\,\frac{1}{2+}\,\frac{1}{6}$.

2. (a) $\frac{B_3}{A_3}=\frac{21}{16}, \frac{B_4}{A_4}=\frac{151}{115}, \frac{B_5}{A_5}=\frac{1320}{991}$.

(b) $\frac{B_2}{A_2}=\frac{3}{2}, \frac{B_3}{A_3}=\frac{10}{7}, \frac{B_4}{A_4}=\frac{23}{16}, \frac{B_5}{A_5}=\frac{56}{39}, \frac{B_6}{A_6}=\frac{168}{124}$.

6. Error is less than 1/4500 in magnitude.

9. $2+\frac{x}{-2+}\,\frac{x-1}{3+}\,\frac{x-2}{0+}\,\frac{x-3}{-5}$. **11.** 38.20.

Exercises 1.5

1. -2, $1\pm 3i$. **2.** 2, $2\pm i$. **3.** 1, 2, $1\pm\sqrt{5}$.

4. $\frac{4}{3}, \pm\sqrt{2}\pm\sqrt{5}$ **5.** $1\pm i$, $-2\pm\sqrt{3}$ **6.** 2, 3, 4.

7. $\frac{2}{3}$, 2, 6. **8.** $-1/5, -1/2, 1, 4$. **9.** 1, 4, 7

10. $-2, 4, -8, 16$. **13.** $-4, 2, 6$. **15.** $-1, -3, 2, 6$.

16. (i) p/r (ii) $-q/r$ (iii) p^2-2q.

(iv) $(r-p)^2+(q-1)^2$ (v) p^3r-q^3. (vi) $3r-pq$.

(vii) $24r-p^3$.

17. (i) $ac-bd$ (ii) $a^2-4a^2b-2b^2+4ac-4d.$

19. $x^3+7x^2-x+1=0.$

22. $qx^3-9p^2x^2-5pqx-2p^2-q^2=0.$

25. $x^4+6x^3+12x^2+11x+1=0.$

26. $2\pm\sqrt{2}, 2\pm\sqrt{3}.$ **27.** 2, 1/2, 2, 1/2.

28. −1, 2, 1/2, −3, −1/3. **29.** $1, (1\pm\sqrt{3})/2, (3\pm\sqrt{5})/2.$

30. −1, 1, 3, 1/3, 2, 1/2. **31.** 2.20 **32.** −1.78

33. 1.520 **34.** 2.279 **35.** −2.047

Exercises 1.6

4. $A^n=\begin{bmatrix}\cos n\theta & \sin n\theta\\ -\sin n\theta & \cos n\theta\end{bmatrix}$

6. (a) $\begin{bmatrix}4&2\\4&6\end{bmatrix}; \begin{bmatrix}-2&2\\2&2\end{bmatrix}; \begin{bmatrix}5&4\\13&8\end{bmatrix}; \begin{bmatrix}3&6\\7&10\end{bmatrix}$

8. $\begin{bmatrix}2&0&1.5\\0&4&1.5\\1.5&1.5&-2\end{bmatrix}+\begin{bmatrix}0&1&1.5\\-1&0&-0.5\\-1.5&0.5&0\end{bmatrix}$

10. $\begin{bmatrix}2&4&2.5\\4&0&2.5\\2.5&2.5&3\end{bmatrix}+\begin{bmatrix}0&-1&1.5\\1&0&1.5\\-1.5&-0.5&0\end{bmatrix}$

13. Orthogonal. **15.** Orthogonal. **16.** Orthogonal.

19. $L=\begin{bmatrix}1&0&0\\1&1&0\\-1&1&1\end{bmatrix}; U=\begin{bmatrix}1&3&1\\0&1&1\\0&0&3\end{bmatrix}; x=1, y=1, z=-1.$

24. $A=\begin{bmatrix}1&2&3\\2&0&5\\0&1&3\end{bmatrix}, X=\begin{bmatrix}x_1\\x_2\\x_3\end{bmatrix}, B=\begin{bmatrix}5\\10\\6\end{bmatrix}$

26. $A=\begin{bmatrix}1&0&1\\7&2&0\\6&0&1\end{bmatrix}, X=\begin{bmatrix}x\\y\\z\end{bmatrix}, B=\begin{bmatrix}1\\0\\2\end{bmatrix}$

29. $\frac{1}{4}\begin{bmatrix} 3 & -10 & -1 \\ -2 & 8 & 2 \\ 2 & -4 & -2 \end{bmatrix}$ **30.** $\begin{bmatrix} 1 & -8 & 10 \\ 0 & 2 & -3 \\ 0 & -1 & 2 \end{bmatrix}$

36. 2. **37.** 2.

38. $x = -2,\ y = 3,\ z = 5.$ **40.** $x = -1,\ y = 4,\ z = 4.$

41. Inconsistent. **47.** (i) No. (ii) Yes.

55. 2. **56.** 3. **57.** 3.

58. 3.

59. (i) $k_1 \neq 3,\ k_2$ arbitrary; (ii) $k_1 = 3,\ k_2 \neq 0$; (iii) $k_1 = 3,\ k_2 = 10$

60. $-1, 3;\ \begin{bmatrix} k_1 \\ -k_1 \end{bmatrix};\ \begin{bmatrix} k_2 \\ -5k_2 \end{bmatrix}$

61. $1, 4, 6;\ \begin{bmatrix} 2k_1 \\ 0 \\ k_1 \end{bmatrix}, \begin{bmatrix} 0 \\ k_2 \\ 0 \end{bmatrix}, \begin{bmatrix} k_3 \\ 0 \\ -2k_3 \end{bmatrix}$

62. $8, 2, 2;\ \begin{bmatrix} 2k_1 \\ -k_1 \\ k_1 \end{bmatrix}, \begin{bmatrix} k_2 \\ 2k_2 \\ 0 \end{bmatrix}, \begin{bmatrix} k_3 \\ 0 \\ 2k_3 \end{bmatrix}$

63. $-2, 4, 6;\ \begin{bmatrix} 0 \\ k_1 \\ 0 \end{bmatrix}, \begin{bmatrix} k_2 \\ 0 \\ -k_2 \end{bmatrix}, \begin{bmatrix} k_3 \\ 0 \\ k_3 \end{bmatrix}$

65. $\begin{bmatrix} -1 & 1 & 2 \\ 0 & 2 & -1 \\ 2 & 0 & 1 \end{bmatrix}, \begin{bmatrix} 2 & 0 & 0 \\ 0 & 2 & 0 \\ 0 & 0 & 8 \end{bmatrix}$

Exercises 2.1

1. (a) 12.5 units, (c) $\frac{1}{2}(ab - bc - 4ac)$. **2.** 20.5 units.

6. $r = 4\cos\theta$.

10. (i) $r^2 = 2\cos\phi$, (ii) $r^2(1 - 4\cos^2\phi) = 0$.

11. (a) $r^2 \sin^2\phi = 5;\ r^2 = 5,\ z = 0.$

(d) $r^2(\sin^2\phi\cos^2\theta - \cos^2\phi) = 4;\quad r^2\cos^2\theta - z^2 = 4.$

13. $r = \cos\theta, z = 0.$

15. $r^2 + 3r^2 \sin 2\theta + z^2 = 1;\ r^2 + 3r^2 \sin^2 \phi \sin 2\theta = 1.$

16. $x^2 + y^2 + (z-1)^2 = 1;\ r^2 + z^2 = 2z.$

Exercises 2.2

1. $2x + y = 4.$

2. $\frac{1}{2}(c_1 - c_2)^2/(m_1 - m_2)^2$

4. $7x + 3y = 33,\ 3x - 7y = -19,\ 3x - 7y = 10,\ 7x + 3y = 4.$

6. $11x + 3y = 30,\ 3x - 11y = 20.$

8. $(\sqrt{5} - 4)\, x + (3 - 2\sqrt{5})\, y = 2 - 4\sqrt{5}.$

9. (7/2, 13/2) or (–3/2, 3/2).

11. 14 units.

15. (1/3, 2/3).

17. $k = -10,\ 3x - 2y + 1 = 0,\ 4x + 5y + 3 = 0;\ \theta = \tan^{-1}(23/2).$

18. $a = 2, c = -3.$

21. $6\lambda^2 - 5\lambda + 52 = 0.$

23. $\sqrt{5}.$

27. $x^2 + y^2 - 4x - 6y = 0.$

29. (100/13, 240/13).

31. $3x^2 + 3y^2 - 7x - 7y + 4 = 0.$

32. $x^2 + y^2 - 14x - 6y - 111 = 0.$

33. $x^2 - y^2 = 4.$

35. Outside.

36. $7x^2 + 7y^2 = 20.$

38. $x^2 + y^2 = a^2 + b^2.$

39. $x^2 + y^2 - 2ax - 2ay + a^2 = 0.$

49. $y = -4a^2/c.$

56. (1, 2), 4/5.

57. 1/59.

58. $e \approx 1/10.$

66. (–4, –1), 5/4.

70. $3x + y + 2 = 0, x - 2y + 5 = 0.$

74. $\frac{x^2}{16} - \frac{y^2}{9} = 1.$

77. $(1 - e^2)x^2 + y^2 - 6x - 9 = 0.$

82. (i) Parabola, (ii) Parabola, (iii) Ellipse.

Exercises 2.3

2. $(4 - 25/\sqrt{2})\,\mathbf{i} + (5 + 5/\sqrt{2} - 4\sqrt{3})\,\mathbf{j}.$

3. $(-5\mathbf{i} + 15\mathbf{j} + 25\mathbf{k})/\sqrt{35}.$

4. $(\mathbf{i} + 4\mathbf{j} + \mathbf{k})/\sqrt{26}.$

8. $\cos(\alpha - \beta).$

9. $\mathbf{a} \times \mathbf{b} = -\mathbf{i} + 7\mathbf{j} + 5\mathbf{k}.$

10. 16 units.

11. 8 units

12. $10\pi(-15\mathbf{i} + 10\mathbf{k})/\sqrt{14}.$

19. (i) −3, (ii) −3, (iii) $22\mathbf{i}-5\mathbf{j}-32\mathbf{k}$, (iv) $12\mathbf{i}+8\mathbf{j}-28\mathbf{k}$.

21. 4 units.

24. Do not intersect.

26. $\sqrt{14}$.

28. $5y-z=3$.

31. $3x-4z+1=0$.

32. $x-3y-6z+8=0$.

36. $(1,-2,7)$.

37. $(7/3, 4/3, 13/3)$.

38. $(9,-6,1)$.

39. $\left(\dfrac{1}{x^2}+\dfrac{1}{y^2}+\dfrac{1}{z^2}\right)\left(x^2+y^2+z^2\right)^2=k^2$.

40. $\left(\dfrac{1}{2},-\dfrac{1}{2},-\dfrac{3}{2}\right)$, $x-2y+z=0$.

44. The given points do not lie in the same plane.

47. Parabolic cylinder with generator parallel to the Y-axis.

48. Two half-planes

49. Cylinder.

50. Circular cylinder.

52. Elliptic cylinder.

53. Hyperbolic cylinder.

56. Real intersecting planes.

57. Elliptic paraboloid.

59. Ellipsoid of revolution.

61. Real parallel planes.

62. Hyperboloid of one sheet.

67. Elliptic paraboloid.

70. Hyperboloid of two sheets.

Exercises 2.4

4. $\left(\sqrt{6}-\sqrt{2}\right)\left(\cos\dfrac{5\pi}{12}+i\sin\dfrac{5\pi}{12}\right)$

6. $\cos\left(-\dfrac{\pi}{2}\right)+i\sin\left(-\dfrac{\pi}{2}\right)$.

9. $2\cos\dfrac{\alpha}{2}\left(\cos\dfrac{\alpha}{2}+i\sin\dfrac{\alpha}{2}\right)$.

10. $2\sin\dfrac{\alpha}{2}\left[\cos\left(\dfrac{\alpha}{2}-\dfrac{\pi}{2}\right)+i\sin\left(\dfrac{\alpha}{2}-\dfrac{\pi}{2}\right)\right]$.

11. $\dfrac{1}{2}\sec\dfrac{\theta}{2}\left(\cos\dfrac{\theta}{2}-i\sin\dfrac{\theta}{2}\right)$.

12. $\dfrac{6}{13}-\dfrac{4}{13}i$.

13. $1+10i$.

15. $-\dfrac{9}{25}+\dfrac{13}{25}i$.

17. $\dfrac{1}{2}-i$.

18. $-8+0i$.

19. $\pm\dfrac{1+i}{\sqrt{2}}$.

20. $-i$.

23. $3+i\sqrt{2}$, $3-i\sqrt{2}$.

27. $x = \cos\frac{(2r+1)\pi}{7} + i\sin\frac{(2r+1)\pi}{7}$, $r = 0, 1, 2, 3, 4, 5, 6.$

28. $x = -1, \frac{1}{2}(1 \pm i\sqrt{3})$; $\cos\frac{(2r+1)\pi}{5} + i\sin\frac{(2r+1)\pi}{5}$; $r = 0, 1, 2, 3, 4.$

32. $-11-2i.$ **33.** $-1/2.$ **37.** $32i.$

38. (i) $\cos\frac{\pi}{6} - i\sin\frac{\pi}{6}$, $\cos\frac{\pi}{2} - i\sin\frac{\pi}{2}$, $\cos\frac{7\pi}{6} - i\sin\frac{7\pi}{6}$.

(iii) $8^{1/6}\left(\cos\frac{2r\pi}{6} + i\sin\frac{2r\pi}{6}\right)$, $r = 0, 1, 2, 3, 4, 5.$

41. $(\sqrt{2})^{1/3}\left(\cos\frac{2r\pi + \pi/4}{3} + i\sin\frac{2r\pi + \pi/4}{3}\right)$, $r = 0, 1, 2.$

42. $\cos\frac{(4r+1)\pi}{8} + i\sin\frac{(4r+1)\pi}{8}$, $r = 0, 1, 2, 3.$

45. $\frac{x}{x^2+y^2} - i\frac{y}{x^2+y^2}$. **46.** $\frac{x-1}{(x-1)^2+y^2} - i\frac{y}{(x-1)^2+y^2}$.

48. $(x^2 - y^2 - x + 1) + i(2xy - y)$. **51.** $\frac{x^2+y^2-1}{x^2+(y-1)^2} + i\frac{2x}{x^2+(y-1)^2}$.

57. (i) Circle, $x^2 + y^2 = 1$, (ii) Circle, $(x+2)^2 + y^2 = 25$.

70. $8\,(16\cos^7\theta - 24\cos^5\theta + 10\cos^3\theta - \cos\theta)$.

72. (i) $\frac{1}{2^4}(\cos 5x + 5\cos 3x + 10\cos x)$,

(ii) $\frac{1}{2^2}(\cos 3x + 3\cos x) - \frac{1}{2^4}(\cos 5x + 5\cos 3x + 10\cos x)$.

73. $\cos 6\theta = \cos^6\theta - 15\cos^4\theta\sin^2\theta + 15\cos^2\theta\sin^4\theta - \sin^6\theta$;

$\sin 6\theta = 6\cos^5\theta\sin\theta - 20\cos^3\theta\sin^3\theta + 6\cos\theta\sin^5\theta$.

75. The vertices of the hexagon are $i-1$ and

$$2 - i + (-3+2i)\left(\cos\frac{r\pi}{3} + i\sin\frac{r\pi}{3}\right),\quad r = 1, 2, 3, 4, 5$$

in the anti-clockwise direction.

81. $\alpha = \frac{1}{2}\log 2$, $\beta = -\pi/4$.

85. $\begin{bmatrix} 3 & -5 \\ 5 & 3 \end{bmatrix}$. **86.** $\begin{bmatrix} 14 & -5 \\ 5 & 14 \end{bmatrix}$.

88. $\begin{bmatrix} -1/2 & -5/2 \\ 5/2 & -1/2 \end{bmatrix}$ **89.** $(2+i)$.

90. $5-3i$. **92.** $1-i\sqrt{2}$.

93. $\begin{bmatrix} 2 & -3 \\ 3 & 2 \end{bmatrix}; \begin{bmatrix} -1 & 1 \\ -1 & -1 \end{bmatrix}$.

Exercises 3.1

9. $3x^2 - 8x^3 + 5x^4$. **11.** $-4x^3/(1+x^2)^2$.

13. $4\tan(2x+3)\sec^2(2x+3)$. **15.** $(-x\sin x - \cos x)/x^2$.

19. $e^{\sin x}\cos x$. **21.** $2/(x^2-1)$.

23. $\dfrac{1}{(1+x^2)\tan^{-1} x}$. **25.** $2/(1+x^2)$.

27. $3(\sinh 3x)e^{1+\cosh 3x}$. **29.** $-\sin x/\sqrt{1+\cos^2 x}$.

33. $-\dfrac{1}{2x\sqrt{x-1}}$. **34.** $2x^{-2/3} - \dfrac{5}{4}x^{-3/4}$.

36. $2x/(x^2+1)^2$. **38.** $-\dfrac{\sin x + 2}{(1+2\sin x)^2}$.

40. $\dfrac{3}{2}(4+3x)^{-1/2}$. **42.** $\dfrac{4xa^2}{a^4 - x^4}$.

43. $-2\sin x$. **46.** $2/[x(1-2x^4)]$.

48. $2e^x/(1-e^x)^2$. **50.** $\tan x/(1+\cos x)$.

51. $3\sec^2 t/(4-\tan^2 t)$. **52.** $2/(1-x^4)$.

53. $\dfrac{2}{\sqrt{x(x-4)}} + \dfrac{2x-5}{2\sqrt{x^2-5x}}$.

57. (i) $-b\cot\theta/a;\ -b/(a^2\sin^3\theta)$, (ii) $-\cot^3\theta;\ -\cot^2\theta\ \text{coses}^5\theta$.

63. (a) $2e^x\cos\left(x+\dfrac{2\pi}{4}\right)$, (b) $2\cos x - 4x\sin x - x^2\cos x$.

64. (c) $x^2\cos\left(x+\dfrac{n\pi}{2}\right) + 2nx\cos\left[x+\dfrac{(n-1)\pi}{2}\right] + n(n-1)\cos\left[x+\dfrac{(n-2)\pi}{2}\right]$.

65. (iii) $\dfrac{1}{2}\left[5^n\sin\left(5x+\dfrac{n\pi}{2}\right) - \sin\left(x+\dfrac{n\pi}{2}\right)\right]$,

(v) $(-1)^n n!\left[\dfrac{1}{(x+1)^{n+1}}-\dfrac{1}{(x+2)^{n+1}}\right]$.

67. (ii) $-(2x+y)/(x+2y)$ (iv) $-y/x$.

68. (i) $-1/(a\sin^3\theta)$, (iii) $-1/(4a\sin^4\theta/2)$, (v) $-1/(4t^3)+3/(4t)$.

69. (iv) $-(2x+y)/(x+2y)$, (vi) $-\left(\dfrac{y}{x}\right)^{1/3}$, (vii) $-(1+y^2)/y^2$,

(ix) $-(e^x\sin y+e^{-y}\sin x)/(e^x\cos y+e^{-y}\cos x)$.

Exercises 3.2

1. $yy_1=2a(x+x_1);\dfrac{x-x_1}{2a}+\dfrac{y-y_1}{y_1}=0$.

4. $\dfrac{xx_1}{a^2}+\dfrac{yy_1}{b^2}=1;\ \dfrac{a^2(x-x_1)}{x_1}=\dfrac{b^2(y-y_1)}{y_1}$.

5. $\dfrac{x}{a}+\dfrac{y}{b}=2$.

6. $\dfrac{x}{a\sin t}+\dfrac{y}{b\cos t}=1$.

15. $p=\dfrac{(a^2-b^2)\sin\alpha\cos\alpha}{\sqrt{a^2\sin^2\alpha+b^2\cos^2\alpha}}$.

16. At the points where $x=\pm a/\sqrt{2}$.

17. $20y-x+7=0;\ 20x+y-140=0$.

28. $\pi/3$.

29. $\pi/2$.

31. $\pi/2$ at (0,0) and $\tan^{-1}\dfrac{3a^{1/3}b^{1/3}}{2(a^{2/3}+b^{2/3})}$ at $(4a^{1/3}b^{2/3}, 4a^{2/3}b^{1/3})$.

33. $\pi/2$.

34. $\dfrac{1}{a}-\dfrac{1}{b}=\dfrac{1}{a_1}-\dfrac{1}{b_1}$.

35. (i) $\sqrt{\dfrac{a+x}{x}}$ (ii) $\left(\dfrac{a}{x}\right)^{1/3}$.

38. (i) $p^2=ar$ (ii) $a^2p=r^3$.

48. $\dfrac{2}{9\pi}$ ft/min.

51. 3/4 ohm/min.

52. $\dfrac{3}{50\pi}$ cm/min.

53. The function is not differentiable at $x=1$; semi-cubical parabola.

58. *Hint:* Apply mean-value theorem to the function $f(x) = \log x$ in $a \le x \le b$.

59. *Hint:* Apply mean value theorem to the function $f(x) = e^x$ *in* $0 \le x \le b$.

66. $\log_e a$.

67. 2.

68. $2k$.

69. a.

70. $3/2$.

74. 1.

75. $-e/2$.

76. 1/3.

77. $-1/3$.

78. 1/6.

79. 1.

80. $(1-\log b)/(1+\log b)$.

82. -1.

83. 1.

84. e.

86. 0.

87. 0.

90. $e^{2/\pi}$

93. $\sqrt{ab}$.

94. 1

95. 1.

96. Maximum value = 1152.

97. $\cos x/(1+\sin x)$.

99. Maximum at $x = 1$ and minimum at $x = 3$.

100. Maximum at $x = \pi/3$.

101. Maximum when $\tan x = \sqrt{p/q}$

102. Maximum at $x = e$.

104. Maximum at $x = \pi/2$ and minimum for $\sin x = 1/e$.

107. $ah/4$.

109. $2\pi\left(1-\sqrt{\dfrac{2}{3}}\right)$.

111. Most economical speed against a current of 2 km/hr. is 3 kms/hr.

113. Breadth $= 2c/\sqrt{3}$, depth $= 2c\sqrt{2/3}$.

117. Concave upwards in $[-2, 0]$ and $[2, \infty]$; Concave downwards in $[-\infty, -2]$ and $[0, 2]$.

118. $(2, -8/3)$.

119. $(\pm\sqrt{3}, \pm\sqrt{3}/2)$; $(0, 0)$.

123. For $x = -2$ and $1 \pm \sqrt{3}$.

128. *Hint:* If $f(x) = \sin x$, then

$$f(x) = \sin\frac{\pi}{2} + \left(x - \frac{\pi}{2}\right) f'\left(\frac{\pi}{2}\right) + \frac{(x-\pi/2)^2}{2!} f''\left(\frac{\pi}{2}\right) + \cdots$$

129. $x + \dfrac{2}{3}x^3 + \dfrac{2}{15}x^5 + \cdots$

131. $1 + x + \dfrac{x^2}{2} - \dfrac{x^4}{8} + \cdots$

133. $\frac{x^2}{2}+\frac{x^4}{12}+\frac{x^6}{45}+\cdots$

134. $1+x+x^2+\frac{2}{3}x^3+\frac{1}{2}x^4+\cdots$

137. (i) $1-\frac{x}{2}+\frac{1}{12}x^2-\frac{1}{720}x^4+\cdots$, (ii) $x-\frac{x^2}{2}+\frac{x^3}{6}-\frac{x^4}{12}+\cdots$

141. $\frac{2}{\sqrt{a}}(a+x)^{3/2}$.

143. y^2/c.

145. $a\sec\frac{x}{a}\cdot$

148. $\frac{a(1+\theta^2)^{3/2}}{2+\theta^2}$.

150. $a/2$.

152. $\frac{(2r)^{3/2}}{\sqrt{a}}$.

155. $\pm\frac{5\sqrt{5}}{18}$.

157. $na/2$.

159. $[a(2,+3t^2)-2at^3]$; $27ay^2=4(x-2a)^3$.

160. $(ax)^{2/3}-(by)^{2/3}=(a^2+b^2)^{2/3}$.

161. $\left(3x+2a,\ -\frac{2}{\sqrt{a}}x^{3/2}\right)$; $4(x-2a)^3=27ay^2$.

162. $(x+y)^{2/3}+(x-y)^{2/3}=2a^{2/3}$.

166. $y=0, y=x-1, y=-x-1$.

167. $y=x,\ y=2x, y=3x$.

168. $y=0,\ y=-x+1,\ y=-x-1$.

169. $y=x, 2y=-x+1, 2y=-x-1$.

Exercises 4.1

7. (a) $(a+b)/2$, (b) $[(a^2+ab+b^2)/3]^{1/2}$, (c) $\sqrt{ab}$.

8. $\frac{1}{2}+\frac{2}{3}(2^{3/2}-1)$.

9. 1.

10. 2.

14. 1.1000; 1.0987.

15. 2.5649.

16. 0.2184; 0.2077.

21. 177.4830.

22. 1.1873.

23. 3.062.

Exercises 4.2

1. $\frac{1}{\sqrt{2}}\log\tan\left(\frac{x}{2}-\frac{\pi}{8}\right)+\text{c}$.

2. $\frac{1}{24}\log\frac{(x+2)^2}{x^2-2x+4}+\frac{1}{4\sqrt{3}}\tan^{-1}\frac{x-1}{\sqrt{3}}+c.$

3. $2\left(\sin\frac{x}{2}-\cos\frac{x}{2}\right)-\sqrt{2}\log\tan\left(\frac{\pi}{8}+\frac{x}{4}\right)+c.$

4. $e^x\tan\frac{x}{2}+\text{c}.$

5. $\frac{1}{4}\log\left|\frac{x-1}{x+1}\right|-\frac{1}{2}\tan^{-1}x+c.$

6. $\frac{1}{8}\left[\frac{(x-3)(3x^2-18x+32)}{(x^2-6x+10)^2}+3\tan^{-1}(x-3)\right]+c.$

7. $\frac{3}{4}\cosh^{-1}\frac{x^4}{\sqrt{5}}+\text{c}.$

8. $-2e^{-\sqrt{x}}(\sqrt{x}+1)+c.$

9. $\frac{1}{16}\log\left|\frac{c(x^2+2x+2)}{x^2-2x+2}\right|+\frac{1}{8}\tan^{-1}\frac{2x}{1-x^2}.$

10. $t\tan^{-1}\sqrt{\frac{t}{a}}-\sqrt{at}+a\tan^{-1}\sqrt{\frac{t}{a}}+c.$

11. $-\frac{t^2}{2}+c.$

12. $x\sec^{-1}x-\log(x+\sqrt{x^2-1})+c.$

13. $-\frac{1}{\sqrt{2}}\sec^{-1}\frac{x^2+1}{x\sqrt{2}}+c.$

14. $\frac{1}{2\sqrt{2}}\left(2\tan^{-1}\frac{\tan x-1}{\sqrt{2\tan x}}+\log\frac{\tan x-\sqrt{2\tan x}+1}{\tan x+\sqrt{2\tan x}+1}\right)+c.$

15. $\sin x-\frac{2}{3}\sin^3 x+\frac{1}{5}\tan^5 x+c.$

16. $\frac{1}{a^4}\left[\frac{x}{\sqrt{x^2+a^2}}-\frac{1}{3}\frac{x^3}{(x^2+a^2)^{3/2}}\right]+c.$

17. $\frac{1}{21}\log\frac{7+3\tan(x/2)}{7-3\tan(x/2)}+c.$

18. $x^2-3x+\frac{13}{9}\log(x-1)-\frac{5}{9}\log(x+2)-\frac{8}{9}\log(2x+1)+c.$

19. $x+\frac{2}{\sqrt{3}}\tan^{-1}\frac{x}{\sqrt{3}}-3\tan^{-1}\frac{x}{2}+c.$

20. $\frac{1}{a^2-b^2}\left(a\tan^{-1}\frac{x}{a}-b\tan^{-1}\frac{x}{b}\right)+c.$

21. $\log(x-1)-\frac{2}{x-1}-\frac{3}{2(x-1)^2}+c.$

22. $\frac{1}{4}\log\frac{x^2-x+1}{x^2+x+1}+2\sqrt{3}\tan^{-1}\frac{2x-1}{\sqrt{3}}-\frac{10}{\sqrt{3}}\tan^{-1}\frac{2x+1}{\sqrt{3}}+c.$

23. $\frac{x}{x+1}\log x-\log(x+1)+c.$

24. $\frac{\pi}{2}-\log 2.$

25. $-\cos\alpha\sin^{-1}\frac{\cos x}{\cos\alpha}-\sin\alpha\cosh^{-1}\frac{\sin x}{\sin\alpha}+c.$

26. $-2\sqrt{1-x}+\sqrt{x(1-x)}+\sin^{-1}\sqrt{1-x}+c.$

27. $\frac{1}{\sqrt{3}}\sin^{-1}\frac{3x+1}{\sqrt{10}}+c.$

28. $2\sqrt{x^2+7x+3}-8\cosh^{-1}\frac{2x+7}{\sqrt{37}}\ c.$

29. $\frac{4}{15}7^{5/4}.$

30. $-\pi.$

31. $\frac{183}{221}x+\frac{18}{221}\log(14\cos x+5\sin x)+c.$

32. $\sin^{-1}\frac{3x+1}{\sqrt{5}(1+x)}+\text{c}.$

33. $\frac{1}{2\sqrt{2}}\left[\sinh^{-1}\frac{1-x}{1+x}-\sinh^{-1}\frac{x+1}{x-1}\right]+c.$

34. $-\cosh^{-1}\frac{x+1}{3x-3}=c.$

35. $\tan x-\frac{x\sec x}{x\sin x+\cos x}+c.$

36. $2x+\log(2\sin x+3\cos x)+c.$

37. $-\frac{2}{\sin\alpha}\left[\frac{\sin(x+2)}{\sin x}\right]^{1/2}+c.$

38. $\sqrt{2}\sin^{-1}(\sin x-\cos x)+c.$

39. $x-4\log x+\frac{5}{4}\log(x-1)+\frac{3}{4}\log(x+1)+\log(x^2+1)-\frac{1}{2}\tan^{-1}x+c.$

40. $\frac{1}{2}\log\frac{5}{8}.$

41. $\frac{1}{3}\sqrt{12+x-6x^2}+\frac{17}{6\sqrt{6}}\sin^{-1}\frac{12x-1}{17}+c.$

42. $3\sin^{-1}\sqrt{\dfrac{x-2}{3}}+\sqrt{(x-2)(5-x)}+c.$

43. $\dfrac{3\pi}{128}.$

44. $\dfrac{\pi}{4}.$

45. $\dfrac{1}{80080}.$

Exercises 4.3

1. 1/2.

2. 16/3.

3. 128/15.

5. $a^2\left(e-\dfrac{1}{e}\right).$

6. $\dfrac{3}{2}a^2.$

7. $\dfrac{1}{3}\left(4\pi+\sqrt{3}\right)a^2.$

8. $\dfrac{3}{2}\pi a^2.$

9. $4\,ab\tan^{-1}\dfrac{b}{a}.$

10. 2.

11. 15/2.

12. 4.

13. 17/4.

14. 18.

15. $\pi ah^2.$

16. $\pi a^3\left(\dfrac{\sinh 2}{2}+1\right).$

17. $2\pi/15.$

18. $\pi/2.$

19. $\dfrac{b}{\pi n};\dfrac{\pi b^2}{4n}.$

20. $\dfrac{4}{3}\pi ab^2.$

21. $\dfrac{1}{3}\pi a^2 b.$

22. $3\pi a^2;5\pi^2 a^3.$

24. $\dfrac{16}{3}r^3.$

26. $2\cdot 4\pi a^2.$

27. $\dfrac{64}{3}\pi a^2.$

31. $6a.$

32. $8a.$

33. (a) $6a$; (b) $3\pi a/2.$

36. $6a.$

40. $\bar{x}=2/3,\ \bar{y}=2.$

41. $\dfrac{16}{15}a^2,\dfrac{4}{7}a.$

42. (i) $(a^2+ab+b^2)/3$, (ii) $2/\pi$, (iii) $\pi/4$.

46. 4mts/sec.

Exercises 5.1

5. $f(1,2)=8/5,\quad f(-2,1)=-8/5,\quad f(0,0)=0.$

8. $f(1,2)=1,\ f(-2,1)=26,\ f(0,0)=5.$

9. 1/3.

12. Does not exist.

13. Does not exist.

21. Continuous.

24. Discontinuous at (0, 0).

25. Discontinuous at (0, 0).

29. Continuous at (0, 0).

30. Discontinuous at (1, 2).

Exercises 5.2

1. $3x(x+2y),\ 3(x^2-y^2).$

3. $\dfrac{2x+3}{y}, -\dfrac{x^2+3x+1}{y^2}.$

4. $\dfrac{2xy(1-y)}{(x^2+y)^2}; \dfrac{x^2(2y-1)+y^2}{(x^2+y)^2}.$

6. $-\dfrac{y}{x^2}+\dfrac{1}{y}; \dfrac{1}{x}-\dfrac{x}{y^2}.$

8. $\dfrac{y^2}{x^2}e^{-y/x}; e^{-y/x}\left(1-\dfrac{y}{x}\right);$

10. $\dfrac{2x}{x^2+y^2}; \dfrac{2y}{x^2+y^2}.$

13. $e^{xy}(y\cos x^2-2x\sin x^2); xe^{xy}\cos x^2.$

15. $\dfrac{y^2}{1+x^2y^4}; \dfrac{2xy}{1+x^2y^4}.$

21. 0.

22. 0.

39. $\dfrac{\partial^2 u}{\partial r^2}+\dfrac{1}{r^2}\dfrac{\partial^2 u}{\partial \theta^2}+\dfrac{1}{r}\dfrac{\partial u}{\partial r}=0.$

Exercises 5.3

1. (i) 1. (ii) $\dfrac{2t(1+2t^2+t^3)}{(1+t)^3}.$

2. (i) $\dfrac{z}{1-z}$ (ii) $2x+8y; 8x.$

7. $r\sin 2\theta; r^2\cos 2\theta.$

9. $\dfrac{\partial z}{\partial u}=\dfrac{x-y}{x^2+y^2}, \dfrac{\partial z}{\partial v}=\dfrac{x}{x^2+y^2}(x\cos v+\sin v).$

14. $\dfrac{2x(4vx^2-1)}{1-4uv}.$

Exercises 5.4

3. $\dfrac{\pi}{4}-\dfrac{1}{2}(x-1)+\dfrac{1}{2}(y-1)+\dfrac{1}{4}(x-1)^2-\dfrac{1}{4}(y-1)^2+\cdots$

5. $-3(x-1)-2(y+1)+6(x-1)^2+6(x-1)(y+1)+(y+1)^2+\cdots$

10. (0, 0); (2, 0).

11. Minimum value is –8 at (2, 1).

12. (0, 0); (*a*, o); (–*a*, 0).

14. Maximum value at (–1, –3, –3).

19. $\frac{8abc}{3\sqrt{3}}$.

20. $x = y = 2z$.

23. If $x_1 + x_2 + x_3 = s$ (where x_1, x_2 and x_3 are positive numbers), then $x_1 x_2 x_3$ is maximum if $x_1 = x_2 = x_3$.

25. $\sqrt{a_1^2 + a_2^2 + \cdots + a_n^2}$.

27. 1/3.

28. $\left(\frac{8}{5}, \frac{3}{5}\right), \left(-\frac{8}{5}, -\frac{3}{5}\right)$.

32. $x = 2, y = 3, z = 1$.

33. $2x \cos x^4 - \cos x^2$.

34. $\frac{2 \sin \alpha^2 - \sin \alpha}{\alpha}$.

Exercises 5.5

1. Straight line.

2. Brachistochrone.

5. (a) *Hint:* Minimize $I = \int_{x_1}^{x_2} [1 + (dy/dx)^2]^{1/2}\, dx$. Here Euler's equation is:

$$\frac{d}{dx}\left(\frac{y'}{1 + y'^2}\right) = 0 \quad \text{where } y' = \frac{dy}{dx}.$$

(b) Great circles; (*Hint:* use spherical polar coordinates.)

7. Circle.

10. $2 \log_e (1 + \sqrt{2})$.

11. *Hint:* $I = \int_{x_1}^{x_2} [\sqrt{1 + y'^2 + z'^2} + \lambda(x^2 + y^2 + z^2)]\, dx$, where $y' = dy/dx$ and $z' = dz/dx$. Write down the two Euler equations and eliminate λ between them.

12. *Hint:* Maximize $I = \int_a^b y\, dx$, subject to the condition $J = \int_a^b ds = L$.

13. (a) Straight line, (b) Catenary,
(c) Circle, (d) Parabola,
(e) Cycolid.

Exercises 6.1

1. 8/3.

2. 24.

3. $ab(a^2 + b^2)/3$.

4. $\log_e \frac{25}{24}$.

5. $\frac{1}{3} \log_e (1 + \sqrt{2})$.

6. 9/2.

7. 15/4.

8. 47/30.

9. 1.

11. 67/60.

15. $3a^4/8$.

17. $a^2 b^2/8$.

19. 3/56.

20. 1/60.

22. $-1025/27$.　　**23.** $\frac{1}{2}\log_e 2$.

24. $\pi a^4/2$.　　**25.** $\pi a^3/6$.

26. $\pi/2$.　　**27.** $\frac{\pi}{4a^2}$.

28. $3\pi a^4/4$.　　**29.** 8π.

30. $\pi a/4$.　　**31.** 10.

33. $\int_0^a \int_0^{\sqrt{ay}} xy\, dx\, dy + \int_0^{2a} \int_0^{2a-y} xy\, dx\, dy = \frac{3a^4}{8}$.

35. $16a^2/3$.　　**36.** $\pi[f(a) - f(0)]$.

37. $\int_0^a \int_0^{\sqrt{4ax}} (x^2 + y^2)\, dy\, dx + \int_a^{3a} \int_0^{3a-x} (x^2 + y^2) dy\, dx = \frac{314a^4}{35}$.

40. $\int_{-1}^{3} \int_{x^2}^{2x+3} dy\, dx$.

Exercises 6.2

1. a^2.　　**2.** πa^2.

3. $3\pi a^2/2$.　　**5.** $a^2(3\pi - 8)/2$.

12. $3\pi a/16$.　　**14.** $ka^2b^2/8$.

15. $8\pi a^3/3$.

Exercises 6.3

1. $a^6/48$.　　**3.** $1/20$.

6. $1/3$.　　**7.** $16a^3/3$.

8. $1024/15$.　　**9.** $\pi a^3(2-\sqrt{2})/3$.

10. Volume of the ellipsoid $(x^2/a^2) + (y^2/b^2) + (z^2/c^2) = 1$. is $4\pi abc/3$.

12. $4\pi \log 3$.　　**13.** $19\pi/6$.

18. $\frac{a^{n+4}}{2(n+4)}$.　　**21.** $2\pi/105$.

22. $\frac{1}{6}[\sqrt{2} + \log(\sqrt{2}+1)]$.　　**25.** $(e-1)/2$.

26. $4[\log(1+\sqrt{2})+(1-\sqrt{2})]$.

Exercises 6.4

4. Divergent.

6. Convergent.

11. Convergent.

18. Divergent.

21. Convergent.

23. Convergent.

24. Converges absolutely.

28. π.

29. $3+3(2^{1/3})$.

30. 0.

31. 1.

32. $\log_e 2$.

33. 0.

34. 6.

35. $\log\sqrt{2}$.

36. $\frac{1}{2}\log 3-2$.

37. $\sqrt{\pi}/2$.

38. $\frac{1}{(m+n)^{n+1}}\Gamma(n+1)$.

39. $\pi/\sqrt{2}$.

40. $\frac{1}{4}\Gamma\left(\frac{5}{4}\right)$.

41. $\Gamma\left(\frac{4}{3}\right)$.

42. $\frac{1}{2}\frac{\Gamma\left(\frac{1}{4}\right)}{\Gamma\left(\frac{3}{4}\right)}\sqrt{\pi}$.

43. $\frac{\sqrt{\pi}\,\Gamma\left(\frac{1}{3}\right)}{\Gamma\left(\frac{5}{6}\right)}$.

44. $\frac{16}{3003}$.

45. $\frac{\pi}{2\sqrt{2}}$.

46. $\frac{\sqrt{\pi}\,\Gamma\left(\frac{1}{n}\right)}{n\Gamma\left(\frac{1}{n}\right)+\frac{1}{2}}$.

52. $\frac{3\pi}{512}$.

64. $\frac{\Gamma(p)\,\Gamma(q)\,\Gamma(r)}{\Gamma(p+q+r+1)}$.

Exercises 7.1

1. 0.002.

2. 2.7586, 2.759, 2.76, 2.8, 3.0, 3.1416, 3.142, 3.14, 3.1, 3.0, 0.5645, 0.564, 0.56, 0.6, 1.0, 4.292, 4.29, 4.3, 4.0, 0.6065, 0.602, 0.60, 0.6, 1.0,

3. 3.142; $\frac{1}{2}\times 10^{-3}$.

4. 563.2.

5. 2.88×10^{-3}.
6. 1.7×10^2.
7. 12.
8. 701.
9. 3.44; 0.0039.
10. 0.1823215; $n = 9$.

Exercises 7.2

2. 14.7.
3. 2.50987.
4. 83.515.
5. 0.
6. $\frac{1}{72}(-x^3 + 29x^2 + 1602x + 47448)$.
7. 0.111; 0.090.
8. 1.913.
9. 194486.
10. 0.
12. 0.5573.
14. 2.4.
15. 93.43.
16. 1.3; 6.04.
17. 0.5580.
18. 162.4.
19. 3250.875.
21. $a_1 = 1.1x + 5.8$.
22. $a_1 = 0.26, a_2 = 1.21$.
23. $a = 0.90, b = 0.55$.
24. $a = -1, b = 14$.
25. $a = 20, b = 1$.
26. $a = 1.42,\ b = -1.07, c = 0.55$.
27. $a = 14.39, b = 0.0826, c = 0.0466$.
28. $y = 0.1 - 0.02x + 0.024x^2$.
29. $a = -0.5,\ b = 1.775,\ c = -0.135$.
30. $a_1 = 0.4645, a_2 = 0.46$.
31. $y = 73.86e^{0.6808x}$.
32. $y = 0.13265(2.9485)^x$.

Exercises 7.3

1. 2.6207 .
2. 1.7963.
3. 2.7065.
4. –2.1049.
6. 2.83.
7. 2.741.
8. 1.05.
9. 2.741.
10. 0.6071.
11. 3.4641.
12. –1.249.
13. 2.911.
14. –2.058.
15. 8.8770.
16. 0.8835849.
18. (a) 1.3247, (b) 1.8393.
19. 3.584428; –1.848127.
20. –1.9266, –1.8533.

Exercises 7.4

1. –0.05.
2. 0.256.
3. 0.12916; –0.00465.
4. 4.054.
5. –1.91675.
6. –16;4
8. 0.5333; –4.56.
9. 0.
10. 2.1509; $x = 0.1388$, $y(0.1388) = -0.2965$.

Exercises 7.5

2. 1.1615.
3. (i) 4.05617, (ii) 4.05106.
4. 66.33.
5. 1.1873.
6. 0.6945; 0.6932; 0.6932.
7. 710 sq.ft.
8. (i) 0.90916, (ii) 90960.
9. 3.1058.
10. 177.4830.
11. 58.00; 53.87; correct value = 53.6.
12. 0.02020; 0.000173.
13. 0.231083.
14. 0.8145.
15. 0.77500; 0.78279; 0.78475; 0.7854.

Index